P9-DCL-958

BIOGRAPHICAL DICTIONARY OF AMERICAN SCIENCE

CONSULTANT EDITORS

BIOGRAPHICAL DICTIONARY OF AMERICAN SCIENCE

The Seventeenth Through the Nineteenth Centuries

CLARK A. ELLIOTT

GREENWOOD PRESS WESTPORT, CONNECTICUT • LONDON, ENGLAND

Library of Congress Cataloging in Publication Data

Elliott, Clark A.
Biographical dictionary of American science.

"A retrospective companion to American men of science."
Includes bibliographical references and index.
1. Scientists—United States—Biography. I. American
men of science. II. Title.
Q141.E37 509'.2'2 [B] 78-4292
ISBN 0-313-20419-5

Library of Congress Catalog Card Number: 78-4292
ISBN: 0-313-20419-5
First published in 1979
Greenwood Press, Inc.
51 Riverside Avenue, Westport, Connecticut 06880
Printed in the United States of America
10 9 8 7 6 5 4 3 2 1

To Priscilla
and to
Andy and Glenn

CONTENTS

PREFACE

GENERAL

This *Dictionary* draws together widely scattered biographical and bibliographical information and, hopefully, will repay its debt to the many other workers on whose research it is based by serving students and scholars not only as a convenient work of reference, but also as a means of promoting further studies in the history of science in America.

RELATION TO OTHER WORKS

This *Dictionary* is deliberately designed as a retrospective companion to *American Men of Science* (*AMS*), which has been recognized as the chief directory of living scientists since its first edition appeared in 1906. The *Dictionary* includes major entries, averaging between 300 and 400 words, for nearly 600 scientists never included in *AMS*. In most instances, therefore, the subjects of the major sketches in this work died before 1906. However, there were a few scientists not recognized or remembered by their contemporaries when *AMS* was first published, and it now seems desirable that these persons should appear in such a reference work.

In addition to these major biographical entries, there are minor (or cross-reference) entries for some 300 other scientists who *were* included in the early editions of *AMS*. These are persons who had reached a significant stage of their careers before 1900 and who therefore were closely identified with science in the late nineteenth century.

SCOPE

As shown in Appendix A, this *Dictionary* includes scientists whose birth dates span the period from 1606 to 1867. Taking the major and minor entries as a whole, the work is intended to be a directory of scientists active in America before the early years of the twentieth century. During that long period—some 250 years—the definition of a scientist changed considerably, and that fact complicates the selection process. In a general sense, the persons represented here were noted for their contributions to scientific knowledge (whether or not that contribution was accepted or proved permanently valid or useful). Individuals noted chiefly as teachers or popularizers usually have not been included, but exceptions have been made for those who were especially important in such roles. Some relatively minor figures, frequently of local significance, have been included, if only to suggest the range of persons involved in the promotion of science in the United States during the period covered by this work.

The term *science* has been somewhat narrowly defined to mean chiefly work in such areas as mathematics, astronomy, physics, chemistry, botany, zoology, geology, and their allied specialties, and including some aspects of applied science. In general, engineers, inventors, physicians, social scientists, cultural anthropologists, ethnologists, explorers, and the like have not been included. Some persons who are known chiefly for work in one of the latter fields will be found here; however, their inclusion usually is based on less well known work that they did in one of the fields of science covered by this volume. For example, H. R. Schoolcraft appears here because of his early geological activities, rather than his more noted ethnological research. A few agriculturists have been included, as have some engineers, but generally for their research or innovative work or for their efforts in generalizing new or established principles, rather than for their applications of such principles or techniques. Some topographical engineers have been included when their work was important in the early exploring and mapping of the American continent. Persons noted for work in medicine are more likely to have been admitted from the colonial period than from later times.

As suggested above, in the historical perspective there is no easy definition of a scientist. This work has relied heavily on the choices made by the editors of the *Dictionary of American Biography* (*DAB*), *Who's Who in America* (1899-1900), and its companion *Who Was Who in America: Historical Volume*, 1607-1896. The starred (that is, most notable) names in *AMS* (1906) were inspected for selection of subjects for the minor sketches. The following works also were used extensively in the selection process: Donald DeB. Beaver, "The American Scientific Community, 1800-1860," (Ph.D. dissertation, Yale University, 1966); Whitfield J. Bell, Jr., *Early American Science: Needs and Opportunities for Study* (Williamsburg, Va., 1955); George H. Daniels, *American Science in the Age of Jackson* (New York, 1968); Clark A. Elliott, "American Scientist, 1800-1863: His Origins, Career, and Interests" (Ph.D. dissertation, Case Western Reserve University, 1970); William H. Goetzmann, *Exploration and Empire: The Explorer and the Scientist in the Winning of the American West* (New York, 1966); Stanley M. Guralnick, *Science and the Ante-Bellum American College* (Philadelphia, 1975); National Academy of Sciences, *A History of the First Half-Century of the National Academy . . .* 1863-1913 (Washington, D.C., 1913); *National Cyclopaedia of American Biography; Notable American Women*, 1607-1950 (Cambridge, Mass., 1971); Raymond P. Stearns, "Colonial Fellows of the Royal Society of London, 1661-1788," *Notes and Records of Royal Society* 8, no. 2 (April 1951): 178-246; Stearns, *Science in the British Colonies of America* (Urbana, Ill., 1970); *World Who's Who in Science* (Chicago, 1968).

Based on an inspection of the works mentioned above, a list of several thousand names was compiled. Selection of a preliminary roster of entries from this much longer list was based upon inclusion in certain of the works or combinations of works mentioned above, as well as some subjective choices based on brief examination of biographical information. This draft list of several hundred names was then circulated among the consultant editors. Subsequently, a few names were dropped and a greater number were added, frequently on the recommendation of the consultant editors.

The *DAB* has proved immensely useful as a source for biographical and bibliographical information. However, it is now more than forty years old. In the preparation of this *Dictionary*, an attempt has been made to locate and to use biographical sources published since the appearance of the *DAB* for those scientists who are not in the recently completed *Dictionary of Scientific Biography* (*DSB*). These more recent sources often add new information or a new perspective that was not available to the authors of the sketches in the DAB. The international *DSB* was an enormous help with the most notable scientists, but could offer no help with the majority of subjects.

MAJOR ENTRIES

The major entries (persons not in *AMS*) are the central core of this *Dictionary*. These sketches typically comprise the following elements:

a. Full name.

b. Date and place of birth and of death.

c. Field(s) of science in which the person was active.

d. Parentage, including father's and mother's names and father's occupation. The mother's maiden name, when definitely known or confirmed in another work of reference, is placed in parentheses. In a few instances the name that most likely is the mother's maiden name is not in parentheses, because this could not be confirmed; parentheses are omitted on the supposition that the name given for the mother may have come from a prior marriage.

   The names of close relatives who are included in this *Dictionary* are listed with family or marital information, with the notation "[q.v.]."

e. Marital status and number of children. When a wife is known to have been married previously and her maiden name is known, this is put in parentheses. Sometimes the number of children is taken from an obituary, and if the total number of children is not known, the entry will read "survived by *x* children."

f. Education. The information given relates chiefly to the individual's highest level(s) of education.

g. Honors. When information on honorary degrees, prizes, medals, and the like is readily available, up to two such honors are listed here. Absence of information regarding honors does *not* mean that the individual never received such honors. It means only that the information was not given in the standard biographical source(s) consulted for that particular person.

h. Career. Whenever available, information is supplied (in general chronological order) about the individual's major appointments, associations, or activities for his or her entire adult life. Both science and non-science occupations or activities are given. While the effort has been made to ensure that dates and other information are accurate, the standard reference sources frequently provide conflicting information. It was not always possible to correct that information here; but when any uncertainty was apparent, that fact generally is indicated by the use of "ca." or other notes of caution.

i. Society memberships. Whenever the information is readily available in the standard biographical sources consulted, membership in, at most, two scientific societies is listed. However, since no extraordinary effort was made to compile society memberships, the absence of such information here does *not* mean that the individual did not belong to a scientific society.

When a person belonged to more than two societies, preference usually is given to the major organizations. Membership information is included partly as an indication of the subject's degree of commitment to science. By an emphasis on national groups, it also suggests the degree of "acceptance" of the person by the scientific community.

j. Scientific contributions. While persons included in the *Dictionary* may have been active in other fields, it is their contributions to science that are emphasized here.

k. Works. Whenever possible, the individual's chief publications are included, listing a maximum of six titles under this heading. However, when more convenient or appropriate, works sometimes are mentioned (with abbreviated citations) in the section discussing "Scientific Contributions"; for reasons of economy, all of these titles are not necessarily repeated in "Works." In some instances, only the general character of the scientist's writings is mentioned, but whenever possible the availability of more complete lists of works is indicated, either as separate bibliographies or as reference to standard bibliographical sources. When the person wrote a number of works, none of which could be considered as his or her chief publication(s), the *Dictionary* sometimes lists representative works, but more frequently refers the reader to the more complete bibliographies published elsewhere.

l. Manuscripts. Whenever the information is available, the location of collections of the scientist's papers is indicated, generally followed by a reference to the source of the information, which may give more complete descriptions. This entry relies heavily on *The National Union Catalog of Manuscript Collections* (*NUCMC*) and Philip M. Hamer, ed., *A Guide to Archives and Manuscripts in the United States* (New Haven, 1961). If no collection was located in such standard sources, inquiries usually were sent to likely repositories. When the source of information on a collection is *not* given here, ordinarily the information came from the repository itself.

Only separate collections of the scientist's papers, or *substantial* groups of letters or other documents in the collections of other persons or institutions, are included here. For example, references to the *NUCMC* include only those which are underlined in its indexes (see that work for further information). In some instances, the indication is given that no collection of manuscripts is known. The intent here is to indicate that no such *separate* collection is known. Of course, researchers may very well find some letters or other manuscripts of the scientist in the collections of papers of other persons, and some of these might be located through *NUCMC*.

m. Works about. This section generally includes only works actually used in the preparation of the *Dictionary*. Both the *DAB* and the *DSB* include bibliographies of works about the scientists; when these dictionaries are cited, the reader should consult them for additional biographical citations. However, when a reference cited in the *DAB* or the *DSB* actually has been consulted in the preparation of this *Dictionary*, the reference is repeated here, along with other sources used.

Sometimes an additional source is cited at the end of an entry. These citations are introduced by "also see" and usually refer to major modern biographies. Ordinarily the works so cited have not been directly consulted for this *Dictionary*, but may have been used by the author of the sketch in the *DAB*, the *DSB*, and so on. The citation is repeated here for the convenience of the user of the *Dictionary*, who seeks definitive or most complete sources on the scientist's life.

MINOR ENTRIES

In addition to the major entries, there are a number of minor (or cross-reference) entries for persons who were active in the nineteenth century, but who lived beyond 1906 and therefore are in the *AMS*. These minor entries supply only the person's name, year of birth and death, scientific fields, and brief identification, in terms of major or representative occupational affiliations or other career descriptions. In addition, where the information is readily available in standard reference works, the location of manuscript collections also is given. Finally, under "Works About," other standard biographical directories or dictionaries are listed as sources for additional information; for these minor sketches, the *AMS* always is included as one of the sources. In instances where a recent major biography has been published and the subject is not included in the *DSB*, the full biography is cited. In other instances, the bibliographies in the *DSB* and the *DAB* are relied on, without repeating the information here.

ACKNOWLEDGMENTS

I am especially indebted to the consultant editors for the advice and encouragement that they supplied throughout this project. They reviewed the list of names and made valuable suggestions to correct my omissions, offered advice on the most appropriate format and coverage for the *Dictionary*, and reviewed the first draft of the major sketches. In addition to the persons listed as consultant editors, I benefited from the help and advice of Dr. Ronald S. Wilkinson, Manuscript Historian at the Library of Congress and Associate in Bibliography at the American Museum of Natural History. No less than the others, Dr. Wilkinson helped to make this project a better one than it might have been without their assistance. While personally assuming full responsibility for all errors of omission or fact, I want to thank each of these colleagues for the time and care they have taken to help ensure that such errors were kept to a minimum.

I found the numerous librarians and manuscript curators to whom I have written to have been uniformly helpful. Within the Harvard University libraries, I particularly want to thank Barbara Dames, Ruth Hoppe, and the other members of the Widener Library interlibrary loan office who responded cheerfully and promptly to my frequent requests for loans from other libraries. A special thanks must go to that great legion of Harvard librarians, living and deceased, who have worked through three centuries to make the Harvard library system the great research collection that it is today. It is upon the availability of those collections that this *Dictionary* rests.

My thanks also go to Dr. Robert Schofield, Professor of the History of Science at Case Western Reserve University, who encouraged and directed my early search for American scientists. Thanks also to Michele Aldrich, Marc Rothenberg, and other friends who have offered their assistance and encouragement. The help that I received from my son Andrew Jordan Elliott, with some of the tiresome work on the appendixes, saved me many hours of work and earned him, I hope, many happy days at summer camp. To the Greenwood Press and its editorial vice-president, James T. Sabin, I acknowledge a special indebtedness. Their commitment to this work has made possible the realization of a project that I, and other students of the history of American science, have long anticipated.

Acknowledgments to one's family are traditional. It was only after I became immersed in this work that I realized how genuine—and how inadequate—such a gesture is. For whatever compensation it might be for thousands of hours that I spent in isolation and for days of ill temper and seeming indifference, I thank my wife, Priscilla, whose love and patience are the essential support for all that I do. To my sons, Andy and Glenn, I say thank you for reassuring me, when I felt discouraged, that perhaps "Daddy's encyclopedia" really was worth doing after all.

ABBREVIATIONS

AMS = *American Men of Science*, 1st ed. (New York, 1906).

Ac. Nat. Scis. Philad. = Academy of Natural Sciences of Philadelphia.

Am. Ac. Arts Scis. = American Academy of Arts and Sciences, Boston.

Am. Ant. Soc. = American Antiquarian Society, Worcester, Mass.

Am. Assoc. Advt. Sci. = American Association for the Advancement of Science.

Am. Chem. Soc. = American Chemical Society.

Am. Inst. Min. Engrs. = American Institute of Mining Engineers.

Am. Journ. Pharm. = *American Journal of Pharmacy.*

Am. Journ. Sci. = *American Journal of Science* ("Silliman's Journal").

Am. Math. Soc. = American Mathematical Society.

Am. Med. Assoc. = American Medical Association.

Am. Mus. Nat. Hist. = American Museum of Natural History, New York.

Am. Ornith. Un. = American Ornithologists' Union.

Am. Phil. Soc. = American Philosophical Society, Philadelphia.

Ann. Med. Hist. = *Annals of Medical History.*

Appleton's CAB = *Appleton's Cyclopaedia of American Biography*, vols. 1-6 (New York, 1887-1889), vol. 7 (New York, 1901), vol. 8 (New York, 1918).

Bell, *Early Am. Sci.* = Whitfield J. Bell, Jr., *Early American Science: Needs and Opportunities for Study* (Williamsburg, Va., 1955).

Biog. Mems. of Natl. Ac. Scis. = *Biographical Memoirs of the National Academy of Sciences.*

Bos. Soc. Nat. Hist. = Boston Society of Natural History, now Boston Museum of Science.

Bull. Hist. Med. = *Bulletin of the History of Medicine.*

Calif. Ac. Scis. = California Academy of Sciences, San Francisco.

Charleston Mus. = Charleston Museum, Charleston, S.C.

Colum. U. Lib. = Columbia University Library, New York.

Corn. U.-CRH = Cornell University Library—Collection of Regional History and University Archives, Ithaca, N.Y.

Cullum, *Biographical Register, West Point* = George W. Cullum, *Biographical Register of Officers and Graduates of the U.S. Military Academy at West Point, New York*, 3d ed., 3 vols. (Boston, 1891).

DAB = *Dictionary of American Biography*, 20 vols. and 4 supplements (New York, 1928-).

DSB = *Dictionary of Scientific Biography*, ed. Charles C. Gillespie, 14 vols. (New York, 1970-1976).

Det. Pub. Lib.-BHC = Detroit Public Library—Burton Historical Collection.

Geol. Soc. Am. = Geological Society of America.

Goetzmann, *Army Exploration* = William H. Goetzmann, *Army Exploration in the American West, 1803-1863* (New Haven, 1959).

Goetzmann, *Exploration and Empire* = William H. Goetzmann, *Exploration and Empire: The Explorer and the Scientist in the Winning of the American West* (New York, 1966).

Guralnick, *Science and American College* = Stanley M. Guralnick, *Science and the Ante-Bellum American College*, Am. Phil. Soc. Memoirs, vol. 109 (Philadelphia, 1975).

Hamer = Philip M. Hamer, ed., *A Guide to Archives and Manuscripts in the United States*, compiled for the National Historical Publications Commission (New Haven, 1961).

Harv. U.-FL = Harvard University—Farlow Reference Library of Cryptogamic Botany, Cambridge, Mass.

Harv. U.-GH = Harvard University—Gray Herbarium.

Harv. U.-HL = Harvard University—Houghton Library.

Harv. U.-MCZ = Harvard University—Museum of Comparative Zoology.

Harv. U.-MS = Harvard University—Medical School Library (Countway).

Harv. U.-UA = Harvard University—University Archives.

Hist. Soc. Penn. = Historical Society of Pennsylvania, Philadelphia.

Hunting. Lib. = Henry E. Huntington Library, San Marino, Calif.

Journ. Chem. Educ. = *Journal of Chemical Education.*

Kelly and Burrage, *Med. Biog.* (1920) = Howard A. Kelly and Walter L. Burrage, *American Medical Biographies* (Baltimore, 1920).

Kelly and Burrage, *Med. Biog.* (1928) = Howard A. Kelly and Walter L. Burrage, *Dictionary of American Medical Biography* (New York, 1928).

L. C. Cat. Prd. Cards 1942 = U.S. Library of Congress, *Catalog of Books Represented by Library of Congress Printed Cards Issued to July 31, 1942* (Ann Arbor, Mich., 1942-1946). Note: This source is cited only if the relevant volumes of *Natl. Un. Cat. Pre-1956 Imp.* have not appeared; the *Natl. Un. Cat. Pre-1956 Imp.*, when completed, will replace this reference.

Lib. Co. Philad. = Library Company of Philadelphia.

Lib. Cong. = U.S. Library of Congress, Washington, D.C.

M.I.T.-Arch. = Massachusetts Institute of Technology—Archives, Cambridge, Mass.

Mallis, *Amer. Entomologists* = Arnold Mallis, *American Entomologists* (New Brunswick, N.J., 1971).

Mass. Hist. Soc. = Massachusetts Historical Society, Boston.

Md. Hist. Soc. = Maryland Historical Society Library, Baltimore.

Meisel, *Bibliog. Amer. Nat. Hist.* = Max Meisel, *A Bibliography of American Natural History: The Pioneer Century, 1769-1865,* 3 vols. (Brooklyn, N.Y., 1924-1929; reprinted, New York, 1967).

Merrill, *Hundred Years of Geology* = George P. Merrill, *The First One Hundred Years of American Geology* (New Haven, Conn., 1924; reprinted, New York, 1964).

Mo. Bot. Gard. = Missouri Botanical Garden, St. Louis.

NCAB = *National Cyclopaedia of American Biography* (New York, 1893-). Note: In some volumes, especially the earlier ones, a date is given on the title page that differs from the copyright date(s) on the verso of the title page. The *Cyclopaedia* includes many persons living at the time of preparation of each volume; ordinarily the biographical data for such entries includes information only prior to the *earliest* date given on the title page or its verso. When volumes of the *Cyclopaedia* are cited, all publication and copyright dates are given. The first date listed is the one that appears on the title page itself. When only one date is given, it is the date of both publication and copyright.

NUCMC = U.S. Library of Congress, *The National Union Catalog of Manuscript Collections* (Ann Arbor, Mich.; Hamden, Conn.; Washington, D.C., 1962-). Volumes with entry dates through 1975 were consulted.

N.Y. Ac. Scis. = New York Academy of Sciences.

N.Y. Bot. Gard. = New York Botanical Garden, New York.

N.Y. Hist. Soc. = New York Historical Society, New York.

N.Y. Lyc. Nat. Hist. = New York Lyceum of Natural History.

N.Y. Pub. Lib. = New York Public Library.

N.Y. St. Lib. = New York State Library, Albany.

Natl. Ac. Scis. = National Academy of Sciences, Washington, D.C.

Natl. Arch. = U.S. National Archives, Washington, D.C.

Natl. Geog. Soc. = National Geographic Society, Washington, D.C.

Natl. Un. Cat. Pre-1956 Imp. = *National Union Catalog: Pre-1956 Imprints* ([Chicago], 1968-).

Nickles, *Geol. Lit. on N. Am.* = John M. Nickles, *Geologic Literature on North America, 1785-1918,* U.S. Geol. Surv. Bulletin no. 746 (Washington, D.C., 1922).

Nota. Am. Wom. = *Notable American Women, 1607-1950: A Biographical Dictionary,* 3 vols. (Cambridge, Mass., 1971).

Phil. Soc. Wash. = Philosophical Society of Washington, D.C.

Phil. Trans. of Roy. Soc. = *Philosophical Transactions of the Royal Society of London.*

Poggendorff, *Biographisch-literarisches Handwörterbuch* = Johann C. Poggendorff, *Poggendorff's Biographisch-literarisches Handwörterbuch zur Geschichte der exacten Wissenschaften* (Leipzig, 1863-1904).

Rens. Poly. Inst.-Arch. = Rensselaer Polytechnic Institute Library—Archives, Troy, N.Y.

Rossiter, *Emergence of Agricultural Science* = Margaret W. Rossiter, *The Emergence of Agricultural Science: Justus Liebig and the Americans, 1840-1880* (New Haven, 1975).

Roy. Soc. = Royal Society of London.

Roy. Soc. Cat. = Royal Society of London, *Catalogue of Scientific Papers,* 19 vols. (London and Cambridge, 1867-1925), covering 1800-1900.

Shipton, *Biog. Sketches, Harvard* = Clifford K. Shipton, *Biographical Sketches of Those Who Attended Harvard College* (Boston).

Smithson. Contr. Knowl. = *Smithsonian Contributions to Knowledge.*

Smithson. Instn. = Smithsonian Institution, Washington, D.C.

Smithson. Misc. Coll. = *Smithsonian Miscellaneous Collections.*

St. Hist. Soc. Wis. = State Historical Society of Wisconsin, Madison.

Stearns, *Sci. in Brit. Colonies* = Raymond P. Stearns, *Science in the British Colonies of America* (Urbana, Ill., 1970).

Torrey Bot. Club = Torrey Botanical Club.

U.S. Geol. Surv. = U.S. Geological Survey.

U.S. Mil. Ac. = U.S. Military Academy, West Point.

U.S. Natl. Mus. = U.S. National Museum, Washington, D.C.

Univ. Calif.-Ban. = University of California, Berkeley—Bancroft Library.

Univ. Mich.-MHC = University of Michigan, Ann Arbor—Michigan Historical Collections.

Univ. N.C.-SHC = University of North Carolina Library, Chapel Hill—Southern Historical Collection.

Univ. S.C.-SCL = University of South Carolina, Columbia—South Caroliniana Library.

Univ. Tex.-TA = University of Texas Library—Texas Archives, Austin.

Univ. Va. Lib. = University of Virginia Library, Charlottesville.

Va. St. Lib. Arch. Div. = Virginia State Library, Archives Division, Richmond.

Who Was Who In America: Historical Volume = *Who Was Who In America: Historical Volume, 1607-1896, A Component Volume of Who's Who in American History* (Chicago, 1963).

Who Was Who In America, vol. 1 = *Who Was Who in America: A Companion Volume to Who's Who in America*, vol. 1, *1897-1942* (Chicago, 1942).

World Who's Who in Science = *World Who's Who in Science: A Biographical Dictionary of Notable Scientists from Antiquity to the Present*, ed. Allen G. Debus (Chicago, 1968).

Yale U.-HMC = Yale University—Historical Manuscript Collection, New Haven, Conn.

Yale U.-RB = Yale University Library—Rare Book Room.

Yale U.-WAC = Yale University—Western Americana Collection.

Youmans, *Pioneers of Sci.* = William Jay Youmans, ed., *Pioneers of Science in America* (New York, 1896).

BIOGRAPHICAL DICTIONARY OF AMERICAN SCIENCE

INTRODUCTION: THE SCIENTIST IN AMERICAN SOCIETY

Although most users of this *Dictionary* may be interested chiefly in the details of a single life, it also should be kept in mind that the career of each scientist was carried out in a larger context. Biographical details differed according to circumstances and the individual's reaction to them, and yet through the course of their lives each of the persons in this work helped to define what role the scientist would play in American society. The professional lives of the persons in this *Dictionary* were played out at a time when the pursuit of knowledge of the physical world was not always recognized by the larger society within which they lived as a necessary or fruitful undertaking. Likewise, would-be scientists at times had to deal with societal pressures that would bend their interests to utilitarian purposes or to the support of traditional values outside of science. In attempting to reconcile these and other societal factors with his or her own career, each person helped to define the social role of the scientist in American society.

It is now more than three and one-half centuries since Europeans first settled in what is now the United States. During these years, vast changes have taken place in the size and character of the population and in the political, economic, and social practices and beliefs of the people. The physical environment itself has been transformed, along with the accommodations of everyday life. If the population at large were asked today to explain these incredible changes, a large number could be expected to attribute them to the effects of science. In a stricter sense, of course, most of these respondents really would be referring to technology, applied science, or the methodology and mentality of science.

The tendency to meld science with its material or popular counterparts is traditional and understandable, since the relations between science and technology are both complex and subtle. Increasingly these relations have come to interest historians.[1] The assumption of this *Dictionary* has been that, in at least a general way, it is possible to separate those who were motivated chiefly by the desire to increase knowledge (the "scientists"[2]) from those who were motivated chiefly by the desire to control the environment (broadly defined to include means of production and household concerns, as well as the natural world itself).

Useful knowledge—technology, applied science, folk practices, and the like—by definition has a social role to fill. This never has been so obvious for science in its more abstracted or detached state. Yet, during the historical period covered by this *Dictionary*, one of the major tasks of the scientist (individually and collectively) was to establish within the American social system a place for abstracted knowledge of the physical world—that is, to create a socially accepted and economically supported role that would allow persons interested in science to devote their lives to its advancement.[3] That the goal was realized by the end of this historical period—1900—was not necessarily inevitable. Indeed, in the colonial period the ideal of a scientific career could scarcely have been envisioned. Nonetheless, during those early years the pursuit of scientific knowledge—both pure and applied— did have certain social sanctions, insofar as its study was considered within the purview of all educated men.[4]

In the early decades after American independence, that supportive relationship began to disintegrate. One of the chief reasons was the increasingly esoteric state of scientific knowledge itself. Science became less accessible to the educated gentleman. At the same time, technology and invention tended to follow a course largely separated from science. In important respects, science moved away from a close relationship with learned culture and technology and began to develop a relatively more independent status in society.[5] Scientists began to erect their own channels of communication and interaction, one of the more important being the *American Journal of Science and Arts* ("Silliman's Journal"), founded in 1818.[6]

The remainder of the nineteenth century showed a gradual movement by American scientists to establish unique and effective avenues of special education and employment, which would allow the individual so motivated to pursue the promotion and development of scientific knowledge on a full-time basis. This was, in conventional terms, a process of "professionalization." As a result, the career-minded scientist came to replace both the devotee of learned culture and the dedicated amateur.[7]

The general outline of this movement toward professionalization is seen in the three tables that follow. They are based on the major sketches in the *Dictionary* (that is, persons generally deceased before 1906 and not listed in *American Men of Science*). The data aggregated in these tables is based on the

Table 1. EDUCATION

Educational Level[1]	BORN BEFORE 1776[2] *Number*	*Percentage*	BORN 1776 TO 1815[3] *Number*	*Percentage*	BORN 1816 TO 1867[4] *Number*	*Percentage*
A.B.[5]	21	25.9	61	27.4	65	27.2
Ph.D, A.B., and B.S., etc.[6]	...	...	4	1.8	35	14.6
M.D. and A.B.	5	6.2	33	14.8	18	7.5
M.D. only	10	12.4	36	16.1	26	10.9
B.S., etc., or military academy, etc.[7]	...	...	16	7.2	25	10.5
College (no degree) and/or other advanced study[8]	19	23.5	26	11.7	36	15.1
Apprenticeship	4	4.9	7	3.1	2	0.8
Public school or academy	14	17.3	35	15.7	28	11.7
Private study, self-taught	8	9.9	5	2.2	4	1.7

1. Individuals are counted only once, under their highest educational level, ordinarily defined in terms of degree(s) earned.

2. Of a total of 98 persons born in this period, the educational level of 81 was determined.

3. Of a total of 240 persons born in this period, the educational level of 223 was determined.

4. Of a total of 243 persons born in this period, the educational level of 239 was determined.

5. Persons in this category also may have engaged in postgraduate studies in law or theology. They also may have pursued advanced studies in science, engineering, or medicine, but did not earn degrees in these latter subjects.

6. This category assumes the individual had an A.B. degree or its equivalent before going on to earn a Ph.D. or professional degree in science or engineering. In several cases, the person also had an M.D. degree.

7. Persons in this category earned professional degrees in science or engineering at the level of the bachelor's degree—such as B.S., Ph.B., C.E., and M.E.—or were graduates of the U.S. Military Academy (West Point) or the U.S. Naval Academy (Annapolis). They did not hold the A.B. degree.

8. Persons in this category attended college but did not graduate, and/or they studied law or theology or pursued non-degree studies in science, medicine, engineering, or military science. Also included are persons who attended technical schools in Europe (including graduates of those schools) and persons who were educated in the Jesuit order.

information used in preparing the appendices to this *Dictionary*.

Table 1 summarizes the data on the education of scientists. Even during the colonial and early republican period the colleges were an important factor in American science (more than 50 percent of the scientists attended college, and more than 30 percent earned the bachelor's degree). During the middle period, an even larger percentage received the bachelor's degree, and this trend continued into the last period. Throughout the years under review, medical studies had an important part in the preparation of scientists. This was most apparent in the middle period. Toward the end of the century, the doctorate of philosophy and the professional degrees in science and engineering became increasingly important. By the time the first edition of *American Men of Science* appeared in 1906, three-quarters of the scientists

Table 2. EMPLOYMENT

	BORN BEFORE 1776[2]		BORN 1776 TO 1815[3]		BORN 1816 TO 1867[4]	
Employment Category[1]	*Number of Individuals*[5]	*Percentage of Fractional Distribution*[6]	*Number of Individuals*[5]	*Percentage of Fractional Distribution*[6]	*Number of Individuals*[5]	*Percentage of Fractional Distribution*[6]
Non-science professions and "white collar" workers[7]	35	24.6	57	14.5	40	9.0
Business[8]	12	7.1	30	7.3	19	4.9
Agriculture	9	4.7	10	2.2	3	0.6
Skilled laborer, craftsman[9]	6	2.3	7	1.4	4	1.2
Inherited wealth, independent study	7	4.1	23	5.7	21	5.5
Medicine[10]	26	20.9	59	17.4	30	7.8
Professor of science[11]	20	12.1	88	23.7	105	29.6
Science and science-related employment: Governmental[12]	6	3.9	64	13.9	84	21.7
Science and science-related employment: Nongovernmental[13]	30	20.3	60	13.8	76	19.8

1. An individual may be counted in as many as four employment categories, although the vast majority appear in no more than three.

2. Of a total of 98 persons born in this period, the occupations of all were determined.

3. Of a total of 240 persons born in this period, reasonably complete occupational data for 237 was determined.

4. Of a total of 243 persons born in this period, reasonably complete occupational data for 241 was determined.

5. Since an individual could be counted in as many as four categories, the total of this column is greater than the number of individuals being studied.

6. For purposes of this column, each individual was assigned to the appropriate employment categories as a fraction of one, based on the total number of categories in which that person was counted. For example, an individual counted in three categories was assigned as one-third to each. The total for this column, therefore, was the same as the total number of individuals being studied.

7. Includes clergy, lawyer, educator-teacher (noncollege), military (except engineer corps), politician, public servant, author, editor, college president, professor (non-science), public lecturer, librarian, publisher, and so on.

8. Includes merchant and manufacturer. Chemical and drug manufacturer is counted in the last category.

9. Includes tradesman and artist. Does not include scientific instrument maker.

10. Includes physician, dentist, professor of medicine.

11. Includes persons who taught science subjects in medical schools, as well as in colleges and universities.

12. Includes science and science-related employment under the auspices of government at all levels. Includes surveyor, curator, army engineer, patent examiner, and employment in such agencies as the U.S. Naval Observatory, Department of Agriculture, Coast Survey, Mint, Assay Office, Smithsonian Institution, and in various geological and natural history surveys. In several instances, an individual was employed in the service of a foreign government.

13. Includes employment in a variety of positions, including actuary, teacher of science (noncollege), editor, astronomer, surveyor, inventor, instrument maker, engineer, miner, patent agent, natural history and geology explorer and collector, curator, horticulturist, gardener, pharmacist, chemist, preparer of almanacs, public lecturer, manufacturing chemist, and so on.

had the bachelor's degree and more than half had the doctorate.[8]

Table 2 shows a similar trend, with the development of special institutions and occupational categories for the employment of scientists. Over a period of two or more centuries, American society slowly developed an occupational support for the scientist, which helped to institutionalize further the role of science in American life.

The importance of occupations that were unrelated to science (Table 2, first five categories) declined, from a high of about 43 percent in the first period to less than 25 percent in the last period. The point should not be missed, however, that even in the colonial and early republican period, occupations that *were* science-related (including medicine) accounted for more than half of the employment total. This finding, of course, tends to dilute somewhat the idea that science had no special occupational base during the earlier period.

Over the years, non-science occupations and medicine were of decreasing importance for American scientists. By the third period, the largest employment category was that of professor of science. By 1910, nearly three-quarters of American scientists were employed as teachers (presumably on the college or university level).[9]

If one considers the importance of the colleges in preparing students of science and in employing them after graduation, it becomes obvious that science in America increasingly became a part of the academic establishment. During the years under study, it is apparent (from Table 2) that government at various levels also played an increasingly significant role in the promotion and institutionalization of the role of the scientist in society. By the time of Cattell's study in 1910, however, the percentage of persons employed in government had declined rapidly, as academic science came more and more to predominate.

Table 2 shows that a number of nongovernmental science and science-related occupations also were available. This general category was larger during the first period than during the last, although some growth is seen between periods two and three. In general, however, during the years under review in Table 2, the areas of most dynamic growth and the greatest sources of support for science were the academic institutions and the government.

It is significant that during these years (and including the first decade of this century, as studied by Cattell), American science had little to do with industry. The first effective relation of science with industry was brought about in the second half of the nineteenth century by scientifically inclined industrial engineers, who borrowed from the original research work done by scientists within the universities. The direct involvement of scientists in industry came about chiefly during the period after 1900, with the establishment of industrial (and government) research laboratories.[10]

One recent study has noted the post-World War II decline of the academic scientist and ascendancy of the scientist in industry and government.[11] To the extent that this is true, American science is undergoing a significant rearrangement of the social pattern of support that developed during the three centuries preceding World War II. If looked at from a very broad perspective, the pattern of movement appears to have been from an alliance with learned culture in the colonial and early republican period, through the era when science was based chiefly in the academic institutions and the government surveys during the later nineteenth and earlier twentieth centuries, to a developing alliance of science with industry and government during recent decades. Of course, the fields of chemistry and physics are especially related to industry. Other sciences have had a closer relation with different economic or social concerns. For example, outside of their connections with academic institutions and government surveys, the biological sciences have been oriented particularly toward agricultural and medical interests. Fields such as meteorology and astronomy have, at various times, had more or less significant relations with navigation and mercantile interests. It is apparent, therefore, that the social relations of science cannot always be viewed as a single phenomenon, but must also be considered from the special perspective of different disciplines.

Table 3 outlines the changing preferences for scientific fields among American scientists during the years before this century. While no dramatic shifts are apparent, throughout these years there was a trend to reduce the predominant interest in the life sciences and to increase the interest in the earth sciences. Such a shift (or leveling process) undoubtedly was due largely to the promotion of geological investigations by federal and state governments during the final three-quarters of the nineteenth century.

The relative position of the physical sciences throughout these years reemphasizes the lack of an effective relationship between science and industry before 1900. If the fields of mathematics, physical sciences, and astronomy are combined, it becomes apparent that there was relatively less interest in these subjects during the middle and last periods than during the first. However, there was some improvement between periods two and three. By the time Cattell produced the first edition of *American Men of Science* in 1906, the position of the physical sciences had improved considerably, chiefly at the expense of the geological sciences.[12] This may be interpreted as an indication of the growing sophistication of laboratory and mathematical science in America and as a prelude to the closer union of science with technology and industry. It also signifies the passing of the time when American science was chiefly concerned with the study of the American continent itself.

The foregoing analysis represents an abstraction of the social environment within which the characters in this *Dictionary* played out their professional lives. It was indeed a giant leap from John Winthrop, Jr.—son of a colonial governor, himself governor of Connecticut, student of alchemy, manufacturer of chemical products, renowned for his medical knowledge, and viewer of the heavens—to that late-nineteenth-century man of science, professor of physics Henry Augustus Rowland. And yet, of course, Winthrop's own relative in the eighteenth century, John Winthrop of Harvard College, was fully as much a "professional" scientist in his day as was Rowland in his. The great difference is that Winthrop of Harvard was the exception for his time, whereas Rowland (considered in terms of his professional connections) increasingly was becoming the rule for science at the outset of the present century.

Until perhaps the last quarter of the nineteenth century, there was no easy road to a scientific career in America. During the period before about 1825, it often was more likely that one would "drift" into a scientific position, rather than actually prepare for and pursue such employment. This is explained largely in terms of the lack of any specific incentive to devote one's life to science, since there was little formal training available for such a career, and there were few jobs to fill even if one was so motivated or prepared. Men such as Benjamin Silliman, Sr., and John Farrar taught in the sciences at Yale and Harvard more because of circumstances than of any design on their part. But once the course of their own lives had been set, they then could promote the study of science in the colleges. Out of this process came a generation of scientifically inclined and liberally educated persons who would deliberately seek out an institutional setting within which their own inclinations could be realized. The actors in this continuing drama of disappointment, conflict, accommodation, and success are to be found in this *Dictionary*. Further detailed studies of the individual and col-

Table 3. FIELDS OF SCIENCE

	BORN BEFORE 1776[2]		BORN 1776 TO 1815[3]		BORN 1816 TO 1867[4]	
Field[1]	*Number of Individuals*[5]	*Percentage of Fractional Distribution*[6]	*Number of Individuals*[5]	*Percentage of Fractional Distribution*[6]	*Number of Individuals*[5]	*Percentage of Fractional Distribution*[6]
Mathematics[7]	20	13.4	21	6.3	23	6.4
Physical sciences[8]	26	16.0	53	14.7	58	19.2
Astronomy and related sciences[9]	25	14.8	41	11.1	36	11.7
Earth sciences[10]	10	5.8	65	20.6	67	23.3
Life sciences[11]	51	43.0	111	40.2	86	31.4
Applied science[12]	13	6.0	21	5.6	19	5.0
Miscellaneous (non-subject)[13]	2	1.0	6	1.5	10	3.0

1. An individual may be counted in as many as four scientific fields, although the vast majority appear in no more than three.
2. There were a total of 98 persons born in this period.
3. There were a total of 240 persons born in this period.
4. There were a total of 243 persons born in this period.
5. Since an individual could be counted in as many as four categories, the total of this column is greater than the number of individuals being studied.
6. For purposes of this column, each individual was assigned to the appropriate categories as a fraction of one, based on the total number of categories in which that person was counted. For example, an individual counted in three categories was assigned as one-third to each. The total for this column, therefore, was the same as the total number of individuals being studied.
7. Includes statistics.
8. Includes natural philosophy, physics, chemistry, metallurgy.
9. Includes astronomy, meteorology, geodesy, climatology, geophysics, astrophysics.
10. Includes geology, mineralogy, paleontology, oceanography, topographical engineering, geography, hydrography, cartography.
11. Includes natural history, zoology, botany, medicine, anatomy, physiology and physiological chemistry, phrenology, psychology.
12. Includes military and civil engineering, agriculture, agricultural chemistry, invention, instrumentation, surveying, exploration and travels.
13. Includes administration, patronage and promotion of science, textbook writer, science education, science and religion, philosophy of science.

lective lives of these pioneers of American science will help to advance our understanding of the complex relations of science to American society.

NOTES

1. According to one study, science and technology operated in essentially separate social channels throughout much of the period covered by this *Dictionary*. See Edwin Layton, "Mirror-Image Twins: The Communities of Science and Technology," in *Nineteenth-Century American Science: A Reappraisal*, ed. George H. Daniels (Evanston, Ill: Northwestern University Press, 1972), pp. 210-30.

2. The term "scientist" will be used in discussing the attributes of the population in the *Dictionary*, because it is both convenient and economical to do so. It should be noted, however, that the word is not entirely accurate, since a number of persons in this study cannot rightly be called scientists in the modern sense of that term. The word is used here only because of the awkwardness of alternatives such as "man of science," "contributor to scientific knowledge," or other essentially descriptive phrases.

3. Much of what is said here relates chiefly to the careers of men only. While numerous obstacles existed for anyone who sought to follow a scientific career, a number of additional discouragements were faced by women, with the result that very few women were involved in the active pursuit of science in America before the current century. See Sally Gregory Kohlstedt, "In from the Periphery: American Women in Science, 1830 to 1880," *Signs: Journal of Women in Culture and Society* 4 (Autumn 1978): 81-96 and Margaret W. Rossiter, "Women Scientists in America before 1920," *American Scientist* 62 (May-June 1974): 312-23.

4. See Alexandra Oleson and Sanborn C. Brown, eds., *The Pursuit of Knowledge in the Early American Republic: American Learned Societies from Colonial Times to the Civil War* (Baltimore: Johns Hopkins University Press, 1976). In John Voss's foreword, he states that one of the chief motivations for the establishment of the early academies was the idea that the educated man can, and should, have access to all areas of knowledge (p. vii). In her introduction, Oleson points out that in the eighteenth century the distinction between pure and applied knowledge did not exist and that the American Philosophical Society, the American Academy of Arts and Sciences, and other early societies were very much interested in practical knowledge (p. xv). These themes recur throughout the work.

5. The earlier involvement of the educated gentleman in the study of science persisted well into the nineteenth century, in the person of the amateur. Significantly, however, by the time of the Civil War the role and function of such persons were defined largely in terms of the ever increasing number of full-time, "professional" scientists. See Sally Gregory Kohlstedt, "The Nineteenth-Century Amateur Tradition: The Case of the Boston Society of Natural History," in *Science and Its Public*, ed. G. Holton and W. A. Blanpied (Dordrecht, Netherlands: D. Reidel Publishing Co., 1976), pp. 173-90. Nathan Reingold reviews the relations of science to "polite learning" in his discussion and definition of the scientific "cultivators," a term that he prefers to the more traditional "amateur"; "Definitions and Speculations: The Professionalization of Science in America in the Nineteenth Century," in Oleson and Brown, eds., *Pursuit of Knowledge*, pp. 38-46. In the national debate over the disposition of the bequest of James Smithson in the 1840s, Howard S. Miller sees evidence that American intellectuals were coming to be divided into two cultural camps, literary versus scientific; *Dollars for Research: Science and Its Patrons in Nineteenth-Century America* (Seattle: University of Washington Press, 1970), p. 15. American inventors frequently came from social classes quite different from those of scientists and were engaged in occupations equally distinct; Clark A. Elliott, "The American Scientist in Antebellum Society: A Quantitative View," *Social Studies of Science* 5 (January 1975): 101, 106.

Also see David F. Noble, *America by Design: Science, Technology, and the Rise of Corporate Capitalism* (New York: Alfred A. Knopf, 1977). Noble sees manufacturing and science as largely unrelated until about the mid-nineteenth century, except insofar as certain upper-class gentlemen had "feet in both worlds." In the later nineteenth century, the increasing union of science with commercial interests came to be represented, not by the scientist, but by the engineer (p. 33). Aside from the concerns of industry, mid-nineteenth-century America was characterized by enthusiastic (if not entirely successful) efforts to apply the science of chemistry to agriculture. These efforts became increasingly important toward the end of the century; Margaret W. Rossiter, *The Emergence of Agricultural Science: Justus Liebig and the Americans, 1840-1880* (New Haven: Yale University Press, 1975). Also see Layton, "Mirror-Image Twins." Arthur P. Molella and Nathan Reingold have wisely asked for a reevaluation of the traditional view that science had little or no effect on technological advance during the early decades of the nineteenth century; "Theorists and Ingenious Mechanics: Joseph Henry Defines Science," *Science Studies* 3 (October 1973): 323-51. Such a closer examination of the historical record is to be encouraged, although it is unlikely to reverse the general conclusion that an effective and continuing relationship between science and technology came about only during the later years of the century.

6. The movement toward a separate and distinct status for scientists was part of an international trend that exhibited local variations depending on conditions in different countries. The beginnings of professionalization in science came first in France after the revolution, as part of the general reform of French society. Professionalization of science came somewhat later in Great Britain than it did in France; Maurice Crosland, "The Development of a Professional Career in Science in France," in *The Emergence of Science in Western Europe*, ed. Crosland (London: Macmillan & Co., 1975), pp. 139-59. Steven Shapin and Arnold Thackray observe that in Great Britain men of science were more involved in general culture, whereas in France and Prussia the governments helped establish distinctive scientific roles at an earlier date; "Prosopography as a Research Tool in History of Science: The British Scientific Community, 1700-1900," *History of Science* 12 (1974) : 4.

7. Nathan Reingold has proposed a useful three-part classification of American scientists of the nineteenth century. He uses the terms "cultivator," "practitioner," and "researcher." By the end of the century, the cultivator (that large group of persons who participated in and supported science, but who generally were not employed as scientists and who usually made few, if any, important contributions to scientific knowledge) had greatly declined, while the group of practitioners and researchers had grown significantly. See "Definitions and Speculations," pp. 33-69. Also see George H. Daniels, "The Process of Professionalization in American Science: The Emergent Period, 1820-1860," *Isis* 58 (Summer 1967) : 160-66.

8. J. McKeen Cattell, "A Further Statistical Study of American Men of Science," *Science*, n.s. 32 (1910) : 643.

9. Ibid., pp. 675-76.

10. Noble, *America by Design*, pp. 110-11; John J. Beer and W. David Lewis, "Aspects of the Professionalization of Science," in *The Professions in America*, ed. Kenneth S. Lynn (Boston: Houghton Mifflin, 1965), pp. 110-30.

11. Roger C. Krohn, *The Social Shaping of Science: Institutions, Ideology, and Careers in Science* (Westport, Conn.: Greenwood Press, 1971), pp. 6, 18.

12. J. McKeen Cattell, "A Statistical Study of American Men of Science," *Science*, n.s. 24 (1906) : 659-60. Cattell's figures yield the following percentages: mathematics, astronomy, physics, and chemistry, 47 percent (chemistry and physics alone, 34 percent) ; life sciences, 41 percent; and geology, 11 percent.

Biographies

A

ABBE, CLEVELAND (1838-1916). *Meteorology; Astronomy.* Meteorologist, U.S. Weather Bureau.

MSS: Lib. Cong. (*NUCMC* 64-1556); Cincinnati Observatory (*NUCMC* 75-1935); also see *DSB.* WORKS ABOUT: See *AMS*; *DAB*; *DSB.*

ABBOT, HENRY LARCOM (1831-1927). *Engineering; Physics.* Officer, U.S. Army Corps of Engineers.

MSS: Oregon Historical Society (*NUCMC* 72-1582); Harv. U.-HL. WORKS ABOUT: See *AMS*; *DAB.*

ABBOT, JOHN (May 31/June 1, 1751, London, England—late 1840/early 1841, Bulloch County, Ga.). *Entomology; Ornithology.* Son of James, wealthy attorney, and Ann Abbot. Married Penelope Warren, before 1779; apparently only one child. EDUCATION: Studied law with father. CAREER: 1773, with support from Dru Drury and other entomologists and from Roy. Soc., which had appointed him "to make researches and collections," emigrated to Virginia; shortly thereafter, settled in Georgia and lived there for remainder of life; in America, collected numerous specimens and prepared colored drawings of arthropods and birds that were sent to British and European workers; supplemented income by farming and teaching, but died in poverty in rural Georgia at an uncertain date. SCIENTIFIC CONTRIBUTIONS: An accomplished scientific illustrator whose earliest surviving watercolors date from 1767. While published nothing on his own, his specimens, observations, and drawings were used in various publications by other persons [see Works]; collected scientific books and was acquainted with taxonomy, but wrote no scientific descriptions. His notes (few of which have been published) show that he was first American naturalist to make really extensive observations of life histories and habits of native insects and arachnids and he gave considerable attention to lesser-known orders; to a somewhat lesser extent, also a pioneer in American ornithology since he observed and painted birds not described until some years later.

WORKS: *The Natural History of the Rarer Lepidopterous Insects of Georgia* (London, 1797), edited by James Edward Smith, includes Abbot's illustrations, his notes on them, and Smith's descriptions of new species, based on Abbot's drawings; J. A. Boisduval and J. E. LeConte [q.v.], *Histoire Générale et Iconographie des Lépidoptères et des Chenilles de l'Amérique Septentrionale* (Paris, 1833), utilized Abbot's illustrations. MSS: Harv. U.-HL and Harv. U.-MCZ; Linnean Society of London; British Museum (Natural History). WORKS ABOUT: This sketch is taken from a biographical outline provided by Ronald S. Wilkinson, who is preparing a life of Abbot that will appear in his *Studies in the History of Entomology.* Also see Abbot's "Notes on My Life," edited by C. L. Remington, in *Lepidopterists' News* 2 (1948): 28-30; Anna Bassett, "Some Georgia Records of John Abbot, Naturalist," *Auk* 55 (1938): 244-54; Elsa G. Allen, "The History of American Ornithology Before Audubon," *Transactions of Am. Phil. Soc.*, n.s. 41 (1951): 543-49.

ABBOTT, CHARLES CONRAD (1843-1919). *Natural history,* especially *Zoology; Archaeology.* Lived near Trenton, N.J.; studied natural history and Indian archaeology of eastern U.S., especially Delaware Valley.

MSS: Ac. Nat. Scis. Philad. (*NUCMC* 66-2); Princeton Univ. Library (Hamer). WORKS ABOUT: See *AMS*; *DAB.*

ABERT, JOHN JAMES (September 17, 1788, probably Shepherdstown, Va.—January 27, 1863, Washington, D.C.). *Topographical engineering.* Son of John and Margarita (Meng) Abert. Married Ellen Matlack Stretch, 1812; six children. EDUCATION: 1808-1811, attended U.S. Military Academy, West Point; ca. 1811-1813, studied law; and 1813, admitted to District of Columbia bar. CAREER: Ca. 1811-1813, assistant to chief clerk in War Office at Washington, D.C.; 1813-1814, law practice in Washington, D.C., and Ohio; 1814, volunteer in District of Columbia militia; 1814, appointed major in Topographical Engineers and was attached to northern division army, serving 1814-1829 on various topographical surveys, including work on East River (N.Y.), Chesapeake and Narragansett bays, Chesapeake and Ohio canal, and Louisville (Ky.) canal, and acting 1816-1818 as assistant to F. R. Hassler [q.v.] on Coast Survey; 1824, brevetted lieutenant colonel; 1829-1861, chief of Topographical Bureau, which in 1831 became separate branch of War Department; 1838, bureau was organized as Corps of Topographical Engineers and that year Abert became colonel in corps; 1832-1834, special agent and commissioner for Indian affairs, charged with relocation of Indians to the West. MEMBERSHIPS: A founder of National Institution for Promotion of Science; member of Geographical Society of Paris. SCIENTIFIC CONTRIBUTIONS: Importance in science closely bound to role of Topographical Engineers at time when corps and its officers played an especially significant part in western explorations; as chief of corps, had difficult task of promotion and guidance of surveys and enforcement of high standards for description of landscape; also assumed responsibility for ensuring presentation of coherent geographical record in textual and cartographic output by various surveys.

WORKS: Writings published mainly as government documents; of particular importance were annual reports to secretary of war, 1831-1861, published with secretary's reports. MSS: See references in *DAB* to official records (now in Natl. Arch.). WORKS ABOUT: *DAB*, supp. 1, pp. 2-3 (Herman R. Friis); William H. Holcombe, "Col. John James Abert," *Professional Memoirs of U.S. Army Corps of Engineers* 7 (March-April 1915): 204-5.

ADAMS, CHARLES BAKER (January 11, 1814, Dorchester, Mass.—January 18, 1853, St. Thomas, V.I.). *Natural history*; *Conchology*. Son of Charles J. Adams, businessman. Married Mary Holmes, 1839; survived by five children. EDUCATION: Attended Yale College; 1834, A.B., Amherst College; 1834-1836, attended Andover Theological Seminary. CAREER: 1836, assistant on Natural History (Geological) Survey of New York state; 1836-1837, tutor and lecturer in geology, Amherst College; 1837, taught without salary in new college at Marion, Mo.; 1838-1847, professor of chemistry and natural history, Middlebury College; 1845, appointed Vermont state geologist; 1843-1844 and 1848-1849 (winters), in Jamaica; 1847-1853, professor of natural history and astronomy, Amherst College; 1849-1850, in Panama; 1852-1853, in St. Thomas. SCIENTIFIC CONTRIBUTIONS: Early work was in geology, and as Vermont state geologist issued four annual reports (1845-1848). Remembered chiefly for work on tropical conchology, an interest that began with first visit to Jamaica in winter of 1843-1844 and continued in later collecting trips to Jamaica, Panama, and St. Thomas. Essentially a descriptive naturalist, but in study of tropical mollusks worked on questions associated with geographical distribution of species; at Amherst compiled zoological cabinet especially strong in conchology.

WORKS: Two chief works were *Contributions to Conchology* (New York, 1849-1852) and *Catalogue of Shells Collected at Panama, with Notes on Their Synonymy, Station and Geographical Distribution* (New York, 1852). Also Vermont geological reports, and, with Alonzo Gray, *Elements of Geology* (New York, 1853). See additional works in *Roy. Soc. Cat.*, vols. 1, 6 (additions), and Meisel, *Bibliog. Amer. Nat. Hist.* MSS: Ac. Nat. Scis. Philad. (Hamer); small collection in Amherst College Archives. WORKS ABOUT: *DAB*, 1: 39-40 (Henry A. Pilsbry); Thomas Bland, "Memoir," *American Journal of Conchology* 1 (1865): 191-204; William S. Tyler, *History of Amherst College* . . . (Springfield, Mass., 1873), pp. 366-68.

ADRAIN, ROBERT (September 30, 1775, Carrickfergus, Ireland—August 10, 1843, New Brunswick, N.J.). *Mathematics*. Of Huguenot ancestry, at least on paternal side. Married Ann Pollock; seven children. EDUCATION: Received adequate education according to parents' financial circumstances, and at age fifteen was on his own. HONORS: M.A., Queen's (Rutgers) College, 1810. CAREER: In Ireland, opened his own school and later was private tutor; 1798, took part in rebellion, and subsequently escaped to America; master at Princeton Academy, and then principal of York County (Pa.) Academy; 1805-1809, principal of academy at Reading, Pa.; 1809-1813, professor of mathematics, Queen's College (New Brunswick, N.J.); 1813-1826, professor at Columbia Univ. and there at various times taught mathematics, astronomy, natural philosophy, and natural history; 1826-1827, returned to Queen's College; 1827, bought farm at New Brunswick, N.J.; 1827-1834, professor at Univ. of Pennsylvania, and in 1828 became vice-provost (while at least part of family remained on New Brunswick farm); 1836-1840, taught at grammar school at Columbia College; 1840-1843, apparently lived on farm. MEMBERSHIPS: Am. Phil. Soc.; Am. Ac. Arts Scis. SCIENTIFIC CONTRIBUTIONS: Shares with N. Bowditch [q.v.] distinction of being America's first creative or original mathematician; contributed to nearly every branch of mathematics, but best-known and most important work was on normal or exponential law of errors [see Works], a law soon thereafter also demonstrated by Gauss and generally bearing latter's name. Author of two articles—on figure of the earth and on mean diameter of the earth (*Transactions of Am. Phil. Soc.*, 1818)—that showed influence of Laplace. Established the *Analyst, or Mathematical Companion*, published 1808 and briefly revived 1814; later established the *Mathematical Diary* (1825-1833) and edited its first six issues.

WORKS: Early papers published in George Baron's *Mathematical Correspondent*, including one on diophantine algebra that ran over fifty pages; "Research Concerning the Probabilities of the Errors Which Appear in Making Observations," *Analyst* 1 (1808): 93-109; also edited American editions of several mathematical works. See bibliography in *DSB*. MSS: A few letters in Am. Phil. Soc. (*DSB*). WORKS ABOUT: *DAB*, 1: 109-10 (David E. Smith); *DSB*, 1: 65-66 (D. J. Struik); J. L. Coolidge, "Robert Adrain and the Beginnings of American Mathematics," *American Mathematical Monthly* 33 (1926): 61-76.

AGASSIZ, ALEXANDER (1835-1910). *Zoology; Oceanography.* Curator and director, Harvard Univ. Museum; investor in coal and copper mines.

MSS: Harv. U.-MCZ (*DSB*). WORKS ABOUT: See *AMS*; *DAB*; *DSB*.

AGASSIZ, JEAN LOUIS RODOLPHE [Louis] (May 28, 1807, Motier-en-Vuly, Switzerland—December 14, 1873, Cambridge, Mass.). *Ichthyology*; *Paleontology*; *Geology*. Son of Rodolphe, Protestant clergyman, and Rose (Mayor) Agassiz. Married Cecile Braun, 1832; Elizabeth Cabot Cary, 1850; three children by first marriage, including Alexander Agassiz [q.v.]. EDUCATION: Studied at College of Lausanne and universities of Zurich and Heidelberg; 1829, Ph.D., Univ. of Erlangen; 1830, M.D., Univ. of Munich. HONORS: Wollaston Medal, Geological Society of London; Copley Medal, Roy. Soc. CAREER: 1830-1832, occasional medical practice, study in Paris; 1832-1846, professor of natural history, College of Neuchâtel (Switzerland); 1846, Lowell Institute lectures (Boston); 1847-1873, professor of natural history, Lawrence Scientific School, Harvard; 1848, conducted geological and biological survey of Lake Superior region; 1865-1866, engaged in explorations in Brazil; 1871, went to California on Coast Survey ship; 1873, helped establish Anderson School of Natural History, Penikese Island. MEMBERSHIPS: Natl. Ac. Scis. and other societies; member, Board of Regents, Smithson. Instn. SCIENTIFIC CONTRIBUTIONS: Prepared under major European scientists, produced important contributions to knowledge of fishes, especially fossil forms, and in glacial studies introduced concept of Ice Age. In U.S. became a focal point for public and professional development of science; trained many outstanding zoologists and geologists of next generation; carried out geological and zoological explorations in Lake Superior region and in Brazil; founded Museum of Comparative Zoology at Harvard. Persistent opponent of evolutionary ideas, including the Darwinian theory.

WORKS: Major works include *Selecta Genera et Species Piscium Quas in Itinere per Brasiliam 1817-1820* . . . (Munich, 1829); *Recherches sur les Poissons Fossiles*, 5 vols. (Neuchâtel, 1833-1844); *Etude sur les Glaciers*, 2 vols. (Neuchâtel, 1840); *Contributions to the Natural History of the United States*, 4 vols. (Boston, 1857-1862), especially vol. 1, pt. 1, *Essay on Classification* (1857); *Lake Superior* (Boston, 1850); "Evolution and Permanence of Type," *Atlantic Monthly* 33 (January 1874): 94-101. Bibliography of works in Jules Marcou, *Life, Letters, and Works of Louis Agassiz* (New York, 1896). MSS: Harv. U.-MCZ; Harv. U.-HL. WORKS ABOUT: *DAB*, 1: 114-22 (David S. Jordan and Jessie K. Jordan); *DSB*, 1: 72-74 (Edward Lurie). Also see Lurie, *Louis Agassiz: A Life in Science* (Chicago, 1960).

AGASSIZ, LOUIS. See AGASSIZ, JEAN LOUIS RODOLPHE.

ALEXANDER, JOHN HENRY (June 26, 1812, Annapolis, Md.—March 2, 1867, Baltimore, Md.). *Applied science*; *Mathematics*. Son of William, merchant, and Mary (Stockett) Alexander. Married Margaret Hammer, 1836; survived by six children. EDUCATION: Ca. 1826 [age fourteen], A.B., St. John's College (Annapolis); studied law for four years. HONORS: LL.D., St. James College. CAREER: Worked on railway survey; 1833, commissioned to conduct preliminary study for topographical survey in connection with a geological survey of Maryland; 1834-1841, state topographical engineer of Maryland, serving 1837-1841 without salary; 1836-1845, a founder and president, George's Creek Coal and Iron Co.; 1845-1867, unofficial adviser to Coast Survey under A. D. Bache [q.v.]; 1857, sent by U.S. government to England on unsuccessful mission to promote unification of coinage; at various times professor of mining and civil engineering at Univ. of Pennsylvania, and professor of natural history at St. James College; at time of death, professor of natural philosophy at Univ. of Maryland; died before plans were completed to appoint him director of U.S. Mint. MEMBERSHIPS: Am. Phil. Soc., Natl. Ac. Scis. SCIENTIFIC CONTRIBUTIONS: Particularly interested in applied science involving mathematics and the physical sciences. As Maryland state topographical engineer, issued annual reports during years 1833-1840, but voluntarily postponed work on accurate survey until U.S. Coast Survey had reached Maryland; in 1841, when accurate survey finally became possible, the state terminated its support. Greatly interested in matter of standardization of weights and measures [see Works]; prepared reports for U.S. Lighthouse Board, and reportedly on occasion was paid as much as $1,000 for scientific opinions. Also an accomplished poet and linguist.

WORKS: *Report on the Standards of Weight and Measure for the State of Maryland* ([Baltimore, 1845]); *A Universal Dictionary of Weights and Measures, Ancient and Modern* (Baltimore, 1850); prepared edition, with extensive additional notes, of F. W. Simms's *Treatise on the Principal Mathematical Instruments Employed in Surveying, Leveling, and Astronomy* (Baltimore, 1836), with 2d ed. (Baltimore, 1848); *Contributions to a History of the Metallurgy of Iron* (Baltimore, 1840, 1842). Full bibliography of works, including number of journal articles, in Hilgard [below], and *Roy. Soc. Cat.*, vol. 1. MSS: Md. Hist. Soc. (*NUCMC* 67-1271). WORKS ABOUT: *DAB*, 1: 168-69 (Joseph S. Ames); J. E. Hilgard, "Memoir," *Biog. Mems. of Natl. Ac. Scis.* 1 (1877): 213-26.

ALEXANDER, STEPHEN (September 1, 1806, Schenectady, N.Y.—June 25, 1883, Princeton, N.J.). *Astronomy.* Son of Alexander Stephenson, successful merchant, and Maria (O'Connor) Alexander; cousin and brother-in-law of Joseph Henry [q.v.]. Married Louisa Meads, 1826; Caroline Forman, 1850; three children by first marriage, and two by second. EDUCATION: 1824, A.B., Union College; 1832-ca. 1833, attended Princeton Theological Seminary. HONORS: LL.D., Columbia Univ., 1852. CAREER: 1824-1830, teacher, including several years at academy at Chittenango, N.Y.; ca. 1830-1832, associated with J. Henry at Albany Academy; at Princeton University, tutor in mathematics (1833-1834), adjunct professor in mathematics (1834-1840), professor of astronomy (with several changes of title) (1840-1877). MEMBERSHIPS: Natl. Ac. Scis. and other societies; president, Am. Assoc. Advt. Sci., 1859. SCIENTIFIC CONTRIBUTIONS: As early as 1830, made observations of star occultations and solar eclipse; in 1834, observed total eclipse in Georgia with new three-and-one-half-inch telescope; made observations of annular eclipses of sun in 1831, 1838, 1854, 1865, and 1875, and during career observed four transits of Mercury as well as 1882 transit of Venus; 1860, led expedition to observe solar eclipse in Labrador, results of which were published in annual report of Coast Survey. Laplace's nebular hypothesis was of particular interest, and this and other philosophical concerns were presented in "Certain Harmonies" [see Works]; interest in nebular hypothesis also instigated drive for large telescope facility at Princeton, which arrived too late for his personal benefit.

WORKS: "Certain Harmonies of the Solar System," *Smithson. Contr. Knowl.* 21 (1875); series of articles entitled "On the Origin of the Forms and the Present Condition of Some of the Clusters of Stars and Several of the Nebulae," in Gould's *Astronomical Journal* 2 (March 13-July 10, 1852); "Fundamental Principles of Mathematics," *Am. Journ. Sci.* 2d ser. 7 (1849): 178-87, 329-43. See works listed in *Roy. Soc. Cat.*, vols. 1, 6 (additions), 9, 13; and in Poggendorff, *Biographisch-literarisches Handwörterbuch*, vol. 3. MSS: Joseph Henry Papers at Smithson. Instn.; miscellaneous items in Princeton Univ. Library—Rare Books and Special Collections. WORKS ABOUT: *DAB*, 1: 174-75 (Frederick E. Brasch); Charles A. Young, "Memoir," *Biog. Mems. of Natl. Ac. Scis.* 2 (1886): 249-59.

ALGER, FRANCIS (March 8, 1807, Bridgewater, Mass.—November 27, 1863, Washington, D.C.). *Mineralogy.* Son of Cyrus, wealthy inventor and iron manufacturer, and Lucy (Willis) Alger. Married and had children. EDUCATION: Attended common schools. HONORS: A.M., Harvard Univ., 1849. CAREER: Engaged in business, probably with father, and took over iron-manufacturing and ordnance business after father's death in 1856. MEMBERSHIPS: Am. Ac. Arts Scis.; Am. Assoc. Advt. Sci.; and other societies. SCIENTIFIC CONTRIBUTIONS: Father a well-known metallurgist, and Francis's interest in mineralogy began as early as 1824; went with father to Nova Scotia in 1826, brought back first collection of minerals from that region to appear in U.S., and subsequently published list [see Works]; 1827 and 1829, returned to Nova Scotia with C. T. Jackson [q.v.] to pursue mineralogical investigations; also conducted number of other mineralogical explorations in New Hampshire and elsewhere; wrote on zinc mines in Sussex County, N.J., and on crystallized gold from California, and gathered important cabinet of minerals. All mineralogical labors were carried out during hours not devoted to normal business pursuits. Also interested in mechanical arts, and at time of death was in Washington to test improved shrapnel that he had invented.

WORKS: List of Nova Scotia minerals in *Am. Journ. Sci.* 12 (1827): 227-32; results of Nova Scotia mineralogical investigations with C. T. Jackson in ibid. 14 (1828): 305-30, in ibid. 15 (1828): 132-60, 201-17, and in *Memoirs of Am. Ac. Arts Scis.* n.s. 1 (1833): 217-330; Alger's Phillips's *Mineralogy* (Boston, 1844) more than doubled size of original work, and had at his disposal the papers of William

Phillips and P. Cleaveland [q.v.] for a planned further edition of this work. See also *Roy. Soc. Cat.*, vols. 1, 7; Meisel, *Bibliog. Amer. Nat. Hist.*; and Jackson [below]. WORKS ABOUT: *Proceedings of Am. Ac. Arts Scis.* 6 (1862-1865): 294-96; C. T. Jackson, in *Proceedings of Bos. Soc. Nat. Hist.* 10 (1865): 2-6; obituary, in *Am. Journ. Sci.*, 2d ser. 38 (1864): 449; also see account of father, in *DAB*, 1:177-78.

ALLEN, HARRISON (April 17, 1841, Philadelphia, Pa.—November 14, 1897, Philadelphia, Pa.). *Anatomy.* Son of Samuel and Elizabeth Justice (Thomas) Allen. Married Julia A. Colton, 1869; survived by two children. EDUCATION: Graduated, Philadelphia Central High School; studied dentistry with Dr. Josiah Foster Flagg; 1861, M.D., Univ. of Pennsylvania. CAREER: Ca. 1861, resident physician, Philadelphia Hospital; 1862-1865, assistant surgeon, U.S. Army, resigning with rank of brevet major; 1865, entered medical practice, Philadelphia, specializing in diseases of nose and throat; at Univ. of Pennsylvania, professor of zoology and comparative anatomy in auxiliary department of medicine (1865-1878), professor of physiology in medical department (1878-1885), professor of comparative anatomy and zoology (1891-1896); 1866-1878, professor of anatomy and surgery, Pennsylvania Dental College; at various times on surgical staffs of Philadelphia, St. Joseph's, and Wills Eye hospitals. MEMBERSHIPS: Ac. Nat. Scis. Philad. and other societies; president, Association of American Anatomists. SCIENTIFIC CONTRIBUTIONS: Chief scientific work done in comparative anatomy; especially noted for study of bats, his first paper on that subject appearing in 1861; also did extensive work on craniology, comparing skulls from Florida mounds, Hawaiian Islands, and other locations. Published papers on joints, muscles, and locomotion and dissected Siamese twins. An early American leader in laryngology.

WORKS: "Monograph of the Bats of North America," *Smithson. Misc. Coll.* 7 (1864; reprinted in 1893); *Outlines of Comparative Anatomy and Medical Zoology* (Philadelphia, 1869; 2d ed., 1877); *Analysis of the Life Form in Art* (Philadelphia, 1875); *System of Human Anatomy, Including Its Medical and Surgical Relations* (Philadelphia, 1884). In addition to number of scientific papers, also published about fifty items on medicine and surgery. See bibliography cited below. MSS: Am. Phil. Soc. (*NUCMC* 68-1459). WORKS ABOUT: *DAB*, 1:191 (Francis R. Packard); [B. G. Wilder], memoir, *Proceedings of Association of American Anatomists*, 10th Annual Session, December 1897, pp. 12-26, with bibliography of works on pp. 19-26.

ALLEN, JOEL ASAPH (1838-1921). *Zoology.* Curator of mammalogy and ornithology, Am. Mus. Nat. Hist. (New York).

MSS: Ac. Nat. Scis. Philad. (*NUCMC* 66-3); Am. Mus. Nat. Hist. (Hamer). WORKS ABOUT: See *AMS*; *DAB*.

ALLEN, OSCAR DANA (February 24, 1836, Hebron, Maine—February 19, 1913, near Ashford, Wash.). *Chemistry.* Son of Alpheus and Hannah (Seabury) Allen. Married Fidelia Totman, 1861; three children. EDUCATION: 1861, Ph.B., Sheffield Scientific School, Yale, and for three years thereafter continued as student at Sheffield; 1871, Ph.D., Yale Univ. CAREER: At Sheffield Scientific School, assistant in chemistry (1860-1864), instructor of analytical chemistry (1870-1871), professor of metallurgy (1871-1874), professor of metallurgy and analytical chemistry (1874-1887); left Yale because of poor health; ca. 1887-ca. 1891, resided in California; ca. 1891, moved to Washington state and established residence at base of Mt. Rainier. MEMBERSHIPS: Said to have been member of number of scientific societies. SCIENTIFIC CONTRIBUTIONS: Work related especially to investigations of cesium and rubidium; at Yale, collaborated with Prof. S. W. Johnson [q.v.] in application of spectral analysis of these elements and helped to establish more accurate determination of atomic weight of cesium. In 1881, prepared American edition of Fresenius's work on quantitative analysis, including update of its nomenclature. Accumulated significant collection of minerals. In retirement devoted some efforts to study of botany in vicinity of Mt. Rainier.

WORKS: "Observations on Caesium and Rubidium," *Am. Journ. Sci.* 2d ser. 34 (1862): 367; "On the Equivalent and Spectrum of Caesium," ibid. 2d ser. 35 (1863): 94-98, with S. W. Johnson; "Chemical Constitution of Hatchettolite and Samarskite from Mitchell County, N.C.," ibid. 3d ser. 14 (1877): 128-31; "Bastuäsite and Tysonite from Colorado," ibid. 3d ser. 19 (1880): 390-93, with W. J. Comstock. MSS: Small collection in Yale Univ. Library—Manuscripts and Archives Division. WORKS ABOUT: *Appleton's CAB*, vol. 1 (1887): *Obituary Record of Graduates of Yale University* (1910-1915); 491-92.

ALLEN, TIMOTHY FIELD (April 24, 1837, Westminster, Vt.—December 5, 1902, New York, N.Y.). *Botany.* Son of Dr. David and Eliza (Graves) Allen. Married Julia Bissell, 1861. EDUCATION: 1858, A.B., Amherst College; 1861, M.D., Univ. of City of New York (New York Univ.). CAREER: 1861, began medical practice, Brooklyn; served as acting assistant surgeon in U.S. Army, then returned to Brooklyn medical practice; 1863-1902, medical practice in New York City; at New York Homeopathic Medical College, professor of materia medica and therapeutics (1867-1902) and dean (1882-1893); 1884, became surgeon in New York Ophthalmic Hospital, and later president of hospital's board of trustees. MEMBERSHIPS: A founder (1871) and vice-president of Torrey Bot. Club. SCIENTIFIC CONTRIBUTIONS: Active botanical interest began after move to New York City in 1863, where met J. Torrey [q.v.]; particularly interested in stoneworts (Characeae), freshwater plants previously neglected by American botanists; published number of papers on these plants and in 1891 donated important collection of specimens to N.Y. Bot. Gard. Chief publication was *Encyclopedia* [see Works]; an important contributor to Torrey Bot. Club's *Flora of New York City* (1870-1876). Also made significant contributions to practice and literature of homeopathic medicine.

WORKS: *Encyclopedia of Pure Materia Medica,* 10 vols. (New York and Philadelphia, 1874-1879); *Ophthalmic Therapeutics* (New York, 1876). Bibliography of botanical writings in memoir by N. L. Britton in *Bulletin of Torrey Bot. Club* 30 (1903): 173-77; see also *Roy. Soc. Cat.*, vols. 9, 12, 13. WORKS ABOUT: *DAB,* 1: 207-8 (Hubert L. Clark); Harry B. Humphrey, *Makers of North American Botany* (New York, 1961), pp. 3-4 [based on *DAB* account].

ALTER, DAVID (December 3, 1807, Allegheny Township, Westmoreland County, Pa.—September 18, 1881, Freeport, Pa.). *Physics.* Son of John, millwright, and Eleanor (Sheetz) Alter. Married Laura Rowley, 1832; Elizabeth A. Rowley, 1844; eleven children by two marriages. EDUCATION: 1831, M.D., Reformed Medical College in New York City. CAREER: 1831, began medical practice in Elderton, Pa., and later moved to Freeport, Pa.; in addition to medical practice carried on series of physical experiments and for a time was engaged in production of coal oil. SCIENTIFIC CONTRIBUTIONS: Interest in electricity began quite early, and physical interests and experiments continued throughout life. 1837, invented electric motor; designed and made electric clock, model for electric locomotive, electrical telegraph and also devised method for purifying bromine; commercial interest in making coal oil from coal thwarted by successful drilling of oil well in Pennsylvania. Lacked advantages of interaction with other scientists, either in person or through substantial library. Made most of own scientific equipment, and among numerous skills was early mastery of techniques of daguerreotype photography. Chiefly remembered for early work on spectrum analysis, publishing two papers [see Works] that presented discovery that individual chemical elements have characteristic spectrum; at about the same time European physicist Angstrom also discovered and published on same phenomenon, but several years elapsed before appearance of work of Kirchhoff and Bunsen, which developed and extended Alter and Angstrom discoveries.

WORKS: "On Certain Physical Properties of Light, Produced by the Combustion of Different Metals in the Electric Spark, Refracted by a Prism," *Am. Journ. Sci.* 2d ser. 18 (1854): 55-57; "On Certain Physical Properties of the Light of the Electric Spark, Within Certain Gases, as Seen Through a Prism," ibid. 2d ser. 19 (1855): 213-14. Both papers are reproduced in Watson [below]. WORKS ABOUT: *DAB,* 1: 230 (Dinsmore Alter); William A. Hamor, "David Alter and the Discovery of Spectrochemical Analysis," *Isis* 22 (1934-1935): 507-10; E. C. Watson, "David Alter on the Light of the Electric Spark," *American Journal of Physics* 25 (1957): 630-32.

ANTHONY, JOHN GOULD (May 17, 1804, Providence, R.I.—October 16, 1877, Cambridge, Mass.). *Conchology.* Son of Joseph and Mary (Gould) Anthony. Married Ann W. Rhodes, 1832; nine children. EDUCATION: At age twelve, left school and went to work. CAREER: 1834, moved to Cincinnati and there engaged in business, including work with silver-plate manufacturers, as independent accountant, and later as partner in bookselling and publishing firm; 1847, contracted scarlet fever, which injured eyesight, and for seven years thereafter was unable to work, but during this period studied and sometimes sold shells; 1853, made walking tour of Kentucky, Tennessee, and Georgia; later, entered partnership as insurance agent and continued this work during last years in Cincinnati; 1863-1877, in charge of mollusks, Museum of Comparative Zoology, Harvard. MEMBERSHIPS: Officer, Western Academy of Natural Sciences (Cincinnati). SCIENTIFIC CONTRIBUTIONS: Early interest in natural history later strengthened through acquaintance with J. P. Kirtland [q.v.] and other Ohio scientists; pursued scientific studies, especially of freshwater shells, while engaged in business, and in 1835 began wide correspondence and exchange with students of mollusks in the East and in Europe. Published occasional pamphlets and made contributions to scientific journals; primarily a descriptive naturalist, and while not a greatly productive author, scientific correspondence contributed to important published works by other persons. Scientific studies attracted attention of J. L. R. Agassiz [q.v.] and in 1863 Anthony took charge of conchological collection in Agassiz's museum at Harvard; 1865, member of scientific corps on Agassiz's Brazilian expedition; later years spent arranging and classifying collections in Museum of Comparative Zoology, to which his own collection was added.

WORKS: Results of some investigations appeared

in British work of L. Reeve, *Conchologia Iconica* 20 vols. (London, 1843-1878). Bibliography in Turner [below], pp. 96-97. MSS: Ac. Nat. Scis. Philad. (Hamer); Harv. U.-MCZ. WORKS ABOUT: *DAB*, 1: 317-18 (Henry A. Pilsbry); Ruth D. Turner, "John Gould Anthony, with a Bibliography and Catalogue of His Species," in Museum of Comparative Zoology (Harvard), Department of Mollusks, *Occasional Papers on Mollusks* 1 (1946): 81-108.

ANTHONY, WILLIAM ARNOLD (1835-1908). *Physics*; *Electrical engineering*. Professor of physics, Cornell Univ.; professor of physics and electrical engineering, School of Science, Cooper Union.

WORKS ABOUT: See *AMS*; *DAB*, supp. 1.

ARTHUR, JOSEPH CHARLES (1850-1942). *Botany*. Professor of vegetable physiology and pathology, Purdue Univ.; botanist, Indiana Experiment Station.

WORKS ABOUT: *See AMS*; *DAB*, supp. 3.

ASHBURNER, CHARLES ALBERT (February 9, 1854, Philadelphia, Pa.—December 24, 1889, Pittsburgh, Pa.). *Geology*. Son of Algernon Eyre and Sarah (Blakiston) Ashburner. Married Roberta M. John, 1881; two children. EDUCATION: 1874, B.S. in civil engineering, Towne Scientific School, Univ. of Pennsylvania. HONORS: D.Sci., Univ. of Pennsylvania, 1889. CAREER: 1872, as undergraduate, participated in survey of raft channel of Delaware River; 1874, worked briefly on lighthouse survey; 1874, under influence of J. P. Lesley [q.v.], state geologist, joined Second Geological Survey of Pennsylvania; during 1874-1880, was assigned to work on surveys of fossil iron-ore belt in Pennsylvania's Juniata Valley, Aughwick Valley, and East Broad Top coal region and of oil district comprising McKean, Elk, Cameron, and Forest counties; 1880-1886, in charge of survey of anthracite field, and after 1885 was first assistant geologist and supervisor of all field and office work; 1886, resigned from Pennsylvania survey; 1886-1889, scientific expert for a Westinghouse company in Pittsburgh; 1888-1889, general manager of New York and Montana Mining and Milling Co.; 1889, vice-president and general manager of Duquesne Mining and Reduction Co.; during these later years also served as private consultant in locating oil and gas deposits. MEMBERSHIPS: Am. Phil. Soc.; Am. Inst. Min. Engrs.; and other societies. SCIENTIFIC CONTRIBUTIONS: Main scientific activity carried out in connection with Second Geological Survey of Pennsylvania, contributing both scientific knowledge and organizational abilities; in work for survey, and in employment thereafter, efforts were directed largely toward economic geology. Became particularly expert in matters related to location of oil deposits; extended the use of contour mapping to indicate underground structure; and prepared more than twenty of Pennsylvania survey's reports, including those on coal, oil, and natural gas.

WORKS: Bibliography in Winslow [below]; also, *Roy. Soc. Cat.*, vols. 9, 13. WORKS ABOUT: *DAB*, 1: 383-84 (George P. Merrill); Arthur Winslow, "Charles Ashburner," *American Geologist* 6 (1890): 69-78.

ASHMEAD, WILLIAM HARRIS (1855-1908). *Entomology*. Assistant curator, U.S. Natl. Mus.

MSS: Ac. Nat. Scis. Philad. (Hamer). WORKS ABOUT: See *AMS*; *DAB*.

ATKINSON, GEORGE FRANCIS (1854-1918). *Botany*. Professor of botany and botanist to Experiment Station, Cornell Univ.

MSS: Corn. U.-CRH (*NUCMC* 62-2338). WORKS ABOUT: See *AMS*; *DAB*.

ATWATER, CALEB (December 25, 1778, North Adams, Mass.—March 13, 1867, Circleville, Ohio). *Natural history*; *Archaeology*. Son of Ebenezer, carpenter, and Rachel (Parker) Atwater. Married Diana Lawrence; Belinda Butler, 1811; nine children. EDUCATION: 1804, M.A., Williams College; later studied theology and law in New York City. CAREER: 1804-1815, in New York City, where in succession he ran school for girls, was minister, was lawyer, and engaged in unsuccessful business venture; 1815-1867, resided in Circleville, Ohio, where early years were devoted to law practice; 1821, elected to Ohio legislature and became leader in promotion of internal improvements and public education; 1829, sent by President Jackson to Wisconsin as member of commission to settle Indian land dispute. MEMBERSHIPS: N.Y. Lyc. Nat. Hist. and other societies. SCIENTIFIC CONTRIBUTIONS: Interested in great range of subjects, natural history of West being only one; published several articles on western natural history, topography, and climate, a few of which appeared in *Am. Journ. Sci.*, and also reported similar observations in *Remarks Made on a Tour* [see Works]. *History of Ohio* [see Works], best-known publication, was twenty years in preparation, and devoted about one-quarter of its content to description of physical and natural history features of Ohio; section on geology emphasized economic aspects of subject, and those parts which dealt with animals and plants included a plea for forestry conservation, argued from economic point of view.

WORKS: Investigation of ancient earthworks and Indian antiquities in Ohio in *Archaeologia Americana* [transactions of Am. Ant. Soc.] 1 (1820); *Remarks Made on a Tour to Prairie du Chien, thence to Washington City, in 1829* (Columbus, Ohio, 1831); these two works, plus additional material, also appeared in *The Writings of Caleb Atwater* (Columbus,

Ohio, 1833); *History of the State of Ohio, Natural and Civil* (Cincinnati, 1838), 403 pp. See additional works in *Roy. Soc. Cat.* vol. 1, and Meisel, *Bibliog. Amer. Nat. Hist.* MSS: Det. Pub. Lib-BHC (*NUCMC* 68-1212); Am. Ant. Soc. (*NUCMC* 75-1882). WORKS ABOUT: *DAB*, 1: 415-16 (W. J. Ghent); Clement L. Martzolff, "Caleb Atwater," *Ohio Archaeological and Historical Quarterly* 14 (1904): 247-71.

ATWATER, WILBUR OLIN (1844-1907). *Physiological chemistry.* Professor of physiological chemistry, Wesleyan Univ.

MSS: Wesleyan Univ. Archives (*DSB*). WORKS ABOUT: See *AMS*; *DAB*; *DSB.*

AUDUBON, JOHN JAMES (April 26, 1785, Les Cayes, Santo Domingo—January 27, 1851, New York, N.Y.). *Ornithology.* Son of Jean Audubon, sea captain and planter, and Mlle. Rabin, a Creole of Santo Domingo. Married Lucy Bakewell, 1808; four children. EDUCATION: Little formal academic education; attended military school for one year. CAREER: 1803-1805, lived on family estate near Philadelphia; 1805, returned to France; 1806, operated lead mine near Philadelphia; 1807, worked in countinghouse in New York; 1808-1819, ran country store in Louisville and Henderson, Ky., and then started grist and lumber mill business, which failed; 1819-1826, worked as itinerant portrait painter, taxidermist in Cincinnati, tutor and drawing teacher in New Orleans and St. Francisville, La.; 1820, decided to publish bird drawings and spent years immediately thereafter in further exploration and preparation; 1826-1839, in Scotland and England, where he painted, worked with publisher, and secured subscribers for *Birds of America*; during same period made several return trips to U.S., conducting field work in the Middle Atlantic states (1829), southeast to the Florida Keys (1831-1832), Labrador (1833), and Galveston, Texas (1837); 1839, settled in New York; 1841, built house on the Hudson. MEMBERSHIPS: Royal Societies of London and of Edinburgh; Am. Ac. Arts Scis. SCIENTIFIC CONTRIBUTIONS: Without advantages of formal training in either science or painting, succeeded at both; above all an artist and naturalist, and as such relied on assistance of others in preparation of formal scientific aspects of work, such as identification of species and Latin names. Greatest work was the folio *Birds of America,* based on years of exploration and observation; as artist, considered superior to A. Wilson [q.v.], but latter produced a more sound and original scientific publication.

WORKS: *Birds of America,* 4 vols. (Edinburgh, 1827-1838), including 435 aquatint plates; text to accompany *Birds* published as *Ornithological Biography,* 5 vols. (Edinburgh, 1831-1839); *Viviparous Quadrupeds of North America,* 3 vols. of plates (New York, 1845-1848) and 3 vols. of text (New York, 1846-1854), prepared in collaboration with J. Bachman [q.v.]; *Synopsis of the Birds of North America* (Edinburgh, 1839). Also see works in *Roy. Soc. Cat.*, vol. 1, and Meisel, *Bibliog. Amer. Nat. Hist.* MSS: various locations; see *NUCMC* and Hamer. WORKS ABOUT: *DAB*, 1: 423-27 (Donald C. Peattie and Eleanor R. Dobson); *DSB*, 1: 329-32 (Robert M. Menzel). Also see Alice E. Ford, *John James Audubon* (Norman, Okla., 1964).

AUSTEN, PETER TOWNSEND (1852-1907). *Chemistry.* Professor of chemistry, Rutgers Univ.; consultant and inventor, especially on dyeing and bleaching processes.

WORKS ABOUT: See *AMS*; *DAB.*

AYRES, WILLIAM ORVILLE (September 11, 1817, New Canaan, Conn.—April 30, 1887, Brooklyn, N.Y.). *Zoology,* especially *Ichthyology.* Son of Jared and Dinah (Benedict) Ayres. Married Maria J. Hildreth, 1847; two children. EDUCATION: 1837, A.B., Yale College; 1854, M.D., Yale. CAREER: 1837-1847, successively, teacher in Berlin, Conn.; Miller's Place, Long Island, N.Y.; East Hartford, Conn.; and Sag Harbor, Long Island, N.Y.; 1847-1852, teacher in Boston; 1854-ca. 1870, engaged in medical practice in San Francisco, and served as professor of theory and practice of medicine and after 1862 as dean of faculty at Tolland Medical College; ca. 1870-ca. 1874, experienced financial reverses and lived for a time in Chicago and in Easthampton, Mass.; ca. 1874-1887, medical practice at New Haven, Conn.; 1879-1887, lecturer on diseases of nervous system in Yale Medical School. MEMBERSHIPS: Bos. Soc. Nat. Hist.; Calif. Ac. Scis. SCIENTIFIC CONTRIBUTIONS: Interest in natural history began at early age, and through friendship with J. J. Audubon [q.v.] his name was given to woodpecker described in Audubon's *Birds of America.* Real interest was in fishes, and during nearly twenty years in California produced number of papers on fishes of that state, particularly during decade 1854-1864, and was an authority on ichthyology of Pacific Coast; occasionally published papers on other classes of animals. All scientific activity carried on in addition to medical practice.

WORKS: Nearly all scientific contributions appeared in proceedings of Bos. Soc. Nat. Hist. and of Calif. Ac. Scis. See bibliographies in *Roy. Soc. Cat.*, vols. 1, 6 (additions), 7, 9, 13; and Bashford Dean, *Bibliography of Fishes* (New York, 1916), 1: 45-46. WORKS ABOUT: Obituaries: W. W. Hawkes, *Proceedings of Connecticut State Medical Society,* n.s. 3 (1887): 174-76; *Obituary Record of Graduates of Yale University* (1880-1890): 370-71; *Science* 10 (July 8, 1887): 24.

B

BABCOCK, JAMES FRANCIS (February 23, 1844, Boston, Mass.—July 19, 1897, Dorchester, Mass.). *Chemistry*; *Invention*. Son of Archibald D. and Fannie (Richards) Babcock. Married Mary P. Crosby, 1869; Marion B. Alden, 1892; five children by first marriage. EDUCATION: 1860, graduated from Boston English High School; 1860-1862, attended Lawrence Scientific School, Harvard. HONORS: Honorary member, Massachusetts College of Pharmacy, 1866. CAREER: Ca. 1862, established private chemical laboratory and office, Boston; 1869-1874, professor of medical chemistry, Massachusetts College of Pharmacy; 1874-1880, professor of chemistry, medical department, Boston Univ.; 1875-1885, Massachusetts state assayer; 1885-1889, head of milk inspection department, city of Boston. SCIENTIFIC CONTRIBUTIONS: Frequently consulted as expert witness in chemical cases, food investigations, and patent suits; particularly noted for work on alcoholic beverages, which was begun in late 1870s and continued during his tenure as state assayer of liquors (1875-1885), and for efforts as Boston milk inspector, in which office he applied effective new methods for detection of coloring matter and other adulterations in dairy products. As state assayer, won legislative endorsement for "three-per-cent limit" as definition of intoxicating liquors. Popular scientific lecturer, inventor of so-called Babcock fire extinguisher, he also patented process for clarifying and bleaching fats and fatty oils and produced procedure for compacting coal dust into usable fuel. Also interested in medicines and drugs and in 1894 was president of Boston Druggists' Association.

WORKS: Publications were mainly official reports on pure foods and sanitation; only other known publications were "Blood Stains," in Allan M. Hamilton, *A System of Legal Medicine* (New York, 1894), and brochure, *Laboratory Talks on Infant Foods* (Boston, 1896). WORKS ABOUT: *DAB*, 1: 457-58 (Hardee Chambliss); *NCAB*, 10 (1909 [copyright 1900]): 445-46; *American Druggist and Pharmaceutical Record* 31 (1897): 86.

BABCOCK, STEPHEN MOULTON (1843-1931). *Agricultural chemistry*. Professor of agricultural chemistry and chemist to Agricultural Experiment Station, Univ. of Wisconsin.

MSS: St. Hist. Soc. Wis. (*NUCMC* 62-2236). WORKS ABOUT: See *AMS*; *DAB*, supp. 1.

BACHE, ALEXANDER DALLAS (July 19, 1806, Philadelphia, Pa.—February 17, 1867, Newport, R.I.). *Physics*, especially *Geophysics*. Son of Richard, postmaster, and Sophia (Dallas) Bache; great-grandson of Benjamin Franklin [q.v.]. Married Nancy Clarke Fowler, 1828; apparently no children. EDUCATION: 1825, graduated from U.S. Military Academy, West Point. HONORS: Recipient of many honors in both U.S. and Europe. CAREER: 1826-1828, with Corps of Engineers at Newport, R.I.; 1828-1836, professor of natural philosophy and chemistry, Univ. of Pennsylvania; 1836, appointed first president, Girard College; 1836-1838, studied education in Europe; 1838-1842, during further delay in opening Girard College, worked at reorganization of public schools in Philadelphia; 1842-1843, resumed professorship at Univ. of Pennsylvania; 1843-1867, superintendent, U.S. Coast Survey, and during this period also member of Lighthouse Board, superintendent of weights and measures, vice-president of U.S. Sanitary Commission, and adviser to President Lincoln during Civil War. MEMBERSHIPS: Roy. Soc. and other societies; first president, Natl. Ac. Scis. SCIENTIFIC CONTRIBUTIONS: While at Univ. of Pennsylvania had charge of research at Franklin Institute, including investigation of steam boiler explosions for federal government. Had general interest in chemistry, physics, and astronomy, but primary research interest was in geophysics; as early as 1830, began study of terrestrial magnetism and in 1840, largely through personal efforts, had available at Girard College the first magnetic observatory in U.S. Scientific and administrative abilities built Coast Survey into firmly established scientific agency, extended its work to

Pacific Coast, helped make astronomical, tidal, and magnetic investigations an integral part of survey's work, and promoted application of telegraphy to longitude determination; while at Coast Survey, continued own investigations on tides, terrestrial magnetism, Gulf Stream, and related subjects. A major force in establishing a self-conscious American scientific community.

WORKS: See bibliography in *Biog. Mems. of Natl. Ac. Scis.* 1 (1877) : 205-12d; see also *Annual Reports of U.S. Coast Survey* (1844-1866). MSS: Natl. Arch. (Coast Survey records), and see additional references in *DSB*. WORKS ABOUT: *DAB*, 1: 461-62 (Louis A. Bauer) ; *DSB*, 1: 363-65 (Nathan Reingold). Also see Merle M. Odgers, *Alexander Dallas Bache, Scientist and Educator* (Philadelphia, 1947).

BACHE, FRANKLIN (October 25, 1792, Philadelphia, Pa.—March 19, 1864, Philadelphia, Pa.). *Chemistry*. Son of Benjamin Franklin, journalist, and Margaret (Markoe) Bache; great-grandson of Benjamin Franklin [q.v.]. Married Aglae Dabadie, 1818; six children. EDUCATION: A.B. 1810 and M.D. in 1814, Univ. of Pennsylvania. CAREER: In U.S. Army, assistant surgeon (1813) and surgeon (1814-1816) ; 1816, began medical practice, Philadelphia; 1824-1836, physician to Walnut Street State Prison; 1829-1836, physician to Eastern Penitentiary; 1826-1832, professor of chemistry, Franklin Institute; 1831-1841, professor of chemistry, Philadelphia College of Pharmacy and Science; 1841-1864, professor of chemistry, Jefferson Medical College (Philadelphia). MEMBERSHIPS: Ac. Nat. Scis. Philad.; president, Am. Phil. Soc., 1853-1855. SCIENTIFIC CONTRIBUTIONS: While engaged in medical practice, interest in chemistry consumed large portion of attention and directly or indirectly contributed to economic well-being. In publication of 1811 [see Works] argued that hydrochloric acid is formed by union of simple substance, chlorine, with hydrogen; became early American advocate of then new concept of chemical equivalents; beginning in 1830, served as important member of committee that prepared various revised editions of U.S. *Pharmacopoeia* and served as chairman for 1863 edition. Beginning in 1833, *U.S. Dispensatory* consumed great deal of time and effort for many years; with George B. Wood, prepared eleven editions of *Dispensatory*, a work that was intended to promote knowledge and use of *Pharmacopoeia*; Bache had responsibility for mineral and chemical aspects. With R. Hare [q.v.], edited *Dictionary of Chemistry* of Andrew Ure (1821) ; also editor of several other chemical works.

WORKS: Article on composition of hydrochloric acid in Philadelphia *Aurora* (1811), a journal founded by father; three articles in *Memoirs of Columbian Chemical Society* (1813) ; *System of Chemistry for the Use of Students in Medicine* (Philadelphia, 1819). MSS: Hist. Soc. Penn. (*NUCMC* 60-2022) ; Am. Phil. Soc. (Hamer). WORKS ABOUT: *DAB*, 1: 463-64 (Edgar F. Smith) ; George B. Wood, *Biographical Memoir of Franklin Bache, M.D., Prepared at the Request of the College of Physicians of Philadelphia* (Philadelphia, 1865).

BACHMAN, JOHN (February 4, 1790, Rhinebeck, N.Y.—February 24, 1874, Columbia, S.C.). *Natural history*, especially *Zoology*. Son of Jacob Bachman, successful farmer. Married Harriet Martin, 1816; Maria Martin, 1848; nine children who lived through infancy, by first marriage. EDUCATION: Reportedly attended Williams College for a time; studied theology with private tutors near Philadelphia; 1813, licensed to preach. HONORS: Ph.D., Univ. of Berlin, 1838; LL.D., Univ. of South Carolina. CAREER: As a boy served as secretary to Johannes Knickerbocker during U.S. exploring expedition and visit to Oneida Indians; taught school in Frankfort, Ellwood, and Philadelphia, Pa.; 1813-ca. 1815, minister to three churches near Rhinebeck, N.Y.; 1815-1874, minister, St. John's Church, Charleston, S.C. MEMBERSHIPS: Corresponding member of Zoological Society of London; member of Ac. Nat. Scis. Philad. and numerous other societies. SCIENTIFIC CONTRIBUTIONS: Interest in natural history began in youth, and at early age provided A. Wilson [q.v.] with specimens of birds from northern New York. In 1831 began close friendship with J. J. Audubon [q.v.], for whom he collected specimens and data on southern birds, and was credited with 134 references in Audubon's *Ornithological Biography;* in 1836 began work with Audubon on quadrupeds [see Works], Bachman having primary responsibility for text and Audubon and sons for drawings as well as participation in collection of specimens and data. Also had active secondary interest in botany and took part in contemporary debate on questions of race.

WORKS: With Audubon, *The Viviparous Quadrupeds of North America*, 6 vols. of plates and text (New York and London, 1845-1854) ; *A Catalogue of the Phaenogamous Plants and Ferns Native and Naturalized, Found Growing in the Vicinity of Charleston, S.C.* (Charleston, 1834) 15 pp.; *The Doctrine of the Human Race, Examined on the Principles of Science* (Charleston, 1850), intended as reply to G. R. Gliddon and J. C. Nott's *Types of Mankind*. See also journal articles in *Roy. Soc. Cat.*, vol. 1, and in Meisel, *Bibliog. Amer. Nat. Hist.* MSS: Charleston Museum (Hamer). WORKS ABOUT: *DAB*, 1: 466-67 (Donald C. Peattie) ; Claude H. Neuffer, ed. *The Christopher Happoldt Journal: His*

European Tour With John Bachman (June-December 1838), Contributions from Charleston Museum 13 (Charleston, 1960), with biography of Bachman on pp. 29-118.

BAILEY, EDGAR HENRY SUMMERFIELD (1848-1933). *Chemistry.* Professor and director of chemical laboratory, Univ. of Kansas.

WORKS ABOUT: See *AMS*; *World Who's Who in Science.*

BAILEY, JACOB WHITMAN (April 29, 1811, Ward [now Auburn], Mass.—February 27, 1857, West Point, N.Y.). *Botany.* Son of Rev. Isaac and Jane (Whitman) Bailey. Married Maria Slaughter, 1835; two children. EDUCATION: At age twelve went to work in circulating library and bookstore in Providence, R.I.; 1832, graduated, U.S. Military Academy, West Point. CAREER: 1833, as second lieutenant in artillery, served as commissary quartermaster at Bellona Arsenal near Richmond, Va.; 1834-1838, assistant professor of chemistry, West Point; 1838-1857, professor of chemistry, mineralogy, geology at West Point. MEMBERSHIPS: Many scientific societies; president-elect of Am. Assoc. Advt. Sci. at time of death. SCIENTIFIC CONTRIBUTIONS: Began collection of shells and insects in teens and at early age procured microscope; most noted as microscopist, in use of which he was an American pioneer, investigating especially so-called infusoria and minute Algae; began studies of Diatomaceae about 1839 and was first person in U.S. to observe and report on diatoms in fossil state. Also made study of deep-sea soundings gathered by U.S. Coast Survey and Wilkes Expedition. Was first American to detect presence of plant structures in anthracite coal ash [see Works]; also contributed to improvement of mechanism of microscope and corresponded with number of important European scientists. Throughout career primary interests were botanical, while professorial duties extended scientific work into other areas and microscopical investigations, especially of fossil forms, contributed to geological knowledge.

WORKS: "On the Crystals Which Occur Spontaneously Formed in the Tissues of Plants," *Am. Journ. Sci.* 45 (1843) : 149-51; "On the Detection of Spirally Dotted or Scalariform Ducts, and Other Vegetable Tissues in Anthracite Coal," ibid. 2d ser. 1 (1846) : 407-10. See bibliography in *Roy. Soc. Cat.*, vols. 1, 6 (additions), and in Meisel, *Bibliog. Amer. Nat. Hist.* Bailey's publications numbered more than fifty, most of which appeared in *Am. Journ. Sci.* MSS: U.S. Mil. Ac. (*NUCMC* 70-1316) ; Bos. Soc. Nat. Hist.—Boston Museum of Science (Hamer) ; N.Y. Pub. Lib.; N.Y. St. Lib. WORKS ABOUT: *DAB*, 1: 498 (Stanley Coulter) ; Harry B. Humphrey, *Makers of North American Botany* (New York, 1961), pp. 9-10.

BAILEY, SOLON IRVING (1854-1931). *Astronomy.* Professor of astronomy, Harvard Univ.

WORKS ABOUT: See *AMS*; *DAB*, supp. 1; *DSB.*

BAIRD, SPENCER FULLERTON (February 3, 1823, Reading, Pa.—August 19, 1887, Woods Hole, Mass.). *Zoology*; *Science administration.* Son of Samuel, successful lawyer, and Lydia M. (Biddle) Baird. Married Mary Helen Churchill, 1846; one child. EDUCATION: B.A. in 1840 and M.A. in 1843, Dickinson College. CAREER: At Dickinson College, professor of natural history (1846-1848) and professor of natural history and chemistry (1848-1850) ; at Smithson. Instn., assistant secretary and "Keeper of the Cabinet" (1850-1878) and secretary (1878-1887) ; 1871-1887, head of U.S. Commission on Fish and Fisheries, without salary. MEMBERSHIPS: Natl. Ac. Scis. and other societies; first permanent secretary of Am. Assoc. Advt. Sci. SCIENTIFIC CONTRIBUTIONS: Teaching at Dickinson College included innovative field studies in zoology and botany; at Smithson. Instn., administrative talents and scientific connections secured cooperation of many persons and agencies who sent natural history specimens to Washington. Personal scientific work, mainly on birds and mammals, was descriptive in nature, but techniques and their detailed accuracy set new and higher standards; article on "Distribution . . . of Birds" [see Works] was one of few theoretical efforts and a contribution to evolutionary thought. Administrative and political abilities were particularly evident during last years when, among other duties, devoted great personal effort to carrying out practical and scientific work of Commission on Fish, especially at Woods Hole, Mass.

WORKS: Publications on mammals and birds issued as vols. 8 and 9 of *Reports of Explorations and Surveys to Ascertain the . . . Route for a Railroad from the Mississippi River to the Pacific Ocean* (Washington, D.C., 1857, 1858) ; "Distribution and Migration of North American Birds," published in abstract in *Am. Journ. Sci.*, 2d ser. 41 (1866) : 78-90, 184-92, 337-47; with T. M. Brewer and R. Ridgway [qq.v.], prepared *A History of North American Birds*, 3 vols. (Boston, 1874) ; G. B. Goode, "The Published Writings of Spencer F. Baird, 1843-1882," *Bulletin of U.S. Natl. Mus.* 20 (1883) ; published over 1,000 items, mainly review articles and government reports. MSS: Smithson. Instn. Archives (*NUCMC* 72-1229) ; Natl. Arch. (records of Commission on Fish) (*DSB*). WORKS ABOUT: *DAB*, 1: 513-15 (David S. Jordan and Jessie K. Jordan) ; *DSB*, 1: 404-6 (Dean C. Allard). Also see William H. Dall, *Spencer Fullerton Baird: A Biography* (Philadelphia, 1915).

BAKER, MARCUS (September 23, 1849, Kalamazoo, Mich.—December 12, 1903, Washington, D.C.). *Geography; Mathematics.* Son of John and Chastina (Fobes) Baker. Married Sarah Eldred, 1874; Marion Una Strong, 1899; survived by two children. EDUCATION: Attended Kalamazoo College; 1870, A.B., Univ. of Michigan; 1896, LL.B. [honorary?], Columbian Univ. (George Washington Univ.). CAREER: 1870-1871, taught mathematics, Albion College; 1871-1873, instructor in mathematics, Univ. of Michigan; 1873-1886, connected with U.S. Coast and Geodetic Survey, including work on survey of Alaskan waters, and after 1882 had charge of Magnetic Observatory at Los Angeles; 1886, joined U.S. Geol. Surv. and directed topographic work of northeastern division, edited survey's topographic maps, and engaged in other research; prepared report for U.S. government on Venezuelan boundary question, and was employed by Venezuela as expert; member, secretary, and editor, Board of Geographic Names; near end of life, became assistant secretary of Carnegie Institution. MEMBERSHIPS: A founder and member of Board of Managers, Natl. Geog. Soc.; secretary, editor, and president of Phil. Soc. Wash. SCIENTIFIC CONTRIBUTIONS: Initial scientific work with W. H. Dall [q.v.] on survey of Alaskan waters included participation in gathering of data and preparing reports, with contributions to *Coast Pilot of Alaska* and to bibliography on Alaskan geography. Later wrote on survey of northwestern boundary of U.S. (1900) and on northwest boundary of Texas (1902). While working for U.S. Geol. Surv., prepared Alaska dictionary [see Works].

WORKS: *Geographic Dictionary of Alaska*, U.S. Geol. Surv. Bulletin no. 187 (Washington, D.C., 1902). See works listed in *Natl. Un. Cat. Pre-1956 Imp.*, and *Roy. Soc. Cat.*, vols. 9, 13. WORKS ABOUT: *DAB*, 1: 525 (Albert P. Bridgman); *World Who's Who in Science*; William H. Dall, biographical sketch, *National Geographic Magazine* 15 (1904): 40-43.

BALDWIN, WILLIAM (March 29, 1779, Newlin, Pa.—August 31, 1819, Franklin, Mo.). *Botany.* Son of Thomas, Quaker minister, and Elizabeth (Garretson) Baldwin. Married Hannah M. Webster, ca. 1808; survived by four children. EDUCATION: 1807, M.D., Univ. of Pennsylvania. CAREER: Worked as teacher; 1805-1806, surgeon on merchant ship to Canton, China; 1807-1811, medical practice, Wilmington, Del.; 1811, moved to Georgia; 1812-1818, naval surgeon stationed at St. Marys and Savannah, Ga.; and 1817-1818, surgeon on voyage of frigate *Congress* to South America; 1818, returned to Wilmington, Del.; 1819, botanist with Maj. Stephen H. Long expedition to Rocky Mountains and died en route. SCIENTIFIC CONTRIBUTIONS: Serious interest in botany stimulated in his early twenties; started collecting plants and after move to Georgia in 1811 began botanical explorations of western part of that state and of East Florida and spent some months with Creek Indians; continued botanical explorations during journey to South America in 1817-1818; after return to Delaware in 1818, planned work on botany of Georgia and East Florida, with an account of botanical observations in Rio de la Plata, and was particularly interested in southern Cyperaceae, but botanical writing was interrupted by appointment to Long expedition. While publishing very little, his botanical study is preserved in correspondence with Rev. G. H. E. Mühlenberg, Dr. W. Darlington [qq.v.], and others, found in Darlington's *Reliquiae Baldwinianae* (1843; reprinted in 1969, with introduction, chronology, and indexes added by Joseph Ewan); also contributed to work of other botanists, and especially to S. Elliott [q.v.], *A Sketch of the Botany of South Carolina and Georgia* (1816-1824); to A. Gray [q.v.], monograph on Rhynchospora, *Annals of N.Y. Lyc. Nat. Hist.* 3 (1828-1836): 191-220; and to J. Torrey [q.v.], monograph on Cyperaceae, ibid. 3: 239-448.

WORKS: See above; also "Account of Two North American Species of Rottbaellia Discovered on the Seacoast of Georgia," *Am. Journ. Sci.* 1 (1819): 355-59; and "Account of Two North American Species of Cyperus, from Georgia, and of four species of Kyllingia, from the Brazilian Coast and from the Rio de la Plata," *Transactions of Am. Phil. Soc.*, n.s. 2 (1825): 167-71. See also Meisel, *Bibliog. Amer. Nat. Hist.* MSS: Harv. U.-GH (Hamer). WORKS ABOUT: *DAB*, 1: 547-48 (Henry R. Viets); Harry B. Humphrey, *Makers of North American Botany* (New York, 1961), pp. 15-16.

BANISTER, JOHN (1650, Twigworth, Gloucestershire, England—May 1692, on Roanoke River, Va.). *Botany; Entomology; Malacology; Anthropology.* Son of John Banister, "pleb." Married, before 1688; one son. EDUCATION: B.A. in 1671 and M.A. in 1674, Magdalen College, Oxford. CAREER: After graduation, served as clerk (and possibly librarian) and, 1676-1678, as chaplain, Magdalen College; 1678, settled in Virginia as Anglican minister (in 1680, no longer listed as minister in Bristol Parish register); natural history was a dominating interest of Banister from the outset; 1690, acquired 1,735 acres near Appomattox River; a founder and trustee of William and Mary College. SCIENTIFIC CONTRIBUTIONS: From first arrival in colonies had ambition to prepare natural history of Virginia and was well acquainted with New World flora at Oxford, but early death prevented completion of projected work. From Virginia, corresponded with John Ray, Bishop Compton, Hans Sloane, Jacob Bobart the

Younger, Martin Lister, and other scientists and sent specimens and drawings; contributed to Linnaeus's understanding of American flora, Lister's iconography of mollusks, and James Petiver's catalogue of insects. After accidental death while on botanizing expedition, originals of catalogues and some drawings, dried plants, seeds, and shells were sent to England and distributed to various persons.

WORKS: Published nothing directly. "Extracts of four letters from Mr. John Banister to Dr. Lister," *Phil. Trans. of Roy. Soc.* 17 (1693): 667-72; James Petiver, "Herbarium Virginianum Banisteri," *Monthly Miscellany or Memoirs for the Curious,* Decad no. 7 (December 1707); Petiver, "Some observations concerning insects made by Mr. John Banister in Virginia, A.D. 1680 . . . ," *Phil. Trans. of Roy. Soc.* 22 (1701): 807-14; catalogue of plants published in John Ray's *Historia Plantarum* and Petiver's *Memoirs; Observations on the Natural Productions of Jamaica; curiosities of Virginia.* MSS: British Library, Bodleian Library at Oxford, and Lambeth Palace; reproduced in Joseph Ewan and Nesta Ewan, *John Banister and His Natural History of Virginia, 1678-1692* (Urbana, Ill., 1970). WORKS ABOUT: *DSB,* 1: 431-32 (Joseph Ewan); *DAB* 1: 575-76 (Armistead C. Gordon, Jr.).

BANNEKER, BENJAMIN (November 9, 1731, Baltimore County, Md.—October 9, 1806, Baltimore County, Md.). *Astronomy*; *Mathematics.* Son of Robert, farmer and former slave, and his wife Mary Banneky, whose name the father adopted. Never married. EDUCATION: Learned reading and writing from maternal grandmother, Molly Walsh, a white English woman; attended country school run by Quakers until old enough to help father on farm. CAREER: 1759, inherited farm from father and remained there all his life, but ca. 1790 sold farm to Ellicott family while retaining right of residence; 1791, assistant to A. Ellicott [q.v.] in survey for federal district of Washington. SCIENTIFIC CONTRIBUTIONS: Showed natural ability in mathematics at early age; in 1753, with borrowed watch as model, completed striking clock, handmade and almost entirely constructed of wood, which continued in operation until his death and which brought him considerable local renown; at age forty-one, Ellicott family moved into region and with assistance and encouragement, especially from George Ellicott, Banneker pursued mathematical and astronomical studies. Appointment as assistant to A. Ellicott in survey for federal district for new city of Washington supported by T. Jefferson [q.v.]; was given charge of astronomical clock and assisted in field and in observatory tent; while he was thus connected with survey, there is no evidence of involvement in plans for actual layout of city and therefore no support for story that Banneker reproduced plans for city from memory after L'Enfant was dismissed and took plans with him. Sent manuscript of early almanac to Jefferson, who sent it to secretary of French Royal Academy of Sciences as evidence of mental abilities of Negroes; for similar reasons, printed almanacs [see Works] received support of antislavery groups in Maryland and Pennsylvania. Produced all calculations for the ephemeris, which was computed annually until 1802, although apparently not published after 1797.

WORKS: Almanacs for 1792 to 1797, the first published by Goddard and Angell in Baltimore and other editions published in both Baltimore and Philadelphia. MSS: Journal and commonplace book extant (in private hands? [see Bedini, below]). WORKS ABOUT: Louis Haber, *Black Pioneers of Science and Invention* (New York, 1970), pp. 1-12; Silvio A. Bedini, *The Life of Benjamin Banneker* (New York, 1972).

BARKER, GEORGE FREDERICK (1835-1910). *Chemistry*; *Physics.* Professor of physics, Univ. of Pennsylvania.

MSS: Univ. of Pennsylvania Library (Hamer). WORKS ABOUT: See *AMS*; *DAB.*

BARNARD, EDWARD EMERSON (1857-1923). *Astronomy.* Professor of practical astronomy, Univ. of Chicago, and astronomer at Yerkes Observatory.

MSS: Joint University Libraries, Nashville, Tenn. (*NUCMC* 66-814). WORKS ABOUT: See *AMS*; *DAB*; *DSB.*

BARNARD, FREDERICK AUGUSTUS PORTER (May 5, 1809, Sheffield, Mass.—April 27, 1889, New York, N.Y.). *Mathematics*; *Physics*; *Astronomy.* Son of Robert Foster, lawyer, and Augusta (Porter) Barnard; brother of John G. Barnard [q.v.]. Married Margaret McMurray, 1847; apparently no childen. EDUCATION: 1828, B.A., Yale College. CAREER: 1828-1837, successively, teacher in Hartford, tutor in mathematics at Yale, teacher in American Institution for Deaf and Dumb at Hartford and in the New York Institution for Deaf and Dumb; at Univ. of Alabama, professor of mathematics and natural history (1837-1848) and professor of chemistry and natural history (1848-1854); at Univ. of Mississippi, professor of mathematics and astronomy (1854-1856), president (1856-1858), and chancellor (1858-1861); ca. 1862-1864, in charge of map and chart depot for U.S. Coast Survey; 1864-1889, president, Columbia Univ. MEMBERSHIPS: Natl. Ac. Scis.; president, Am. Assoc. Advt. Sci., 1866. SCIENTIFIC CONTRIBUTIONS: Best known as college president and was so involved in wide range of other interests that scientific work was in a sense avocational. Astronomer for Alabama and Florida Boundary Commission; established astronomical observatories at universities of Alabama and Mississippi, and accompanied eclipse expedition to Labrador in 1860. Credited with invention of stereoscopic photography and made improvements in daguerreotype process. Wrote on such diverse topics as expenditure of heat in hot air furnace, aurora borealis and zodiacal light, and explosive force of gunpowder; a leading American proponent of metric system and published on that topic and on standardization of such things as coinage, time, weights and measures; also interested in theory of magic squares. In addition to scientific interests, wrote on politics, economics, religion, and, above all, education, especially matters relating to academic administration.

WORKS: Bibliography of scientific works in Davenport [below], pp. 268-72; major writings mentioned in *DAB*. MSS: Colum. U. Lib. (*NUCMC* 61-3232). WORKS ABOUT: *DAB*, 1: 619-21 (Charles F. Thwing); Charles B. Davenport, "Memoir," *Biog. Mems. of Natl. Ac. Scis.* 20 (1939): 259-72.

BARNARD, JOHN GROSS (May 19, 1815, Sheffield, Mass.—May 14, 1882, Detroit, Mich.). *Engineering*; *Mathematics*; *Physics*. Son of Robert Foster, lawyer, and Augusta (Porter) Barnard; brother of Frederick A. P. Barnard [q.v.]. Married Jane Elizabeth Brand (died 1853); Mrs. Anna E. (Hall) Boyd; four children by first marriage, three by second. EDUCATION: 1833, graduated from U.S. Military Academy, West Point. HONORS: LL.D., Yale Univ., 1864. CAREER: 1833-1881, officer in U.S. Army Corps of Engineers and in 1865 achieved rank of colonel; 1855-1856, superintendent, U.S. Military Academy; 1867-1881, member and, at time of retirement, president of Permanent Board of Engineers for Fortifications and River and Harbor Improvements; member of a number of boards and commissions, including Lighthouse Board. MEMBERSHIPS: Natl. Ac. Scis. SCIENTIFIC CONTRIBUTIONS: As army engineer, oversaw engineering of fortifications on East and West coasts and on Gulf coast; published several works on coastal defense, and in 1870 government sent him to Europe to study progress in utilization of iron for purposes of defense. Also engaged in important work on improvement of rivers and harbors, including acting as superintending engineer for construction of Delaware breakwater and improvements of Hudson River and New Jersey harbors; made particularly significant contribution to work on mouth of Mississippi River; in 1850 was chosen to carry out survey for projected railroad across Isthmus of Tehuantepec in Mexico. During Civil War, served as chief engineer for construction of defenses of Washington, D.C.; in 1864, was proposed by President Lincoln as head of Army Corps of Engineers, but declined appointment. In addition to engineering works, also carried out studies on mathematical and physical topics, and among publications were works on gyroscope and related questions.

WORKS: "Problems of Rotary Motion Presented by the Gyroscope, the Precession of the Equinoxes, and the Pendulum," *Smithson. Contr. Knowl.* 19 (1872). See *Natl. Un. Cat. Pre-1956 Imp.*, and *Roy. Soc. Cat.*, vols. 1, 7, 9. WORKS ABOUT: DAB, 1:626-27 (Gustave J. Fiebeger); Henry L. Abbot, "Memoir," *Biog. Mems. of Natl. Ac. Scis.* 5 (1905): 219-29.

BARNES, CHARLES REID (1858-1910). *Plant physiology*. Professor of botany and plant physiology, universities of Wisconsin and Chicago.

WORKS ABOUT: See *AMS*; *DAB*.

BARTLETT, WILLIAM HOLMS CHAMBERS (September 4, 1804, Lancaster County, Pa.—February 11, 1893, Yonkers, N.Y.). *Mathematics*; *Astronomy*. No information on parentage. Married Harriet Whiteborne, 1829; eight children. EDUCATION: 1826, graduated from U.S. Military Academy, West Point. HONORS: A.M., College of New Jersey (Princeton), 1837; LL.D., Hobart College, 1847. CAREER: 1826, made second lieutenant in U.S. Army Corps of Engineers; 1826-1829, assistant professor of engineering, U.S. Military Academy; 1828-1834, engineering duty at Fort Monroe, Va., Fort Adams in Newport, R.I., and Washington, D.C.; at U.S. Military Academy, acting professor (1834-1836) and professor of natural and experimental philosophy (1836-1871); 1871-1888, actuary at Mutual Life Insurance Co. of New York. MEMBERSHIPS: Am. Phil. Soc.; Natl. Ac. Scis.; and other societies. SCIENTIFIC CONTRIBUTIONS: Interested in number of mathematical, astronomical, and physical questions; early investigation of building stones [see Works] resulted from engineering work at Fort Adams. In 1840, visited European observatories and upon return directed erection of observatory at U.S. Military Academy, at which much of his research work was done; report on solar eclipse of 1854 [see Works] said to have made first use of photographic plates in astronomical measurement. Prepared series of textbooks, beginning with *Treatise on Optics* (1839), with later works on acoustics, astronomy, spherical astronomy, mechanics, molecular physics. Best-known textbook, that on analytical mechanics [see Works], and certain aspects of physical research were characterized by a mathematical approach to physical and astronomical questions. In later life, became actively interested in actuarial science and prepared several actuarial publications for Mutual Life Insurance Co. of New York.

WORKS: "Experiments on the Expansion and

Contraction of Building Stones at Variations of Temperature," *Am. Journ. Sci.* 22 (1832) : 136-40; "On the Solar Eclipse of 1854, May 26," *Astronomical Journal* 4 (1856) : 33-35; *Elements of Analytical Mechanics* (New York, 1853), which had a total of nine editions. Bibliography of principal writings in Holden [below], pp. 192-93. WORKS ABOUT: *DAB*, supp. 1, pp. 54-55 (Evan Thomas); Edward S. Holden, "Memoir," *Biog. Mems. of Natl. Ac. Scis.* 7 (1911) : 170-93.

BARTON, BENJAMIN SMITH (February 10, 1766, Lancaster, Pa.—December 19, 1815, Philadelphia, Pa.). *Botany*; *Zoology.* Son of Thomas, Episcopal rector, and Esther (Rittenhouse) Barton; nephew of David Rittenhouse [q.v.] and uncle of William P. C. Barton [q.v.]. Married Mary Pennington, 1797; two children. EDUCATION: Studied literature, science, and medicine in College of Philadelphia; 1786-1789, studied medicine at Univ. of Edinburgh and in London. CAREER: 1785, assisted uncle, D. Rittenhouse, in survey of western boundary of Pennsylvania; 1789, began medical practice in Philadelphia; at College of Philadelphia [after 1791 known as Univ. of Pennsylvania], professor of natural history and botany (1790-1795), transferred to professorship of materia medica (1795), professor of theory and practice of medicine (1813-1815); 1798, became physician to Pennsylvania Hospital. MEMBERSHIPS: Royal Society of Edinburgh; Linnean Society of London; and other societies. SCIENTIFIC CONTRIBUTIONS: Wrote on numerous botanical, zoological, medical, and ethnographical topics, including work on natural history of Lewis and Clark Expedition; during years 1804-1809, published on many topics in his own *Philadelphia Medical and Physical Journal*, and in 1808 began preparation of American editions of various European works. *Elements* [see Works] was first botanical textbook written by an American, and *Collection for . . . Materia Medica* [see Works] was based largely on original investigations. Diversity of interests and projects partly accounted for lack of substantive achievement commensurate with training and ability; nevertheless, career was influential stimulus to natural history investigation in U.S. Had largest natural history library and herbarium of native plants then available in this country.

WORKS: *New Views of the Origin of the Tribes and Nations of America* (Philadelphia, 1797) ; *Elements of Botany* (Philadelphia, 1803), with plates by W. Bartram [q.v.]; *Collections for an Essay towards a Materia Medica of the U.S.*, 2 vols. (Philadelphia, 1798, 1804). See bibliographies in Meisel, *Bibliog. Amer. Nat. Hist.*, and in *Elements of Botany*, 6th ed., edited by W. P. C. Barton (Philadelphia, 1836), pp. 27-30. Also, see bibliographical references in *DSB.* MSS: Hist. Soc. Penn. (*NUCMC* 60-2224 and 61-316) ; Am. Phil. Soc. (*NUCMC* 61-721). WORKS ABOUT: *DAB*, 2: 17-18 (George Blumer) ; *DSB*, 1: 484-86 (Joseph Ewan).

BARTON, WILLIAM PAUL CRILLON (November 17, 1786, Philadelphia, Pa.—February 29, 1856, Philadelphia, Pa.). *Botany.* Son of William, lawyer and judge, and Elizabeth (Rhea) Barton; nephew of Benjamin S. Barton [q.v.]. Married Esther Sergeant, 1814; no information on children. EDUCATION: 1805, A.B., Princeton Univ.; 1808, M.D., Univ. of Pennsylvania. CAREER: 1808, began medical practice, Philadelphia; 1809, appointed navy surgeon and served on frigates *United States*, *Essex*, and *Brandywine* and in marine hospitals at Philadelphia, Norfolk, and Pensacola; 1815, made professor of botany, Univ. of Pennsylvania; for three years, instructor in materia medica, Jefferson Medical School (Philadelphia) ; 1842-1844, under orders of U.S. Navy Department, served as first chief of Bureau of Medicine and Surgery, in New York; 1844-1856, inactive naval duty and in 1852 was president of Board of Medical Examiners. SCIENTIFIC CONTRIBUTIONS: Studied with uncle, B. S. Barton, at Univ. of Pennsylvania. Botanical interests were both general and medical; major works on botany were produced during earlier years and were usually catalogues and compilations or popularizations based on works of other persons. Prepared drawings for *Flora of North America* [see Works], which were colored by wife; although title of latter publication was overly ambitious, it was an effective popularization of efforts by earlier botanists to systematize American flora.

WORKS: *Flora Philadelphicae Prodromus* (Philadelphia, 1815), 100 pp.; *Compendium Florae Philadelphicae*, 2 vols. (Philadelphia, 1824) ; *Syllabus of the Lectures Delivered on Vegetable Materia Medica and Botany in the University of Pennsylvania* (Philadelphia, 1819), 12 pp.; *Vegetable Materia Medica of the U.S.*, 2 vols. (Philadelphia, 1817-1818) ; *Flora of North America*, 3 vols. (Philadelphia, 1821-1823). See bibliography in Harshberger [below]. WORKS ABOUT: *DAB*, 2: 25-26 (Donald C. Peattie) ; John W. Harshberger, *Botanists of Philadelphia and Their Work* (Philadelphia, 1899), pp. 159-63.

BARTRAM, JOHN (May 23, 1699, Marple, Pa.—September 22, 1777, Kingsessing, Pa.). *Botany.* Son of William, farmer, and Elizabeth (Hunt) Bartram; cousin of Humphry Marshall [q.v.]. Married Mary Maris, 1723; Ann Mendenhall, 1729; two children by first marriage, nine children by second, including William Bartram [q.v.]. EDUCATION: Attended country school and apparently later had Latin tutor in order to learn to read Linnaeus. CAREER: Inherited farm from an uncle; 1728, sold that farm and bought one of 102 acres at Kingsessing, Pa., four miles from Philadelphia; by use of crop rotation and fertilizers Bartram prospered as farmer; ca. 1734, introduced to Peter Collinson of London and became a supplier of American flora; 1736, made first botanical journey to Schuylkill River; 1738, to Virginia and the Blue Ridge; 1742, to Catskill Mountains; 1743, through Pennsylvania and New York; 1755, again to the Catskills, with son William; 1760, to Carolinas; 1761, to forks of Ohio River; 1765, appointed king's botanist with annual grant of £50 pounds and went to Florida with William; remaining years were spent at Kingsessing. MEMBERSHIPS: Am. Phil. Soc.; Royal Academy of Sciences of Sweden. SCIENTIFIC CONTRIBUTIONS: First native American botanist and, according to Linnaeus, "the greatest contemporary 'natural botanist' in the world." By 1730 was cultivating plants, trees, and shrubs from various parts of U.S. in garden at Kingsessing, and there performed what perhaps were first American experiments in plant hybridization; sent more than 100 new botanical species to Europe. Also collected other natural history specimens, but was not systematic botanist or zoologist and wrote little intended for publication.

WORKS: *Observations on the Inhabitants, Climate, Soil, Rivers, Productions, Animals, and Other Matters . . . from Pensilvania to Onondago, Oswego and the Lake Ontario in Canada* (London, 1751); "Diary of a Journey Through the Carolinas, Georgia, and Florida," edited by Francis Harper, *Transactions of Am. Phil. Soc.* n.s. 33, pt. 1 (1942-1944): 1-10; *Description of East Florida, With a Journal by John Bartram* (London, 1769). Also, see Bartram bibliography in *Bartonia, Proceedings of the Philadelphia Botanical Club*, Supplement to vol. 12 (December 31, 1931). MSS: Hist. Soc. Penn. (*NUCMC* 60-1467); Am. Phil. Soc. (*NUCMC* 61-729); Ac. Nat. Scis. Philad. (*NUCMC* 66-11). WORKS ABOUT: *DAB*, 2: 26-28 (Donald C. Peattie); *DSB* 1: 486-88 (Whitfield J. Bell, Jr.). Also see Ernest Earnest, *John and William Bartram: Botanists and Explorers* (Philadelphia, 1940).

BARTRAM, WILLIAM (April 9, 1739, Kingsessing, Pa.—July 22, 1823, Kingsessing, Pa.). *Botany*; *Ornithology*. Son of John [q.v.], well-known botanist, and Ann (Mendenhall) Bartram. Never married. EDUCATION: 1752-1756, attended Academy of Philadelphia. CAREER: 1757-1761, apprenticed to Philadelphia merchant; 1761, began unsuccessful operation of store at Cape Fear, N.C.; 1765-1766, explored Florida with father; ca. 1766, established briefly as rice and indigo planter on St. Johns River, Fla., but a year later was back in Pennsylvania; ca. 1767-1771, worked as farmer and did natural history drawings for London physician, Dr. John Fothergill; 1771, again in the South; 1773-1777, sponsored by Fothergill, explored in Carolinas, Georgia, and Florida; 1778-1823, at Kingsessing, where he worked in family's botanical garden; 1782, declined appointment as professor of botany, Univ. of Pennsylvania. MEMBERSHIPS: Am. Phil. Soc., but did not attend meetings; Ac. Nat. Scis. Philad.; foreign societies. SCIENTIFIC CONTRIBUTIONS: Showed natural ability and interest in botany and drawing and before 1756 was providing natural history illustrations for father's friend and patron in London, Peter Collinson, and for others; real career as naturalist began in 1773 with explorations in southeastern U.S., during which, in exchange for financial support, he sent seeds, specimens, drawings, and descriptions to John Fothergill. Reputation rests on *Travels* [see Works], which included sound descriptions of Indians and natural history, including list of 215 native bird species, and was the most complete such list before work of A. Wilson [q.v.], whose own work Bartram encouraged; the elaborate literary style of the *Travels* has been criticized, but was a major influence on the writings of European romantic authors.

WORKS: *Travels Through North and South Carolina, Georgia, East and West Florida, the Cherokee Country, the Extensive Territories of the Muscogulges, or Creek Confederacy, and the Country of the Choctaws* (Philadelphia, 1791), which within a a decade was republished in London and Dublin and translated into German, Dutch, and French. For other works see *DSB*, and Meisel, *Bibliog. Amer. Nat. Hist.* MSS: Hist. Soc. Penn. (*NUCMC* 60-1467); Ac. Nat. Scis. Philad. (Hamer); British Museum (Natural History) Department of Botany (*DAB*). WORKS ABOUT: *DAB*, 2: 28-29 (Lane Cooper); *DSB*, 1: 488-90 (Whitfield J. Bell, Jr.). Also see Ernest Earnest, *John and William Bartram: Botanists and Explorers* (Philadelphia, 1940).

BARUS, CARL (1856-1935). *Physics*. Professor of physics and dean of graduate department, Brown Univ.

MSS: Brown Univ. Library (*NUCMC* 73-41). WORKS ABOUT: See *AMS*; *DAB*, supp. 1; *DSB*.

BATCHELDER, JOHN MONTGOMERY (October 13, 1811, New Ipswich, N.H.—July 3, 1892, Cambridge, Mass.). *Engineering; Physics; Scientific apparatus*. Son of Samuel, merchant and cotton goods manufacturer, and Mary (Montgomery) Batchelder. Married Mary E. Wood, 1843; Elizabeth C. (Bird) Beardsley (died 1898); three children. EDUCATION: 1831, student at Brunswick [Bowdoin College?]; said to have studied civil engineering with Professor [James?] Hayward, Harvard Univ. CA-

REER: For number of years, worked as civil engineer at York Mills, Maine; also practiced civil engineering at Lawrence, Mass.; had charge of mill at Ipswich, Mass.; employed by A. D. Bache [q.v.] on Coast Survey and carried out observations to test base-line equipment; 1858, went from Coast Survey, as assistant to B. A. Gould [q.v.] at Dudley Observatory (Albany); thereafter, apparently resided at Cambridge, Mass. MEMBERSHIPS: Am. Ac. Arts Scis.; Bos. Soc. Nat. Hist.; and other societies. SCIENTIFIC CONTRIBUTIONS: During period of association with Coast Survey, carried out numerous experimental studies on variety of topics, including the compressibility of seawater and several other liquids by pressure and the effects of temperature in compression as related to sounding instruments developed by J. Saxton [q.v.]. During this time also worked on compressibility of rubber, use of vulcanized india rubber in compression-sounding apparatus, and other studies on use of rubber, and on apparatus for sounding as well as determination of speed of vessels and currents of water. At Dudley Observatory worked chiefly at task of making operable a calculating machine and also on telegraphic connections and apparatus. Interested in wide range of scientific subjects and in notebooks kept observations of temperature in well, aurora borealis, temperature of Saco River, and other topics; interested in invention and problems in mechanics, devised deep-sea sounding apparatus as well as tide meter, and prior to 1853 made independent invention of Bunsen burner. Also interested in electricity; inventor of a dynamometer and of other devices.

WORKS: See *Roy. Soc. Cat.*, vols. 1, 6 (additions), 7, 9, 13. WORKS ABOUT: John Trowbridge, memoir, *Proceedings of Am. Ac. Arts Scis.* 28 (1892-1893): 305-10; Charles Chandler and S. F. Lee, *History of New Ipswich, N.H.* (Fitchburg, Mass., 1914), pp. 222-24.

BAYMA, JOSEPH (November 9, 1816, Cirie [near Turin], Italy—February 7, 1892, Santa Clara, Calif.). *Mathematics; Physics.* Father was physician. EDUCATION: Graduated from Jesuit college in Turin; matriculated at Royal Academy of Turin, apparently intending to study medicine; 1832, entered Jesuit novitiate at Chieri; 1847, ordained as Jesuit priest. CAREER: After ordination, served one year as missionary in Algiers; thereafter, assistant to Father Angelo Secchi, director of Osservatorio del Collegio Romano; worked with Father Enrico Vasco, editing *Il Ratio Studiorum adattato ai tempi presenti*; 1852, became rector, Episcopal Seminary of Bertinoro in Romagna; 1858-1869, held chair of philosophy, Stonyhurst College (Lancashire, England); 1869-ca. 1872, president and, later, professor of higher mathematics, St. Ignatius College (now Univ. of San Francisco); 1880-1892, at Santa Clara College, Calif. SCIENTIFIC CONTRIBUTIONS: Remembered mainly for book on molecular mechanics [see Works]. Essentially a philosopher and mathematician writing in tradition of Ruggiero Boscovich's dynamism; contended that force was more fundamental than volume, mass, or extension in explanation of physical phenomena; though contrary to Boscovich, he held view that attractive or repulsive character is inherent in any elementary point-particle of which matter is composed and that such attraction or repulsion is not determined by distance. Produced no permanently useful ideas on molecular configuration, but work is of interest because written at time when few chemists were working in terms of three-dimensional atomic and molecular shapes. In U.S. turned to development of new work on "Cycloidal functions," left uncompleted. At St. Ignatius College built up what was called finest collection of scientific apparatus in U.S. at the time.

WORKS: *Elements of Molecular Mechanics* (London, 1866); related works in *Proceedings of Roy. Soc.* 13 (1864): 126-35, and in *Philosophical Magazine* 37 (1869): 182-88, 275-87, 348-58, and 431-32; about 1880 published several elementary mathematical textbooks. MSS: Georgetown Univ. Archives (*DAB*). WORKS ABOUT: *DAB*, 2: 79-80 (Francis A. Tondorf); Mel Gorman, "Stereochemical Concepts in the Molecular System of Joseph Bayma," in *Proceedings of Tenth International Congress of History of Science* (Paris, 1964), pp. 899-901.

BEAL, WILLIAM JAMES (1833-1924). *Botany.* Professor of botany, Michigan Agricultural College. (Michigan State Univ.).

WORKS ABOUT: See *AMS*; *DAB*.

BEAN, TARLETON HOFFMAN (1846-1916). *Zoology; Ichthyology.* Affiliated with U.S. Natl. Mus., U.S. Commission on Fish, and New York Aquarium; New York state fish culturist.

WORKS ABOUT: See *AMS*; *DAB*.

BEAUMONT, WILLIAM (November 21, 1785, Lebanon, Conn.—April 25, 1853, St. Louis, Mo.). *Physiology.* Son of Samuel, farmer, and Lucretia (Abel) Beaumont. Married Mrs. Deborah Platt, 1821; survived by three children. EDUCATION: Probably not more than common school education; studied medicine privately; 1812, medical license, Third Medical Society of Vermont. CAREER: 1807-1810, taught school and studied medicine, Champlain, N.Y.; 1810-ca. 1812, apprenticed to Dr. Benjamin Chandler, St. Albans, Vt.; 1812-1815, surgeon's mate, Sixth Infantry, Plattsburgh, N.Y.; ca. 1815-1820, medical practice, Plattsburgh, N.Y.; 1820-1839, U.S.

Army surgeon, assigned to Fort Mackinac (now in Michigan) in 1820, Fort Niagara in ca. 1825, Fort Howard on Green Bay in 1826, Fort Crawford on upper Mississippi in 1828, six months' leave in Washington, D.C., in 1832, New York City in ca. 1833, Jefferson Barracks near St. Louis in 1834 and, later, St. Louis Arsenal, and was allowed to engage in private medical practice in St. Louis; 1839-1853, medical practice, St. Louis. SCIENTIFIC CONTRIBUTIONS: June 1822, began treatment of nineteen-year-old Canadian trapper, Alexis St. Martin, for gunshot wound in stomach that healed, but did not close; this left opening to stomach that made it possible for Beaumont to carry out series of experiments on digestion, beginning about 1825; early investigations included study of digestibility of certain foods and effectiveness of digestive juices when removed from stomach; St. Martin left Beaumont about 1825 and went to Canada, but returned during years 1829-1831 and 1832-1833 for further experiments. Beaumont had support of surgeon general and secretary of war, and later studies were conducted in Washington, where he received valuable advice and assistance from R. Dunglison [q.v.]; experiments summarized in *Experiments and Observations* [see Works], published at personal expense. Results of nearly 250 experiments were first major American contributions to physiology and one of most important ever made to study of gastric digestion, including study of both physical and chemical properties of gastric juice.

WORKS: Early reports on studies in *Medical Recorder* 8 (1825-1826); *Experiments and Observations on the Gastric Juice and the Physiology of Digestion* (Plattsburgh, N.Y., 1833), 280 pp., with 2d ed. (1847) and German translation (Leipzig, 1834). MSS: Washington Univ. Medical Library, St. Louis (*NUCMC* 60-1967); Univ. of Chicago Library (*NUCMC* 64-39); St. Hist. Soc. Wis. (*NUCMC* 68-2114). WORKS ABOUT: *DAB*, 2: 104-10 (Victor C. Vaughan); *DSB*, 1: 542-45 (George Rosen). Also see Jesse S. Myer, *Life and Letters of Dr. William Beaumont* . . . (St. Louis, 1912; 2d ed., 1939).

BECK, LEWIS CALEB (October 4, 1798, Schenectady, N.Y.—April 20, 1853, Albany, N.Y.). *Chemistry*; *Mineralogy*. Son of Caleb, lawyer, and Catherine (Romeyn) Beck. Married Hannah Maria Smith, 1825; apparently seven children. EDUCATION: 1817, A.B., Union College; studied medicine at College of Physicians and Surgeons, N.Y., which later conferred M.D. degree; 1818, licensed to practice medicine. CAREER: 1819, began medical practice in Schenectary; 1819-ca. 1822, in Albany and in St. Louis and other parts of the West preparing gazeteer; ca. 1822, returned to Albany and to medical practice there; 1824, teacher of botany at Berkshire Medical Institution; 1824-1829, junior professor of botany, mineralogy, and zoology, Rensselaer Polytechnic Institute; 1826-1832, professor of botany and chemistry, Vermont Academy of Medicine; 1830-1837 and 1838-1853, professor of chemistry and natural history, Rutgers College; 1836, professor at Univ. of City of New York (New York Univ.); 1836-1842, mineralogist with Natural History Survey of New York; 1840-1853, professor of chemistry and pharmacy, Albany Medical College. MEMBERSHIPS: Numerous literary and scientific societies in Europe and the United States. SCIENTIFIC CONTRIBUTIONS: During travels to St. Louis gathered information on botany, mineralogy, geology, climate, and civic matters that was presented in *Gazeteer* [see Works]. In addition to mineralogical work for New York state survey, also was employed by that state to investigate potash, and in 1836 his report on latter subject was issued as New York state document. For the first time, in 1848, Congress made appropriation for chemical examination of foods; Beck was chosen to carry out this work and prepared reports dealing mainly with breadstuffs, issued with 1848 and 1849 reports of U.S. commissioner of patents. In addition to other work, also published on topics of medical interest.

WORKS: *Gazeteer of the States of Illinois and Missouri* (Albany, 1823), 352 pp.; *Botany of the Northern and Middle States* (Albany, 1833), 471 pp.; *Manual of Chemistry* (Albany, 1831), 458 pp.; *Mineralogy of New York* (Albany, 1842), 536 pp., being the final report on state survey. See also *Roy. Soc. Cat.*, vols. 1, 12, and Gross [below]. MSS: Rutgers Univ.—Alexander Library. WORKS ABOUT: *DAB*, 2: 116 (Lyman F. Kebler); Samuel D. Gross, *Lives of Eminent American Physicians and Surgeons of the Nineteenth Century* (Philadelphia, 1861), pp. 679-96.

BECKER, GEORGE FERDINAND (1847-1919). *Geology*; *Mathematics*; *Physics*. Geologist in charge, U.S. Geol. Surv.

MSS: Lib. Cong. (*NUCMC* 60-24). WORKS ABOUT: See *AMS*; *DAB*; *DSB*.

BEECHER, CHARLES EMERSON (October 9, 1856, Dunkirk, N.Y.—February 14, 1904, New Haven, Conn.). *Paleontology*. Son of Moses, banker, and Emily (Downer) Beecher. Married Mary Salome Galligan, 1894; two children. EDUCATION: 1878, B.S., Univ. of Michigan; 1889, Ph.D., Yale Univ. CAREER: 1878-1888, assistant to paleontologist J. Hall [q.v.] at New York State Museum (Albany); 1888, became assistant, in charge of invertebrate fossils, at Peabody Museum, Yale Univ.; at Yale Univ., instructor in paleontology (1891-

1892), assistant professor of paleontology (1892-1897), professor of historical geology (1897-1902), University professor of paleontology (1902-1904); 1897, became member of governing board of Sheffield Scientific School, Yale; 1899-1904, curator of geological collections, Peabody Museum, and member and secretary of museum's board of trustees. MEMBERSHIPS: Natl. Ac. Scis.; Geol. Soc. Am.; and other societies. SCIENTIFIC CONTRIBUTIONS: Began collection of shells at age twelve and published first scientific paper, a list of freshwater shells found in vicinity of Ann Arbor, Mich., at age twenty; early publications, on recent mollusks, were written with other persons; while at Albany assisted J. Hall with latter's monographs on invertebrate fossils of New York. First paleontological paper published by Second Pennsylvania Geological Survey (1884) and dealt with new genera and species from Devonian period; skilled in preparation of drawings, and facility in physical manipulation made possible uncovering of structures such as limbs and antennae of trilobites (1893). Became supporter of developmental and evolutionary ideas of A. Hyatt [q.v.] and is best remembered for work in application of these ideas, producing a natural classification of Brachiopoda and Trilobita.

WORKS: One of important works was "The Origin and Significance of Spines: A Study in Evolution," *Am. Journ. Sci.*, 4th ser. 6 (1898). Most major papers reprinted in Beecher, *Studies in Evolution*, Yale Bicentennial Series (New York, 1901), 638 pp. Full bibliography in C. Schuchert, *Am. Journ. Sci.* 4th ser. 17 (1904): 421-22; also see Dall [below]. MSS: Some letters in O. C. Marsh Papers, Yale Univ. Library—Manuscripts and Archives Division. WORKS ABOUT: *DAB*, 2: 127-28 (George P. Merrill); William H. Dall, "Memoir," *Biog. Mems. of Natl. Ac. Scis.* 6 (1909), 57-70.

BEHR, HANS HERMAN (August 18, 1818, Cöthen, Germany—March 6, 1904, San Francisco, Calif.). *Entomology; Botany.* For centuries family was involved in administration of duchy at Anhalt, Germany. Married Agnes Omylska, 1853; three children. EDUCATION: Classical education in the Prince's College at Zebst; 1843, M.D., Univ. of Berlin, also having studied medicine and natural sciences at universities of Halle and Würzburg. CAREER: 1844-1847, on recommendation of Alexander von Humboldt, studied aborigines and natural history in Australia, Java, Straits Settlements, East India, and South Africa; 1848-1851, left Germany for Brazil and practiced medicine two years in Philippines; 1851-ca. 1892, medical practice in San Francisco, and for number of years professor of botany at California College of Pharmacy, San Francisco; ca. 1892-1904, curator of entomology, Calif. Ac. Scis. MEMBERSHIPS: Vice-president, Calif. Ac. Scis., 1864-1904. SCIENTIFIC CONTRIBUTIONS: Had number of friends among important European scientists, and breadth of interests and scientific background made him best trained scientist in Calif. Ac. Scis. for many years. At first, interests were mainly botanical, and *Flora* [see Works], prepared in connection with teaching at College of Pharmacy, demonstrated a familiarity with current state of botanical science and a comprehension of its various orders. Later became interested chiefly in entomological studies, and is remembered especially for descriptions of number of western butterflies; credited with sixteen entomological papers, especially on Lepidoptera, published chiefly in *Proceedings of Calif. Ac. Scis.*; gave large collection of Lepidoptera to academy about 1893, later destroyed in earthquake of 1906. Also wrote poetry and fiction; after-dinner speeches prepared for San Francisco's Bohemian Club contained entomological knowledge and were published posthumously as *The Hoot* [see Works].

WORKS: *The Flora of San Francisco* (San Francisco, 1888); *The Hoot of the Owl* (San Francisco, 1904), 227 pp. Also see Meisel, *Bibliog. Amer. Nat. Hist.*, and *Roy. Soc. Cat.*, vols. 7, 9, 12, 13. MSS: No collection is known (papers not in Calif. Ac. Scis.). WORKS ABOUT: Mallis, *Amer. Entomologists*, pp. 285-88; E. O. Essig, *A History of Entomology* (New York, 1931), pp. 553-56.

BELL, ALEXANDER GRAHAM (1847-1922). *Physics; Invention; Education of deaf.* Inventor of telephone.

MSS: Natl. Geog. Soc. and Lib. Cong. (*NUCMC* 68-2006 and *DSB*). WORKS ABOUT: See *AMS; DAB; DSB.*

BEMAN, WOOSTER WOODRUFF (1850-1922). *Mathematics.* Professor of mathematics, Univ. of Michigan.

MSS: Univ. Mich.-MHC (*NUCMC* 64-1437). WORKS ABOUT: See *AMS; Who Was Who in America*, vol. 1.

BENT, SILAS (October 10, 1820, South St. Louis, Mo.—August 26, 1887, Shelter Island, Long Island, N.Y.). *Oceanography.* Son of Judge Silas and Martha (Kerr) Bent. Married Ann Eliza Tyler, 1857; no information on children. EDUCATION: No information on early education; 1836, appointed midshipman, U.S. Navy. CAREER: 1836-1861, U.S. Navy, engaged in extensive sea duty and achieved rank of lieutenant; 1849, under Cdr. James Glynn of U.S. brig *Preble*, accompanied successful mission to obtain release of eighteen American prisoners held in Japan; 1852-1854, flag lieutenant aboard *Mississippi* during Com. M. C. Perry's Japan expedition; 1860-1861, assigned to Hydrographic Division of U.S. Coast Survey; 1861, resigned from U.S. Navy on account of southern loyalties, returned to St. Louis, and took over management of wife's family estate. SCIENTIFIC CONTRIBUTIONS: While serving on *Preble*, invented card device for use on dry compass to make automatic the conversion of magnetic observation to a true bearing; device was praised by M. F. Maury [q.v.], but apparently never extensively used. Best known for hydrographic surveys of Japanese waters, published as part of government's *Sailing Directions and Nautical Remarks by Officers of the Late U.S. Naval Expedition to Japan* (1857). Published first scientific study of Kuro Siwo, so-called Japanese Gulf Stream; this investigation also was carried out during Perry expedition, when Bent was only officer having sound training in hydrography and familiarity with oceans of that area; after completion of Perry mission ordered to Washington and there worked on charts and data for publication, full completion of which was prevented by Civil War. After leaving navy, continued interest in oceanography and, in publications that appeared in 1869 and 1872, argued that Atlantic Gulf Stream and Kuro Siwo in Pacific created open sea about North Pole; while this thesis was not generally accepted, it nonetheless helped foster interest in polar studies.

WORKS: "The Japanese Gulf Stream," *Bulletin of American Geographical and Statistical Society* 2 (1856) : 203-13. See also *Natl. Un. Cat. Pre-1956 Imp.* and *Roy. Soc. Cat.*, vol. 13. WORKS ABOUT: *DAB*, 2: 206 (Harry A. Marmer) ; Edward L. Towle, "Lt. Silas Bent's Device to Eliminate Variation Correction in the Magnetic Compass, 1849," *American Neptune* 25 (1965) : 93-98.

BESSEY, CHARLES EDWIN (1845-1915). *Botany.* Professor of agriculture and botany, dean, and chancellor, Univ. of Nebraska; state botanist of Nebraska.

MSS: Univ. of Nebraska Archives (*NUCMC* 70-316). WORKS ABOUT: See *AMS; DAB; DSB.*

BEYER, HENRY GUSTAV (1850-1918). *Physiology; Hygiene.* Surgeon and medical director, U.S. Navy.

WORKS ABOUT: See *AMS; NCAB, 14* (1910) : 250; *Who Was Who in America,* vol. 1.

BIDDLE, OWEN (1737, apparently Philadelphia, Pa.—March 10, 1799, probably Philadelphia, Pa.). *Astronomy.* Son of John, shipping and import merchant, and Sarah (Owen) Biddle. Married Sarah Parke, 1760; at least seven children. EDUCATION: No information on education. CAREER: For a time, engaged in clock and watch-making business; several years before beginning of American Revolution, entered shipping and importing business; 1775, appointed delegate to Provincial Conference, Philadelphia, and same year became member of Pennsylvania Committee of Safety; 1776, a delegate to Pennsylvania Constitutional Convention; among other offices held during this period, in 1777 appointed deputy commissioner of forage; 1782, appointed to establish boundary line between Pennsylvania and Virginia; as merchant, suffered serious losses during war period. MEMBERSHIPS: Am. Phil. Soc. SCIENTIFIC CONTRIBUTIONS: Participated in Am. Phil. Soc.'s observations of transit of Venus in 1769, during which he was given charge of operations at Cape Henlopen, Del., and was assigned use of reflecting telescope belonging to Philadelphia Library Company and once used by Mason and Dixon in their survey work [see Works]. A manager of Society for Encouraging the Culture of Silk in Pennsylvania, formed in 1770 with other members of Am. Phil. Soc., and in 1781 Biddle reported to Am. Phil. Soc. that silk society had not succeeded. During Revolution, conducted observations of solar eclipse, in company with D. Rittenhouse [q.v.] and others; also served on other scientific committees, including one on determination of distance and geographical locations of New Castle courthouse and State House observatory and another on experiments on electric eel. During later years, turned attentions from science to concern with youth in Quaker community and played important role in establishment of Quaker preparatory school at Westtown, Pa. (opened in 1799).

WORKS: "An Account of the Transit of Venus . . . 1769," *Transactions of Am. Phil. Soc.* 1 (1771) : 89-96. MSS: Hist. Soc. Penn., Lib. Cong., and Swarthmore College (Bell [below]). WORKS ABOUT: Bell, *Early Am. Sci.*, p. 48; Henry D. Biddle, "Owen Biddle," *Pennsylvania Magazine of History and Biography* 16 (1892) : 299-329.

BIGELOW, JACOB (February 27, 1786, Sudbury, Mass.—January 10, 1879, Boston, Mass.). *Botany; Technology.* Son of Jacob, Congregational minister, and Elizabeth (Wells) Bigelow. Married Mary Scollay, 1817; five children. EDUCATION: 1806, A.B., Harvard College; 1810, M.D., Univ. of Pennsylvania. CAREER: 1811, began medical practice with Dr. James Jackson, Boston; 1812, lectured with W. D. Peck [q.v.] on botany, Harvard; 1815-1855, professor of materia medica, Harvard Medical School; 1816-1827, Rumford professor of application of science to useful arts, Harvard; visiting physician, Massachusetts General Hospital. MEMBERSHIPS: President, Am. Ac. Arts. Scis., 1847-1863; member of Am. Assoc. Advt. Sci. SCIENTIFIC CONTRIBUTIONS: Had widely varied interests and as scientist is known especially for botanical work, done principally in period from about 1814 to 1824; *Florula Bostoniensis* (1814), with expanded editions in 1824 and 1840 covering all of New England, made available for first time a work by native American botanist that effectively described and systematized flora of the area; third edition of *Florula* was last work in U.S. to be arranged by artificial system, which Bigelow continued for sake of unity of work while appreciating advantages of natural system. Chief botanical publication, *Medical Botany* [see Works], grew out of professional interests, and for this work he invented process for reproducing plates. Appointment to Rumford professorship grew out of interest in mechanics, and he promoted use of the word *technology*. Played important role in preparation of first U.S. *Pharmacopoeia* (1820), and in 1822 published *Treatise on the Materia Medica* as sequel to the *Pharmacopoeia*; wrote extensively on medical subjects. Largely responsible for establishment of Mt. Auburn Cemetery in Cambridge, Mass., in 1831.

WORKS: *Florula Bostoniensis* (Boston, 1814); *American Medical Botany*, 3 vols. (Boston, 1817-1820), with 60 colored plates; *Elements of Technology* (Boston, 1829) comprised his Rumford lectures, issued in expanded form as *The Useful Arts*, 2 vols. (Boston, Mass., 1840; reprinted, New York, 1972); *Discourse on Self-Limited Diseases* (Boston, 1835). MSS: Harv. U.-MS; Mass. Hist. Soc. (Hamer). WORKS ABOUT: *DAB*, 2: 257-58 (Donald C. Peattie and John F. Fulton); L. H. Bailey, Jr., "Some North American Botanists: V. Jacob Bigelow." *Botanical Gazette* 8 (May 1883): 217-22.

BINNEY, AMOS (October 18, 1803, Boston, Mass.—February 18, 1847, Rome, Italy). *Zoology*. Son of of Col. Amos, businessman and merchant, and Hannah (Dolliver) Binney. Married cousin, Mary Ann Binney, 1827; five children, including William G. Binney [q.v.]. EDUCATION: 1821, A.B., Brown Univ.; 1826, M.D., Harvard Univ. CAREER: 1823, journeyed by horseback to Cincinnati and in 1824-1825 went to Europe, both for reasons of health; 1826, received M.D. degree, but with no interest in practice of medicine, began work as merchant; thereafter joined father in real estate and mining operations and inherited business interests following father's death; 1836-1837, member of Massachusetts state legislature; 1842, retired from business. MEMBERSHIPS: Am. Phil. Soc.; Am. Ac. Arts Scis. SCIENTIFIC CONTRIBUTIONS: Serious interest in natural sciences and collection of shells began while in college, and studied medicine not to practice profession, but as means of securing scientific training. While actively engaged in business, scientific studies consumed leisure hours; and as soon as practicable, retired and devoted fuller attentions to science. A founder and president, 1843-1847, of Bos. Soc. Nat. Hist.; one of the society's chief financial and scientific patrons, contributing papers on zoological and paleontological topics; also helped to establish museum and conducted effort to acquire building for society. About 1835, began work on *Terrestrial . . . Mollusks*, chief contribution to science, which was nearly completed at time of death and on which no expense had been spared in collection of necessary specimens and in employment of superior illustrators; work was completed through editorial labors of A. A. Gould [q.v.]; it became model for other zoologists and established Binney as authority on subject, a reputation that later passed to his son William.

WORKS: *The Terrestrial Air-breathing Mollusks of the United States, and the Adjacent Territories of North America* (Boston, 1851); most papers published by Bos. Soc. Nat. Hist., in whose *Boston Journal of Natural History* early aspects of *Terrestrial . . . Mollusks* appeared. See *Roy. Soc. Cat.*, vol 1, and Meisel, *Bibliog. Amer. Nat. Hist.* WORKS ABOUT: *DAB*, 2: 279-80 (Henry A. Pilsbry); Augustus A. Gould, "Memoir," in *Terrestrial . . . Mollusks*, I: xi-xxix.

BINNEY, WILLIAM GREENE (October 22, 1833, Boston, Mass.—August 3, 1909, Burlington, N.J.). *Conchology*. Son of Amos [q.v.], businessman and zoologist, and Mary Ann (Binney) Binney. Married Maria Louisa Chamberlin, 1855; two children. EDUCATION: 1850-1852, attended Harvard College and received A.B. in 1854 (out-of-course, 1857); probably studied in Europe. HONORS: M.A., Harvard, 1884. CAREER: Inherited wealth from his family and that of his wife; for some years, director of several financial institutions. MEMBERSHIPS: Kaiserlich Königlichen Zoologisch-botanischen Gesellschaft, Vienna, and other scientific societies; one of ten honorary members of Conchological Society of Great Britain. SCIENTIFIC CONTRIBUTIONS: Succeeded father as major authority on American land mollusks and continued his work; supplement to father's *Terrestrial . . . Mollusks* appeared in *Boston Journal of Natural History* (1863) and is referred to as volume 4 of the paternal work; also prepared volume 5 for publication in *Bulletin of Museum of Comparative Zoology* (Harvard) in 1878, and later additions appeared in that bulletin during period 1883-1892. Prepared edition of conchological writings of T. Say [q.v.] (1858); and with G. W. Tryon, Jr. [q.v.] in 1864 edited similar work on C. S. Rafinesque [q.v.]; under commission from state of Massachusetts, prepared new edition of earlier work by A. A. Gould [q.v.] on invertebrates of the state, Binney's edition appearing in 1870; collaborated with T. Bland [q.v.] on *Land and Fresh Water Shells* [see Works] and on number of other publications.

WORKS: "Bibliography of North American Conchology Previous to the Year 1860," *Smithson. Misc. Coll.* 5, 9 (1863-1864); "Land and Freshwater Shells of North America," *Smithson. Misc. Coll.* 7, 8, 16 (in 4 vols.) (1865-1873), with T. Bland and G. W. Tryon, Jr. See also Arthur F. Gray, "Bibliography of W. G. Binney, 1856-1892," *Nautilus* 50 (1937): 98-100, 135-40. WORKS ABOUT: Charles J. F. Binney, *Genealogy of the Binney Family in the U.S.* (Albany, 1886), pp. 173-74; *Harvard Graduates' Magazine* 18 (1909): 142-43.

BLAIR, ANDREW ALEXANDER (1848-1932). *Chemistry.* Consulting chemist, especially for analysis of metals; associated with Booth, Garrett & Blair, Philadelphia.

WORKS ABOUT: See *AMS*; *NCAB*, 25 (1936): 131.

BLAKE, FRANCIS (1850-1913). *Physics*; *Invention.* Associated for a time with U.S. Coast Survey; inventor; worked with Bell Telephone Co.

WORKS ABOUT: See *AMS*; *DAB.*

BLAKE, WILLIAM PHIPPS (1825-1910). *Mineralogy*; *Geology*; *Metallurgy.* Mineralogist of California state board of agriculture; professor of mineralogy and geology, College of California (Berkeley); professor of geology and mining and director of School of Mines, Univ. of Arizona.

MSS: Arizona Pioneers' Historical Society, Tucson (*NUCMC* 62-2173). WORKS ABOUT: See *AMS*; *DAB.*

BLAND, THOMAS (October 4, 1809, Newark, Nottinghamshire, England—August 20, 1885, Brooklyn, N.Y.). *Natural history*, especially *Conchology.* Son of Thomas Bland, physician. No information on marital status. EDUCATION: Educated at the Charterhouse School in London and studied law. CAREER: Practiced law for a while; 1842, emigrated to Barbados and later to Jamaica; 1850, became superintendent of gold mine at Marmato, New Granada (Colombia); 1852, moved to New York and lived there for most of remainder of life, in later years beset by poverty. MEMBERSHIPS: Am. Phil. Soc. and other societies; fellow, Royal Geological Society of London: SCIENTIFIC CONTRIBUTIONS: Mother interested in natural history and her uncle, Richard Shepard, was conchologist. Active interest in mollusks resulted from visit of C. B. Adams [q.v.] to Jamaica in 1848-1849; in that period, began investigation of animal life of West Indies, and interest in shells continued after move to New Granada; met W. G. Binney [q.v.] in New York and with him produced number of works on terrestrial mollusks, most noteworthy of which was part 1 of "Land and Freshwater Shells" [see Works], important contribution to growth and development of knowledge in field and for many years a recognized authority; wide-ranging experience in conchology gave Bland particular interest in questions of geographical distribution of fauna, and work on distribution of shells of Antillean region and U.S. in effect continued efforts of C. B. Adams. Bland's many papers on mollusks dealt with anatomy, classification, distribution, and development; published over seventy papers on mollusks, most of which appeared in *Annals of N.Y. Lyc. Nat. Hist.* and in *American Journal of Conchology.*

WORKS: With W. G. Binney, part 1 of "Land and Freshwater Shells of North America," *Smithson. Misc. Coll.* 8 (1869); "On the Geographical Distribution of the Genera and Species of Land Shells of the West India Islands, and a Catalogue of the Species of Each Island," *Annals of N.Y. Lyc. Nat. Hist.* 7 (1862): 335-61. See *Roy. Soc. Cat.*, vols. 1, 7, 9, 13. Arthur F. Gray produced *A Complete List of the Scientific Papers of Thomas Bland . . . from 1852 to 1883* (Salem, Mass., 1884), which was not seen during research for this volume and presumably is rare. WORKS ABOUT: *DAB*, 2: 357 (Frank E. Ross); [W. G. Binney], Obituary, *Am. Journ. Sci.*, 3d ser. 30 (1885): 407-8.

BLODGET, LORIN (May 25, 1823, near Jamestown, Chautauqua County, N.Y.—March 24, 1901, Philadelphia, Pa.). *Climatology.* Son of Arba, probably farmer, and Bebe (Bullock) Blodget. Married Mary Elizabeth Gibbs, 1856; no information on children. EDUCATION: Went to Jamestown (N.Y). Academy; attended Geneva (Hobart) College, but left without graduating. CAREER: Age seventeen, taught school in Chautauqua County, N.Y.; as early as 1848, involved in politics; 1851, put in charge of climatological research at Smithson. Instn.; 1852, joined Pacific Railroad Survey and worked on barometric determination of altitudes and gradients; 1854-1856, at War Department Office in Washington, while continuing association with Pacific Railroad Survey; 1857-1864, associate editor of Philadelphia newspaper, *North American*; 1858-1865, secretary of Philadelphia Board of Trade; 1863-1877, with U.S. Treasury Department, in charge of financial and statistical reports (1863-1865) and as U.S. appraiser of merchandise at Philadelphia (1865-1877); after 1877, held number of positions in Philadelphia. MEMBERSHIPS: Am. Phil. Soc.; Am. Assoc. Advt. Sci. SCIENTIFIC CONTRIBUTIONS: Became one of Smithsonian's corps of volunteer meteorological observers at about age twenty; presentation in 1853 to Am. Assoc. Advt. Sci. was one of earliest papers on atmospheric physics produced in U.S. *Climatology of the U.S.* [see Works], based on all available meteorological data at his disposal, was produced while serving as "assistant professor" at Smithsonian; it was first successful attempt systematically to examine the climate of any part of America, and major conclusions have generally been upheld by subsequent data. Credited with producing over 150 volumes, 350 pamphlets, and other papers, most of which related to financial and legislative matters.

WORKS: *Climatology of the U.S. and the Temperate Latitudes of the North American Continent* (Philadelphia, 1857), 536 pp.; "Climatology of Pennsylvania," published with *Annual Report of Secretary of Internal Affairs of Pennsylvania . . . 1888* (Harrisburg, Pa., 1889), pp. 107-310. See *Roy. Soc. Cat.*, vols. 1, 7, 9, 13, and *Natl. Un. Cat. Pre-1956 Imp.* WORKS ABOUT: *DAB*, 2: 379 (William J. Humphreys); *NCAB*, 4 (1897 [copyright 1891 and 1902]): 530.

BODLEY, RACHEL LITTLER (December 7, 1831, Cincinnati, Ohio—June 15, 1888, Philadelphia, Pa.). *Botany*; *Chemistry*. Daughter of Anthony Prichard, carpenter and pattern maker, and Rebecca Wilson (Talbott) Bodley. Never married. EDUCATION: Until age twelve attended private school conducted by mother; 1849, A.B., Wesleyan Female College, Cincinnati; 1860-ca. 1861, studied advanced chemistry and physics at Polytechnic College and practical anatomy and physiology at Woman's Medical College, Philadelphia. HONORS: A.M., Wesleyan Female College, 1871; M.D., Woman's Medical College, 1879. CAREER: 1849-1860, teacher at Wesleyan Female College, in later years of appointment holding rank as preceptress in higher college studies; 1862-1865, teacher of natural sciences, Cincinnati Female Seminary; at Woman's Medical College, Philadelphia, appointed professor of chemistry (1865) and served as dean (1874-1888); 1882-1885 and 1887-1888, elected director of twenty-ninth school section in Philadelphia. MEMBERSHIPS: Ac. Nat. Scis. Philad.; Am. Chem. Soc.; and other societies. SCIENTIFIC CONTRIBUTIONS: Became interested in chemistry and botany while teaching at Wesleyan Female College; at Cincinnati Female Seminary arranged and catalogued according to latest methods a large collection of plants [see Works]. First female chemist on staff of Woman's Medical College, and there also studied sea plants, collected botanical specimens on summer trips, and gave summer lectures on botany and other subjects; during early 1880s, gave six lectures on "household chemistry" at Franklin Institute. In Philadelphia, scientific interests were directed by practical concerns of the Woman's Medical College, and she became particularly interested in work of medical missionaries.

WORKS: Meisel, *Bibliog. Amer. Nat. Hist.*, 3: 493, lists: "Catalogue of Plants Contained in the Herbarium of Joseph Clark, Arranged According to the Natural System. Cincinnati, R. P. Thompson, 1865. 47 p. In Library of Surgeon General's office. Not pub." No works listed in *Roy. Soc. Cat.* MSS: Library of Woman's Medical College (*Nota. Am. Wom.*). WORKS ABOUT: *Nota. Am. Wom.*, 1: 186-87 (Gulielma F. Alsop); Sarah K. Bolton, *Successful Women* (Boston, 1888), pp. 149-74.

BOLL, JACOB (May 29 1828, Dieticon, Canton Aargau, Switzerland—September 29, 1880, Wilbarger County, Tex.). *Natural history*; *Geology*. Son of Henry and Magdalena (Peier) Boll, a family of moderate wealth. Married Henriette Humbel, 1854; three children. EDUCATION: Attended gymnasium in Switzerland and apparently at that time met J. L. R. Aggassiz [q.v.]; attended Univ. of Jena, but left in 1853 without attaining degree. CAREER: 1854-1869, operated apothecary shop in Bremgarten; 1869 went to U.S., stopping in Cambridge; spent 1869-1870 in Texas with relatives, collecting specimens for Agassiz; winter of 1870-1871, assistant to entomologist H. A. Hagen [q.v.] at Museum of Comparative Zoology, Harvard; 1871, after brief trip to Switzerland, returned to Cambridge and collected New England insects for Harvard museum; 1872-1873, returned to Switzerland; 1874, returned to Cambridge expecting appointment in museum, but learning of Agassiz's death, went to Texas; 1874-1880, studied geology and natural history of Texas; and 1878-1880, employed as collector of fossils for E. D. Cope [q.v.]. MEMBERSHIPS: Bos. Soc. Nat. Hist.; Academia Caesaraea Leopoldino—Carolina Naturae Curiosorum of Germany. SCIENTIFIC CONTRIBUTIONS: Produced small flora for Bremgarten region (1869); while in Switzerland, 1872-1873, prepared several publications with friend Heinrich Frey at Univ. of Zurich, based on his American collections. In U.S. was above all a natural history explorer-collector, most notably of series of fossil vertebrates (thirty-two of fifty-seven new Texas Permian vertebrate species and genera described by Cope were Boll's discoveries); explorations aided in development of Texas mineralogical resources, and "Geological Explorations" [see Works] was first informed investigation of Permian rocks of the state; at time of death plans were under way for state geological survey under Boll's direction. Also contributed report on locusts of Texas region for 1878 report of U.S. Entomological Commission.

WORKS: "Geological Examinations in Texas," *American Naturalist* 14 (1880): 684-86. Personal publications were not numerous; see list of works in Geiser, *Naturalists* [below], p. 265. MSS: Cope Collection at Am. Mus. Nat. Hist.; Harv. U.-MCZ. (Geiser, *Naturalists*). WORKS ABOUT: *DAB*, 2: 419-20 (Samuel W. Geiser); S. W. Geiser, *Naturalists of the Frontier*, 2d ed., rev. and enl. ([Dallas], 1948), pp. 19-29.

BOLTON, HENRY CARRINGTON (January 28, 1843, New York, N.Y.—November 19, 1903, Washington, D.C.). *Chemistry*. Son of Jackson, successful physician, and Anna Hinman (North) Bolton. Married Henrietta Irving, 1893; no information on children. EDUCATION: 1862, A.B., Columbia Univ., 1866, Ph.D., Univ. of Göttingen, having previously studied chemistry in Paris and at universities of Heidelberg, Berlin, and Göttingen. CAREER: 1866-1867, traveled in Europe; ca. 1868, opened private laboratory in N.Y.; 1872-1877, assistant in analytical chemistry, head of quantitative laboratory, and lecturer, School of Mines, Columbia Univ.; 1875-1877, professor of chemistry, Woman's Medical College of New York Infirmary; 1877-1887, professor of chemistry, Trinity College (Hartford); 1887, with an adequate income, retired and pursued private literary and scientific work, living first in New York City and later in Washington, D.C. MEMBERSHIPS: Vice-president of Am. Assoc. Advt. Sci., 1882; elected president of N.Y. Ac. Scis., 1893; member of other societies. SCIENTIFIC CONTRIBUTIONS: Active as researcher, bibliographer, and historian of science. Most successful research was on compounds of uranium, with results published especially during period 1866-1870; also investigated action of organic acids on minerals, with results published in three papers in *Annals of N.Y. Ac. Scis.* (1877-1882). Probably best known for bibliographical work, which began with bibliography of literature on uranium (published in *Annals of N.Y. Lyc. Nat. Hist.*, 1870); most important bibliographical effort was "Select Bibliography" [see Works]. In 1882, became interested in "singing beach" at Manchester, Mass., and traveled widely studying similar phenomena elsewhere in the world. Published several works on history of chemistry. Also interested in folklore; published several works in that field and was a founder of American Folklore Society (1887).

WORKS: "Select Bibliography of Chemistry, 1492-1892," *Smithson. Misc. Coll.* 36 (1893), with three additional parts published by Smithsonian between 1899 and 1904, which extended coverage to 1902; "Catalogue of Scientific and Technical Periodicals, 1665-1882," *Smithson. Misc. Coll.* 29 (1885). Also see *Roy. Soc. Cat.*, vols. 7, 9, 13, and *Popular Science Monthly* [below]. WORKS ABOUT: *DAB*, 2: 422-23 (Edward Hart); sketch of life to 1893, with abbreviated bibliographic references to works, in *Popular Science Monthly* 43 (1893): 688-95.

BONAPARTE, CHARLES LUCIEN (May 24, 1803, Paris, France—July 29, 1857, Paris, France). *Zoology*, especially *Ornithology*. Son of Lucien, diplomat and brother of Napoleon I, and Alexandrine (de Bleschamp) Bonaparte. Married Zénaïde C. J. Bonaparte (cousin), 1822; twelve children. EDUCATION: Educated in Italy. HONORS: M.A., Princeton Univ., 1825. CAREER: Ca. 1822-1828, lived in Philadelphia, where father-in-law, Joseph Bonaparte, resided; 1828-1850, resided in Italy and engaged in scientific and political activity; 1840, on death of father became prince of Canino and Musignano; 1850, returned to Paris, after initial opposition from Louis Napoleon; 1850-1857, resided in France and devoted time to scientific work, with special interest in the Muséum d'histoire naturelle. MEMBERSHIPS: Ac. Nat. Scis. Philad.; Academy of Sciences (Paris); and other societies. SCIENTIFIC CONTRIBUTIONS: In U.S. for only short time, but made great contribution to development of American ornithology, and also wrote on comparison of American and European birds. Of the early triumvirate of American ornithologists—A. Wilson, J. J. Audubon [qq.v.], and Bonaparte—latter was most informed and most able systematist and placed American studies on firm scientific foundation; *American Ornithology* [see Works] was prepared as supplement to work of Wilson; "Genera" [see Works] was first important classified arrangement of American birds, while *Conspectus* [see Works] was most significant ornithological publication. Work in Europe dealt not only with ornithology, but also with broad questions of classification in zoology, including consideration of both morphological characters and physiological factors.

WORKS: *American Ornithology*, 4 vols. (Philadelphia, 1825-1833); *Observations on the Nomenclature of Wilson's "Ornithology"* (Philadelphia, 1826), 25 pp., originally published in *Journal of Ac. Nat. Scis. Philad.* 3-5 (1823-1825); "The Genera of North American Birds, and a Synopsis of the Species Found Within the Territory of the U.S., Systematically Arranged in Orders and Families," *Annals of N.Y. Lyc. Nat. Hist.* 2 (1828): 1-128, 293-451; *Conspectus Generum Avium* (Leyden, 1850). Also see *DSB* and Meisel, *Bibliog. Amer. Nat. Hist.* MSS: Am. Phil. Soc. (Hamer); Muséum d'histoire naturelle (*DSB*). WORKS ABOUT: *DSB*, 2: 281-83 (G. Petit); Spencer Trotten, sketch, *Cassinia: Proceedings of the Delaware Valley Ornithological Club* 9 (1905): 1-5.

BOND, GEORGE PHILLIPS (May 20, 1825, Dorchester, Mass.—February 17, 1865, Cambridge, Mass.). *Astronomy*. Son of William Cranch [q.v.], director of Harvard College Observatory, and Selina (Cranch) Bond. Married Harriet Gardner Harris; three children. EDUCATION: 1845, A.B., Harvard

College. CAREER: At Harvard College Observatory, assistant observer (1845-1859) and director (1859-1865). MEMBERSHIPS: Am. Ac. Arts Scis.; foreign associate of Royal Astronomical Society, London. SCIENTIFIC CONTRIBUTIONS: Worked so closely with father at observatory that it frequently is difficult to isolate their individual contributions. At age twenty-three discovered satellite Hyperion while working with father on observations of Saturn; two years later discovered crepe ring of Saturn; and in 1851 produced theory that rings of Saturn were in fluid rather than solid state; work on Donati comet of 1858 [see Works] widely praised for unsurpassed completeness, and for that work was first American recipient of Royal Astronomical Society's gold medal (1865). Most important contribution to science was pioneering advocacy and use of photography as astronomical tool; with father experimented with use of daguerreotype, 1847-1851, and produced first photograph of star; in 1857, conducted further experiments using wet collodion photography; used photography to map heavens, to measure double stars and brightness of stars, and to determine stellar parallax, and in 1860 produced important work on comparative brightness of sun, moon, and Jupiter. At time of death, working on Orion nebula, an investigation begun by father.

WORKS: Monograph on Donati comet of 1858 in *Annals of Harvard College Observatory* 3 (1862); "Observations Upon the Great Nebula of Orion," in ibid. 5 (1867). Bibliography in Edward S. Holden, *Memorials of William Cranch Bond and His Son* . . . (San Francisco, 1897). Also see *Roy. Soc. Cat.*, vols. 1, 6 (additions), 12. MSS: Harv. U.-UA (*NUCMC* 65-1227); Lick Observatory, Mt. Hamilton, Calif. (*NUCMC* 62-3730). WORKS ABOUT: *DAB*, 2: 430-31 (Raymond S. Dugan); *DSB*, 2: 284-85 (Owen Gingerich). Also see Bessie Z. Jones and Lyle Boyd, *The Harvard College Observatory: The First Four Directorships* . . . (Cambridge, Mass., 1971).

BOND, WILLIAM CRANCH (September 9, 1789, Falmouth [now Portland], Maine—January 29, 1859, Cambridge, Mass.). *Astronomy.* Son of William—lumber exporter, silversmith, and clockmaker—and Hannah (Cranch) Bond. Married cousin, Selina Cranch, 1819; after her death in 1831, married her sister, Mary Roope Cranch; six children, including George P. Bond [q.v.]. EDUCATION: Left public schools early to help support family; apprenticed to father as silversmith and clockmaker. HONORS: A.M., Harvard Univ., 1842. CAREER: Watchmaker; 1815, during plans for trip to Europe, commissioned by Harvard Univ. to study European observatories and astronomical equipment; 1839-1859, director, Harvard College Observatory, before 1846 without salary; 1848-1859, Phillips professor of astronomy, Harvard. MEMBERSHIPS: Am. Ac. Arts Scis. and other societies; first American foreign associate of Royal Astronomical Society. SCIENTIFIC CONTRIBUTIONS: By time of War of 1812, his instruments were used to rate chronometers of most ships sailing from Boston and performed similar service for number of expeditions to determine longitude in eastern U.S. Self-taught in astronomy and made independent discovery of comet of 1811; built astronomical observatory in home at Dorchester, Mass., and pioneered in rating of chronometers and in meteorological and magnetic observations. In 1838, appointed by U.S. government to cooperate with exploring expedition led by C. Wilkes [q.v.], and in 1839 asked by Harvard to move family and equipment to Cambridge and to become astronomical observer for university. Comet of 1843 elicited interest and generosity of citizenry, and in 1847 a fifteen-inch refractor telescope, as large as any then in existence, was mounted at Cambridge, for which Bond designed dome and observing chair; with Harvard telescope Bond and son George undertook studies of Saturn, and daguerreotype process was first used for astronomical purposes at Harvard Observatory; also conducted detailed studies of Orion nebula and was responsible for device that made chronograph an instrument of great accuracy. Work intimately associated with that of son George. *Roy. Soc. Cat.* lists over ninety contributions of W. C. Bond, most relating to work at Harvard.

WORKS: Bibliography in Edward S. Holden, *Memorials of William Cranch Bond and His Son* . . . (San Francisco, 1897). MSS: Harv. U.-UA (*NUCMC* 65-1228). WORKS ABOUT: *DAB*, 2: 434-35 (Raymond S. Dugan); *DSB*, 2: 285 (Owen Gingerich). Also see Bessie Z. Jones and Lyle Boyd, *The Harvard College Observatory: The First Four Directorships* . . . (Cambridge, Mass., 1971).

BOOTH, JAMES CURTIS (July 28, 1810, Philadelphia, Pa.—March 21, 1888, Haverford, Pa.). *Chemistry.* Son of George and Ann (Bolton) Booth. Married Margaret M. Cardoza, 1853; no information on children. EDUCATION: 1829, A.B., Univ. of Pennsylvania; for year or more, studied chemistry at Rensselaer Polytechnic Institute; ca. 1832-1836, studied chemistry with Frederick Wöhler and Gustav Magnus in Germany and attended lectures in Vienna. HONORS: LL.D., Univ. of Lewisburg (Bucknell Univ.), 1867; Ph.D., Rensselaer Polytechnic Institute, 1884. CAREER: 1831-1832, taught chemistry in Flushing, N.Y., before going to Germany; 1836-1888, conducted student chemical laboratory in Philadelphia, which in 1878 became known as Booth, Garrett & Blair; 1836-1837, assistant in Pennsylvania Geological Survey; 1836-1845, professor of chemistry applied to arts, Franklin Institute; 1837-1838, Delaware state geologist; 1842-1845, teacher of chemistry, Philadelphia Central High School; 1849-1888, melter and refiner, Philadelphia Mint; 1851-1855, professor of chemistry applied to arts, Univ. of Pennsylvania. MEMBERSHIPS: Am. Phil. Soc. and other societies; president of Am. Chem. Soc., 1883 and 1884. SCIENTIFIC CONTRIBUTIONS: An accomplished analytical chemist whose unique instructional laboratory became a major American center of chemical education; especially interested in application of informed chemical investigation to solution of practical problems. In 1841, published official report on geology of Delaware. In mid-1840s pioneered in America in use of polariscope to analyze sugar and molasses; particularly skillful in dealing with metallurgical processes, a talent used in work at Philadelphia Mint. *Encyclopaedia* [see Works] contained over 1,000 pages and incorporated chemical knowledge from Europe as well as from Great Britain; paper "On Conversion of Benzoic Acid" [see Works] said to be one of early American contributions to organic chemistry.

WORKS: *Encyclopaedia of Chemistry, Practical and Theoretical* (Philadelphia, 1850), with C. Morfit [q.v.]; "On the Conversion of Benzoic Acid into Hippuric," *Proceedings of Am. Phil. Soc.* 3 (1843): 129, with M. Boyé [q.v.]. See *Roy Soc. Cat.*, vols 1, 13, and Meisel, *Bibliog. Amer. Nat. Hist.* WORKS ABOUT: *DAB*, 2: 447-48 (Lyman C. Newell); Edgar F. Smith, *James Curtis Booth* ([Philadelphia?], 1922), 17 pp.

BOSS, LEWIS (1846-1912). *Astronomy.* Director of Dudley Observatory, Albany.

WORKS ABOUT: See *AMS*; *DAB*; *DSB*.

BOWDITCH, HENRY PICKERING (1840-1911). *Physiology.* Professor of physiology and dean, Harvard Medical School.

MSS: Harv. U.-MS (*NUCMC* 62-3533). WORKS ABOUT: See *AMS*; *DAB*; *DSB*.

BOWDITCH, NATHANIEL (March 26, 1773, Salem, Mass—March 16, 1838, Boston, Mass.). *Astronomy*; *Mathematics.* Son of Habakkuk, shipmaster and cooper, and Mary (Ingersoll) Bowditch. Married Elizabeth Boardman, 1798; Mary Ingersoll (cousin), 1800; eight children by second marriage. EDUCATION: At age ten, left school to work in father's cooper shop; ca. 1785-1795, apprenticed to ship chandlers, and at same time engaged in intensive self-education. HONORS: A.M., Harvard Univ., 1802. CAREER: 1795-1803, five sea voyages, first as clerk, three as supercargo, and last as master and supercargo; 1804-1823, president, Essex Fire and Marine Insurance Co.; 1823-1838, actuary for Massachusetts Hospital Life Insurance Co., Boston. MEMBERSHIPS: Roy. Soc.; president, Am. Ac. Arts Scis., 1829-1838. SCIENTIFIC CONTRIBUTIONS: Began study of algebra at age fourteen and computed almanac when only fifteen; at age seventeen began study of Latin in order to read Newton's *Principia* and made much use of library of Irish chemist, Dr. Richard Kirwan, in Philosophical Library Company of Salem. As scientist, characterized not by original discoveries, but by useful criticisms of work of others; nearly all scientific work done in Salem, prior to removal to Boston in 1823. In 1799, prepared American edition of John H. Moore's *Practical Navigator*, with ten editions during his lifetime and many after death; contributed mathematical solutions to R. Adrain [q.v.] for *Analyst* (1808, 1814), and *Mathematical Diary* (1825-1828); between 1804 and 1820 published twenty-three papers on astronomy, mathematics, and physics in *Memoirs of Am. Ac. Arts Scis.*; that on meteor explosion over Weston, Conn., in 1807 (first published in *Nicholson's Journal* 28 [1811]: 89-98, 206-19) was of particular interest, but of greater scientific importance was paper on pendulum [see Works]. Major scientific project was translation, with notes, of first four volumes of Laplace's *Mécanique Céleste*, which was published at own expense, 1829-1839, and was in English-speaking world frequently preferred edition.

WORKS: "On the Motion of a Pendulum Suspended from Two Points," *Memoirs of Am. Ac. Arts Scis.* 3 (1815), pt. 2, pp. 413-36. Best bibliography in Peabody Museum's *Catalogue of a Special Exhibition of Manuscripts, Books, Portraits and Personal Relics of Nathaniel Bowditch . . .* (Salem, Mass., 1937); also see *Roy. Soc. Cat.*, vol. 1. MSS: Bost. Pub. Lib. (Hamer); Essex Institute (Salem, Mass.). WORKS ABOUT: *DAB*, 2: 496-98 (Raymond C. Archibald); *DSB*, 2: 368-69 (Nathan Reingold). Also see Robert E. Berry, *Yankee Stargazer* (New York, 1941), a popularized biography.

BOWDOIN, JAMES (August 7, 1726, Boston, Mass. —November 6, 1790, Boston, Mass.). *Physics*; *Astronomy.* Son of James, wealthy merchant, and Hannah (Portage) Bowdoin. Married Elizabeth Erving, 1748; two children. EDUCATION: 1745, A.B., Harvard. HONORS: LL.D., Harvard, 1782, and Univ. of Edinburgh. CAREER: Merchant and capitalist; 1753-1756, elected to Massachusetts General Court; 1757-1769 and 1770-1773, member of Council, siding with colony against Great Britain; chosen member of Continental Congress, but unable to accept because of health; 1775-1777, appointed member of executive council by Massachusetts Provincial Congress; 1779, elected president of Massachusetts Constitutional Convention; 1785-1787, governor of

Massachusetts; 1788, delegate to Massachusetts Convention on adoption of federal Constitution. MEMBERSHIPS: Roy. Soc. and other societies; first president of Am. Ac. Arts Scis. SCIENTIFIC CONTRIBUTIONS: In midst of active business and public career, indulged an interest in science that began in 1750 with visit to B. Franklin [q.v.]; carried on correspondence with Franklin on electricity and other topics, and when Franklin was in England, he presented Bowdoin's letters before Roy. Soc. Turned attention to astronomy at Franklin's suggestion, and in 1780s became interested in subject of theory of light, Newton's theory of gravity, and related philosophical matters. Achieved considerable reputation among Americans for scientific work.

WORKS: Essentials of early correspondence with Franklin included in latter's *Experiments and Observations on Electricity* (London, 1751); "An Improvement Proposed for Telescopes," *London Magazine*, November 1761, pp. 602-4; "Observations Upon a Hypothesis for Solving the Phenomena of Light . . . ," *Memoirs of Am. Ac. Arts Scis.* 1 (1785): 187-94; "Observations on Light, and the Waste of Matter in the Sun and fixt Stars . . . ," ibid., pp. 195-207; "Observations, Tending to Prove by Phaenomena and Scripture, the Existence of an Orb Which Surrounds the Whole Visible Material System, . . . ," ibid., pp. 208-33. Bibliography in Shipton [below]. MSS: Mass. Hist. Soc. (Shipton). WORKS ABOUT: *DAB*, 2: 498-501 (William A. Robinson); Shipton, *Biog. Sketches, Harvard* 11 (1960): 514-50; evaluation of Bowdoin's scientific work by Frederick E. Brasch, *Scientific Monthly* 33 (1931): 461-63.

BOWSER, EDWARD ALBERT (June 18, 1837, Sackville, New Brunswick, Canada—February 19, 1910, Honolulu, Hawaii). *Mathematics.* No information on parentage or marital status. EDUCATION: Attended Santa Clara College (California); 1865, graduated, Albany (N.Y.) Normal School; B.S. in 1868, M.S. and C.E. in 1871, Rutgers College. HONORS: LL.D., Lafayette College in 1881 and Rutgers in 1905. CAREER: After college, taught school for two years in the West; ca. 1865/1866, taught at Brooklyn Polytechnic Institute; at Rutgers College, tutor in engineering and mathematics (1868-1870), adjunct professor of mathematics (1870-1871), and professor of mathematics and engineering (1871-1904); 1875-1895, assistant for U.S. Coast and Geodetic Survey in New Jersey; 1901-1904, on leave from Rutgers, traveled in Europe, Asia, Middle East. MEMBERSHIPS: Am. Assoc. Advt. Sci.; Am. Math. Soc. SCIENTIFIC CONTRIBUTIONS: Prepared number of very good textbooks on mathematics and engineering used in many educational institutions. In work for Coast Survey, determinations of geographic points were so well executed that they were used for all other maps of state of New Jersey; 1882 survey of border of New York and New Jersey, prepared for joint commission of the two states, was later accepted by both states.

WORKS: Among textbooks were those on *Analytic Geometry* (1880), *Differential and Integral Calculus* (1880), *Analytic Mechanics* (1884), *Hydromechanics* (1885), *College Algebra* (1888), *Treatise on Roofs and Bridges* (1898), all published in New York. Does not appear to have prepared any research papers in science and mathematics and has no entries in *Roy. Soc. Cat.* See *Natl. Un. Cat. Pre-1956 Imp.* for full titles of works and for additional titles. MSS: Rutgers Univ. Library (*NUCMC* 65-1573). WORKS ABOUT: *NCAB*, 19 (1926): 415; *Who Was Who in America*, vol. 1.

BOYÉ, MARTIN HANS (1812-1909). *Chemistry; Physics; Geology.* Geologist and chemist with First Pennsylvania Geological Survey; professor of chemistry and natural philosophy, Philadelphia Central High School.

WORKS ABOUT: See *AMS*; *DAB.*

BOYLSTON, ZABDIEL (March 9, 1680, Brookline, Mass.—March 1, 1776, Brookline, Mass.). *Medicine* (smallpox inoculation). Son of Thomas, physician, and Mary (Gardner) Boylston. Married Jerusha Minot, 1705; eight children. EDUCATION: Studied medicine with father and with Dr. John Cutter of Boston. CAREER: Successfully operated apothecary shop in Boston, and practiced medicine, accumulating considerable wealth; 1721, began administering inoculations for smallpox; 1724-1726, in England; ca. 1740, retired from medical practice and raised horses on his Brookline farm. MEMBERSHIPS: Roy. Soc. SCIENTIFIC CONTRIBUTIONS: Remembered primarily as first American physician to inoculate for smallpox, beginning June 26, 1721, following suggestion made earlier by C. Mather [q.v.]; after consulting evidence provided by Mather from *Phil. Trans. of Roy. Soc.* and after interviewing blacks in Boston who had been inoculated in Africa, Boylston inoculated 241 persons in a little more than six months. Practice of inoculation aroused bitter public opposition, which for a time threatened lives of Mather and Boylston, and during this period series of letters and pamphlets appeared, probably written by Mather with Boylston's aid; first publication on inoculation issued under Boylston's own name was *An Historical Account* [see Works], which showed author to have been careful observer and efficient record keeper. While in London, lectured to Roy. Soc. and Royal College of Physicians on observations regarding inoculation. Interested in general natural history and collected plants and animals; sent communications to Roy. Soc., including observations on rattlesnakes in 1735, and final communication was on cure for rattlesnake bite and on balsam taken from bark of white cedar.

WORKS: *Some Account of What is Said of Inoculating or Transplanting the Small Pox By the Learned Dr. Emanuel Timonius, and Jacobus Pylarinus . . . Answer to the scruples of many about the Lawfulness of this method* (Boston, 1721), with pts. 1 and 3 by Mather and pt. 2 mainly by Boylston; *An Historical Account of the Small-Pox inoculated in New England* (London, 1726; 2d ed., Boston, 1730); "Ambergris in Whales," *Phil. Trans. of Roy. Soc.* 33 (1724): 193. Publications are mentioned in *DAB*. MSS: British Museum, Roy. Soc., and Mass. Hist. Soc. (*DAB*). WORKS ABOUT: *DAB*, 2: 535-36 (John F. Fulton); Stearns, *Sci. in Brit. Colonies*, pp. 435-42. Also see Gerald M. Mager, "Zabdiel Boylston: Medical Pioneer of Colonial Boston," *Dissertation Abstracts International* 36 (1976): 6270-A.

BRACE, De WITT BRISTOL (January 5, 1859, Wilson, N.Y.—October 2, 1905, Lincoln, Neb.). *Optics.* Son of Lusk and Emily (Bristol) Brace. Married in 1901; no information on children. EDUCATION: 1881, bachelor's degree, Boston Univ.; 1881-1883, graduate study at Massachusetts Institute of Technology and Johns Hopkins Univ.; 1885, Ph.D., Univ. of Berlin. CAREER: 1886, assistant professor of physics, Univ. of Michigan; at Univ. of Nebraska, instructor in physics and astronomy (1887-1888) and professor of physics (1888-1905), until 1896, as sole member of department. MEMBERSHIPS: British and Am. Assocs. Advt. Sci. SCIENTIFIC CONTRIBUTIONS: During early years at Nebraska, had little time for personal research, but during that period built up physics department and began what became department of electrical engineering; after 1896, was able to do more personal research, especially study of factors (such as magnetism and electric field, pressure, strain, and "ether drift") that affect velocity of light propagation through matter; in course of this research, invented several new instruments, including spectropolariscope, and spectrophotometer which bears his name. Best remembered for experiments testing Lorentz-Fitzgerald contraction hypothesis in 1904, in which he repeated Lord Rayleigh's experiments in more accurate manner and definitely established that double refraction was not caused by movement of refracting medium through ether; at first mistakenly thought his experiments disproved contraction hypothesis, but Joseph Larmor later showed that double refraction need not occur. In addition to works by students, personally published about fifteen articles.

WORKS: "Observations on Light Propagated in a Dielectric Normal to the Lines of Force," *Philosophical Magazine* 44 (1897): 342-49; "Double Refraction in Matter Passing Through the Ether," ibid., n.s. 7 (1904): 317-28; "Observation on the Circular Components in the 'Faraday Effect,'" *Nature* 62 (1900): 368-69. Also see *Roy. Soc. Cat.*, vol. 12, and Poggendorff, *Biographisch-literarisches Handwörterbuch*, 5: 157. MSS: No collection is known; some letters in Board of Regents papers, Univ. of Nebraska (Lincoln) Library—Archives. WORKS ABOUT: *DAB*, 2: 540-41 (Clarence A. Skinner); *DSB*, 2: 382-83 (Eugene Frankel).

BRACKENRIDGE, WILLIAM DUNLOP (June 10, 1810, Ayr, Scotland—February 3, 1893, Baltimore, Md.). *Botany.* No information on parentage or marital status. EDUCATION: Trained as gardener; student of Friedrich Otto, garden director at Berlin. CAREER: Started as gardener's boy; became head gardener for Dr. Patrick Neill, Edinburgh, and then for several years in Europe, especially in Poland and with Friedrich Otto; ca. 1837, arrived in U.S. as employee of Robert Buist, Philadelphia nurseryman; 1838-1842, horticulturist with U.S. Exploring Expedition to the Pacific (Wilkes Expedition); 1842-1854, was in charge of living plants and prepared report on ferns for the expedition; 1855, purchased thirty acres near Baltimore, Md., where he worked as nurseryman and landscape architect; for several years horticultural editor for *American Farmer*. SCIENTIFIC CONTRIBUTIONS: Reputation rests on work done for Wilkes Expedition, a position secured after resignation of A. Gray [q.v.]; most of time on expedition spent working with naturalist C. Pickering [q.v.], and in October 1841, on Mt. Shasta in California, collected specimen of California pitcher plant not definitely classified until flower was found ten years later; had responsibility for some 250 living plants and seeds gathered by expedition, as well as for report on ferns. Had reputation as good field botanist and horticulturist, but lacked training in systematics of botany and in Latin. Labors with living plants and seeds, first in greenhouse behind Patent Office and after 1850 in new National Botanic Garden near Capitol, consumed much of his effort; J. Torrey [q.v.], and especially Gray, helped with report on ferns and probably did much of work on Latin description and nomenclature, but despite this aid, final report is considered as Brackenridge's and is his one important botanical publication [see Works]. Fire in 1856 destroyed most of the work, and it now is quite rare.

WORKS: *Filices, including Lycopodiaceae and Hydropterides*, vol. 16 of publications of Wilkes Expedition, issued in 2 pts., text in 1854 and folio atlas of 46 plates in 1855 (Philadelphia). See Meisel, *Bibliog. Amer. Nat. Hist.* MSS: Md. Hist. Soc. (*NUCMC* 67-1336); Smithson. Instn. Library

(Hamer). WORKS ABOUT: *DAB*, 2: 545-46 (Donald C. Peattie); John H. Barnhart, "Brackenridge and His Book on Ferns," *Journal of N.Y. Bot. Gard.* 20 (1919): 117-24.

BRACKETT, CYRUS FOGG (1833-1915). *Physics; Electrical engineering.* Professor of physics, Princeton Univ.

WORKS ABOUT: See *AMS*; *Who Was Who in America*, vol. 1.

BRADLEY, FRANK HOWE (September 20, 1938, New Haven, Conn.—March 27, 1879, near Nacoochee, Ga.). *Geology.* Son of Abijah and Eliza Collis (Townsend) Bradley. Married Sarah M. Bolles, 1867; four children. EDUCATION: 1863, A.B., Yale College; ca. 1864/65, spent year as student in chemistry laboratory, Sheffield Scientific School, Yale. CAREER: 1863-1864, teacher in Hartford, Conn.; 1865-1866, in Panama area, collected coral and other zoological specimens, some for Yale Museum; 1867-1868, assistant geologist for survey of Illinois; 1868-1869, professor of natural sciences, Hanover College (Indiana); 1869, assistant geologist for survey of Indiana; 1869-1875, professor of geology and mineralogy, Univ. of Tennessee; 1872, attached to U.S. Geol. Surv. under F. V. Hayden [q.v.] and assigned as chief geologist to Snake River division, Idaho; 1875-1879, engaged in private mining operations and died in mine accident. SCIENTIFIC CONTRIBUTIONS: Early interest in geology led to discovery of new trilobite species in Potsdam sandstone at Keesville, N.Y., at age nineteen, announcement of which was made at 1857 Montreal meeting of Am. Assoc. Advt. Sci.; also proved existence of crinoids in same layer. Prepared report on carboniferous rocks of Vermillion County for Indiana survey; report on Snake River work with U.S. Geol. Surv. included identification of fossils of Quebec group in Idaho and at base of Teton Mountains. In 1876, published small geological chart of the U.S.

WORKS: "Description of a New Trilobite from the Potsdam Sandstone," with note by Elkanah Billings, *Proceedings of Am. Assoc. Advt. Sci.* 14 (1860): 161-66 (also in *Am. Journ. Sci.* 2d ser. 30 [1860]); "On the Silurian Age of the Southern Appalachians," *Am. Journ. Sci.* 3d ser. 9 (1875): 279-88, 370-83. See *Natl. Un. Cat. Pre-1956 Imp.* and *Roy. Soc. Cat.*, vols. 1, 7, 9, 12. WORKS ABOUT: *DAB*, 2: 570 (George P. Merrill); James D. Dana, obituary, *Am. Journ. Sci.*, 3d ser. 17 (1879): 415-16; *Obituary Record of Graduates of Yale University* (1870-1880), pp. 369-70.

BRANDEGEE, MARY KATHARINE LAYNE CURRAN (October 28, 1844, western Tennessee—April 3, 1920, Berkeley, Calif.). *Botany.* Daughter of Marshall Bolling, pioneering miller and farmer, and Mary (Morris) Layne. Married Hugh Curran, 1866; Townshend S. Brandegee [q.v.], 1889; apparently no children. EDUCATION: Early education uncertain; 1878, M.D., Univ. of California, San Francisco. CAREER: Said to have taught school, probably near Folsom, Calif., where part of childhood was spent; 1878, entered medical practice, but had little success; 1883-ca. 1894, curator of herbarium, Calif. Ac. Scis.; 1894-1906, lived in San Diego, while continuing botanical studies; 1906-1920, worked, without salary or position, in herbarium at Univ. of California, Berkeley, where husband served as honorary curator. MEMBERSHIPS: Calif. Ac. Scis. SCIENTIFIC CONTRIBUTIONS: Began botanical studies as outgrowth of study of materia medica, though initially more interested in insects; active botanical collecting began in 1882 and subsequently botanized throughout the state, becoming a leading authority on California plants, though a projected flora of state never materialized. In various studies was particularly attentive to variation and stages of development of plants. In early days, lack of other comparable collections in region made her position as curator at Calif. Ac. Scis. one of particular importance for botanical studies. Responsible for origin of academy's series of *Bulletins* and served as acting editor; 1900-1908, edited *Zoe*, publication established in 1890 with aid of president of academy and husband to record her own botanical observations and criticisms, along with those of other persons. In 1891, organized a botanical club associated with academy; was actively engaged in number of botanical projects, few being completed, and was particularly noted for critical reviews of work of others.

WORKS: Bibliography in Setchell (below), pp. 174-77; nearly all publications (1884-1914) in *Proceedings* and *Bulletins* of Calif. Ac. Scis. or in *Zoe*. MSS (and herbarium): Herbarium of Univ. of California, Berkeley (*Nota. Am. Wom.*). WORKS ABOUT: *Nota. Am. Wom.*, 1: 228-29 (Hunter Dupree and Marian L. Gade); William A. Setchell, "T. S. Brandegee and M. K. L. C. Brandegee," Univ. of California, *Publications in Botany* 13 (1926): 156-78.

BRANDEGEE, TOWNSHEND STITH (February 16, 1843, Berlin, Conn.—April 7, 1925, Berkeley, Calif.). *Botany.* Son of Elishama, physician, and Florence (Stith) Brandegee. Married Mary Katharine (Layne) Curran [Brandegee] [q.v.], 1889; apparently no children. EDUCATION: 1870, Ph.B., Sheffield Scientific School, Yale. CAREER: ca. 1862, enlisted for two years in Company G, First Connecticut Volunteers; 1871, became county surveyor and city engineer, Cañon City, Colo.; 1875, appointed assistant topographer and botanical collector, Hayden Exploring Expedition of Southwest Colorado; subsequently, served as engineer for various railway surveys in the West; and took part in survey for production of forest map of Adirondacks in New York and similar work with Northern Transcontinental Survey; during all this time, also engaged in botanical collecting; beginning in late 1880s, abandoned engineering in favor of botanical interests; ca. 1889-1894, closely associated with Calif. Ac. Scis., where wife was curator of herbarium; 1894-1906, resided at San Diego; 1906, became honorary curator of botany, Univ. of California, Berkeley. MEMBERSHIPS: Botanical Society of America and other societies. SCIENTIFIC CONTRIBUTIONS: From earliest days in Colorado collected ferns for John H. Redfield, eventually collecting plants of all kinds, with encouragement of Redfield and A. Gray [q.v.]; later, for C. S. Sargent [q.v.], collected timber specimens for U.S. Census and for Jesup Collection at Am. Mus. Nat. Hist.; turned attention to botany of California, especially of its islands, and eventually became noted authority on flora of Lower California and adjacent and southern parts of Mexico. During early years, was essentially collector, results of which were worked up by others; but after marriage and association with Calif. Ac. Scis. at San Francisco, began publishing results of own work. Especially noted as botanical explorer; chief work on Mexican flora based on collections of Dr. C. A. Purpus [see Works].

WORKS: *Plantae Mexicanae Purpusianae,* I-XII, in vols. 3, 4, 6, 7, 10 of Univ. of California, *Publications in Botany* (1909-1924). Bibliography of works (1876-1924) in Setchell [below], pp. 169-73. MSS: Herbarium of Univ. of California, Berkeley [see account of wife, in *Nota. Am. Wom.*]. WORKS ABOUT: *DAB,* 2:599-600 (Donald C. Peattie); William A. Setchell, "T. S. Brandegee and M.K.L. C. Brandegee," Univ. of California, *Publications in Botany* 13 (1926) : 156-78.

BRANNER, JOHN CASPER (1850-1922). *Geology.* Professor of geology and president, Stanford Univ.

MSS. Hunting. Lib. (*NUCMC* 61-2199); Stanford Univ. Archives (*NUCMC* 67-2095). WORKS ABOUT: See *AMS; DAB.*

BRASHEAR, JOHN ALFRED (1840-1920). *Astronomy; Physics; Mechanics.* Manufacturer of astronomical and physical apparatus.

MSS: See *DAB.* WORKS ABOUT: See *AMS; DAB; DSB.*

BRATTLE, THOMAS (June 20, 1658, Boston, Mass. —May 18, 1713, Boston, Mass.). *Astronomy.* Son of Thomas, wealthy trader and landowner, and Elizabeth (Tyng) Brattle. Never married. EDUCATION: 1676, A.B., Harvard College; 1689, returned from period of travel and study abroad. CAREER: Inherited wealth from father (died 1683) and engaged in business; 1693-1713, treasurer, Harvard College. MEMBERSHIPS: Probably member of Philosophical Society formed in Boston, 1683. SCIENTIFIC CONTRIBUTIONS: Observed comet of 1680 (called "Newton's comet"), in which he made one of earliest recorded uses of three-and-one-half-foot telescope and other astronomical apparatus that had been donated to Harvard College, ca. 1671, by J. Winthrop (d. 1676) [q.v.]; results of Brattle's observations of comet were published in local almanac for 1681 and also made known to John Flamsteed, astronomer royal at Greenwich, who sent them to Newton; and in *Principia Mathematica* Newton referred favorably to contributions of "the observer in New England" who took "the position of the comet with reference to the fixed stars." Made additional astronomical observations up to time of death, reporting to Roy. Soc. on solar eclipses in 1694 and 1703 and on lunar eclipses in 1700, 1703, and 1707; visited with Flamsteed at Greenwich in 1689 and also corresponded with the British astronomer. In 1711, sent to an English correspondent his account of smallpox epidemics in Boston, which was read before Roy. Soc.; on basis of previous epidemics in Boston since 1666, Brattle projected in this paper a twelve-year cycle and predicted next outbreak would occur 1714 (which, however, did not occur until 1721).

WORKS: "An Account of Some Eclipses of the Sun and Moon, observed by Mr. Tho. Brattle, at Cambridge, about four miles from Boston in New-England, when the Difference of Longitude between Cambridge and London in determin'd, from an Observation made of one of them at London," by J. Hodgson, *Phil. Trans. of Roy. Soc.* 24 (1704-1705) : 1630-38; "Observatio Eclipsis Lunaris peracta Bostonij Nov. Anglorum, die quinto Aprilis vespere, A.D. 1707," ibid. 25 (1706-1707) : 2471-72. WORKS ABOUT: *DAB,* 2:606-7 (George H. Genzmer); Stearns, *Sci. in Brit. Colonies,* pp. 153-56, 401-2, 487.

BREWER, THOMAS MAYO (November 21, 1814, Boston, Mass.—January 23, 1880, Boston, Mass.). *Ornithology; Oology.* Son of Col. James Brewer, Revolutionary War patriot. Married Sally R. Coffin, 1849; survived by one child. EDUCATION: 1835, A. B., Harvard College; 1838, M.D., Harvard Medical Scool. CAREER: 1838-ca. 1840, medical practice in Boston; ca. 1840-1857, contributor and later editor, *Boston Atlas;* 1844-1880, member, Boston School Committee; 1857-1875, member of publishing firm of Swan and Tileston, later Brewer and Tileston, and in 1875 became president; 1875, retired. MEMBERSHIPS: Bos. Soc. Nat. Hist. SCIENTIFIC CONTRIBUTIONS: In midst of active business career, developed international reputation as ornithologist and was first American noted for oological work, which included study of breeding habits and life histories of birds. Catalogue of Massachusetts birds [see Works] added forty-five species and increased list compiled by E. Hitchcock [q.v.] by one-quarter; received number of acknowledgments from J. J. Audubon [q.v.] for information provided for *Birds in America;* in 1840, prepared edition of *American Ornithology* by A. Wilson [q.v.], which for first time made that work available to anyone interested in subject and included Brewer's synopsis of all then-known North American birds. First part of *Oology* [see Works] appeared in 1857, including life histories and accounts of geographical distribution of birds and featuring colored illustrations of eggs; but cost of printing plates was so high that further volumes never appeared. Material on bird biographies eventually published in *History* [see Works], done in collaboration with S. F. Baird and R. Ridgway [qq.v.], with estimated two-thirds of text by Brewer; also left manuscripts used extensively in two additional volumes on *Water Birds* (1884). Published many papers and notes in *Proceedings of Bos. Soc. Nat. Hist.*

WORKS: Supplement to Hitchcock's catalogue of Massachusetts birds, in *Boston Journal of Natural History* 1 (1837): 435-39; "North American Oology," pt. 1, *Smithson. Contr. Knowl.* 11 (1857); *A History of North American Birds,* 3 vols. (Boston, 1874) with S. F. Baird and R. Ridgway. See *Roy. Soc. Cat.,* vols. 1, 7, 9, 12, 13. MSS: Letters to Baird in Smithson. Instn. (*DAB*); other papers in Harv. U.-MCZ. WORKS ABOUT: *DAB,* 3: 24-25 (Witmer Stone); W. L. McAtee, "Thomas Mayo Brewer," *Nature Magazine* 46 (November 1953): 468.

BREWER, WILLIAM HENRY (1828-1910). *Agriculture; Botany; Geology; Chemistry.* Professor of agriculture, Sheffield Scientific School, Yale.

MSS: Yale U.-HMC (*NUCMC* 71-2011). WORKS ABOUT: See *AMS; DAB.*

BREWSTER, WILLIAM (1851-1919). *Ornithology.* Curator of mammals and birds, Museum of Comparative Zoology, Harvard Univ.; accumulated private ornithological museum.

MSS: Ac. Nat. Scis. Philad. (Hamer); Harv. U.-MCZ. WORKS ABOUT: See *AMS; DAB.*

BRIDGES, ROBERT (March 5, 1806, Philadelphia, Pa.—February 20, 1882, Philadelphia, Pa.). *Botany; Chemistry.* Son of Culpeper, merchant [?], and Sarah (Cliffton) Bridges. Never married. EDUCATION: 1824, A.B., Dickinson College; 1828, M.D., Univ. of Pennsylvania. CAREER: 1826-1864, private assistant to F. Bache [q.v.] at Bache's chemical lectures at Franklin Institute, at Phildelphia College of Physicians, and finally at Jefferson Medical School; 1828, began several years of medical practice; 1830-1840, vaccine physician in southwestern district, Philadelphia, and district physician during cholera outbreak in 1832; 1839-1846, assistant editor, *Am. Journ. Pharm.;* 1842-1879, professor of general and pharmaceutical chemistry, Philadelphia College of Pharmacy, and 1860-1882, chairman of its board of trustees; 1846-1848, professor of chemistry, Franklin Medical College; 1842-1860, taught chemistry, Philadelphia Association for Medical Instruction. MEMBERSHIPS: Am. Phil. Soc.; president and vice-president, Ac. Nat. Scis. Philad. SCIENTIFIC CONTRIBUTIONS: During early years in Bache's laboratory, discovered residuum of nitre, reported by Bache [see Works]. Became effective teacher of chemistry; 1845, published American edition of George Fownes's *Elementary Chemistry,* in which he made additions to original work and later prepared several other editions, the last in 1878. Published several papers in *Proceedings of Ac. Nat. Scis. Philad.* during 1840s; 1839-1865, published papers of chemical-pharmaceutical nature in *Am. Journ. Pharm.;* assisted George B. Wood with editions of *U. S. Dispensatory* in 1865, 1870, and 1877, and was member of Committee on Revision of U.S. *Pharmacopoeia* in 1840 and 1870. Active member of several committees of Ac. Nat. Scis. Philad., including botany, entomology, crustacea, and herpetology. Chief botanical effort was index of the genera in the herbarium of the Ac. Nat. Scis., prepared by Bridges and Dr. Paul B. Goddard and presented to academy in 1835; in 1843 presented, for the botanical committee, a new index and also one for Menke Herbarium.

WORKS: "Residuum of Nitre, After Exposure to a Red Heat," *North American Medical and Surgical Journal* 5 (1828): 241-42, reported by F. Bache. See works either mentioned or listed in Ruschenberger [below]. WORKS ABOUT: *DAB,* 3:35-36 (Francis R. Packard); W. S. W. Ruschenberger, "Obituary Notice," *Proceedings of Am. Phil. Soc.* 21 (1884): 427-47.

BRIGHAM, ALBERT PERRY (1855-1932). *Geology; Physical geography.* Clergyman; professor of geology, Colgate Univ.

MSS: See *DAB.* WORKS ABOUT: See *AMS; DAB,* supp. 1.

BRITTON, ELIZABETH GERTRUDE KNIGHT (1858-1934). *Botany,* especially *Bryology.* Associated with husband, N. L. Britton [q.v.], director of N.Y. Bot. Gard.; a founder and active officer of Wild Flower Preservation Society of America.

MSS: In N.Y. Bot. Gard. and Harv. U.-FL (*Nota. Am. Wom.).* WORKS ABOUT: See *AMS* (Mrs. N. L.) ; *Nota. Am. Wom.*

BRITTON, NATHANIEL LORD (1859-1934). *Botany.* Director, N.Y. Bot. Gard.

MSS: Ac. Nat. Scis. Philad. (Hamer) ; N.Y. Bot. Gard. WORKS ABOUT: *AMS; DAB,* supp. 1; *DSB.*

BROADHEAD, GARLAND CARR (1827-1912). *Geology.* Professor of geology, Univ. of Missouri.

MSS: State Historical Society of Missouri, Columbia (*NUCMC* 64-266). WORKS ABOUT: See *AMS; DAB.*

BROCKLESBY, JOHN (October 8, 1811, West Bromwich, England [1820 came to U.S., settled with family at Avon, Conn.]—June 21, 1889, Hartford, Conn.). *Meteorology; Microscopy.* Son of John and Ann (Brooks) Brocklesby. Married Mary Louisa Kain, 1842; three children. EDUCATION: 1835, A.B., Yale College; 1836, began study of law, Hartford; 1838-1839, studied at Yale Law School. HONORS: LL.D., Hobart College, 1868. CAREER: 1835-1836, taught school, Avon, Conn.; 1838-1840, tutor in mathematics, Yale College; 1840-1842, law practice, Hartford; at Trinity College (Hartford), professor of mathematics and natural philosophy (1842-1873), professor of astronomy and natural philosophy (1873-1882), and acting president (1860, 1864, 1866-1867, 1874). MEMBERSHIPS: Am. Assoc. Advt. Sci. SCIENTIFIC CONTRIBUTIONS: Especially interested in meteorology and microscopy. Active contributor to meetings of Am. Assoc. Advt. Sci., and during 1850s published papers in *Proceedings* on topics such as range of temperature at Hartford, meteor viewed from that city (1850), swelling of springs before rain, tornado in Connecticut (1851), experiments on visual direction. In 1848, published paper on influence of color on dew *(Am. Journ. Sci.),* and in *Proceedings of Am. Assoc.* in 1874 published paper on relation of periodicity of rainfall in U.S. to periodicity of solar spots. Also wrote secondary school textbooks on meteorology (1848), microscopic world (1851), astronomy 1855), and physical geography (1868) and a handbook on microscopy for amateurs (1871).

WORKS: See *Roy. Soc. Cat.,* vols. 1, 9, and *Natl. Un. Cat. Pre-1956 Imp.* MSS: Meteorological journal (1864-1876) in Trinity College Library. WORKS ABOUT: *NCAB* 12 (1904) : 287; *Obituary Record of Graduates of Yale University* (1880-1890) : 540-41.

BROOKE, JOHN MERCER (December 18, 1826, near Tampa, Fla.—December 14, 1906, Lexington, Va.). *Invention; Astronomy; Physics.* Son of Bvt. Maj. George Mercer, U.S. Army, and Lucy (Thomas) Brooke. Married Elizabeth Selden Garnett; Mrs. Kate Corbin Pendleton; six children. EDUCATION: 1847, graduated from U.S. Naval Academy, Annapolis. CAREER: 1841-1845, midshipman, U.S. Navy, with service under Cdr. David G. Farragut off Brazil and with cruise around Cape Horn to Pacific; 1845-1847, at Annapolis; 1847-1849, routine naval duty; 1849-1850, hydrographic service with Coast Survey, under Lt. Samuel P. Lee; 1851-1853, service with M. F. Maury [q.v.] at Naval Observatory; 1854-1858, served on *Vincennes* during North Pacific Surveying and Exploring Expedition and then in Washington with Cdr. John Rodgers preparing expedition's charts and records for publication; 1855, promoted to lieutenant; 1858-1860, engaged in survey of route from California to China; 1861, resigned from U.S. Navy and soon thereafter attached to Confederate navy; 1863-1865, chief, Confederate Bureau of Ordnance and Hydrography; 1865-1899, professor of physics and astronomy, Virginia Military Institute. SCIENTIFIC CONTRIBUTIONS: At Naval Observatory invented deep-sea sounding apparatus that for first time made possible study of ocean specimens and mapping of topography of ocean bottom; for this the King of Prussia in 1860 bestowed on him gold medal in science from Academy of Berlin. During connection with North Pacific exploring party put in charge of astronomical work, with assignment to fix geographical location of primary points and to measure with chronometer the differences of longitude; in surveying route from San Francisco to China, made deep-sea soundings and surveyed several Pacific islands and east coast of Japan. In 1861, produced plans for converting U.S.S. *Merrimac* (later, C.S.S. *Virginia)* to ironclad. Invented "Brooke" gun, most powerful weapon produced by Confederacy.

WORKS: Said to have written on ordnance and other naval matters for the *United Service, U.S. Nautical Magazine,* and others. No science publications under his name have been found. WORKS ABOUT: *DAB,* 3:69-70 (Charles L. Lewis) ; *NCAB,* 22 (1932) : 29.

BROOKS, WILLIAM KEITH (1848-1908). *Zoology.* Professor of zoology, Johns Hopkins Univ.

MSS: Johns Hopkins Univ. (see this and other references in *DSB*). WORKS ABOUT: See *AMS; DAB; DSB.*

BROOKS, WILLIAM ROBERT (1844-1921). *Astronomy; Physics.* Professor of astronomy and director of Smith Observatory, Hobart College.

WORKS ABOUT: See *AMS; DAB; DSB.*

BROWN, JOSEPH (December 3/14, 1733, Providence, R. I.—December 3, 1785, Providence, R. I.). *Physics; Astronomy.* Son of James, wealthy merchant and West Indies trader, and Hope (Power) Brown. Married Elizabeth Power, 1759; four children. EDUCATION: No information. HONORS: A. M., Rhode Island College, 1770. CAREER: Merchant and manufacturer in partnership with brothers, retiring at an early age; maintained connections with family iron manufacturing facility, "Furnace Hope," at Scituate, R.I., until his death and served there as technical adviser; served several years in Rhode Island General Assembly; 1769-1785, trustee, Rhode Island College (Brown Univ.); 1784-1785, professor of natural philosophy, Rhode Island College, without pay. MEMBERSHIPS: Am. Ac. Arts Scis. SCIENTIFIC CONTRIBUTIONS: Gave up business activities as soon as financially possible and turned attention to science and related matters. Involved in planning and construction of several notable Providence buildings; mechanical interests led to his supervising construction of fire engine and also of steam engine used at "Furnace Hope"; also engaged in electrical experiments. Most noted scientific work done in relation to observations of transit of Venus in 1769, a project that he apparently formulated while reading account by J. Winthrop [q.v.] of transit of 1761; having seen list of apparatus compiled by Am. Phil. Soc., and following consultation with B. West [q.v.], ordered telescope and other apparatus costing him nearly £100; observations of transit were made with West, whose astronomical and mathematical skills were necessary component in success of venture, Brown's contribution being enthusiasm, skill in use of apparatus, and financial backing. West published *An Account of the Observation of Venus Upon the Sun* (1769), the first account to appear other than brief notices in newspapers.

WORKS: "An Observation of a Solar Eclipse, Oct. 27, 1780, at Providence," *Memoirs of Am. Ac. Arts Scis.* 1 (1785): 149-50. WORKS ABOUT: *DAB*, 3: 141 (Lawrence C. Wroth); J. Walter Wilson, "Joseph Brown, Scientist and Architect," *Rhode Island History* 4 (1945): 67-79, 121-28.

BROWNE, DANIEL JAY (December 4, 1804, Fremont, N.H.—1867). *Agriculture.* Son of Isaac, probably farmer, and Mary Browne. No information on marital status. EDUCATION: Little is known regarding education; reportedly attended some courses at Harvard Univ. CAREER: Engaged in farming; 1830-1832, published a monthly, *Naturalist* (Boston); 1833-1835, spent mainly in traveling in West Indies, Europe, and South America; 1836-1842, took part in several engineering projects in New York state and in Canada; thereafter, spent several years engaged in literary pursuits and served as corresponding secretary for American Agicultural Association and on Board of Agriculture of American Institute of New York; 1845-1851, while employed in agricultural warehouse of R. L. Allen and Co., N.Y., also served as an editor of their paper, *American Agriculturalist;* 1852, attached to U.S. Census Office, especially concerned with agricultural statistics; 1853-1859, agricultural clerk, U.S. Patent Office; 1861, sent to Europe by Patent Office to investigate cultivation and manufacture of flax. SCIENTIFIC CONTRIBUTIONS: Before age thirty, published *Naturalist,* as well as works on trees and on technical words and phrases used in arts and sciences; later published another work on American trees (1846), works on cage birds and poultry (1850), and *American Muck Book* (1851). As clerk in Patent Office prepared number of papers and reports on agriculture; 1854 and 1855 was sent to Europe to collect information on agricultural questions and to arrange for supplies of seeds and cuttings, and in this capacity was first government agricultural explorer; tenure in Patent Office, and especially activities in distribution of seeds, created controversy in agricultural press. Results of study of flax appeared in U.S. Patent Office's *Report for 1861.* Not an original author, but nonetheless an effective compiler and popularizer.

WORKS: See major works mentioned in *DAB;* see also *Natl. Un. Cat. Pre-1956 Imp.* WORKS ABOUT: *DAB*, 3: 164-65 (Claribel R. Barnett); Foster W. Russell, *Mt. Auburn Biographies* ([Wakefield, Mass. 1953]), p. 28.

BRUCE, ARCHIBALD (February 1777, New York, N.Y.—February 22, 1818, New York, N.Y.). *Mineralogy.* Son of Dr. William, with medical department of British army in New York, and Judith (Bayard) Van Rensselaer Bruce. Married ca. 1803; no information on children. EDUCATION: 1797, A.B., Columbia College; studied medicine in New York; 1800, M.D., Univ. of Edinburgh; during two years after 1800, studied mineralogy in Europe. CAREER: 1803, began medical practice, New York City; at College of Physicians and Surgeons of New York, professor of mineralogy (1807-1808), professor of materia medica (1808-1811), trustee and registrar, (1807-1811); 1812-1818, professor of materia medica and mineralogy, Queen's (Rutgers) College; 1810-1814, founder and editor, *American Mineralogical Journal.* MEMBERSHIPS: Literary and Philosophical Society of New York; natural history department, N.Y. Hist. Soc. SCIENTIFIC CONTRIBUTIONS: Interest in mineralogy apparently arose while studying medicine with Dr. David Hosack in New York, before going abroad in 1798; during years in Europe, brought together valuable mineralogical collection and became acquainted with leading European mineralogists. In U.S. made collection freely available; this generosity, wide correspondence with American students of mineralogy, and publication of *American Mineralogical Journal* all contributed greatly toward promotion of study of American mineralogical wealth. Journal was pioneering venture as first such American publication devoted exclusively to scientific subjects, although only one volume was published. In New Jersey, discovered metal known as brucite, as well as deposits of zinc oxide in Sussex County. Not greatly productive as an author; accounts of chief investigations appeared in his *American Mineralogical Journal, American Medical and Philosophical Register* 1 (1810): 117, and *Medical Repository* 11 (1808): 441-42.

WORKS: "Description of Some Combinations of Titanium Occurring Within the U.S.," *American Mineralogical Journal* 1 (1810-1814): 233-43, was one of longer papers. See Meisel, *Bibliog. Amer. Nat. Hist.* MSS: No collection is known. WORKS ABOUT: *DAB*, 3: 179-80 (James M. Phalen); *Am. Journ. Sci.* 1 (1819): 299-304.

BRUNNOW, FRANZ FRIEDRICH ERNST (November 18, 1821, Berlin, Germany—August 20, 1891, Heidelberg, Germany). *Astronomy.* Son of Johann, German privy councillor of state, and Wilhelmine (Weppler) Brunnow. Married Rebecca Lloyd Tappan, 1857; at least one child. EDUCATION: 1843, Ph.D., Univ. of Berlin. HONORS: Amsterdam Academy gold medal, 1848. CAREER: 1847, became head of private observatory, Bilk, Germany; 1851-1854, assistant director, observatory in Berlin; 1854-1860 and 1861-1863, professor of astronomy and director of observatory, Univ. of Michigan, Ann Arbor; 1860-1861, associate director, Dudley Observatory, Albany, N.Y.; 1865-1874, Andrews professor of astronomy, Univ. of Dublin, and royal astronomer for Ireland; 1874-1891, retired, living in Basle (1874-1880), Vevey (1880-1889), and Heidelberg (1889-1891). MEMBERSHIPS: Fellow, Royal Astronomical Society. SCIENTIFIC CONTRIBUTIONS: While at Univ. of Berlin, wrote doctoral thesis entitled "De Attractione Moleculari" and became associated with Johann F. Encke, director of observatory; at Bilk, wrote notable paper on De Vico's comet and received gold medal from Amsterdam Academy. *Lehrbuch* [see Works] reached four editions and several translations and is prime source of Brunnow's astronomical reputation. In U.S. for only eleven years; at Ann Arbor and at Albany was involved in determination of physical and astronomical constants for those observatories and their instruments; contacts with students at Ann Arbor were limited, and for a time J. C. Watson [q.v.], successor at that observatory, was his only student; nevertheless, Brunnow played a central role in introducing German astronomical methods into U.S.; during period 1858-1862, while at Ann Arbor, published *Astronomical Notices,* which contained number of his own papers. In Ireland, did important work in organizing observatory and in publishing results of investigations. During career published number of papers on comets, small planets, and stellar parallax.

WORKS: *Lehrbuch der sphärischen Astronomie* (Berlin, 1851); "Examination of the Division of the Ann Arbor Meridian Circle," *Astronomical Notices*, nos. 10-12 (1859), pp. 77-80, 86-88, 94-96; *Astronomical Observations and Researches Made at Dunsink*, in 3 pts. (Dublin, 1870, 1873, 1879). See *Roy. Soc. Cat.*, vols. 1, 7, 9, and *Natl. Un. Cat. Pre-1956 Imp.* WORKS ABOUT: *NCAB*, 13 (1906): 78; obituary, *Monthly Notices of Royal Astronomical Society* 52 (1891-1892): 230-33. Also see Marc Rothenberg, "The Educational and Intellectual Background of American Astronomers, 1825-1875" (Ph.D. diss., Bryn Mawr College, 1974), pp. 139-42.

BRUSH, CHARLES FRANCIS (1849-1929). *Physics; Invention.* Inventor of Brush arc light; founder of Brush Electric Co. and Linde Air Products Co.

MSS: Case Western Reserve Univ.—Archive of Contemporary Science (*NUCMC* 68-904). WORKS ABOUT: See *AMS; DAB,* supp. 1; Henry J. Eisenman II, "Charles Francis Brush: Pioneer Innovator in Electrical Technology," *Dissertation Abstracts* 28 (1967): 2159-A.

BRUSH, GEORGE JARVIS (1831-1912). *Mineralogy.* Professor of mineralogy and director, Sheffield Scientific School, Yale.

MSS: Yale U.-HMC (Hamer). WORKS ABOUT: See *AMS; DAB.*

BUCKLEY, SAMUEL BOTSFORD (May 9, 1809, Torrey, N.Y.—February 18, 1884, Austin, Tex.). *Botany; Natural history.* No information on parentage. Married Charlotte Sullivan, 1852; Sarah Porter, 1855; Libbie Myers, 1864; three children. EDUCATION: 1836, A.B., Wesleyan Univ.; 1842-1843, studied at College of Physicians and Surgeons, New York. HONORS: Ph. D., Waco Univ. (Texas), 1872. CAREER: Taught in Illinois and Alabama, two years as principal of Allenton Academy, Wilcox, Ala.; 1842-1843, collected plants and other natural history objects in Alabama, Tennessee, North and South Carolina, Florida; 1843-1855, lived on family farm, Torrey, N.Y.; 1855-1856, worked in bookstore, Yellow Springs, Ohio; ca. 1857-1860, traveled through South conducting various natural history studies; 1860-1861, assistant geologist and naturalist, Texas Geological Survey; 1862-1865, chief examiner, Statistical Department, U.S. Sanitary Commission; 1866-1867 and 1874-1877, Texas state geologist; 1871-1872, agricultural and scientific editor of *State Gazette*, Austin, Tex. MEMBERSHIPS: Ac. Nat. Scis. Philad. and other societies. SCIENTIFIC CONTRIBUTIONS: Primarily field naturalist, traveled widely in South from Carolinas to Texas and studied and collected flora and fauna of region; knowledgeable concerning southern plants in natural setting, but less able to deal with subject from more formal scientific view. In 1841 discovered skeleton of Zeuglodon in Alabama; brought to light number of new species of plants, insects, shells, and other groups. Paper entitled "Description of . . . Plants" [see Works] in 1843, one of his earliest and most notable works, described new species of plants in southwestern region of Appalachian Mountains, including shrub named Buckleya distichophylla by Torrey; engaged in works on trees and shrubs of America and on geology and natural history of Texas, which were never published. In 1858, investigated elevation of some higher mountains in North Carolina and Tennessee, and Buckley's Peak in Tennessee bears his name. Publications appeared in *Am. Journ. Sci.*, *Proceedings of Am. Phil. Soc.*, and elsewhere; also issued several reports as Texas geologist.

WORKS: "Description of Some New Species of Plants," *Am. Journ. Sci.* 45 (1843): 170-177. See *Roy. Soc. Cat.*, vols. 1, 7, 9, 12, and Meisel, *Bibliog. Amer. Nat. Hist.* MSS: Harv. U.-GH (Hamer). WORKS ABOUT: *DAB*, 3: 232-33 (Theodore S. Palmer); obituary in *Am. Journ. Sci.* 3d ser. 29 (1885): 171-72.

BURGESS, EDWARD (June 30, 1848, West Sandwich, Mass.—June 12, 1891, Boston, Mass.). *Entomology*. Son of Benjamin F., wealthy merchant in West Indian trade, and Cordelia (Ellis) Burgess. Married Caroline L. Sullivant; survived by two sons. EDUCATION: 1871, A.B., Harvard College. CAREER: 1879, deprived of wealth through failure of father's business; 1879-1882 and 1883-1884, instructor in entomology, Bussey Institution, Harvard; 1883-1891, engaged in business as yacht designer; 1885, designed his first America's Cup winner; 1887, appointed member of U.S. Naval Board awarding prizes for design of cruisers and battleships; 1888, became permanent chairman, board of life-saving appliances, U.S. Life-Saving Service. MEMBERSHIPS: Bos. Soc. Nat. Hist.; president of Cambridge Entomological Club. SCIENTIFIC CONTRIBUTIONS: For number of years, worked as volunteer assistant to S. H. Scudder [q.v.]; particularly interested in insect anatomy, and first work was published with Scudder while working with him on abdominal appendages of New England butterflies. Burgess's most important paper was that on milkweed butterfly [see Works], which dealt with entire anatomy of insect, both internal and external, and through text and drawings gave more accurate and detailed picture than had ever been rendered for any perfect insect. Adept in use of microscope and in 1876 was called upon to investigate source of odor in Boston water supply. Had vast knowledge of native American diptera, though publishing little on the subject, and collection of insects of that order (amounting to 14,000 specimens) was sent to U.S. Natl. Mus. Entomological publication ceased about 1885 when attention turned to yacht design.

WORKS: "On Symmetry in the Appendages of Hexapod Insects Especially as Illustrated in the Lepidopterous Genus Nisoniades," with S. H. Scudder, in *Proceedings of Bos. Soc. Nat. Hist.* 13 (1870): 282-306; "Contributions to the Anatomy of the Milk Weed Butterfly, Danais archippus," in Bos. Soc. Nat. Hist., *Anniversary Memoirs* (Boston, 1880). See bibliography of works in Scudder [below], pp. 362-64. WORKS ABOUT: *DAB*, 3: 275 (W. J. Ghent); Samuel H. Scudder, "The Services of Edward Burgess to Natural Science," *Proceedings of Bos. Soc. Nat. Hist.* 25 (1892): 358-64.

BURNETT, WALDO IRVING (July 12, 1827, Southborough, Mass.—July 1, 1854, Boston, Mass.). *Biology*, especially *Histology*. Son of Joel Burnett, physician. No information on marital status. EDUCATION: Studied medicine privately and at Tremont Medical School (Boston); 1849, M.D., Harvard Medical School. CAREER: 1845, after father's death, began teaching school; 1849-1854, following several months in Europe, engaged in scientific investigation and may have practiced medicine. MEMBERSHIPS: Am Assoc. Advt. Sci. and, 1849, secretary to zoology section; Am. Ac. Arts Scis. SCIENTIFIC CONTRIBUTIONS: After 1849 graduation, went to Europe to study, but returned after four months because of health and died of tuberculosis at age twenty-seven. During short life, wrote sixty or more articles and was one of first Americans to pursue special studies in microanatomy; while not credited with any important new discoveries in his special field of histology, was familiar with wide range of European literature, repeated and confirmed results of European investigators, and made small contributions of his own; paper entitled "The Cell" [see Works] was first work by American to treat cell in comprehensive manner and in some respects was more complete and systematic than contemporary European works. Also extended microscopical investigations to study of infusoria, embryology, and other subjects, including entomology. Range of interests is represented by papers on relation of distribution of lice to different faunas, relation of embryology and spermatology to animal classification, reproduction of lost parts in reptiles, formation and function of Allantois, physical features of Florida, and other subjects.

WORKS: "The Cell: Its Physiology, Pathology, and Philosophy; as Deduced from Original Investigations, to Which are Added Its History and Criticism," *Transactions of Am. Med. Assoc.* 6 (1853): 645-832. See *Roy. Soc. Cat.*, vol. 1; also, partial bibliography in Cornfield [below]. WORKS ABOUT: *Appleton's CAB*, 1: 459; Jerome R. Cornfield, "W. I. Burnett, Early American Histologist," *Bull. Hist. Med.* 26 (1952): 431-51.

BURNHAM, SHERBURNE WESLEY (1838-1921). *Astronomy.* Professor of practical astronomy, Univ. of Chicago, and astronomer, Yerkes Observatory.

WORKS ABOUT: See *AMS; DAB; DSB.*

BURRILL, THOMAS JONATHAN (1839-1916). *Botany; Bacteriology.* Professor of botany and horticulture, vice-president, and dean, Univ. of Illinois.

MSS: Univ. of Illinois Archives (*NUCMC* 65-1306). WORKS ABOUT: See *AMS; DAB.*

BURRITT, ELIJAH HINSDALE (April 20, 1794, New Britain, Conn.—January 3, 1838, Houston, Tex.). *Astronomy.* Son of Elihu—farmer, shoemaker, and miller—and Elizabeth (Hinsdale) Burritt. Married Ann W. Watson, 1819; five children. EDUCATION: For about two years apprenticed to blacksmith at Simsbury, Conn.; 1816, entered Williams College, but did not graduate. CAREER: 1817, became teacher in Sanderson Academy, Ashfield, Mass.; 1819-1829, settled in Georgia, at Milledgeville, and there taught school, engaged in civil engineering, and for several years owned and edited weekly newspaper until forced to leave state because of supposed abolitionist sympathies; 1829, returned to New Britain and for several years conducted boarding and day school; 1837, organized and directed group of colonists to Texas; died there of yellow fever. SCIENTIFIC CONTRIBUTIONS: While in Georgia, appointed by state to conduct survey of Chattahoochie River, and in 1826 submitted report. Had interest in study of mathematics and astronomy from early age, and while student at Williams College published *Logarithmick Arithmetick* (1818), with astronomical tables; while in Georgia, published pamphlet, *Astronomia, or Directions for the Ready Finding of All the Principal Stars in the Heavens Which Are Named on Carey's Celestial Globe* (1821). Later, established observatory in building that housed boarding and day school in New Britain. In 1830, published pamphlet, *Universal Multipliers for Computing Interest, Simple and Compound.* Remembered chiefly for *Geography of the Heavens* [see Works], with an accompanying "celestial atlas" based on Burritt's drawings; intended for schools and colleges, this work probably ranked as best American publication of its sort up to that time and had wide circulation, selling over 300,000 copies and sixteen editions by 1876. Older brother of Elihu Burritt, "The Learned Blacksmith."

WORKS: See above; *Geography of the Heavens . . . Accompanied by a Celestial Atlas* (Hartford, Conn., 1833). WORKS ABOUT: Albert J. Brooks, Biography, *Popular Astronomy* 44 (1936): 293-99.

BYERLY, WILLIAM ELWOOD (1849-1935). *Mathematics.* Professor of mathematics, Harvard Univ.

WORKS ABOUT: See *AMS; DAB*, supp. 1.

C

CABOT, SAMUEL, JR. (September 20, 1815, Boston, Mass.—April 13, 1885, Boston, Mass.). *Ornithology.* Son of Samuel, influential merchant, and Eliza (Perkins) Cabot. Married Hannah Lowell Jackson, 1844; nine children. EDUCATION: 1836, A.B., Harvard College; 1839, M.D., Harvard Medical School; two years of medical study in Paris. CAREER: 1841-1842, ornithologist with Stephens-Catherwood expedition to Yucatan; ca. 1843-1885, practiced medicine, and particularly surgery, in Boston; 1853-1884, visiting surgeon, Massachusetts General Hospital; president of Boston children's hospital at time of death. MEMBERSHIPS: Member of Bos. Soc. Nat. Hist. and curator of birds; Am. Ac. Arts Scis. SCIENTIFIC CONTRIBUTIONS: Published contributions to ornithology said to number nearly fifty and dealt most notably with birds of Yucatan; prepared material on birds for John Lloyd Stephens's *Incidents of Travel in Yucatan* (1843) and published several articles on birds of Yucatan in *Boston Journal of Natural History* and *Proceedings of Bos. Soc. Nat. Hist.,* where most of his other ornithological writings also appeared; work on ornithology of Yucatan resulted in descriptions of seven new species as well as more adequate descriptions of thirty previously known species. Work on birds of U.S. related especially to rarer species of New England; ornithological writings included some anatomical aspects. Also pioneered in surgery as one of earliest Americans to use disinfectants and anaesthesia and performed important innovative surgical operations. Active ornithological work ceased about 1850 because of press of medical interests; as ornithologist was well known in Europe and reportedly corresponded with leading scientists there.

WORKS: Abbreviated bibliographical references to ornithological works in *Proceedings of Am. Ac. Arts Scis.* [below], pp. 519-20; also see Meisel, *Bibliog. Amer. Nat. Hist.* Notes and ornithological collection went to Bos. Soc. Nat. Hist. [*Auk,* below]. WORKS ABOUT: *NCAB,* 25 (1936) : 262-63; obituaries in *Proceedings of Am. Ac. Arts Scis.* 21 (1885-1886) : 517-20, and in *Auk* 3 (1886) : 144.

CALDWELL, GEORGE CHAPMAN (1834-1907). *Chemistry.* Professor of chemistry, Cornell Univ.

MSS: Corn. U.-CRH (*NUCMC* 62-3892 and 66-766). WORKS ABOUT: See *AMS; NCAB,* 26 (1937) : 142-43.

CALVIN, SAMUEL (1840-1911). *Geology; Paleontology.* Professor of geology, Iowa State Univ.; state geologist of Iowa.

WORKS ABOUT: See *AMS; DAB.*

CAMPBELL, MARIUS ROBINSON (1858-1940). *Gelogy; Physiography.* Geologist with U.S. Geol. Surv.

WORKS ABOUT: See *AMS; DAB,* supp. 2.

CARHART, HENRY SMITH (1844-1920). *Physics.* Professor of physics, Northwestern Univ. and Univ. of Michigan.

WORKS ABOUT: See *AMS*; *NCAB,* 4 (1897 [copyright 1891 and 1902]) : 455-56; *World Who's Who in Science.*

CARLL, JOHN FRANKLIN (May 7, 1828, Bushwick [now Brooklyn], N.Y.—March 13, 1904, Waldron, Ark.). *Geology.* Son of John, farmer, and Margaret (Walters) Carll. Married Hannah A. Burtis, 1853; Martha Tappan, 1868; no information on children. EDUCATION: Educated at Union Hall Academy, Flushing, N.Y. CAREER: Ca. 1846-ca. 1849, worked with father on farm; 1849-1853, engaged in publishing with brother-in-law, assisted with editing and printing of *Daily Eagle,* Newark, N. J.; 1853-ca. 1863, civil engineer and surveyor, Flushing, N.Y.; 1864, moved to Pleasantville, Pa., and there became involved in oil development; 1874-1885, assistant geologist in Second Geological Survey of Pennsylvania, at head of petroleum and natural gas surveys; 1885, began private practice as geological consultant. SCIENTIFIC CONTRIBUTIONS: While involved with development of oil resources in Pennsylvania, introduced several related inventions, including static-pressure sand pump, removable pump chamber, and adjustable sleeves for piston rods. So ably did he expand, systematize, and communicate great practical knowledge gained in oil fields after 1864, that in 1883 J. P. Lesley [q.v.], head of second Pennsylvania survey, labeled him as virtual founder of geology of petroleum. Work for Pennsylvania survey was of pioneering character inasmuch as he was first to understand and describe adequately characteristic features of the region surveyed [see Works], and his reports became standard sources for oilmen; also effective in communicating great store of growing knowledge on day-to-day basis; reports included comments on oil drilling equipment and presented comparison of geology of northwestern Pennsylvania with that of adjacent parts of Ohio and New York.

WORKS: Prepared seven reports for Second Geological Survey of Pennsylvania, covering oil regions of Warren, Venango, Clarion, and Butler counties and also including some geological work on Crawford County, as well as Chautauqua County in New York. See *Roy. Soc. Cat.*, vols. 9, 14. WORKS ABOUT: *DAB*, 3: 496 (Katherine W. Clendinning); *Who's Who in America, 1903-1905*.

CARPENTER, GEORGE WASHINGTON (July 31, 1802, Germantown, Pa.—June 7, 1860, Germantown, Pa.). *Mineralogy; Chemistry (pharmacology)*. Son of Conrad and Ann (Adams) Carpenter. Married Annabella Wilbank, 1836; Ellen Douglas, 1841; one child by first marriage, six children by second. EDUCATION: Educated at Germantown collegiate academy. CAREER: 1820-1828, assistant in Charles Marshall, Jr., wholesale drug firm, Philadelphia; 1828-1860, operated his own drug business and engaged in real estate and other business ventures. MEMBERSHIPS: Ac. Nat. Scis. Philad. and other American and European societies. SCIENTIFIC CONTRIBUTIONS: While assistant in Marshall drug firm, met T. Nuttall [q.v.], and interest in natural sciences, which apparently developed from this acquaintance, occupied leisure time for remainder of life. Particularly interested in mineralogy, and collections and descriptions were made available to mineralogist P. Cleaveland [q.v.]. Contributed several papers to *Am. Journ. Sci.* during period 1825-1832, dealing mainly with pharmacological matters; also published papers in *American Journal of Medical Sciences*. On Germantown estate, had museum with large collections of specimens relating to all aspects of natural history, deposited by widow with Ac. Nat. Scis.; also had greenhouses containing rare plants collected by Nuttall and given in 1893 by widow to city of Philadelphia.

WORKS: "On the Mineralogy of Chester County, With an Account of Some of the Minerals of Delaware, Maryland, and Other Localities," *Am. Journ. Sci.* 14 (1828): 1-10; "On the Muriate of Soda, or Common Salt, with an Account of the Salt Springs in the U.S.," ibid. 15 (1829): 1-6; *Carpenter's Essays on Some of the Most Important Articles of the Materia Medica* (Philadelphia, 1831). See also *Roy. Soc. Cat.*, vol. 1. WORKS ABOUT: S. B. Winslow, *Biographies of Successful Philadelphia Merchants* (Philadelphia, 1864), pp. 124-29; *NCAB*, 10 (1909 [copyright 1900]): 235.

CASSIN, JOHN (September 6, 1813, Upper Providence Township, Delaware, Pa.—January 10, 1869, Philadelphia, Pa.). *Ornithology*. Son of Thomas Cassin, probably farmer. No information on marital status. EDUCATION: Educated at Quaker school, Westtown, Pa. CAREER: Ca. 1834, entered mercantile business in Philadelphia; attained position in U.S. Custom-House; later, manager of important engraving and lithographic firm in Philadelphia; served on Philadelphia City Council. MEMBERSHIPS: Vice-president of Ac. Nat. Scis. Philad., and curator 1842-1869. SCIENTIFIC CONTRIBUTIONS: During period 1846-1850, Dr. Thomas B. Wilson began assembling large collection of 26,000 birds and related library, and Cassin assumed responsibility for its proper arrangement and identification at Ac. Nat. Scis.; work on Wilson collection and resultant publications established Cassin as leading American ornithologist of the time. Unlike earlier American ornithologists, great interest was in systematic zoology and publications were technical in nature, emphasizing matters of proper description and classification and dealing with problems of synonymy and nomenclature. Published descriptions of 194 new bird species, mainly from West African specimens collected by Mr. P. B. Du Chaillu. Most papers appeared in *Proceedings of Ac. Nat. Scis. Philad.*; also prepared ornithological sections for reports of several government-sponsored expeditions, including Wilkes Expedition and Pacific Railroad Survey, and of expeditions to China seas, Japan, and Southern Hemisphere. *Illustrated Birds* [see Works] prepared as supplement and updating of *Birds of America* by J. J. Audubon [q.v.]. Distinguished as first American thoroughly familiar with ornithology of both Old and New Worlds.

WORKS: Ornithological work for government surveys included his *Mammalogy and Ornithology*, 2d ed., Vol. 8 of U.S. Exploring Expedition report (Philadelphia, 1858); and contributions to the Report of Pacific Railroad Survey, vol. 9, pt. 2 *(Senate Executive Document no. 78* and *House Executive*

Document no. 91, 33d Congress, 2d session); *Illustrated Birds of California, Texas, Oregon, British and Russian America* (Philadelphia, 1856). See Meisel, *Bibliog. Amer. Nat. Hist.* WORKS ABOUT: *DAB*, 3: 568-69 (Witmer Stone); Stone, "John Cassin," *Cassinia: Proceedings of the Delaware Valley Ornithological Club* 5 (1901): 1-7.

CASWELL, ALEXIS (January 29, 1799, Taunton, Mass.—January 8, 1877, Providence, R. I.). *Physical sciences*, especially *Astronomy*. Twin son of Samuel, probably farmer, and Polly (Seaver) Caswell. Married Esther L. Thompson, 1830; Elizabeth B. Edmands, 1855; six children by first marriage. EDUCATION: 1822, A.B., Brown Univ.; ca. 1822-ca. 1827, studied theology with president of Columbian College (later, George Washington Univ.). HONORS: D.D. in 1841 and LL.D. in 1865, Brown Univ. CAREER: 1822-1827, taught at Columbian College, last two years as professor of ancient languages; 1827, minister, Halifax, Nova Scotia; 1828-1863, professor of mathematics, natural philosophy, and astronomy, Brown Univ.; ca. 1863, became president of National Exchange Bank and of American Screw Co., Providence, R.I.; 1868-1872, president of Brown Univ.; 1875-1877, president of Rhode Island Hospital. MEMBERSHIPS: Vice-president, Am. Assoc. Advt. Sci., 1855; member of Natl. Ac. Scis. and other societies. SCIENTIFIC CONTRIBUTIONS: Made no important original contributions to scientific knowledge, but was above all dedicated and effective science educator; during career, called on to teach wide spectrum of mathematical and physical sciences, which tended to prevent development of specialization. Particularly interested in meteorology and astronomy, and in 1858 delivered four lectures on latter subject at Smithson. Instn., published in its *Annual Report* for that year; carried on meteorological observations at Providence, 1831-1876, and publication of these observations for period 1831-1860 [see Works] probably constituted most noteworthy scientific publication. Frequent attendant at meetings of Am. Assoc. Advt. Sci. and, while making no formal presentations to that body, was highly enough regarded to be chosen vice-president.

WORKS: "Meteorological Observations Made at Providence, R.I. . . . from December 1831 to May 1860," *Smithson. Contr. Knowl.* 12 (1860); Reviews of scientific works for *Christian Review* (June 1836 and December 1841); "On Zinc, as a Covering for Buildings," *Am. Journ. Sci.* 31 (1837): 248-52. See *Roy. Soc. Cat.*, vol. 1. WORKS ABOUT: *DAB*, 3: 570-71 (Walter C. Bronson); Joseph Lovering, "Memoir," *Biog. Mems. of Natl. Ac. Scis.* 6 (1909): 363-72 (reprinted from *Proceedings of Am. Ac. Arts Scis.*, 1877).

CASWELL, JOHN HENRY (December 27, 1846, New York, N.Y.—October 16, 1909, apparently New York, N.Y.). *Mineralogy*. No information on parentage. Married Mary B. Curtiss, 1872; no information on children. EDUCATION: 1865, A.B., Columbia Univ.; 1865-1868, studied at Mining Academy, Freiberg, Germany. CAREER: 1868-1871 and 1874-1877, assistant in mineralogy, School of Mines, Columbia Univ.; 1871-1874, involved in concerns of parental estate (father died in 1871); 1877, gave up position at Columbia and thereafter devoted attentions to business affairs; in later years, held number of positions of public responsibility in New York. MEMBERSHIPS: N.Y. Ac. Scis. SCIENTIFIC CONTRIBUTIONS: During period as assistant in mineralogy at Columbia, spent time during summers in mines of Colorado, Nevada, and California. Remembered for microscopic work on metamorphic and eruptive rocks, collected in Black Hills in 1875 by Walter P. Jenney and H. Newton [q.v.]; results of study, published in Jenney and Newton's *Report* (1880) [see Works], ranked as second government-issued American contribution to micropetrography (Ferdinand Zirkel's report for Fortieth Parallel Survey in 1876 was first such contribution); Black Hills report included two colored plates and was noteworthy as well for first identification in America of the rock phonolite. Thereafter, apparently ceased active work in mineralogy, but gathered admirable collection of minerals and for years was active in N.Y. Ac. Scis.

WORKS: "Microscopic Petrography of the Black Hills of Dakota," in H. Newton and W. P. Jenney, *Report on the Geology and Resources of the Black Hills of Dakota* (Washington, D.C., 1880), pp. 469-527. WORKS ABOUT: James F. Kemp, "Memoir," *Annals of N.Y. Ac. Scis.* 19 (1909): 353-56; Merrill, *Hundred Years of Geology*, pp. 484, 647.

CATESBY, MARK (April 3, 1683, Castle Hedingham, Essex, England—December 23, 1749, London, England). *Natural history*, especially *Botany*. Son of John—lawyer, landowner, and mayor of Sudbury, Suffolk—and Anne (Jekyll) Catesby. Married in 1747; two children. EDUCATION: Probably educated in Castle Hedingham; Catesby knew Latin. CAREER: Inherited modest property from father; 1712, went to Virginia, where sister was residing, and while there explored the colony, visited Bermuda and Jamaica, and sent plants back to England; 1719, returned to England; 1722-1726, sent to South Carolina by William Sherard to collect plants, and explored South Carolina, Georgia, and Florida; 1725-1726, passed parts of this period in Bahamas; 1726, returned to England, settled at Hoxton and, having learned art of etching, devoted remainder of life to writing and illustrating his *Natural History,* the sale of which earned a meager living. MEMBERSHIPS: Roy Soc. SCIENTIFIC CONTRIBUTIONS: While a young man, met John Ray, but scientific work did not really begin before going to Virginia, where brother-in-law introduced him to various persons having botanical interests; sent botanical specimens back to England. Main activity during much of adult life was preparation of *Natural History,* of which first volume was largely devoted to ornithology with botanical backgrounds, while second volume had elements of cartography, geology, anthropology, and zoology, as well as botany; although chief interest was in botany, most lasting contributions were to ornithology. Work was used by Linnaeus as source for classifying number of American birds and lesser number of plants and other animals; Catesby employed accepted theories and modes of description current in his day.

WORKS: *Natural History of Carolina, Florida, and the Bahama Islands,* 2 vols., folio, with text in French and English and with 220 illustrations (London, 1731-1743, 1729-1747, with new London eds. in 1754 and 1771; reprinted in 1974 with introduction by George Frick and notes by Joseph Ewan); "Of Birds of Passage," *Phil. Trans. of Roy. Soc.* 44 (1747): 435-44; and, posthumously, *Hortus Brittanno-Americanus* (London, 1763). MSS: Univ. Va. Lib. (Hamer). WORKS ABOUT: *DAB,* 3: 571-72 (W. H. B. Court); *DSB,* 3: 129-30 (George F. Frick). Also see Frick and Raymond P. Stearns, *Mark Catesby: The Colonial Audubon* (Urbana, Ill., 1961).

CHALMERS, LIONEL (1715, Campbellton [Cambleton?], Argyllshire, Scotland—May [?], 1777, probably Charleston, S.C.). *Medicine*; *Meteorology.* No information on parentage. Married Martha Logan, ca. 1739; had children. EDUCATION: Apparently attended St. Andrews Univ., Scotland; 1756, awarded M.D. degree, St. Andrews, on recommendation of Robert Whytt at Edinburgh. CAREER: Ca. 1737-1777, medical practice, Charleston, S.C.; 1740, dissolved partnership in what apparently was drug establishment begun year or two before; while apparently residing in Charleston, owned plantation that was near the city and in 1742 was advertised for sale; 1754, medical partnership with Dr. J. Lining [q.v.] came to end after number of years' duration; served as justice of peace. MEMBERSHIPS: American Society (later Am. Phil. Soc.); probably member of Philosophical Society of Edinburgh (later, Royal Society of Edinburgh). SCIENTIFIC CONTRIBUTIONS: Following retirement of Dr. Lining, ranked as leading Charleston physician; as did Lining, took up study of weather and disease, and during period 1750-1759 carried out meteorological observations. In 1755 communicated meteorological record for Charleston during 1752 to Roy. Soc., along with some record of diseases and with conclusions as to relations of the two; results of these studies were anticipated in series of papers communicated to American Society and published (1769) in *American Magazine*; emphasized importance of temperature while minimizing Lining's concern with barometric pressure and humidity; investigations culminated in chief work, *An Account* [see Works], published shortly before death. Has been ranked high among colonial scientists [Stearns, *Sci. in Brit. Colonies,* p. 680] for attempts at establishment of scientific principles based on correlation of data on disease and climate.

WORKS: "Of the Opisthotonos and Tetanus," *Medical Observations and Inquiries* (London) 1 (1757); *Essay on Fevers* (Charleston, 1767; German translation, Riga, 1773); *An Account of the Weather and Diseases of South Carolina,* 2 vols. (London, 1776). MSS: See references (footnotes) in Waring [below]. WORKS ABOUT: Joseph I. Waring, "Lionel Chalmers, Medical Author," *Bull. Hist. Med.* 32 (1958): 349-55, Stearns, *Sci. in Brit. Colonies,* pp. 587-98, passim.

CHAMBERLIN, THOMAS CHROWDER (1843-1928). *Geology.* Chief geologist, Glacial Division, U.S. Geol. Surv.; president, Univ. of Wisconsin; professor of geology, Univ. of Chicago; founder and editor, *Journal of Geology.*

MSS: St. Hist. Sc. Wis. (*NUCMC* 62-1830); Univ. of Chicago Library (*NUCMC* 64-51). WORKS ABOUT: See *AMS; DAB; DSB.*

CHANDLER, CHARLES FREDERICK (1836-1925). *Industrial chemistry.* A founder and dean, School of Mines, Columbia Univ.; professor of chemistry, Columbia Univ.

MSS: Colum. U. Lib. (Hamer). WORKS ABOUT: See *AMS; DAB.*

CHANDLER, SETH CARLO (1846-1913). *Astronomy.* Editor of *Astronomical Journal.*

WORKS ABOUT: See *AMS; DAB; DSB.*

CHAPMAN, ALVAN WENTWORTH (September 23, 1809, Southampton, Mass.—April 6, 1899, Apalachicola, Fla.). *Botany.* Son of Paul, tanner, and Ruth (Pomeroy) Chapman. Married Mary Ann Simmons Hancock, 1839; one child who died in infancy. EDUCATION: 1830, B.A., Amherst College; 1833-1835, studied medicine with Albert Reese in Washington, Ga. (date and place of M.D. not known). HONORS: M.D., Univ. of Louisville, 1846; LL.D., Univ. of North Carolina, 1886. CAREER: 1831-1833, tutor on Whitemarsh Island, Savannah,

Ga.; 1833-ca. 1835, principal of academy at Washington, Ga.; 1835, began medical practice and moved that year to Quincy, Fla.; later, moved to Marianna, Fla.; 1847, moved to Apalachicola, Fla., where he continued practice of medicine and surgery and lived for rest of life. MEMBERSHIPS: Ac. Nat. Scis. Philad. and other societies; associate of Am. Ac. Arts Scis. SCIENTIFIC CONTRIBUTIONS: For half a century, was leader of botanical studies in the South and played role in relation to flora of that region comparable to A. Gray [q.v.] in the North. As early as 1837, botanized on Apalachicola River and about same time began sending Southern botanical specimens to Gray and J. Torrey [q.v.]; first botanical publication appeared in 1845; attracted by largely unknown semitropical vegetation of Florida. Reputation as botanist rests on *Flora* [see Works], for which Torrey read proof for first edition; *Flora* first appeared without Chapman's knowledge and was published by same company responsible for Gray's *Manual*, the model for Chapman's work; *Flora* had three editions during author's lifetime and for many years was standard authority for southern botany. Collected three herbaria. Also interested in entomology and had large collection of southern Lepidoptera.

WORKS: *Flora of the Southern United States* (New York, 1860; 3d ed., 1897, was complete revision); "List of the Plants Growing Spontaneously in the Vicinity of Quincy, Fla.," *Western Journal of Medicine and Surgery* 3 (1845): 461-83. MSS: N.Y. Bot. Gard., Harv. U.-GH, and Mo. Bot. Gard. (*DSB*). WORKS ABOUT: *DAB*, 4: 16-17 (Donald C. Peattie); *DSB*, 3: 196-97 (Joseph Ewan); Charles Mohr, "Alvan Wentworth Chapman," *Botanical Gazette* 27 (June 1899): 473-78.

CHASE, PLINY EARLE (August 18, 1820, Worcester, Mass.—December 17, 1886, Haverford, Pa.). *Physical sciences*. Son of Anthony, newspaper owner and treasurer of Worcester County, and Lydia (Earle) Chase. Married Elizabeth Brown Oliver, 1843; six children. EDUCATION: 1839, A.B., Harvard College. HONORS: LL.D., Haverford College, 1876; Magellanic Medal of Am. Phil. Soc., 1864. CAREER: Ca. 1839-ca. 1841, teacher in Leicester and Worcester, Mass., and in Providence, R.I.; ca. 1841-1844, 1845-1848, and 1861-1871, teacher in Philadelphia; 1844-1845, in New England; ca. 1848-1861, partner in stove and foundry business, Philadelphia; at Haverford College, professor of natural sciences (1871-1875) and professor of philosophy and logic (1875-1886); 1884, appointed lecturer in psychology and logic, Bryn Mawr College. MEMBERSHIPS: Am. Phil. Soc. and other societies. SCIENTIFIC CONTRIBUTIONS: Showed special abilities in mathematics while in college; wrote several school textbooks on mathematics. Productive and diversified author; wrote at early date on philology, but later wrote mainly on meteorological, physical, and cosmical subjects; published more than 150 papers, including over 120 in *Proceedings of Am. Phil. Soc.*, beginning in 1858; for some years, prepared summaries of scientific information for *Journal of Franklin Institute*. Awarded Magellanic Medal for paper entitled "Numerical Relations" [see Works]. In later years became increasingly interested in attempt to prove universal law that "all physical phenomena are due to an Omnipresent Power, acting in ways which may be represented by harmonic or cyclical undulations in an elastic medium"; began publication on this subject in 1863, and his efforts attracted interest of physicists in England. Also interested in botany.

WORKS: "Numerical Relations of Gravity and Magnetism," *Proceedings of Am. Phil. Soc.* 9 (1862-1864): 425-40, and *Transactions of Am. Phil. Soc.*, n.s. 13 (1869): 117-36; *Elements of Meteorology for Schools and Households* (Philadelphia, 1884). See *Roy. Soc. Cat.*, vols. 1, 7, 9, 14; also, bibliography of contributions to Am. Phil. Soc., through 1880, in its *Proceedings* 19 (1880-1881): 184-90. WORKS ABOUT: *DAB*, 4: 27 (Marjory H. Davis); Samuel S. Green, biography, *Proceedings of Am. Ant. Soc.*, n.s. 4 (1885-1887): 316-21.

CHATARD, THOMAS MAREAN (1848-1927). *Chemistry*. Chemist with U.S. Geol. Surv.; consulting chemist; expert special agent with Twelfth Census.

WORKS ABOUT: See *AMS*.

CHAUVENET, WILLIAM (May 24, 1820, Milford, Pa.—December 13, 1870, St. Paul, Minn.). *Astronomy; Mathematics*. Son of William Marc, businessman, and Mary B. (Kerr) Chauvenet. Married Catherine Hemple, 1841; survived by five children. EDUCATION: 1840, A.B., Yale College. HONORS: LL.D., St. John's College, 1860. CAREER: 1841, appointed professor of mathematics, U.S. Navy, serving on U.S.S. *Mississippi;* 1842-1845, in charge of school for midshipmen, Naval Asylum, Philadelphia; 1845-1859, professor of mathematics and astronomy, later of astronomy, navigation, and surveying, at U.S. Naval Academy, Annapolis, which he helped to establish; at Washington Univ. (St. Louis), professor of mathematics and natural philosophy (1859-1869) and chancellor (1862-1869). MEMBERSHIPS: President of Am. Assoc. Advt. Sci. at time of death; member of Natl. Ac. Scis. and other societies. SCIENTIFIC CONTRIBUTIONS: Began scientific career with work for A. D. Bache [q.v.] on magnetic observations at Girard College. Played conspicuous role in establishing U.S. Naval Academy, and was largely responsible through his teaching for setting academy on strong and scientific foundation; during years at Annapolis (after 1853, in charge of department of astronomy and navigation) was academy's most distinguished professor. In 1855 and 1859 offered professorships at Yale, but declined. Wrote on mathematics and astronomy, and textbooks were particularly important; *Treatise on . . . Trigonometry* [see Works] said to have been more complete than any other such work in English language; *Manual* [see Works], most important book, was held in highest regard in both Europe and America, and a number of investigations reported therein were original, in whole or in part, with Chauvenet. Most noteworthy journal article was on lunar distances, later incorporated into *Manual.* In 1925, Mathematical Association of America established in his memory the Chauvenet Prize for Mathematical Exposition.

WORKS: *Treatise on Plane and Spherical Trigonometry* (Philadelphia, 1850); *Manual for Spherical and Practical Astronomy,* 2 vols. (Philadelphia, 1863); *Treatise on Elementary Geometry, With Appendices Containing a Collection of Exercises for Students and an Introduction to Modern Geometry* (Philadelphia, 1870). See bibliography in Coffin [below], pp. 243-44. WORKS ABOUT: *DAB,* 4: 43-44 (William R. Roever); J. H. C. Coffin, "Memoir," *Biog. Mems. of Natl. Ac. Scis.* 1 (1877): 227-44.

CHESTER, ALBERT HUNTINGTON (November 22, 1843, Saratoga Springs, N.Y.—April 13, 1903, New Brunswick, N. J.). *Chemistry; Mineralogy.* Son of Albert Tracy and Elizabeth (Stanley) Chester. Married Alethea S. Rudd, 1869; Georgiana Waldron Jenks, 1898; no information on children. EDUCATION: Two years at Union College; at Columbia Univ., M.E. in 1868 and Ph.D. in course in 1878. HONORS: Sc.D., Hamilton College, 1891. CAREER: 1870-1891, professor of chemistry, mineralogy, and metallurgy, Hamilton College; 1891-1903, professor of chemistry and mineralogy, Rutgers College; 1882, chemist to New York state board of health; practicing mining engineer. MEMBERSHIPS: Am. Assoc. Advt. Soc. SCIENTIFIC CONTRIBUTIONS: Published numerous works and, in addition to teaching duties, also engaged in large mining engineering practice.

WORKS: "The Iron Region of Northern Minnesota," *Annual Report of Minnesota Geological Survey,* no. 11 (1884), pp. 154-67; *A Catalogue of Minerals Alphabetically Arranged, With Their Chemical Compositions and Synonyms* (New York, 3d ed., 1897, 56 pp.); *A Dictionary of the Names of Minerals* (New York, 1896), 320 pp. See Poggendorff, *Biographisch-literarisches Handwörterbuch,* 3: 266, 4: 244; and *Roy. Soc. Cat.,* vols. 7, 9, 14. WORKS ABOUT: *NCAB,* 11 (1909 [copyright 1901 and 1909]): 422-23; *Who Was Who in America,* vol. 1.

CHITTENDEN, RUSSELL HENRY (1856-1943). *Physiological chemistry; Physiology.* Professor of physiological chemistry and director of Sheffield Scientific School, Yale Univ.

MSS: Yale Univ. Library (*DAB* and *DSB*). WORKS ABOUT: See *AMS; DAB,* supp. 3; *DSB.*

CHRISTY, DAVID (1802-ca. 1868 [?]). *Geology.* No information on parentage, marital status, or education. CAREER: 1824-1836, journalist; as agent for American Colonization Society, in 1848 appointed to Ohio, where he lectured to state legislature and solicited funds from individuals for purchase of land known as "Ohio in Africa"; geologist for Nantahala and Tuckasege Land and Mineral Co. of North Carolina. MEMBERSHIPS: May have been member of Western Academy of Natural Sciences, Cincinnati. SCIENTIFIC CONTRIBUTIONS: On extensive travels in work for American Colonization Society, made geological observations and collections and by 1855 had sizable cabinet. Published observations in form of letters to *Cincinnati Gazette,* which were later reprinted in pamphlet addressed to J. Locke [q.v.], assistant to Ohio's chief geologist; while difficult to identify portions of the pamphlet that were based on Christy's own field work, Locke credits him in introduction as first to have "actually drawn approximate sections of the strata from the Atlantic to Iowa and from Lake Erie to the Gulf of Mexico" and claims most of work resulted from Christy's firsthand observations. In 1847 wrote letter, later published, to M. De Verneuil of Paris; in it, while doubting validity of glacial theory, in fact gave first delineation of southern margin of drift. *De Bow's Review* of November 1867 reported Christy working on book to be entitled "Geology Attesting Christianity"; published several pamphlets for American Colonization Society; and most important work was *Cotton Is King; or, The Economical Relations of Slavery* (1855).

WORKS: *Letters on Geology: Being a Series of Communications Originally Addressed to Dr. John Locke, of Cincinnati, Giving an Outline of the Geology of the West and Southwest, Together With an Essay on the Erratic Rocks of North America, Addressed to M. De Verneuil, Illustrated by Geological Sections and Engravings of Some Rare Fossils* (Rossville, Ohio, 1848), 81 pp., 5 plates. See Meisel, *Bibliog. Amer. Nat. Hist.,* which lists several geological articles, and two reports of Nantahala and Tuckasege Land and Mineral Co. (1856, 1858). MSS: A few items at Cincinnati Historical Society. WORKS ABOUT: *DAB,* 4: 97-98 (Reginald C. Mc-

Grane); Merrill, *Hundred Years of Geology,* pp. 255-56, 629.

CIST, JACOB (March 13, 1782, Philadelphia, Pa.—December 30, 1825, Wilkes-Barre, Pa.). *Geology; Natural history.* Son of Charles, wealthy printer and publisher, and Mary (Weiss) Cist. Married Sarah Hollenback, 1807; survived by five children. EDUCATION: Attended Philadelphia public schools; 1794-1795, Moravian boarding school, Nazareth, Pa. CAREER: 1797-1800, worked in father's Philadelphia printing business; 1800, managed short-lived printing firm established by father in Washington, D.C.; 1800-1808, postal clerk, Washington, D.C.; 1808-1825, postmaster, Wilkes-Barre, Pa.; 1808, and for some years thereafter, business partner with father-in-law; 1808, made unsuccessful attempt to establish glassworks and pottery business; 1813-1815, engaged in mining and selling of anthracite coal; 1816, treasurer of Luzerne County, Pa.; 1816-1818, a founder and treasurer of Wilkes-Barre Bridge Co.; 1817, first cashier of Wilkes-Barre's first bank; 1820, tried unsuccessfully to organize ironworks; owned large areas of iron lands. MEMBERSHIPS: A founder and secretary, Luzerne County Agricultural Society. SCIENTIFIC CONTRIBUTIONS: Remembered especially for work on geology and promotion of anthracite coal, his father having been involved in 1792 formation of Lehigh Coal Mine Co.; in 1813, with others, Jacob leased land of old Lehigh Coal Mine Co., but was no more successful in marketing anthracite than old company. Article in *Am. Journ. Sci.* (1822) [see Works], summarizing years of study and constituting first important work on anthracite in U.S., included map, three sections of strata, and discussion of mining procedures as well as various descriptions of geology, mineralogy, and fossils of region. Worked for some years on manuscript on American entomology, including great many original drawings, but work was not published and manuscript was lost.

WORKS: "An Account of the Mines of Anthracite, in the Region about Wilkes-Barre, Penn.," *Am. Journ. Sci.* 4 (1822): 1-16, with author given as "Zachariah Cist," though most certainly Jacob was intended. Also see other works under "Zachariah Cist" in Meisel, *Bibliog. Amer. Nat. Hist.* MSS. Ac. Nat. Scis. Philad. (*NUCMC* 66-24). WORKS ABOUT: *DAB,* 4: 109-10 (Carl W. Mitman); George B. Kulp, *Families of the Wyoming Valley . . .* (Wilkes-Barre, Pa., 1890), 3: 1342-54.

CLAP, THOMAS (June 26, 1703, Scituate, Mass.—January 7, 1767, New Haven, Conn.). *Astronomy.* Son of Deacon Stephen and Temperance (Gorham) Clap. Married Mary Whiting, 1727; Mary (Haynes) Lord Saltonstall, twice a widow; five children. EDUCATION: 1722, A.B., Harvard College. CAREER: 1726-1739, minister, Windham, Conn.; 1739-1766, president (until 1745 called rector), Yale College. SCIENTIFIC CONTRIBUTIONS: While theologically conservative, was well versed in and accepted most advanced scientific thought of day, and participated in and actively promoted teaching of science at Yale. Constructed crude orrery, first to be made in colonies, and apparently had mastered Newtonian physics; after J. Winthrop (died 1779) [q.v.], most able of colonial mathematicians. Actively interested in wide range of scientific subjects, but special interest was astronomy; observed transit of Mercury in 1742 and planned to observe transit of 1753, but cloud cover prevented success. Played especially important role in motivating others to make such observations of transits and to appreciate importance of such studies. Investigations of meteors began at Windham, Conn.; in 1750s made known to interested American scientists his theory that certain meteors shared characteristics with comets (had ellipical orbits and did not fall to earth or dissipate in gas); in 1755, prompted by anonymous article in *Gentleman's Magazine* that presented similar theory, Clap prepared manuscript for distribution to American astronomers. Collected data on meteor of 1759 and results appeared in *Boston News Letter* (May 31, 1759) in form of letter to J. Winthrop, who later published similar views in *Phil. Trans. of Roy. Soc.* (1761); in 1763, Clap revised his manuscript and sent it to Roy. Soc., and also prepared several other accounts of meteors, published in 1765 and 1766. Even in own time his meteor theory was recognized to have serious flaws.

WORKS: Account of observations of Halley's comet in *Connecticut Gazette* (May 1759); *Conjectures Upon the Nature and Motion of Meteors Which are Above the Horizon* (Norwich, Conn., 1781), published posthumously by Ezra Stiles. MSS: Yale Univ. Library. WORKS ABOUT: *DAB* 4: 116-17 (Harris E. Starr); Leonard Tucker, "Pres. Thomas Clap of Yale College: Another 'Founding Father' of American Science," *Isis* 52 (1961): 55-57. Also see Tucker, *Puritan Protagonist: President Thomas Clap of Yale College* (Chapel Hill, N.C., 1962).

CLAPP, CORNELIA MARIA (1849-1935). *Zoology.* Professor of zoology, Mt. Holyoke College.

MSS: See *Nota. Am. Wom.* WORKS ABOUT: See *AMS*; *Nota Am. Wom.*

CLARK, ALVAN (March 8, 1804, Ashfield, Mass.—August 19, 1887, Cambridgeport, Mass.). *Astronomy; Telescope construction.* Son of Abram, farmer and miller, and Mary (Bassett) Clark. Married Maria Pease, 1826; four children, including Alvan G. Clark [q.v.]. EDUCATION: Until ca. 1821, attended school built on father's farm. HONORS: M.A., Amherst in 1854, Princeton in 1865, old Univ. of Chicago in 1866, and Harvard in 1874; Rumford Medal of Am. Ac. Arts Scis., 1867. CAREER: Ca. 1821-ca. 1823, employed in wagon maker's shop; 1824-1825, engraver and itinerant portrait painter; 1825-1836, engraver of textile printing rolls at Lowell, Mass., Providence, R.I., New York City, and Fall River, Mass.; 1836-1860, portrait painter with studio in Boston; 1844, began study of optics; 1850, founding date of Alvan Clark & Sons, telescope makers; 1860-1887, devoted full efforts to telescope business. MEMBERSHIPS: Am. Ac. Arts Scis.; Am. Assoc. Advt. Sci. SCIENTIFIC CONTRIBUTIONS: Interest in optics attributed to efforts of son George Bassett Clark at casting and grinding mirrors, and out of these beginnings arose firm of Alvan Clark & Sons (George Bassett Clark and Alvan Graham Clark), master lens makers; unexcelled quality of refracting lenses was due mainly to care exercised in grinding and polishing glass. First Clark telescope, five inches, sold in 1848 and before Civil War produced several twelve-inch objectives; greatest achievement of firm during Alvan Clark's lifetime was thirty-six-inch lens produced for Lick Observatory (California). In testing telescopes, discovered several double stars, reports of which elicted endorsement of telescopes by W. R. Dawes in England; during years 1862-1863, three Clarks conducted certain unique photometrical experiments, published by father [see Works, two articles on the sun].

WORKS: W. R. Dawes, "New Double Stars Discovered by Mr. Alvan Clark . . . ," *Monthly Notices of Royal Astronomical Society* 17 (1857): 257-59; "The Sun, a Small Star," *Memoirs of Am. Ac. Arts Scis.* n.s. 8 (1863): 569-72, and "The Sun and Stars Photometrically Compared," *Am. Journ. Sci.* 2d ser. 36 (1863): 76-82; "On a New Method of Measuring Celestial Arcs," *Proceedings of Am. Assoc. Advt. Sci.* 10 (1856): 108-11. WORKS ABOUT: *DAB,* 4: 119-20 (Raymond S. Dugan); *DSB,* 3: 228 (Deborah J. Warner); Warner, *Alvan Clark & Sons: Artists in Optics,* U.S. Natl. Mus. Bulletin No. 274 (Washington, D.C., 1968), 120 pp. (pp. 39-113 constitute catalogue of instruments made by Alvan Clark & Sons).

CLARK, ALVAN GRAHAM (July 10, 1832, Fall River, Mass.—June 9, 1897, Cambridgeport, Mass.). *Astronomy; Telescope construction.* Son of Alvan [q.v]—painter, engraver and telescope maker—and Maria (Pease) Clark. Married Mary Mitchell Willard, 1865; had children. EDUCATION: Attended public schools, Cambridge, Mass.; ca. 1848, began four or five years' training as machinist in machine shop. HONORS: Lalande Medal, French Academy of Sciences, 1862. CAREER: Ca. 1852-1897, partner in firm of Alvan Clark & Sons, telescope makers. MEMBERSHIPS: Am. Ac. Arts Scis.; Astronomical Society of France; and other societies. SCIENTIFIC CONTRIBUTIONS: Showed interest and ability in astronomy and its apparatus as a schoolboy, and in 1847-1848, at age fifteen, prepared seven-and-one-half-inch speculum and employed it in preparation of map of stars of Orion nebula. In firm of Alvan Clark & Sons, did the optical work with his father, while brother George Bassett Clark had charge of mechanical work. While testing instruments made by family firm, discovered sixteen double stars; with eighteen-inch lens that went to Dearborn Observatory (Evanston, Ill.), discovered companion star to Sirius, previously known only theoretically by influence on Sirius itself, and for this discovery received Lalande Medal. Went on astronomical expeditions to observe eclipses in Shelbyville Tenn. (1869), and in Spain (1870), both sponsored by U.S. Coast Survey and directed by J. Winlock [q.v.]; in 1878, accompanied U.S. Naval Observatory expedition to observe eclipse in Creston, Wyo. Did much of work on thirty-six-inch Lick Observatory lens, and while head of firm the great forty-inch lens was prepared for Yerkes Observatory (Wisconsin), an achievement still unrivaled; while achievements of firm of Alvan Clark & Sons were epoch-making, apparently the Clarks were not able to envision that large telescopes of future would be reflective rather than refractive.

WORKS: S. W. Burnham, "Double Stars Discovered by Mr. Alvan G. Clark," *Am. Journ. Sci.* 3d ser. 17 (1879): 283-89. WORKS ABOUT: *DAB,* 4: 120 (Raymond S. Dugan); *DSB* 3: 288 (Deborah J. Warner); also see sketch of Alvan Clark [above], including Works About.

CLARK, HENRY JAMES (June 22, 1826, Easton, Mass.—July 1, 1873, Amherst, Mass.). *Zoology; Botany.* Son of Henry Porter, Swedenborgian clergyman, and Abigail Jackson (Orton) Clark. Married Mary Young Holbrook, 1854; eight children. EDUCATION: 1848, A.B., Univ. of City of New York (New York Univ.); 1854, S.B., Lawrence Scientific School, Harvard. CAREER: 1848-1850, teacher, White Plains, N.Y.; 1854, became assistant to J. L. R. Agassiz [q.v.]; 1860-1865, assistant professor of zoology, Lawrence Scientific School, Harvard; 1866-1869, professor of botany, zoology, and geology, Pennsylvania State College; 1869-1872, professor of natural history, Univ. of Kentucky; 1872-1873, professor of veterinary science, Massachusetts Agricultural College. MEMBERSHIPS: Natl. Ac. Scis. and number of other societies. SCIENTIFIC CONTRIBUTIONS: Regarded as most eminent American histologist and microscopist of his day. Early interests were botanical, and published several works restricted to that subject, the last appearing in 1859; under influence of Agassiz, focus of interest changed from botany to zoology, while training in both areas gave useful view of entire field of biology. Worked 1856-1863 with Agassiz on *Contributions to the Natural History of the U.S.* (1857-1862), preparing sections on anatomy and histology, but dispute over authorship [see Works, *A Claim*] led to eventual departure from Cambridge. In 1864 presented Lowell lectures, published as *Mind in Nature* [Works], which included much original research and as general work on zoology set new mark for American scientists. Various publications dealt with great range of subjects, including Protozoa, infusoria and lower plants, Coelenterata, spontaneous generation, and the cell and the nature of protoplasm; most notable work probably was that on sponges, and research proved that flagellated cells of sponges that he discovered were zoological in character.

WORKS: *A Claim for Scientific Property* (Cambridge, Mass., 1863); *Mind in Nature; or, The Origins of Life, and the Mode of Development of Animals* (New York, 1864); "On the Spongiae Ciliatae as Infusoria Flagellata; or, Observations on the Structure, Animality, and Relationship of Leucosolenia botryoides, Bowerbank," *Memoirs of Bos. Soc. Nat. Hist.* 1 (1867): 305-10. See *Roy. Soc. Cat.*, vols. 1, 6 (additions), 7, 9; also, bibliography in Packard [below], pp. 327-28. MSS: Harv. U.-MCZ. WORKS ABOUT: *DAB* 4: 131-32 (Frederick Tuckerman); A. S. Packard, "Memoir," *Biog. Mems. of Natl. Ac. Scis.* 1 (1877): 317-28.

CLARK, WILLIAM BULLOCK (1860-1917). *Geology.* Professor of geology, Johns Hopkins Univ.; Maryland state geologist; geologist with U.S. Geol. Surv.

WORKS ABOUT: See *AMS; DAB.*

CLARKE, FRANK WIGGLESWORTH (1847-1931). *Geological chemistry.* Chief chemist, U.S. Geol. Surv.

MSS: Lib. Cong. (Hamer). WORKS ABOUT: See *AMS; DAB,* supp. 1; *DSB.*

CLARKE, JOHN MASON (1857-1925). *Paleontology; Geology.* New York state paleontologist and director of state museum; professor of geology, Rensselaer Polytechnic Institute.

WORKS ABOUT: See *AMS; DAB.*

CLAYPOLE, EDWARD WALLER (June 1, 1835, Ross, Herefordshire, England—August 17, 1901, Long Beach, Calif.). *Geology.* Son of Edward Angell, Baptist minister, and Elizabeth (Blunt) Claypole. Married Jane Trotter, 1865; Katharine Benedicta Trotter, 1879; three children by first marriage. EDUCATION: At Univ. of London, matriculated in 1854, A.B. in 1862, and B.S. in 1864. HONORS: D.Sc., Univ. of London, 1888. CAREER: 1852, began teaching, Abingdon; 1866-1872, tutor in classics and mathematics, Stokescroft College (Bristol); 1872, went to U.S.; 1873-1881, professor of natural history, Antioch College; 1881-1883, on staff of Pennsylvania Geological Survey; 1883-1898, professor of natural sciences, Buchtel College (now Univ. of Akron); 1888, a founder and an editor of *American Geologist;* 1898-1901, professor of geology and biology, Throop Polytechnic Institute (now California Institute of Technology). MEMBERSHIPS: Geological societies of London, Edinburgh, and America; Am. Phil. Soc.; and other societies. SCIENTIFIC CONTRIBUTIONS: First intent was to become civil engineer, but instead became teacher, and during career taught large variety of subjects. Geological work also was wide ranging, but carefully done, and dealt especially with questions of stratigraphy and paleontology; one of most important contributions was discovery of fish remains in Silurian rocks of Pennsylvania, a find that he continued to study and report in detail even after terminating connection with Pennsylvania survey; also wrote on glacial matters and produced several works on physical characteristics and changes in the earth. Among miscellaneous papers were those on migration of plants and animals (1877-1879); bibliography is extensive, and later papers tended to be philosophical in nature; unpublished paper entitled "The Devonian Formation of the Ohio Basin" won Walker Prize of Massachusetts Institute of Technology, 1895.

WORKS: Earliest known work represented in paper "On Some Evidences in Favor of Subsidence in the Southwest Counties of England During the Recent Period" (abstract), *Proceedings of Bristol Naturalists' Society* 5, pt. 2 (1870), pp. 34-36. Bibliography of works in Comstock [below], pp. 491-97. WORKS ABOUT: *DAB,* 4: 181-82 (George P. Merrill); Theodore B. Comstock, "Memoir," *Bulletin of Geol. Soc. Am.* 13 (1901-1902): 487-97.

CLAYTON, JOHN (Autumn 1694, Fulham, England—week preceding January 6, 1774, Gloucester County, Va.). *Botany.* Son of John, lawyer and attorney general of Virginia, and Lucy Clayton. Married Elizabeth Whiting, 1723; eight children. EDUCATION: May have attended Cambridge University, and studied law. CAREER: Ca. 1715, came to Virginia (where father had lived since ca. 1705); 1720-1773, clerk of Gloucester County Court, Va. MEMBERSHIPS: Am. Phil. Soc.; Swedish Royal Academy of Science. SCIENTIFIC CONTRIBUTIONS: Correspondence with M. Catesby [q.v.] led to introduction to John Frederick Gronovius, and by 1735 Clayton was sending plant specimens, largely by way of Catesby, to Gronovius for identification. Sent Gronovius "A Catalogue of Plants, Fruits, and

Trees Native to Virginia," which was prepared for publication without Clayton's knowledge; Clayton's familiarity with Ray's system made possible his determination of plant genera; Gronovius added species names and other technical contributions and published work as *Flora Virginica* [see Works], which credited Clayton's "Catalogue" in introduction, but most recognition went to Gronovius, who had had assistance of Linnaeus; *Flora* was first such work of importance on plants of British North America. Continued sending specimens to Gronovius and had correspondence with J. Bartram [q.v.] and Peter Collinson, the latter attempting to persuade Gronovius to produce second edition of *Flora*; Clayton sent enlarged and improved manuscript of "Flora" to Collinson along with specimens to be drawn by best-known botanical artist of the day, but manuscript was not published and is lost; second edition of *Flora* was prepared for publication by Gronovius's son, and reputation of work rests on this edition. Clayton made trip to Canada in 1746 and may have traveled as far as the Mississippi.

WORKS: John Frederick Gronovius, *Flora Virginica,* pt. 1, 128 pp., about 600 species (Leyden, 1739); pt. 2, 77 pp., about 300 species (Leyden, 1743); 2d ed. prepared by Laurens Theodore Gronovius (Leyden, 1762). MSS: See references in notes to Berkeley and Berkeley [below]. WORKS ABOUT: *DAB,* 4: 184-85 (Donald C. Peattie); Edmund Berkeley and Dorothy Berkeley, *John Clayton, Pioneer of American Botany* (Chapel Hill, N.C., 1963).

CLEAVELAND, PARKER (January 15, 1780, Rowley, Byfield Parish, Mass.—October 15, 1858, Brunswick, Maine). *Mineralogy.* Son of Parker, physician, and Elizabeth (Jackman) Cleaveland. Married; had children. EDUCATION: 1799, A.B., Harvard College; private study of law and theology. CAREER: 1799-1803, teacher at Haverhill, Mass., and three years at York, Maine; 1803-1805, tutor in mathematics and natural philosophy, Harvard; 1805-1858, professor of mathematics and natural philosophy and of chemistry, Bowdoin College; 1820, began teaching materia medica, and was involved in administration, Medical School of Maine (associated with Bowdoin College). MEMBERSHIPS: Numerous scientific societies, including European. SCIENTIFIC CONTRIBUTIONS: Interest in mineralogy began in 1807 when Maine lumbermen asked him to examine possibly valuable rocks. In 1808 added chemistry and mineralogy to teaching duties, and in 1812 began wide correspondence with American scientists, collecting data on localities and kinds of minerals in various parts of country; this mass of new information was incorporated in *Treatise* [see Works], the first important American manual on mineralogy, a work of wide popularity and usefulness; for theoretical portions of the work, depended on European authors, but use of Brongniart's classification by chemical makeup, rather than crystal form, cast work in modern light. For twenty years, planned third edition of *Treatise,* but other duties prevented realization of plan, and works of others came to supplant his as major authority on American mineralogy. For many years maintained record of weather, prepared for publication by C. A. Schott [q.v.] [see Works].

WORKS: *Elementary Treatise on Mineralogy and Geology* (Boston, 1816), 688 pp., with 2d ed. in 2 vols. (Boston, 1822); Charles A. Schott, "Results of Meteorological Observations Made at Brunswick, Maine, Between 1807 and 1859," *Smithson. Contr. Knowl.* 16 (1867); accounts of meteorological observations and of fossil shells in *Memoirs of Am. Ac. Arts Scis.* 3 (1809): 119-21 and 155-58, respectively; observations on solar eclipse of 1811, ibid. 3, pt. 2 (1815): 247-48; description of halos and parhelia, ibid, 4 (1818): 120-28. MSS: Bowdoin College Library (*NUCMC* 71-34); N.Y. Pub. Lib. (*NUCMC* 72-1012). WORKS ABOUT: *DAB,* 4: 189-90 (Marshall P. Cram); *DSB,* 3: 313 (John G. Burke).

CLEMSON, THOMAS GREEN (July 1, 1807, Philadelphia, Pa.—April 6, 1888, "Fort Hill," Oconee County, S.C.). *Mining engineering; Chemistry; Geology.* Son of Thomas Clemson, wealthy merchant. Married Anna Maria Calhoun, 1838; three children. EDUCATION: 1826, went to Paris and entered practical laboratory, attending chemical lectures at Sorbonne; 1828-1832, student at Ecole des mines royale; received diploma as assayer after examination at French Royal Mint. CAREER: Ca. 1832-1838, consulting mining engineer in Paris (1832-1836) and in Philadelphia and Washington (1837-1838); 1838-1844, lived on South Carolina plantations, engaged in mining engineering, and became associated with father-in-law, John C. Calhoun, in agriculture and gold mining; 1844-1851, U.S. chargé d'affaires in Belgium; 1851-1853, at residences on Long Island, N.Y., and in South Carolina; 1853-1861, resided at "The Home" near Bladensburg, Md., involved in planting and assaying; 1859-1861, U.S. superintendent of agriculture; during Civil War, served with Confederate army and became supervisor of mines and metal works in Trans-Mississippi Department; 1865-1888, resided in South Carolina, including "Fort Hill," the Calhoun family plantation. SCIENTIFIC CONTRIBUTIONS: Chief scientific work was as chemical analyst; while in France, 1830-1836, produced sixteen scientific papers, most on chemical subjects, especially assays and analyses of minerals and iron and copper ore of U.S.; during period 1856-1858, published several papers on agricultural chemistry in *American Farmer.* Said to have discovered phosphoric acid in the air [see Works, "Sources of

Ammonia"], which he explained as originating from infusoria and animalculae that are everywhere, although French chemist, M. Barral, reported to French Academy of Sciences in 1860 his own, but later, discovery of the acid in the air. Interest in agricultural and scientific education resulted in bequest of estate to South Carolina, which established Clemson College.

WORKS: "Sources of Ammonia," *American Farmer* 11 (May 1856); two articles on fertilizer in *Annual Report of U.S. Patent Office, Agricultural Division*, 1859 and 1860. See *Roy. Soc. Cat.*, vol. 1; also bibliography in Brackett [below], pp. 584-85. MSS: Clemson College Library (*NUCMC* 61-2106). WORKS ABOUT: *DAB*, 4: 200-1 (E. W. Sikes); R. N. Brackett, "Thomas Green Clemson, LL.D., the Chemist," *Journ. Chem. Educ.* 5 (1928): 433-44, 576-85.

COAN, TITUS (February 1, 1801, Killingworth, Conn.—December 1, 1882, Hilo, Hawaii). *Geology*. Son of Gaylord, farmer, and Tamza (Nettleton) Coan. Married Fidelia Church, 1834; Lydia Bingham, 1873; had children. EDUCATION: Graduated from East Guilford (now Madison, Conn.) Academy; 1831-1833, student at Auburn Theological Seminary (Auburn, N.Y.); 1833, licensed as minister. CAREER: From time of academy graduation until 1826, taught in Killingworth, Conn., and surrounding towns; 1826-1831, taught in western New York state, in charge of school at Riga; 1833, ordained missionary of American Board; 1833-1834, went on mission to Patagonia; 1835-1882, missionary at Hilo, Hawaii; 1870-1871, on leave in U.S. MEMBERSHIPS: Am. Assoc. Advt. Sci. SCIENTIFIC CONTRIBUTIONS: Achievements only incidental to missionary work, relating mainly to volcanoes of Hawaii. Observed volcanoes for more than forty years, including the great volcano Kilauea, which was situated in his district; in *Life* [see Works], gave fairly detailed accounts of volcanoes based on own and natives' observations and told how Wilkes Expedition spent three months in exploration at Hilo, during which time life-long friendship with J. D. Dana [q.v.] developed; also told of later unsuccessful attempts to measure lava's heat with instrument sent by Dana. Undoubtedly was through contacts with Dana that scientific articles were published, a series of twenty-three papers on Hawaiian volcanoes that appeared in *Am. Journ. Sci.* during years 1852-1881, portions of which also were included in *Life*.

WORKS: *Adventures in Patagonia* (New York, 1880), with observations on natural phenomena; *Life in Hawaii: An Autobiographical Sketch of Mission Life and Labors, 1835-1881* (New York, 1882), with observations on volcanoes, especially pp. 262-335; "Coral Reefs of Hawaii," *Am. Journ. Sci.* 3d ser. 8 (1874): 466. Also see references to articles on Hawaiian volcanoes in *Roy. Soc. Cat.*, vols. 1, 7, 9, 14. MSS: N.Y. Hist. Soc. (*NUCMC* 60-1869); Hawaiian Historical Society (Hamer). WORKS ABOUT: *DAB*, 4: 236-37 (John C. Archer); *Life in Hawaii* [above]; Obituary in *Science* 1 (1883): 27.

COFFIN, JAMES HENRY (September 6, 1806, Williamsburgh, Mass.—February 6, 1873, Easton, Pa.). *Mathematics*; *Meteorology*. Son of Matthew and Betsy (Allen) Coffin. Married Aurelia Medici Jennings, 1833; Abby Elizabeth Young, 1851; no information on children. EDUCATION: 1828, A.B., Amherst College. CAREER: 1823-1828, tutor at college and teacher during summer; 1829-1837, operated school for boys in Greenfield, Mass.; ca. 1830, added to school a manual labor department that became Fellenberg Manual Labor Institution; 1837-1839, head of academy, Ogdenburg, N.Y.; 1840-1843, tutor, Williams College; 1843-1846, principal of academy at South Norwalk, Conn.; 1846-1873, professor of mathematics and natural philosophy, Lafayette College. MEMBERSHIPS: Natl. Ac. Scis.; Am. Assoc. Advt. Sci. SCIENTIFIC CONTRIBUTIONS: Active research in meteorology began about 1838, and at Williams College designed and constructed instrument, installed on Mt. Greylock, to record automatically wind direction and velocity. For over thirty years collected data on winds from both published works and through correspondence; also had available data from Smithsonian's observers and from army, and in 1846 became collaborator of Smithson. Instn. Most important works were those on winds of Northern Hemisphere and of the globe [see Works]; in these studies, drew upon enormous body of meteorological data; his unprecedented labors firmly established ideas about general flow of winds on globe and in some cases confirmed general suppositions, or theoretical projections, regarding general movement of winds. Also wrote several papers on mathematical and astronomical subjects.

WORKS: Description of wind instrument in *Proceedings of Am. Assoc. Advt. Sci.* 2 (1849): 386-89; "On the Winds of the Southern Hemisphere," ibid. 13 (1859): 284-91; "Winds of the Northern Hemisphere," *Smithson. Contr. Knowl.* 6 (1853), with over 150 pages of tables; "Winds of the Globe; or, The Laws of Atmospheric Circulation Over the Surface of the Earth," ibid. 20 (1875). See also *Roy. Soc. Cat.*, vols. 2, 7, 9. MSS: Smithson. Instn. Archives; also a few items in Lafayette College Library. WORKS ABOUT: *DAB*, 4: 267-68 (W. F. Humphreys); A. Guyot "Memoir," *Biog. Mems. of Natl. Ac. Scis.* 1 (1877): 257-64.

COFFIN, JOHN HUNTINGTON CRANE (September 14, 1815, Wiscasset, Maine—January 8, 1890, Washington, D.C.). *Mathematics.* Son of Nathanael and Mary (Porter) Coffin. Married Louisa Harrison, 1845; survived by five children. EDUCATION: 1834, A.B., Bowdoin College. HONORS: LL.D., Bowdoin College, 1884. CAREER: 1834, went on extensive sea voyage with uncle, learning navigation and practical seamanship; 1836, appointed professor of mathematics in U.S. Navy and, 1836-1843, served on *Vandalia* and *Constitution* in West India squadron, at Norfolk Navy Yard, and in Florida surveys; 1844-1853, at Naval Observatory, Washington, D.C., most of time in charge of mural circle; 1853-ca. 1866, head of department of mathematics and later of astronomy and navigation, U.S. Naval Academy, Annapolis; 1866-1877, director, *Nautical Almanac;* 1877, retired from navy. MEMBERSHIPS: Am. Phil. Soc.; Natl. Ac. Scis.; and other societies. SCIENTIFIC CONTRIBUTIONS: Official connections led to chief achievements, and published little on his own. Conducted extensive studies of mural circle at Naval Observatory, results of which appeared in first five volumes of its *Washington Observations;* this probably was most sophisticated and complete of his scientific efforts, even though the work had been closely supervised by M. F. Maury [q.v.] and was treated as part of official work of the observatory and not as personal research. At Annapolis, prepared textbook, *Navigation and Nautical Astronomy,* published by Naval Academy in 1863; it was based on manuscript produced by W. Chauvenet [q.v.] for use of students and became a standard manual at academy, with many subsequent editions. As superintendent of *American Ephemeris and Nautical Almanac,* responsible mainly for issues of years 1869-1880.

WORKS: See abbreviated references to publications in *Proceedings of Am. Ac. Arts Scis.* 25 (1889-1890) : 312-13; also see *Natl. Un. Cat. Pre-1956 Imp.* WORKS ABOUT: *NCAB,* 5 (1907 [copyright 1891 and 1907]) : 530; George C. Comstock, "Memoir," *Biog. Mems. of Natl. Ac. Scis.* 8 (1919) : 1-7; *Obituary Record of the Graduates of Bowdoin College . . . for the Decade Ending 1 July 1899* (Brunswick, Maine, 1899), p. 12.

COLDEN, CADWALLADER (February 7, 1688, Ireland [raised in Berwickshire, Scotland]—September 20, 1776, Flushing, N.Y.). *Medicine; Botany; Physics.* Son of Alexander Colden, minister in Church of Scotland. Married Alice Christy, 1715; ten children, including Jane Colden [q.v.]. EDUCATION: 1705, A.B., Univ. of Edinburgh; studied medicine in London. CAREER: 1710-1718, engaged in business and medical practice in Philadelphia; 1718, gave up medicine and moved to New York; 1720, appointed surveyor general of New York; 1721-1776, member of New York Governor's Council; 1739-1762, spent most of time at Coldengham, country seat west of Newburgh, N.Y.; 1761-1776, lieutenant governor of New York, and during much of this period virtual head of state government; 1762, moved to Spring Hill estate, Flushing, N.Y. MEMBERSHIPS: Am. Phil. Soc. SCIENTIFIC CONTRIBUTIONS: Corresponded with and sent plant specimens to European scientists; catalogue that accompanied a collection sent to Linnaeus was published as "Plantae Coldenghamiae" [see Works]; mastered Linnaeus's classification, but also felt need for more natural system. Wrote on medical topics, interested in astronomy and cartography, and was competent mathematician. Most lasting contributions made to botany, but was engrossed for last thirty years of life with work in physics, and *First Causes of Action in Matter* [see Works], in which he tried to explain cause of gravity, was most ambitious scientific undertaking in American colonial period; not interested in experimentation, but favored wholly rational and mathematical approach to fundamental questions with which he was not fully prepared to deal. Most influential work was perhaps that on *Dr. Berkley's Treatise* [see Works], with comments appended; also wrote well-known *History of Five Indian Nations* (1727).

WORKS: "Plantae Coldenghamiae," *Acta societatis regiae scientiarum Upsaliensis* 4 (1743) : 81-136 and 5 (1744-1750) : 47-82; *An Abstract from Dr. Berkley's* [sic] *Treatise on Tar-Water* (New York, 1745) ; *Explication of the First Causes of Action in Matter* (New York, 1745; London, 1746), with German (1748) and French (1751) translations; *The Principles of Action in Matter,* 2d ed., expanded (London, 1751). MSS: N.Y. Hist. Soc. (*NUCMC* 61-546), most of which have been published in society's *Collections;* reference to other repositories in *DSB.* WORKS ABOUT: *DAB,* 4: 286-87 (Alice M. Keys) ; *DSB,* 3: 343-45 (Brooke Hindle) ; Stearns, *Sci. in Brit. Colonies,* pp. 559-75. Also see Alice M. Keys, *Cadwallader Colden* (New York, 1906).

COLDEN, JANE (March 27, 1724, New York, N.Y.—March 10, 1766, probably New York, N.Y.). *Botany.* Daughter of Cadwallader [q.v.], scientist and lieutenant governor of New York, and Alice (Christy) Colden. Married William Farquhar, 1759; one child. EDUCATION: Educated by mother and father. SCIENTIFIC CONTRIBUTIONS: Father provided her with botanical education and prepared for her a manuscript that explained fundamentals of botany, including Linnaean system; in letter to Linnaeus on May 12, 1756, Peter Collinson praised her as perhaps first woman to master Linnaeus's system. Learned to take botanical impressions and to prepare descriptions in English, but never learned Latin; by 1757, had prepared catalogue of over 300 specimens of local flora and exchanged specimens and seeds with several colonial and European botanists; as father's interests turned to other subjects, he planned that she take over his botanical activities, including correspondence. The first American woman to win distinction as botanist, but work seems never to have been accepted on par with that of father and other American male botanists. No evidence that botanical interests continued after marriage in 1759. Only one contemporary publication, that by A. Garden [q.v.] [see Works], is known.

WORKS: Alexander Garden, "The Description of a New Plant," *Essays and Queries of Edinburgh Philosophical Society* 2 (1757): 1-7, of which pp. 5-7 are Colden's description of plant sent by her to Garden in 1754; H. W. Ricketts and Elizabeth Hall, eds., *Botanic Manuscript of Jane Colden, 1724-1766* (Garden Club of Orange and Duchess Counties of New York, 1963), a selection of fifty-seven descriptions as given in her manuscript on New York flora, which included some 340 illustrations and was purchased by Sir Joseph Banks and deposited in British Museum. MSS: See Ricketts and Hall [above]; other manuscripts with family papers in N.Y. Hist. Soc., published in society's *Collections*. WORKS ABOUT: *DAB*, 4: 288-89 (Marion P. Smith); *Nota. Am. Wom.*, 1: 357-58 (Brooke Hindle); Stearns, *Sci. in Brit. Colonies*, pp. 565-67.

COLLETT, JOHN (January 6, 1828, Eugene, Ind.—March 15, 1899, Indianapolis, Ind.). *Geology*. Son of Stephen S., businessman and public official, and Sarah (Groenendyke) Collett. Never married. EDUCATION: 1847, A.B., Wabash College; 1882, M.D., Central College of Physicians and Surgeons, Indianapolis. HONORS: M.A. in ca. 1852 and Ph.D. in 1879, Wabash College. CAREER: Ca. 1847, entered upon successful career in farming and related activities; 1870-1878, assistant with Indiana geological survey; 1871-1873, member of Indiana state senate; 1878-1879, member of commission to build Indiana statehouse; 1879, appointed chief, Indiana Bureau of Statistics and Geology; 1881-1885, Indiana state geologist. MEMBERSHIPS: Am. Assoc. Advt. Sci. SCIENTIFIC CONTRIBUTIONS: Began collecting specimens and fossils at age eight; these interests continued, but did not have opportunity until middle years to use geological interest and knowledge in official capacity. Primary scientific contributions appear in various state reports on Indiana geology, in study of which he was an acknowledged leader; known particularly for stratigraphic work; one notable discovery was evidence of preglacial river in Indiana and western Kentucky. Contributions to Indiana reports are extensive and dedication was such that when legislature failed to make appropriation in 1884, Collett provided funds from personal resources to continue survey and was later reimbursed.

WORKS: See list in Nickles, *Geol. Lit. on N. Am.* WORKS ABOUT: *NCAB*, 9 (1907 [copyright 1899 and 1907]): 548 (includes some errors); George P. Merrill, *Contributions to a History of American State Geological and Natural History Surveys*, U.S. Natl. Mus. Bulletin no. 109 (Washington, D.C., 1920), pp. 81-83; *Biographical History of Eminent and Self-Made Men of the State of Indiana* (Cincinnati, 1880), vol. 1, 7th district, pp. 23-25.

COLLIER, PETER (August 17, 1835, Chittenango, N.Y.—June 29, 1896, Ann Arbor, Mich.). *Agricultural chemistry*. Son of Jacob and Mary Elizabeth Collier. Married Caroline F. Angell, 1871; one child. EDUCATION: 1861, A.B., Yale College; 1866, Ph.D., Sheffield Scientific School, Yale. HONORS: M.D., Univ. of Vermont, 1870. CAREER: 1862-1866, assistant in chemistry, Sheffield Scientific School; 1867-1877, professor of chemistry, mineralogy, and metallurgy, Univ. of Vermont, and professor of toxicology and chemistry in medical school; 1871 made dean of medical faculty; 1871, elected secretary, Vermont state board of agriculture, mining, and manufacture; 1877-1883, chief chemist, U.S. Department of Agriculture; 1883-1887, in Washington, D.C., where worked on sorghum report; 1887-1895, director, New York State Agricultural Experiment Station (Geneva, N.Y.); 1895, moved to Ann Arbor, Mich. SCIENTIFIC CONTRIBUTIONS: A special student of Prof. S. W. Johnson [q.v.] at Sheffield School, and in subsequent career was able to combine interest in chemistry with practical agricultural considerations. In Vermont, responsible for number of farm institutes held throughout the state; most notable work, on sorghum [see Works], was done mainly while at U.S. Department of Agriculture, and investigations behind the report are significant as being first important chemical studies done under department's auspices; at Department of Agriculture also studied grasses and forage crops with departmental botanist. Ability to combine demands of science and practical agriculture continued after going to New York Experiment Station.

WORKS: *Sorghum, Its Culture and Manufacture, Economically Considered as a Source of Sugar, Syrup and Fodder* (Cincinnati, 1884), 570 pp.; also see annual reports of U.S. Commissioner of Agriculture, 1877-1883, and of New York State Agricultural Experiment Station, 1888-1895. See *Natl. Un. Cat. Pre-1956 Imp.* MSS: Univ. of Wisconsin Library, Rare Books Department (*NUCMC* 70-471). WORKS ABOUT: *DAB*, 4: 304 (E. H. Jenkins); *Obituary Records of Graduates of Yale University* (1890-1900): 468-69.

COLLINS, FRANK SHIPLEY (1848-1920). *Botany.* Factory manager; authority on algae.

MSS: Am. Phil. Soc. (*NUCMC* 62-747); Harv. U-FL (Hamer). WORKS ABOUT: See *AMS, DAB.*

COLMAN, BENJAMIN (October 19, 1673, Boston, Mass.—August 29, 1747, Boston, Mass.). *Medicine* (smallpox inoculation); *Natural history.* Son of William, wealthy shopkeeper, and Elizabeth Colman. Married Jane Clark, 1700; Sarah (Crisp) Clark, 1732; Mary (Pepperell) Frost, 1745; three children by first marriage. EDUCATION: 1692, A.B., Harvard College; 1695, A.M., following study of theology as resident student, Harvard. HONORS: D.D., Glasgow Univ., 1731. CAREER: 1692, minister, Medford, Mass., for six months; 1695-1699, in England, where he preached at Bath; 1699-1747, minister, Brattle Street Church, Boston; 1717-1728, fellow, Harvard College; 1724, declined offer of presidency of Harvard College. SCIENTIFIC CONTRIBUTIONS: Made no noteworthy contribution to scientific knowledge, but was identifield with scientific community in colonies, and prominence as clergyman tended to support and promote acceptance and study of science. Defended smallpox inoculation during controversy in Boston, 1721, and in *Some Observations* [see Works] suggested that smallpox may be transmitted by "animalcules," which had been discovered in pustules, an idea appearing in more developed form in unpublished work, "Angel of Bethesda," by C. Mather [q.v.]. In 1723, sent description of hummingbird and sea horse to secretary of Roy. Soc., subsequently presented to the society itself; in 1728, sent account of New England earthquake of October 29, 1727, to bishop of Peterborough [see Works], in response to bishop's request for Colman's meteorological observations; while report contained no original thoughts on nature and cause of earthquakes, it did confirm other accounts of location and effects of the disturbance.

WORKS: *Some Observations on the New Method of Receiving the Small-pox, by Ingrafting or Inoculation* (Boston, 1721); report on earthquake of October 1727 in *Phil. Trans. of Roy. Soc.* 36, no. 409 (May-June 1729), pp. 124-27. See Shipton, *Biog. Sketches, Harvard,* 4 (1933): 129-37, for complete list of religious and other works. MSS: Mass. Hist. Soc. (Hamer); Harv. U.-HL (Shipton [above]). WORKS ABOUT: *DAB,* 4: 311 (James T. Adams); Stearns, *Sci. in Brit. Colonies,* pp. 442-46.

COMPTON, ALFRED GEORGE (1835-1913). *Astronomy; Physics.* Professor of applied mathematics and physics, College of City of New York (City College).

WORKS ABOUT: See *AMS; Who Was Who in America,* vol. 1.

COMSTOCK, ANNA BOTSFORD (1854-1930). *Entomology; Nature study; Science illustration and wood engraving.* Lecturer and professor in nature study, Cornell Univ.

MSS: Corn. U.-CRH (*NUCMC* 70-1067); also see *Nota. Am. Wom.* WORKS ABOUT: See *AMS* (Mrs. J. H.); *Nota. Am. Wom.*

COMSTOCK, GEORGE CARY (1855-1934). *Astronomy.* Professor of astronomy, director of Washburn Observatory, Univ. of Wisconsin.

WORKS ABOUT: See *AMS; DAB,* supp. 1; *DSB.*

COMSTOCK, JOHN HENRY (1849-1931). *Entomology.* Professor of entomology, Cornell Univ.

MSS: Corn. U.-CRH (*NUCMC* 62-2323, 66-1031, 70-1067); Ac. Nat. Scis. Philad. (Hamer). WORKS ABOUT: See *AMS; DAB,* supp. 1.

COMSTOCK, THEODORE BRYANT (1849-1915). *Geology; Mining engineering.* Founder and director, Arizona School of Mines; president, Univ. of Arizona; consulting engineer; associated with several mining companies.

MSS: Univ. of Arizona Library—Special Collections (Hamer). WORKS ABOUT: See *AMS; NCAB,* 13 (1906): 450; *Who Was Who in America,* vol. 1.

CONRAD, TIMOTHY ABBOTT (June 21, 1803, near Trenton, N.J.—August 8, 1877, Trenton, N.J.). *Paleontology; Malacology.* Son of Solomon White, printer and professor of botany at Univ. of Pennsylvania, and Elizabeth (Abbott) Conrad. Never married. EDUCATION: Attended Quaker school at Westtown, Pa.; self-taught in advanced studies. CAREER: Connected with father's printing and publishing business and ran shop for short time after father's death in 1831; after giving up shop, turned full attention to natural history explorations and study, and often experienced poverty; after a time, through fortunate investments, became financially self-supporting; 1833-1834, in Alabama; for New York Natural History Survey, geologist (1837) and paleontologist (1838-1842), this being his first salaried position; 1854-1857, part-time paleontologist for Smithson. Instn.; 1870-1871, assistant in invertebrate paleontology, North Carolina Geological Survey. MEMBERSHIPS: An organizer of Association of American Geologists; member of Am. Phil. Soc. and other societies. SCIENTIFIC CONTRIBUTIONS: Published number of works on his own, including preparation of illustrations. For some years was active as chief investigator of American Tertiary strata and was one of first Americans to study interregional and intercontinental geology on basis of comparison of fossil content. Also studied modern freshwater and saltwater mollusks, and in addition to own field work prepared reports on fossils collected by Wilkes Expedition and by other exploring parties. Did much important work, which won well-deserved reputation, but some publications, especially systematic portions, were at times carelessly executed; in addition to works listed below, also published number of shorter papers.

WORKS: *Fossil Shells of the Tertiary Formations of North America* (Philadelphia, 1832-1835), republished by G. D. Harris (Washington, D.C., 1893); *Eocene Fossils of Claiborne, with Observations on This Formation in the U.S., and a Geological Map of Alabama* (Philadelphia, 1835); *Fossils of the Medial Tertiary of the U.S.* (Philadelphia, 1838-1861), republished by W. H. Dall (Philadelphia, 1893); *American Marine Conchology* (Philadelphia, 1831). See references to bibliographies of works in *DSB*. WORKS ABOUT: *DSB*, 3: 391-93 (Ellen J. Moore); Youmans, *Pioneers of Sci.*, pp. 385-93.

COOK, GEORGE HAMMELL (January 5, 1818, Hanover, N.J.—September 22, 1889, New Brunswick, N.J.). *Geology*. Son of John and Sarah (Munn) Cook. Married Mary Halsey Thomas, 1846; five children. EDUCATION: 1839, C.E. and, later, B.N.S. from Rensselaer Polytechnic Institute. HONORS: Ph.D., Univ. of New York; LL.D., Union College. CAREER: 1836-1838, assistant, Morris and Essex Railroad and Catskill and Canojoharie Railroad; at Rensselaer, appointed tutor (1839), then senior professor (1842-1846); 1846-1848, engaged in glassmaking business, Albany; at Albany Academy, appointed professor of mathematics and natural philosophy (1848) and principal (1851-1853); 1852, sent to Europe by state of New York to study salt manufacture; 1853-1889, professor of chemistry and natural science, Rutgers College; 1854-1856, assistant geologist, New Jersey survey; 1864, took charge of agricultural work at Rutgers scientific school; 1864-1889, New Jersey state geologist; 1880-1889, director, New Jersey State Agricultural Experiment Station; 1886, made director of New Jersey weather service. MEMBERSHIPS: Natl. Ac. Scis.; vice-president of Am. Assoc. Advt. Sci., 1888. SCIENTIFIC CONTRIBUTIONS: Much of career was devoted to practical science, especially in regard to geology, agriculture, and water supply; reports as New Jersey state geologist stressed practical and economic concerns, and through various activities made important contribution to development of agriculture in New Jersey. Most published work appeared as official reports of New Jersey Geological Survey, Rutgers scientific school, and State Agricultural Experiment Station.

WORKS: *Geology of New Jersey* (Newark, 1868), 900 pp.; *Report on the Clay Deposits of Woodbridge, South Amboy, and Other Places in New Jersey* (Trenton, N.J., 1878), with J. C. Smock. Two significant contributions to general—rather than practical—geology were "On a Subsidence of the Land on the Sea-Coast of New Jersey and Long Island," *Am. Journ. Sci.* 2d ser. 24 (1857): 341-54, and "The Southern Limits of the Last Glacial Drift Across New Jersey and the Adjacent Parts of N.Y. and Penn.," *Transactions of Am. Inst. Min. Engrs.* 6 (May 1877): 467-70. See bibliography in Gilbert [below], pp. 143-44. MSS: Rutgers Univ. Library (*NUCMC* 65-1596); Ac. Nat. Scis. Philad. (Hamer). WORKS ABOUT: *DAB*, 4: 373-74 (George P. Merrill); G. K. Gilbert, "Memoir," *Biog. Mems. of Natl. Ac. Scis.* 4 (1902): 135-44. Also see Jean W. Sidar, *George Hammell Cook* (New Brunswick, N.J., 1976).

COOKE, JOSIAH PARSONS (October 12, 1827, Boston, Mass.— September 3, 1894, Newport, R.I.). *Chemistry*. Son of Josiah Parsons, successful lawyer, and Mary (Pratt) Cooke. Married Mary Hinckley Huntington, 1860; no children. EDUCATION: 1848, A.B., Harvard College; ca. 1850, spent year in Europe and attended chemistry lectures in France. HONORS: LL.D., Cambridge University in 1882 and Harvard in 1889. CAREER: 1849, tutor in mathematics, Harvard; 1850-1894, professor of chemistry and mineralogy, Harvard. MEMBERSHIPS: Natl. Ac. Scis. and other societies; foreign honorary member of Chemical Society of London. SCIENTIFIC CONTRIBUTIONS: Largely self-taught in chemistry. Credited with establishment of Harvard's department of chemistry, based on introduction of meaningful chemical instruction into curriculum and purchase of equipment for laboratory largely out of personal resources. In 1854, published paper [see Works] that was pioneering effort in classification of elements by atomic weights and was widely noticed; during years 1877-1882, prepared series of significant papers on atomic weight of antimony, and 1887 and 1888 published two papers with Theodore W. Richards on relative values of atomic weights of hydrogen and oxygen; during years 1855-1893, presented eight lecture courses for Lowell Institute. *Chemical Philosophy* [see Works] had four editions and was greatly influential.

WORKS: "The Numerical Relation Between the Atomic Weights, with Some Thoughts on the Classification of the Chemical Elements," *Am. Journ. Sci.*, 2d ser. 17 (1854): 387-407; papers on atomic weight of antimony and of hydrogen and oxygen [see above] appeared in *Proceedings of Am. Ac. Arts Scis.*; *Principles of Chemical Philosophy* (Cambridge, Mass., 1868); *Religion and Chemistry as Proof of God's Plan in the Atmosphere and the Elements* (New York, 1864; new ed., 1880), originally prepared as lectures for Brooklyn Institute of Arts and Sciences. Major works mentioned in *DSB*; fuller bibliography in *Proceedings of Am. Ac. Arts Scis.* 30 (1894-1895): 544-47. MSS: Harv. U.-UA (*NUCMC* 65-1230). WORKS ABOUT: *DAB*, 4: 387-88 (Lyman C. Newell); *DSB*, 3: 397-99 (George S. Forbes).

COOLEY, LEROY CLARK (1833-1916). *Physics.* Professor of physics and chemistry, Vassar College.

WORKS ABOUT: See *AMS*; *NCAB*, 11 (1909 [copyright 1901 and 1909]): 263; *Who Was Who in America*, vol. 1.

COOPER, JAMES GRAHAM (June 19, 1830, New York, N.Y.—July 19, 1902, probably Hayward, Calif). *Zoology*; *Botany*; *Geology.* Son of William [q.v.], wealthy heir and naturalist, and Mary (Wilson) Cooper. Married Rosa M. Wells, 1866; no information on children. EDUCATION: 1851, M.D., College of Physicians and Surgeons, N.Y. CAREER: 1851-1853, worked in New York hospitals; 1853, contracted as surgeon with governor of Washington Territory; 1853-1854, surgeon-naturalist in western division of Pacific Railroad Survey; 1854-1857, personal explorations in Pacific Coast region, for a time with Washington governor on Indian treaty commission, returning to New York, ca. 1856, by way of Panama; 1857, appointed surgeon on wagon expedition and went as far west as Wyoming; 1858-1859, probably in New York; 1859, in Florida; 1860-1861, contract surgeon with army to Oregon and other parts of the West; 1861, became zoologist with California Geological Survey: 1864-1865, surgeon with California Volunteers; 1866-1871, medical practice in Santa Cruz, Calif.; 1871-1875, gave up medical practice, removed to Ventura County, Calif., and continued natural history explorations and study; 1875-1902, at Hayward, Calif. MEMBERSHIPS: N.Y. Lyc. Nat. Hist.; active early member of Calif. Ac. Scis. SCIENTIFIC CONTRIBUTIONS: During period 1853-1866, carried out extensive explorations of natural history of the West, especially on Pacific Coast. Interests included botany, ornithology, mammalogy, geology, and paleontology, but was perhaps best known as conchologist, and more than half of his eighty or so publications were on that subject. Also developed an early ecological perspective as result of information gathered in wide travels and wrote on geographical distribution of plants and later of animals, especially birds and mollusks; presented observations on distribution of forests in relation to climate, in *Annual Report of Smithson. Instn.* 13 (1858): 246-80, the first such work published by the government. After about 1865, became more of a systematist than a collector. Cooper Ornithological Club, Santa Clara, Calif., named for him (1893).

WORKS: See *Roy. Soc. Cat.*, vols. 2, 7, 9, 12, 14; additional bibliographical references in Max Arnim, *Internationale Personalbibliographie, 1800-1943* (Leipzig, 1944), p. 242. MSS: Univ. of California General Library, Berkeley (Hamer); Smithson. Instn. Library (Hamer). WORKS ABOUT: *DAB*, 4: 406-7 (Joseph Grinnell); W. O. Emerson, "Dr. James G. Cooper," *Bulletin of Cooper Ornithological Club* 1 (1899): 1-5; *Memorial and Biographical History of Northern California* (Chicago, 1891), pp. 395-96. (The last source includes some biographical data that conflicts with that given in this sketch.)

COOPER, THOMAS (October 22, 1759, London, England—May 11, 1839, Columbia, S.C.). *Chemistry.* Son of Thomas Cooper, person of money and position. Married Alice Greenwood, 1779; Elizabeth P. Hemming, ca. 1811; five children by first marriage, three by second. EDUCATION: 1779, matriculated at Oxford Univ., but did not receive degree; studied medicine and law. HONORS: M.D., Univ. of New York, ca. 1817. CAREER: 1787, became barrister in England, but apparently practiced little; ca. 1790-1793, partner in calico-printing business that failed; 1793, emigrated to U.S., settled near J. Priestley [q.v.] in Pennsylvania, took up farming and began law practice, and acted as consulting physician; by 1799, living with Priestley; 1801-1804, commissioner, Luzerne County, Pa.; 1804-1811, Pennsylvania state judge; 1811-1815, professor of chemistry, Dickinson College; 1815-1819, professor of applied chemistry and mineralogy, Univ. of Pennsylvania; at South Carolina College (Univ. of South Carolina), professor of chemistry (1819-1834), president (1821-1834), and teacher of political economy; 1834-1839, edited statutes of South Carolina for publication. MEMBERSHIPS: Manchester [England] Literary and Philosophical Society; Am. Phil. Soc. SCIENTIFIC CONTRIBUTIONS: In England, undertook advanced work in use of chlorine in bleaching, used in his calico-printing firm; in 1810, prepared potassium, probably for first time in U.S. After political removal from judgeship in 1811, became disillusioned with politics and turned attention more fully to science. During period 1813-1815, edited *Emporium* [see Works]. Prepared American editions of several English textbooks on chemistry, with his own annotations; also wrote several theoretical works, among more significant of which were descriptions of Priestley's scientific discoveries, published as appendix 1 of Priestley's *Memoirs.* Made no great contribution to science, but was important as disseminator of practical scientific information and played significant part in promotion of science in U.S.

WORKS: Bleaching procedure presented in *Emporium of Arts and Sciences*, n.s. 1 (1813): 158-61, and in *Transactions of Am. Phil. Soc.*, n.s. 1 (1818):

317-24. Major works given in *DSB*, which also gives references to more complete bibliographies. MSS: Univ. S.C.-SCL (*NUCMC* 66-524). WORKS ABOUT: *DAB*, 4: 414-16 (Dumas Malone); *DSB*, 3: 399-400 (E. L. Scott). Also see D. Malone, *The Public Life of Thomas Cooper, 1783-1839* (New Haven, 1926; reprinted, New York, 1961).

COOPER, WILLIAM (1797, New York [City?]—April 20, 1864). *Natural history*. Son of James Cooper, wealthy merchant, and Frances Graham Cooper. Married Mary Wilson; six children, including James G. Cooper [q.v.]. EDUCATION: 1821-ca. 1824, in Europe, where he studied zoology. CAREER: Inherited wealth and spent life in study of natural history; 1838-1853, in retirement on farm at Guttenberg, N.J.; 1853, moved to Hoboken, N.J. MEMBERSHIPS: Founding member of the N.Y. Lyc. Nat. Hist.; first American member of London Zoological Society. SCIENTIFIC CONTRIBUTIONS: Information on life sparse and uncertain. Only about nineteen years old when participated in founding N.Y. Lyc. Nat. Hist., and number of his papers appeared in its *Annals*. Earliest publication was that on Georgia Megatherium [see Works]. While apparently reasonably active as naturalist, did not write great deal, and modesty sometimes led him to permit others to make use of discoveries; aided number of other naturalists, including J. Torrey, T. Nuttall, J. J. Audubon [qq.v.], and made natural history materials available to J. E. DeKay [q.v.] for New York state reports on zoology; helped edit two volumes of works of C. L. Bonaparte [q.v.], and much of Cooper's own ornithological work appeared there. State of health prevented extensive field work, but at one time visited Big Bone Lick in Kentucky. In mid-1850s, began to collect in conchology and for that purpose made dredging excursions along Atlantic Coast, from Nova Scotia to Bahamas; last paper, on shells of West Coast, appeared in reports of Pacific Railroad Survey.

WORKS: "On the Remains of the Megatherium Recently Discovered in Georgia," *Annals of N.Y. Lyc. Nat. Hist.* 1 (1824): 114-24. See Meisel, *Bibliog. Amer. Nat. Hist.* WORKS ABOUT: Herman L. Fairchild, *A History of the New York Academy of Sciences . . .* (New York, 1887), pp. 70-73; William H. Dall, obituary of James G. Cooper, *Science*, n.s. 16 (1902): 268.

COPE, EDWARD DRINKER (July 28, 1840, Philadelphia, Pa.—April 12, 1897, Philadelphia, Pa.). *Zoology; Paleontology*. Son of Alfred, heir to wealth who lived in retirement on estate, and Hannah (Edge) Cope. Married Annie Pim, 1865; one child. EDUCATION: Went to Friends' School, Westtown, Pa.; 1860-1861, studied comparative anatomy with J. Leidy [q.v.], Univ. of Pennsylvania; studied reptiles with S. F. Baird [q.v.] at Smithson. Instn.; 1863-1864, several months study abroad. HONORS: Ph.D., Univ. of Heidelberg, ca. 1886; Gold Medal, Geological Society of London, 1879. CAREER: 1864-1867, professor of comparative zoology and botany, Haverford College; 1867-1889, engaged in exploration and research, until 1871 entirely self-financed; 1872 went to Wyoming and 1873 to Colorado, both in association with U.S. geological survey under F. V. Hayden [q.v.]; 1874-1875 and 1877, paleontologist, U.S. surveys under G. M. Wheeler [q.v.]; 1878-1897, part owner and senior editor of *American Naturalist*; 1879, terminated connection with U.S. geological surveys and his most active period of western exploration came to an end; ca. 1886, lost much of inherited fortune and ceased most field exploration; at Univ. of Pennsylvania, professor of geology and mineralogy (1889-1895) and professor of zoology and comparative anatomy (1895-1897); 1892, began association with Texas Geological Survey; at times connected with Indiana and Canadian geological surveys. MEMBERSHIPS: Natl. Ac. Scis. and other societies; president, Am. Assoc. Advt. Sci., 1896. SCIENTIFIC CONTRIBUTIONS: Early work was on reptiles, in study of which he was one of country's leading experts by age twenty-two. Attention shifted to vertebrate paleontology in 1865, and began work in 1870 with Leidy on Wyoming fossils collected by Hayden survey. An enormously productive scientist, he eventually introduced over 600 species and a number of genera, mainly of extinct vertebrates. Engaged in famous controversy with O. C. Marsh [q.v.], his great rival in collection and study of American fossil forms. Wrote number of works on evolution and related questions, at times carrying views into works on social questions; a major advocate of neo-Lamarckian evolutionary views.

WORKS: Bibliography includes over 1,300 titles; *The Vertebrata of the Tertiary Formations of the West*, Reports of U.S. Geological Survey of the Territories (Hayden), vol. 3 (1884), 1009 pp., known as "Cope's Bible." See Osborn [below]. MSS: Haverford College Library—Quaker Collection (*NUCMC* 62-4336); Ac. Nat. Scis. Philad. (*NUCMC* 66-26); Am. Mus. Nat. Hist. WORKS ABOUT: *DAB* 4: 421-22 (George P. Merrill); Henry F. Osborn, *Cope: Master Naturalist. The Life and Letters . . . with a Bibliography of His Writings Classified by Subject* (Princeton, 1931).

CORNWALL, HENRY BEDINGER (1844-1917). *Chemistry; Mineralogy*. Professor of applied chemistry and mineralogy, Princeton Univ.

WORKS ABOUT: See *AMS*; *Who Was Who in America*, vol. 1.

CORY, CHARLES BARNEY (January 31, 1857, Boston, Mass.—July 29, 1921, Ashland, Wis.). *Ornithology.* Son of Barney, well-to-do importer, and Eliza Ann Bell (Glynn) Cory. Married Harriet W. Peterson, 1883; two children. EDUCATION: 1876, entered Lawrence Scientific School, Harvard, but did not graduate; 1878, several months study at Boston Law School. CAREER: 1877-1906, wealth allowed him to travel, especially to Florida, but also to other parts of America and Europe; 1906, lost fortune; 1906, became salaried curator of zoology, Field Museum of Natural History, Chicago. MEMBERSHIPS: Bos. Soc. Nat. Hist.; a founder of Am. Ornith. Un. and president, 1903-1905. SCIENTIFIC CONTRIBUTIONS: Until 1906, inherited wealth made possible extensive travel and pursuit of ornithological and other interests; on estate at Hyannis on Cape Cod, established one of first bird sanctuaries in U.S., and in Florida established Florida Museum of Natural History at Palm Beach. In 1878, visited Bahamas, and through subsequent studies became acknowledged expert on birds of West Indies, in 1880s publishing several books on birds of that region. In 1893, when Field Museum was organized in Chicago, presented entire bird collection and was made curator of ornithology for life, although until 1906 assistant did much of work. Greatest ornithological effort and one that occupied him during later years at Field Museum was *Birds of the Americas* [see Works], of which later volumes were published by museum after his death; in addition to other works, wrote number of papers published in *Bulletin of Nuttall Ornithological Club* and in *Auk.* During much of life, involved in numerous other activities, including athletics and writing of opera librettos and short stories.

WORKS: *Beautiful and Curious Birds of the World* (Boston, 1880-1883), an illustrated folio work; *Birds of Eastern North America*, 2 pts. (Boston, 1899, 1900), a popular manual on bird identification; *Birds of Illinois and Wisconsin* (Chicago, 1909); *Mammals of Illinois and Wisconsin* (Chicago, 1912); *Catalogue of Birds of the Americas*, 15 pts. (Chicago, 1918-1949), of which Cory was sole author of only two. Major books mentioned in *DAB.* MSS: Ac. Nat. Scis. Philad. (Hamer). WORKS ABOUT: *DAB*, 4: 458-59 (Witmer Stone); Wilfred H. Osgood, "In Memoriam," *Auk* 39 (1922): 151-66.

COTTING, JOHN RUGGLES [born Cutting] (November 16, 1778, Acton, Mass.—October 13, 1867, Milledgeville, Ga.). *Chemistry; Geology.* No information on parentage. Married; no information on children. EDUCATION: Attended Harvard College and studied medicine at Dartmouth College; 1802, A.B., Dartmouth College; studied theology. CAREER: 1807-1812, minister, Waldoboro, Maine; during War of 1812, employed as chemist with Boston firm; apparently continued as minister and also at times was a teacher in academies in Boston, Worcester, and Greenfield, Mass.; 1824-1825, lecturer in chemistry, Berkshire Medical Institution, Pittsfield, Mass.; ca. 1829, teacher of chemistry, Amherst Academy; 1835, moved to Augusta, Ga.; 1836-1840, state geologist of Georgia; ca. 1840, began operation of school with his wife at Milledgeville, Ga. SCIENTIFIC CONTRIBUTIONS: Facts of Cotting's life very uncertain; remembered chiefly for work with Georgia Geological Survey, which began in 1836 when certain persons in Burke and Richmond counties wanted geological and agricultural survey and engaged Cotting for that purpose. Later, Georgia legislature appointed him as state geologist, but legislative support terminated in 1840 and Cotting's efforts in this capacity had no lasting importance.

WORKS: *Report of a Geological and Agricultural Survey, of Burke and Richmond Counties, Georgia* (Augusta, Ga., 1836), 198 pp.; *An Introduction to Chemistry, With Practical Questions; Designed for Beginners in the Science* (Boston, 1822), 420 pp.; *A Synopsis of Lectures on Geology, Comprising the Principles of the Science. Designed as a Textbook* (Taunton, Mass., 1835), 120 pp.; *An Essay on the Soils and Available Manures of the State of Georgia . . . Founded on a Geological and Agricultural Survey* (Milledgeville, Ga., 1843), 121 pp.; edited the one volume (of twenty numbers) of *Chemist and Meteorological Journal* (Amherst, Mass., 1826). No works listed in *Roy. Soc. Cat.* WORKS ABOUT: *NCAB*, 5 (1907 [copyright 1891 and 1907]): 185-86; obituary, *Am. Journ. Sci.*, 2d ser. 45 (1868): 141-42; Merrill, *Hundred Years of Geology*, pp. 174-76; also biographical notes by Samuel W. Geiser in Harvard University Archives.

COUES, ELLIOTT (September 9, 1842, Portsmouth, N.H.—December 25, 1899, Baltimore, Md.). *Ornithology.* Son of Samuel Elliott, merchant and U.S. patent officer, and Charlotte Haven (Ladd) Coues. Married Jane Augusta McKenney, 1867; Mrs. Mary Emily Bates, 1887; five children by first marriage. EDUCATION: A.B. in 1861 and M.D. in 1863, Columbian College (now George Washington Univ.). HONORS: Ph.D., Columbian College, 1869. CAREER: 1862, enlisted as medical cadet, U.S. Army; 1864-1881, assistant surgeon, U.S. Army, with assignments in Arizona, North Carolina, Dakota Territory, and elsewhere; 1873-1876, surgeon-naturalist, U.S. Northern Boundary Commission; 1876-1880, secretary-naturalist, U.S. geological survey under F. V. Hayden [q.v.]; 1877-1886, professor of anatomy, Columbian College; 1881, resigned from army and devoted remaining years to research and writing; 1884-1891, biology editor, *Century Dictionary*; 1897-1899, connected with ornithology magazine,

Osprey. MEMBERSHIPS: Natl. Ac. Scis.; a founder of Am. Ornith. Un. and president, 1893-1895. SCIENTIFIC CONTRIBUTIONS: As surgeon in U.S. Army, studied and collected birds in various parts of country where he was stationed. Western explorations by Coues and others greatly increased knowledge of New World birds; *Key* [see Works], perhaps most important work over time, was produced to facilitate study of these native birds; also made alterations in classification and introduced artificial key to ornithology. Among publications of Hayden survey, Coues wrote on fur-bearing animals, rodents, birds of Northwest, and birds of the Colorado Valley, the latter an uncompleted work that included important bibliography of North American ornithology. Published nearly 1,000 scientific and popular works. After about 1885, interests turned largely away from ornithology to psychical research, theosophy, and historical studies of early western exploration; and between 1893 and 1900, published fifteen volumes on early travelers in the West.

WORKS: *Key to North American Birds* (Salem, Mass., and New York, 1872; 5th ed., 1903). For Hayden survey, wrote *Fur-Bearing Animals* (Washington, D.C., 1877), *Monographs of North American Rodentia* (Washington, D.C., 1877), *Birds of the Northwest* (Washington, D.C., 1874), and *Birds of the Colorado Valley* (Washington, D.C., 1878). Chief technical works listed in "Memoir," *Biog. Mems. of Natl. Ac. Scis.* 6 (1909): 395-446. MSS: St. Hist. Soc. Wis. (*NUCMC* 68-2159); Ac. Nat. Scis. Philad. (Hamer); Smithson. Instn. Archives (various collections). WORKS ABOUT: *DAB*, 4: 465-66 (Witmer Stone); *DSB*, 3: 438-39 (Elizabeth N. Shor).

COULTER, JOHN MERLE (1851-1928). *Botany.* Professor of botany, Univ. of Chicago; founder and editor, *Botanical Gazette.*

MSS: See Rodgers [below], especially pp. 308-9. WORKS ABOUT: See *AMS*; *DAB*; Andrew D. Rodgers III, *John Merle Coulter* (Princeton, 1944).

COUPER, JAMES HAMILTON (March 4, 1794, Sunbury, Liberty County, Ga.—June 3, 1866, Glynn County [?], Ga. [buried at Frederika, St. Simon's Island]). *Agriculture*; *Natural history.* Son of John, large landowner and planter, and Rebecca (Maxwell) Couper. Married Caroline Wylly; at least eight children. EDUCATION: 1814, A.B., Yale College. CAREER: Lived all of adult life in Glynn County, Ga.; became manager of father's plantation, called Hopeton; subsequently purchased and inherited land of his own, and also managed property of other persons; 1829, erected sugar mill at Hopeton; 1834, operated mills for production of cottonseed oil at Natchez, Miss., and at Mobile, Ala., later abandoned for lack of capital. MEMBERSHIPS: American Ethnological Society. SCIENTIFIC CONTRIBUTIONS: After leaving Yale, went to Europe to study water control methods in Holland. As planter in Georgia, was distinguished for efforts at experimentation and application of scientific procedures, including improved dike and drainage system. Experimented with many new crops and substituted sugar for cotton as chief crop; 1829, constructed first up-to-date and complete mill in South for processing sugar cane; after 1838, along with most Georgia planters, turned to rice as chief crop. Especially remembered for early production of cottonseed oil, the first to manufacture that product in the South; also carried out number of experiments on production of olive oil. Noted as geologist and microscopist, and for a period ranked as a leading conchologist of the South; accumulated large library, including number of works on natural history. In 1861 presented large collection of fossils and other natural history specimens to College of Charleston, S.C.

WORKS: "Detailed Description of the Strata in Which the Fossil Bones and Shells from the Brunswick Canal Were Found," *Proceedings of Ac. Nat. Scis. Philad.* 1 (1842): 216-17; "Fossils at Chatahoochie River, Ga.," *Proceedings of Bos. Soc. Nat. Hist.* 2 (1846): 123-24; "On the Geology of a Part of the Seacoast of Georgia," in W. B. Hodgson, *Memoir on the Megatherium . . .* (New York, 1846), pp. 31-47. MSS: See reference in *DAB* to material in private hands; plantation records in Univ. NC.-SHC. WORKS ABOUT: *DAB*, 4: 468-69 (R. P. Brooks); W. J. Northen, *Men of Mark in Georgia* (Atlanta, Ga., 1910 [reprinted, Spartanburg, S.C., 1974]), 2: 215-22.

COURTENAY, EDWARD HENRY (November 19, 1803, Baltimore, Md.—December 21, 1853, Charlottesville, Va.). *Mathematics.* No information on parentage. Married; had children. EDUCATION: 1821, graduated, U.S. Military Academy, West Point. HONORS: M.A., Univ. of Pennsylvania, 1834; LL.D., Hampden-Sydney College (Virginia), 1846. CAREER: 1821, second lieutenant, assigned to Army Corps of Engineers; at U.S. Military Academy, assistant professor of natural and experimental philosophy (1821-1822), principal assistant professor of engineering (1822-1824), acting professor (1828-1829) and professor (1829-1834) of natural and experimental philosophy; 1824-1826, assistant engineer in construction of Fort Adams, R.I.; 1826-1828, assistant to chief engineer, Washington, D.C.; 1834, resigned from U.S. Army; 1834-1836, professor of mathematics, Univ. of Pennsylvania; 1836-1837, division engineer, New York and Erie Railroad; 1837-1841, U.S. civil engineer on construction of Fort Independence in Boston; 1841-1842, chief engineer of dry dock, Brooklyn Navy Yard, N.Y.; at

Univ. of Virginia, professor of mathematics (1842-1853) and chairman of faculty (1845-1846). SCIENTIFIC CONTRIBUTIONS: Noted for mathematical abilities, but devoted chief efforts to teaching. Edited and translated *Elementary Treatise on Mechanics*, based on French work of Jean Louis Boucharlat, adapting it to needs of West Point cadets (1833). Work on differential and integral calculus was published after death.

WORKS: *Treatise on the Differential and Integral Calculus, and the Calculus of Variations* (New York, 1855). See also *Natl. Un. Cat. Pre-1956 Imp.* and *Roy. Soc. Cat.*, vol. 2. MSS: Duke Univ. Library (*NUCMC* 62-1084). WORKS ABOUT: Cullum, *Biographical Register, West Point,* 1 (1891) : 263-66; *NCAB*, 5 (1907 [copyright 1891 and 1907]) : 519-20.

CRAFTS, JAMES MASON (1839-1917). *Chemistry.* Professor of chemistry and president, Massachusetts Institute of Technology.

WORKS ABOUT: See *AMS*; *DAB*.

CRAIG, THOMAS (December 20, 1855, Pittston, Pa.—May 8, 1900, Baltimore, Md.). *Mathematics.* Son of Alexander, mining engineer, and Mary (Hall) Craig. Married Louise Alvord, 1880; survived by two children. EDUCATION: 1875, C.E., Lafayette College; 1878, Ph.D. in mathematics, Johns Hopkins Univ. CAREER: 1875-1876, teacher, Newton, N.J.; 1876-1879, graduate student and fellow, Johns Hopkins Univ.; 1879-1881, connected with U.S. Coast Guard and Geodetic Survey; at John Hopkins Univ., instructor (1879-1880), associate (1880-1883), associate professor (1883-1892), and professor of mathematics (1892-1900) ; 1894-1899, chief editor and earlier member of editorial staff, *American Journal of Mathematics.* MEMBERSHIPS: A number of mathematical and scientific societies. SCIENTIFIC CONTRIBUTIONS: One of first to hold fellowship in mathematics at Johns Hopkins Univ.; studied with Prof. J. J. Sylvester [q.v.] and began lecturing there before receipt of degree. *Projections* [see Works] produced while working for Coast Survey. Special interests included theory of functions, differential equations, mechanics, and hydrodynamics, and scope of mathematical interests made him particularly effective in editorial role with *American Journal of Mathematics*; publication of *Linear Differential Equations* [see Works] grew out of university teaching and was one of most important efforts. Praised by contemporaries for facility in comprehending advanced work of the great geometers, but showed less ability in original investigation. Papers appeared in *American Journal of Mathematics* and other American and foreign journals; a work on theory of surfaces left unfinished at time of death.

WORKS: *Treatise on Projections* (Washington, D.C., 1882) ; *A Treatise on Linear Differential Equations,* vol. 1, *Equations with Uniform Coefficients* (New York, 1889) ; *Elements of the Mathematical Theory of Fluid Motion* (New York, 1879), 178 pp., reprinted from *Van Nostrand's Magazine.* See lists of works in Poggendorff, *Biographisch-literarisches Handwörterbuch,* 4: 279, and *Roy. Soc. Cat.*, vols. 9, 12, 14. MSS: Letters to Daniel C. Gilman, and several other items, in Johns Hopkins Univ. Library —Archives. WORKS ABOUT: *DAB*, 4: 496 (David E. Smith) ; F. P. Matz, "Prof. Thomas Craig, C.E., Ph.D.," *American Mathematical Monthly* 8 (1901) : 183-87.

CRESSON, EZRA TOWNSEND (1838-1926). *Entomology.* Curator and editor, American Entomological Society (founded as Entomological Society of Philadelphia) ; insurance company official.

MSS: Ac. Nat. Scis. Philad. (Hamer). WORKS ABOUT: See *AMS*, 2d ed. (1910) ; *DAB*.

CROCKER, LUCRETIA (December 31, 1829, Barnstable, Mass.—October 9, 1886, Boston, Mass.). *Science education.* Daughter of Henry, businessman in insurance and county sheriff, and Lydia E. (Farris) Crocker. Never married. EDUCATION: 1850, graduated, Massachusetts State Normal School, West Newton; attended lectures of J. L. R. Agassiz [q.v.] at Harvard Univ. CAREER: 1850-1854, instructor in geography, mathematics, and natural science, State Normal School, West Newton; thereafter, spent some time at home with family; 1857-1859, professor of mathematics and astronomy, Antioch College; 1859, returned to Boston, assumed responsibility for care of parents, and soon began involvement in number of educational concerns; during period of years after 1865, assisted in selection of Sunday school books for American Unitarian Association; 1866-1875, member of Committee on Teaching, New England Freedman's Aid Society, and in 1869 made tour of freedmen's schools in South; during same period, taught botany and mathematics in private school of friend; 1873-ca. 1876, in charge of science department, Society to Encourage Studies at Home (correspondence school) ; 1873, elected to Boston School Committee; 1876-1886, elected by Boston School Committee to serve on board of school supervisors. MEMBERSHIPS: Am. Assoc. Advt. Sci. SCIENTIFIC CONTRIBUTIONS: As classroom teacher, particularly noted for work in mathematics. Among other activities after return to Boston from Antioch, assisted Mary L. Hall in preparation of geography textbook, *Our World* (1865), and later (1883) wrote *Methods of Teaching Geography: Notes on Lessons*; chief contribution to science was in promotion of that subject in public education, especially in Boston where, as member of board of school supervisors, took special responsibility for natural sciences, actively promoted their importance in school curriculum, and was particularly successful in promotion of study of zoology and botany at high school level. With E. H. S. Richards [q.v.], 1881-1882, developed course in mineralogy and helped introduce it into curriculum; worked with and promoted Teachers' School of Science, conducted by A. Hyatt [q.v.] under sponsorship of Bos. Soc. Nat. Hist. Strongly influenced by friend J. L. R. Agassiz.

WORKS: See above; also *Natl. Un. Cat. Pre-1956 Imp.* WORKS ABOUT: *Nota. Am. Wom.*, 1: 407-9 (Norma K. Green); Ednah D. Cheney, *Memoirs of Lucretia Crocker and Abby W. May* (Boston, 1893).

CROSBY, WILLIAM OTIS (1850-1925). *Geology.* Professor of geology, Massachusetts Institute of Technology; consulting engineer for dam construction.

WORKS ABOUT: See *AMS; DAB.*

CROSS, CHARLES ROBERT (1848-1921). *Experimental physics.* Professor of physics and director of laboratory, Massachusetts Institute of Technology.

WORKS ABOUT: See *AMS; NCAB*, 11 (1909 [copyright 1901 and 1909]) : 183; *Who Was Who in America*, vol. 1.

CURLEY, JAMES (October 25, 1796, County Roscommon, Ireland—July 24, 1889, Georgetown, Washington, D.C.). *Astronomy.* No information on parentage; father died when son was quite young. EDUCATION: Early education in Ireland was not extensive, but received encouragement in mathematical studies; 1826, began study of Latin at Catholic Seminary, Washington, D.C., and soon thereafter entered Society of Jesus. CAREER: 1817, left Ireland for U.S.; 1817-ca. 1819, bookkeeper, Philadelphia; thereafter, taught mathematics at Frederick, Md.; 1826, began study of Latin and also taught course, Catholic Seminary, Washington, D.C.; 1827, entered Society of Jesus at Georgetown and in 1833 was ordained as priest; 1829, sent to teach at Frederick; 1831-1879, teacher of natural philosophy, mathematics, and astronomy, Georgetown College (now Georgetown Univ.); for fifty years, chaplain to Visitation Convent, Georgetown; 1850-1860, socius and procurator of the province. SCIENTIFIC CONTRIBUTIONS: At Georgetown Univ., prepared plans and supervised building construction for observatory, completed about 1844 and intended especially as instructional facility. In 1847, determined latitude of city of Washington; not accepted by astronomers at the time, but during 1860s advanced methods allowed government astronomers to substantiate his earlier results. During early years of Georgetown observatory, published several short papers on apparatus, latitude and longitude of Washington, comets, and eclipse in *Monthly Notices of Astronomical Society* (1849-1850) and in B. A. Gould's *Astronomical Journal* (1851-1854). At one time, worked with C. G. Page [q.v.] on experiments on electric currents. After Civil War, lack of funds and other factors led to cessation of research at observatory, and thereafter devoted chief efforts to teaching; also interested in botany and established greenhouse at Georgetown Univ.

WORKS: See *Roy. Soc. Cat.*, vol. 2. MSS: Georgetown Univ. Archives (Durkin [below]). WORKS ABOUT: Obituary, *Woodstock Letters* 18 (1889) : 381-84; Joseph T. Durkin, *Georgetown University: The Middle Years, 1840-1900* (Washington, D.C., 1963).

CURTIS, MOSES ASHLEY (May 11, 1808, Stockbridge, Mass.—April 10, 1872, Hillsboro, N.C.). *Botany.* Son of Jared—merchant, teacher and prison chaplain—and Thankful (Ashley) Curtis. Married Mary de Rosset, 1834; ten children. EDUCATION: 1827, A.B., Williams College; studied theology. HONORS: D.D., Univ. of North Carolina, 1852. CAREER: 1827-1830, teacher in Walden and Watertown, Mass., and perhaps elsewhere; 1830-1833, tutor, Wilmington, N.C.; 1833-1834, theological study in Massachusetts; 1835-1837, missionary work in western North Carolina; 1837-1840, teacher and head of Episcopal school, Raleigh, N.C.; 1840-1841, missionary in Washington, N.C.; 1841-1847 and 1856-1872, Episcopal priest, Hillsboro, N.C., and oftentimes teacher; 1847-1856, priest, Society Hill, S.C.; 1860-1862, botanist and zoologist, North Carolina Geological Survey. MEMBERSHIPS: Am. Soc. Advt. Sci. SCIENTIFIC CONTRIBUTIONS: Began active botanical studies around 1830, and when first publication, "Enumeration" [see Works], appeared in 1834, it included descriptions of 1,031 species. About 1846 began to study fungi; became particularly close collaborator with English botanist, Rev. Miles Joseph Berkeley, with whom he wrote series of works on mycology, and is recognized as having been, in effect, coproducer of Berkeley's *Notices of North American Fungi*, published in London (1872-1876). Catalogue of plants of North Carolina, published in 1867 as his second important botanical contribution to geological and natural history survey of the state [see Works], was pioneering effort that enumerated both flowering and flowerless plants.

WORKS: "Enumeration of Plants Growing Spontaneously Around Wilmington, N.C., with Remarks on Some New and Obscure Species," *Boston Journal of Natural History* 1 (1834) : 82-140; *Geological and Natural History Survey of North Carolina*, pt. 3, *Botany; Containing a Catalogue of the Plants of the State, with Descriptions and History of the Trees,*

Shrubs, and Woody Vines (Raleigh, 1860), 124 pp., and its sequel, . . . *Botany; Containing a Catalogue of the Indigenous and Naturalized Plants of the State* (Raleigh, 1867), 158 pp.; works written with M. J. Berkeley included "Contributions to the Mycology of North America," *Am. Journ. Sci.* 2d ser. 8 (1849) : 401-3, ibid. 2d ser. 9 (1850) : 171-75, ibid. 2d ser. 10 (1850) : 185-88, and ibid. 2d ser. 11 (1851) : 93-95, the latter a description of fungi collected by Wilkes Expedition. See bibliography in Powell [below]. MSS: Univ. N.C.-SHC (*NUCMC* 64-477); Harv. U.-GH (Hamer). WORKS ABOUT: *DAB,* 4: 617-18 (Donald C. Peattie); William S. Powell, *Moses Ashley Curtis . . .* (Chapel Hill, N.C., 1958), 26 pp.

CUTBUSH, JAMES (1788, Pennsylvania [probably Philadelphia]—December 15, 1823, West Point, N.Y.). *Chemistry.* Son of Edward, stonecutter, and Anne (Marriat) Cutbush. Married; no information on children. EDUCATION: Uncertain; was referred to as an M.D., but date and source of degree are not determined. CAREER: Ca. 1811-1813, public lecturer on chemistry and related subjects, Philadelphia; professor of chemistry, mineralogy, and natural philosophy in St. John's College, affiliated with St. John's (Lutheran) Church, Philadelphia; 1814-ca. 1819, assistant apothecary general, U.S. Army; 1819, identified in newspaper advertisement as manufacturing chemist and apothecary; 1820, appointed chief medical officer, West Point; 1820-1823, acting professor of chemistry and mineralogy, West Point. MEMBERSHIPS: A founder (1811) and first president, Columbian Chemical Society; member of Am. Phil. Soc. and other societies. SCIENTIFIC CONTRIBUTIONS: What little is known about Cutbush is based chiefly on his publications. Scientific work characterized by interest in application of chemistry to manufacturing and practical problems; *Philosophy* [see Works] was one of first chemistry textbooks by American-born author, and *Artist's Manual* [see Works] was, in effect, popular dictionary on chemical technology. Most important work carried out at West Point, where he continued earlier studies on gunpowder and also conducted investigations of higher explosives; experimental results of these studies appeared in *Pyrotechny* [see Works], published posthumously by wife and former students.

WORKS: "The Application of Chemistry to Arts and Manufactures," a series of fifteen articles in the Philadelphia *Aurora* (1808); "Prognostic Signs of the Weather" and "On the Oxyacetite of Iron as a Test or Reagent for the Discovery of Arsenic" in *Memoirs of Columbian Chemical Society* (1813); *Philosophy of Experimental Chemistry,* 2 vols. (Philadelphia, 1813); *American Artist's Manual,* 2 vols. (Philadelphia, 1814); *System of Pyrotechny* (Philadelphia, 1825). These and other works mentioned in sources listed below. WORKS ABOUT: *DAB,* 5: 10 (Edgar F. Smith); E. F. Smith, *James Cutbush: An American Chemist* (Philadelphia, 1919), 94 pp.

CUTLER, MANASSEH (May 13, 1742, Killingly, Conn.—July 28, 1823, Hamilton, Mass.). *Botany.* Son of Hezekiah, farmer, and Susanna (Clark) Cutler. Married Mary Balch, 1766; at least four children. EDUCATION: 1765, A.B., Yale College; later studied law, theology, medicine. HONORS: LL.D., Yale, 1791. CAREER: Ca. 1766, teacher, Dedham, Mass.; 1766-1769, operated store on Martha's Vineyard and while there was admitted to practice as attorney; 1769-1770, studied theology at Dedham, Mass.; 1771-1823, minister at Ipswich Hamlet (now Hamilton), Mass.; 1775, 1776, and 1778, army chaplain; 1779, began occasional practice of medicine; 1782, opened private boarding school that operated for more than twenty-five years; 1787, elected one of three directors of the Ohio Company, negotiated land contract with U.S. Congress, and contributed to draft of Ordinance of 1787; 1788-1789, in Ohio; 1800, member of Massachusetts General Court; 1800-1804, U.S. congressman. MEMBERSHIPS: Am. Ac. Arts Scis.; Am. Phil. Soc.; and other societies. SCIENTIFIC CONTRIBUTIONS: While actively involved in numerous other activities, managed to find time to pursue scientific studies as well. Interested in astronomy and meteorology, and physical studies included repetition of Franklin's electrical experiments. Explored White Mountains in 1784, reached summit of Mt. Washington, and made rough estimate of its elevation. As man of science, best remembered for botanical work; used Linnaean system in "Account of . . . Vegetable Productions" [see Works], the first systematic publication on botany of New England, which won for him wide reputation as botanist of merit.

WORKS: "Observations of the Transit of Mercury Over the Sun, November 12, 1782, at Ipswich," "Observations of an Eclipse of the Moon, March 29, 1782, and an eclipse of the Sun, on the 12th of April following, at Ipswich . . . ," and "Meteorological Observations at Ipswich, in 1781, 1782, and 1783," all in *Memoirs of Am. Ac. Arts Scis.* 1 (1785) : 128, 162-64, 336-71, respectively; "An Account of Some of the Vegetable Productions, Naturally Growing in this Part of America, Botanically Arranged," ibid., pp. 396-493. See Meisel, *Bibliog. Amer. Nat. Hist.* MSS: Northwestern Univ. Library (*NUCMC* 60-2228); see additional references in *NUCMC* and Hamer. WORKS ABOUT: *DAB,* 5: 12-14 (Claude M. Fuess); Shipton, *Biog. Sketches, Harvard,* 16 (1972) : 138-52.

CUTTING, HIRAM ADOLPHUS (December 23, 1832, Concord, Vt.—April 18, 1892, Lunenburg, Vt.). *Natural history; Geology.* Son of Stephen Church and

Eliza Reed (Darling) Cutting. Married Miranda E. Haskell, 1856; adopted one child. EDUCATION: At age fifteen, began study of medicine; attended St. Johnsbury Academy. HONORS: A.M. in 1868 and Ph.D. in 1879, Norwich Univ.; M.D., Dartmouth College, 1870. CAREER: Before about age twenty-one, taught school and worked as surveyor and in itinerant daguerreotype car; 1854-1892, owned dry goods store at Lunenburg; ca. 1865, licensed as claim agent for veterans' benefits; 1870, received honorary M.D. degree and thereafter turned increasingly to practice of medicine; 1870-1886, Vermont state geologist and curator of state Natural History Cabinet; 1880-1886, lecturer in natural sciences, Norwich Univ.; 1881-1886, secretary, state board of agriculture; 1881-1886, state fish commissioner; lecturer at several Vermont institutions. MEMBERSHIPS: Am. Assoc. Advt. Sci.; American Microscopical Society [?]. SCIENTIFIC CONTRIBUTIONS: Scientific work was of diversified nature; amassed two large natural history collections. With self-made instrument, took up study of microscopy and for some years studied miscroscopic anatomy and disease and carried out such examinations for number of physicians, demonstrating particular skill in devising improved methods for mounting and preserving specimens. Maintained observatory at store and for many years collected meteorological data for Smithson. Instn. Reports as secretary of Vermont Board of Agriculture were models of their kind, and of particular note were investigations on nutrition of plants, in which he concluded that only modest amounts of nitrogen were needed in most cases; as state geologist, tests of building stones proved to have economic significance. Among other fields touched on in writings were mining, agricultural pests, geology, and ornithology.

WORKS: See *Roy. Soc. Cat.*, vol. 7, and *Natl. Un. Cat. Pre-1956 Imp.* WORKS ABOUT: *NCAB,* 10 (1909 [copyright 1900]): 204; *Biographical Encyclopaedia of Vermont of the Nineteenth Century* (Boston, 1885), pp. 320-27; T. D. Seymour Bassett, *A History of the Vermont Geological Surveys and State Geologists* (Burlington, Vt., 1976), pp. 15-18.

D

DABOLL, NATHAN (April 24, 1750, Groton, Conn.—March 9, 1818, Groton, Conn.). *Mathematics; Navigation.* Son of Nathan and Anna (Lynn) Daboll. Married Elizabeth Daboll (cousin); Elizabeth Brown (widow); at least one child. EDUCATION: Educated in town school and with local Congregational minister; self-taught in advanced subjects, especially mathematics. CAREER: Worked as cooper; 1770, employed to revise calculations for almanac; 1773-1775 and 1793-1818, almanac published under Daboll's name; during Revolutionary War and perhaps before that, worked as instructor to seamen; 1783-1788, teacher of mathematics and astronomy, Plainfield (Conn.) Academy; 1788, returned to Groton, reestablished his schools of navigation there and at New London, instructing members of merchant marine and navy; 1811, invited by Com. John Rodgers to teach class on board frigate *President* in New London harbor. SCIENTIFIC CONTRIBUTIONS: Demonstrated natural talent for mathematics and before age twenty-one was employed by Timothy Green to revise calculations for almanacs; Daboll's name appeared on Green's almanacs for 1773 to 1775; for number of years thereafter Daboll continued to revise calculations for Green's *New England Almanac,* and beginning in 1793 his name reappeared on the publication, which remained in hands of his descendants into twentieth century. *Schoolmaster's Assistant* [see Works], for many years a standard arithmetic textbook, was particularly noted for early introduction of calculations in terms of new federal money. With N. Pike [q.v.] and Daniel Adams, recognized as one of three major early American arithmeticians.

WORKS: *Daboll's Complete Schoolmaster's Assistant* (New London, Conn., 1800); *Daboll's Practical Navigator . . .* (New London, Conn., 1820). WORKS ABOUT: *DAB,* 5: 23 (Harris E. Starr); Florian Cajori, *The Teaching and History of Mathematics in the U.S.* (Washington, D.C., 1890), pp. 46-47.

DALL, WILLIAM HEALEY (1845-1927). *Malacology; Paleontology.* Paleontologist, U.S. Geol. Surv.

MSS: Smithson. Instn. Archives (*NUCMC* 71-1872); Stanford Univ. Libraries (*NUCMC* 72-624); Ac. Nat. Scis. Philad. (Hamer). WORKS ABOUT: See *AMS*; *DAB*; *DSB.*

DALTON, JOHN CALL, Jr. (February 2, 1825, Chelmsford, Mass— February 12, 1889, New York, N.Y.). *Physiology.* Son of John Call, physician, and Julia Ann (Spalding) Dalton. Never married. EDUCATION: 1844, A.B., Harvard College; 1847, M.D., Harvard Medical School; 1847, went to Paris and was associated with Claude Bernard. CAREER: 1851-1854, professor of physiology, Univ. of Buffalo; 1854-1856, professor of physiology, Univ. of Vermont; at College of Physicians and Surgeons, New York, professor of physiology and microscopical anatomy (1855-1883) and president (1884-1889); 1859-1861, connected with Long Island College Hospital; during Civil War, surgeon in volunteer army, with rank of brigadier general. MEMBERSHIPS: Natl. Ac. Scis. and other American and foreign societies. SCIENTIFIC CONTRIBUTIONS. Came under influence of Claude Bernard in Paris in 1847, turned away from medical practice, and became first American to devote life to teaching and research in experimental physiology. Recent discovery of ether made possible pioneering efforts in use of living animal processes as illustrations for lectures. Publications also based on own experimental observations; essay on corpus luteum won Am. Med. Assoc. prize, 1851; single most important experimental paper probably was that on sugar formation [see Works], which confirmed discovery by Bernard that had seemingly been disproved by researchers in England. Later life devoted largely to preparation of *Anatomy of the Brain* [see Works], a work that included number of photographs of specimens prepared by Dalton himself. Expended much effort in defense of work against criticisms by antivivisectionists. Made no great discoveries, but unprecedented efforts in experimental physiology, in country where such work was practically unknown, gave contributions special significance.

WORKS: *Treatise on Human Physiology* (Philadelphia, 1859; 7th ed., 1882); "Sugar Formation in the Liver," *Transactions of New York Academy of Medicine* 14 (1871): 15-33; *Topographical Anatomy of the Brain* (Philadelphia, 1885); *The Experimental Method in Medical Science* (New York, 1882), 108 pp. See *Roy. Soc. Cat.*, vols. 2, 7, 9, 12, 14, and *Natl. Un. Cat. Pre-1956 Imp.* MSS: Letters in several collections in Harv. U.-MS. WORKS ABOUT: *DAB*, 5: 40 (Russell H. Chittenden); S. Weir Mitchell, "Memoir," *Biog. Mems. of Natl. Ac. Scis.* 3 (1895): 177-85.

DANA, EDWARD SALISBURY (1849-1935). *Mineralogy*; *Crystallography*. Professor of physics and curator of mineralogy, Yale Univ.; editor of *Am. Journ. Sci.*

WORKS ABOUT: See *AMS*; *DAB*, supp. 1.

DANA, JAMES DWIGHT (February 12, 1813, Utica, N.Y.—April 14, 1895, New Haven, Conn.). *Geology*; *Zoology*. Son of James, saddler and hardware merchant, and Harriet (Dwight) Dana. Married Henrietta F. Silliman, daughter of B. Silliman [q.v.], 1844; four children. EDUCATION: 1833, A.B., Yale College. HONORS: Received number of honorary degrees; Wollaston Medal, Geological Society of London, 1872; Copley Medal, Roy. Soc., 1877. CAREER: 1833-1835, instructor in U.S. Navy and accompanied cruise to Mediterranean; 1836, assistant to Silliman in Yale chemistry laboratory; 1838-1842, geologist and mineralogist, Wilkes Expedition; 1842-1856, in Washington and New Haven, preparing expedition reports; 1840, became editor of *Am. Journ. Sci.*; 1849, appointed professor of natural history, Yale, but did not assume duties of professorship until 1856; 1864-1890, professor of geology and mineralogy, Yale. MEMBERSHIPS: Roy. Soc. and other foreign and American societies; president, Am. Assoc. Advt. Sci., 1854. SCIENTIFIC CONTRIBUTIONS: First scientific paper appeared in 1835, on volcanoes, and long-standing research interest in that subject later was summarized in publication of 1890 [see Works]. Reputation as one of world's leading geologists grew out of work on Wilkes Expedition, which gave him globular view of geology; initially appointed as mineralogist and geologist, but later also took on responsibility for marine zoology when Joseph P. Couthouy left expedition. Three great reports resulting from work on expedition [see Works] made number of highly significant contributions to science, including studies of corals and related animals considered in both zoological and geological aspects; work on crustaceans led to his idea of zoological progress, called "cephalization." Also made important contributions to theory of mountain building.

WORKS: *System of Mineralogy* (New Haven, 1837); *Manual of Geology* (Philadelphia and London, 1862); *Characteristics of Volcanoes* (New York, 1890); reports for Wilkes Expedition: *Zoophytes* (Philadelphia, 1846), 741 pp. with atlas of 61 plates; *Geology* (Philadelphia, 1849), 756 pp. with atlas of 21 plates; and *Crustacea*, 2 vols. (Philadelphia, 1852-1854), with atlas of 96 plates. Fairly complete bibliography in Daniel C. Gilman, *Life of James Dwight Dana* (New York and London, 1899). MSS: Yale U.-RB (Hamer); Harv. U.-GH (Hamer); Ac. Nat. Scis. Philad. (Hamer). WORKS ABOUT: *DAB*, 5: 55-56 (George P. Merrill); *DSB*, 3: 549-54 (William Stanton).

DANA, JAMES FREEMAN [born Jonathan] (September 23, 1793, Amherst, N.H.—April 14, 1827, New York, N.Y.). *Chemistry*; *Mineralogy*. Son of Luther, merchant and sea captain, and Lucy (Giddings) Dana; brother of Samuel L. Dana [q.v.]. Married Matilda Webber, 1818; one child who died in childhood. EDUCATION: Attended Phillips Exeter Academy; 1813, A.B., Harvard College; 1817, M.D., Harvard Medical School; 1815, studied chemistry with Friedrich Accum, London. CAREER: Ca. 1813, as college undergraduate, helped prepare lecture experiments; 1815, sent to England to purchase chemical equipment for Harvard; ca. 1816-1820, assistant to chemistry professor, and acted as instructor in experimental chemistry, Harvard; 1817-1820, medical practice, Cambridge, Mass.; 1817, appointed lecturer in chemistry for medical students, Dartmouth College; 1820-1826, professor of chemistry and mineralogy and sometimes of botany, Dartmouth; ca. 1825, elected to New Hampshire General Court; 1826-1827, professor of chemistry, College of Physicians and Surgeons, New York; 1826, appointed by secretary of war as visitor to U.S. Military Academy at West Point; occasionally employed as consulting chemist. MEMBERSHIPS: Founding member of Linnaean Society of New England (1814); N.Y. Lyc. Nat. Hist. SCIENTIFIC CONTRIBUTIONS: As undergraduate, won Boylston Prize for paper entitled "Tests for Arsenic," and in 1817 won second Boylston Prize for essay entitled "Composition of Oxymuriatic Acid"; published chemical papers in *Am. Journ. Sci.* on variety of topics, including electrical battery, analysis of wax myrtle (berries), examination of urinary and other calculi, analysis of nitrous gas, and analyses and experiments with other substances. Work on vapor of flame [see Works] was suggestive of significant work, not pursued beyond this one short paper; most important original publication was *Epitome* [see Works], based on syllabus for lectures at Dartmouth.

WORKS: "Outlines of the Mineralogy and Geology of Boston and Its Vicinity, with a Geological Map," *Memoirs of Am. Ac. Arts. Scis.* 4 (1818): 129-224, with S. L. Dana; "On the Effect of Vapour on Flame," *Am. Journ. Sci.* 1 (1818): 401-2; *Epitome of Chemical Philosophy* (Concord, N.H., 1825). See *Roy. Soc. Cat.*, vol. 2. WORKS ABOUT: *DAB*, 5: 56 (Lyman C. Newell); "Memoir," *Collections of New Hampshire Historical Society* 2 (1827): 290-300.

DANA, SAMUEL LUTHER (July 11, 1795, Amherst, N.H.—March 11, 1868, Lowell, Mass.). *Chemistry*; *Mineralogy*. Son of Luther, merchant and sea captain, and Lucy (Giddings) Dana; brother of James F. Dana [q.v.]. Married Ann T. Willard, 1820; later, married Augusta Willard; survived by four children. EDUCATION: 1813, A.B., Harvard College; 1818, M.D., Harvard Medical School; studied law for time after college. CAREER: 1813-1815, served in U.S. Army and received commission as first lieutenant; 1818-ca. 1820, medical practice, Gloucester, Mass.; ca. 1820-1826, medical practice, Waltham, Mass.; 1826-1833, gave up medicine and turned to manufacture of sulphuric acid and bleaching salts, soon merging with Newton Chemical Co. and serving there as superintendent and chemist; 1830-1833 and 1837-1840, associated with geological surveys of Massachusetts under E. Hitchcock [q.v.]; 1833, visited Europe; 1834-1868, chemist, Merrimac print works, Lowell, Mass., with later years spent largely on a farm. MEMBERSHIPS: Linnaean Society of New England. SCIENTIFIC CONTRIBUTIONS: Early American authority on technical and agricultural chemistry; as result of studies of bleaching of cotton, introduced ca. 1838 what came to be known as "American system" of bleaching; investigations in regard to calico printing led to discovery that sodium phosphate was the ingredient in cow dung that made that substance useful in dyeing process. Interest in study of manures resulted in *Muck Manual* [see Works], one of earliest scientific works on agriculture to be produced by an American investigator; in this work were presented views that conflicted with those of Justus Liebig, yet some aspects of Dana's ideas later were proved to have been correct. Also studied problem of lead in Lowell water supply, carried on extensive studies of madder and its uses as dye, and conducted other studies. Much of work was done for employer and resulted in improvement of technical processes, but published relatively little.

WORKS: "Outlines of the Mineralogy and Geology of Boston . . ." [see J. F. Dana]; *A Muck Manual for Farmers* (Lowell, Mass., 1842; 5th ed. 1855). See *Roy. Soc. Cat.*, vol 2, and *Natl. Un. Cat. Pre-1956 Imp.* WORKS ABOUT: *DAB*, 5: 61 (Lyman C. Newell); Youmans, *Pioneers of Sci.*, pp. 311-18; Rossiter, *Emergence of Agricultural Science*, especially chap. 3, passim.

DARLINGTON, WILLIAM (April 28, 1782, Dilworthtown, Pa.—April 23, 1863, West Chester, Pa.). *Botany*. Son of Edward, probably farmer, and Hannah (Townsend) Darlington. Married Catharine Lacey, 1808; eight children. EDUCATION: Studied medicine with John Vaughan, Wilmington, Del.; 1804, M.D., Univ. of Pennsylvania, where he studied medical botany with B. S. Barton [q.v.]. HONORS: LL.D., Yale, 1848; Sc.D., Dickinson College, 1856. CAREER: 1804-1806, engaged in limited medical practice, Birmingham, Pa.; 1806-1808, surgeon on ship to Calcutta; 1808, began medical practice in West Chester, Pa.; in War of 1812, served with state volunteers and attained rank of major; 1815-1817 and 1819-1823, member, U.S. House of Representatives; 1830, elected president of Bank of Chester County; held various public offices, including prothonotary and clerk of courts and canal commissioner; director and president, West Chester Railroad. MEMBERSHIPS: A number of societies; an organizer of Chester County Cabinet of Natural Sciences. SCIENTIFIC CONTRIBUTIONS: While actively engaged in professional and public career, devoted leisure hours to botanical study and related correspondence, and these activities seem always to have been chief interest. Labeled as "Nestor of American botany" by A. Gray [q.v.], a title earned as much by Darlington's personal influence and enthusiasm as by his direct scientific contributions. Published two works on plants in region of West Chester, *Florula* and much enlarged *Flora* [see Works]; best remembered for two historical-biographical publications [see Works].

WORKS: *Florula Cestrica* (West Chester, Pa., 1826); *Flora Cestrica* (West Chester, Pa., 1837); *Agricultural Botany* (Philadelphia, 1847). *Reliquiae Baldwinianae: Selections from the Correspondence of the Late William Baldwin . . .* (Philadelphia, 1843) and *Memorials of John Bartram and Humphry Marshall; with Notices of Their Botanical Contemporaries* (Philadelphia, 1849) have both been reprinted (New York, N.Y., 1969 and New York, N.Y., 1967, respectively) with notes and commentary by Joseph Ewan. See Meisel, *Bibliog. Amer. Nat. Hist.* MSS: Various repositories (see *DSB*). WORKS ABOUT: *DAB*, 5: 78-79 (Donald C. Peattie); *DSB*, 3: 562-63 (Joseph Ewan).

DAVENPORT, GEORGE EDWARD (1833-1907). *Botany*. Resided in Medford, Mass.; engaged in botanical research; member of Medford School Committee.

WORKS ABOUT: See *AMS*; *World Who's Who in Science.*

DAVIDSON, GEORGE (1825-1911). *Geodesy; Geography*; *Astronomy.* Associated with U.S Coast and Geodetic Survey, chiefly on Pacific Coast; regent, honorary professor of geodesy and astronomy, and professor of geography, Univ. of California. MSS: Univ. Calif.-Ban. (*NUCMC* 71-723); also see Lewis [below]. WORKS ABOUT: See *AMS; DAB*; Oscar Lewis, *George Davidson: Pioneer West Coast Scientist* (Berkeley, Calif., 1954).

DAVIS, CHARLES HENRY (January 16, 1807, Boston, Mass.—February 18, 1877, Washington, D.C.). *Hydrography*; *Astronomy.* Son of Daniel, judge and solicitor general of Massachusetts, and Lois (Freeman) Davis. Married Harriette Blake Mills, 1842; six children. EDUCATION: 1821-1823, attended Harvard College; later studied there with Prof. B. Peirce [q.v.] and in 1841 took A.B. degree (listed with class of 1825); 1823, appointed midshipman in U.S. Navy, and in 1827 passed examination for lieutenant. HONORS: LL.D., Harvard, 1868. CAREER: 1824-1840, periods of sea duty to South America, South Seas, Mediterranean, Russia, and England, with intervening years and period 1840-1842 at Boston and Cambridge, Mass.; 1842-1849, assistant on Coast Survey; 1849-1855 and 1859-1861, in charge of *American Ephemeris and Nautical Almanac*, at Cambridge, Mass.; 1854, made commander; 1856-1859, sea duty in Pacific; 1861, with U.S. Navy Bureau of Detail, Washington, D.C.; 1862, made commodore and engaged in active duty on upper Mississippi; 1862-1865, head of Bureau of Navigation, including Bureau of Detail and all naval scientific work; 1863, made rear admiral; 1865-1867, head of Naval Observatory; 1867-1869, in command of Brazilian squadron; 1870-1873, command at Norfolk Navy Yard; 1873-1877, superintendent of Naval Observatory. MEMBERSHIPS: Natl. Ac. Scis. SCIENTIFIC CONTRIBUTIONS: Scientific work began with appointment to Coast Survey, taking charge of hydrographic work from Rhode Island northward; studied especially Nantucket shoals and tides of New York harbor and Long Island Sound. Chief scientific work was generalized study of laws of tidal action and alluvial deposits. Played important role in establishing *American Ephemeris.* In 1863, with A. D. Bache and J. Henry [qq.v.] appointed by secretary of navy to government technical and scientific advisory committee, which was formed at Davis's suggestion and led to formation of Natl. Ac. Scis.

WORKS: "Upon the Geological Action of the Tidal and Other Currents of the Ocean," *Memoirs of Am. Ac. Arts Scis.*, n.s. 4 (1849): 117-56; "On the Law of the Deposit of the Flood Tide," *Smithson. Contr. Knowl.* 3 (1852). See *Roy. Soc. Cat.*, vols. 2, 9. MSS: Greenfield Village and Henry Ford Museum (Dearborn, Mich.) [access is restricted]; Harv. U.-HL. WORKS ABOUT: *DAB*, 5: 106-7 (Allen Wescott); Charles H. Davis [son], *Life of Charles Henry Davis* . . . (Boston and New York, 1899); Don Groves, "The Unsung Sailor of Science," *Naval Engineers Journal* 81 (October 1969): 27-34.

DAVIS, WILLIAM MORRIS (1850-1934). *Physical geography*; *Geology.* Professor of geology, Harvard Univ.

MSS: Harv. U.-HL. WORKS ABOUT: See *AMS*; *DAB*, supp. 1; *DSB.*

DAY, WILLIAM CATHCART (May 30, 1857, Urbana, Ohio—August 5, 1905, Swarthmore, Pa.). *Chemistry.* Son of William G. and Caroline (Cathcart) Day. Married Jennie Leamy, 1884; survived by five children. EDUCATION: A.B. in 1880 and Ph.D. in chemistry in 1883, Johns Hopkins Univ. CAREER: 1883-1884, professor of chemistry and physics, St. John's College (Maryland); 1884-1887, professor of chemistry, Univ. of Nashville; at Swarthmore College, professor of chemistry and physics (1887-1888) and professor of chemistry (1888-1901). MEMBERSHIPS: German Chemical Society (Berlin); Am. Phil. Soc.; and other societies. SCIENTIFIC CONTRIBUTIONS: Worked on analysis of chrome iron ore, oxidation of brom. cymene, and action of carbon dioxide on sodium aluminate. Succeeded in producing substances that to varying degrees had characteristic properties of asphalts, through distillation of animal and vegetable products under ordinary atmospheric pressure, and thereby solved question of the nature of asphalt. Also prepared various reports for the U.S. Geol. Surv. and for the Eleventh U.S. Census, especially on coal, minerals, and building stone, and produced most complete report published up to that time on American stone industry.

WORKS: "The Laboratory Production of Asphalts from Animal and Vegetable Materials," *Journal of Franklin Institute* whole ser. 148 (1899): 205-26. See *Roy. Soc. Cat.*, vols. 9, 14, and *Natl. Un. Cat. Pre-1956 Imp.* WORKS ABOUT: *World Who's Who in Science; Who's Who in America, 1903-1905; Johns Hopkins [University] Half-Century Directory* (Baltimore, 1926); Marcus Benjamin, "Some American Contributions to Technical Chemistry," *Science*, n.s. 21 (1905): 880; obituary, *Swarthmorean* 1 (1905): 4-5.

DEANE, JAMES (February 14, 1801, Colrain, Mass.—June 8, 1858, Greenfield, Mass.). *Geology.* Father probably a farmer. No information on marital status. EDUCATION: Probably no more than common school education; ca. 1826-1831, studied medicine; 1831, M.D., Univ. of State of New York. CAREER: 1822-ca. 1826, copyist in law office; 1831-1858, medical practice, Greenfield, Mass. MEMBERSHIPS: Corresponding member, Bos. Soc. Nat. Hist. SCIENTIFIC CONTRIBUTIONS: Remembered chiefly for 1835 discovery of fossil footprints in red sandstone of Connecticut Valley, which he brought to attention of E. Hitchcock [q.v.]; Hitchcock in turn undertook detailed study and began publication of series of papers on subject; question arose whether tracks were made by extinct reptiles or birds. In 1843, Deane's correspondence with English geologist G. A. Mantell was published [see Works]; thus Deane, who originally believed tracks were those of birds but later modified his views, belatedly took up himself the subject earlier left to Hitchcock and over next fifteen years published number of additional papers. Scientific work summarized in *Ichnographs* [see Works], which appeared after death; apparently did not publish on any scientific topic other than fossil footprints, but is said to have contributed to *Boston Medical and Surgical Journal.*

WORKS: Correspondence with G. A. Mantell regarding fossil footprints in *Am. Journ. Sci.* 45 (1843): 177-88; *Ichnographs from the Sandstone of the Connecticut Valley* (Boston, 1861), 61 pp. and 46 plates. See Meisel, *Bibliog. Amer. Nat. Hist.* WORKS ABOUT: *Appleton's CAB* 2: 116; T. T. Bouvé, biographical sketch in *Proceedings of Bos. Soc. Nat. Hist.* 6 (1856-1859): 391-94; Merrill, *Hundred Years of Geology,* pp. 553-63.

DeBRAHM, WILLIAM GERARD (1717—ca. 1799). *Surveying; Military engineering; Hydrography.* Parentage and early life not known. Married, first wife died 1774; Mary (Drayton) Fenwick, 1776; at least one child, apparently by first marriage. EDUCATION: Not known, but seemingly received some training as military engineer. CAREER: Captain of engineers, army of Emperor Charles VI; 1751, with 160 Protestants from Salzburg, founded Bethany, Ga.; 1754, appointed surveyor of Georgia; in years thereafter engaged in various surveying and engineering projects, including fortification of Charleston, S.C. (1755), defense of Savannah, Ga. (1757), and planning and defense of Ebenezer, Ga. (1757), and took up land at Ebenezer, on Savannah River; 1764, appointed by king as surveyor general of Southern District; 1765, began survey of Florida coast and there received large land grant; 1770, suspended from Florida survey; 1771-1775, in England; 1775, returned to Charleston and again given charge of Florida survey, residing at Charleston and St. Augustine; apparently spent later years in Philadelphia. SCIENTIFIC CONTRIBUTIONS: In 1757, in combination with surveys of Lt. Gov. William Bull of South Carolina, produced first full map of that state and Georgia; also prepared map of East Florida. As surveyor general, prepared four folio manuscript reports on South Carolina, Georgia, and East Florida, including maps, coastal observations, comments on soil, trees, shrubs, and wildlife, especially from utilitarian view, as well as local history and other matters. Studied Florida coast, and during 1771 trip to England investigated Gulf Stream, which was first described, with chart, in *Atlantic Pilot* [see Works]; observations on stream were thought by some to be superior to, and apparently begun before, those of B. Franklin [q.v.], although De Brahm studied only movements of currents and not temperatures; 1775 voyage from England to Charleston was ostensibly to study what De Brahm then called George Stream.

WORKS: *The Atlantic Pilot* (London, 1772); *The Leveling Balance and Counter-Balance; or, The Method of Observing, by the Weight and Height of Mercury . . .* (London, 1774); *De Brahm's Zonical Tables for the 25 Northern and Southern Climates* (London, 1774); parts of manuscript reports later published (see *DAB*). MSS: British Museum and Harv. U.-HL (Brown [below]). WORKS ABOUT: *DAB,* 5: 182-83 (Wilbur H. Siebert); Ralph H. Brown, "The De Brahm Charts of the Atlantic Ocean, 1772-1776," *Geographical Review* 28 (1938): 124-32.

De FOREST, ERASTUS LYMAN (June 27, 1834, Watertown, Conn.—June 6, 1888, probably Watertown, Conn.). *Mathematics.* Son of John, wealthy physician, and Lucy Starr (Lyman) De Forest. Never married. EDUCATION: 1854, A.B., Yale College; 1856, Ph.B., Sheffield Scientific School, Yale. CAREER: Inherited wealth; 1857-ca. 1858, in California, worked in gold mines and engaged in teaching; ca. 1858-ca. 1860, in Australia, tutor at Univ. of Melbourne; traveled in India and Europe; subsequently returned to Connecticut and there spent remainder of life, devoted to care of father and to pursuit of mathematical studies. SCIENTIFIC CONTRIBUTIONS: First publication, "Method of Estimating" [see Works], appeared in 1866, and mathematical work throughout life dealt with questions related to correcting irregular series of numbers, theory of errors and probability, and similar matters. Work little noticed by contemporaries, but in 1920s was recognized and appreciated for its importance in actuarial science [see Wolfenden, below].

WORKS: "A Method of Estimating and Correcting the Error Caused by the Unequal Length of the Calendar Months, in Reducing Observations of Temperature," *Am. Journ. Sci.* 2d ser. 41 (1866) : 371-79; "On Some Methods of Interpolation Applicable to the Graduation of Irregular Series, Such as Tables of Mortality, Etc.," *Annual Report of Smithson. Instn.* (1871) : 275-339, and ibid. (1873) : 319-53; *Interpolation and Adjustment of Series* (New Haven, 1876), 52 pp. See bibliography in Wolfenden [below], pp. 120-21 (based on *Roy. Soc. Cat.*, vols. 7, 9, 14, listing twenty-three papers, 1866-1888). WORKS ABOUT: *DAB*, 5: 197-98 (David E. Smith); *Obituary Record of Graduates of Yale University* (1880-1890) : 457; Hugh H. Wolfenden, "On the Development of Formulae for Graduation by Linear Compounding, with Special Reference to the Work of Erastus L. Deforest," *Transactions of Actuarial Society of America* 26 (1925) : 81-121, a review of De Forest's mathematical work, with little biographical information.

De KAY, JAMES ELLSWORTH (October 12, 1792, probably at Lisbon, Portugal [brought to New York at age two]—November 21, 1851, Oyster Bay, Long Island, N.Y.). *Zoology.* Son of George, sea captain, and Catherine (Colman) De Kay. Married Janet Eckford, 1821; survived by three children. EDUCATION: 1807-1812, attended Yale College; ca. 1812, began study of medicine, probably in New York City; 1819, M.D., Univ. of Edinburgh. CAREER: 1819, returned to New York; apparently devoted little time to medical practice, was active in affairs of N.Y. Lyc. Nat. Hist. as editor and librarian, and promoted work of its museum; throughout life, maintained close ties with literary community, and for several years apparently had career as writer in mind; 1831-1832, accompanied father-in-law, Henry Eckford, to Turkey, the latter having been given responsibility for Turkish naval yard; 1836-1844, engaged in zoological survey for state of New York; ca. 1844-1851, resided at Oyster Bay, Long Island. MEMBERSHIPS: N.Y. Lyc. Nat. Hist.; Am. Assoc. Advt. Sci. SCIENTIFIC CONTRIBUTIONS: Served as editor for volumes 1 (1825) and 2 (1828) of *Annals of N.Y. Lyc. Nat. Hist.*, and there published several articles on natural history, especially fossil remains. Chief scientific work done in connection with New York Natural History Survey, and report, based on extensive travel and study, incorporated descriptions of 1,600 species, including mammals, birds, reptiles, amphibians, fishes, mollusks, and crustaceans.

WORKS: *Zoology of New York* (report for New York Natural History Survey), 6 pts. in 5 vols. (Albany, 1842-1844) ; *Anniversary Address on the Progress of the Natural Sciences in the U.S.* (New York, 1826; reprinted New York, 1970), 78 pp. Also, see Meisel, *Bibliog. Amer. Nat. Hist.* MSS: No collection is known. WORKS ABOUT: *DAB*, 5: 203-4 (William B. Shaw); *NCAB*, 9 (1907 [copyright 1899 and 1907]) : 204-5; and information received from Yale Univ. Archives.

DELAFIELD, JOSEPH (August 22, 1790, New York, N.Y.—February 12, 1875, New York, N.Y.). *Mineralogy; Geodesy.* Son of John, wealthy merchant, and Ann (Hallett) Delafield. Married Julia Livingston, 1833; three children. EDUCATION: 1808, A.B., Yale College; studied law in a New York law office. CAREER: 1810, commissioned lieutenant, Fifth New York State Militia; 1811, admitted to bar as partner of former law teacher; 1812, made captain of drafted militia and captain in regular army; 1814, achieved rank of major, Forty-sixth U.S. Infantry; 1815, honorably discharged from army; appointed secretary to U.S. agent (1816), later acting agent, and finally agent (1821-1828) for joint commission to settle U.S.-Canadian boundary under articles VI and VII of Treaty of Ghent; during later years not actively engaged in business, but devoted efforts to scientific and related interests; after marriage in 1833, divided time between residence in New York and estate in Yonkers on the Hudson, and at latter location constructed innovative and profitable limekiln. MEMBERSHIPS: President of N.Y. Lyc. Nat. Hist., 1827-1866. SCIENTIFIC CONTRIBUTIONS: During period of connection with joint commission, directed surveying operations in field, charged with interpretation of previously indefinite boundary as outlined in treaty; in this largely unexplored region, extending from St. Regis, Quebec, to Lake of the Woods (Minnesota), made notes of fossils, shells, animals, rocks; *Diary* [see McElroy and Riggs, below] includes great many notes on natural history of region, in addition to references to problems regarding boundary determination. Particularly interested in mineralogy; during field work on boundary survey gathered minerals and in subsequent years devoted much effort to study of mineralogical collection, one of best private collections in U.S. during that time. Published papers on geology and mineralogy, including early accounts of Great Lakes region, in *Am. Journ. Sci.* and *Annals of N.Y. Lyc. Nat. Hist.*, especially during 1820s.

WORKS: See Meisel, *Bibliog. Amer. Nat. Hist.* MSS: See references in introduction and notes to *Diary* [below]. WORKS ABOUT: Franklin B. Dexter, *Biographical Sketches of the Graduates of Yale College . . .* (New Haven, 1912), 6: 187-88; Robert McElroy and T. Riggs, eds., *The Unfortified Boundary: A Diary of the First Survey of the Canadian Boundary Line from St. Regis to the Lake of the Woods by Major Joseph Delafield* (New York, 1943).

DEWEY, CHESTER (October 25, 1784, Sheffield, Mass.—December 15, 1867, Rochester, N.Y.). *Botany.* Son of Stephen, probably farmer, and Elizabeth

(Owen) Dewey. Married Sarah Dewey, 1810; Olivia Hart Pomeroy, 1825; five children by first marriage, ten by second. EDUCATION: 1806, A.B., Williams College; studied for ministry and in 1807 was licensed by Berkshire Congregationalist Association. HONORS: Said to have held D.D. and LL.D. degrees. CAREER: Ca. 1807-1808, taught and preached at West Stockbridge and Tyringham, Mass.; at Williams College, tutor (1808-1810) and professor of mathematics and natural philosophy (1810-1827); 1822, became lecturer on chemistry and medical botany, Berkshire Medical Institution; 1827-1836, principal of Berkshire Gymnasium (Pittsfield, Mass.); 1836-1850, principal of high school (later collegiate institute) in Rochester, N.Y.; 1837, appointed to prepare botanical report for Zoological and Botanical Survey of Massachusetts; 1842-1849, lecturer on chemistry, medical school at Woodstock, Vt.; 1850-1861, professor of chemistry and natural sciences, Univ. of Rochester. MEMBERSHIPS: Am. Assoc. Advt. Sci.; Am. Phil. Soc. SCIENTIFIC CONTRIBUTIONS: Best known for botanical investigations, especially sedges (Carices), which he studied over forty-year period; results of these studies were highly regarded by A. Gray [q.v.]. While teaching at Williams College, gathered geological and botanical specimens for museum. In 1818, published report of meteorological observations made at Williamstown, and while at Rochester kept daily record of weather over long period of time; also wrote on geology, mineralogy, and other subjects.

WORKS: "Caricography," series of articles on sedges, *Am. Journ. Sci.*, 1824-1866; *Reports on the Herbaceous Plants and on the Quadrupeds of Massachusetts published . . . by the Commissioners on the Zoological and Botanical Survey of the State* (Cambridge, Mass., 1840), with section on quadrupeds by E. Emmons [q.v.]. See Meisel, *Bibliog. Amer. Nat. Hist.*, and *Roy. Soc. Cat.*, vols 2, 7. MSS: Univ of Rochester Library (*NUCMC* 62-4133). *WORKS* ABOUT: *DAB*, 5: 267-68 (Charles W. Dodge); Martin B. Anderson, "Sketch of the Life of Prof. Chester Dewey, D.D., LL.D.," *Annual Report of Smithson. Instn.* (1870): 231-40.

DILLER, JOSEPH SILAS (1850-1928). *Geology; Petrography.* Geologist with U.S. Geol. Surv.

WORKS ABOUT: See *AMS; DAB*, supp. 1.

DOLBEAR, AMOS EMERSON (1837-1910). *Physics; Astronomy.* Professor of physics and astronomy, Tufts Univ.

MSS: Tufts Univ. Library (*NUCMC* 70-2020). WORKS ABOUT: See *AMS; NCAB*, 9 (1907 [copyright 1899 and 1907]): 414-15; *World Who's Who in Science.*

DONALDSON, HENRY HERBERT (1857-1938). *Neurology.* Professor of neurology, Univ. of Chicago; professor of neurology and director of research, Wistar Institute, Philadelphia.

MSS: Am. Phil. Soc. (*NUCMC* 60-2732). WORKS ABOUT: See *AMS; DAB*, supp. 2; *DSB*.

DOOLITTLE, CHARLES LEANDER (1843-1919). *Astronomy; Mathematics.* Professor of astronomy and director of Flower Observatory, Univ. of Pennsylvania.

WORKS ABOUT: See *AMS; DAB*.

DOREMUS, ROBERT OGDEN (1824-1906). *Chemistry; Physics; Medicine.* Professor of chemistry and physics, College of City of New York (City College).

WORKS ABOUT: See *AMS*; *DAB*.

DOUGLAS, SILAS HAMILTON (October 16, 1816, Fredonia, N.Y.—August 26, 1890, Ann Arbor, Mich.). *Chemistry.* Son of Benjamin and Lucy (Townsend) Douglass [*sic*]. Married Helen Wells, 1845; seven children. EDUCATION: attended Univ. of Vermont; studied medicine with Dr. Z. Pitcher [q.v.], Detroit; 1841-1842, attended Medical Department, Univ. of Maryland; 1842, licensed by Censors of Michigan State Medical Society. HONORS: A.M., Univ. of Vermont, 1847. CAREER: 1838, settled in Detroit, joined D. Houghton [q.v.] on geological surveys of Michigan, and served as physician on staff of H. R. Schoolcraft [q.v.]; 1843, began medical practice at Ann Arbor; at Univ. of Michigan, assistant to professor of chemistry (1844-1845), lecturer in chemistry and geology (1845-1846), professor of chemistry, mineralogy, and geology (1846-1851), professor of chemistry, pharmacy, medical jurisprudence, geology, and mineralogy (1851-1855), professor of chemistry, mineralogy, pharmacy, and toxicology (1855-1870), professor of chemistry (1870-1875), professor of metallurgy and chemical technology (1875-1877), and director of chemical laboratory (1870-1877). MEMBERSHIPS: Several scientific societies. SCIENTIFIC CONTRIBUTIONS: Chief scientific work was establishment of chemical laboratory and related instruction at Univ. of Michigan; in 1848, took part in establishment of medical department; and in 1856 separate laboratory building was constructed for analytical chemistry, serving all schools in university, the first such laboratory erected by a state institution. In 1875, controversy arose over deficits in accounts related to student laboratory fees; legislature mounted inquiry; and in 1877 Douglas was dismissed, although no criminal conduct was involved. Took part in organizing Ann Arbor water and gas works. Published works on ventilation, poison detection, and analysis of waters and coal.

WORKS: *Guide to a Systematic Course of Qualitative Analysis Prepared for the Chemical Laboratory at the University of Michigan* (Ann Arbor, Mich., 1864) 28 pp.; with A. B. Prescott [q.v.], *Qualitative Chemical Analysis: A Guide in the Practical Study of Chemistry and in the Work of Analysis* (Ann Arbor, Mich., 1874) 259 pp. See references to works in Kelly and Burrage [below]. No items listed in *Roy. Soc. Cat.* MSS: Univ. Mich.-MHC. WORKS ABOUT: Kelly and Burrage, *Med. Biog.* (1928); Burke A. Hinsdale, *History of the University of Michigan* (Ann Arbor, Mich., 1906); Wilfred B. Shaw, *The University of Michigan: An Encyclopedic Survey* (Ann Arbor, Mich., 1951).

DOUGLASS, WILLIAM (ca. 1691, Gifford, Haddington County, Scotland—October 21, 1752, Boston, Mass.). *Medicine; Natural history.* Son of George Douglass, "portioner" of Gifford and factor for marquis of Tweeddale. Never married. EDUCATION: Studied medicine in Edinburgh, Leyden, and Paris; 1712, M.D., Univ. of Utrecht. HONORS: M.A., Univ. of Edinburgh, 1750. CAREER: Ca. 1717, traveled in West Indies; 1718-1752, medical practice, Boston. SCIENTIFIC CONTRIBUTIONS: At beginning of practice was sole physician in Boston area with earned medical degree; in 1721, opposed smallpox inoculation in Boston and wrote several pamphlets on subject, but later came to favor inoculation procedure; this changing attitude was reflected in *Practical Essay* [see Works]. Chief medical publication, *Practical History* [see Works], was earliest effective clinical account of scarlet fever. Pioneered in colonies in medical applications of mercuric substances. By 1721, collected more than 700 Boston-area plants and eventually had some 1,100 specimens; *Summary* [see Works] included botanical data, particularly on trees of New England, and showed Douglass to have been accomplished in botanical studies. Exchanged botanical, meteorological, and astronomical data with C. Colden [q.v.] and published almanac [see Works]. Interested in establishing correct latitude and longitude for colonial locations and traveled over New England to gather data for projected map.

WORKS: *A Practical Essay Concerning the Small Pox* (London, 1730); *The Practical History of a New Epidemical Eruptive Miliary Fever, with an Angina Ulcusculosa, which prevailed in Boston, New England, in the Years 1735 and 1736* (Boston, 1736), 20 pp.; *Mercurius Nov-Anglicanus; or, An Almanack,* by "William Nadir" (Boston, 1743); *A Summary, Historical and Political, of the First Planting, Progressive Improvements and Present State of the British Settlements in North America,* 2 vols. (Boston, 1749 and 1751; reprinted, Boston in 1755 and London in 1755, 1760). Writings listed in G. H. Weaver, "Life and Writings of William Douglass," *Bulletin of Society of Medical History of Chicago* 2 (April 1921): 229-59. WORKS ABOUT: *DAB*, 5: 407-8 (John F. Fulton); Stearns, *Sci. in Brit. Colonies*, pp. 477-84; Kelly and Burrage, *Med. Biog.* (1928).

DRAKE, DANIEL (October 20, 1785, near Bound Brook, N.J.—November 5, 1852, Cincinnati, Ohio). *Medicine; Natural history.* Son of Isaac, farmer, and Elizabeth (Shotwell) Drake. Married Harriet Sisson, 1807; at least four children. EDUCATION: 1816, M.D., Univ. of Pennsylvania. CAREER: 1807, took over medical practice of Dr. William Goforth, Cincinnati, with whom he had previously studied; 1810, established store, operated by father and brother, that dealt at various times in groceries, drugs, books, stationery, etc.; 1813, organized Cincinnati Manufacturing Co.; at Transylvania Univ. (Lexington, Ky.), professor of materia medica and medical botany (1817-1818 and 1823-1824), professor of theory and practice of medicine and dean of medical faculty (1824-1827); at Ohio Medical College (Cincinnati), founder, president, and professor of medicine (1819-1822), professor of clinical medicine (1831-1832), professor of pathology, practice, and clinical medicine (1849-1850), and professor of theory and practice of medicine (1852); founder and editor (1828-1838), *Western Journal of the Medical and Physical Sciences;* 1830-1831, professor of theory and practice of medicine, Jefferson Medical College (Philadelphia); 1835-1839, professor of theory and practice of medicine and dean, Medical Faculty of Cincinnati College; 1839-1849 and 1850-1852, professor of pathology and practice of medicine, Louisville Medical Institute. MEMBERSHIPS: Wernerian Academy of Natural Sciences (Edinburgh); Am. Phil. Soc.; and other societies. SCIENTIFIC CONTRIBUTIONS: Played pivotal role in early medical education in the West. Compiled enormous amount and variety of information on western natural history, topography, geology, meteorology, climate, population, and customs, presented in relation to diseases and regional living conditions, in his *Systematic Treatise* [see Works], which culminated some forty years of medical and scientific work.

WORKS: *Natural and Statistical View, or Picture of Cincinnati and the Miami Country, Illustrated by Maps, with an Appendix Containing Observations on the Late Earth Quakes, the Aurora Borealis and the Southwest Winds* (Cincinnati, 1815); *A Systematic Treatise, Historical, Etiological and Practical, on the Principal Diseases of the Interior Valley of North America, as They Appear in the Caucasian, African, Indian, and Esquimoux Varieties of Its Population,* 2 vols. (Cincinnati, 1850 and 1854). List of writings in Shapiro and Miller [below]. MSS: Cincinnati Historical Society (*NUCMC* 71-1534); also see references in Shapiro and Miller. WORKS ABOUT: *DAB*, 5: 426-27 (Albert P. Mathews); Henry D. Shapiro and Zane L. Miller, *Physician to the West: Selected Writings of Daniel Drake on Science and Society* (Lexington, Ky., 1970). Also see Emmet F. Horine, *Daniel Drake . . . Pioneer Physician of the Midwest* (Philadelphia, 1961).

DRAPER, DANIEL (1841-1931). *Meteorology; Mechanics.* Director, New York Meteorological Observatory.

WORKS ABOUT: See *AMS*; *NCAB*, 6 (1929 [copyright 1892 and 1929]): 172-73; *Who Was Who in America*, vol. 1.

DRAPER, HENRY (March 7, 1837, Prince Edward County, Va.—November 20, 1882, New York, N.Y.). *Astronomy.* Son of John William [q.v.], physician and chemist, and Antonia Coetana de Paiva Pereira (Gardner) Draper. Married Anna M. Palmer, 1867; no information on children. EDUCATION: 1858, M.D., Univ. of City of New York (New York Univ.), having also studied earlier in the academic department. HONORS: LL.D., Univ. of City of New York and Univ. of Wisconsin, 1882. CAREER: 1858-1859, on staff of Bellevue Hospital; 1860-1882, professor of natural science in undergraduate department of Univ. of City of New York; 1862, surgeon of Twelfth Regiment of New York Militia; 1866-1873, professor of physiology and dean of medical faculty; 1870, appointed professor of analytical chemistry in academic department; 1873, father-in-law, Courtlandt Palmer, died and Draper took over management of estate; 1882, resigned joint professorship in chemistry and physics, to which he had been appointed following the death of his father earlier that year. MEMBERSHIPS: Astronomische Gesellschaft; Natl. Ac. Scis.; and other societies. SCIENTIFIC CONTRIBUTIONS: About 1857 began to develop idea of photography applied to astronomy, a field in which he was a pioneer. In 1861, completed first of his glass mirrors and soon began making daguerreotypes of sun and moon; in 1872, with twenty-eight-inch reflecting telescope, produced first photograph of stellar spectrum lines; also photographed spectra of sun, moon, Jupiter, and Venus. Published extensively both on techniques for applying photography to astronomy and on actual results of applications. In 1874, directed photographic department of U.S. commission to observe transit of Venus, and in 1878 organized solar eclipse expedition to Wyoming.

WORKS: "On the Construction of a Silvered Glass Telescope 15½ Inches in Aperture, and Its Use in Celestial Photography," *Smithson. Contr. Knowl.* 14, pt. 2 (1864), became standard manual on telescope making; *A Textbook on Chemistry* (New York, 1866). Bibliographies in *DSB* and Barker [below]. MSS: N.Y. Pub. Lib. (*NUCMC* 68-1698). WORKS ABOUT: *DAB*, 5: 435-37 (Raymond S. Dugan); *DSB*, 4: 178-81 (Charles A. Whitney); George F. Barker, "Memoir," *Biog. Mems. of Natl. Ac. Scis.* 3 (1895): 81-139.

DRAPER, JOHN WILLIAM (May 5, 1811, St. Helen's, Lancashire, England—January 4, 1882, Hastings-on-Hudson, N.Y.). *Chemistry*; *Science and religion.* Son of John Christopher, itinerant Methodist minister, and Sarah (Ripley) Draper. Married Antonia Coetana de Paiva Pereira Gardner, 1831; five children, including Henry Draper [q.v.]. EDUCATION: 1829, entered Univ. of London, and received honors certificate in chemistry; 1836, M.D., Univ. of Pennsylvania, also having studied chemistry and physics under R. Hare [q.v.]. HONORS: Rumford Medal, Am. Ac. Arts Scis. CAREER: 1832, emigrated to U.S. and established laboratory in family farmhouse in Virginia; 1836-1839, professor of chemistry and natural philosophy, Hampden-Sydney College; at Univ. of City of New York (New York Univ.), professor of chemistry, undergraduate department (1839-1882), helped organize school of medicine and was made professor of chemistry and physiology (1841), and became president of medical school (1850). MEMBERSHIPS: Natl. Ac. Scis. and other societies; first president, Am. Chem. Soc. SCIENTIFIC CONTRIBUTIONS: Coauthored three articles on geology before arrival in U.S. Known especially for work on radiant energy, and interest in chemical effects involved him in pioneering work in photography, including portraiture and microphotography; did important early work on diffraction spectrum; in 1841 formulated law regarding chemical action of absorbed rays; studied phenomenon of light from heat. Also studied osmosis in relation to physiology and prepared leading textbook on physiology. During later period of life turned to historical studies, including relations of science and religion, and *Intellectual Development of Europe* [see Works] is probably his major publication.

WORKS: Among most significant works were *Human Physiology Statical and Dynamical; or, The Conditions and Course of the Life of Man* (New York, 1856); a collection of his papers, *Scientific Memoirs, Being Experimental Contributions to a Knowledge of Radiant Energy* (New York, 1878; reprinted, New York, 1973); *A History of the Intellectual Development of Europe* (New York, 1863); *History of the Conflict Between Religion and Science* (New York, 1874). Bibliography in Fleming [below]. MSS: Lib. Cong. WORKS ABOUT: *DAB*, 5: 438-41 (Ellwood Hendrick); *DSB*, 4: 181-83 (Donald Fleming); Fleming, *John William Draper and the Religion of Science* (Philadelphia, 1950; reprinted, New York, 1972).

DROWN, THOMAS MESSINGER (March 19, 1842, Philadelphia, Pa.—November 16, 1904, South Bethlehem, Pa.). *Chemistry*; *Metallurgy*. Son of William Appleton, umbrella manufacturer, and Mary (Peirce) Drown. Married Helen Leighton, 1868; no information on children. EDUCATION: 1862, M.D., Univ. of Pennsylvania; ca. 1862-1865, studied chemistry at Sheffield (Yale) and Lawrence (Harvard) scientific schools; 1865-1868, studied at School of Mines, Freiberg, Saxony, and with Robert von Bunsen at Heidelberg Univ. HONORS: LL.D., Columbia Univ., 1895. CAREER: 1862, made single voyage as ship's surgeon to Europe; 1869-1871, instructor in metallurgy, Lawrence Scientific School; 1871-ca. 1874, operated private analytical and consulting practice, Philadelphia; for Am. Inst. Min. Engrs., a manager (1871-1873) and secretary and editor (1873-1883); 1874-1881, professor of chemistry, Lafayette College; 1881-ca. 1885, involved in settlement of father's business affairs; at Massachusetts Institute of Technology, was professor of chemistry (1885-1895), then became head of chemistry department (1888) and also took charge of chemical engineering program (1893); 1887-1895, in charge of chemical investigation of inland waters and of sewage for Massachusetts Board of Health, and after 1895 continued to act as consulting chemist for the state; 1895-1904, president of Lehigh Univ. MEMBERSHIPS: President, Am. Inst. Min. Engrs. SCIENTIFIC CONTRIBUTIONS: Extensive training in chemistry and metallurgy, in both U.S. and Europe, made him one of the best prepared American analytical chemists of his time. Developed in his students of analytical chemistry an interest in making original contributions and, as editor of *Transactions of Am. Inst. Min. Engrs.*, made that journal an outlet for such work. Developed and published number of analytical processes that were related to metallurgy and had commercial or industrial applicability. Extent and thoroughness of work for Massachusetts Board of Health, especially development of "map of normal chlorine" in Massachusetts surface waters, made his efforts model for others; results of investigations for Massachusetts board given chiefly in its annual reports.

WORKS: List of his papers published in *Transactions of Am. Inst. Min. Engrs.* is given in Raymond [below], pp. 292-93; see also *Roy. Soc. Cat.*, vols. 9, 12, 14. WORKS ABOUT: *DAB*, 5: 460-62 (Henry P. Talbot); R. W. Raymond, "Biographical Notice," *Transactions of Am. Inst. Min. Engrs.* 36 (1906): 288-304.

DUDLEY, CHARLES BENJAMIN (1842-1909). *Chemistry*; *General technology*. Chemist to Pennsylvania Railroad Co.

WORKS ABOUT: See *AMS*; *DAB*.

DUDLEY, PAUL (September 3, 1675, Roxbury, Mass.—January 25, 1751, Roxbury, Mass.). *Natural history*. Son of Joseph, governor of Massachusetts, and Rebecca (Tyng) Dudley. Married Lucy Wainwright, 1703; six children, all died in infancy. EDUCATION: 1690, A.B., Harvard College; studied law at Inner Temple, London. CAREER: 1702, appointed attorney general of Massachusetts; 1718, appointed judge of superior court of judicature; at times served in Massachusetts House and Executive Council; 1745-1751, chief justice of Massachusetts. MEMBERSHIPS: Roy. Soc. SCIENTIFIC CONTRIBUTIONS: Sent natural history specimens and papers on various subjects to Roy. Soc. between 1719 and 1736, the longest period of continuous communication with the society of any colonial before him. Earliest communication was on making maple syrup and this and number of other contributions were published in *Phil. Trans.*; one item not published was 1733 account of locusts in New England, later shown to be cicadas; in paper on whales answered question of derivation of ambergris; also wrote on moose, molasses made from apples, rattlesnakes, earthquakes in New England, height of Niagara Falls, and other topics, including medical subjects. One of most important papers, on plants of New England, referred to hybridization of Indian corn and included useful comments on role of wind, but these observations were little noticed by contemporaries; last communication with Roy. Soc. was on evergreens, including distinction between spruce and hemlock.

WORKS: Paper on Niagara Falls, *Phil. Trans. of Roy. Soc.* 32 (April-May 1722): 69-72; "Observations on Some of the Plants of New England, with Remarkable Instances of the Nature and Power of Vegetation," ibid. 33 (October-December 1724): 194-200; "An Essay Upon the Natural History of Whales, with a Particular Account of the Ambergreese, found in the Sperma Caeti Whale," ibid. 33 (March-April 1725): 256-69; "An Account of the Several Earthquakes Which Have Happen'd in New-England, Since the First Settlements of the English in That Country, especially of the Last, Which Happen'd on Octob. 29, 1727," ibid. 39 (April-June 1735): 63-73. Bibliography of works in Shipton, *Biog. Sketches, Harvard*, 4 (1933): 42-54. MSS: Bos. Pub. Lib. (Shipton [above]). WORKS ABOUT: *DAB*, 5: 483-84 (H. W. Howard Knott); Stearns, *Sci. in Brit. Colonies*, pp. 455-72.

DUDLEY, WILLIAM RUSSEL (1849-1911). *Botany*. Professor of botany, Stanford Univ.

WORKS ABOUT: See *AMS*; *DAB*.

DUNBAR, WILLIAM (1749, near Elgin, Morayshire, Scotland—October 1810, plantation near Natchez, Miss.). *Mathematics*; *Astronomy*; *Mete-*

orology. Son of Sir Archibald Dunbar, Scottish earl. Married; several children. EDUCATION: Educated in Glasgow, Scotland; later studied mathematics and astronomy in London. CAREER: 1771, arrived at Fort Pitt (Pittsburgh) with goods to trade with Indians in exchange for furs; 1773, established plantation in West Florida, as partner of Philadelphia merchant; 1792, with partner established second plantation ("The Forest") near Natchez, and Dunbar later acquired sole ownership; 1798, appointed surveyor general of District of Natchez and acted as representative of Spanish government in establishing boundary between U.S. and Spanish possessions east of Mississippi; 1804, appointed by President Jefferson to explore territory of Ouachita River; 1805, appointed to explore region about Red River; served in Mississippi territorial legislature. MEMBERSHIPS: Am. Phil. Soc. SCIENTIFIC CONTRIBUTIONS: Put scientific knowledge to work on plantation, including design of plow and harrow, better cotton gin, mechanism for making square cotton bales, and introduction of idea for manufacturing cottonseed oil. Conducted meteorological studies in 1799, including temperature, barometer, wind, and weather, which resulted in best account of climate of his region to be produced up to that time [see Works]; had private observatory; among other achievements, made first observation of elliptical rainbow and gave adequate explanation of it. Also studied natural history of Mississippi region, and included such observations in manuscript report on boundary survey (1798), submitted to Spanish government; in correspondence with Jefferson, and as result of Ouachita explorations commissioned by the president, gave first scientific account of Hot Springs, including analysis of water there. Also wrote on fossil bones found in Louisiana, on Indian sign language, and other topics.

WORKS: Scientific work found chiefly in *Transactions of Am. Phil. Soc.*, including "Meteorological Observations for One Entire Year, Ending the 31st of January 1800 . . . ," ibid. 6 (1809): 9-24; "Description of the River Mississippi and its Delta, with That of the Adjacent Parts of Louisiana," ibid., pp. 165-87 and appendix, pp. 191-200. See *Roy. Soc. Cat.*, vol. 2, and Meisel, *Bibliog. Amer. Nat. Hist.* MSS: Mississippi Department of Archives and History (*NUCMC* 60-2578). WORKS ABOUT: *DAB*, 5: 507-8 (Franklin L. Riley); Riley, "Sir William Dunbar: The Pioneer Scientist of Mississippi," *Publications of Mississippi Historical Society* 2 (1899): 85-111.

DUNGLISON, ROBLEY (January 4, 1798, Keswick, Cumberland, England—April 1, 1869, Philadelphia, Pa.). *Medical education*; *Physiology.* Son of William, wool manufacturer, and Elizabeth (Jackson) Dunglison. Married Harriet Leadam, 1824; seven children. EDUCATION: Attended Green Row Academy, Abbey Holme; apprenticed to surgeon; attended medical lectures in Edinburgh, Paris, and London; 1818, medical degree, Royal College of Surgeons (London); diploma, Society of Apothecaries; 1823, M.D., Univ. of Erlangen. HONORS: M.D., Yale Univ., 1825. CAREER: 1819, began medical practice, London, and there was appointed physician-accoucheur to Eastern Dispensary; 1825-1833, professor of medicine, Univ. of Virginia; 1833-1836, professor of medicine, Univ. of Maryland; 1836-1868, professor of institutes of medicine and sometimes dean, Jefferson Medical College (Philadelphia). MEMBERSHIPS: Vice-president, Am. Phil. Soc. SCIENTIFIC CONTRIBUTIONS: While in England, issued several medical publications, including compilation on childhood diseases, *Commentaries on the Diseases of the Stomach and Bowels of Children* (1824), which helped win appointment to Univ. of Virginia. In U.S., played important role in medical education, through both teaching and publications, contributing significantly to early success of medical school at Univ. of Virginia and of Jefferson Medical School. At Univ. of Virginia became first in America to devote full efforts to medical teaching, wrote textbooks for medical education, was first American to write book on physiology, and ranks as pioneer in teaching of that subject; also wrote medical dictionary and works on hygiene (public health) and other medical topics. Knowledge of medicine derived largely from publications of other authors rather than personal observation or research; edited several medical journals, in England and U.S., and also translated and edited number of foreign works. Gave significant advice and assistance to W. Beaumont [q.v.] in his study of gastric juice.

WORKS: *Human Physiology*, 2 vols. (Philadelphia, 1832); *A New Dictionary of Medical Science and Literature*, 2 vols. (Boston, 1833). List of chief works in *DSB*; bibliography of works and text of autobiographical manuscript at College of Physicians of Philadelphia published in *Transactions of Am. Phil. Soc.*, n.s. 53, pt. 8 (December 1963): 212 pp. WORKS ABOUT: *DAB*, 5: 512-13 (Russell H. Chittenden); *DSB*, 4: 251-53 (Samuel X. Radbill); Jerome J. Bylebyl, "William Beaumont, Robley Dunglison, and the 'Philadelphia Physiologists,'" *Journal of the History of Medicine and Allied Sciences* 25 (1970): 3-21.

DURAND, ELIAS [ELIE MAGLOIRE] (January 25, 1794, Mayenne, France—August 14, 1873, Philadelphia, Pa.). *Botany.* Son of André Durand, recorder of deeds at Mayenne. Married Polymnia Rose Ducatel, 1820; Marie Antoinette Berauld, 1825; one child by first marriage, four by second. EDUCA-

TION: 1808-1812, apprenticed to pharmacist-chemist in Mayenne; studied for one year in Paris. CAREER: 1813-1814, assistant pharmacist in French army; 1814-1816, employed as druggist in Nantes, France, returning to army during Napoleon's Hundred Days; 1816-ca. 1817, went to New York, worked in Boston as superintendent in chemical laboratory, and later found similar employment in Philadelphia; 1817-1824, chief clerk and then partner in Baltimore pharmacy of Edme Ducatel; 1824, returned to France to buy apparatus and supplies; 1825, established pharmaceutical shop in Philadelphia; 1852, retired. MEMBERSHIPS: Ac. Nat. Scis. Philad.; Am. Phil. Soc. SCIENTIFIC CONTRIBUTIONS: Study of American plants began while in Baltimore, but evidence suggests active interest in botany before leaving France and he is said to have been commissioned at that time by wealthy Frenchman to provide collection of North American flora. Shop in Philadelphia became center for physicians and botanists; as pharmacist introduced number of new items to American medical practice and was first in U.S. to bottle mineral water. Encouraged and supported botanical collectors and procured herbaria of T. Nuttall, C. S. Rafinesque [qq.v.], and others; in 1837, explored Dismal Swamp in Virginia, and after early retirement devoted full efforts to botanical studies and collecting. Gave specimens to Ac. Nat. Scis. Philad., gathered collections for sale, and in 1868 presented Jardins des Plantes in Paris with 100,000 specimens of North American flora, which together with later donations amounted to 15,000 species, designated by recipient as Herbier Durand.

WORKS: Works on chemical and botanical topics, totaling fifteen items, listed in *Roy. Soc. Cat.*, vols. 2, 7, 9. MSS: Harv. U.-GH (Hamer). WORKS ABOUT: *DAB*, 5: 538-40 (George H. Genzmer); William J. Robbins, "French Botanists and the Flora of the Northeastern U.S.: J. G. Gilbert and Elias Durand," *Proceedings of Am. Phil. Soc.* 101 (1957): 362-68.

DUTTON, CLARENCE EDWARD (1841-1912). *Geology*; *Physics*. U.S. Army officer; assigned to U.S. Geol. Surv.

MSS: Univ. Tex.-TA (*NUCMC* 69-2002). WORKS ABOUT: See *AMS*; *DAB*; *DSB*.

DWIGHT, THOMAS (1843-1911). *Anatomy*. Professor of anatomy, Harvard Univ. Medical School.

WORKS ABOUT: See *AMS*; *DAB*.

E

EASTMAN, JOHN ROBIE (1836-1913). *Astronomy; Mathematics.* Astronomer, U.S. Naval Observatory.

WORKS ABOUT: See *AMS*; *DAB.*

EATON, AMOS (May 17, 1776, Chatham, N.Y.—May 10, 1842, Troy, N.Y.). *Geology; Botany.* Son of Capt. Abel, farmer, and Azuba (Hurd) Eaton. Married Polly Thomas, 1799; Sally Cady, 1803; Anne Bradley, 1816; Alice Johnson, 1827; ten children; grandfather of Daniel C. Eaton [q.v.]. EDUCATION: 1799, A.B., Williams College; studied law and in 1802 admitted to New York state bar; 1816-1817, studied science at Yale, under B. Silliman [q.v.] and Eli Ives. CAREER: 1802-1810, law practice, land agent, surveyor, in Catskill, N.Y.; 1811-1815, imprisoned in Greenwich jail, New York City, on charge of forgery in land dispute, although Eaton and many others always argued for his innocence; 1817, began public lectures on science in Williamstown, Mass., and later in other towns in western Massachusetts, Vermont, and the Hudson Valley; 1820, appointed professor of natural history, medical school, Castleton, Vt.; 1820-1821, conducted geological and agricultural surveys of Albany and Rensselaer counties, and in 1824 surveyed region crossed by Erie Canal, under sponsorship of Stephen Van Rensselaer; 1824-1842, senior professor, Rensselaer School (later Rensselaer Polytechnic Institute), established at Eaton's suggestion by Stephen Van Rensselaer. MEMBERSHIPS: Founder of Troy, (N.Y.) Lyceum of Natural History. SCIENTIFIC CONTRIBUTIONS: A scientist of wide interests with noted ability to arouse public interest; his popular lectures and textbooks covered botany, zoology, chemistry, geology, and other subjects. Lectures by invitation before New York state legislature in 1818 said to have prepared way for 1836 New York Natural History Survey. As result of work on Van Rensselaeer-sponsored surveys, was perhaps leading American geologist in 1820s, working especially on stratigraphy, but produced theories not supported by sufficient data. Also did work in botany and *Manual* [see Works] had many editions, but was later criticized by A. Gray [q.v.]. Most enduring work was formulation of Rensselaer School, in which science was viewed in relation to daily life, and students learned by laboratory and field work; the school was operated by Eaton almost single-handedly, and number of influential scientists of next generation were educated there.

WORKS: *Manual of Botany for the Northern States* (Albany, 1817); *Geological and Agricultural Survey of the District Adjoining the Erie Canal* (Albany, 1824); *Geological Nomenclature for North America* (Albany, 1828). Representative works listed in *DSB.* Full bibliography, and citation of manuscript sources, in Ethel M. McAllister, *Amos Eaton, Scientist and Educator* (Philadelphia, 1941). WORKS ABOUT: *DAB,* 5: 605-6 (George P. Merrill); *DSB,* 4: 273-75 (Samuel Rezneck).

EATON, DANIEL CADY (September 12, 1834, Fort Gratiot [now part of Port Huron], Mich.—June 29, 1895, New Haven, Conn.). *Botany.* Son of General Amos B., army officer, and Elizabeth (Selden) Eaton; grandson of Amos Eaton [q.v.]. Married Caroline Ketcham, 1866; three children. EDUCATION: 1857, B.A., Yale College; 1860, B.S., Lawrence Scientific School, Harvard. CAREER: Ca. 1861-1864, clerk and inspector of stores, Commissary Department, U.S. Army; 1864-1895, professor of botany, Yale Univ., attached especially to Sheffield Scientific School. MEMBERSHIPS: Torrey Bot. Club. SCIENTIFIC CONTRIBUTIONS: Remembered chiefly for work with ferns, beginning as undergraduate with paper entitled "Three New Ferns" [see Works]; continued preparatory studies during three years with A. Gray [q.v.] at Harvard, while enrolled in Lawrence Scientific School. Worked on ferns collected by C. Wright [q.v.] of Rodgers's exploring expedition in Japan and in Cuba, and also contributed to sections on ferns for report by J. Torrey [q.v.] on botany of Mexican boundary survey and to A. W. Chapman [q.v.], *Flora of the Southern United States*; also botanized in eastern U.S. and in 1869 explored and collected in Utah. Most important work was *The Ferns* [see Works], which included descriptions of all known species north of Mexico. After about 1883, chief botanical interest was in mosses and liverworts, including Hawaiian species, and also was interested in study of Algae. Contributed botanical definitions to *Webster's International Dictionary*.

WORKS: "On Three New Ferns from California and Oregon," *Am. Journ. Sci.* 2d ser. 22 (1856): 138; *The Ferns of North America*, 2 vols. (Salem, Mass., and Boston, 1879, 1880), including 81 colored plates and descriptions of 149 species; *A Check-list of North American Sphagna* . . . (New Haven, 1893), 10 pp. Bibliography in Setchell [below]. MSS. Yale U. Lib. (*NUCMC* 71-2016); Harv. U.-GH (Hamer). WORKS ABOUT: *DAB*, 5: 606-7 (William R. Maxon); William A. Setchell, "Daniel Cady Eaton," *Bulletin of Torrey Bot. Club* 22 (August 31, 1895): 341-51; *Obituary Record of Graduates of Yale University* (1890-1900): 396.

EDDY, HENRY TURNER (1844-1921). *Mathematics*; *Mechanics*; *Engineering*. Professor of mathematics, astronomy, and civil engineering, Univ. of Cincinnati; president, Rose Polytechnic Institute; professor of engineering and mechanics, Univ. of Minnesota.

WORKS ABOUT: See *AMS*; *DAB*.

EDISON, THOMAS ALVA (1847-1931). *Physics; Invention*. Inventor.

MSS: Edison National Historic Site, West Orange, N.J. (*NUCMC* 66-797). WORKS ABOUT: See *AMS*; *DAB*, supp. 1; *DSB*.

EDWARDS, WILLIAM HENRY (1822-1909). *Entomology*. Lawyer; businessman; author of *Butterflies of North America*.

MSS: West Virginia Department of Archives and History (*NUCMC* 62-2588; also see entry 66-652); Ac. Nat. Scis. Philad. (*NUCMC* 66-34); Am. Mus. Nat. Hist. (Hamer). WORKS ABOUT: See *AMS*; *DAB*.

EIGENMANN, ROSA SMITH (October 7, 1858, Monmouth, Ill.—January 12, 1947, San Diego, Calif.). *Ichthyology*. Daughter of Charles Kendall, printer, and clerk of San Diego school board, and Lucretia (Gray) Smith. Married Carl H. Eigenmann, 1887; five children. EDUCATION: Attended business college in San Francisco; 1880 (summer), studied fishes on Pacific Coast with D. S. Jordan [q.v.]; 1880-1882, studied at Indiana Univ. and in summer of 1881, with thirty-three other students, accompanied Jordan to Europe; 1887-1888, studied cryptogamic botany with W. G. Farlow [q.v.], Harvard Univ. CAREER: Ca. 1886, reporter for *San Diego Union* (owned by brother and brother-in-law); 1887, went with husband, a zoologist, to Harvard Univ. to study South American fishes in Museum of Comparative Zoology; 1888, returned to San Diego and continued researches with husband at their small biological facility; 1891, moved to Bloomington, Ind., where husband was made professor of zoology, and continued joint researches for several years; 1893, retired from research to care for children; 1926, returned to San Diego. MEMBERSHIPS: First woman member, San Diego Society of Natural History; member of Calif. Ac. Scis. SCIENTIFIC CONTRIBUTIONS: First published in 1880; in same year, paper read at meeting of San Diego society won favorable comment from D. S. Jordan (leading American ichthyologist), who was in attendance at society meeting and who encouraged her to attend Indiana Univ. In period 1888-1893, with husband wrote fifteen papers, including significant works on fishes of South America and western North America, and dealt with questions of embryology and evolution and with classification and related matters. Personal research interest was fishes of San Diego region and as individual author published twenty scientific papers. After 1893, ceased active research but continued to assist husband in editing his research work; despite relatively short research career, is recognized as first American woman to attain distinction in ichthyology.

WORKS: Most papers listed in Bashford Dean, *Bibliography of Fishes* 1 (New York, 1916): 365-67, 2 (1917): 463, 3 (1923): 55. WORKS ABOUT: *Nota. Am. Wom.*, 1: 565-66 (Carl L. Hubbs); Leonard Stejneger, "Memoir of Carl H. Eigenmann," *Biog. Mems. of Natl. Ac. Scis.* 18 (1938): 305-36.

EIGHTS, JAMES (1798, Albany, N.Y.—1882, Ballston, N.Y.). *Natural history*; *Geology*. Son of Jonathan Eights, physician. Never married. EDUCATION: Apparently studied medicine; known in adult years as "Doctor." CAREER: Details of life uncertain; apparently never practiced medicine; 1829-1830, sailed as naturalist on Capt. Edmund Fanning's voyage of discovery to South Sea Islands; 1835-1853, generally in Albany; 1837, appointed geologist, Wilkes Expedition, but was eliminated from scientific corps before sailing; at times assisted L. Vanuxem and E. Emmons [qq.v.] on geological work for New York Natural History Survey (1836-1842); 1853, made last appearance in Albany directory, listed as draftsman and geologist; during late 1850s, may have been in North Carolina; in later years lived with sister at Ballston. MEMBERSHIPS: Albany Institute. SCIENTIFIC CONTRIBUTIONS: Reputation derived chiefly from work on collections and observations gathered while on Fanning expedition, which resulted in several papers on zoology, botany, and geology of South Shetland Islands in Antarctic region, including contribution of first geological observations ever made of Antarctic. In 1837, published first account of a ten-legged pycnogonid (sea spider), later confirmed by other Antarctic explorers; 1835-1840, wrote anonymously for Albany magazine, *Zodiac*, on natural history, geology, meteorology, and other topics; also published papers in other journals on geology of Albany, the Gulf Stream, North Carolina geology (1854, 1858), and other geological and natural history subjects. An accomplished artist, produced illustrations to accompany science publications, and in 1815 drew sketches of Albany that have been extensively reproduced.

WORKS: "Description of a New Crustaceous Animal (Brongniartia trilobitoides) Found on the Shores of the South Shetland Islands, with Remarks on Their Natural History" [1833], *Transactions of Albany Institute* 2 (1852): 53-69. See also Meisel, *Bibliog. Amer. Nat. Hist.* WORKS ABOUT: John M. Clarke, "The Reincarnation of James Eights, Antarctic Explorer," *Scientific Monthly* 2 (1916): 189-202; Kelly and Burrage, *Med. Biog.* (1928).

EIMBECK, WILLIAM (January 29, 1841, Brunswick, Germany—March 27, 1909, Washington, D.C.). *Geodesy.* Son of Frederick and Henrietta Eimbeck. Never married. EDUCATION: Attended public and private schools at Brunswick, and later attended the Polytechnical and Agricultural College. CAREER: 1857, came to U.S., settled at St. Louis, and became draftsman with locomotive builders; 1860-1869, civil engineer associated with various public projects in and around St. Louis; ca. 1867-1869, professor of engineering and practical astronomy, Washington Univ.; 1870, participated in U.S. Coast Survey eclipse expedition to southern Europe; 1871-1906, attached to engineering force, U.S. Coast and Geodetic Survey; 1871-1878, took part in astronomical and triangulation work in Midwest, on Pacific Coast (1872-1877), and in the East; 1878, undertook eighteen years of triangulation work from Nevada eastward. MEMBERSHIPS: Am. Assoc. Advt. Sci.; Natl. Geog. Soc.; and other societies. SCIENTIFIC CONTRIBUTIONS: During thirty-five years of service with Coast Survey, participated especially in triangulation work, and became well acquainted with various aspects of geodetic field work. Chief scientific labor began in 1878 with involvement in extension eastward from Nevada of primary scheme of triangulation, following approximately along thirty-ninth parallel and connecting on continental divide with second party. In 1885, presented plans for what came to be called duplex base apparatus, description of which appeared in superintendent of survey's report for 1897, along with article on its accurate use in measuring base along Great Salt Lake. Interested in wider aspects of geodesy, in 1900 commenced experiments on seasonal range in value of coefficient of refraction (never published); also studied detectable tides in earth's surface and other related questions, without completing plans for publication.

WORKS: See above. No papers listed in *Roy. Soc. Cat.*; paper on duplex base apparatus is only publication listed in *Natl. Un. Cat. Pre-1956 Imp.* WORKS ABOUT: *DAB*, 6: 64-65 (H. A. Marmer); Edwin Smith, memoir, *Science* n.s. 30 (1909): 48-50.

ELKIN, WILLIAM LEWIS (1855-1933). *Astronomy.* Director, Yale Univ. Observatory.

WORKS ABOUT: See *AMS*; *DAB*, supp.1; *DSB*.

ELLICOTT, ANDREW (January 24, 1754, Solebury Township, Pa.—August 28, 1820, West Point, N.Y.). *Surveying*; *Mathematics.* Son of Joseph, successful miller, and Judith (Bleaker) Ellicott. Married Sarah Brown, 1775; nine children. EDUCATION: Probably attended Quaker school in Solebury, Pa.; later studied in Philadelphia, including time with R. Patterson [q.v.]. HONORS: M.A., College of William and Mary, 1784. CAREER: Ca. 1775, joined Maryland militia, in course of the war attaining rank of major, and probably began work as surveyor about same time; 1784-1785, a representative for Virginia to extend Virginia-Pennsylvania boundary survey begun by Mason and Dixon; 1785, moved to Baltimore and taught mathematics at academy; 1785-1786, employed on Pennsylvania-Ohio boundary survey; 1786, one term in Maryland legislature; 1786-1787, survey of Pennsylvania-New York boundary; 1788, survey of islands in Ohio and Allegheny rivers; 1789, moved to Philadelphia; 1789-1790, survey of Pennsylvania-New York western boundary; 1791-1793, survey for District of Columbia; 1792, survey in western New York; 1793-1795, survey of road from Reading, Pa., to Presqu'Isle (Erie), and in 1794 appointed one of commissioners to lay out town of Erie; 1796-1800, survey of U.S.-Florida boundary; 1801-1808, secretary to Pennsylvania Land Office; 1811-1812, survey of Georgia-South Carolina boundary; 1813-1820, professor of mathematics, U.S. Military Academy, West Point; ca. 1817-1819, astronomer for U.S.-Canada boundary commission to effect terms of Treaty of Ghent. MEMBERSHIPS: Am. Phil. Soc. SCIENTIFIC CONTRIBUTIONS: Most important work in science was surveys outlined above, results of which remained permanently valid. Particularly remembered for work on survey of federal district, and after

L'Enfant was dismissed, redrew latter's plan for the city, but did not make original contribution to design. During 1789-1790 survey of Pennsylvania-New York boundary for federal government, carried out first topographical survey of Niagara River and measured Niagara Falls.

WORKS: Published *U.S. Almanac* during period of the Revolution; "Description of the Falls" (letter to Benjamin Rush), *European Magazine and London Review* (October 1793); sixteen articles on astronomical observations and surveying activities in *Transactions of Am. Phil. Soc.* 3-6 (1793-1809), and n.s. 1 (1818). See *Roy. Soc. Cat.*, vol. 2. MSS: Lib. Cong. (Hamer). WORKS ABOUT: *DAB*, 6: 89-90 (Harrison G. Dwight); Catharine V. C. Mathews, *Andrew Ellicott: His Life and Letters* (New York, 1908).

ELLIOT, DANIEL GIRAUD (1835-1915). *Zoology.* Curator of zoology, Field Museum, Chicago; a founder of Am. Ornith. Un.

MSS: Ac. Nat. Scis. Philad. (Hamer). WORKS ABOUT: See *AMS*; *DAB*.

ELLIOTT, EZEKIEL BROWN (July 16, 1823, Sweden, Monroe County, N.Y.—May 24, 1888, Washington, D.C.). *Statistics*; *Electricity.* Son of John Brown, physician, and Joanna Balch Elliott. Never married. EDUCATION: 1844, graduated, Hamilton College. CAREER: 1844-1849, taught in Michigan, New York, and Maine; 1849-1854, engaged in telegraphic work in Boston, associated with House printing telegraph lines between Boston, New York, and Albany; 1854, took up actuarial work for life insurance company and later for state of Massachusetts; 1861, became actuary with U.S. Sanitary Commission, Washington, D.C.; 1865, appointed secretary to commission for revision of U.S. revenue laws, and until 1870 remained as actuary in office of U.S. secretary of treasury; 1870, appointed chief clerk, U.S. Bureau of Statistics, and in 1878 transferred to Bureau of Mint; 1871-1875, member, U.S. Civil Service Commission; 1881-1888, held office of U.S. actuary. MEMBERSHIPS: Am. Ac. Arts Scis.; American Metrological Society; and others. SCIENTIFIC CONTRIBUTIONS: While engaged in telegraph work in Boston, produced several inventions, including dynamos and motors, and in 1853 received medal for invention of telegraph insulator. Early reputation for abilities as actuary gained in computation of life insurance tables and in work for U.S. Sanitary Commission; in 1863, went as delegate of American Statistical Association to International Statistical Congress at Berlin and there presented paper, "On the Military Statistics of the United States of America." During period 1867-1888, despite changing titles, was in effect an actuary in U.S. Treasury Department and prepared statistical works on number of topics, including mortality and birth, financial matters, coinage, congressional apportionment, population, and other subjects. Presented scientific papers before various societies, one of most significant being mathematical theory that made possible easier computation of life tables, presented in 1866 to Am. Assoc. Advt. Sci. and published as "Remarks Upon the Statistics of Mortality," *U.S. Bureau of the Census, Ninth Census, 1870*, vol. 2 (Washington, D.C., 1872), pp. ix-xvi.

WORKS: See above; also *Natl. Un. Cat. Pre-1956 Imp.* and *Roy. Soc. Cat.*, vols. 2, 7, 9 [see erratum slip], 14. WORKS ABOUT: *NCAB*, 2 (1899 [copyright 1891 and 1899]): 255; William Harkness, obituary, *Bulletin of Phil. Soc. Wash.* 11 (1892): 470-73.

ELLIOTT, STEPHEN (November 11, 1771, Beaufort, S.C.—March 28, 1830, Charleston, S.C.). *Botany.* Son of William, merchant, and Mary (Barnwell) Elliott. Married Esther Habersham, 1796; no information on children. EDUCATION: 1791, A.B., Yale College. HONORS: LL.D., Yale in 1819, Harvard in 1822, and Columbia in 1825; Elliott Society of Natural History (Charleston, S.C., 1853) was named for him. CAREER: 1792, became established as planter near Beaufort, S.C.; 1796, elected to South Carolina state legislature; 1808-1812, member of South Carolina state senate; 1812, moved to Charleston, S.C.; 1812-1830, president, Bank of the State of South Carolina; 1820, elected president of Univ. of South Carolina, but did not serve; 1824, appointed professor of natural history, Medical College of South Carolina, which he helped to establish, and served without pay; 1828, joined Hugh Swinton Legaré in establishing short-lived *Southern Review.* MEMBERSHIPS: A founder and president, 1814-1830, Literary and Philosophical Society of South Carolina. SCIENTIFIC CONTRIBUTIONS: While life was devoted to number of activities, is now remembered chiefly for botanical work and especially for *Sketch* [see Works], which was begun while residing on Beaufort plantation during period 1800-1808; *Sketch*, which began to appear in parts in 1816, added great many genera and species to those presented in T. Walter [q.v.], *Flora Caroliniana* (London, 1788), and also included number of botanical discoveries by W. Baldwin [q.v.] that might have been lost to view for many years because of Baldwin's early death. Also interested in insects and collected shells. During years 1828-1830, prepared number of contributions on wide range of subjects for *Southern Review.*

WORKS: *A Sketch of the Botany of South Carolina and Georgia*, 2 vols. (Charleston, 1816-1824; with facsimile reprint, New York, 1971, in 2 vols.

with introduction by Joseph Ewan and chronology of Elliott's life); "Observations on the Genus Glycine, and Some of Its Kindred Genera," *Journal of Ac. Nat. Scis. Philad.* 1 (1818): 320-26, 371-73. MSS: Charleston Museum; also correspondence with G. H. E. Mühlenberg [q.v.] at Harv. U.-GH and Hist. Soc. Penn. (Ewan [above]). WORKS ABOUT: *DAB*, 6: 99 (Arney C. Childs); also see Ewan [above].

ELLIS, JOB BICKNELL (1829-1905). *Botany; Mycology.* Teacher in Pennsylvania, New York, and Georgia; established residence in New Jersey; collector of fungi.

MSS: Harv. U.-FL (Hamer); Ac. Nat. Scis. Philad. (Hamer). WORKS ABOUT: See *AMS*; *DAB*.

EMERSON, BENJAMIN KENDALL (1843-1932). *Geology.* Professor of geology, Amherst and Smith colleges; geologist with U.S. Geol. Surv.

WORKS ABOUT: See *AMS*; *DAB*, supp. 1; *DSB*.

EMERSON, GEORGE BARRELL (September 12, 1797, Wells, Maine—March 4, 1881, Newton, Mass.). *Botany.* Son of Samuel, physician, and Sarah (Barrell) Emerson. Married Olivia Buckminster, 1823; Mrs. Mary (Rotch) Fleming, 1834; three children by first marriage. EDUCATION: 1817, A.B., Harvard College. HONORS: LL.D., Harvard Univ., 1859. CAREER: 1817-1819, principal of private school, Lancaster, Mass.; 1819-1821, tutor in mathematics, Harvard College; 1821, became principal of Boston English Classical School; 1823, established private school for girls, Boston; 1837, appointed chairman of commission for Zoological and Botanical Survey of Massachusetts; 1855, retired from teaching. MEMBERSHIPS: Am. Ac. Arts Scis.; president, Bos. Soc. Nat. Hist., 1837-1843. SCIENTIFIC CONTRIBUTIONS: Had interest in wide range of sciences, and in 1827 participated in foundation of Boston Mechanics' Institution. Came to concentrate scientific interests in botany; during period of presidency of Bos. Soc. Nat. Hist., Zoological and Botanical Survey of Massachusetts was established by legislature and Emerson assumed responsibility for study of trees and shrubs, devoting nine summers to botanical explorations of state; issued *Report* in 1846 [see Works] and in 1878 issued third edition in two volumes; *Report* was characterized by A. Gray [q.v.] as classic of New England botany and praised for its popular appeal as well as for its scientific reputation. At one time, reportedly was offered professorship in natural history and position as head of botanical garden at Harvard; was said to have been influential in obtaining bequest that founded Arnold Arboretum at Harvard. Interested in and published extensively on subject of education.

WORKS: *Report on the Trees and Shrubs Growing Naturally in the Forests of Massachusetts . . .* (Boston, 1846), 547 pp.; *Manual of Agriculture for the School, the Farm, and the Fireside* (Boston, 1862), 306 pp., with Charles L. Flint. MSS: Mass Hist. Soc. (Hamer). WORKS ABOUT: *DAB*, 6: 127-28 (Edward H. Jenkins); Robert C. Waterston, *Memoir of George Barrell Emerson . . .* (Cambridge, Mass., 1884).

EMERTON, JAMES HENRY (1847-1930). *Zoology; Arachnology.* Zoological illustrator; prepared zoological and anatomical models for museums and medical schools.

WORKS ABOUT: See *AMS*; *DAB*, supp. 1.

EMMET, JOHN PATTEN (April 8, 1796, Dublin, Ireland—August 15, 1842, New York, N.Y.). *Chemistry.* Son of Thomas Addis, lawyer and one-time physician, and Jane (Patten) Emmet. Married Mary B. F. Tucker, 1827; three children. EDUCATION: 1814-1817, attended U.S. Military Academy at West Point, but for reasons of health did not graduate; ca. 1818-1819, in Europe, studied languages and art; 1822, M.D., College of Physicians and Surgeons, New York. CAREER: While at West Point, appointed acting assistant professor of mathematics; 1822, began medical practice, Charleston, S.C.; 1824, offered series of public lectures on chemistry in Charleston; at Univ. of Virginia, professor of natural history, including zoology, botany, mineralogy, chemistry, and geology (1825-1827), professor of chemistry and materia medica (1827-1842). SCIENTIFIC CONTRIBUTIONS: Made no important original contributions to science, but during decade of 1830s prepared several papers, published especially in *Am. Journ. Sci.*, on chemical, pharmaceutical, and physical topics, including articles on magnetism and electricity. In 1834, began private pursuit of horticulture, experimenting with various fruits and flowers and raising of silkworms. About 1840, as result of observations of colored edge of shadow on white paper, performed investigations the outcome of which was said to contradict Newtonian refraction theory, but work was never published. With R. Dunglison [q.v.] performed analyses of gastric juice for study by W. Beaumont [q.v.] on physiology of digestion.

WORKS: Two representative papers are "The Iodide of Potassium . . . as a Test for Arsenic, with Remarks Upon the Nature and Properties of the Compound Formed," *Am. Journ. Sci.* 18 (1830): 58-63, and "An Inquiry Into the Cause of the Voltaic Currents Produced by the Action of Magnets and Electro-dynamic Cylinders Upon Coils and Revolving

Plates," ibid. 26 (1834): 23-44. Eight works listed in *Roy. Soc. Cat.*, vol. 2. MSS: Notebooks in Univ. of Virginia (see *Natl. Un. Cat. Pre-1956 Imp.*). WORKS ABOUT: Kelly and Burrage, *Med. Biog.* (1928); Thomas Addis Emmet, *The Emmet Family With . . . a Biographical Sketch of Prof. John Patten Emmet M.D. and Other Members* (New York, 1898), especially pp. 271-318.

EMMONS, EBENEZER (May 16, 1799, Middlefield, Mass.—October 1, 1863, Brunswick County, N.C.). *Geology*. Son of Ebenezer, farmer, and Mary (Mack) Emmons. Married Maria Cone, 1818; had children. EDUCATION: 1818, A.B., Williams College; 1826, graduated from Rensselaer Polytechnic Institute; took course at Berkshire Medical School. CAREER: Student of geology and assistant to A. Eaton [q.v.], Rensselaer Institute, and, ca. 1826, lecturer on chemistry, Albany Medical College; ca. 1827, began medical career in Chester, Mass., and continued practice, especially obstetrics, throughout life; 1828-1834, lecturer in chemistry, Williams College; 1830, appointed junior professor at Rensselaer Institute and lecturer at Castleton, Vt., medical school; 1833-1859, professor of natural history, Williams College; 1836-1842, geologist on New York Natural History Survey; 1838, appointed professor of chemistry, Albany Medical School, and moved to that city; 1842, appointed custodian of New York state collections at Albany and became involved in survey of state agricultural resources; 1851, appointed state geologist of North Carolina and soon thereafter moved south; 1859-1863, professor of geology and mineralogy, Williams College. MEMBERSHIPS: Association of American Geologists was planned at meeting in his Albany home, 1838; member, Am. Assoc. Advt. Sci. SCIENTIFIC CONTRIBUTIONS: Played a central role in geological survey of New York, out of which developed a system of geological classification that—together with Emmons's nomenclature based on geographic reference—strongly influenced future views of American geology. Advocacy of his so-called Taconic system, which referred to formations underlying the Potsdam, led to bitter controversy within geological community, with most important geologists opposed to Emmons's views. Work on geology of North Carolina has been judged as well executed.

WORKS: *Manual of Mineralogy and Geology* (Albany, 1826); most important work is the now rare report on *Geology . . . of the Second Geological District*, issued as part of division 4 of the *Natural History of New York* (Albany, 1842); also notable is report on *Agriculture of New York*, 5 vols., issued as part 5 of same series (Albany, 1846-1854) and including summary of New York geology; *American Geology*, 3 vols. (Albany, 1855-1857), was heavily criticized. Bibliography in Nickles, *Geol. Lit. on N. Am.* MSS: Personal papers apparently lost (*DSB*). WORKS ABOUT: *DAB*, 6: 147 (George P. Merrill); *DSB*, 4: 363-65 (Cecil J. Schneer).

EMMONS, SAMUEL FRANKLIN (1841-1911). *Geology; Mining engineering*. Geologist with U.S. Geological Survey of Fortieth Parallel and with U.S. Geol. Surv.

MSS: Lib. Cong. (Hamer). WORKS ABOUT: See *AMS*; *DAB*; *DSB*.

EMORY, WILLIAM HEMSLEY (September 7, 1811, Queen Annes County, Md.—December 1, 1887, Washington, D.C.). *Topographical engineering; Astronomy*. Son of Thomas and Anna Maria (Hemsley) Emory. Married Matilda Wilkins Bache, 1838; no information on children. EDUCATION: 1831, graduated, U.S. Military Academy, West Point. CAREER: 1831-1836, assigned to military duty at various stations; 1836-1838, resigned from army and engaged in civil engineering; 1838, appointed first lieutenant, Corps of Topographical Engineers; 1844-1846, principal assistant, northeastern boundary survey between U.S. and Canada; 1846-1848, duty during Mexican War included chief topographical engineer and acting adjutant general, Army of the West; 1848-1853, chief astronomer, boundary survey between California and Mexico; 1854-1857, commissioner and astronomer to survey boundaries outlined in Gadsden Treaty; 1855, became major of cavalry; during Civil War, served first in frontier posts and in 1862 appointed brigadier general of volunteers; 1866-1876, various military commands; 1876, retired as brigadier general. SCIENTIFIC CONTRIBUTIONS: Chief scientific labor done in West during period between Mexican War and Civil War; in *Notes* [see Works] and other publications gave earliest reliable scientific account of Southwest. In addition to accurate mapping of geographical points, also compiled much useful astronomical, meteorological, and ethnographical data and collected botanical and other natural history specimens. Besides *Notes*, report on Mexican boundary survey [see Works], and several scientific papers, also published *Observations, Astronomical, Magnetic and Meteorological* (1850), made in Panama.

WORKS: *Notes of a Military Reconnaissance from Fort Leavenworth in Missouri, to San Diego in California . . . Made in 1846-1847*, U.S. Senate document (Washington, D.C., 1848; reprinted, Albuquerque, N. Mex., 1951, minus scientific data); *Report on the U.S. and Mexican Boundary Survey*, U.S. House Document (Washington, D.C., 1857-1859). See also *Roy. Soc. Cat.*, vol. 2. MSS: Yale U.-WAC (Hamer). WORKS ABOUT: *DAB*, 6: 153-54 (Charles F. Carey); Cullum, *Biographical Register, West Point*, 1 (1891): 481-83.

ENGELMANN, GEORGE (February 2, 1809, Frankfurt am Main, Germany—February 4, 1884, St. Louis, Mo.). *Botany*; *Meteorology*. Son of Julius Bernhardt [in some sources given as George], schoolmaster, and Julia (May) Engelmann; brother of Henry Engelmann [q.v.]. Married Dorothea Horstmann, 1840; at least one child. EDUCATION: 1827-1828, attended Univ. of Heidelberg; studied two years at Univ. of Berlin; 1831, M.D., Univ. of Würzburg. CAREER: 1832, went to Paris and then to U.S.; 1833, arrived in St. Louis; 1833-1835, lived on farm in Illinois (near St. Louis), prospected, explored natural history, and made journey to Southwest; 1835, established successful medical practice in St. Louis and continued to pursue botanical interests; after settling in St. Louis, also was active in German-American concerns, including publication of periodical and daily newspaper; 1869, terminated most active medical practice and turned attentions more fully to other activities. MEMBERSHIPS: Natl. Ac. Scis. and other societies; established St. Louis Academy of Science (1856). SCIENTIFIC CONTRIBUTIONS: Made early botanical collecting trips to Louisiana (1835) and to Arkansas (1837), but for some time thereafter large medical practice prevented extensive field work; over period of years, had number of persons who collected western flora for him, including friend Ferdinand Lindheimer [q.v.], who began as early as 1839. In 1842, began cooperation with A. Gray [q.v.] in collection and study of American flora of the West, Engelmann becoming focal point for activities of collectors and movement of specimens from west to east. 1857, commissioned by Henry Shaw, St. Louis businessman, to gather books and herbaria specimens in Europe for collection that was foundation of Missouri Botanical Garden. Formed first full herbarium of American grapes and 1861 paper on grape diseases was important early contribution to plant pathology in U.S.; wrote important works on oaks, grapes, rushes, cacti, euphorbias, and conifers; also conducted zoological and anatomical investigations and made meteorological observations from 1836 to 1884.

WORKS: *Botanical Works of the Late George Engelmann Collected for Henry Shaw* (Cambridge, Mass., 1887); also see bibliography of writings in *Missouri Historical Review* 23 (1928-1929): 189-95. MSS: Mo. Bot. Gard. (Hamer); also Harv. U.-GH and N.Y. Bot. Gard. (Hamer). WORKS ABOUT: *DAB*, 6: 159-60 (George T. Moore); Richard G. Beidelman, "George Engelmann, Botanical Gatekeeper of the West," *Horticulture* 48 (April 1970): 42.

ENGELMANN, HENRY (October 1, 1831, Frankfurt am Main, Germany—March 30, 1899, LaSalle, Ill.). *Geology*. Son of Julius Bernhardt Engelmann [in some sources given as George], schoolmaster; brother of George Engelmann [q.v.]. No information on marital status. EDUCATION: Graduated from Gymnasium of Creuznach and Univ. of Berlin; studied at mining school at Freiberg, Saxony; for one year before coming to U.S., visited and studied mines and metallurgical works. CAREER: 1856, came to U.S. and for a time resided in St. Louis; 1856, geologist with Lt. F. T. Bryan expedition from Kansas to Utah; 1857, assistant to G. C. Swallow [q.v.], Geological Survey of Missouri; 1858-1859, geologist, meteorologist, and botanical collector on Utah expedition led by Capt. J. H. Simpson; ca. 1861, began several years service as assistant geologist, State Geological Survey of Illinois; 1869-1899, superintendent of ore department and general consulting engineer, Matthiesen & Hegeler Zinc Co., LaSalle, Ill., and spent two years during this period in Utah engaged in study of coal deposits and manufacturing coke. MEMBERSHIPS: Am. Inst. Min. Engrs. SCIENTIFIC CONTRIBUTIONS: Chief geological work grew out of explorations with Bryan and Simpson expeditions, the latter across Great Basin in Utah. Results of these studies, as belatedly published in 1876, brought together geological picture of region from Sierra Nevadas across Great Basin and eastward to Mississippi and, in association with work of J. S. Newberry [q.v.] on stratigraphy of Grand Canyon and that of F. V. Hayden and F. B. Meek [qq.v.] in Kansas and Dakotas, gave geological overview of large portion of West. Also collected botanical specimens that were described by brother George and located vertebrate fossil sites later exploited by O. C. Marsh [q.v.]. Work for Illinois survey chiefly stratigraphical. Later took important role in development of American zinc-smelting industry.

WORKS: Appendixes E, I, and M of James H. Simpson, *Report of Explorations Across the Great Basin of the Territory of Utah . . .* (Washington, D.C., 1876). See also *Roy. Soc. Cat.*, vols. 2, 7, 14, and Meisel, *Bibliog. Amer. Nat. Hist.* WORKS ABOUT: Memoir, *Transactions of Am. Inst. Min. Engrs.* 30 (1900); Goetzmann, *Army Exploration*, pp. 391-423, passim.

ESPY, JAMES POLLARD (May 9, 1785, Washington County, Pa.—January 24, 1860, Cincinnati, Ohio). *Meteorology*. Son of Josiah, pioneer who moved from one location to another and probably farmer, and Elizabeth (Patterson) Espy. Married Margaret Pollard, 1812; no children. EDUCATION: Early years in Kentucky, where taken as infant; 1808, A.B., Transylvania Univ. (Lexington, Ky.); studied law. HONORS: Magellanic Prize, Am. Phil. Soc., 1836. CAREER: Ca. 1808, began teaching school in Xenia, Ohio, and studied law; 1812-1817, principal, academy at Cumberland, Md.; 1817-1836, taught mathematics and classics in Philadelphia, including association with Franklin Institute; 1836, gave up teaching to devote full efforts to study and lectures on meteorology; 1842-1859, appointed by Congress as U.S. meteorologist, with connections first with U.S. War Department; 1848, became associated also with U.S. Navy Department, and after 1852 was connected with Smithson. Instn.; 1859-1860, visited in Pennsylvania and Ohio. MEMBERSHIPS: Am. Phil. Soc.; Franklin Institute. SCIENTIFIC CONTRIBUTIONS: Work in meteorology begun about 1825 while teaching at Franklin Institute; developed theory of storms later promoted in lectures, including presentation at 1840 meeting of British Association for Advancement of Science at Glasgow and communication to French Academy of Sciences. In 1836, established system of meteorological observations in Pennsylvania, and later worked to effect national network of weather observers, subsequently connected by telegraph and making possible beginnings of weather forecasting. Unlike other Americans working in meteorology at that time, Espy engaged in experiments, especially on heat effects; involved in controversy with W. C. Redfield [q.v.] on movement of storms, and Espy's views eventually were proven incorrect; but his experimental and theoretical work that related to explanation of precipitation in terms of rising and subsequent expansion and cooling of moist air was important contribution.

WORKS: *Philosophy of Storms* (Boston, 1841); *Report on Meteorology,* 4 vols. (Washington, D.C., 1843-1857) [see *DSB* for bibliographic details]). Also see *Roy. Soc. Cat.,* vol. 2. MSS: Am. Phil. Soc. and Franklin Institute (also see additional references in *DSB*). WORKS ABOUT: *DAB,* 6: 185-86 (William J. Humphreys); *DSB,* 4: 410-11 (Nathan Reingold).

EUSTIS, HENRY LAWRENCE (February 1, 1819, Fort Independence, Boston, Mass.—January 11, 1885, Cambridge, Mass.). *Engineering.* Son of Brig. Gen. Abraham, soldier, and Rebecca (Sprague) Eustis. Married Sarah Augusta Eckley, 1844; Caroline Bartlett Hall, 1856; four children by first marriage, two by second. EDUCATION: 1838, A.B., Harvard College; 1842, graduated, U.S. Military Academy, West Point. CAREER: 1842, commissioned second lieutenant, Corps of Engineers; 1843-1845, assistant engineer, construction of Fort Warren and Lovell's Island sea wall, Boston harbor; 1845-1847, in charge of engineering operations, Newport, R.I.; 1847-1849, assistant professor of engineering, West Point; at Lawrence Scientific School, Harvard Univ., professor of engineering (1849-1885) and dean (1862-1863 and 1871-1885); 1862-1864, served with Massachusetts Volunteers, and achieved rank of brigadier general of volunteers. MEMBERSHIPS: Am. Ac. Arts Scis. and other societies. SCIENTIFIC CONTRIBUTIONS: Said to have published number of articles on scientific and technical matters, but specific bibliographic references have not been located. Prepared paper on tornado in Boston area, 1851 [see Works], and reportedly presented paper in 1860 to Am. Assoc. Advt. Sci. on hurricane that passed through vicinity of Boston. Particularly noted as teacher, and in that capacity strongly influenced number of successful scientists and engineers.

WORKS: "The Tornado of August 22d, 1851, in Waltham, West Cambridge, and Medford, Middlesex County, Mass.," with map, *Memoirs of Am. Ac. Arts Scis.* n.s. 5 (1853): 169-78. WORKS ABOUT: *DAB,* 6: 192-93 (Claude M. Fuess); memoir, *Proceedings of Am. Ac. Arts Scis.* 20 (1884-1885): 513-19.

EVANS, JOHN (February 14, 1812, Portsmouth, N.H.—April 13, 1861, Washington, D.C.). *Geology.* Son of Richard Evans, associate justice of New Hampshire supreme court. Married Sarah Zane Mills, 1835; four children. EDUCATION: Attended schools at Andover, Mass.; M.D., St. Louis Univ. Medical Department. HONORS: Honorary degree, St. Louis Univ. Medical Dept., 1854. CAREER: For several years, pursued mercantile interests in Massachusetts; ca. 1831-1839, clerk in post office department, Washington, D.C.; 1839, moved to St. Louis and there pursued medical studies; 1847, chosen as assistant to D. D. Owen [q.v.] in U.S. geological survey of Wisconsin, Iowa, Minnesota, and Nebraska; 1851-1852, engaged in geological survey in Oregon for U.S. Department of Interior; 1852, returned to Washington, D.C.; 1853, appointed geologist of railroad survey along northern route from Minnesota to Washington territory; remained in Washington and Oregon territories and continued earlier geological survey until 1856, and then returned to Washington, D.C.; 1860. U.S. geologist with Chiriqui exploration expedition to Isthmus of Panama. MEMBERSHIPS: Bos. Soc. Nat. Hist.; Ac. Nat. Scis. Philad.; and other societies. SCIENTIFIC CONTRIBUTIONS: In work with Owen survey, was first scientist to explore fossil remains in Badlands of South Dakota, and gathered fossil bones later described by J. Leidy [q.v.], as well as Cretaceous fossils written up by Owen and B. F. Shumard [q.v.]. Official instructions for explorations in Oregon were based largely on Evans's own ideas and called for a geological and mineralogical survey and barometric determination of mountain elevations, as well as geographical determination for certain localities; details of survey were forwarded in reports to General Land Office; also sent geological samples, and in 1858 B. F. Shumard published paper on the fossils collected; reports on geology of northern railroad route were lost before publication, but

Evans and Shumard published two papers on fossil collections (1854). During period after completion of surveys of Washington and Oregon in 1856, worked on report for Congress, encompassing the region's geology, topography, geography, and natural history, that probably was Evans's most important work, but was not published and now is lost. Geological report on Chiriqui expedition published as U.S. House Document (1861).

WORKS: References to publications and to manuscripts are given in text and footnotes to Lange [below]. WORKS ABOUT: *Lamb's Biographical Dictionary of the U.S.* (Boston, 1900), vol. 3: 10-11; Erwin F. Lange, "Dr. John Evans . . . ," *Proceedings of Am. Phil. Soc.* 103 (1959) : 476-84.

EWELL, ERVIN EDGAR (October 22, 1867, Washington, Mich.—February 7, 1904, New Orleans, La.). *Chemistry.* Son of Samuel D. E. Ewell. Married Alice Priest, 1896; survived by one child. EDUCATION: Ph.C. in 1888 and B.S. in pharmacy in 1900, College of Pharmacy, Univ. of Michigan; 1888-1889, attended College of Literature, Science, and Arts at Univ. of Michigan. CAREER: 1888-1890, assistant in qualitative analysis, Univ. of Michigan; 1890-1893, assistant chemist, Bureau of Chemistry, U.S. Department of Agriculture; 1893-1894, chemist, Magnolia Sugar and Railroad Co., Lawrence, La.; for Bureau of Chemistry, U.S. Department of Agriculture, assistant chemist (1894-1897) and assistant chief (1897-1903); 1903-1904, manager of Atlanta, Ga., office of German Kali Works. MEMBERSHIPS: Am. Chem. Soc. SCIENTIFIC CONTRIBUTIONS: In Bureau of Chemistry, was given charge of greatly varied work done for other departments of government; personal research interests related especially to physiology and chemisty; and despite handicaps imposed by official duties, was able to produce number of worthy research papers. One of most significant investigations dealt with intoxicating qualities of mescal button, a cactus used by Indians in religious ceremonies. Keenly interested in standards of measurement, and through chairmanship of committee of Am. Chem. Soc. and other means, played role in formation of U.S. Bureau of Standards. Work for German Kali Works related to question of importance of potash salts as crop nutriment.

WORKS: See *Roy. Soc. Cat.*, vol. 14; *Natl. Un. Cat. Pre-1956 Imp.* WORKS ABOUT: Obituary, *Science* n.s. 19 (May 6, 1904) : 741; *Who Was Who in America,* vol. 1; University of Michigan, *Catalogue of Graduates, Non-Graduates, Officers, and Members of the Faculties* (Ann Arbor, Mich., 1923) ; obituary from unidentified alumni publication, including evaluation by Harvey W. Wiley [q.v.] of work done in Bureau of Chemistry (photocopy of clipping provided by Alumni Records Office, Univ. of Michigan).

F

FAIRCHILD, HERMAN LEROY (1850-1943). *Geology.* Professor of geology, Univ. of Rochester.
MSS: Univ. of Rochester Library (*NUCMC* 62-4385). WORKS ABOUT: See *AMS; NCAB,* 33 (1947) : 291-92.

FARLOW, WILLIAM GILSON (1844-1919). *Cryptogamic botany.* Professor of cryptogamic botany, Harvard Univ.
MSS: Harv. U.-FL (Hamer). WORKS ABOUT: See *AMS*; *DAB.*

FARRAR, JOHN (July 1, 1779, Lincoln, Mass.—May 8, 1853, Cambridge, Mass.). *Mathematics; Physics; Astronomy.* Son of Deacon Samuel and Mary (Hoar) Farrar. Married Lucy Maria Buckminster; Eliza Ware Rotch; no information on childen. EDUCATION: 1803, A.B., Harvard College; ca. 1803-1805, student at Andover Theological Seminary. HONORS: LL.D., Brown Univ., 1833. CAREER: 1805-1807, tutor in Greek, Harvard College; 1807-1836, Hollis professor of mathematics and natural philosophy, Harvard. MEMBERSHIPS: Am. Ac. Arts Scis. SCIENTIFIC CONTRIBUTIONS: Noted for reform and update of science and mathematics curriculum at Harvard; made available to American students for first time translations of important French and other European works on mathematics, electricity, astronomy, and mechanics; credited with change in America from Newtonian fluxional notations to Leibniz's algorithms for the calculus. Translations during years 1818-1829 constituted two series, the Cambridge Mathematics and the Cambridge Natural Philosophy; carefully chosen, adapted, and arranged to be of greatest benefit to American students, they were used not only at Harvard, but at other colleges as well. Published little that was original, and most of such works were meteorological and astronomical observations or were related to instrumentation and technique.
WORKS: Among works translated or otherwise made available to American students were Euler's and Lacroix's works on algebra (1818 and 1825), the spherical trigonometry of Lacroix and Bezout (1820), works on electricity (1826) and optics (1826) from Biot's work on elementary physics, and Biot's work on astronomy; also prepared *An Elementary Treatise on the Application of Trigonometry to Projection, Mensuration, Navigation, and Surveying* (Cambridge, Mass., 1822), and *Elementary Treatise on Mechanics* (Cambridge, Mass., 1825). See bibliography in *DSB.* MSS: Harv. U-UA (*NUCMC* 65-1237) ; also Mass. Hist. Soc. and Boston Public Library (*DSB*). WORKS ABOUT: *DAB,* 6: 292-93 (David E. Smith) ; *DSB,* 4: 546-47 (Brooke Hindle).

FERGUSON, JAMES (August 31, 1797, Perthside, Scotland [ca. 1800, brought to U.S., early life spent in Albany, N.Y.]—September 26, 1867, Washington, D.C.). *Astronomy.* Parentage, marital status, and education not known. HONORS: Astronomical prize medals, French Academy of Sciences, 1854 and 1860. CAREER: 1817-1819, assistant civil engineer, Erie Canal; for Northwestern Boundary Survey under Treaty of Ghent, assistant surveyor (1819-1822) and astronomical surveyor (1822-1827) ; 1827-1832, civil engineer for Pennsylvania; 1833-1847, first assistant in U.S. Coast Survey; 1848-1867, assistant astronomer, U.S. Naval Observatory. MEMBERSHIPS: Albany Institute. SCIENTIFIC CONTRIBUTIONS: During years that F. R. Hassler [q.v.] directed Coast Survey, held number two position, but after A. D. Bache [q.v.] took command in 1843, forced out on grounds that work on Chesapeake and Delaware bays was incorrect, although rivalry with Bache also played part in removal. At Naval Observatory had charge of equatorial instrument, engaged particularly in observation of asteroids, and discovered Euphrosyne, Virginia, and Echo. Total number of scientific publications surpassed those of any other antebellum astronomer, although writings were mainly reports on observations, with little mathematical or theoretical content. Was at one time considered as candidate to succeed C. H. Davis [q.v.] as head of Naval Observatory.

WORKS: Published nearly ninety papers, chiefly in Gould's *Astronomical Journal* and in *Astronomische Nachrichten*; influential article on science and politics of Coast Survey, written by Ferguson but published unsigned, appeared in *North American Review* (April 1842): 446-57. See *Roy. Soc. Cat.*, vols. 2, 9. MSS: No collection of papers is known; see references to several items in Reingold [below]. WORKS ABOUT: *Appleton's CAB*, 2: 433; B. T. Sands, "The Late Mr. James Ferguson . . . ," *Astronomische Nachrichten* 71 (1868): 101-2 (short sketch, in English); Nathan Reingold, ed., *The Papers of Joseph Henry* (Washington, D.C., 1975), 2: 15-16.

FERNALD, CHARLES HENRY (1838-1921). *Entomology.* Professor of zoology, Massachusetts Agricultural College (Univ. of Massachusetts).

WORKS ABOUT: See *AMS; DAB.*

FERREL, WILLIAM (January 29, 1817, Fulton County, Pa.—September 18, 1891, Maywood, Kans.). *Meteorology.* Son of Benjamin Ferrel, farmer and sawmill operator, and his wife, a Miss Miller. Never married. EDUCATION: 1839-1841, attended Marshall College (Pennsylvania); 1844, graduated, Bethany College (West Virginia). HONORS: Said to have received honorary A.M. and Ph.D. degrees. CAREER: 1844-1846, teacher, Liberty, Mo., having taught elsewhere both before and between years in college; 1846-1854, teacher, Allensville, Todd County, Ky.; 1854-1858, operated private school in Nashville, Tenn.; 1857 and 1858-1867, employed in *Nautical Almanac* office, Cambridge, Mass.; 1867, began work with Coast and Geodetic Survey; 1882-1886, worked with U.S. Signal Service; 1886-1891, lived with family in Kansas, and spent 1889-1890 in Martinsburg, W.Va. MEMBERSHIPS: Natl. Ac. Scis. and other societies; honorary member, Royal Meteorological Society. SCIENTIFIC CONTRIBUTIONS: Began scientific work about 1850 with study of tidal theory and was first to view quantitatively the question of tidal friction and its effect on earth's rate of rotation; while with Coast Survey, made contributions to means for prediction of tides. Most significant work related to effects of earth's rotation on movements of bodies near its surface, especially in regard to movement of winds and ocean currents, and most important publication was "The Motions of Fluids" [see Works]; gave nonmathematical version of work in 1856 "Essay" [see Works], and later developed mathematical detail, presenting Ferrel's law: "if a body is moving in any direction, there is a force, arising from the earth's rotation, which always deflects it to the right in the northern hemisphere, and to the left in the southern." Outcome of work was to make meteorology part of mathematical physics, and with Laplace was founder of so-called geophysical fluid dynamics.

WORKS: "Essay on the Winds and Currents of the Ocean," *Nashville Journal of Medicine and Surgery* (October 1856); "The Motions of Fluids and Solids Relative to the Earth's Surface," *Mathematical Monthly* 1 (1859): 140-48, 210-16, 300-307, 366-73, 397-406, and 2 (1860): 89-97, 339-46, 374-90, reprinted as *Professional Papers of U.S. Signal Service*, no. 8 (Washington, D.C., 1882). Major works mentioned in *DSB*; full bibliography in *Biog. Mems. of Natl. Ac. Scis.* 3 (1895): 300-309. MSS: Duke Univ. Libraries (Hamer); autobiography in Harv. U.-HL (*DSB*). WORKS ABOUT: *DAB*, 6: 338 (William J. Humphreys); *DSB*, 4: 590-93 (Harold L. Burstyn).

FINE, HENRY BURCHARD (1858-1928). *Mathematics.* Professor of mathematics and dean, Princeton Univ.

WORKS ABOUT: See *AMS; DAB; DSB.*

FITCH, ASA (February 24, 1809, Salem, N.Y.—April 8, 1879, Salem, N.Y.). *Entomology.* Son of Asa, physician and farmer, and Abigail (Martin) Fitch. Married Elizabeth McNeil, 1832; at least one child. EDUCATION: 1827, graduated, Rensselaer School (later Rensselaer Polytechnic Institute); attended medical schools in New York City and Albany; 1829, M.D., Vermont Academy of Medicine, Castleton, Vt. CAREER: 1830, assistant professor of natural history, Rensselaer School; 1830-1831 (winter), medical practice and scientific activities at Greenville, Ill.; 1831, established medical practice at Fort Miller, N.Y.; 1838, gave up medical practice, returned to Salem, N.Y., and engaged in agricultural and scientific activities; ca. 1845, engaged to collect and identify native insects for New York State Cabinet of Natural History; 1854-1870, state entomologist of New York. SCIENTIFIC CONTRIBUTIONS: Interests in insects and in agriculture led to investigations of insect pests, and beginning in 1845 made helpful recommendations to farmers based on personal study of insect life histories in relation to agricultural crops; one notable investigation was the study of grain insects. Service as New York entomologist was first such recognition by any of the states of the importance of entomology for agriculture; annual reports as state entomologist became standard sources, and influence was spread as well through wide correspondence.

WORKS: Prepared fourteen "Reports on the Noxious, Beneficial, and Other Insects of the State of New York," *Transactions of New York State Agricultural Society* 14-30 (1855-1872). Bibliography in

J. A. Lintner, *First Annual Report on the Injurious and Other Insects of the State of New York* (Albany, 1882). MSS: Corn. U.-CRH (*NUCMC* 62-4215); Smithson. Instn. Archives (*NUCMC* 74-975); Yale U.-HMC (*NUCMC* 74-1190). WORKS ABOUT: *DAB,* 6: 424 (Leland O. Howard); *DSB,* 5: 11-12 (Samuel Rezneck).

FITZ, HENRY (December 31, 1808, Newburyport, Mass.—October 31, 1863, New York, N.Y.). *Telescope construction.* Son of Henry, hatter, and Susan Bradley (Page) Fitz. Married Julia Ann Wells, 1844; survived by six children. EDUCATION: No information regarding formal education; learned printing, and then pursued locksmith's trade in shop of William Day, New York City. CAREER: For a time worked as printer; ca. 1827, gave up printing and began work as locksmith; 1830-1839, traveled between New York, Philadelphia, Baltimore, and New Orleans, as practicing locksmith; 1839, went to Europe to learn about Daguerre's photographic procedures and about optics and optical glass; 1840-ca. 1845, operated photographic studio in Baltimore; 1845, moved to New York and turned full attentions to making telescopes. MEMBERSHIPS: A founder, American Photographical Society. SCIENTIFIC CONTRIBUTIONS: His first telescope, made in 1838, was a reflector; after settling in Baltimore began to make refractors, and in meantime (1839) made what was thought to be first camera portrait. In 1845, won gold medal at fair of American Institute for six-inch refracting telescope, and this success attracted several astronomers. Made six-inch telescope that Lt. J. M. Gilliss [q.v.] took to Chile, several telescopes for L. M. Rutherfurd [q.v.], telescopes for various other persons and institutions, including Vassar College, Univ. of Michigan, and Dudley Observatory (Albany), and produced sixteen-inch instrument for a Mr. Van Duzee of Buffalo; several of his productions were used in important astronomical work. Invented process for correcting curves of objective glass; reportedly used local polishing some fifteen years prior to its development in Europe, and constantly worked to perfect lenses. Began work on telescopes somewhat before A. Clark [q.v.], but early death prevented full development of potential.

WORKS: No works are listed in either *Roy. Soc. Cat.* or *Natl. Un. Cat. Pre-1956 Imp.* MSS: Papers and shop in Smithson. Museum of History and Technology (Multhauf [below], which also includes references to manuscripts in private hands). WORKS ABOUT: *DAB,* 6: 433 (George H. Genzmer); "Holcomb, Fitz, and Peate: Three Nineteeth-Century American Telescope Makers," introduction by Robert P. Multhauf, Contributions from Museum of History and Technology, paper 26, *Bulletin of U.S. Natl. Mus.,* no. 228 (1962), pp. 156-83.

FLEMING, WILLIAMINA PATON STEVENS (1857-1911). *Astronomy.* Curator of astronomical photographs, Harvard College Observatory.

WORKS ABOUT: See *AMS; DAB; DSB; Nota. Am. Wom.*

FLINT, ALBERT STOWELL (1853-1923). *Astronomy.* Astronomer, Washburn Observatory, Univ. of Wisconsin.

WORKS ABOUT: See *AMS; DAB.*

FOLGER, WALTER (June 12, 1765, Nantucket, Mass.—September 8, 1849, Nantucket, Mass.). *Mathematics; Astronomy; Invention and scientific apparatus.* Son of Walter, successful whale oil merchant and candle maker, and Elizabeth (Starbuck) Folger. Married Anna Ray, 1785; ten children. EDUCATION: Formal education did not proceed beyond elementary levels; self-taught in science, mathematics, law, and medicine. CAREER: Lived entire life on Nantucket; worked with father in whale oil business, and became known as expert watch and clock maker; for number of years, practiced law; 1808, elected for one term to Massachusetts General Court; 1808-1814, chief justice of Court of Common Pleas (Nantucket); 1809-1814 and 1822, served in Massachusetts senate; 1817-1821, served in U.S. Congress; at times worked as land surveyor, taught navigation and nautical astronomy to shipmasters and others, participated in various business enterprises, and held several local political offices. MEMBERSHIPS: President, Nantucket Philosophical Institute, 1826-1832. SCIENTIFIC CONTRIBUTIONS: Showed natural mechanical ability, and in 1790 completed astronomical clock of unique design, which included time, day and year, as well as rising and setting of sun and moon, the movements and positions of which were demonstrated by movable balls. Calculated and published almanac for 1790 and 1791; made several telescopes, including casting of speculum (mirror) for instrument completed about 1821, which at time was one of best telescopes in U.S.; observed comet of 1811. About 1823, made thermometers and, 1827-1848, kept record of temperature, barometer, winds, and weather. An accomplished chemist, was knowledgeable concerning coloring, and in 1812 discovered process for annealing wire, antedating by some years announcement of similar procedure in Europe. Said to have published in newspapers and elsewhere on mathematical problems and solutions.

WORKS: "A Topographical Description of Nantucket [May 21, 1791]," *Collections of Mass. Hist. Soc.* 3 (1794): 153-55; "Observations of the Solar Eclipse of 17 Sept. 1811, made at Nantucket," *Memoirs of Am. Ac. Arts Scis.* 3, pt. 2 (1815): 252-54. MSS: Small collection in Peter Foulger Museum (Nantucket). WORKS ABOUT: *DAB,* 6: 489 (H.

W. Howard Knott); Will Gardner, *The Clock That Talks and What It Tells: A Portrait Story of the Maker, Walter Folger, Jr.* (Nantucket, Mass., 1954).

FONTAINE, WILLIAM MORRIS (1835-1913). *Paleobotany.* Professor of natural history and geology, Univ. of Virginia.

WORKS ABOUT: See *AMS; NCAB,* 19 (1926): 187-88.

FORBES, STEPHEN ALFRED (1844-1930). *Zoology; Entomology.* Professor of zoology and entomology and dean of science, Univ. of Illinois; state entomologist of Illinois.

MSS: Univ. of Illinois—Illinois Historical Survey (*NUCMC* 62-1221 and 69-1592); Univ. of Illinois Archives (*NUCMC* 65-1919). WORKS ABOUT: See *AMS,* 2d ed. (1910); *DAB; DSB.*

FOSTER, JOHN WELLS (March 4, 1815, Petersham, Mass.—June 28 [?], 1873, Chicago, Ill.). *Geology.* Son of Festus, Unitarian minister, and Patience Foster. Married Lydia Conerse [Converse?], 1838; no information on children. EDUCATION: 1834, completed scientific course at Wesleyan Univ.; 1835, admitted to bar. CAREER: Ca. 1834, went to Zanesville, Ohio, and in 1835 began law practice there; 1837-1839, assistant on Ohio Geological Survey; 1839, resumed law practice at Zanesville; 1844, returned to Massachusetts and engaged in civil engineering; 1845-1847, with W. W. Mather [q.v.], explored Point Kewenaw, Mich., for several copper companies; 1847, with J. D. Whitney [q.v.], appointed to assist C. T. Jackson [q.v.] on U.S. geological survey of Lake Superior region, completion of which later was delegated to Foster and Whitney; 1852, returned again to Massachusetts, engaged in politics, and became an organizer of Native American party; later helped found Republican party in Massachusetts and served on governor's executive council; 1858, returned to the West, settling in Chicago; 1858-1863, land commissioner for Illinois Central Railway; 1863, resigned and devoted time to scientific work and for some two years lectured at Chicago Univ. MEMBERSHIPS: President, Am. Assoc. Advt. Sci., 1869; president, Chicago Academy of Sciences. SCIENTIFIC CONTRIBUTIONS: While connected with Ohio survey, prepared report on great central coal beds of that state; reports on Michigan survey [see Works], prepared jointly with J. D. Whitney, were first authoritative works on copper and iron deposits of that region; in conjunction with work for Illinois Central Railway, investigated coal fields of Indiana and Illinois. Also interested in mound builders in the West.

WORKS: With Whitney, *Report on the Geology and Topography of a Portion of the Lake Superior Land District . . . ,* pt. 1, *Copper Lands,* and pt. 2, *The Iron Region Together With the General Geology* (Washington, D.C., 1850 and 1852), both published by order of U.S. Congress; *Prehistoric Races of the U.S.A.* (Chicago and London, 1873). See *Roy. Soc. Cat.,* vols 2, 7, and *Natl. Un. Cat. Pre-1956 Imp.* WORKS ABOUT: *NCAB,* 10 (1909 [copyright 1900]): 169-70; Charles Whittlesey, "Personnel of the First Geological Survey of Ohio," *Magazine of Western History* 2 (1885): 85-87; biographical sketch, *Science* n.s. 10 (November 17, 1899): 707.

FOWLER, LORENZO NILES (June 23, 1811, Cohocton, N.Y.—September 2, 1896, West Orange, N.J.). *Phrenology.* Son of Deacon Horace, farmer and pioneer, and Martha (Howe) Fowler; brother of Orson S. Fowler [q.v.]. Married Lydia Folger, 1844; three children. EDUCATION: Attended Amherst (Mass.) Academy. CAREER: At early age, joined brother Orson as itinerant lecturer and practitioner of phrenology; 1836, established phrenological office in New York City and in 1845, with Orson and S. R. Wells [q.v.], created firm of Fowlers and Wells; at one time, lecturer on phrenology and mental science in New York. Hydropathic and Physiological School (founded in 1853); 1858-1860, with wife, Lydia, and with S. R. Wells conducted extensive phrenological lecture tour throughout U.S. and Canada; 1860-1862, lectured in England; 1863-1896, lived in London, there engaged in lecturing and practicing phrenology; 1880, established *Phrenological Magazine* in England; died on visit to U.S. SCIENTIFIC CONTRIBUTIONS: Attracted to phrenology through influence of brother Orson, and won wide fame as lecturer and practitioner; described as straightforward and practical in presentation and analysis, and promoted practical ethics. Not as prolific as Orson, but nonetheless produced several important and widely read books on phrenology and was actively interested in number of reform movements. Wife, Lydia Folger Fowler, in 1850 became second woman in U.S. to earn M.D. degree and took active part in phrenological work of her husband, as well as teaching medicine; daughter, Jessie Allen Fowler, continued father's phrenological work in England.

WORKS: *Principles of Phrenology and Physiology Applied to Man's Social Relations* (New York, 1842); *Marriage: Its History and Ceremonies* (New York, 1847); *Lectures on Man* (London, 1864). See *Natl. Un. Cat. Pre-1956 Imp.* MSS: Fowler Family Papers at Corn. U.-CRH (Stern [below]). WORKS ABOUT: Madeleine B. Stern, *Heads & Headlines: The Phrenological Fowlers* (Norman, Okla., 1971); *Appleton's CAB,* 2:518, and 7:304 (list of deaths).

FOWLER, ORSON SQUIRE (October 11, 1809, Cohocton, N.Y.—August 18, 1887, Sharon Station,

Dutchess County, N.Y.). *Phrenology*. Son of Deacon Horace, farmer and pioneer, and Martha (Howe) Fowler; brother of Lorenzo N. Fowler [q.v.]. Married Mrs. Martha (Brevoort) Chevalier, 1835; Mrs. Mary (Aiken) Poole, 1865; Abbie L. Ayres, 1882; two children by first marriage, three children by third. EDUCATION: 1834, A.B., Amherst College. CAREER: Soon after leaving college, became itinerant lecturer and practitioner of phrenology; 1838, established Phrenological Museum or Athenaeum, Philadelphia; 1838, sponsored establishment of *American Phrenological Journal and Miscellany* and in 1842, with brother Lorenzo, became editor; in partnership with Lorenzo, founded firm of O. S. and L. N. Fowler; in 1842 moved to New York City and in 1845, with S. R. Wells [q.v.] and Lorenzo, established firm of Fowlers and Wells; 1850-1853, chief attentions given to planning and construction of octagonal house at Fishkill, N.Y.; ca. 1855, withdrew from Fowlers and Wells, but continued phrenological lecturing and writing independently; 1863, established home at Manchester, Mass., and phrenological practice in Boston; ca. 1882, moved to Sharon Station, N.Y. MEMBERSHIPS: A founder, American Phrenological Society (1849). SCIENTIFIC CONTRIBUTIONS: Took leading role in bringing about widespread influence of phrenology in American life during nineteenth century, among ultimate byproducts of which were the promotion of ideas of brain as organ of mind and of localization of functions in brain and research on physiological psychology. Produced first work, *Phrenology Proved, Illustrated and Applied*, in 1836, with Samuel Kirkham and brother Lorenzo; subsequently published great many books and articles on phrenology (which Fowler family saw as related to all aspects of life of individual and society), promoting application of ideas of phrenology and physiology to wide range of social reforms. After mid-1850s Orson devoted chief attention to questions of sex and sexual guidance.

WORKS: *Creative and Sexual Science* (New York, 1875). See several other works listed in *DAB*; also see *Natl. Un. Cat. Pre-1956 Imp.* MSS: Fowler Family Papers at Corn. U.-CRH (Stern [below]). WORKS ABOUT: *DAB*, 6: 565-66 (Ernest S. Bates); Madeleine B. Stern, *Heads & Headlines: The Phrenological Fowlers* (Norman, Okla., 1971).

FRANKFORTER, GEORGE BELL (1859-1947). *Chemistry*. Professor of chemistry, dean of School of Chemistry, and director of chemistry laboratory, Univ. of Minnesota.

WORKS ABOUT: See *AMS*; *NCAB*, 42 (1958): 269-70 [includes some inaccuracies]; *Who Was Who in America*, vol. 3, *1951-1960*.

FRANKLIN, BENJAMIN (January 17, 1706, Boston, Mass.—April 17, 1790, Philadelphia, Pa.). *Physics*, especially *Electricity*; *Oceanography*; *Meteorology*. Son of Josiah, tallow chandler and soap boiler, and his second wife, Abiah (Folger) Franklin. Common law marriage to Deborah Read, 1730; four children, two illegitimate. EDUCATION: At age eight, attended grammar school and then went to school for writing and arithmetic; at age ten began working for father; at age twelve apprenticed to brother as printer. HONORS: Honorary degrees fom Harvard (1753), Oxford (1762), and others. CAREER: 1723, went to Philadelphia; 1724-1726, printer in London; 1728, formed printing partnership in Philadelphia and published *Pennsylvania Gazette* and *Poor Richard's Almanack*; 1736-1751, clerk of state assembly; 1751-1764, member of state assembly; 1737-1753, deputy postmaster at Philadelphia; 1753-1774, deputy postmaster general for the colonies; 1748, gave up active involvement in business; 1757-1762 and 1764-1775, Pennsylvania's agent in England; 1775, member of Second Continental Congress and elected postmaster general by Congress; 1776-1785, went to France to negotiate alliance; later, minister to French court and a negotiator of peace with England; 1785-1787, president of Pennsylvania Executive Council; 1787, member of the Constitutional Convention. MEMBERSHIPS: Roy. Soc., recipient of its Copley Medal, 1753; founder, Am. Phil. Soc.; member of other societies. SCIENTIFIC CONTRIBUTIONS. Experimental and theoretical work in electricity done mainly during period 1743-1752 and resulted in his conception of electricity as single "fluid"; introduced idea of conservation of charge and words such as *positive* and *negative*, *charge* and *battery*, and suggested experiment that demonstrated that lightning is electrical in nature. Reputation as important scientific thinker preceded first trip to Europe in 1757. Also studied and printed first chart of Gulf Stream, investigated nature of atmospheric currents and direction of storms, wrote on medical subjects, invented or improved number of common items, and made experimental apparatus. Played important role in promotion of science in colonies.

WORKS: *Experiments and Observations on Electricity* (London, 1751); *The Papers of Benjamin Franklin* (New Haven, 1959-) includes all his writings and correspondence. MSS: Lib. Cong. and Am. Phil. Soc. (*DSB*); additional references in *NUCMC* and Hamer. WORKS ABOUT: *DAB*, 6: 585-98 (Carl L. Becker); *DSB*, 5: 129-39 (I. Bernard Cohen). Also see Carl Van Doren, *Benjamin Franklin* (New York, 1938).

FRAZER, JOHN FRIES (July 8, 1812, Philadelphia,

Pa.—October 12, 1872, Philadelphia, Pa.). *Geophysics; Chemistry.* Son of Robert, lawyer, and Elizabeth (Fries) Frazer. Married Charlotte Jeffers Cave, 1838; three children, including Persifor Frazer [q.v.]. EDUCATION: 1830, A.B., Univ. of Pennsylvania; studied law with John M. Scott and, ca. 1836, admitted to bar. HONORS: LL.D., Harvard Univ., 1857. CAREER: While in college, and for some time afterward, served as laboratory assistant to Prof. A.D. Bache [q.v.]; ca. 1836, began practice of law, but found it not to his liking; 1836, assistant to H. D. Rogers [q.v.] on Geological Survey of Pennsylvania; 1842-1844, professor of chemistry and natural philosophy, Philadelphia Central High School; 1844-1872, professor of chemistry and natural philosophy and, for number of years, also vice-provost, Univ. of Pennsylvania; was lecturer at Franklin Institute and, 1850-1866, edited its journal. MEMBERSHIPS: Am. Phil. Soc.; Natl. Ac. Scis. SCIENTIFIC CONTRIBUTIONS: Noted as science educator, editor, and critic, but made no contributions to scientific knowledge on his own, and chief literary contributions were made in capacity as editor of *Journal of Franklin Institute.* While acting as assistant to A. D. Bache, participated in first accurate determination in this country of periods of daily variations of magnetic needle, and also was associated with work on connection of aurora borealis with magnetic forces. While chief teaching area was chemistry, main interest was mechanics; also well acquainted with study of history of science. In teaching, and as editor of *Journal of Franklin Institute,* played important role in critical analysis and communication of discoveries by other scientists.

WORKS: The only article by Frazer cited in the *Roy. Soc. Cat.* is "Digest of the Results of Experiments on Friction Performed by Capt. Morin," *Journal of Franklin Institute* 3d ser. 5 (1843): 412-19. MSS: Am. Phil. Soc. (*NUCMC* 60-2734). WORKS ABOUT: *DAB,* 7: 3 (Thomas D. McCormick); John L. Le Conte, "Memoir," *Biog. Mems. of Natl. Ac. Scis.* 1 (1877): 245-56.

FRAZER, PERSIFOR (1844-1909). *Geology; Chemistry.* Mineralogist and metallurgist with Hayden's U.S. geological survey; assistant to geologist, Second Geological Survey of Pennsylvania; professor of chemistry, Franklin Institute; consulting and reporting geologist, metallurgist, and mining engineer.

MSS: Am. Phil. Soc. (*NUCMC* 60-2982); Ac. Nat. Scis. Philad. (Hamer). WORKS ABOUT: See *AMS*; *DAB.*

FREAR, WILLIAM (1860-1922). *Agricultural chemistry.* Professor of agricultural chemistry, vice-director, and chemist of Experiment Station, Pennsylvania State Univ,; chemist, Pennsylvania State Board of Agriculture.

WORKS ABOUT: See *AMS*; *DAB.*

FREEMAN, THOMAS ([?], Ireland—November 8, 1821, Huntsville, Ala.). *Civil engineering; Astronomy; Exploration.* No information on parentage or marriage. EDUCATION: No specific information available, but career accomplishments suggest sound scientific training. CAREER: 1784, came to U.S.; 1794-1796, employed as a surveyor for federal district of Washington; 1796-1798, appointed surveyor of boundary line between U.S. and Spanish territories, but difficulties with Commissioner A. Ellicott [q.v.] led to discharge from duties; 1799, enlisted by Gen. Wilkinson in planning and supervision of construction of Fort Adams, Miss.; 1804, appointed by President Jefferson to explore Red and Arkansas rivers and, after delay, in 1806 undertook exploration; 1808, appointed register of land office in Madison County, Mississippi Territory, and was given responsibility for organizing county government; 1811-1821, surveyor of public lands south of Tennessee, and in this position was generally known as surveyor general. SCIENTIFIC CONTRIBUTIONS: Achievements summarized in activities as surveyor [above]; in work on U.S. capital district, surveyed northern part and before resignation began first topographical survey of city. In 1806, exploration of Red and Arkansas rivers with naturalist Peter Custus, proceeded as far as place where Arkansas, Oklahoma, and Texas now meet, but turned back at insistence of Spanish; on this expedition, collected measurements and observations, accurately mapping lower Red River for first time, and astronomical observations rechecked in 1914 were found to be nearly exact. Ca. 1807, mapped portion of boundary between Tennessee and Alabama.

WORKS: With Custus, *An Account of the Red River in Louisiana* (Washington, 1806), 63 pp. MSS: See references in *DAB.* WORKS ABOUT: *DAB,* 7: 13-14 (Albert T. Witbeck); Dunbar Rowland, ed., *Encyclopedia of Mississippi History* (Madison, Wis., 1907), 1: 749-50.

FRISBY, EDGAR (1837-1927). *Mathematics; Astronomy.* Professor of mathematics, U.S. Navy.

WORKS ABOUT: See *AMS*; *World Who's Who in Science.*

FRY, JOSHUA (ca. 1700, Crewkerne, Somerset, England—May 31, 1754, camp at Wills's Creek, site of Cumberland, Md.). *Surveying; Mathematics.* Son of Joshua Fry. Married Mary (Micou) Hill; survived by five children. EDUCATION: 1718, matriculated at Wadham College, Oxford Univ. CAREER: Arrived in Virginia before 1720; served as vestryman and magistrate, Essex County, Va.; 1729, became master of grammar school attached to College of William and Mary; 1731, appointed professor of mathematics and natural philosophy, William and Mary, this connection with college continuing until 1737; ca. 1737 [?], removed to back settlements of Virginia with hopes of gaining landed fortune and in 1744 was living on Hardware River, Goochland County; 1745, with establishment of Albermarle out of Goochland County, became Albemarle presiding justice of peace, surveyor, representative in House of Burgesses, and county lieutenant, latter being an office of first importance; 1746, served as commissioner of king in establishing boundaries of Lord Fairfax's grant in Northern Neck; 1749, with Peter Jefferson, commissioned to continue boundary line between Virginia and North Carolina; 1752, appointed one of four commissioners to negotiate treaty with Six Nations and other Indian tribes; 1754, appointed commander in chief of Virginia militia to deal with French in West, but died in camp and was succeeded by George Washington. SCIENTIFIC CONTRIBUTIONS: Most lasting contribution was map of Virginia [see Works], prepared with Peter Jefferson (father of Thomas Jefferson [q.v.]) and said to be first map of region based on actual surveys; remained best Virginia map produced in eighteenth century.

WORKS: *A Map of the Inhabited Part of Virginia Containing the Whole Province of Maryland with Part of Pensilvania, New Jersey and North Carolina Drawn by Joshua Fry and Peter Jefferson in 1751* (probably first printed in 1754). MSS. See references in *DAB*. WORKS ABOUT: *DAB*, 7: 48-49 (Richard L. Morton); P. Slaughter, *Memoir of Col. Joshua Fry* . . . [Richmond, Va., 1880]; *The Fry and Jefferson Map of Virginia and Maryland* . . . (Charlottesville, Va., 1966), with brochure containing introduction by Dumas Malone.

FURBISH, KATE (May 19, 1834, Exeter, N.H.—December 6, 1931, Brunswick, Maine). *Botany*. Daughter of Benjamin, owner of hardware store and manufacturer of tinware and stoves, and Mary A. Lane Furbish. Never married. EDUCATION: Attended schools in Brunswick, Maine, where had been taken at about age one; studied drawing in Portland and Boston, and French literature in Paris. CAREER: Following father's death in 1873, inherited income allowing self-sufficiency. MEMBERSHIPS: A founding member (1895), Josselyn Botanical Society of Maine. SCIENTIFIC CONTRIBUTIONS: In 1870, began serious study of flora of Maine, collecting and classifying botany and preparing watercolor illustrations; traveled systematically over state, including largely unexplored remote areas such as Aroostook County, and was notably successful in ambition to prepare drawings of all native plants of Maine, excepting grasses, sedges, trees, and ferns. Admired by professional botanists such as A. Gray [q.v.]; published several articles and occasionally lectured on botany.

WORKS: See *American Naturalist* 15 (June 1881): 469-70; ibid., May 1882, pp. 397-99; ibid., 16 (December 1882): 1004; *Rhodora* 3 (June 1901): 185; and Asa Gray, *Synoptical Flora of North America*, 1 (1884): 189. Sixteen folio volumes of watercolor drawings at Bowdoin College; collection of dried plants, totaling nearly 4,000 sheets, in Harv. U.-GH. Other information on collections and manuscripts in *Nota. Am. Wom.* WORKS ABOUT: *Nota. Am. Wom.*, 1: 686-87 (Lazella Schwarten); Louis H. Coburn, "Kate Furbish, Botanist," *Maine Naturalist* 4 (November 15, 1924): 106-8.

G

GABB, WILLIAM MORE (January 20, 1839, Philadelphia, Pa.—May 30, 1878, Philadelphia, Pa.). *Geology*; *Paleontology*. Son of Joseph, salesman, and J. H. (More) Gabb. Apparently never married. EDUCATION: 1857, graduated, Central High School, Philadelphia; 1857-1860, student and assistant to paleontologist J. Hall [q.v.] in Albany. CAREER: 1860, spent some time at Smithson. Instn.; 1862, appointed paleontologist, California Geological Survey, and for six years traveled throughout California and elsewhere on Pacific Coast, including 1867 exploration in Lower California; 1869-1872, carried out officially sponsored topographical and geological survey of Santo Domingo; 1873-1876, conducted similar government survey of province of Talamanca, Costa Rica; 1876, returned to U.S.; later started for Santo Domingo to pursue mining claim, but returned to Philadelphia and soon thereafter died of tuberculosis. MEMBERSHIPS: Ac. Nat. Scis. Philad.; Natl. Ac. Scis.; and other societies. SCIENTIFIC CONTRIBUTIONS: When joining California survey in his early twenties, was considered foremost American authority on Cretaceous marine fossils, having already published some fifteen papers on that subject; map and report on Lower California [see Works] were first to present correct geological structure of that region; also wrote on Upper Mesozoic and Tertiary rocks of California in connection with state geological survey. Wrote on geology and geography of Santo Domingo, but planned work on Costa Rican paleontology was not finished. In course of explorations, gathered many geological, ethnological, entomological, and other specimens for Smithsonian and other collections; as an accomplished field worker, his paleontological publications were mainly descriptive and included some 474 Cretaceous fossils.

WORKS: Wrote parts of vol. 1 (1864) and all of vol. 2 (1869) of *Geological Survey of California: Paleontology*; geological map and report on Lower California, *Petermanns Geographische Mitteilungen* 14 (1868); "On the Topology and Geology of Santo Domingo," *Transactions of Am. Phil. Soc.* n.s. 15 (1881): 49-259; geographical work on Costa Rica appeared in publication by that government (1875); "On the Indian Tribes and Languages of Costa Rica," *Proceedings of Am. Phil. Soc.* 14 (1874-1875): 483-602. Bibliography in *Biog. Mems. of Natl. Ac. Scis.* 6 (1909): 356-61. WORKS ABOUT: *DAB*, 7: 81-82 (Charles Schuchert); *DSB*, 5: 214 (Elizabeth N. Shor).

GAGE, SIMON HENRY (1851-1944). *Histology*; *Embryology*. Professor of histology and embryology, Cornell Univ.

MSS: Corn. U.-CRH (*NUCMC* 62-2335 and 66-428). WORKS ABOUT: See *AMS*; *NCAB*, 36 (1950): 127-28.

GAMBEL, WILLIAM (June 1823, Philadelphia, Pa.—December 13, 1849, on Feather River, Calif.). *Ornithology*. Son of William and Elizabeth Richardson Gambel, a teacher. Married Catherine M. Towson, ca. 1848; no information on children. EDUCATION: 1848, M.D., Univ. of Pennsylvania. CAREER: 1838-1839, accompanied T. Nuttall [q.v.] on southern trip; 1840, went to Cambridge, Boston, and Maine with Nuttall; 1841-1842 traveled in West to Rio Grande, New Mexico, and California, collecting specimens for Nuttall; 1842-ca. 1845, served as clerk to Com. Thomas Ap Catesby Jones of Pacific fleet and collected specimens on west coasts of North and South America and on Sandwich Islands; 1845, returned to Philadelphia and began study of medicine; 1849, left for California, apparently to establish medical practice, but died en route. MEMBERSHIPS: Ac. Nat. Scis. Philad. SCIENTIFIC CONTRIBUTIONS: By about 1838, had collection of minerals, and in travels in West collected botanical specimens as well as birds. Prepared illustrations for Nuttall's supplements to F. A. Michaux [q.v.], *North American Sylva*, and, ca. 1846, sent collection of some 350 botanical specimens to Nuttall, who had returned to England. Most significant contributions made in ornithology and especially exploration and study of birds of the West, producing first significant work on California birds. Also discovered and named kangaroo rat and parasitic mouse of California.

WORKS: Publications appeared chiefly in *Proceedings of Ac. Nat. Scis. Philad.*; of particular importance was "Remarks on the Birds Observed in Upper California, with Descriptions of New Species," *Journal of Ac. Nat. Scis. Philad.* n.s. 1 (1847-1850): 25-56, 215-29; also, T. Nuttall, "Description of Plants Collected by William Gambel, M.D., in the Rocky Mountains and Upper California," ibid., pp. 148-89. See *Roy. Soc. Cat.*, vol. 2. MSS: Hist. Soc. Penn. (Graustein [below]). WORKS ABOUT: Witmer Stone, "William Gambel, M.D.," *Cassinia: Proceedings of Delaware Valley Ornithological Club* 14 (1910): 1-8; Jeannette E. Graustein, *Thomas Nuttall*... (Cambridge, Mass., 1967), pp. 342-90, passim.

GANNETT, HENRY (1846-1914). *Geography; Statistics.* Chief geographer, U.S. Geol. Surv.

WORKS ABOUT: See *AMS*; *DAB*.

GARDEN, ALEXANDER (January 1730, Birse, Scotland—April 15, 1791, London, England). *Botany; Natural history.* Son of Alexander Garden, clergyman. Married Elizabeth Peronneau, 1755; three children who lived beyond childhood. EDUCATION: Ca. 1743-ca. 1746, apprenticed to Dr. James Gordon and studied medicine at Marischal College, Aberdeen; 1750, studied medicine at Univ. of Edinburgh; 1753, M.D. and A.M., Marischal College. CAREER: 1746, went to London, qualified as surgeon's second mate in British navy, but was not appointed; probably returned to Aberdeen and worked with Dr. Gordon; 1748-1750, surgeon's first mate in navy; 1752, arrived in South Carolina and began medical practice in Prince William Parish; 1755, established medical practice in Charleston and acquired large and profitable following; 1782, as loyalist during Revolution, returned to England and was active in Roy. Soc. and in pursuit of efforts to regain confiscated property, which was eventually restored. MEMBERSHIPS: Am. Phil. Soc.; Roy. Soc.; and other societies. SCIENTIFIC CONTRIBUTIONS: Interested in natural history of South Carolina from outset; entered into correspondence with British and European scientists, including Linnaeus, and his 1754 trip northward brought contact with other American scientists. Scientific activities, at least in part, related to medical concerns, and early work on Indian Pink [see Works], prepared in 1757 but not published until 1771, included botanical and medical aspects. Main activity was as botanist, but also discovered several new animals, and description of mud iguana that was sent to Linnaeus in 1765 necessitated erection of entirely new class of Amphibia. Made some attempts at designating new species and genera and at times disagreed with Linnaeus and other contemporary authorities, but published little under his own name; main contributions were specimens and descriptions provided for Linnaeus and other scientists. Also interested in other scientific subjects, including mineralogy and meteorology. Gardenia named for him by British naturalist John Ellis.

WORKS: "An Account of the Indian Pink . . . ," *Essays and Observations, Physical and Literary* 3 (1771): 145-53. Bibliography of works and locations of manuscripts in Berkeley and Berkeley [below]. WORKS ABOUT: *DAB*, 7: 132-33 (Donald C. Peattie); Edmund Berkeley and Dorothy S. Berkeley, *Dr. Alexander Garden* . . . (Chapel Hill, N.C., 1969).

GARDINER, JAMES TERRY (May 6, 1842, Troy, N.Y.—September 10, 1912, Northeast Harbor, Maine). *Topographical engineering.* Son of Daniel and Ann (Terry) Gardiner. Married Josephine Rogers, 1868; Eliza Greene Doane, 1881; no information on children. EDUCATION: Said to have had little formal education; for brief periods, studied at Rensselaer Polytechnic Institute and at Sheffield Scientific School, Yale. HONORS: Ph.B., Yale Univ., 1868. CAREER: At young age, was made subassistant engineer, Brooklyn Water Works; 1861-1862, inspector, U.S. Ordnance Corps; during Civil War, employed on construction of earthworks around San Francisco harbor; 1864-1867, topographical assistant, Geological Survey of California; 1867-1873, chief topographer, U.S. Geological Survey of Fortieth Parallel (King survey); 1873-1875, topographer with Hayden Survey of the Territories; 1876-1886, director, New York state survey; 1880-1886, member of New York state board of health; engaged in practice as consulting engineer; 1892-1895, president, Street Railroad & Lighting Co., St. Joseph, Mo.; 1895, became vice-president of coal companies of Erie Railroad Co.; 1899, elected president, Mexican Coke & Coal Co.; director of several coal, railroad, and related enterprises; during later years, lived in New York. SCIENTIFIC CONTRIBUTIONS: A boyhood friend of C. R. King [q.v.], and as one of most accomplished American field topographers of time, declined appointment in geodesy at Harvard Univ. to join King survey; as topographer on this survey, and on that of F. V. Hayden [q.v.], made important contribution to exploration of West. Drew atlas of Colorado for Hayden survey, as well as atlas for King survey; latter was done in some haste and therefore did not achieve degree of excellence that it might have. Later, became interested in public health, and helped promote adequate sewerage systems in state of New York. Despite early report on coal and iron fields of Colorado (1875), apparently made no significant technical contributions during years of involvement in coal industry.

WORKS: See above; also *Natl. Un. Cat. Pre-1956 Imp.* MSS: N.Y. St. Lib. (*NUCMC* 62-937). WORKS ABOUT: DAB, 7: 139 (Ray P. Baker); Goetzmann, *Exploration and Empire*, pp. 362-521, passim.

GARMAN, SAMUEL (1843-1927). *Zoology*. Curator of fishes, Museum of Comparative Zoology, Harvard Univ.

MSS: Harv. U-MCZ. WORKS ABOUT: See *AMS*, 3d ed. (1921); *DAB*.

GATTINGER, AUGUSTIN (February 3, 1825, Munich, Germany—July 18, 1903, Nashville, Tenn.). *Botany*. Son of August, important official of German government, and Rosa Gattinger. Married Josephine Dury, 1849; survived by three children. EDUCATION: 1849, having previously completed medical course, expelled from Univ. of Munich for political reasons. CAREER: 1849, left Germany and established medical practice at Cave Spring, Tenn.; later, practiced medicine at Charleston, Tenn.; 1858-1864, resident surgeon at copper mines, Ducktown, Tenn.; 1864, enlisted in U.S. Army as assistant surgeon; ca. 1865, took up residence in Nashville; ca. 1865-1869, Tennessee state librarian; thereafter, apparently chief activity was medical practice in Nashville; 1883, made assistant in office of state commissioner of agriculture to collect minerals and plants for Louisville Exposition and other exhibits; 1897, worked with Tennessee Centennial Exposition in Department of Mineralogy and Mines. MEMBERSHIPS: Am Assoc. Advt. Sci. SCIENTIFIC CONTRIBUTIONS: Devoted spare time to study of botany and geology, and engaged in number of government-related scientific activities, including assistance in 1878 to state commissioner of agriculture in preparation of work on grasses and forage plants. In same year produced for Nashville Board of Health a publication on trees and shrubs suitable to area; in 1880 collected specimens of Tennessee timber and stones for census; in 1894 wrote work on medical plants for commissioner of agriculture, and is referred to as first state botanist of Tennessee. Years as librarian gave opportunity for closer contact with latest scientific literature, and extensive collecting activities made him authority on botany of middle Tennessee, an area previously largely unexplored. Collected second largest herbarium in the South; life's work in botany was summarized in *Flora* [see Works], originally produced at personal expense in 1887; its 1901 edition included more than 2,100 species.

WORKS: *The Flora of Tennessee and a Philosophy of Botany* (Nashville, Tenn., 1901), this edition published by authority of the state through the Bureau of Agriculture. Bibliography in *Amer. Hist. Mag.* [below]. WORKS ABOUT: *NCAB*, 15 (1916 [copyright 1914 and 1916]): 91; "Dr. Augustin Gattinger," *American Historical Magazine and Tennessee Historical Society Quarterly* 9 (1904): 153-78.

GENTH, FREDERICK AUGUSTUS, Sr. (May 17, 1820, Wächtersbach, Hesse-Cassel, Germany—February 2, 1893, Philadelphia, Pa.). *Chemistry*; *Mineralogy*. Son of George Friedrich, forest warden, and Karoline Amalie (Freyin von Swartzenau) Genth. Married Karoline Jäger, 1847; Minna Paulina Fischer, 1852; three children by first marriage, nine by second. EDUCATION: 1839-1845, studied chemistry, geology, and mineralogy at universities of Heidelberg, Giessen, and Marburg; 1845, Ph.D., Univ. of Marburg. CAREER: 1845-1848, privatdocent and assistant to Bunsen, Univ. of Marburg; 1848, came to U.S. and opened analytical laboratory, Philadelphia; ca. 1849, engaged in mining work, North Carolina; 1850-1872, reestablished Philadelphia laboratory and engaged in research, commercial analytical work, and teaching of special students; 1872-1888, professor of chemistry, Univ. of Pennsylvania; 1888-1893, again reopened private laboratory; held positions as chemist to Second Geological Survey of Pennsylvania and to Pennsylvania Board of Agriculture. MEMBERSHIPS: Natl. Ac. Scis. and other societies; president Am. Chem. Soc., 1880. SCIENTIFIC CONTRIBUTIONS: Training under noted German teachers made him one of best prepared U.S. scientists of time. A mineral chemist of first rank who was particularly adept at analytical work; published 102 works in American and German periodicals, of which over 70 were on minerals; described some 215 species and is credited with discovery of 24 new minerals. Carried out analysis of fertilizers and thereby aided development of agricultural industry in Pennsylvania. Best known for work on ammonia-cobalt bases, begun in 1847 and later developed jointly with O. W. Gibbs [q.v.]; Genth's original paper (1851) [see Works] gave first clear evidence of existence of salts of cobalt ammines.

WORKS: Earliest paper on cobalt ammines in *Keller-Tiedemann's Nordamerikan Monatsbericht* 2 (1851): 8-12; with O. W. Gibbs, "Researches on the Ammonia-Cobalt Bases," *Smithson. Contr. Knowl.* 9 (1857), which also appeared in other publications. See bibliography in *Biog. Mems. of Natl. Ac. Scis.* 4 (1902): 222-31. MSS: Am. Phil. Soc. WORKS ABOUT: *DAB*, 7: 209-10 (Lyman C. Newell); *DSB*, 5: 349-50 (George B. Kauffman).

GIBBES, LEWIS REEVE (August 14, 1810, Charleston, S.C.—November 21, 1894, Charleston, S.C.). *Natural history; Astronomy; Chemistry*. Son of Lewis Ladson and Maria Henrietta Drayton

Gibbes. Married Anna Barnwell Gibbes, 1848; nine children. EDUCATION: 1829, A.B., South Carolina College (Univ. of South Carolina); 1836, M.D., Medical College of State of South Carolina; ca. 1836-1837, in Paris, studied at Sorbonne and Jardin des Plantes and attended lectures of noted physicians and surgeons. CAREER: At South Carolina College, tutor in mathematics (1831-1834) and acting professor of mathematics (1834-1835); 1838-1892, professor of science, College of Charleston, at various times teaching mathematics, astronomy, physics, chemistry, and mineralogy. MEMBERSHIPS: Ac. Nat. Scis. Philad. and other societies; president, Elliott Society of Natural History (Charleston), 1856-1889. SCIENTIFIC CONTRIBUTIONS: Among most notable early contributions was 1850 revision of species of Crustacea in several U.S. collections, together with notes and descriptions of new species [see Works]. Also wrote on other zoological subjects and on astronomical and physical topics, as well as on geology, botany, and mathematics. Belated recognition has been given for work in classification of chemical elements, prepared as lecture illustration during period 1870-1875 and presented in latter year to Elliott Society; publication was delayed for ten years, by which time arrangements of elements by Lothar Meyer and Mendeleev were well known; Gibbes's table was original with him, and demonstrated awareness of systematic, if not periodic, relationships between elements; felt need to leave blank spaces in table, which he thought might be filled by elements not known at time.

WORKS: "On the Carcinological Collections of the U.S., and an Enumeration of Species Contained in Them, with Notes on the Most Remarkable, and Descriptions of New Species," *Proceedings of Am. Assoc. Advt. Sci.* 3 (1850): 167-201; "Synoptical Table of the Chemical Elements," *Proceedings of Elliott Society of Natural History* 2 (October 1875): 77-90, published in March 1886. See *Roy. Soc. Cat.*, vols. 2, 7, 15. MSS: Lib. Cong. (Hamer); Charleston Museum (Hamer). WORKS ABOUT: Kelly and Burrage, *Med. Biog.* (1920); Wendell H. Taylor, "Lewis Reeve Gibbes and the Classification of the Elements," *Journ. Chem. Educ.* 18 (1941): 403-7.

GIBBES, ROBERT WILSON (July 8, 1809, Charleston, S.C.—October 15, 1866, Columbia, S.C.). *Paleontology*; *Geology*. Son of William Hasell, lawyer, and Mary Philp (Wilson) Gibbes. Married Carolina Elizabeth Guignard, 1827; twelve children. EDUCATION: 1827, A.B., South Carolina College (later Univ. of South Carolina); 1827-1828, attended medical lectures, Philadelphia; 1830, M.D., Medical College of State of South Carolina. CAREER: 1827-1834, assistant professor of chemistry, geology, and mineralogy, South Carolina College; 1834-1866, medical practice, Columbia, S.C.; owner (1852-1865) and editor (1852-1858) of *South Carolinian*, newspaper and publishing business acquired in mortgage settlement; twice mayor of Columbia; during Civil War, served as surgeon general of South Carolina; also had business interests, including manufacture of cotton shirting. SCIENTIFIC CONTRIBUTIONS: Interested in geology, paleontology, mineralogy, conchology, and ornithology, but collections relating to these subjects, along with all other property, were destroyed during Civil War. Chief scientific interest was paleontology, especially Tertiary period; in study of South Carolina fossil forms, named one specimen Dorudon, an assertion refuted by English anatomist Richard Owen, who claimed it was not a separate genus, but a small Zeuglodon previously named by himself. Defender of geological studies against attack by religious interests. Close friend of S. G. Morton [q.v.] and J. C. Nott and supported their ideas on separate creation of races. Also wrote on medical and historical subjects.

WORKS: "Description of the Teeth of a New Fossil Animal Found in the Green Sand of S.C.," *Proceedings of Ac. Nat. Scis. Philad.*, 2 (1845): 254-56; "On the Fossil Genus Basilosaurus, Harlan, (Zeuglodon, Owen), with a Notice of Specimens from the Eocene Green Sand of South Carolina," *Journal of Ac. Nat. Scis. Philad.* n.s. 1 (1847-1850): 5-15; "Monograph of the Fossil Squalidae of the United States," ibid., pp. 139-47, 191-206; *The Documentary History of the American Revolution*, 3 vols. (New York, 1853-1857). See Meisel, *Bibliog. Amer. Nat. Hist.* MSS: Letters to S. G. Morton in Lib. Co. Philad. (Jellison and Swartz [below]). WORKS ABOUT: *DAB*, 7: 235-36 (Arney R. Childs); Richard M. Jellison and Phillip S. Swartz, "The Scientific Interests of Robert W. Gibbes," *South Carolina Historical Magazine* 66 (1965): 77-97.

GIBBS, GEORGE (January 7, 1776, Newport, R.I.—August 5, 1833, estate near Astoria, Long Island, N.Y.). *Mineralogy*; *Patronage of science.* Son of George, wealthy merchant, and Mary (Channing) Gibbs. Married Laura Wolcott; three children, including O. Wolcott Gibbs [q.v.]. EDUCATION: Apparently did not attend college. HONORS: M.A., Rhode Island College (Brown Univ.) in 1800 and Yale in 1808. CAREER: Inherited wealth; during early life visited Europe; 1814, purchased country estate on Long Island and latter part of life devoted to rural pursuits. MEMBERSHIPS: American Geological Society (organized in New Haven, 1819); N.Y. Lyc. Nat. Hist. SCIENTIFIC CONTRIBUTIONS: In 1800, went to Europe and while there became acquainted with number of leading mineralogists. In 1805, brought to Newport, R.I., his collection of more than 12,000 mineralogical specimens purchased abroad, largest such collection then in U.S.; in 1810, through friendship with B. Silliman [q.v.], collection was deposited at Yale, and during years 1810-1812 Gibbs helped to organize it according to Haüy's method; collection was available for use by college and public, and in 1825 Yale purchased it for $20,000. Supported *American Mineralogical Journal* of A. Bruce [q.v.] and was credited with having persuaded Silliman to undertake *Am. Journ. Sci.*; through prizes and other efforts, attempted to promote study of mineralogy in U.S. His relatively few publications dealt with minerals and mineral deposits, notices regarding ironworks, gunpowder, magnetism and light, and dry rot.

WORKS: Appeared in *American Mineralogical Journal* and *Am. Journ. Sci.* See Meisel, *Bibliog. Amer. Nat. Hist.* MSS: St. Hist. Soc. Wis. (*NUCMC* 62-2787); Newport Historical Society. WORKS ABOUT: *DAB* 7: 244-45 (Charles Schuchert); John C. Greene, "Die Anfänge der Entwicklung der Mineralogie in den Vereinigten Staaten" [text in English], *Geologie* 20 (1971): 485-90.

GIBBS, JOSIAH WILLARD (February 11, 1839, New Haven, Conn.—April 28, 1903, New Haven, Conn.). *Mathematics*; *Physics.* Son of Josiah Willard, philologist and professor of sacred literature at Yale, and Mary Anna (van Cleve) Gibbs. Never married. EDUCATION: 1858, A.B., Yale; 1863, Ph.D. in engineering, Yale; 1866-1869, a year's study each at universities of Paris, Berlin, and Heidelberg. HONORS: Copley Medal, Roy. Soc., 1901; Rumford Medal, Am. Ac. Arts Scis.; several honorary degrees. CAREER: 1863-1866, tutor for Latin and natural philosophy at Yale; 1871-1903, professor of mathematical physics at Yale, without salary until 1880. MEMBERSHIPS: Natl. Ac. Scis. and other societies; foreign honorary member, Roy. Soc. SCIENTIFIC CONTRIBUTIONS: Early work was in engineering, while postgraduate study in Europe determined direction of later career. Began work in thermodynamics in early 1870s; in publications on that subject assumed concept of entropy to be as important as those of energy, temperature, pressure, and volume in thermodynamic system; in dealing with their interrelations clarified entropy concept as originally stated by Rudolf Clausius; in paper entitled "Heterogeneous Substances" [see Works], combined chemical, elastic, surface, electromagnetic, and electrochemical phenomena into one integrated system, and in this paper included new concept of chemical potential, the foundation for his "phase rule"; with this work helped to establish theoretical basis for studies in physical chemistry. Sent copies of his papers to various persons, and early support came chiefly from James Clerk Maxwell in England. In 1880s, Gibbs worked on optics, especially Maxwell's electromagnetic theory of light. Later book on statistical mechanics [see Works] came to dominate that subject.

WORKS: Fundamental papers on thermodynamics published in *Transactions of Connecticut Academy of Arts and Sciences,* beginning in 1873 and including 300-page memoir, "On the Equilibrium of Heterogeneous Substances," vol. 3 (1875-1878); *Elementary Principles in Statistical Mechanics Developed with Special Reference to the Rational Foundation of Thermodynamics* (New Haven, 1902); Edwin B. Wilson, *Vector Analysis Founded Upon the Lectures of Josiah Willard Gibbs* (New York, 1901); H. A. Bumstead and R. G. Van Name, *The Scientific Papers of Josiah Willard Gibbs* (New York, 1906; reprinted, New York, 1961). MSS: Yale U.—Beinecke Library (*DSB*). WORKS ABOUT: *DAB*, 7: 248-51 (Edwin B. Wilson); *DSB*, 5: 386-93 (Martin J. Klein). Also see Lynde P. Wheeler, *Josiah Willard Gibbs: The History of a Great Mind* (New Haven, 1962).

GIBBS, [OLIVER] WOLCOTT (1822-1908). *Chemistry.* Professor of chemistry and physics, Free Academy of New York (City College); Rumford professor of applied science, Harvard Univ.

MSS: Franklin Institute, Philadelphia (*NUCMC* 64-963). WORKS ABOUT: See *AMS*; *DAB*; *DSB*.

GILBERT, GROVE KARL (1843-1918). *Geology; Physical geography.* Geologist with Wheeler and Powell surveys and with U.S. Geol. Surv.

WORKS ABOUT: See *AMS*; *DAB*; *DSB*.

GILL, THEODORE NICHOLAS (1837-1914). *Zoology*; *Comparative anatomy.* Professor of zoology, George Washington Univ.; associated with Smithson. Instn.

MSS: Ac. Nat. Scis. Philad. (Hamer). WORKS ABOUT: See *AMS*; *DAB*; *DSB*.

GILLISS, JAMES MELVILLE (September 6, 1811, Georgetown, Washington, D.C.—February 9, 1865, Washington, D.C.). *Astronomy.* Son of George, employee of U.S. government, and Mary (Melville) Gilliss. Married Rebecca Roberts, 1837; had children. EDUCATION: Ca. 1826 [age fifteen], entered navy as midshipman; 1831, achieved grade of passed midshipman; 1833, studied at Univ. of Virginia; 1835, studied in Paris. CAREER: Ca. 1826, entered navy; 1835, completed fourth cruise; 1836, assigned to Depot of Charts and Instruments, Washington, D.C.; 1837, made head of depot; 1842-1844, given responsibility for planning the building and procuring the instruments for recently authorized Naval Observatory; for that purpose, 1842-1843, visited Europe; 1845-1846, prepared his astronomical observations for publication; ca. 1846-1848, assigned to duty with Coast Survey; 1848, given charge of expedition to

South America to observe Venus and Mars; 1849-1852, at Santiago, Chile, observatory; 1852, commenced work on report and reduction of South American observations; 1855, placed on navy reserved list, but ordered by secretary of navy to continue work on Santiago observations; 1858, observed eclipse in Peru; 1860, observed eclipse in Washington Territory under sponsorship of Coast Survey; 1861-1865, superintendent, Naval Observatory. MEMBERSHIPS: Natl. Ac. Scis. SCIENTIFIC CONTRIBUTIONS: At Depot of Charts and Instruments, made astronomical, magnetic, and meteorological observations for Wilkes Expedition; publication of studies in 1846 [see Works] was first volume of astronomical observations to appear in U.S.; during early 1840s, was only American giving full time to practical astronomy; played major role in transformation of depot into Naval Observatory, but directorship went to M. F. Maury [q.v.]. Expedition to Santiago resulted in great many observations of Venus and Mars, as well as of stars, and also resulted in permanent observatory in Chile, but failure of coordinate observations in Northern Hemisphere nullified value for determination of solar parallax. Published four reports on Santiago observations, including astronomy and meteorology, and also on Chile's geography, climate, earthquakes, minerals, and social and political conditions.

WORKS: *Astronomical Observations Made at the Naval Observatory . . .* , (U.S. Senate Document no. 172, pt. 1, 28th Congress, 2d session, 1846). See also *Roy. Soc. Cat.*, vol. 2, and *Natl. Un. Cat. Pre-1956 Imp.* MSS: Natl. Arch.; also in G. P. Marsh Papers, Univ. of Vermont Library. WORKS ABOUT: *DAB*, 7: 292-93 (Raymond S. Dugan); Benjamin A. Gould, "Memoir," *Biog. Mems. of Natl. Ac. Scis.* 1 (1877): 135-79.

GIRARD, CHARLES FREDERIC (March 9, 1822, Mulhouse, France—January 29, 1895, Neuilly-sur-Seine, near Paris, France). *Zoology*. No information on parentage. Never married. EDUCATION: Educated at Neuchâtel, Switzerland, including study with J. L. R. Agassiz [q.v.]; 1856, M.D., Georgetown College. HONORS: Cuvier Prize, Institute of France, 1861. CAREER: Assistant to Agassiz at Neuchâtel; 1847-1850, student and assistant to Agassiz at Cambridge, Mass.; 1850-1860, connected with Smithson. Instn. as assistant to S. F. Baird [q.v.]; 1860-1863, in France, contracted to supply drugs and surgical apparatus to Confederate army; 1863-1865, in the southern states; 1865, returned to Paris and practiced medicine for some twenty years. SCIENTIFIC CONTRIBUTIONS: Main interest was study of fishes and reptiles, but also wrote on other subjects, including flatworms, echinoderms, quadrupeds, spiders, centipedes, and insects, as well as medicine, bibliography of science, and related topics. At Smithsonian, worked on reports on fishes and reptiles collected by Mexican boundary survey, Pacific Railroad Survey, and other explorations. While in U.S., played important role as descriptive naturalist, made especially important contributions to study of North American herpetology and ichthyology, and in latter field is particularly noted for contributions on Cyprinidae of North America and fauna of Pacific Coast.

WORKS: "Contributions to the Natural History of Freshwater Fishes of North America. I. Monograph of the Cottoids," *Smithson. Contr. Knowl.* 3 (1852); "Researches Upon the Cyprinoid Fishes Inhabiting the Fresh Waters of the U.S.A., West of the Mississippi Valley," *Proceedings of Ac. Nat. Scis. Philad.* 8 (1856-1857): 165-213; U.S. Exploring Expedition (Wilkes), *Herpetology* (Philadelphia, 1858); *Reports of Exploration and Surveys . . . for a Railroad from the Mississippi River to the Pacific Ocean*, vol. 10, pt. 4: *Fishes* (Washington, D.C., 1859); with S. F. Baird, "Catalogue of North American Reptiles in the Museum of the Smithsonian Institution," *Smithson. Misc. Coll.* 2 (1862). See Goode [below] for elaborate bibliography covering period 1849-1891. MSS: Smithson. Instn. Archives. WORKS ABOUT: *DAB*, 7: 319 (Hubert L. Clark); George B. Goode, "The Published Writings of Dr. Charles Girard," *Bulletin of U.S. Natl. Mus.*, no. 41 (1891).

GLOVER, TOWNEND (February 20, 1813, Rio de Janeiro, Brazil—September 7, 1883, Baltimore, Md.). *Entomology*. Son of Henry, wealthy businessman, and Mary (Townend) Glover. Married Sarah T. Byrnes, 1840; at least one adopted child. EDUCATION: Educated by relatives in England; 1834-1836, studied art in Germany. HONORS: Gold medal for entomological work from Napoleon III, 1865. CAREER: As infant, was sent to live with relatives in England, following mother's death; when about six years old, father died, and in 1834 [age twenty-one], inherited remains of father's wealth; 1836, went to U.S.; 1836-1838, traveled around U.S., especially in South; 1838, settled in New Rochelle, N.Y., and after marriage moved to Fishkill, N.Y., and pursued horticultural, natural history, and artistic interests; 1846, purchased father-in-law's estate, followed life of country gentleman, studied entomology, and produced models of fruits; 1854-1859, held appointment in U.S. Bureau of Agriculture, collecting statistics and other information; ca. 1859-ca. 1863, professor of natural sciences, Maryland Agricultural College (later, part of Univ. of Maryland); 1863-1878, U.S. entomologist, Department of Agriculture. SCIENTIFIC CONTRIBUTIONS: Winter of 1853-1854, went to Washington with large collection of fruit models; this effort culminated in appointment in Bureau of Agriculture; thereafter traveled extensively in South and to South America, investigating insects, plants and plant diseases, soil, natural history, and other subjects. Became first official entomologist in employ of U.S. government and as such worked on Department of Agriculture museum. More interested in illustrating insects than in assembling entomological collection; after 1859 devoted much time and effort to preparation of ambitious illustrated work on U.S. insects that was never fully completed, although twelve copies were printed in 1878, including 273 copper etchings, with 6,179 figures [see Works]. Published relatively little else, but did circulate portion of his great accumulation of entomological knowledge in annual reports. Ca. 1865, he warned of need to guard against accidental importation of insect pests, but advice was not heeded.

WORKS: *Illustrations of North American Entomology in the Orders of Coleoptera, Orthoptera, Neuroptera, Hymenoptera, Lepidoptera, and Diptera* (Washington, D.C., 1878). See bibliography in *Bulletin of U.S. Department of Agriculture, Division of Entomology,* no. 18 (1888), pp. 63-68. MSS: Smithson. Instn. Archives (*NUCMC* 74-976). WORKS ABOUT: *DAB,* 7: 333-34 (Leland O. Howard); Mallis, *Amer. Entomologists,* pp. 61-69.

GODFREY, THOMAS (1704, Bristol Township, Pa. —December 1749, Philadelphia, Pa.). *Mathematics; Technology.* Son of Joseph Godfrey, farmer and maltster who died while Thomas was an infant. Married; no information on children. EDUCATION: Little or no formal education; apprenticed to glazier. CAREER: Worked as glazier on Pennsylvania Statehouse (now Independence Hall) and on home of Gov. J. Logan [q.v.], who recognized and encouraged his mathematical talents; 1729-1736, published almanacs; 1740, advertised instruction in navigation, astronomy, and mathematics. MEMBERSHIPS: Philadelphia Junto, for a short time; mathematician of Am. Phil. Soc. (as founded in 1743). SCIENTIFIC CONTRIBUTIONS: Had natural mathematical ability. In 1730 devised double reflecting quadrant for determining latitude and tested it in Delaware Bay and on voyage to Jamaica; sometime afterward, learned that Roy. Soc. was to grant award to James Hadley, its vice-president, for similar quadrant; through Godfrey's own solicitations and active intervention of J. Logan on his behalf, Roy. Soc. published account of Godfrey's quadrant and sent him award in form of household furniture; the invention, generally known as Hadley's quadrant, soon replaced other devices for measuring elevations, and conflict appears to have been case of simultaneous, uncoordinated invention. Contributed to mathematical queries, as well as astronomical data and essays, in the *Pennsylvania Gazette* and *Pennsylvania Journal*; Lewis Evans used astronomical data gathered by himself and Godfrey to fix longitude of Philadelphia in Evans's *General Map of the Middle British Colonies* (1755). Godfrey published series of almanacs in Philadelphia for years 1730-1737.

WORKS: J. Logan, "An Account of Mr. Thomas Godfrey's Improvement of Davis's Quadrant, transferred to the Mariner's-Bow," *Phil. Trans. of Roy. Soc.* 38 (1733-1734): 441-50 (reprinted with Godfrey's letter of November 9, 1732, to Roy. Soc. in *American Magazine* 1 [1757-1758]: 475-80, 529-34). MSS: Logan papers related to Godfrey are in Roy. Soc. Library and in Logan Papers at Hist. Soc. Penn. (*DSB*). WORKS ABOUT: *DAB,* 7: 345-46 (Joseph Jackson); *DSB,* 5: 434 (Brooke Hindle).

GODMAN, JOHN DAVIDSON (December 20, 1794, Annapolis, Md.—April 17, 1830, Germantown, Pa.). *Anatomy; Natural history.* Son of Capt. Samuel, Revolutionary War officer, and Anna (Henderson[?]) Godman. Married Angelica Kauffman Peale, 1821; apparently had children. EDUCATION: 1811-1812, printer's apprentice; 1815, began study of medicine with physician in Elizabethtown, Pa., and later continued studies with physician in Baltimore; 1818, M.D., Univ. of Maryland. CAREER: While student at Univ. of Maryland, acted as demonstrator and sometimes as lecturer in anatomy; ca. 1818, began medical practice in New Holland, Pa.; soon thereafter moved to location near Baltimore, expecting appointment at Univ. of Maryland that did not materialize; removed to Philadelphia and lectured on anatomy and physiology; 1821-1822, professor of surgery, Medical College of Ohio (Cincinnati), and editor of *Western Quarterly Reporter of Medical, Surgical and Natural Science*; 1822-1823, returned to Philadelphia and probably engaged in medical practice; 1823, took charge of Philadelphia School of Anatomy; 1826-1827, professor of anatomy, Rutgers Medical College (New York City); 1828, in West Indies for part of year, seeking cure for tuberculosis; last years passed at Germantown, Pa., engaged in study and writing. MEMBERSHIPS: Am. Phil. Soc. SCIENTIFIC CONTRIBUTIONS: Made wide reputation as teacher of anatomy, and also published results of anatomical studies. Edited first trans-Allegheny medical journal (*Western Quarterly Reporter*); in 1825 became member of editorial board of *Philadelphia Journal of Medical and Physical Sciences* and was chiefly responsible for its transition in 1827 to *American Journal of Medical Sciences.* Chief work was *Natural History* [see Works], which dealt with mammals and ranks as first original American publication in its field. While in Baltimore and in Germantown, gathered many observations and impressions during nature walks and in final illness prepared these for serial publication, later gathered together in *Rambles* [see Works].

WORKS: *Anatomical Investigations, Comprising Descriptions of the Various Fasciae of the Human Body* (Philadelphia, 1824); *American Natural History,* 3 vols. (Philadelphia, 1826-1828); *Rambles of a Naturalist* (Philadelphia, 1833). See *Natl. Un. Cat. Pre-1956 Imp.* WORKS ABOUT: *DAB,* 7: 350-51 (William S. Miller); Miller, "John Davidson Godman," *Ann. Med. Hist.* 2d ser. 9 (1937): 293-303.

GOESSMANN, CHARLES ANTHONY (1827-1910). *Chemistry.* Professor of chemistry, Massachusetts Agricultural College (Univ. of Massachusetts).
WORKS ABOUT: See *AMS; DAB.*

GOOCH, FRANK AUSTIN (1852-1929). *Chemistry.* Professor of chemistry and director of chemical laboratory, Yale Univ.
WORKS ABOUT: See *AMS; NCAB,* 12 (1904): 329-30.

GOOD, ADOLPHUS CLEMENS (December 19, 1856, West Mahoning, Pa.—December 13, 1894, Efulen, German territory north of French Congo, Africa). *Natural history.* Son of Abram, farmer, and Hannah (Irwin) Good. Married Lydia B. Walker, 1883; survived by one child. EDUCATION: 1879, A.B., Washington and Jefferson College; 1882, graduated, Western Theological Seminary (Allegheny, Pa.). HONORS: Ph.D., Washington and Jefferson College, ca. 1890. CAREER: 1882, ordained by presbytery of Kittanning, Pa., and went to Africa under sponsorship of Presbyterian Board of Foreign Missions; 1882-1885, with mission at Baraka on Gaboon River, French Congo; 1885-1889 and 1890-1892, with new mission at Kangive, 150 miles inland; 1889-1890, in United States; 1892-1894, missionary work in Bulu country, German territory of Africa. SCIENTIFIC CONTRIBUTIONS: Before going to Africa, asked by W. J. Holland [q.v.], chancellor of Western Univ. of Pennsylvania, to collect insects; Good collected thousands of specimens, especially Lepidoptera and Coleoptera, and was credited with having contributed more to knowledge of African insects than any person before him. Apparently wrote nothing on subject himself, but Holland and others published papers based on his collections that contained hundreds of new species and many new genera; at Holland's request, Good provided notes on life histories of West African Lepidoptera, including observations of butterfly migrations. Also collected birds and mammals. In missionary work, went deeper inland than any previous white explorer.
WORKS: Holland [see below], pp. 292-95, lists twenty-nine papers, most by himself, published in 1886-1897 and based on Good's collections of Lepidoptera. WORKS ABOUT: *DAB,* 7: 375 (Frederick T. Persons); Ellen C. Parsons, *A Life for Africa: Rev. Adolphus Clemens Good* (Edinburgh and London, 1899), especially appendix A, pp. 291-98, W. J. Holland, "The Scientific Labors of Rev. A. C. Good."

GOODALE, GEORGE LINCOLN (1839-1923). *Botany.* Professor of botany, Harvard Univ.
WORKS ABOUT: See *AMS; DAB.*

GOODE, GEORGE BROWN (February 13, 1851, New Albany, Ind.—September 6, 1896, Washington, D.C.). *Ichthyology; Science administration.* Son of Francis Collier, merchant, and Sarah Woodruff (Crane) Goode. Married Sarah Lamson Ford Judd; survived by four children. EDUCATION: 1870, A.B., Wesleyan College (Wesleyan Univ.); 1870-1871, studied natural history with J. L. R. Agassiz [q.v.], Harvard Univ. CAREER: 1871-1877, curator, Orange Judd Museum of Natural History, Wesleyan College; 1873-1877, part-time work for U.S. Commission on Fish and U.S. Natl. Mus., both under S. F. Baird [q.v.]; 1877-1887, museum curator, Smithson. Instn.; ca. 1887-1888, U.S. fish commissioner; 1887-1896, assistant secretary, Smithson. Instn., in charge of U.S. Natl. Mus. MEMBERSHIPS: Natl. Ac. Scis; Am. Phil. Soc.; and other societies. SCIENTIFIC CONTRIBUTIONS: Added to science as administrator, research ichthyologist, and historian; wrote on fundamentals of museum administration, and except for early death, perhaps would have become secretary of Smithsonian. Publications on ichthyology began in 1871; during connection with Commission on Fish, collected fishes along East Coast, became familiar with economic problems of fish industry, and conducted survey of American fisheries for Tenth U.S. Census (1880); most important scientific work published as *Oceanic Ichthyology* [see Works], dealing with all then-known deep-sea fishes; it drew on collections of number of explorations and added 156 new species and 55 new genera to Atlantic fauna.
WORKS: "Catalogue of the Fishes of Bermuda," *Bulletin of U.S. Natl. Mus.,* no. 5 (1876); *The Fisheries and Fishery Industries of the U.S.,* 7 vols. (Washington, D.C., 1884-1887); *Oceanic Ichthyology,* written with T. F. Bean [q.v.], issued as *Special Bulletin of U.S. Natl. Mus.,* 2 vols. (1895); *Smithsonian Institution, 1846-1896* (Washington, D.C., 1897). See bibliography in *Annual Report of U.S. Natl. Mus., 1897,* pt. 2, pp. 479-500. MSS: Va. St. Lib. Arch. Div. (*NUCMC* 61-2419); Smithson. Instn. Archives (*NUCMC* 72-1237). WORKS ABOUT: *DAB,* 7: 381-82 (David S. Jordan); Paul H. Oehser, *Sons of Science: The Story of the Smithsonian Institution and Its Leaders* (New York, 1949), pp. 92-109.

GORHAM, JOHN (February 24, 1783, Boston,

Mass.—March 27, 1829, Boston, Mass.). *Chemistry.* Son of Stephen, merchant, and Molly (White) Gorham. Married Mary Warren, 1808; no information on children. EDUCATION: A.B. in 1801 and M.B. [medicine] in 1804, Harvard Univ.; ca. 1804-1806, studied medicine and chemistry in Paris, London, and Edinburgh; 1811, M.D., Harvard. CAREER: 1806-1829, medical practice, Boston; at Harvard, adjunct professor of chemistry and materia medica (1809-1816) and Erving professor of chemistry and mineralogy (1816-1827). MEMBERSHIPS: Am. Ac. Arts Scis. SCIENTIFIC CONTRIBUTIONS: While teaching chemistry, primary concern was medical practice and finally gave up chemistry to devote more time to duties as physician; during active years, had reputation as best chemist in Boston area. Chief work was *Elements* [see Works], which ranks as one of first systematic treatments of chemistry written by American author and published in U.S.; it grew out of chemical lectures and was attempt to give overview of chemistry, without involving unnecessary details; *Elements* was avowedly compilation from other sources, but showed thorough knowledge of subject and was standard text for number of years. Teaching and writing helped to promote study of chemistry in American educational institutions. A founder and for fifteen years an editor of *New England Journal of Medicine and Surgery.*

WORKS: *The Elements of Chemical Science,* 2 vols. (Boston, 1819-1820); "An Analysis of Sulphate of Barytes from Hatfield, Mass.," *Memoirs of Am. Ac. Arts Scis.* 3, pt. 2 (1815): 237-40; "Contributions to Chemistry. No. 1. Indigogene," *New England Journal of Medicine* 6 (1817): 169-76; "Chemical Analysis of Indian Corn," ibid. 9 (1820): 320-28. MSS: Harv. U-UA (*NUCMC* 65-1245). WORKS ABOUT: *DAB,* 7: 433 (Lyman C. Newell); review of *Elements* in *Am. Journ. Sci.* 3 (1821): 330-41; James Jackson, *An Address Delivered at the Funeral of Dr. John Gorham . . .* (Boston, 1829).

GOULD, AUGUSTUS ADDISON (April 23, 1805, New Ipswich, N.H.—September 15, 1866, Boston, Mass.). *Conchology.* Son of Nathaniel Duren—farmer, music teacher, and conductor—and Sally Andrews (Prichard) Gould. Married Harriet Cushing Sheafe, 1833; ten children. EDUCATION: A.B. in 1825 and M.D. in 1830, Harvard Univ. CAREER: Ca. 1825/1826, private tutor in Maryland for one year; ca. 1826, began study of medicine, including internship at Massachusetts General Hospital; 1830, commenced medical practice and became one of chief physicians in Massachusetts; 1837, engaged to prepare report on invertebrates of Massachusetts, in connection with state survey; also catalogued pamphlets for Boston Athenaeum; 1835 and 1836, taught botany and zoology at Harvard. MEMBERSHIPS: President, Bos. Soc. Nat. Hist., 1865; member of Natl. Ac. Scis. and other societies. SCIENTIFIC CONTRIBUTIONS: Was major figure in early conchological study in America. One of his most important works was done in connection with Zoological and Botanical Survey of Massachusetts; survey *Report* [see Works], illustrated with drawings done by author, was first attempt in U.S. to study mollusks of a region, including freshwater, marine, and terrestrial forms, and long remained standard work. Prepared reports on mollusks gathered by number of government-sponsored explorations, especially shells collected by Wilkes Expedition and North Pacific Surveying and Exploring Expedition of 1853-1855; in addition to describing new species, works were pioneering contributions to study of geographical distribution of mollusks, their relations to geology, study of their anatomy, and other related subjects. In addition to other work, edited for publication *Terrestrial Air-breathing Mollusks* by A. Binney [q.v.]. All scientific work done during time not devoted to medical practice.

WORKS: *Report on the Invertebrata of Massachusetts* [not including insects] (Cambridge, Mass., 1841); *Principles of Zoology* (Boston, 1848), with J. L. R. Agassiz [q.v.]; *Mollusca and Shells,* Report for Wilkes Expedition, vol. 12 (Philadelphia, 1852); "Descriptions of Shells Collected in the North Pacific Exploring Expedition," *Proceedings of Bos. Soc. Nat. Hist.* 6 (1856-1859), 7 (1859-1861), and 8 (1861-1862). See bibliography in *Biog. Mems. of Natl. Ac. Scis.* 5 (1905): 106-13. MSS: Several locations including Boston Museum Science, Harv. U-HL, and other repositories (*DSB*). WORKS ABOUT: *DAB,* 7:446-47 (Frank C. Baker); *DSB,* 5:477-79 (George E. Gifford, Jr.).

GOULD, BENJAMIN APTHORP, Jr. (September 27, 1824, Boston, Mass.—November 26, 1896, Cambridge, Mass.). *Astronomy.* Son of Benjamin Apthorp, educator and merchant, and Lucretia Dana (Goddard) Gould. Married Mary Apthorp Quincy, 1861; five children. EDUCATION: 1844, A.B., Harvard College; studied at universities of Berlin and Göttingen; 1848, Ph.D., Univ. of Göttingen. CAREER: 1844-1845, taught at Boston Latin School; 1848, after returning to Cambridge from Germany, taught French, German, and mathematics for two years; 1849-1861 and 1886-1896, published and edited *Astronomical Journal*; 1852-1867, head of longtitude department, U.S. Coast Survey; 1855-1859, executive officer of scientific council and later director, Dudley Observatory, Albany; 1858, moved to Albany; 1859, returned to Cambridge; 1859-1864, involved in caring for late father's mercantile business; 1864, became actuary with U.S. Sanitary Commission; 1870-1885, at invitation of Pres. Domingo Sarmiento, established Argentine National Observatory at Cordoba; 1885-1896, returned to Massachusetts and worked on astronomical data gathered while in South America. MEMBERSHIPS: Roy. Soc.; Natl. Ac. Scis.; and other societies. SCIENTIFIC CONTRIBUTIONS: After return from Germany in 1848, set out to establish more firmly the study of astronomy in U.S., and made important contribution through publication of his *Astronomical Journal.* In work with Coast Survey, early recognized usefulness of telegraph for longitude studies, and prepared *Standard Mean* catalogue [see Works], earliest effort at correcting and combining star positions as determined at various observatories; after marriage in 1861, established private observatory near Cambridge and investigated star positions near north celestial pole. In 1866, became interested in use of photography in astronomy and collaborated with L. M. Rutherfurd [q.v.] in study of Pleiades and Praesepe star clusters. Most important work was stellar observations in South America, where thousands of observations and great many photographs were made and later measured and reduced, and general and zone catalogues were published as part of *Resultados* [see Works].

WORKS: *Standard Mean Right Ascensions of Circumpolar and Time Stars Prepared for the Use of U.S. Coast Survey* (Washington, D.C., 1862); *Resultados del Observatorio Nacional Argentino en Córdoba,* vols. 1-15 (Buenos Aires, 1879-1896). Bibliography in *Memoirs of Natl. Ac. Scis.* 17 (1924): 171-80. MSS: No collection is known; some items in Yale Univ. Library and in E. D. Cope Collection at Haverford College Library. WORKS ABOUT: *DAB,* 7: 447-49 (Raymond S. Dugan); *DSB,* 5: 479-80 (Brian G. Marsden).

GRAHAM, JAMES DUNCAN (April 4, 1799, Prince William County, Va.—December 28, 1865, Boston, Mass.). *Topographical engineering.* Son of Dr. William and Mary (Campbell) Graham. Married Charlotte Hustler Meade, 1828; later, Frances Wickham; had children. EDUCATION: 1817, graduated, U.S. Military Academy, West Point. CAREER: 1817, made third lieutenant, assigned to artillery; 1819-1821, assistant to Maj. Stephen H. Long on expedition to Rocky Mountains; 1822, assigned to topographical duty; 1838, commissioned major, U.S. Corps of Topographical Engineers; 1839-1840, astronomer in survey of U.S.-Texas border; 1840-1843, commissioner for survey and exploration of northeast boundary of U.S. (Maine); 1843-1847, head of scientific corps and principal U.S. astronomer for demarcation of boundary between U.S. and British Provinces, under Treaty of Washington; 1849-1850, in charge of resurvey of Mason-Dixon line; 1850-1851, principal astronomer and head of scientific corps, survey of portion of U.S. and Mexican boundary; 1854-1864, duty on Great Lakes, especially director of harbor improvements; 1863, promoted to colonel, Corps of Engineers; 1864-1865, in charge of harbor repair and construction on Atlantic Coast. MEMBERSHIPS: Am. Phil. Soc. SCIENTIFIC CONTRIBUTIONS: In addition to labors as topographical engineer with important surveying projects [outlined above], also published several scientific papers on problems in geodesy, including observations of magnetic dip in northeastern and southwestern frontiers of U.S. (1844), transit of Mercury in 1845, and latitude and longitude of various points, the latter prepared especially during period of duty in Great Lakes region. During period 1858-1859 carried out significant studies leading to discovery of lunar tide on Great Lakes.

WORKS: "Geographical Determinations and the Discovery of a Lunar Tidal Wave in Lake Michigan," *Proceedings of Am. Phil. Soc.* 7 (1859-1860): 378-384; "Investigation of the Problem Regarding the Existence of a Lunar Tidal Wave on the Great Freshwater Lakes of North America," *Proceedings of Am. Assoc. Advt. Sci.* 14 (1860): 52-60. See *DAB* (which also includes reference to manuscripts in private hands) and *Roy. Soc. Cat.,* vols. 2, 6 (additions), 7. WORKS ABOUT: *DAB,* 7: 476 (Carroll S. Alden); Cullum, *Biographical Register, West Point,* 1 (1891): 157-58.

GRAY, ASA (November 18, 1810, Sauquoit, Oneida County, N.Y.—January 30, 1888, Cambridge, Mass.). *Botany.* Son of Moses, farmer and tanner, and Roxana (Howard) Gray. Married Jane Lathrop Loring, 1848; no children. EDUCATION: 1825, entered Fairfield Academy (Herkimer County, N.Y.); 1831, M.D., Fairfield College of Physicians and Surgeons of Western District of New York. HONORS: Degrees from universities of Oxford, Cambridge, and Aberdeen. CAREER: Ca. 1831-1832, medical practice, Bridgewater, N.Y.; 1832-1835, taught science, especially at Bartlett's High School, Utica, N.Y., and during summers botanized in New York, Pennsylvania, and New Jersey; 1835, moved to New York City, became full botanical partner with J. Torrey [q.v.], and worked as curator, N.Y. Lyc. Nat. Hist.; 1836-1838, held appointment as botanist with planned Wilkes Expedition; 1838, resigned as Wilkes's botanist and accepted appointment as professor of botany at new Univ. of Michigan; 1838-1839, in Europe to buy books for university and to study type-specimens for Torrey-Gray *Flora;* 1839-1842, worked on *Flora* while waiting in vain for commencement of duties at Michigan; 1842-1888, Fisher professor of natural history, Harvard Univ., directing attention chiefly to botany and the botanic garden; 1873, retired from teaching, but retained professorship; while at Harvard, made several trips to Europe and to Florida, California, and Mexico. MEMBERSHIPS: Natl. Ac. Scis.; Roy. Soc.; and many other societies. SCIENTIFIC CONTRIBUTIONS: The leading American botanist of his generation. Collaboration with Torrey firmly established reliance on type-specimens and on the natural over the Linnaean classification in U.S. botany. Most work was in descriptive and taxonomic botany; described and classified specimens collected by government explorers and by explorers sponsored by himself; did important work on plant geography, especially on Japan and North America. Letter from Darwin (1857) later used as evidence of latter's priority over A. R. Wallace; became major defender of Darwinism while maintaining personal religious and scientific views. Wrote influential botany textbooks; firmly established Botanic Garden and Herbarium at Harvard.

WORKS: Among major works were *Flora of North America*, 2 vols. (New York, 1838-1843), with Torrey; *Manual of the Botany of the Northern U.S.* (Boston, 1848; with 5 eds. in his lifetime); *Darwiniana* (New York, 1876), a collection of Gray's essays on evolution; monograph on relation of botany of Japan and North America in *Memoirs of Am. Ac. Arts Scis.* n. s. 6 (1859): 377-452. Bibliography of works (780) in *Am. Journ. Sci.* 3d ser. 36 (1888), appendix 3-67. MSS: Harv. U.-GH (*NUCMC* 75-1978); see additional references in *NUCMC* and Hamer. WORKS ABOUT: *DAB*, 7: 511-14 (George H. Genzmer); *DSB*, 5: 511-15 (A. Hunter Dupree). Also see Dupree, *Asa Gray* (Cambridge, Mass., 1959).

GREEN, FRANCIS MATHEWS (February 23, 1835, Boston, Mass.—December 19, 1902, Albany, N.Y., while on visit from home in Boston). *Hydrography.* Son of Mathews Wylly and Margaret Augusta (Gilchrist) Green. Married Elizabeth S. Cushing, 1870; no information on children. EDUCATION: 1854, graduated from English High School, Boston. CAREER: 1854-1861, seaman; 1861, joined volunteer navy as acting master; 1864, became acting volunteer lieutenant; 1868, appointed lieutenant commander in regular navy; 1872, attached to U.S. Hydrographic Office to prepare publication on sailing directions for West Indies; 1877, headed expedition to determine longitudes on east coast of South America; 1881, led expedition for similar purposes to China, Japan, and East Indies; 1883, promoted to commander; in later years, spent time as captain of Pennsylvania schoolship *Saratoga*; 1897, retired from U.S. Navy; in years just before death, engaged as editorial contributor to *Century Dictionary*. SCIENTIFIC CONTRIBUTIONS: Scientific work done with U.S. Navy, and over years various expeditions made important additions to more accurate knowledge of surface of earth. In preparation of publication on West Indies sailing, took part (1874) in survey of coast of Mexico. Through personal effort, became accomplished astronomical observer, and when West Indies submarine cables were completed, was put in charge of Hydrographic Office expedition to determine longitude in West Indies and Central America by exchange of telegraphic time signals; 1877 expedition to determine longitudes of chief locations on east coast of South America involved exchange of time signals on transatlantic cables and submarine cables from Brazil to Buenos Aires; similar work later done in Pacific area.

WORKS: *The Navigation of the Caribbean Sea and Gulf of Mexico*, vol. 1 (Washington, D.C., 1877); *Report on the Telegraphic Determination of Differences of Longitude in the West Indies and Central America* (Washington, D.C., 1877); also, with C. H. Davis and Lt. N. A. Norris, *Telegraphic Determination of Longitudes on the East Coast of South America* (Washington, D.C., 1880) and *Telegraphic Determination of Longitudes in Japan, China, and the East Indies* (Washington, D.C., 1883). Major works mentioned in *DAB*. WORKS ABOUT: *DAB*, 7: 543-44 (George W. Littlehales); Edward W. Callahan, *List of Officers of the Navy of the U.S. and the Marine Corps from 1775 to 1900* (New York, 1901).

GREEN, JACOB (July 26, 1790, Philadelphia, Pa.—February 1, 1841, Philadelphia, Pa.). *Chemistry; Zoology; Botany.* Son of Ashbel, minister and president of Princeton Univ., and Elizabeth (Stockton) Green. Married; two children. EDUCATION: 1807, A.B., Univ. of Pennsylvania; studied medicine, law, and theology; admitted to New York bar. HONORS: M.D., Yale Univ., 1827. CAREER: For several years, bookseller in Albany, N.Y.; practiced law; 1816, moved to Princeton, N.J., to reside with father and study theology; 1818-1822, professor of chemistry, experimental philosophy, and natural history, Princeton Univ.; ca. 1822, moved to Philadelphia; 1822-1825, pursued scientific and literary interests and gave course of public lectures on chemistry; 1825-1841, professor of chemistry, Jefferson Medical College (Philadelphia), which he had helped to found. MEMBERSHIPS: Apparently member of Ac. Nat. Scis. Philad. SCIENTIFIC CONTRIBUTIONS: During course of career was actively interested in wide range of scientific fields; had early interest in electricity and in botany and astronomy. Work on trilobites [see Works] most important contribution to biology (paleontology); also wrote on mollusks, salamanders, and lizards, and had some interest in ethnology. Chief interest was chemistry and published three textbooks on subject. While writing a great deal, his original contributions to knowledge were relatively minor, perhaps because of diversity of interests; most effective as teacher and as disseminator of scientific knowledge; even chemical works were essentially compilations.

WORKS: *A Catalogue of the Plants Indigenous to the State of New York* (Albany, 1814); *Astronomical Recreations* (Philadelphia, 1824); *Electro-Magnetism* (Philadelphia, 1827); *Textbook of Chemical Philosophy on the Basis of Turner's Elements* (Philadelphia, 1829); *Monograph on the Trilobites of North America* (Philadelphia, 1832; with supplement, Philadelphia, 1835). Chief works listed in *DSB*; also see *Roy. Soc. Cat.*, vol. 3. WORKS ABOUT: *DAB*, 7: 548-49 (Horace B. Baker); *DSB*, 5: 517-18 (Wyndham D. Miles).

GREENE, EDWARD LEE (1843-1915). *Botany.* Professor of botany, Univ. of California and Catholic Univ.; associate in botany, Smithson. Instn.

MSS: Univ. of Notre Dame Archives (*NUCMC* 67-521); Harv. U.-GH (Hamer). WORKS ABOUT: See *AMS; DAB.*

GREENWOOD, ISAAC (May 11, 1702, Boston, Mass.—October 12, 1745, Charleston, S.C.). *Mathematics; Natural philosophy.* Son of Samuel, well-to-do merchant and shipbuilder, and Elizabeth (Bronsdon) Greenwood. Married Sarah Shrimpton Clarke, 1729; five children. EDUCATION: 1721, A.B., Harvard College; 1723-1726, went to London and pursued theological studies, audited lectures on experimental philosophy by John T. Desaguliers, and also may have spent some time in study of medicine. CAREER: While in London, began preaching; 1727, offered public experimental course in mechanical philosophy at Boston; 1727, appointed as first incumbent in Hollis professorship of mathematics and natural philosophy at Harvard, having been nominated by English benefactor, Thomas Hollis; 1728-1738, was Hollis professor and during this period also delivered series of popular lectures on astronomy in Boston; 1738, removed from professorship because of excessive drinking; 1738-1845, attempted to establish school of experimental philosophy in Boston, lectured in Philadelphia, and served as tutor at sea. SCIENTIFIC CONTRIBUTIONS: Noteworthy for teaching and especially for role in updating and more firmly establishing place of science in Harvard College; used experiment and demonstration in teaching, having brought from England fine collection of scientific apparatus, the gift of Thomas Hollis. Was familiar with work of Newton, which no doubt formed basis for much of his teaching, and among other subjects apparently taught Newtonian fluxions. *Arithmetick* [see Works] was first such textbook in English to be written by native American; published several articles dealing with meteors, damps, and aurora borealis in *Phil. Trans. of Roy. Soc.*

WORKS: *Experimental Course of Mechanical Philosophy* (Boston, 1726); *Arithmetick, Vulgar and Decimal: With the Application thereof to a Variety of Cases in Trade and Commerce* (Boston, 1729); *Prospectus of Explanatory Lectures on the Orrery* (Boston, 1734); *A Philosophical Discourse Concerning the Mutability and Changes of the Material World* (Boston, 1731). MSS: Harv. U.-UA and Mass. Hist. Soc. (*DSB*). WORKS ABOUT: *DAB*, 7: 591-92 (David E. Smith); *DSB*, 5: 519-20 (Brooke Hindle); Shipton, *Biog. Sketches, Harvard*, 6 (1942): 471-82.

GREW, THEOPHILUS (died 1759, Philadelphia, Pa.). *Mathematics; Astronomy.* No information on parentage. Married Elizabeth Cosins, 1735; Frances Bowen, 1739; Rebecca Richards, 1747; no information on children. EDUCATION: No information. HONORS: M.A., College and Academy of Philadelphia (later, Univ. of Pennsylvania), 1757. CAREER: 1732, published almanac, Annapolis, Md.; as early as 1734, teacher of mathematics, surveying, navigation, and astronomy in Philadelphia; 1740-ca. 1742, headmaster, Public School of Kent County, Md.; ca. 1742, returned to Philadelphia and reopened mathematical school; 1750-1759, first professor of mathematics, College and Academy of Philadelphia, serving as master of its mathematical school and also teacher in the college; 1753, given permission to conduct private evening school; consulted in relation to various city and colonial land surveys. SCIENTIFIC CONTRIBUTIONS: Details of life are uncertain. In 1732, published almanac in Annapolis and in later years published almanacs there and in Philadelphia, New York, and Williamsburg, Va., and achieved wide reputation for astronomical calculations; in newspapers, published solutions to mathematical questions relating to surveying and navigation. Consulted in regard to protracted boundary dispute between Pennsylvania and Maryland, and served as one of Pennsylvania commissioners at meeting at New Castle, Del., in November 1750.

WORKS: *The Maryland Almanack for the Year of our Lord God, 1733 . . .* (Annapolis, Md., 1732); *The Description and Use of the Globes, celestial and terrestrial; with variety of examples for the learner's exercise* (Germantown, Pa., 1753), prepared as text for students at academy and college. See Charles Evans, *American Bibliography* (Chicago, 1903-). WORKS ABOUT: *DAB*, 7: 609-10 (Robert F. Seybolt); Edward P. Cheyney, *History of the University of Pennsylvania, 1740-1940* (Philadelphia, 1940).

GRIFFITH, ROBERT EGLESFELD (February 13, 1798, Philadelphia, Pa.—June 26, 1850, Philadelphia, Pa.). *Botany; Conchology.* Son of Robert Eglesfeld and Maria Thong (Patterson) Griffith. Married Mary Eyre, 1829; three children. EDUCATION: 1820, M.D., Univ. of Pennsylvania. CAREER: 1820, began medical practice, Philadelphia; 1831-1835, editor, *Journal of Philadelphia College of Pharmacy;* 1833-1836, physician to city board of health; 1835, professor of materia medica, Philadelphia College of Pharmacy; 1835-1836, editor of *Am. Journ. Pharm.;* 1836-1838, professor of materia medica, therapeutics, hygiene, and medical jurisprudence, Univ. of Maryland; 1837-1839, professor of practice, obstetrics, and medical jurisprudence, Univ. of Virginia; 1839, returned to Philadelphia. MEMBERSHIPS: Am. Phil. Soc.; Franklin Institute; and other societies. SCIENTIFIC CONTRIBUTIONS: Was known as botanist and conchologist; botanical works published especially in pharmaceutical journals, and also prepared *Medical Botany.* At time of death had begun plans for a work on "Botany of the Bible." Gave large collection of shells to Ac. Nat. Scis. Philad., and had started preparation of work on conchology before death, but no bibliographic references to publications on that subject have been found in standard sources.

WORKS: *Medical Botany* (Philadelphia, 1847); *Universal Formulary* (Philadelphia, 1848). See *Roy. Soc. Cat.,* vol. 3, for references to medical-botanical works in *Journal of Philadelphia College of Pharmacy* and *Am. Journ. Pharm.;* also see *Natl. Un. Cat. Pre-1956 Imp.* WORKS ABOUT: Kelly and Burrage, *Med. Biog.* (1928); *NCAB,* 12 (1904): 552.

GRIMES, JAMES STANLEY (May 10, 1807, Boston, Mass.—September 27, 1903, Evanston, Ill.). *Natural philosophy; Phrenology.* Son either of Andrew and Polly Robbins Grimes or of Joseph and Sally Robbins Grimes. No information on marital status. EDUCATION: No information. CAREER: Practiced law in Boston and New York City; then professor of medical jurisprudence, Castleton Medical College; also taught at Willard Institute for women; finally went to Evanston, Ill., and lived there for remainder of life; lectured and debated throughout eastern states, and in 1860 gave series of eight lectures on spiritualism in Boston, his last public appearance in New England; shortly thereafter, retired to Evanston, and last forty years of life were passed out of public view. MEMBERSHIPS: Am. Assoc. Advt. Sci. SCIENTIFIC CONTRIBUTIONS: Details of life uncertain. Attempted to deal with science in widest philosophical sense and to expand its realm of concern to include study of phenomena of mind, spirit, and the like; lacked sufficient background for magnitude of studies addressed, but was opposed to superstition and produced some original and interesting ideas. Early interest was in phrenology and, beginning in 1832, ardently promoted its study; in *New System* [see Works] proposed classification for phrenology that differed from accepted system of Spurzheim, but was opposed by O. S. Fowler [q.v.]. Considered one of first American evolutionists, and claimed his *Phreno-Geology* [see Works] was earliest work ever published on theistic evolution. Also studied mental healing, and was one of first Americans to contribute constructively to study of mesmerism. *Geonomy* [see Works] dealt with problems of formation of planetary system and earth.

WORKS: *A New System of Phrenology* (Buffalo and New York, 1839); *Phreno-Geology: The progressive creation of man, indicated by natural history, and confirmed by discoveries which connect the organization and functions of the brain with the successive geological periods* (Boston and Cambridge, Mass., 1851); *Outlines of Geonomy* (Boston, 1858); "Laws of the Ocean Currents," *Proceedings of Am. Assoc. Advt. Sci.* 17 (1868): 106-13. Major works mentioned in *DAB;* see also *Natl. Un. Cat. Pre-1956 Imp.* WORKS ABOUT: *DAB,* 7: 630-31 (Ernest S. Bates); *World Who's Who in Science.*

GRISCOM, JOHN (September 27, 1774, Hancock's Bridge, N.J.—February 26, 1852, Burlington, N.J.). *Chemistry.* Son of William, farmer and saddle and harness maker, and Rachel (Denn) Griscom. Married Abigail Hoskins, 1800; Rachel Denn, 1843; eight children. EDUCATION: Attended country schools; 1793, studied for several months in Friends' Academy, Philadelphia; self-taught in chemistry. HONORS: LL.D., Union College, 1824. CAREER: 1791-1793, teacher near Salem, N.J.; 1793-1794, teacher; 1794-1807, taught in Burlington, N.J., and began lecturing on chemistry; 1807-1831, operated school in New York City, which in 1825 became New York School for boys; 1818-1819, in Europe; while in New York, gave public lectures on chemistry and in 1813-1820 lectured on chemistry at Columbia College and Rutgers Medical School (New York City); 1832-1834, principal, Friends' School, Providence, R.I.; ca. 1834-1840, lived with daughters in West Haverford, Pa., and for three or four years arranged and supervised proof sheets for *Journal of Franklin Institute;* 1841-1852, returned to Burlington, N.J., lectured, served as superintendent of schools until death, although almost blind during last two years. MEMBERSHIPS: Literary and Philosophical Society of New York. SCIENTIFIC CONTRIBUTIONS: Did little original work in chemistry, but is known as teacher and as disseminator of scientific discoveries from Europe; school in New York was well equipped for teaching science; for nearly thirty years was leading teacher of chemistry in that city, and helped prepare great number of chemical students. Well acquainted with European chemical literature and during 1818-1819 trip to Europe met number of leading scientists, giving impressions in *Year in Europe* [see Works]. For some years, prepared abstracts, selections, and translations of foreign scientific publications for *Am. Journ. Sci.* and *Journal of Franklin*

Institute, and for some years was an editor of latter publication.

WORKS: *Year in Europe* (New York and Philadelphia, 1823). See *Roy. Soc. Cat.*, vol. 3. MSS: N.Y. Pub. Lib. (Hamer). WORKS ABOUT: *DAB*, 8: 7 (Edgar F. Smith); E. F. Smith, *John Griscom* (Philadelphia, 1925), 27 pp.

GROTE, AUGUSTUS RADCLIFFE (February 7, 1841, Aigburth, near Liverpool, England—September 12, 1903, Hildesheim, Germany). *Entomology.* Son of Friedrich Rudolf, farmer who engaged in real estate and other business ventures, and Anna (Radcliffe) Grote. Married first in 1880; second marriage after removal to Germany; at least two children by first marriage. EDUCATION: After 1846, lived with family in Staten Island, N.Y.; prepared for Harvard College, but family's financial reverses made attendance impossible; later studied in Europe. HONORS: M.A., Lafayette College, 1874. CAREER: Spent several years at Demopolis, Ala.; associated with Buffalo (N.Y.) Society of Natural Sciences and served for some years as society's curator; 1879-1880, edited *North American Entomologist* at Buffalo; ca. 1880, after death of father, returned to Staten Island; 1884-ca. 1894, resided in Bremen, Germany; ca. 1894, removed to Hildesheim, Germany, becoming honorary curator in Roemer-Museum. MEMBERSHIPS: Am. Assoc. Advt. Sci. SCIENTIFIC CONTRIBUTIONS: Publications in entomology began in 1862, and during career described in excess of 2,000 new species of American lepidopterous insects. Number of early papers written with Coleman T. Robinson (also associated with Buffalo Society), their earliest joint paper appearing in 1865. Grote came to rank as one of most important American entomologists of nineteenth century, and was first to investigate in detail the noctuid moths; while chief contributions were to descriptions of species of Lepidoptera, also wrote on other entomological subjects; published more than 200 works in entomology, and continued to contribute to American literature after removal to Germany. Also wrote on philosophical topics, including relations of science and religion, and published poetry.

WORKS: *An Illustrated Essay on the Noctuidae of North America* (London, 1882; reprinted, Hampton, England, 1971, with biographical summary by Ronald S. Wilkinson). See references to published bibliographies of works in *DAB* and *DSB*. WORKS ABOUT: *DAB*, 8: 27 (Leland O. Howard); *DSB*, 5: 555-56 (Ronald S. Wilkinson). Also Wilkinson [Works, above].

GUMMERE, JOHN (1784, near Willow Grove, Pa.—May 31, 1845, Burlington, N.J.) *Mathematics.* Son of Samuel, minister of Society of Friends and at one time postmaster, and Rachel (James) Gummere. Married Elizabeth Buzby, 1808; had children. EDUCATION: Attended common schools; at early age, began independent study of mathematics; ca. 1804, studied six months at Friends' Boarding School, Westtown, Pa. HONORS: M.A., Princeton Univ., 1825. CAREER: Ca. 1803/1804, teacher, Horsham, Pa.; ca. 1805-1811, teacher, Rancocas, N.J.; 1811-ca. 1814, teacher, Westtown, Pa.; 1814-1833, operated boarding school, Burlington, N.J.; 1833-1843, teacher of mathematics and for some years also principal, Haverford College; 1843-1845, with son, reestablished school at Burlington, N.J. MEMBERSHIPS: Am. Phil. Soc. SCIENTIFIC CONTRIBUTIONS: Largely through self-study, became one of top mathematicians in U.S., having knowledge of mathematical works in French, and later also studied German language as well; correspondent of R. Adrain and N. Bowditch [qq.v.]. School at Burlington said to have had better stock of scientific apparatus than any other private institution in U.S. Textbook on surveying [see Works] was chief American publication on subject and had twenty-two editions; work on astronomy [see Works] was used as textbook at West Point and was recognized in its day as unique among American works as aid to carrying out astronomical calculations. Said to have been offered, but declined, professorship of mathematics at Univ. of Pennsylvania.

WORKS: *Treatise on Surveying* (Philadelphia, 1814); *Elementary Treatise on Astronomy* (Philadelphia, 1822); "On the Construction of Eclipses of the Sun," *Transactions of Am. Phil. Soc.* n.s. 3 (1830): 467-71; "Practical Rule for Calculating, from the Elements in the 'Nautical Almanac,' the Circumstances of an Eclipse of the Sun for a Particular Place," ibid. n.s. 5 (1837): 297-318. WORKS ABOUT: *DAB*, 8: 49 (George H. Genzmer); William J. Allinson, *Memorials of the Life and Character of John Gummere* [1845].

GUTHRIE, SAMUEL (1782, Brimfield, Mass.—October 19, 1848, Sacketts Harbor, N.Y.) *Chemistry.* Son of Samuel, physician, and Sarah Guthrie. Married Sybil Sexton, 1804; four children. EDUCATION: Little formal education; studied medicine with father; 1810-1811, course of lectures, College of Physicians and Surgeons, New York; 1815, course of lectures, Univ. of Pennsylvania. CAREER: 1802-1811, medical practice, Smyrna, N.Y.; 1811, moved to Sherbourne, N.Y.; during War of 1812, examination surgeon; after war, resumed medical practice; 1817, moved to Sacketts Harbor, N.Y., engaged in farming, for a time practiced medicine, and was active as manufacturer and chemist, producing vinegar, alcohol, molasses from potato starch, and percussion pill and punch lock; held several local offices. SCIENTIFIC CONTRIBUTIONS: Best known to contemporaries for invention and production of priming powder, or percussion pill, and punch lock that exploded it, replacing flash-in-pan powder and flintlock musket; commenced manufacture of these "pellets" in 1826. At Sacketts Harbor, had laboratory near home and carried out various experiments, and in 1830 devised means for rapid conversion of potato starch to molasses, communicating discovery and sample of product to B. Silliman [q.v.] at *Am. Journ. Sci.* [see Works]. Other correspondence with Silliman related to work on crystallized potassium chlorate, several kinds of powder, oil of turpentine, and chloric ether, and in 1832 these communications also appeared in *Am. Journ. Sci.*; chloric ether produced in 1831 by distillation of chloride of lime with alcohol in copper still was actually chloroform, and Guthrie is recognized as original discoverer of that substance, antedating very slightly independent discoveries in Europe.

WORKS: "New Mode of Preparing a Spirituous Solution of Chloric Ether," *Am. Journ. Sci.* 21 (1832): 64-65; "On Sugar from Potato Starch," ibid., pp. 284-88; "On Pure Chloric Ether," ibid. 22 (1832): 105-6. Pawling [below], appendix, pp. 94-122, reproduces the several articles by Guthrie that appeared in *Am. Journ. Sci.* in 1832. WORKS ABOUT: *DAB*, 8: 62 (Lyman C. Newell); Jesse R. Pawling, *Dr. Samuel Guthrie: Discoverer of Chloroform, Manufacturer of Percussion Pellets, Industrial Chemist* (Watertown, N.Y., 1947).

GUYOT, ARNOLD HENRY (September 28, 1807, Boudevilliers, Switzerland—February 8, 1884, Princeton, N.J.). *Geography*; *Glaciology.* Son of David Pierre and Constance (Favarger) Guyot. Married Sarah Doremus Haines, 1867; no information on children. EDUCATION: 1821, began studies at Academy of Neuchâtel; 1825-1827, studied in Germany, especially in Karlsruhe and Stuttgart; 1827, returned to Neuchâtel to pursue preparation for ministry; 1829, went to Berlin to continue study for ministry, but later changed to science; 1835, Ph.D., Univ. of Berlin. CAREER: Ca. 1835-1839, tutor in Paris; 1839-1848, professor of history and physical geography, Academy of Neuchâtel; 1848, joined J. L. R. Agassiz [q.v.] at Cambridge, Mass.; 1849, gave Lowell lectures in Boston; ca. 1849-ca. 1854, lectured on geography and methods of teaching that subject for Massachusetts Board of Education; 1854-1884, professor of physical geography and geology, Princeton Univ.; under sponsorship of Smithson. Instn., undertook task of locating and equipping weather observation stations, especially in Massachusetts and New York. MEMBERSHIPS: Natl. Ac. Scis.; honorary member of geographical societies of London and Paris. SCIENTIFIC CONTRIBUTIONS: Prepared doctoral dissertation on natural classification of lakes; at Agassiz's suggestion, in 1838 began field work in Switzerland, investigating concept of glacial age; during number of trips to that country, made important contributions to study of glacial structures and their movement and to knowledge of moraines, but published little because of projected joint work with Agassiz that never appeared; in this work Guyot was to be responsible for section on erratic boulders of plains of Switzerland. *Earth and Man*, based on Lowell Institute lectures, was most significant work; during period 1861-1875, also prepared series of textbooks on geography. In work for Smithsonian, laid groundwork for system of weather stations throughout U.S.; and in establishing best locations for stations, carried out altitude surveys of Appalachians from Vermont to North Carolina.

WORKS: *Earth and Man; or, Lectures on Comparative Physical Geography in Its Relation to the History of Mankind* (Boston, 1849). Chief works listed in *DSB*; see bibliography in *Biog. Mems. of Natl. Ac. Scis.* 2 (1886): 345-47. MSS: Miscellaneous items and references in Princeton Univ. Library—Rare Books and Special Collections. WORKS ABOUT: *DAB*, 8: 63-64 (Robert M. Brown); *DSB*, 5: 599-600 (Albert V. Carozzi).

H

HAGEN, HERMANN AUGUST (May 30, 1817, Königsberg, Germany—November 9, 1893, Cambridge, Mass.). *Entomology.* Son of Carl Heinrich, royal councillor and professor of political economy, technology, and agriculture, and Anna Dorothea Linck Hagen. Married Johanna Maria Elise Gerhards, 1851; no information on children. EDUCATION: 1840, M.D., Univ. of Königsberg; ca. 1840-1843, studied medicine in Berlin, Vienna, and Paris; 1843, returned to Königsberg. CAREER: 1843-1867, practiced medicine and surgery, Königsberg, where served on school board and city council; 1867, at invitation of J. L. R. Agassiz [q.v.], took charge of entomology department, Museum of Comparative Zoology, Harvard; 1870-1893, professor of entomology, Harvard Univ., but continued to devote chief attention to duties in museum; 1882, took part in survey of injurious insects along Northern Pacific Railway. MEMBERSHIPS: Honorary member of most European entomological societies. SCIENTIFIC CONTRIBUTIONS: In 1839, published first entomological paper, on dragonflies, and during years in Europe wrote number of other entomological works, especially on neuropteroid group, including dragonflies, termites, lacewings, ant lions, psocids, and others, as well as on fossil and amber-encapsulated insects; of work done in Europe, bibliography of entomology, covering world literature to 1862, was most widely known [see Works]. At suggestion of C. R. R. Osten-Sacken [q.v.] undertook study of North American Neuroptera, and "Synopsis" (1861) [see Works] effectively began study of those insects in this country. In U.S., was first incumbent in professorship restricted to entomology, ranked as one of chief entomologists in world during lifetime, was teacher of several eminent entomologists, and also did notable work at Museum of Comparative Zoology, to which he gave personal collections in 1875.

WORKS: "Monographie der Termiten," *Linnaea Entomologica* 10 (1855) and 14 (1860); "Synopsis of the Neuroptera of North America," *Smithson. Misc. Coll.* 4 (1861); *Bibliotheca Entomologica,* 2 vols. (Leipzig, 1862-1863). Wrote in excess of 400 works; see W. Horn and S. Schenkling, *Index Litteraturae Entomologicae: Serie I . . . bis inklusive 1863* (Berlin, 1928-1929); W. Derksen and U. Scheiding-Göllner, *Index Litteraturae Entomologicae: Serie II . . . von 1864 bis 1900* (Berlin, 1963-). MSS. Harv. U.-MCZ. WORKS ABOUT: *DAB,* 8: 82 (Leland O. Howard); Mallis, *Amer. Entomologists,* pp. 119-26.

HAGEN, JOHN GEORGE (1847-1930). *Mathematics; Astronomy.* Member of Society of Jesus; director of observatory, Georgetown Univ.; director of Vatican observatory.

WORKS ABOUT: See *AMS; World Who's Who in Science.*

HAGER, ALBERT DAVID (November 1, 1817, Chester, Vt.—July 29, 1888, Chicago, Ill.). *Geology.* Son of David, probably carpenter, and Hannah (Caryl) Hager. Married three times; names of two wives are known: Julia A. Wheeler, ca. 1844; Rosa F. Blood, 1872 (third); three children by second marriage, one child by third. EDUCATION: Attended public schools in Vermont; learned carpentry from father. HONORS: A.M., Amherst College. CAREER: 1836, employed by map publisher and traveled in Ohio and Kentucky in pursuit of business; on return home, again took up carpenter's trade, and studied geology; 1838-ca. 1844, conducted school in Kentucky; ca. 1844-ca. 1849, farmer in Vermont; ca. 1849, returned to work as carpenter; 1854, became general agent for Cyrus H. McCormick; 1856-1857, Vermont assistant state naturalist; in late 1850s and until 1861, school superintendent at Cavendish, Vt.; 1857-1861, assistant geologist of Vermont; 1864-1870, Vermont state geologist and curator of cabinet of natural history; 1865-1869, Vermont fish commissioner; during late 1860s, worked as geological and mining consultant; 1870-1871, state geologist of

Missouri; 1872, moved to Chicago; 1873-1874, superintendent of Washingtonian Home (Chicago), an employment that grew out of interest in reform of alcoholics; 1877-1887, secretary-librarian, Chicago Historical Society. MEMBERSHIPS: Am. Assoc. Advt. Sci. SCIENTIFIC CONTRIBUTIONS: During period 1857-1861, served with Vermont Geological Survey under E. Hitchcock [q.v.], with responsibility for economic aspects, and had charge of preparation of final report. Only other known geological work was as head of Missouri survey, resulting in one progress report; work in Missouri was not successful, in either geological or administrative sense, and thereafter apparently ceased geological activity.

WORKS: *Report on the Economical Geology, Physical Geography and Scenery of Vermont; Being a Portion of the Geological Report of the State made by Prof. Hitchcock* . . . (Claremont, N.H., 1862). See *Natl. Un. Cat. Pre-1956 Imp.* MSS: Correspondence during later years in Chicago Historical Society. WORKS ABOUT: *NCAB*, 3 (1893 [copyright 1891]): 224; *Chicago Tribune*, July 30, 1888, p. 2; Merrill, *Hundred Years of Geology*; information from Chicago Historical Society; T. D. Seymour Bassett, *A History of the Vermont Geological Surveys and State Geologists* (Burlington, Vt., 1976), pp. 12-15.

HAGUE, ARNOLD (1840-1917). *Geology.* Assistant under C. R. King [q.v.] on U.S. Geological Survey of Fortieth Parallel; geologist, U.S. Geol. Surv.

WORKS ABOUT: See *AMS*; *DAB*; *DSB*.

HAGUE, JAMES DUNCAN (1836-1908). *Geology; Mining.* First assistant geologist under C. R. King [q.v.] on U.S. Geological Survey of Fortieth Parallel; consulting engineer and expert on mining, New York City.

MSS: Hunting. Lib. (*NUCMC* 61-3003). WORKS ABOUT: See *AMS*; *DAB*.

HALDEMAN, SAMUEL STEMAN (August 12, 1812, Locust Grove, Lancaster County, Pa.—September 10, 1880, Chickies, Pa.). *Zoology*; *Philology.* Son of Henry, businessman, and Frances (Steman) Haldeman. Married Mary A. Hough, 1835; survived by four children. EDUCATION: 1828-1830, attended Dickinson College. CAREER: 1830-1835, studied at home and operated father's sawmill; 1836-1837, assistant, Geological Survey of New Jersey; 1837, became assistant on Geological Survey of Pennsylvania, maintaining official connection for only one year, but assisted geologist H. D. Rogers [q.v.] until 1842; 1837-1880, resided at Chickies, Pa., was inactive partner, with brothers, in ironworks, and pursued various studies; during early 1840s, traveled to Tennessee, Carolinas, and elsewhere to collect shells; 1842-1843, lecturer in zoology, Franklin Institute; 1847, 1859, 1861-1862, 1866, and 1875, study trips to Europe; 1851-1855, professor of natural history, Univ. of Pennsylvania; 1851-1852, editor of *Pennsylvania Farmer's Journal*; 1852, became chemist and geologist to Pennsylvania State Agricultural Society; 1855-1858, professor of natural history, Delaware College; 1868-1880, professor of comparative philology, Univ. of Pennsylvania. MEMBERSHIPS: Am. Phil. Soc.; Natl. Ac. Scis.; and other societies. SCIENTIFIC CONTRIBUTIONS: Early reputation in science based on *Freshwater . . . Mollusca* [see Works], first part of which appeared in 1840; later became noted for entomological studies. During period of work at father's sawmill, became interested in vocal sounds, and by late 1850s these studies eclipsed work in zoology; had finely tuned hearing and was able to differentiate wide range of sounds, including those produced by insects [see Works]; studied speech patterns in U.S. and Europe, and became authority on dialects of American Indians.

WORKS: *Monograph of the Freshwater Univalve Mollusca of the U.S.*, 8 pts. (Philadelphia, 1840-1845; reprinted, [Philadelphia], 1871); "A New Organ of Sound in Lepidoptera," *Am. Journ. Sci.* 2d ser. 5 (1848): 435; *Analytic Orthography: An Investigation of the Sounds of the Voice, and Their Alphabetic Notation* (Philadelphia, 1860). References to bibliographies of works in *DAB*. MSS: Am. Phil. Soc. (*NUCMC* 61-789); Ac. Nat. Scis. Philad. (*NUCMC* 66-49); Hist. Soc. Penn. WORKS ABOUT: *DAB*, 8: 94-95 (George C. Harvey); J. P. Lesley, "Memoir," *Biog. Mems. of Natl. Ac. Scis.* 2 (1886): 139-72.

HALL, ASAPH (1829-1907). *Astronomy.* Professor, U.S. Naval Observatory.

MSS: Lib. Cong. (*NUCMC* 72-1723); also see *DSB*. WORKS ABOUT: See *AMS*; *DAB*; *DSB*.

HALL, CHRISTOPHER WEBBER (1845-1911). *Geology; Mineralogy.* Professor of geology and mineralogy, Univ. of Minnesota; associated with U.S. Geol. Surv.

WORKS ABOUT: See *AMS*; *NCAB*, 9 (1907) [copyright 1899 and 1907]): 502; *Who Was Who in America*, vol. 1.

HALL, EDWIN HERBERT (1855-1938). *Physics.* Professor of physics, Harvard Univ.

MSS: Harv. U.-HL (*NUCMC* 71-107). WORKS ABOUT: See *AMS*; *DAB*, supp. 2; DSB.

HALL, GRANVILLE STANLEY (1846-1924). *Psychology.* Professor of psychology and president, Clark Univ.

MSS: Univ. of Akron—Archives of American Psychology (*NUCMC* 70-1425); Clark Univ. Library

(*NUCMC* 70-1499) ; Clark Univ. Archives (*NUCMC* 73-320). WORKS ABOUT: See *AMS*; *DAB*; *DSB*.

HALL, JAMES (September 12, 1811, Hingham, Mass.—August 7, 1898, Bethlehem, N.H.). *Paleontology*; *Geology*. Son of James, woolen mill superintendent, and Susan (Dourdain) Hall. Married Sarah Aikin, 1838; four children. EDUCATION: B.N.S. in 1832 and M.A. in 1833, Rensselaer Polytechnic Institute. HONORS: Wollaston Medal, Geological Society of London, 1858; also other honors, including several honorary degrees. CAREER: 1832, appointed librarian, Rensselaer; 1835-1841, professor of chemistry, Rensselaer, and for number of years thereafter held title of professor of mineralogy and geology; 1836-1837, assistant to E. Emmons [q.v.] in New York Natural History [Geological] Survey; 1837-1843, chief geologist, fourth (western) district, New York survey; 1843-1898, New York state paleontologist; 1855-1858, Iowa state geologist; 1857-1860, Wisconsin state geologist; at New York State Museum, appointed curator (1865) and director (1871) ; 1893, appointed New York state geologist. MEMBERSHIPS: Natl. Ac. Scis. and other societies; first president, Geol. Soc. Am., 1889. SCIENTIFIC CONTRIBUTIONS: Nearly all geological work related to paleontology and stratigraphic questions. Was unrivaled as collector of geological specimens and, through scientific work, persistence when appropriations failed, and force of personality, became chief American invertebrate paleontologist of his time. Chief work done in New York state, beginning with report of 1843 [see Works] and continuing with unprecedented series of thirteen reports on paleontology. In addition to work in Iowa and Wisconsin, also aided in establishment of geological surveys in California, Missouri, New Jersey, and Ohio; trained number of assistants who later contributed to advance of paleontological and geological studies. Presidential address to Am. Assoc. Advt. Sci. in 1857 [published in 1883; see Works] indicates one of few diversions from stratigraphical and paleontological interests.

WORKS: *Geology of New York: Part IV, Comprising the Survey of the Fourth Geological District* (Albany, 1843) ; *New York State Natural History Survey: Paleontology*, 8 vols. in 13 (Albany, 1847-1894) ; "Contributions to the Geological History of the American Continent," *Proceedings of Am. Assoc. Advt. Sci.* 31 (1883) : 29-71. Wrote as single author and as collaborator a total of 302 scientific works; major publications listed in *DSB*; full bibliography in *Bulletin of Geol. Soc. Am.* 10 (1899). MSS: N.Y. St. Lib. (Hamer) ; Ac. Nat. Scis. Philad. (Hamer) ; also some items in E. N. Horsford Papers at Rens. Poly. Inst.-Arch. WORKS ABOUT: *DAB*, 8: 135-37 (George P. Merrill) ; DSB, 6: 56-58 (Donald W. Fisher). Also see John M. Clarke, *James Hall of Albany* (Albany, 1921).

HALLOWELL, EDWARD (September 14, 1808, Philadelphia, Pa.—February 20, 1860, Philadelphia, Pa.). *Herpetology*. Son of Edward Hallowell, proprietor of coal lands in Schuylkill County, Pa. No information on marital status. EDUCATION: B.A. in 1827 and M.D. in 1830, Univ. of Pennsylvania. CAREER: Engaged in medical practice, apparently in Philadelphia; an invalid for several years before death. MEMBERSHIPS: Ac. Nat. Scis. Philad. SCIENTIFIC CONTRIBUTIONS: Little biographical information available. Made large number of contributions to *Journal* and *Proceedings* of Ac. Nat. Scis., nearly all on reptiles, describing some fifty-five new species of lizards, snakes, and batrachians from Africa, the American West, and elsewhere; also prepared papers on anatomy of various specimens, including harpy eagle, gorilla, and bat; report on reptiles of North Pacific Surveying and Exploring Expedition was completed by E. D. Cope [q.v.] after Hallowell's death. Said to have made notable contributions to pathology.

WORKS: "Report Upon the Reptilia of the North Pacific Exploring Expedition, Under the Command of Capt. Rodgers," *Proceedings of Ac. Nat. Scis. Philad.* 12 (1860) : 480-510. See Meisel, *Bibliog. Amer. Nat. Hist.* MSS: Ac. Nat. Scis. Philad. (*NUCMC* 66-43). WORKS ABOUT: Kelly and Burrage, *Med. Biog.* (1928) ; communication from Univ. of Pennsylvania Archives.

HALSTED, GEORGE BRUCE (1853-1922). *Mathematics; Logic*. Professor of mathematics, Univ. of Texas and, later, Kenyon College and Colorado State Teachers' College.

MSS: See *DAB*. WORKS ABOUT: See *AMS*; *DAB*; *DSB*.

HAMMOND, WILLIAM ALEXANDER (August 28, 1828, Annapolis, Md.—January 5, 1900, Washington, D.C.). *Neurology; Physiology*. Son of Dr. John W. and Sarah (Pinkney) Hammond. Married Helen Nisbet, 1849; Esther D. Chapin, 1886; no information on children. EDUCATION: 1848, M.D., Univ. of City of New York (New York Univ.). CAREER: Ca. 1849, medical practice, Saco, Maine, for several months; 1849-1859, assistant surgeon, U.S. Army, stationed in New Mexico, Kansas, Florida, and West Point; 1860-1861, served as professor of anatomy and physiology, Univ. of Maryland, and pursued medical practice, Baltimore; 1861, returned to duties as assistant surgeon, U.S. Army; 1862-ca. 1864, surgeon general, with rank of brigadier general; 1864, court-martialed on charge of irregularities in granting of contracts and dismissed from army, but in 1878 exonerated by review board; 1864, began medical practice in New York City; 1866-1867, lecturer on nervous diseases, College of Physicians and Surgeons, New York; 1867-1874, professor of nervous and mental diseases, Bellevue Hospital Medical College; 1867-1883, involved in founding and editing several medical journals; 1873-1882, professor of nervous and mental diseases, Univ. of City of New York; 1882-1888, a founder and professor, Post-Graduate Medical School of New York; for a time, on faculty of Univ. of Vermont (Burlington); 1888-1900, medical practice, Washington, D.C. SCIENTIFIC CONTRIBUTIONS: During years as army surgeon, before Civil War, was interested in botany, but especially in physiological chemistry and carried out various experiments on himself; chief publication during these early years was "Experimental Researches" [see Works], which won Am. Med. Assoc. prize. Also during antebellum period carried out studies on urine and body's metabolism, which were published with some other papers as *Physiological Memoirs* [see Works]. Later career was devoted to more strictly medical concerns, and won renown for work on neurological diseases.

WORKS: "Experimental Researches Relative to the Nutritive Value and Physiological Effects of Albumen, Starch, and Gum, When Singly and Exclusively Used as Food," *Transactions of Am. Med. Assoc.* 10 (1857): 511-87; *Physiological Memoirs* (Philadelphia, 1863). Bibliography of works in *Post-Graduate* [below], pp. 611-19. WORKS ABOUT: *DAB*, 8:210-11 (James M. Phalen); "Hammond Memorial Meeting," *Post-Graduate* (New York Post-Graduate Medical School and Hospital) 15 (May 1900): 594-643.

HANKS, HENRY GARBER (May 12, 1826, Cleveland, Ohio—June 19, 1907, Alameda, Calif.). *Geology; Chemistry.* No information on parentage or marital status. EDUCATION: Attended common schools, Cleveland. CAREER: 1842, went to New Orleans; ca. 1842, seaman on ship from Boston to Calcutta; remained in India about one year; ca. 1843/1844, shipped as seaman to New York; engaged in scientific studies and extensive travel throughout U.S.; 1852-1856, worked in California, in mines and engaged in several business activities in Sacramento and San Francisco; 1856, returned to New York via Cape Horn; ca. 1858, returned by Cape Horn to California; 1866, established Pacific Chemical Works; 1880-1886, state mineralogist of California and established state mining bureau. MEMBERSHIPS: Calif. Ac. Scis.; Geol. Soc. Am.; and other societies. SCIENTIFIC CONTRIBUTIONS: On voyage around Cape Horn, ca. 1858, searched for gold, carried out geological study in Straits of Magellan, and investigated coal measures and effects of earthquakes on western coast of South America; later, as member of California Academy, undertook systematic study of geology of gold, earthquakes, and volcanoes. A founder of San Francisco Microscopical Society (1870), he served as president and read papers before that body. During trip to Europe in 1872, studied glaciers in Switzerland. Later, founded California State Geological Society, served as California commissioner to Paris Exposition of 1878, and at that time attended first international geological congress. In 1885, investigated and reported on iron and coal mines of Alabama and gold mines in Georgia.

WORKS: See *Natl. Un. Cat. Pre-1956 Imp.* and *Roy. Soc. Cat.*, vols. 7, 10, 15. WORKS ABOUT: *NCAB*, 13 (1906): 129; notice of death in *Science* n.s. 25 (1907): 1015; *Who Was Who in America*, vol. 1.

HARE, ROBERT (January 17, 1781, Philadelphia, Pa.—May 15, 1858, Philadelphia, Pa.). *Chemistry.* Son of Robert, businessman and brewery owner, and Margaret (Willing) Hare. Married Harriet Clark, 1811; six children. EDUCATION: Privately tutored; attended chemical lectures by J. Woodhouse [q.v.] at Univ. of Pennsylvania. HONORS: Honorary degrees from Yale in 1806 and Harvard in 1816; Rumford Medal, Am. Ac. Arts Scis., 1839. CAREER: Until ca. 1816, managed father's brewery in Philadelphia; made unsuccessful attempt at manufacture of illuminating gas, in New York City; 1810-1812, professor of natural philosophy, Univ. of Pennsylvania Medical School; 1818, for several months, professor of natural philosophy and chemistry, College of William and Mary; 1818-1847, professor of chemistry, Univ. of Pennsylvania Medical School. MEMBERSHIPS: Am. Phil. Soc. SCIENTIFIC CONTRIBUTIONS: Chemical researches began early during years of management of brewery. In 1801 made most important contribution to science with invention of oxyhydrogen blowpipe that gave higher temperature than was possible by any other means; blowpipe made possible melting of platinum and was used in limelight and other forms of illumination. Chief scientific interest was in investigations in electricity, and here also produced several useful devices, especially calorimotor (1819), deflagrator (1821) that produced high electric current, and electric furnace (1839) in which phosphorus, calcium carbide, artificial graphite from carbon and other products were produced; introduced idea of mercury cathode in electrolysis of aqueous solutions of metallic salts; also did notable work on constitution of salts, use of tar in lighting, on eudiometers for study of gases, and other subjects. Published large number of papers, and also prepared editions of several standard chemical works; had talent for preparing experimental and demonstration apparatus, and laboratory perhaps was best stocked of any in the country; numerous students learned chemistry under his direction. After retirement, wrote novel and published and lectured on spiritualism.

WORKS: *Memoir of the Supply and Application of the Blow-Pipe* . . . (Philadelphia, 1802) ; *Compendium of the Course of Chemical Instruction* ... (Philadelphia, 1828; 4th ed., 1840). See *Roy. Soc. Cat.*, vol. 3. MSS: Am. Phil. Soc. (*NUCMC* 62-932). WORKS ABOUT: *DAB*, 8: 263-64 (Edgar F. Smith and Helen C. Boatfield) ; *DSB*, 6: 114-15 (Wyndham D. Miles). Also see E. F. Smith, *The Life of Robert Hare, an American Chemist* (Philadelphia, 1917).

HARKNESS, WILLIAM (December 17, 1837, Ecclefechan, Scotland—February 28, 1903, Jersey City, N.J.). *Astronomy.* Son of James, Presbyterian clergyman and physician, and Jane (Weild) Harkness. Never married. EDUCATION: 1854-1856, attended Lafayette College; A.B. in 1858 and M.A. in 1861, Univ. of Rochester; 1862, graduated from New York Homeopathic Medical College. CAREER: 1858 and 1860, newspaper reporter in New York and Pennsylvania state legislatures; 1862, appointed aide in U.S. Naval Observatory; 1863-1865, professor of mathematics at Naval Observatory; 1865-1866, on cruise of monitor *Monadnock;* 1866-1867, at Hydrographic Office; 1867-1899, at Naval Observatory as civilian astronomical director (1892-1899) and as director of *American Ephemeris and Nautical Almanac* (1897-1899) ; 1899, retired with rank of rear admiral. MEMBERSHIPS: President, Am. Assoc. Advt. Sci., 1893; a founder of Phil. Soc. Wash. SCIENTIFIC CONTRIBUTIONS: While on *Monadnock* cruise, made observations of terrestrial magnetism and effects of ship's armor on compass. During total solar eclipse in 1869 discovered coronal line K1474 (λ 5303), and in 1870 observed eclipse in Sicily. Chief work began in 1871 with appointment as transit of Venus commissioner, and much of time thereafter was concerned with preparations, observations, and discussion of results of transits in 1874 and 1882; in 1874 had charge of observations in Tasmania. Devised measuring machine, the spherometer caliper, to determine accurately positions of Venus and sun in photographs, and was in charge of reduction of all American transit observations; interests in astronomy were chiefly of practical nature, and he took central role in devising various pieces of apparatus used at Naval Observatory. Chief theoretical paper was "The Solar Parallax" [see Works].

WORKS: "The Solar Parallax and Its Related Constants," *Washington Observations . . . 1885*, appendix 3 (Washington, D.C., 1891). See *Roy. Soc. Cat.*, vols. 7, 10, 15. MSS: Documents relating to service with Naval Observatory are with its records in Natl. Arch. *(DSB)*; Univ. of Rochester Library (Hamer). WORKS ABOUT: *DAB*, 8: 266-67 (Raymond S. Dugan) ; *DSB*, 6: 119 (Nathan Reingold).

HARLAN, RICHARD (September 19, 1796, Philadelphia, Pa.—September 30, 1843, New Orleans, La.). *Comparative anatomy; Zoology.* Son of Joshua, wholesale grocer and merchant, and Sarah (Hinchman) Harlan. Married Margaret Hart (Simmons) Howell, 1833; survived by four children. EDUCATION: Studied medicine with Dr. Joseph Parrish in Philadelphia; 1818, M.D., Univ. of Pennsylvania. CAREER: 1816-1817, surgeon on ship to Calcutta; 1818, began medical practice, Philadelphia; ca. 1818, demonstrator in school of anatomy established by Dr. Parrish; 1820, appointed a physician to Philadelphia Dispensary; 1821, became professor of comparative anatomy, Peale's Philadelphia Museum; 1822-1838, physician to Philadelphia Almshouse; 1833 and 1838-1840, visited Europe; 1843, established medical practice in New Orleans, but died soon thereafter. MEMBERSHIPS: Ac. Nat. Scis. Philad.; Am. Phil. Soc.; and other societies. SCIENTIFIC CONTRIBUTIONS: Among earliest works was paper on animal heat. First paleontological work appeared 1823, and he became first American to make specialty of study of vertebrate paleontology; worked on specimens collected by Lewis and Clark, Stephen Long, T. Nuttall [q.v.], T. R. Peale [q.v.], J. J. Audubon [q.v.], and others; personally searched for fossil remains in New Jersey, Ohio River area, and Virginia, and also purchased collections. Early studies published in *Fauna* [see Works], which included both living and extinct forms and ranks as first American attempt at systematic presentation of mammals, but was heavily criticized, especially by J. D. Godman [q.v.], as being essentially translation and adaptation of A. G. Desmarest's *Mammalogie;* second part of *Fauna,* on reptiles, never appeared, but published *Herpetology* [see Works]. A productive author, he collected valuable information and made notable contributions to taxonomy, including new species, but chief contribution was less in original research than in collection and systematization of zoological and paleontological work of earlier writers.

WORKS: *Fauna Americana* (Philadelphia, 1825) ; *American Herpetology* . . . (Philadelphia, 1827), 87 pp.; "Critical Notices of Various Organic Remains Hitherto Discovered in North America," *Transactions of Geological Society of Pennsylvania* 1 (1835) : 46-112; *Medical and Scientific Researches* (Philadelphia, 1835), a collection of Harlan's earlier writings. Chief works mentioned in *DAB;* also see *Roy. Soc. Cat.*, vol. 3. WORKS ABOUT: *DAB*, 8: 273-74 (Daniel M. Fisk) ; *DSB*, 6: 119-21 (Whitfield J. Bell, Jr.).

HARRINGTON, MARK WALROD (August 18, 1848, Sycamore, Ill.—October 9, 1926). *Astronomy; Meteorology.* Son of James, physician, and Charlotte (Walrod) Harrington. Married Rose Smith; at least one child. EDUCATION: Attended Northwestern Univ.; A.B. in 1868 and A.M. in 1871, Univ. of Michigan; 1876-1877, studied in Leipzig. HONORS: LL.D., Univ. of Michigan, 1894. CAREER: 1868-1876, at Univ. of Michigan, served as assistant curator in museum, instructor, and assistant professor, teaching mathematics, geology, zoology, and botany; 1871 (summer), astronomical assistant, U.S. Coast and Geodetic Survey, Alaska; 1877-1878, professor of astronomy, Cadet School, Foreign Office, Peking, China; 1878-1879, professor, Louisiana State Univ.; 1879-1891, professor of astronomy and director of observatory, Univ. of Michigan; 1884-1892, established and edited *American Meteorological Journal*; 1891-1895, chief, U.S. Weather Bureau; 1895-1897, president, Univ. of Washington; 1898-1899, resumed work with U.S. Weather Bureau, situated at Puerto Rico and then New York; 1899-1907, traveled widely to China and elsewhere, engaging in manual labor, his whereabouts unknown to family; 1907-1926, institutionalized in asylum for mentally ill. MEMBERSHIPS: Am. Assoc. Advt. Sci. and other societies. SCIENTIFIC CONTRIBUTIONS: Advanced the study of meteorology through publication of *Journal* and as head of Weather Bureau; promoted collection and publication of data on rainfall by bureau, and his work on climate and meteorology of Death Valley, Calif., appeared as bureau's first bulletin. Chief work was *About Weather* [see Works], based on information collected at Weather Bureau; as first civilian head of Weather Bureau, experienced problems that made term of office a difficult one.

WORKS: *About the Weather* (New York, 1899). Also see *Roy. Soc. Cat.*, vols. 10, 12, 15. WORKS ABOUT: *DAB*, 8: 301-2 (William J. Humphreys); *Who's Who in America, 1901-1902.*

HARRIS, THADDEUS WILLIAM (November 12, 1795, Dorchester, Mass.—January 16, 1856, Cambridge, Mass.). *Entomology.* Son of Thaddeus Mason, Unitarian minister, and Mary (Dix) Harris. Married Catherine Holbrook, 1824; twelve children. EDUCATION: A.B. in 1815 and M.D. in 1820, Harvard Univ. CAREER: 1820, began medical practice, Milton, Mass., and later moved to Dorchester; 1831-1856, librarian, Harvard College; 1834 and 1837-1842, lecturer on natural history, Harvard; 1837, commissioned to prepare report on insects of Massachusetts, in connection with state survey. MEMBERSHIPS: Am. Ac. Arts Scis. and other societies; corresponding member, London Entomological Society. SCIENTIFIC CONTRIBUTIONS: Student of insect taxonomy and life histories; best known as economic entomologist, and played central role in directing attention in U.S. to study of insects in relation to agriculture; reputation as entomologist rests on *Treatise* [see Works], standard work containing much information on life histories of insects, and he played important part in demonstrating need for this kind of information in dealing with problem of insect damage. Catalogue included in *Report* of 1833 [see Works] by E. Hitchcock [q.v.] was classified list of 2,300 species of insects of Massachusetts. Gathered large collection of insects. Hopes for permanent position as professor of natural history at Harvard frustrated by appointment of A. Gray [q.v.] in 1842, and during later years duties in library consumed much attention formerly given to entomological studies. In addition to entomology, had interest in botany and produced work on squash-gourd group.

WORKS: "Insects," in Edward Hitchcock, *Report on the Geology, Mineralogy, Botany, and Zoology of Massachusetts* . . . (Amherst, Mass., 1833), pp. 566-95; *Treatise on Some of the Insects of New England Which Are Injurious to Vegetation* (Cambridge, Mass., 1842), a private reprinting of report prepared as commissioner for Zoological and Botanical Survey of Massachusetts, with revised edition in 1852 and illustrated edition in 1862, published posthumously. Bibliography of works in *Entomological Correspondence of Thaddeus William Harris, M.D.*, ed. S. H. Scudder, Bos. Soc. Nat. Hist. Occasional Papers, no. 1 (1869). MSS: Bost. Soc. Nat. Hist.—Boston Museum of Science (Hamer); also Harv. U.-UA and Harv. U.-MCZ. WORKS ABOUT: *DAB*, 8:321-22 (Leland O. Howard); Mallis, *Amer. Entomologists*, pp. 25-33.

HARRISON, JOHN (December 17, 1773, Philadelphia, Pa.—July 19, 1833, Philadelphia, Pa.). *Chemical manufacturing.* Son of Thomas, merchant, and Sarah (Richards) Harrison. Married Lydia Leib, 1802; eight children. EDUCATION: Apprenticed to druggist; went to Europe for two years, where he studied chemical manufacturing business and also studied chemistry with J. Priestley [q.v.]. CAREER: 1793-ca. 1801, engaged in wholesale and retail trade in chemicals and drugs, Philadelphia, in partnership with Samuel Betton; 1801-1833, operated chemical manufacturing business, after 1831 known as John Harrison and Sons; 1821-1824, recorder, city and county of Philadelphia. MEMBERSHIPS: Member, board of managers, Franklin Institute. SCIENTIFIC CONTRIBUTIONS: In 1801, became first American manufacturing chemist, and subsequently built up large business. As early as 1793, began experimentation with practical methods for manufacture of sulphuric, nitric, and hydrochloric acids; by

1806 was firmly established in manufacture of sulphuric acid and other chemicals; that year began production of white lead and later added other chemicals and colors; in 1807 had built for his use large-capacity chamber that made possible annual production of almost half-million pounds of sulphuric acid. Has been credited as American originator of platinum still for concentration of sulphuric acid, at time when procedure was known to only very few in Europe; Harrison's still was made by Dr. Erick Bollman in Philadelphia.

WORKS: No references to publications appear in *Natl. Un. Cat. Pre-1956 Imp.* or *Roy. Soc. Cat.* WORKS ABOUT: *DAB*, 8: 344-45 (John H. Frederick); Edgar F. Smith, *Chemistry in Old Philadelphia* ([Philadelphia], 1919), pp. 103-4.

HART, EDWARD (1854-1931). *Chemistry.* Professor of chemistry and dean of science department, Lafayette College, Easton, Pa.

WORKS ABOUT: See *AMS*; *DAB*, supp. 1.

HARTT, CHARLES FREDERICK (August 23, 1840, Frederickton, New Brunswick—March 18, 1878, Rio de Janeiro, Brazil). *Geology.* Son of Jarvis William, educator, and Prudence (Brown) Hartt. Married Lucy Lynde, 1869; two children. EDUCATION: 1860, B.A., Acadia College; 1861, became student of J. L. R. Agassiz [q.v.], Museum of Comparative Zoology, Harvard; 1863-1866, enrolled in Lawrence Scientific School, Harvard. CAREER: 1860-1861, went to St. John, New Brunswick, to teach in high school established by father, and began geological investigation of region with G. F. Mathew; 1861-1864, while student at Cambridge, Mass., pursued geological studies during vacations; 1864, participant in provincial survey of southern portion of New Brunswick; 1865-1866, geologist, Agassiz's expedition to Brazil; 1866-1867, scientific teacher and lecturer, New York City and elsewhere; 1867-1868, second expedition to Brazil; 1868, professor of natural history, Vassar College; 1868-1878, professor of economic, general, and agricultural geology, Cornell Univ.; 1874, went again to Brazil; 1875-1878, served as head of Geological Survey of Brazil; 1876, appointed director of geological department, Brazilian National Museum, Rio de Janeiro, but resigned shortly thereafter. MEMBERSHIPS: Bos. Soc. Nat. Hist.; Am. Assoc. Advt. Sci.; and other societies. SCIENTIFIC CONTRIBUTIONS: Began explorations in Nova Scotia before leaving college, and work with Mathew in New Brunswick made geology of that region understandable for first time; numerous references to Hartt's work in John William Dawson's *Acadian Geology* (1855), including description by S. H. Scudder [q.v.] of Hartt's discovery of oldest known fossil insects. Careful geological studies in Brazil refuted Agassiz's initial conjecture that glaciers had reached Amazon Valley. Majority of papers written in Brazil are in Portuguese. In addition to geological work, also wrote on ethnology.

WORKS: *The Geology and Physical Geography of Brazil* (Boston, 1870) gave results of first two Brazilian expeditions; "Geological Survey of Brazil," *Am. Journ. Sci.* 3d ser. 11 (June 1876): 466-72. Chief works in geology and ethnology listed in *NCAB*. Lists of works in Rathbun and in Simonds [below]. MSS: Duke Univ. Library (*NUCMC* 61-2509); Corn. U.-CRH (*NUCMC* 70-1088). WORKS ABOUT: *NCAB*, 11 (1909 [copyright 1901 and 1909]): 260-61; Richard Rathbun, "Sketch," *Proceedings of Bos. Soc. Nat. Hist.* 19 (1878): 338-64; Frederic W. Simonds, "Tribute," *American Geologist* 19 (1897): 69-90.

HASSLER, FERDINAND RUDOLPH (October 7, 1770, Aarau, Switzerland—November 20, 1843, Philadelphia, Pa.). *Geodesy*; *Mathematics.* Son of Jakob Hassler, successful clock manufacturer. Married Marianne Gaillard, 1798; nine children. EDUCATION: Studied at Univ. of Bern. CAREER: Geodetic field worker in Switzerland; 1805, joined land company and emigrated to U.S. with intention of taking up farming, but venture failed because of company's economic problems; 1807, Hassler's plan for a coast survey was accepted by a federal commission and he was nominated for the work, but political conditions prevented start of survey; 1807-1809, acting professor of mathematics, U.S. Military Academy, West Point; 1809-1811, professor of natural philosophy and mathematics, Union College; 1811-1815, in London at request of secretary of treasury to purchase surveying instruments; 1816-1818, engaged in field work as superintendent of Coast Survey; 1819-1830, employed as an astronomer in settlement of northeastern boundary dispute and engaged in farming in New York state, teaching, writing mathematical textbooks, and working as gauger in New York Custom-House; 1830, appointed superintendent of weights and measures; 1832-1843, superintendent of revived Coast Survey. MEMBERSHIPS: Am. Phil. Soc. SCIENTIFIC CONTRIBUTIONS: To position as superintendent of Coast Survey brought a familiarity with best methods of geodetic surveying in Europe, and applications of this knowledge resulted in survey of high quality and permanent usefulness; his conception of survey would extend it to include investigation of geophysical facets such as terrestrial magnetism and tides. Combination of Hassler's uncompromising scientific standards, his personality, and the requirements of practical politics led to series of administrative disputes, but ultimately he was able to launch the first federal scientific agency on path to which it continued to adhere more than century later.

WORKS: Bibliography in G. A. Weber, *The Coast and Geodetic Survey* (Baltimore, 1923), and *The Bureau of Standards* (Baltimore, 1925). MSS: N.Y. Pub. Lib. (*NUCMC* 71-1221); also Am. Phil. Soc. and Natl. Arch. *(DSB)*. WORKS ABOUT: *DAB*, 8: 385-86 (H. A. Marmer); *DSB*, 6: 165-66 (Nathan Reingold). Also see Florian Cajori, *The Chequered Career of Ferdinand Rudolph Hassler* (Boston, 1929).

HASTINGS, CHARLES SHELDON (1848-1932). *Physics*. Professor of physics, Sheffield Scientific School, Yale.

WORKS ABOUT: See *AMS*; *DAB*, supp. 1.

HATCHER, JOHN BELL (October 11, 1861, Cooperstown, Ill.—July 3, 1904, Pittsburgh, Pa.). *Paleontology*. Son of John, farmer and businessman, and Margaret Columbia (Lanning) Hatcher. Married Anna M. Peterson, 1887; four children. EDUCATION: 1881, attended Grinnell College (Iowa); 1884, Ph.B., Sheffield Scientific School, Yale. CAREER: 1884-1893, assistant to O. C. Marsh [q.v.] at Yale Univ., and collected fossils in West and South; 1891-1892, assistant in geology, Yale; 1893-1900, curator of vertebrate paleontology and assistant in geology, Princeton Univ., and during these years made several collecting expeditions, including three trips to Patagonia (1896-1899); 1900-1904, served as curator of paleontology and osteology, Carnegie Institute (Pittsburgh), and continued collecting in West. MEMBERSHIPS: Am. Phil. Soc.; Geol. Soc. Am.; and other societies. SCIENTIFIC CONTRIBUTIONS: Through extensive field work, became chief American collector of fossil animals and helped transform that work from activity of untrained amateurs to highly skilled technique; began collection of vertebrate fossils in Kansas, Nebraska, Texas, North Carolina, Virginia, and District of Columbia while at Yale, and in Wyoming in 1889 made one of most important discoveries, the first fossil remains of Triceratops. Major work done while at Princeton was exploration of Patagonia, a project he instigated to collect mammalian fossils and to examine stratigraphy of that area; with aid of J. Pierpont Morgan, undertook preparation of series of reports on Patagonian investigations, and among results was conjecture (on basis of similarity of fauna) that Patagonia and Australia once were connected. In addition to collecting, also wrote on fossil mammals and reptiles, as well as on other subjects, but did not publish many of his original conceptions on dynamical and stratigraphical geology.

WORKS: *Narrative of the Expeditions: Geography of Southern Patagonia*, Princeton Univ. Expeditions to Patagonia, 1896-1899, Reports, vol. 1 (Princeton, 1903); *The Ceratopsia . . . Based on Preliminary Studies by Othniel Marsh, ed. and completed by Richard S. Lull*, U.S. Geol. Surv., Monograph 49 (Washington, D.C., 1907). Bibliography in *American Geologist* 35 (1905): 139-41. WORKS ABOUT: *NCAB*, 21 (1931): 212-13; W. B. Scott, obituary, *Science* n.s. 20 (1904): 139-42.

HAY, OLIVER PERRY (1846-1930). *Paleontology*. Professor of zoology and geology, Butler College (Indiana); associated with Field Museum of Natural History (Chicago), Am. Mus. Nat. Hist., and Carnegie Institution (Washington).

MSS: Univ. of Florida Library, Gainesville (*NUCMC* 69-223); Smithson. Instn. Library (Hamer). WORKS ABOUT: See *AMS; DAB*.

HAYDEN, EDWARD EVERETT (April 14, 1858, Boston, Mass.—November 17, 1932, Baltimore, Md.). *Meteorology*. Son of William and Louise Anna (Dorr) Hayden. Married Kate Reynolds, 1882; six children. EDUCATION: 1879, graduated, U.S. Naval Academy; 1884, studied at Harvard College Observatory. CAREER: 1879-ca. 1882, sea duty; ca. 1882, assigned to special duty at Smithson. Instn. and to field work with U.S. Geol. Surv.; 1883, while attached to Geol. Surv., suffered accident resulting in loss of leg; 1885, retired from active naval duty; 1885-1886, assistant geologist, U.S. Geol. Surv.; 1887-1893, marine meteorologist and editor of pilot charts, Hydrographic Office; 1892-1896, assistant editor, *American Meteorological Journal;* 1898, head of Naval Observatory, Mare Island, Calif.; 1899, assigned to branch of Hydrographic Office at Manila; 1901, restored to active duty in U.S. Navy; 1902-1910, in charge of chronometers and time service, U.S. Naval Observatory, Washington, D.C.; 1910-1915, commandant, Key West (Fla.) Naval Station; 1915-1921, president of General Court Martial, Norfolk (Va.) Navy Yard; 1921, retired as rear admiral. MEMBERSHIPS: Vice-president (1890-1893) and secretary (1895-1897), Natl. Geog. Soc. SCIENTIFIC CONTRIBUTIONS: During period 1885-ca. 1895, became noted for work in meteorology, especially hurricanes. While at Mare Island during war with Spain, produced work on rate of clocks and barometric pressure, and while at Naval Observatory at Washington applied this work to regulation of observatory's time-signal transmission clock by means of barometric pressure and temperature, thereby making possible accurate standard time-service in U.S. Also contributed to knowledge regarding ship routings.

WORKS: "The Great Storm Off the Atlantic Coast . . . March 11-14, 1888," *Nautical Monographs of*

Hydrographic Office, no. 5 (Washington, D.C., 1888); "The Law of Storms," *National Geographic Magazine* 2 (1890): 199-211; "Tropical Cyclones," *United Service* (June 1889); "Clock Rates and Barometric Pressure . . . ," *Publications of Astronomical Society of the Pacific* 11 (1899): 101-14. See *Natl. Un. Cat. Pre-1956 Imp.* and *Roy. Soc. Cat.*, vol. 15. WORKS ABOUT: *DAB*, supp. 1: 387-88 (Allan Westcott); *NCAB*, 27 (1939): 73; obituary, *New York Times*, November 18, 1932, p. 19.

HAYDEN, FERDINAND VANDIVEER (September 7, 1829, Westfield, Mass.—December 22, 1887, Philadelphia, Pa.). *Geology.* Son of Asa and Melinda (Hawley) Hayden. Married Emma C. Woodruff, 1871; no children. EDUCATION: 1850, A.B., Oberlin College; 1853, M.D., Albany Medical College; and during this time studied paleontology with J. Hall [q.v.]. CAREER: 1853, along with F. B. Meek [q.v.], was sent by Hall to collect fossils in Dakotas; 1854-1855, carried out geological explorations in region of Missouri and Yellowstone rivers, part of time with cooperation from American Fur Company; 1856-1857, geologist with Lt. G. K. Warren [q.v.] in exploration of Dakotas and Black Hills; 1858, with Meek in Kansas Territory; 1859-1860, joined Capt. William F. Raynolds in exploration of northern Rocky Mountains; 1862-1865, surgeon, U.S. Army; 1865-1872, professor of geology, Univ. of Pennsylvania; 1866, under sponsorship of Ac. Nat. Scis. Philad., again explored Badlands; 1867-1879, director, U.S. Geological and Geographical Survey of Territories, carrying out explorations especially in Colorado, Idaho, Montana, Wyoming, and Utah; 1879-1886, geologist, U.S. Geol. Surv. under C. R. King and J. W. Powell [qq.v.]. MEMBERSHIPS: Geological societies of London and Edinburgh; Natl. Ac. Scis.; and other societies. SCIENTIFIC CONTRIBUTIONS: Explorations included large area of the West, although by necessity much was covered at rapid pace; surveys gave employment to large number of American scientists. In course of travels Hayden gathered great many fossils, sent especially to Meek and to J. Leidy and J. S. Newberry [qq.v.], while Hayden himself dealt with geology of territory crossed; frequently published jointly with Meek, creating detailed vertical section of western Cretaceous and Tertiary formations; in 1871-1872, investigated Yellowstone Park area, presented findings of recent volcanic action and importance of horizontal pressure in mountain building, and also lobbied for establishment of Yellowstone Park. In other geological work, demonstrated importance of glacial action in the West.

WORKS: Publications included government reports as well as journal articles. See partial bibliography in *Biog. Mems. of Natl. Ac. Scis.* 3 (1895): 409-13; see additional references to bibliographies in *DSB*. MSS: Natl. Arch. (U.S. Geological and Geographical Survey of Territories); see this and other references in *DSB*. WORKS ABOUT: *DAB*, 8: 438-40 (George P. Merrill); *DSB*, 6: 186-88 (Michele L. Aldrich).

HAYDEN, HORACE H. (October 13, 1769, Windsor, Conn.—January 26, 1844, Baltimore, Md.). *Geology; Dentistry.* Son of Thomas—carpenter, joiner, and architect—and Abigail (Parsons) Hayden. Married Marie Antoinette Robinson, 1805; six children. EDUCATION: Attended school in Windsor, Conn.; served as cabin boy on several voyages to West Indies; for five years, studied and practiced architecture with father; 1792, removed to New York City, where studied dentistry with aid of John Greenwood; 1810, received dental license, Medical and Chirurgical Faculty of Maryland; studied medicine at Univ. of Maryland while practicing dentistry; 1840, as member of American Society of Dental Surgeons, received D.D.S. degree. HONORS: M.D., Univ. of Maryland, 1840. CAREER: Ca. 1800, began dental practice, Baltimore, Md., and also practiced in other localities in that state; taught dentistry in office; ca. 1825, lectured on dental physiology and pathology, Univ. of Maryland; 1840-1844, as a founder and first president, Baltimore College of Dental Surgery, served there first as professor of principles of dental science and then as professor of dental physiology and pathology. MEMBERSHIPS: Baltimore Physical Association; Maryland Academy of Sciences and Literature. SCIENTIFIC CONTRIBUTIONS: Dentistry was chief activity, and worked to improve its practice and prestige; Baltimore College of Dental Surgery, which he helped to found, was first such institution in the world; studied dentistry in broadest sense, probing functions of thyroid, salivary, lacrimal, and other glands, and published in medical journals. Pursued science as avocation, studying botany, culture of silkworms, and especially geology. *Geological Essays* [see Works], an early American work in its field, dealt with Atlantic and Gulf coast alluvial deposits (coastal plain) and presented global theory to explain their formation; this work was reviewed with favor by B. Silliman [q.v.] in *Am. Journ. Sci.*, but was criticized by reviewer in *North American Review.*

WORKS: *Geological Essays; or, An Inquiry Into Some of the Geological Phenomena to Be Found in Various Parts of America and Elsewhere* (Baltimore, 1820), 412 pp. Also see Meisel, *Bibliog. Amer. Nat. Hist.* WORKS ABOUT: *DAB*, 8: 440-42 (L. Parmly Brown); J. A. Taylor, *History of Dentistry* (Philadelphia and New York, 1922), pp. 80-82; Merrill, *Hundred Years of Geology*, pp. 81-85.

HAYES, AUGUSTUS ALLEN (February 28, 1806, Windsor, Vt.—June 21, 1882, Brookline, Mass.). *Chemistry.* Son of Capt. Thomas Allen and Sophia (West) Hayes. Married Henrietta Bridge Dana, 1836; three children. EDUCATION: 1823, graduated from military academy in Norwich, Vt.; ca. 1823, began study of chemistry with J. F. Dana [q.v.], professor at Dartmouth College. HONORS: M.D., Dartmouth College, 1846. CAREER: Assistant professor of chemistry, New Hampshire Medical College; 1828, moved to Boston and resided in that area for remainder of life, engaged in chemical research, consulting, and management; served in turn as director of Roxbury plant that produced colors and other chemicals and as consulting chemist or director of several important New England manufacturing concerns engaged in dyeing, bleaching, gas producing, and smelting; for number of years, Massachusetts state assayer. MEMBERSHIPS: Am. Ac. Arts. Scis. and other societies. SCIENTIFIC CONTRIBUTIONS: Chemical work was of diverse nature; while at Dartmouth, undertook proximate analysis of various medicinal plants, and isolated organic alkaloid sanguinaria; this study was followed by investigation of compounds of chromium. In 1837, commenced study of efficient use of fuels in production of steam, results of which made important contribution to improved furnaces and arrangement of steam boilers. Also contributed to improvements in processing iron and copper; studied formation, composition, and kinds of guano; and researches during Civil War led to American production of pure saltpeter for use by Navy. Under commission from U.S. Navy (ca. 1850) to study copper and copper sheathing on ships, investigated difference in chemical composition and action of ocean, at and below its surface and in entrance to rivers; while examining public water supply for Charlestown, Mass. (1859-1860), originated electrical means of detecting limits of impurity in body of water. Published more than fifty scientific papers on chemical and related topics.

WORKS: Bibliography of works in *Proceedings of Am. Ac. Arts Scis.* [below]. MSS: Several items in J. D. Dana and Silliman Family Papers in Yale Univ. Library—Manuscripts and Archives Division and in E. N. Horsford Papers at Rens. Poly. Inst.-Arch. WORKS ABOUT: *DAB,* 8:443-44 (Lyman C. Newell); memoir, in *Proceedings of Am. Ac. Arts Scis.,* 18 (1882-1883): 422-27.

HAZEN, HENRY ALLEN (January 12, 1849, Sirur, India—January 23, 1900, Washington, D.C.). *Meteorology.* Son of Rev. Allen, Congregational missionary, and Martha (Chapin) Hazen. Never married. EDUCATION: Attended school in St. Johnsbury, Vt.; 1871, A.B., Dartmouth College; attended one year at Thayer School of Civil Engineering, Dartmouth. CAREER: 1873-1876, instructor in drawing, Sheffield Scientific School, Yale; 1877-1881, assistant to Prof. E. Loomis [q.v.] in physics and meteorology, Yale; 1881, appointed to meteorological section, U.S. Signal Service; 1891, appointed professor of meteorology, forecast division, U.S. Weather Bureau, on occasion of transfer of meteorological service to Department of Agriculture. SCIENTIFIC CONTRIBUTIONS: While assistant to Prof. Loomis, participated in preparation of several meteorological papers. In U.S. meteorological service, began as computer in scientific division; in 1887 began preparation of official weather forecasts, and in 1888 commenced editing *Monthly Weather Review* and also did work in records division. During these years with meteorological service, engaged in experimentation and preparation of theoretical works; one of early works dealt with problem of reduction of barometric readings to equivalents under similar conditions at sea level; also experimented with measurement of humidity and determination of dew point, best means of exposing thermometers, and other problems of instrumentation and made five balloon ascensions to investigate vertical distribution of temperature and humidity; also studied thunderstorms and tornadoes, theories of cyclones, effects of sunspots and the moon, and other topics. Productive in both experimental and theoretical work, although some theoretical contributions met with criticism by researchers in the field.

WORKS: *Reduction of Air Pressure to Sea Level, at Elevated Stations West of the Mississippi River,* U.S. Signal Corps, Professional Papers, no. 6 (Washington, D.C., 1882), 42 pp.; *The Climate of Chicago,* U.S. Weather Bureau Bulletin, no. 10 (Washington, D.C., 1893); *Handbook of Meteorological Tables* (Washington, D.C., 1888); *The Tornado* (New York, 1890). See Poggendorff, *Biographisch-literarisches Handwörterbuch,* vols. 3, 4. WORKS ABOUT: *DAB,* 8:477 (William J. Humphreys); Cleveland Abbe, biographical sketch, *Monthly Weather Review,* January 1900, pp. 14-15.

HEILPRIN, ANGELO (1853-1907). *Geology; Paleontology; Geography.* Professor of invertebrate paleontology and geology, Ac. Nat. Scis. Philad.; professor of geology, Wagner Free Institute of Science, Philadelphia; explorer in Mexico, Alaska, British Guiana, North Africa, West Indies, Arctic with Robert E. Peary.

MSS: Ac. Nat. Scis. Philad. (*NUCMC* 66-46). WORKS ABOUT: See *AMS; DAB.*

HENRY, JOSEPH (December 17, 1797, Albany, N.Y.—May 13, 1878, Washington, D.C.). *Physics; Science administration.* Son of William, laborer, and

Ann (Alexander) Henry; cousin and brother-in-law of Stephen Alexander [q.v.]. Married Harriet L. Alexander, 1830; six children. EDUCATION: Attended school in Galway and Albany, N.Y.; apprenticed to jeweler and watchmaker; 1819-1822, attended Albany Academy and taught school for one year. HONORS: Received numerous honorary degrees. CAREER: 1823-1824 and later, assistant in chemistry to T. Romeyn Beck, Albany Academy; 1822-1826, private tutor in home of Stephen van Rensselaer, teacher to elder Henry James, employed as surveyor, directed survey of road from Hudson River to Lake Erie (1825); 1826-1832, professor of mathematics and natural philosophy, Albany Academy; 1832-1848, professor of natural philosophy, College of New Jersey (Princeton Univ.); 1846-1878, secretary (director), Smithson. Instn.; 1852, became member of Lighthouse Board, and from 1871 on, served as its president. MEMBERSHIPS: Am. Phil. Soc. and other societies; president, Natl. Ac. Scis., 1868-1878. SCIENTIFIC CONTRIBUTIONS: After early interest in dramatics, began experimentation in electricity and magnetism in 1827 while teaching at Albany Academy; one of early achievements was improved electromagnet having greater strength than those before his. In 1831 devised first electromagnetic telegraph and later invented electromagnetic relay; research in electromagnetism led to discovery of self-induction in 1832, independently of Michael Faraday, and unit of self-inductance or mutual inductance is called *henry;* Henry (as did Faraday) later carried out experiments on numerous aspects of electromagnetism. Also did experimental work on capillarity, phosphorescence, and other physical phenomena, and with S. Alexander studied solar radiation and heat of sunspots. After becoming head of Smithsonian, had little time for original research, but made that institution a prime force in promotion and disseminaton of scientific investigations, including research in meteorology, ethnology, and anthropology.

WORKS: *Scientific Writings of Joseph Henry,* 2 vols. (Washington, D.C., 1886). MSS: Smithson. Instn. Archives; also see *DSB, NUCMC,* and Hamer. WORKS ABOUT: *DAB,* 8: 550-53 (William F. Magie); *DSB,* 6: 277-81 (Nathan Reingold); Reingold, ed., *The Papers of Joseph Henry* (Washington, D.C.; vol. 1, 1972; 15 vols. are planned). Also see Thomas Coulson, *Joseph Henry: His Life and Work* (Princeton, 1950).

HENTZ, NICHOLAS MARCELLUS (July 25, 1797, Versailles, France—November 4, 1856, Mariana, Fla.. *Entomology*; *Arachnology.* Father, a lawyer, engaged in politics and was forced to leave France after defeat of Napoleon. Married Caroline Lee Whiting, 1824; had children. EDUCATION: Between ages twelve and fourteen, undertook study of miniature painting; 1813-1815, student at Hospital Val-de-Grâce;1820-1821, attended medical lectures, Harvard Univ. CAREER: 1816, settled with parents at Wilkes-Barre, Pa.; for several years, resided in Boston and Philadelphia as instructor in French and miniature painting, and for some time was private tutor near Charleston, S.C.; 1824-1825, teacher at Round Hill School (Northampton, Mass.) in association with George Bancroft; 1826-1831, professor of modern languages and belles lettres, Univ. of North Carolina; thereafter, with wife conducted female seminaries or other schools in various locations in the South, going to Covington, Ky. (1830), Cincinnati, Ohio, (ca. 1832), Florence, Ala. (1834), Tuscaloosa, Ala. (1842), Tuskegee, Ala. (1846), Columbus, Ga. (1847); ca. 1850, in ill health, removed to Mariana, Fla. SCIENTIFIC CONTRIBUTIONS: Chief scientific interests, pursued during leisure hours, were study of beetles and spiders; in early work on beetles described several new species from Massachusetts and Pennsylvania. Later work was chiefly on spiders, and became pioneer American authority on them; by early 1840s had described nearly 150 species of spiders; prepared illustrations of great number of them and studied their life histories and behavior; Bos. Soc. Nat. Hist. undertook publication of work on *Spiders* [see Works], with twenty-one plates, which ranks as earliest significant publication on American specimens. A friend and correspondent of T. W. Harris [q.v.].

WORKS: Edward Burgess and J. H. Emerton, eds., *The Spiders of the United States: A Collection of the Arachnological Writings of . . . Hentz,* Bost. Soc. Nat. Hist. Occasional Papers no. 2 (Boston, 1875); Letters to T. W. Harris, in S. H. Scudder, ed., *Entomological Correspondence of Thaddeus William Harris, M.D.,* Bos. Soc. Nat. Hist. Occasional Papers no. 1 (Boston, 1869). Bibliography of works in Burgess and Emerton, eds. [above]. MSS: Correspondence with T. W. Harris in Bos. Soc. Nat. Hist.—Boston Museum of Science and Harv. U.-MCZ. WORKS ABOUT: NCAB, 9 (1907 [copyright 1899 and 1907]): 428; Mallis, *Amer. Entomologists,* pp. 405-8.

HERING, DANIEL WEBSTER (1850-1938). *Physics; Mechanics.* Professor of physics and applied mechanics and dean, New York Univ.

WORKS ABOUT: See *AMS; NCAB,* 42 (1958): 184-85.

HERRICK, EDWARD CLAUDIUS (February 24, 1811, New Haven, Conn.—June 11, 1861, New Haven, Conn.). *Astronomy; Entomology.* Son of Claudius, clergyman and educator, and Hannah (Pierpont) Herrick. Never married. EDUCATION: Had early classical education, but because of difficulties with eyes did not attend college. HONORS: M.A., Yale College, 1838. CAREER: At age sixteen, became clerk in a leading New Haven bookstore; 1835-ca. 1838, proprietor of same bookstore; ca. 1838-ca. 1841, served as New Haven city clerk and in office of *Am. Journ. Sci.*; at Yale Univ., librarian (1843-1858) and treasurer (1852-1861). MEMBERSHIPS: Am. Assoc. Advt. Sci. SCIENTIFIC CONTRIBUTIONS: Chief entomological work was investigation of Hessian fly, begun in 1832 and continuing over period of years; other entomological work included discovery and first description of parasites of eggs of spring cankerworm moth and investigation of seventeen-year locusts. Known among contemporaries especially for work in astronomy and meteorology, an interest initiated by meteor showers in 1833; astronomical papers were chiefly on so-called shooting stars, and in 1837 presented theory of periodic recurrence of number of such meteors in month of August, published without awareness of similar theory propounded by M. Quetelet of Brussels. Also prepared thorough catalogue of meteoric showers of the past, and conducted observations and study of aurora borealis. An early collaborator with J. D. Dana [q.v.] in zoological studies.

WORKS: "A Brief Preliminary Account of the Hessian Fly and Its Parasites," *Am. Journ. Sci.* 41 (1841): 153-58; "Further Proof of an Annual Meteoric Shower in August, with Remarks on Shooting Stars in General," ibid. 33 (1838): 176, 354-64, 401-2; catalogue of past meteoric showers, in ibid., 35 (1839), 39 (1840), 40 (1841), 43 (1842). Nearly all papers published in *Am. Journ. Sci.;* see bibliography in *Roy. Soc. Cat.*, vols. 3, 7. MSS: Yale Univ. Library—Manuscripts and Archives Division; also letters in J. D. Dana Papers at Yale—Beinecke Library. WORKS ABOUT: *DAB,* 8 586-87 (Harris E. Starr); obituary in *Am. Journ. Sci.,* 2d ser. 34 (1862): 159-60.

HILDRETH, SAMUEL PRESCOTT (September 30, 1783, Methuen, Mass.—July 24, 1863, Marietta, Ohio). *Natural history.* Son of Samuel—physician, farmer, and merchant—and Abigail (Bodwell) Hildreth. Married Rhoda Cook, 1807; six children. EDUCATION: Attended Phillips Andover and Franklin academies and Methuen public schools; studied medicine with father and for two years with another physician; attended series of lectures in Harvard College; 1805, diploma, Medical Society of Massachusetts. CAREER: 1803-1804, taught school while studying medicine; ca. 1805-1806, medical practice, Hampstead, N.H.; 1806, went to Marietta, Ohio; ca. 1807-1808, medical practice, Belpre, Ohio; 1808-1861, medical practice in Marietta; 1810-1811, member of Ohio state legislature; 1836, appointed chairman of committee to plan geological survey of Ohio; 1836-1837, first assistant geologist and paleontologist, Ohio Geological Survey under W. W. Mather [q.v.]. SCIENTIFIC CONTRIBUTIONS: As naturalist in broadest sense, collected insects, shells, fossils, and plants; investigated geological phenomena; kept records of flowering of plants and of weather; was interested in relations of climate and health; wrote on archaeology and later on early history of Ohio; and also published on medical subjects. Meteorological record [see Works]was one of longest continuous records of weather maintained during those years. Has been praised as early authority on Ohio geology, and work on coal deposits [see Works] was considered to be of great importance by British geologists. Probably best remembered as historian, contributing much toward preservation of early history of Ohio.

WORKS: Scientific work appeared especially in *Am. Journ. Sci.;* "Observations on the Bituminous Coal Deposits of the Valley of the Ohio, and the Accompanying Rock Strata . . . ," ibid. 29 (1836): 1-154; Charles A. Schott [q.v.], "Results of Meteorological Observations Made at Marietta, Ohio, Between 1826 and 1859, Inclusive by S. P. Hildreth . . . To Which Are Added Results of Observations Made at Marietta by Mr. Joseph Wood Between 1817 and 1823," *Smithson. Contr. Knowl.* 16 (1870). See *Roy. Soc. Cat.,* vol. 3, and *Natl. Un. Cat. Pre-1956 Imp.* MSS: Ohio Historical Society (*NUCMC* 75-1031); Campus Martius Museum, Marietta, Ohio (Hamer); Marietta College Library (Waller [below]). WORKS ABOUT: *DAB,* 9: 21-22 (Albert P. Mathews); A. E. Waller, "Dr. Samuel Prescott Hildreth," *Ohio Archaeological and Historical Quarterly* 53 (1944): 313-38.

HILGARD, EUGENE WOLDEMAR (1833-1916). *Geology; Agricultural chemistry and physics.* Professor of agriculture and director of Agricultural Experiment Station, Univ. of California, Berkeley.

MSS: Univ. Calif.-Ban. (Hamer). WORKS ABOUT: See *AMS; DAB.*

HILGARD, JULIUS ERASMUS (January 7, 1825, Zweibrücken, Bavaria—May 8, 1891, Washington, D.C.). *Geodesy.* Son of Theodor Erasmus—lawyer, author, and horticulturist—and Margaretha (Pauli) Hilgard; brother of Eugene W. Hilgard [q.v.]. Married Katherine Clements, 1848; four children. EDUCATION: In Germany, attended gymnasium through third grade; 1836, family emigrated to U.S. and settled on farm at Belleville, Ill.; thereafter father taught children at home; 1843, went to Philadelphia to study civil engineering. CAREER: Ca. 1843, participated in preliminary surveys for Bear Mountain Railroad; 1844, entered computing division, U.S. Coast Survey, the beginning of forty years of association with that organization; ca. 1844-1851, engaged in field work for Coast Survey on Gulf and Atlantic coasts and also worked on computations and investigations in survey office; 1851, put in charge of survey's computing department; 1853, continued as supervisor of computing department, but after this date also was given number of special assignments; 1855-1860, continued some field work for survey, but chief responsibility was publication of its records and results; 1860-1862, severed connection with Coast Survey, and engaged in business at Paterson, N. J.; 1862, rejoined survey, put in charge of office, and during the period 1864-1867 performed duties of acting superintendent; as part of duties in Coast Survey, also conducted Office of Weights and Measures; 1881-1885, superintendent, U.S. Coast Survey. MEMBERSHIPS: Natl. Ac. Scis.; president, Am. Assoc. Advt. Sci., 1875. SCIENTIFIC CONTRIBUTIONS: While studying civil engineering in Philadelphia, made positive impression on A. D. Bache [q.v.], who took him into Coast Survey; chief scientific work done in geodesy and geophysics as part of activities of survey, and for many years before becoming superintendent played major role in conduct and standards of the organization. Special tasks within survey related, for example, to measurement of bases and investigation of instrumental errors; was noted for knowledge of precision instruments, and took leadership role in matters relating to standard weights and measures.

WORKS: Writings appeared especially in *Annual Reports of U.S. Coast Survey.* See *Roy. Soc. Cat.,* vols. 3, 6 (additions), 7, 10, 12, 15. MSS: Hilgard Family Papers at Univ. Calif.-Ban. WORKS ABOUT: *DAB,* 9: 23 (George W. Littlehales); E. W. Hilgard, "Memoir," *Biog. Mems. of Natl. Ac. Scis.* 3 (1895): 327-38.

HILL, GEORGE WILLIAM (1838-1914). *Mathematics; Astronomy.* Associated with *American Ephemeris and Nautical Almanac;* resided at West Nyack, N.Y.

WORKS ABOUT: See *AMS; DAB; DSB.*

HILL, HENRY BARKER (April 27, 1849, Waltham, Mass,—April 6, 1903, Cambridge, Mass.). *Chemistry.* Son of Thomas [q.v.]—minister, college president, and scientist—and Ann Foster (Bellows) Hill. Married Ellen Grace Shepard, 1871; at least one child. EDUCATION: Attended Antioch College for less than one year; 1869, A.B., Harvard College; ca. 1869-ca. 1870, studied at Univ. of Berlin. CAREER: At Harvard Univ., assistant in chemistry (1870-1874), assistant professor of chemistry (1874-1884), professor of chemistry (1884-1903), director of chemical laboratory (1894-1903); 1891, lecturer on organic chemistry, Massachusetts Institute of Technology; in addition to teaching, also conducted chemical investigations for various commercial and governmental organizations. MEMBERSHIPS: Am. Ac. Arts Scis.; Natl. Ac. Scis.; and other societies. SCIENTIFIC CONTRIBUTIONS: At outset of teaching career, had little formal training in chemistry; teaching responsibilities related chiefly to qualitative analysis and organic chemistry, and notes for former subject [see Works] constituted sole separate publication. Original contributions to chemistry largely grew out of consulting and commercial engagements, which included work on adulterations of food for state board of health and chemical investigations for bleaching establishment; first scientific papers were on uric acid. For period of years, acted as consulting chemist for Carter Ink Company; in this capacity visited Squibb pharmaceutical factory and was induced to begin investigation of waste products created in making acetic acid from distillations of oak wood; in this waste he found furaldehyde (furfurol), then largely uninvestigated. Remainder of life was devoted to study of this substance and its derivatives, resulting in more than thirty scientific papers.

WORKS: *Lecture Notes on Qualitative Analysis* (New York, 1874); "On the Ethers of Uric Acid," *Proceedings of Am. Ac. Arts Scis.* 12 (1876-1877): 26-36; most papers appeared in *Proceedings of Am. Ac. Arts Scis.* and *American Chemical Journal.* See bibliography in Jackson [below], pp. 264-66. WORKS ABOUT: *DAB,* 9: 33-34 (Walter L. Jennings); Charles L. Jackson, "Memoir," *Biog. Mems. of Natl. Ac. Scis.* 5 (1905): 255-66.

HILL, ROBERT THOMAS (1858-1941). *Geology; Geography.* Geologist, U.S. Geol. Surv.; engaged in private consulting practice.

MSS: Univ. Tex.-TA (*NUCMC* 70-811). WORKS ABOUT: See *AMS; NCAB,* 40 (1955): 159.

HILL, THOMAS (January 7, 1818, New Brunswick, N. J.—November 21, 1891, Waltham, Mass.). *Mathematics; Astronomy; Science and religion.* Son of Thomas, tanner and judge of court of common pleas, and Henrietta (Barker) Hill. Married Ann Foster Bellows, 1845; Lucy E. Shepard, 1866; survived by seven children, including Henry B. Hill [q.v.]. EDUCATION: 1830-1838, in turn, apprenticed to printer, spent eighteen months under oldest brother at academy in Holmesburg, Pa., and apprenticed to apothecary; 1843, A.B., Harvard College; 1845, graduated, Harvard Divinity School. HONORS: S.T.D., Harvard, 1860; LL.D., Yale Univ., 1863; Scott Prize, Franklin Institute, ca. 1843. CAREER: 1845-1859, Unitarian minister,

Waltham, Mass.; 1859, gave series of Lowell Institute lectures; 1859-1862, president and professor of metaphysics, ethics, and natural religion, Antioch College; 1862-1868, president, Harvard Univ.; 1868-1873, moved to Waltham, preached at various churches, established cranberry business in Massachusetts and New Jersey; 1869, traveled in West and elsewhere in U.S.; 1870, gave Lowell lectures; 1871, served in state legislature; 1871-1872, botanist, physicist, and photographer on Coast Survey expedition around Cape Horn led by J. L. R. Agassiz [q.v.]; 1873-1891, minister, Portland, Maine; ca. 1881-1891, nonresident professor of natural theology and ethics, Meadville Theological School. MEMBERSHIPS: Am. Ac. Arts Scis.; Am. Phil. Soc.; and other societies. SCIENTIFIC CONTRIBUTIONS: While student at Harvard, demonstrated unusual ability in mathematics; invented "occultator" to calculate times and paths of eclipses, which won Scott Prize; and on graduation was recommended by B. Peirce [q.v.] to be director of national observatory in Washington, but Hill declined offer. Later, invented devices for adding and subtracting and "nautrigon" for determination of latitude and longitude at sea. Mathematical papers demonstrated lasting influence of Peirce and dealt especially with curves; wrote several mathematical textbooks; despite great ability and interest in mathematics, these studies always remained recreational. Made early attempt in U.S. to reconcile science and religion.

WORKS: *Geometry and Faith* (New York, 1849; rev. eds., 1874 and 1882), 48 pp.; "Systems of Coordinates in One Plane," *Proceedings of Am. Assoc. Advt. Sci.* 11 (1857): 42-45. See *Natl. Un. Cat. Pre-1956 Imp.* and *Roy. Soc. Cat.*, vols. 3, 7, 10, 15. MSS: Harv. U.-UA (*NUCMC* 65-1251). WORKS ABOUT: *DAB*, 9:45-46 (William G. Land); Land, *Thomas Hill* (Cambridge, Mass., 1933).

HILLEBRAND, WILLIAM FRANCIS (1853-1925). *Chemistry.* Chemist, U.S. Geol. Surv.; chief chemist, U.S. Bureau of Standards.

MSS: See *DAB*. WORKS ABOUT: See *AMS; DAB.*

HINRICHS, GUSTAVUS DETLEF (1836-1923). *Chemistry.* Professor of physical science, State Univ. of Iowa; professor of chemistry, St. Louis College of Pharmacy; professor of chemistry, Medical Department, St. Louis Univ.; consulting chemist in St. Louis.

WORKS ABOUT: See *AMS; NCAB*, 21 (1931): 390-91.

HITCHCOCK, CHARLES HENRY (1836-1919). *Geology.* Professor of geology, Dartmouth College; head of New Hampshire Geological Survey.

WORKS ABOUT: See *AMS; DAB.*

HITCHCOCK, EDWARD (May 24, 1793, Deerfield, Mass.—February 27, 1864, Amherst, Mass.). *Geology.* Son of Justin, farmer and hatter, and Mercy (Hoyt) Hitchcock. Married Orra White, 1821; six children. EDUCATION: Attended Deerfield Academy; ca. 1820, studied theology, Yale Univ.; 1825-1826, audited courses taught by B. Silliman [q.v.] at Yale. HONORS: M.A., Yale, 1818; LL.D., Harvard, 1840. CAREER: 1815-1819, principal, Deerfield Academy; 1821-1825, Congregational minister, Conway, Mass.; at Amherst College, professor of chemistry and natural history (1825-1845), president (1844-1854), professor of geology and natural theology (1845-1864); 1830-1833 and 1837-1841, Massachusetts state geologist; 1836, appointed, but soon resigned as geologist for first district, New York Natural History Survey; 1856-1861, Vermont state geologist. MEMBERSHIPS: First chairman, Association of American Geologists and Naturalists, 1840; member of Natl. Ac. Scis. SCIENTIFIC CONTRIBUTIONS: Research in science began with observations of 1811 comet, and corrected errors in *Blunt's Nautical Almanac*, but difficulties with eyesight soon terminated activities as astronomer. Was largely responsible for establishment of Massachusetts Geological Surveys in 1830 and 1837, and chief scientific work grew out of geological appointments in Massachusetts and Vermont; one noteworthy geological work was study of terraces in Connecticut River Valley, which was presented in *Surface Geology* [see Works] and included study of actions of rivers and glaciers; also wrote on metamorphosis of sediments; played central role in study of large tracks (originally discovered by J. Deane [q.v.]) in red sandstone of Connecticut Valley, and became convinced these were produced by early birds, although they have since been judged as of dinosaur origin. Textbook, *Elementary Geology*, had number of editions and was widely used. Also wrote on science and religion, motivated by conviction that they were parts of single subject.

WORKS: *Report on the Geology, Mineralogy, Botany, and Zoology of Massachusetts* (Amherst, Mass., 1833), with *Final Report*, 2 vols., published in 1841; *Elementary Geology* (Amherst, Mass., 1840); *The Religion of Geology and Its Connected Sciences* (Boston, 1851); "Illustrations of Surface Geology," *Smithson. Contr. Knowl.* 9 (1857). Bibliography of works in *American Geologist* 16 (1895): 139-49. MSS: Amherst College Library (*NUCMC* 75-114). WORKS ABOUT: *DAB*, 9: 70-71 (George P. Merrill); *DSB*, 6: 437-38 (Michele L. Aldrich); Stanley M. Guralnick, "Geology and Religion Before Darwin: The Case of Edward Hitchcock . . . ," *Isis* 63 (1972) : 529-43.

HOLBROOK, JOHN EDWARDS (December 30, 1794, Beaufort, S.C. [removed soon thereafter to paternal home at Wrentham, Mass.]—September 8, 1871, Norfolk, Mass.). *Zoology.* Son of Silas and Mary (Edwards) Holbrook. Married Harriott Pinckney Rutledge, 1827; no children. EDUCATION: 1815, A.B., Brown Univ.; 1818, M.D., Univ. of Pennsylvania; 1818-1822, traveled and studied in Europe, especially Edinburgh and Paris. CAREER: 1822, began medical practice, Charleston, S.C.; 1824, a founder of Medical College of South Carolina, and for more than thirty years served there as professor of anatomy; during Civil War, served as medical officer in Confederate army, serving as head of examining board of surgeons, South Carolina; died in Massachusetts while vacationing with sister. MEMBERSHIPS: Natl. Ac. Scis. SCIENTIFIC CONTRIBUTIONS: While studying in Europe, became associated with zoologists at Jardin des Plantes in Paris and became interested in study of reptiles; after return to U.S., undertook study of American reptiles. Resultant work, *Herpetology* [see Works], was one of first major zoological publications by American investigator and won for author international reputation as the chief U.S. zoologist of the time; begun in serial form in 1836, *Herpetology* was furnished with nearly 150 color plates made by Italian artist employed by Holbrook, working from live specimens, and parts of publication are now quite rare. After work on reptiles, turned to study of fishes and before Civil War terminated his scientific work, produced *Ichthyology* [see Works]. In addition to aforementioned works only one scientific paper is known to have been written by Holbrook.

WORKS: *North American Herpetology,* 5 vols. (Philadelphia, 1842) ; *Ichthyology of South Carolina* (Charleston, 1860; superseding earlier version issued 1855-1857) ; "An Account of Several Species of Fish Observed in Florida, Georgia, etc.," *Journal of Ac. Nat. Scis. Philad.* n.s. 3 (1855) : 47-58. See bibliography, with lists of species, in Theodore Gill, "Memoir," *Biog. Mems. of Natl. Ac. Scis.* 5 (1905) : 67-77. Manuscripts and drawings said to have been lost during Civil War (Mansueti [below]). WORKS ABOUT: *DAB,* 9: 129-30 (Hubert L. Clark) ; Romeo Mansueti, "Father of American Herpetology," *Nature Magazine* 43 (1950) : 16-20, 52.

HOLDEN, EDWARD SINGLETON (1846-1914). *Astronomy.* President of Univ. of California and director of Lick Observatory; librarian, U.S. Military Academy, West Point.

WORKS ABOUT: See *AMS; DAB; DSB.*

HOLDER, JOSEPH BASSETT (October 26, 1824, Lynn, Mass.—February 27, 1888, New York, N.Y.). *Zoology.* Son of Aaron Lummus, wholesale and retail druggist, and Rachael (Bassett) Holder. Married Emily Augusta Gove, before 1851; apparently only one child. EDUCATION: Graduated, Friends' School, Providence, R.I.; said to have entered Harvard Medical School and to have served there as demonstrator in anatomy. CAREER: 1846, began medical practice in Swampscott, Mass., and later moved to Lynn, Mass., where became city physician; 1859, at suggestion of J. L. R. Agassiz and S. F. Baird [qq.v.], became surgeon in chief with government engineers, Florida reef, and engaged in scientific studies of the reef; ca. 1861, entered U.S. Army, assigned as health officer and surgeon, military prison at Fort Jefferson, Dry Tortugas, Fla.; 1869, transferred to Fortress Monroe, Va.; at Am. Mus. Nat. Hist. (New York), assistant director (1871-1881) and curator of zoology (1881-1888). MEMBERSHIPS: N.Y. Ac. Scis.; Am. Ornith. Un.; and other societies. SCIENTIFIC CONTRIBUTIONS: Prepared first list of birds and plants of Essex County, Mass. Became friend of J. L. R. Agassiz, who maintained laboratory at Nahant, near Lynn, Mass., and the two dredged bay and collected specimens together. In Florida, Holder undertook extensive studies of reef creation and its plant and animal life, and shipped specimens and data to Agassiz's and other museums; studies of reef resulted in important information on corals, especially determination of their relatively rapid growth, in contrast to contemporary opinion that corals grew very slowly; also experimented with other marine specimens and studied whales and flights of birds. Interested in, and frequently wrote on, popularization of scientific knowledge and promotion of interest in nature.

WORKS: *History of the American Fauna,* 5 vols. (New York, [1877]). See several other works listed in *DAB.* See also *Roy. Soc. Cat.,* vols. 12, 15. WORKS ABOUT: DAB, 9: 140-41 (Mary B. Hart) ; Charles F. Holder, *The Holders of Holderness* [Pasadena, Calif., 1902], pp. 226-31, written by the son of J. B. Holder and largely based on the father's genealogical researches; memoir, *Transactions of N.Y. Ac. Scis.* 7 (1887-1888) : 251-55.

HOLLAND, WILLIAM JACOB (1848-1932). *Zoology; Paleontology.* Director, Carnegie Museum of Pittsburgh.

MSS: Historical Society of Western Pennsylvania, Pittsburgh (*NUCMC* 70-2131) ; Ac. Nat. Scis. Philad. (Hamer). WORKS ABOUT: See *AMS; DAB,* supp. 1.

HOLLICK, CHARLES ARTHUR (1857-1933). *Ge-*

ology; Paleobotany. Curator, department of fossil botany, and paleobotanist, N.Y. Bot. Gard.
WORKS ABOUT: See *AMS; DAB,* supp. 1.

HOLMES, FRANCIS SIMMONS (December 9, 1815, Charleston, S.C.—October 19, 1882, Charleston, S.C.). *Geology; Natural history.* Son of John and Anna Glover Holmes. Married Miss Toomer; later, married Sarah Hazzard; seven children by first marriage, six by second. EDUCATION: Ca. 1830 [at about age fourteen], left school and shortly thereafter entered countinghouse; apparently largely self-educated. HONORS: Referred to as A.M. CAREER: At young age, entered upon mercantile career, and for time engaged in dry goods business in partnership with brother-in-law, as Trenholm & Holmes; thereafter, turned to agriculture and established plantation in St. Andrew's Parish, S.C.; 1850, appointed curator of museum of natural history, College of Charleston, and shortly thereafter also became professor of geology and natural history; ca. 1862, granted leave from college to become chief of Confederate Mining and Nitre Bureau in Georgia and South Carolina; after war, became a founder and president of Charleston Mining and Manufacturing Co.; 1869, resigned position with Charleston museum (and presumably also gave up professorship); in later years, retired to rice plantation near Charleston. MEMBERSHIPS: Ac. Nat. Scis. Philad. and other societies. SCIENTIFIC CONTRIBUTIONS: In 1842, published *Southern Farmer and Market Gardener,* which won some popularity. Discovered fossils in vicinity of Charleston, accumulated large collection, and in 1847 was given a room at College of Charleston, later to become museum of natural history; in 1849 published paper on geology of Charleston *(Am. Journ. Sci.),* in which he pointed out similarities with geological formations of Paris and London; in 1850s, under auspices of state legislature, undertook study of fossils with M. Tuomey [q.v.], which resulted in two volumes completed after Tuomey's death. After Civil War, came to realize value of fossils as source of phosphate and established mining company for commerical production of fertilizer.

WORKS: *Pleiocene Fossils of South Carolina* (Charleston, 1858), with M. Tuomey; *Post-Pleiocene Fossils of South Carolina* (Charleston, 1860), with J. Leidy [q.v.]; *Phosphate Rocks of South Carolina and the "Great Carolina Marl Bed"* . . . (Charleston, 1870), 87 pp. See also Meisel, *Bibliog. Amer. Nat. Hist.* WORKS ABOUT: [City of] Charleston, *Yearbook—1882,* pp. 335-38; J. H. Easterby, *A History of the College of Charleston* ([Charleston], 1935).

HOLMES, WILLIAM HENRY (1846-1933). *Anthropology; Geology.* Associated with Hayden survey of territories and U.S. Geol. Surv.; chief, U.S. Bureau of American Ethnology; curator of anthropology, U.S. Natl. Mus.; director, National Gallery of Art; artist.

MSS: Smithson. Instn. Archives (*NUCMC* 72-1240); Yellowstone National Park, Wyoming (Hamer). WORKS ABOUT: See *AMS; DAB,* supp. 1.

HORN, GEORGE HENRY (April 7, 1840, Philadelphia, Pa.—November 24, 1897, Beesley's Point, N.J.). *Entomology.* Son of Philip Henry, druggist and businessman, and Frances Isabella (Brock) Horn. Never married. EDUCATION: Graduated, Philadelphia High School; 1861, M.D., Univ. of Pennsylvania. CAREER: 1862-1866, surgeon, California Volunteers; ca. 1866-1897, practiced medicine, especially obstetrics, Philadelphia; 1874, 1882, and 1888, visited European museums; 1889, appointed professor of entomology, Univ. of Pennsylvania, an honorary title. MEMBERSHIPS: Entomological Society of France; Ac. Nat. Scis. Philad.; and other societies. SCIENTIFIC CONTRIBUTIONS: Confined entomological work almost exclusively to Coleoptera, and became associated with J. L. LeConte [q.v.], at that time the leading American coleopterist; while in West during Civil War, collected large number of coleopterous insects. Horn's first publication appeared in 1860; wrote extensively, described 1,583 species and varieties of Coleoptera, and designated some 150 genera; prepared section on Otiorhynchidae for joint publication with LeConte, "Rhynchophora" [see Works], and worked with LeConte on their joint publication entitled "Classification of the Coleoptera" [see Works], an achievement of great importance. After death of LeConte in 1883, Horn became leading American authority on beetles, and succeeded LeConte as director of entomological division at Ac. Nat. Scis. Philad. Descriptive work and keys to classification of Coleoptera remained authoritative for many years.

WORKS: "Rhynchophora of America, North of Mexico," *Proceedings of Am. Phil. Soc.* 15 (1876): 1-455, with J. L. LeConte; "Classification of the Coleoptera of North America," *Smithson. Misc. Coll.* 26 (1883), 567 pp., also with LeConte; "Synopsis of the Silphidae of the U.S. with Reference to the Genera of other Countries," *Transactions of American Entomological Society* 8 (1880): 219-322; "On the Genera of Carabidae with Special Reference to the Fauna of Boreal America," ibid. 9 (1881): 91-96. See bibliography by Samuel Henshaw in ibid. 25 (1898), appendix 25-72. MSS: Ac. Nat. Scis. Philad. (Hamer). WORKS ABOUT: *DAB,* 9: 229-30 (Leland O. Howard); *DSB,* 6: 506-7 (Melville H. Hatch).

HORNER, WILLIAM EDMONDS (June 3, 1793, Warrenton, Fauquier County, Va.—March 13, 1853, Philadelphia, Pa.). *Anatomy.* Son of William, merchant, and Mary (Edmonds) Horner. Married Elizabeth Welsh, 1820; ten children. EDUCATION: 1809, completed academy education; 1814, M.D., Univ. of Pennsylvania; 1821, studied surgery in Europe. CAREER: 1813-1815, as surgeon's mate, U.S. Army, was granted furlough in winter of 1813-1814 to complete medical studies; 1815, medical practice for short period in Warrenton, Va.; toward end of that year, moved to Philadelphia and there developed lucrative medical and surgical practice; 1816, at invitation of professor of anatomy C. Wistar [q.v.], became prosector in Univ. of Pennsylvania; at Univ. of Pennsylvania, appointed adjunct professor of anatomy (1819), served as professor of anatomy (1831-1853), and for some thirty years served as dean of medical school; 1823, began what was said to be twenty-five years' service on staff of Blockley, the almshouse hospital; at St. Joseph's Hospital, became one of physicians (1849), thereafter surgeon and president (1850). MEMBERSHIPS: Am. Phil. Soc. SCIENTIFIC CONTRIBUTIONS: A noted anatomist and an early American microscopist, he prepared first American work on pathological anatomy [see Works]. Studied sweat glands, muscles of rectum and larynx; best remembered for investigation of tensor tarsi, muscle of the eye related to tear mechanism, now called muscle of Horner; also did important work in describing cartilages at bronchial subdivisions. *Special . . . Anatomy* [see Works] included original microscopical studies and was Horner's most ambitious publication; also wrote on cases for medical journals and prepared paper on Asiatic cholera based on 1832 Philadelphia epidemic.

WORKS: "Description of a Small Muscle at the Internal Commissure of the Eyelids," *Philadelphia Journal of Medical and Physical Sciences* 8 (1824): 70-80; *A Treatise on Special and General Anatomy*, 2 vols (Philadelphia, 1826); *A Treatise on Pathological Anatomy* (Philadelphia, 1829). Incomplete bibliography in Middleton [below]. WORKS ABOUT: *DAB*, 9: 233-34 (William S. Miller); William S. Middleton, "William Edmonds Horner," *Ann. Med. Hist.* 5 (1923): 33-34.

HORSFIELD, THOMAS (May 12, 1773, near Bethlehem, Pa.—July 24, 1859, London, England). *Natural history.* Son of Timothy (Jr.), apparently farmer, and Juliana Sarah (Parsons) Horsfield. Married; two children. EDUCATION: Studied pharmacy with a Dr. Otto in Bethlehem, Pa.; 1798, M.D., Univ. of Pennsylvania. CAREER: 1799-1800, surgeon on merchant ship to Java; 1800, returned to Philadelphia and prepared for future exploring and collecting in Java; 1801-1819, resided in Java, before 1811 as surgeon in Dutch colonial army; after British occupation in 1811, pursued natural history explorations under auspices of East India Co.; 1819, left East Indies for England; 1820-1859, curator, East India Co. museum, London. SCIENTIFIC CONTRIBUTIONS: At Univ. of Pennsylvania, prepared graduation thesis on effects of sumac and poison ivy, a ground-breaking American effort in experimental pharmacology. Noted chiefly for extensive botanical and zoological explorations in Java and the several volumes that resulted from those studies; during initial journey to Java, attracted by its abundant unexplored flora and by drugs derived by natives from certain plants. Chief writings were prepared after leaving Java, having taken extensive collections to London for study; major work, *Plantae* [see Works], included descriptions of more than 2,000 species, all collected by Horsfield, and was prepared with assistance of Robert Brown and J. J. Bennett, botanists. Also prepared publications, on his own or in collaboration with other naturalists, on Lepidoptera and other insects, mammals, and birds, based on collections of East India Co. museum.

WORKS: *Plantae Javanicae Rariores* (London, 1838-1852). See *DAB* and *Dictionary of National Biography* [below], for references to other works; also *Roy. Soc. Cat.*, vols. 3, 6 (additions), and *Natl. Un. Cat. Pre-1956 Imp.* WORKS ABOUT: *DAB*, 9: 236 (John F. Fulton); *Dictionary of National Biography*, 9: 1273-74.

HORSFORD, EBEN NORTON (July 27, 1818, Moscow, [now Leicester], N.Y.—January 1, 1893, Cambridge, Mass.). *Chemistry.* Son of Jerediah, farmer and Indian missionary, and Charity Maria (Norton) Horsford. Married Mary L'Hommedieu Gardiner, 1847; Phoebe Dayton Gardiner, 1857; four children by first marriage, and one by second. EDUCATION: 1838, graduated in civil engineering, Rensselaer Polytechnic Institute; 1844-1847, in Europe, where studied chemistry two years with Justus Liebig at Giessen, Germany. CAREER: After leaving Rensselaer, worked for year or so on New York State Natural History [Geological] Survey; 1840-1844, was professor of mathematics and natural sciences, Albany Female Academy, and gave annual chemistry lectures in Newark College (Delaware); at Lawrence Scientific School, Harvard Univ., Rumford professor and lecturer on application of science to useful arts (1847-1863) and dean (1861-1862); 1863, resigned Harvard appointment and thereafter directed full efforts to industrial chemistry, making fortune from Rumford Chemical Co. in Rhode Island; served as president of board of visitors, Wellesley College. MEMBERSHIPS: Early member, Am. Chem. Soc. SCIENTIFIC CONTRIBUTIONS: As second American to study chemistry with Liebig, Horsford played significant role in introduction of European chemical methods and knowledge into U.S.; and as professor in Lawrence Scientific School, developed first U.S. laboratory for teaching analytical chemistry, based on Liebig's practice; published number of papers in American and German journals. Chemical investigations were varied and frequently had practical orientation, including use of lead pipes for Boston water supply, study of condensed milk, fermentation, acid phosphates in medicine, vulcanization of rubber, and other subjects; particularly interested in food chemistry and nutrition, and development of phosphatic baking powder as substitute for yeast was one of basic products of his Rumford Chemical Co. In later years, studied and wrote about Indian languages and early Viking explorations in North America.

WORKS: *The Theory and Art of Bread-Making* (Cambridge, Mass., 1861). See several works listed in *DSB* and in *Roy. Soc. Cat.*, vols. 3, 7. MSS: Rens. Poly. Inst.-Arch. (*DSB*, which also includes reference to other manuscripts); also Univ. of Delaware Archives (*NUCMC* 70-1513). WORKS ABOUT: *DAB* 9: 236-37 (Lyman C. Newell); *DSB*, 6: 517-18 (Samuel Rezneck); Rossiter, *Emergence of Agricultural Science*, chaps. 4 and 5.

HOUGH, FRANKLIN BENJAMIN (July 22, 1822, Martinsburg, Lewis County, N.Y.—June 11, 1885, Lowville, N.Y.). *Forestry*; *Botany*. Son of Horatio G., physician, and Martha (Pitcher) Hough. Married Maria S. Eggleston, 1845; Mariah E. Kilham, 1849; one child by first marriage, and eight by second. EDUCATION: 1843, A.B., Union College; 1848, M.D., Western Reserve Medical College. CAREER: Ca. 1843-1846, taught one year in academy at Champion, N.Y., and then was principal, Gustavus Academy, Ohio; ca. 1848, established medical practice, Somerville, N.Y.; 1854, appointed director of New York state census at Albany, and continued medical practice; ca. 1861, inspector, U.S. Sanitary Commission; 1862-1863, regimental surgeon, New York Volunteers; ca. 1865 (after Civil War), settled at Lowville, N.Y.; 1865, director, New York state census; prepared manual and other documents for 1867 New York constitutional convention; 1867, superintendent, census of District of Columbia; 1870, superintendent, U.S. Census; 1876, appointed forestry agent, U.S. Department of Agriculture; 1881, commissioned head of newly established Division of Forestry in Department of Agriculture; ca. 1881/1882, studied forestry situation in Europe; 1882, established short-lived *American Journal of Forestry*. MEMBERSHIPS: Am. Assoc. Advt. Sci. and other societies; an original member of American Forestry Association. SCIENTIFIC CONTRIBUTIONS: Pioneer advocate of preservation and planning in use of American forests, having become aware of problems during work for U.S. Census, and largely through personal efforts was appointed first U.S. forestry agent; in that capacity, produced three reports during years 1877-1882 and performed important role in arousing public interest in forest problems. Had earlier interest in botany, and also wrote on geology, meteorology, and climatology, in addition to several historical works.

WORKS: *A Catalogue of Indigenous, Naturalized and Filicoid Plants of Lewis County, N.Y.* (Albany, [1846]); "On the Duty of Governments in the Preservation of Forests," *Proceedings of Am. Assoc. Advt. Sci.* 22, pt. 2 (1873), sec. B, pp. 1-10; *Elements of Forestry* (Cincinnati, 1882). See bibliography of writings in *Annual Report of Regents of Univ. of State of New York* 99 (1886): 321-47. MSS: N.Y. St. Lib. (Hamer); Det. Pub. Lib.-BHC (*NUCMC* 68-1246). WORKS ABOUT: *DAB*, 9: 250-52 (Henry S. Graves); Charles E. Randall, "Hough . . . ," *American Forests* 67 (May 1961): 10-11, 41-42, 44.

HOUGH, GEORGE WASHINGTON (1836-1909). *Astronomy*. Professor of astronomy and director of Dearborn Observatory, Northwestern Univ.

WORKS ABOUT: See *AMS*; *DAB*; *DSB*.

HOUGHTON, DOUGLASS (September 21, 1809, Troy, N.Y.—October 13, 1845, Eagle River, Mich.). *Geology*. Son of Jacob, lawyer and judge, and Mary Lydia (Douglas) Houghton. Married Harriet Stevens, 1833; two children. EDUCATION: 1829, A.B., Rensselaer Polytechnic Institute; 1831, completed requirements for practice of medicine. CAREER: 1829-1830, assistant professor of chemistry and natural history, Rensselaer Polytechnic Institute; 1830, delivered course of science lectures in Detroit, having been recommended by A. Eaton [q.v.]; 1831, appointed surgeon and botanist on expedition led by H. R. Schoolcraft [q.v.] to find sources of Mississippi River; 1832-1837, practiced medicine and surgery and engaged in real estate dealings, Detroit; 1837-1845, state geologist, and professor of geology, mineralogy, and chemistry at Univ. of Michigan; 1842-1843, mayor of Detroit; 1844-1845, engaged in field work for projected federal geological and geophysical survey of the West in conjunction with linear survey of public lands already planned by government. MEMBERSHIPS: Ac. Nat. Scis. Philad.; Bos. Soc. Nat. Hist.; and other societies. SCIENTIFIC CONTRIBUTIONS: Chief work done in conjunction with early Geological Surveys of Michigan, undertaken largely through his initiative; explorations of Upper Peninsula were particularly extensive and important because of economic value of mineral content.

WORKS: Chief publications were reports on Michigan Geological Surveys. See Nickles, *Geol. Lit. on N. Am.*; also George N. Fuller, ed., *Geological Re-*

ports of Douglass Houghton, 1837-1845 (Lansing, Mich., 1928). MSS: Univ. Mich.-MHC (*NUCMC* 65-327); also Det. Pub. Lib. (*DSB*). WORKS ABOUT: *DAB*, 9: 254-55 (George P. Merrill); *DSB*, 6: 523-24 (Samuel Rezneck). Also see Edsel K. Rintala, *Douglass Houghton, Michigan's Pioneer Geologist* (Detroit, 1954).

HOWARD, LELAND OSSIAN (1857-1950). *Entomology.* Chief, Bureau of Entomology, U.S. Department of Agriculture.

MSS: Ac. Nat. Scis. Philad. (Hamer). WORKS ABOUT: See *AMS*; *DAB*, supp. 4; *DSB*.

HOWE, HERBERT ALONZO (1858-1926). *Astronomy.* Professor of mathematics and astronomy, dean, and director of Chamberlin Observatory, Univ. of Denver.

MSS: See *DAB*. WORKS ABOUT: See *AMS*; *DAB*.

HOWE, JAMES LEWIS (1859-1955). *Chemistry.* Professor of chemistry, Washington and Lee Univ.

WORKS ABOUT: See *AMS*; *DSB*.

HUBBARD, HENRY GUERNSEY (May 6, 1850, Detroit, Mich.—January 18, 1899, Crescent City, Fla.). *Entomology.* Son of Bela, wealthy and influential citizen of Detroit, and Sarah (Baughman) Hubbard. Married Kate Lasier, 1887; four children. EDUCATION: Preparatory education included several years under private tutors in Europe; 1873, A.B., Harvard College; 1873-1874 (winter), studied entomology with H. A. Hagen [q.v.], at Harvard. CAREER: 1874, began private museum in Detroit and undertook entomological collecting with E. A. Schwarz [q.v.] near Detroit and in Florida; 1876-1878, carried out several expeditions, exploring Lake Superior region, partly with Schwarz; 1879, appointed and served for short time as naturalist with Geological Survey of Kentucky; 1879, following accidental death of brother, took charge of property at Crescent City, Fla., and resided there for many years; 1880-ca. 1885, occupied with studies of cotton worm for U. S. Entomological Commission and of insects affecting oranges for U.S. Department of Agriculture; ca. 1885, began several years work in horticulture, established well-known semitropical garden at Crescent City; 1889-1892, visited Lake Superior region and various places in West, part of time with Schwarz; 1894, again began work as special agent for Department of Agriculture, especially in revision of work on orange insects, but progress was interrupted by tuberculosis, which finally took life. MEMBERSHIPS: President, Entomological Society of Washington. SCIENTIFIC CONTRIBUTIONS: His father had strongly pronounced scientific interests and was a founder of Am. Assoc. Advt. Sci.; son's entomological interests developed at Harvard. Did notable collecting work with E. A. Schwarz and was conspicuous among contemporaries for field studies. In work for Kentucky survey conducted early studies of cave insects; as entomological agent for U.S. government, did important work in economic entomology, especially on orange insects, and in these studies developed useful kerosene-soap emulsion.

WORKS: "The Coleoptera of Michigan," *Proceedings of Am. Phil. Soc.* 17 (1877-1878): 593-669, with E. A. Schwarz (descriptions of new species by J. L. Le Conte [q.v.]); *Report on the Insects Affecting the Orange* . . . (Washington, D.C., 1885). Bibliography of entomological writings in *Proc. Ent. Soc.* [below], pp. 356-60. MSS: Smithson. Instn. Archives (*NUCMC* 74-977). WORKS ABOUT: *DAB*, 9: 327-28 (Leland O. Howard); memoir, *Proceedings of Entomological Society of Washington* 4 (1901): 350-60.

HUBBARD, JOSEPH STILLMAN (September 7, 1823, New Haven, Conn.—August 16, 1863, New Haven, Conn.). *Astronomy.* Son of Ezra Stiles and Eliza (Church) Hubbard. Married Sarah E. L. Handy, 1848; one child, who died young. EDUCATION: 1843, A.B., Yale Univ. CAREER: 1843-1844 (winter), taught at Southington, Conn.; 1844, acted as assistant to S. C. Walker [q.v.], Philadelphia High School observatory, and, later that year, employed in Washington as computer of observations for latitude and longitude gathered by Lt. John C. Frémont in West; 1845-1863, professor of mathematics, U.S. Navy, assigned to Naval Observatory; performed duties of acting editor, *Astronomical Journal*, when B. A. Gould [q.v.] was out of country. MEMBERSHIPS: Am. Phil. Soc.; Natl. Ac. Scis.; and other societies. SCIENTIFIC CONTRIBUTIONS: Astronomical labors began before entering college, with construction of telescope. At Naval Observatory, worked with J. H. C. Coffin [q.v.] in arranging for zone observations, which were carried out between 1846 and 1850 and involved use of three instruments at one time; in 1848, officially put in charge of observatory's prime-vertical transit instrument, with which most important work was done, and was particularly concerned with problem of parallax of Alpha Lyrae. Most significant investigations were reported in *Astronomical Journal*, and among more important were those relating to computations on zodiacs of all known asteroids, orbit of comet of 1843, Biela's comet, and fourth comet of 1825; lesser papers dealt with elements and ephemerides of comets and asteroids, and also wrote on techniques for computation, and on labor-saving devices. B. A. Gould gave Hubbard large measure of credit for encouragement and success of *Astronomical Journal*. Thought to have been considering abandonment of scientific work in later years for career in ministry.

WORKS: Major works referred to in *DAB*. Also see *Roy. Soc. Cat.*, vol. 3. WORKS ABOUT: *DAB*, 9: 329-30 (Raymond S. Dugan); B. A. Gould, "Memoir," *Biog. Mems. of Natl. Ac. Scis.* 1 (1877): 1-34.

HUBBARD, OLIVER PAYSON (March 31, 1809, Pomfret, Conn.—March 9, 1900, New York, N.Y.). *Geology*; *Mineralogy*. Son of Stephen, merchant, and Zeruiah (Grosvenor) Hubbard. Married Faith Wadsworth Silliman (daughter of B. Silliman and sister of B. Silliman, Jr. [qq.v.]), 1837; three children. EDUCATION: 1825-1826, attended Hamilton College; 1828, A.B., Yale College; 1837, M.D. (honorary [?]), South Carolina Medical College. HONORS: LL.D., Hamilton College, 1861. CAREER: Ca. 1828-1831, taught at Geneva, N.Y., and Richmond, Va.; 1831-1836, assistant to B. Silliman, Yale chemistry laboratory; at Dartmouth College, professor of chemistry, pharmacy, mineralogy, and geology (1836-1866) and lecturer on same subjects (1866-1871); 1871-1883, professor of chemistry and pharmacy, Dartmouth medical school; 1852-1865, with wife, conducted school for girls, Hanover, N.H.; 1865-1873, ran comparable establishment at New Haven; 1863-1864, member of New Hampshire legislature; 1871-1895, an overseer, Thayer School of Engineering, Dartmouth. MEMBERSHIPS: A founding member, Association of American Geologists; member of N.Y. Ac. Scis. and other societies. SCIENTIFIC CONTRIBUTIONS: While in Silliman's laboratory, aided Charles Goodyear in experiments eventually leading to vulcanization of rubber. Scientific work done especially during years prior to about 1855. Wrote on geology of White Mountains of New Hampshire, especially descriptions of trap dikes and New Hampshire drift; in 1841 at meeting of Association of American Geologists, presented paper on slates at Waterville, Maine, in which he suggested that markings were of organic origin; also wrote on minerals found in New Hampshire and northern New York; prepared reviews on several reports of New York and other geological surveys, and wrote other miscellaneous papers. Scientific work was not extensive, perhaps due to burden of teaching duties. At Dartmouth, in 1853 built Shattuck Observatory, and later wrote several historical works on aspects of that college.

WORKS: Bibliography of works on geology and mineralogy, which appeared chiefly in *Am. Journ. Sci.* and *Transactions of N.Y. Ac. Scis.*, in Hovey [below], pp. 362-63; also see *Roy. Soc. Cat.*, vols. 3, 15. MSS: Dartmouth College Library. WORKS ABOUT: *NCAB*, 9 (1907 [copyright 1899 and 1907]): 557; J. J. Stevenson and Alexis A. Julien, obituary, *Science* n.s. 11 (1900): 742-43; E. O. Hovey, memoir, *American Geologist* 25 (1900): 360-63.

HUMPHREYS, ANDREW ATKINSON (November 2, 1810, Philadelphia, Pa.—December 27, 1883, Washington, D.C.). *Hydraulic engineering; Promotion of science*. Son of Samuel, chief naval constructor, and Letitia (Atkinson) Humphreys. Married Rebecca Hollingsworth, 1839; four children. EDUCATION: 1831, graduated, U.S. Military Academy, West Point. HONORS: LL.D., Harvard Univ., 1868. CAREER: 1831-1836, commissioned lieutenant, U.S. artillery, and thereafter served with U.S. Topographical Engineers, stationed primarily in South, where participated in Seminole War; 1836, resigned commission; 1837-1838, served as civil engineer with U.S. Topographical Engineers and worked especially in Delaware Bay; 1838, appointed lieutenant, U.S. Corps of Topographical Engineers; 1838-1844, worked on various surveys and harbor and bridge projects and as assistant in Bureau of Topographical Engineers at Washington; 1844-1850, assigned to Coast Survey; 1848, commissioned captain; 1850-1851 and 1857-1861, directed topographic and hydrographic survey of delta of Mississippi River, during latter period assisted by Lt. H. L. Abbot [q.v.]; 1851-1854, recuperated from disabling sunstroke, and then went to Europe to study means of protection of river deltas on that continent; 1854-ca. 1857, in charge of explorations for railroad route from Mississippi River to Pacific; 1856, appointed member of Lighthouse Board; 1861-1866, held various military commands in U.S. Army; 1866-1879, brigadier general and head of Corps of Engineers; 1879, retired. MEMBERSHIPS: Am. Phil. Soc.; Natl. Ac. Scis.; and other societies. SCIENTIFIC CONTRIBUTIONS: Achievements in science and engineering made in official capacity rather than as personal investigations. Study of Mississippi River established lasting reputation as hydraulic engineer, and final report [see Works] brought together large amount of data collected by investigators under Humphrey's direction, as well as from other sources; as director of Pacific Railroad Surveys, came to have general responsibility for various trans-Mississippi surveys and explorations. During this period and later as chief of Corps of Engineers did much to promote geological, zoological, and botanical surveys under auspices of U.S. government.

WORKS: *The Report Upon the Physics and Hydraulics of the Mississippi River* (Philadelphia, 1861), with H. L. Abbot. MSS: Hist. Soc. Penn. (*NUCMC* 60-2611). WORKS ABOUT: *DAB*, 9: 371-72 (Gustav J. Fiebeger); Henry L. Abbot, "Memoir," *Biog. Mems. of Natl. Ac. Scis.* 2 (1886): 203-15. Also see Henry H. Humphreys, *Andrew Atkinson Humphreys* (Philadelphia, 1924).

HUNT, EDWARD BISSELL (June 15, 1822, Portage, Livingston County, N.Y.—October 2, 1863, Brooklyn, N.Y.). *Physics.* Son of Sanford and Fanny (Rose) Hunt. Married Helen Maria Fiske (noted as author under name H. H.—Helen Hunt—and later as Helen Hunt Jackson), 1852; two sons. EDUCATION: 1845, graduated, U.S. Military Academy, West Point. CAREER: 1845, appointed second lieutenant, Corps of Engineers; 1846-1849, assistant professor of engineering, West Point; 1851-1855, assigned to U.S. Coast Survey, especially at Washington, D.C., having charge of engraving of charts; 1855-1857, engineering duty in Rhode Island; 1857, assigned duty with Corps of Engineers and until 1862 stationed at Fort Taylor, Key West, Fla., where involved in construction of fortifications and lighthouses; 1859, promoted to captain; 1862, chief engineer, department of Shenandoah; 1862-1863, stationed at New York and engaged in military construction and testing so-called sea-miner; 1863, promoted to major. MEMBERSHIPS: Am. Ac. Arts Scis.; Natl. Ac. Scis. SCIENTIFIC CONTRIBUTIONS: During and after period of service with Coast Survey, contributed number of items to *Annual Reports,* covering topics such as map making and printing and descriptions of apparatus devised by himself and others; also prepared indexes to the *Annual Reports.* Contributions to scientific journals were on variety of topics; article in 1850 on Mariotte's law of gaseous pressure gave early evidence of a theory of molecular physics that later became Hunt's primary scientific concern; his work in this area was given most fully in 1854 paper, "Forces" [see Works] (which disputed work of Boscovich), but his ideas were never fully presented. Also wrote on other physical topics, proposed New York trigonometrical and geographical survey, problems of scientific publications, and other matters. Activities at Key West, Fla., included study of ocean currents and Florida reef. In early 1860s, conceived idea of sea-miner, a submarine projectile for destruction of vessel at distance; in testing this apparatus at New York there occurred an accident in which Hunt lost his life.

WORKS: "On the Nature of Forces," *Am. Journ. Sci.* 2d ser. 18 (1854) : 237-49. Published number of works in *Proceedings of Am. Assoc. Advt. Sci., Am. Journ. Sci.,* and *Annual Reports of U.S. Coast Survey.* See bibliography of some thirty works in *Roy. Soc. Cat.,* vols. 3, 6 (additions). MSS: Some items in E. N. Horsford Papers at Rens. Poly. Inst. WORKS ABOUT: NCAB, 11 (1909 [copyright 1901 and 1909]) : 440; F. A. P. Barnard, "Memoir," *Biog. Mems. of Natl. Ac. Scis.* 3 (1895) : 31-41.

HUNT, THOMAS STERRY (September 5, 1826, Norwich, Conn.—February 12, 1892, New York, N.Y.). *Chemistry; Geology.* Son of Peleg and Jane Elizabeth (Sterry) Hunt. Married Anna Gale, 1877; no children. EDUCATION: Attended public schools and held number of jobs to help in support of family; 1845, became student assistant to chemist B. Silliman, Jr. [q.v.], at Yale Univ. CAREER: 1846-1872, chemist and mineralogist, Geological Survey of Canada; 1856-1862, taught chemistry, Univ. of Laval (Quebec) ; 1862-1868, taught chemistry, McGill Univ. (Montreal) ; 1872-1878, professor of geology, Massachusetts Institute of Technology; 1874-1878, chemist in Second Geologic Survey of Pennsylvania; after 1878, devoted efforts to writing and geological consulting work; made number of trips to Europe. MEMBERSHIPS: Roy. Soc.; Natl. Ac. Scis.; and other societies. SCIENTIFIC CONTRIBUTIONS: While attached to Canadian survey, produced numerous works of geological-chemical nature that were of considerable importance. Greatly interested in theoretical and speculative questions, and insistence on personal views and priority often led to controversy. Introduced to America chemical ideas of C. F. Gerhardt and A. Laurent on atoms and molecules, and also was strongly influenced by writings of *naturphilosoph* J. B. Stallo [q.v.]; early appreciated significance of Laurent's suggestion of "water-type" for organic chemistry; probably first to view silica as central element in mineralogy, just as Gerhardt had viewed carbon as fundamental to organic chemistry, a field in which Hunt also was interested. Prepared "natural system" for classification of minerals, which was not adopted; in geology, especially interested in Paleozoic rocks, and attempted to devise chemical means for determining history of rocks not containing fossil or stratigraphic evidence.

WORKS: Wrote large number of papers, and reprinted them in the following: *Chemical and Geological Essays* (Boston and London, 1875; 4th ed., New York, 1891) ; *Mineral Physiology and Physiography: A Second Series of Chemical and Geological Essays* (Boston, 1886; 2d ed., New York, 1890) ; and *A New Basis for Chemistry: A Chemical Philosophy* (Boston, 1887; 3d ed., New York, 1891). Bibliography in *Biog. Mems. of Natl. Ac. Scis.,* 15 (1934). Major publications given in *DSB.* MSS: Edinburgh Univ. Library; Roy. Soc.; Colum. U. Lib.; and Smithson. Instn. (*DSB*). WORKS ABOUT: *DAB,* 9: 393-94 (Lyman C. Newell) ; *DSB* 6: 564-66 (W. H. Brock).

HUTCHINS, THOMAS (1730, Monmouth County, N.J.—April 28, 1789, Pittsburgh, Pa.). *Geography; Military engineering.* No information on parentage or marital status; orphaned before age sixteen. EDUCATION: Said to have spent youth in "Western country," but no evidence available regarding education; engineering and astronomy probably learned through military service. CAREER: 1757-1759, officer in Pennsylvania colonial regiment during

French and Indian War; ca. 1760-1780, in British army; during early years situated chiefly at Fort Pitt and made several excursions to other locations in West, including Detroit (1762) and New Orleans and Cuba (1766); 1768-1770, stationed at Fort Chartres, Illinois country; 1772-ca. 1777, attached to Southern Department of army, involved in various engineering projects in West Florida; 1776, promoted to captain; ca. 1777-1780, in London, and during years 1779-1780 imprisoned on charge of passing information to friends of United States in France; 1780, resigned army commission, went to France, and with recommendation of B. Franklin [q.v.] sailed for Charleston, S.C., where joined colonial southern army; 1781-1789, geographer to United States; 1783, employed in several surveying projects for Pennsylvania and was chosen a Pennsylvania commissioner to determine western extension of Pennsylvania-Virginia boundary; 1785-1789, as U.S. geographer had charge of all surveys, called for by Ordinance of 1785, of western lands ceded by states to Congress; 1787, determined boundary between New York and Massachusetts. MEMBERSHIPS: Am. Phil. Soc. SCIENTIFIC CONTRIBUTIONS: As military engineer, devised plans for fortifications at Fort Pitt and Pensacola, Fla. While on expeditions, kept journals and prepared maps to accompany them, and several have been published; chief publication was map and description of Virginia, etc. [see Works]. 1783-1785, assignment for commission on Pennsylvania-Virginia boundary (of which Hutchins was member) was to complete Mason-Dixon line; before death, completed more than half of basic surveys for western lands.

WORKS: *A Topographical Description of Virginia, Pennsylvania, Maryland and North Carolina* (London, 1778), reprinted with biographical sketch and list of publications and manuscripts by Frederick C. Hicks (Cleveland, 1904). MSS: Hist. Soc. Penn. (*NUCMC* 60-2226). WORKS ABOUT: *DAB,* 9: 435-36 (F. C. Hicks); F. C. Hicks [above]; William L. Jenks, "The 'Hutchins' Map of Michigan," *Michigan History Magazine* 10 (1926): 358-73.

HYATT, ALPHEUS (April 5, 1838, Washington, D.C.—January 15, 1902, Cambridge, Mass.). *Zoology; Invertebrate paleontology.* Son of Alpheus, wealthy merchant, and Harriet R. (King) Hyatt. Married Ardella Beebe, 1867; three children. EDUCATION: 1856-1857, attended Yale College; 1862, B.S., Lawrence Scientific School, Harvard. CAREER: 1861, collected fossils and marine animals on Island of Anticosti in Gulf of St. Lawrence; during Civil War, captain in Forty-seventh Massachusetts; 1865-1902, in charge of fossil cephalopods in Museum of Comparative Zoology, Harvard; 1867-ca. 1870, at Salem, Mass., connected with Essex Institute, was a founder of both Peabody Academy of Sciences and *American Naturalist* and from 1867 to 1871 served as an editor of *Naturalist;* at Bos. Soc. Nat. Hist., custodian (1870-1902) and curator (1881-1902); 1870-1888, professor of zoology and paleontology, Massachusetts Institute of Technology; 1870-1902, director, Teachers' School of Science at Bos. Soc. Nat. Hist.; 1877-1902, taught zoology and paleontology at Boston Univ.; 1879, founder of marine laboratory at Annisquam, Mass., and later first president of Woods Hole laboratory. MEMBERSHIPS: Natl. Ac. Scis.; cofounder of American Society of Naturalists. SCIENTIFIC CONTRIBUTIONS: Although a student of J. L. R. Agassiz [q.v.], accepted theory of evolution shortly after 1859. Chief work during earlier career was in freshwater Polyzoa and sponges, while classification and evolution of cephalopods occupied much of final twenty-five years of life. Along with E. D. Cope [q.v.], was cofounder of neo-Lamarckian theory; is best remembered for theoretical work on evolution, and above all was interested in laws governing individual development and evolution of groups; his theory of aging of species leading to extinction was particularly important.

WORKS: Series of papers on freshwater Polyzoa in *American Naturalist* (April-June 1867) and in *Proceedings of Essex Institute* 4-5 (1866-1868); "Revision of North American Poriferae [Sponges]," *Memoirs of Bos. Soc. Nat. Hist.* 2 pt. 4, nos. 2 and 5 (1875-1877); "The Genesis of the Tertiary Species of Planorbis at Steinheim," Bos. Soc. Natl. Hist., *Anniversary Memoirs* (Boston, 1880); "Genesis of the Arietidae [Cephalopods]," *Smithson. Contr. Knowl.* 26 (1889); "Phylogeny of an Acquired Characteristic," *Proceedings of Am. Phil. Soc.* 32 (1893): 349-647. MSS: Princeton Univ. Library (*NUCMC* 61-1277); Md. Hist. Soc. (*NUCMC* 67-1523); Harv. U.-MCZ; Boston Museum of Science. WORKS ABOUT: *DAB,* 9: 446-47 (Hubert L. Clark); *DSB,* 6: 613-14 (Stephen J. Gould).

HYDE, EDWARD WYLLYS (1843-1930). *Mathematics.* Professor of mathematics, Univ. of Cincinnati.

WORKS ABOUT: See *AMS; Who Was Who in America,* vol. 1.

I

IDDINGS, JOSEPH PAXSON (1857-1920). *Geology; Petrology.* Professor of petrology, Univ. of Chicago; geologist, U.S. Geol. Surv.

WORKS ABOUT: See *AMS; DAB; DSB.*

IRVING, ROLAND DUER (April 29, 1847, New York, N.Y.—May 27, 1888, Madison, Wis.). *Geology; Mining engineering.* Son of Pierre Paris, Episcopal clergyman, and Anna Henrietta (Duer) Irving. Married Abby Louise McCulloh, 1872; three children. EDUCATION: 1863-1864, pursued classical course, Columbia Univ.; 1869, M.E., School of Mines, Columbia Univ. CAREER: 1869-1870, was superintendent of smelting works in New Jersey and served with Ohio Geological Survey; 1870-1888, professor of geology and mineralogy, Univ. of Wisconsin; 1873-1879, assistant geologist, Wisconsin Geological Survey; 1880-1888, conducted investigations of Lake Superior region for U.S. Geol. Surv. MEMBERSHIPS: President, Wisconsin Academy of Sciences, Arts and Letters. SCIENTIFIC CONTRIBUTIONS: Most significant work was done in structural geology and in genetic petrography, and in latter field ranks as one of first American investigators to demonstrate usefulness of such studies. Work for U.S. Geol. Surv. began with investigation of copper-bearing rocks that touched upon parts of Michigan, Wisconsin, Minnesota, and Canada; resulting publication was the first attempt at overall view of Lake Superior copper region [see Works]; after copper, began similar studies of iron ores and continued these investigations until death. Single most important geological contribution probably was explanation of origin of iron ores in Lake Superior region.

WORKS: Work on Wisconsin survey included in *Geology of Wisconsin, Survey of 1873-1879,* 4 vols. (Madison, Wis., 1882-1883); *Copper-Bearing Rocks of Lake Superior,* U.S. Geol. Surv. Monograph no. 5 (1883). Chief works listed in text and footnotes in Chamberlin [below]; also see bibliography in Nickles, *Geol. Lit. on N. Am.* WORKS ABOUT: *DAB,* 9: 505 (George P. Merrill); T. C. Chamberlin, memoir, *Transactions of Wisconsin Academy of Sciences, Arts and Letters* 8 (1888-1891): 432-37 (first published in *American Geologist,* January 1889).

J

JACKSON, CHARLES LORING (1847-1935). *Chemistry.* Professor of chemistry, Harvard Univ.
WORKS ABOUT: See *AMS; NCAB,* 42 (1958): 556-57.

JACKSON, CHARLES THOMAS (June 21, 1805, Plymouth, Mass.—August 28, 1880, Somerville, Mass.). *Chemistry; Mineralogy; Geology.* Son of Charles, merchant, and Lucy (Cotton) Jackson. Married Susan Bridge, 1834; survived by five children. EDUCATION: Studied medicine under James Jackson and Walter Channing; 1829, M.D., Harvard Medical School; 1829-1832, in Europe, studied medicine at Univ. of Paris and geology and mineralogy at Ecole des Mines (Paris). CAREER: At early age, made walking tour through New York and New Jersey with several geologist-naturalists; 1827 and 1829, visited Nova Scotia with F. Alger [q.v.]; ca. 1832-1836, medical practice, Boston, Mass.; 1836, established laboratory for teaching analytical chemistry, Boston; 1837-1839, geological surveyor of public lands of Maine and Massachusetts and also Maine state geologist; 1839-1840, conducted Geological Survey of Rhode Island; 1839-1844, engaged in Geological Survey of New Hampshire; 1847, appointed by U.S. government to conduct mineral survey of Lake Superior region, but dismissed after two seasons' work because of conflicts with J. W. Foster and J. D. Whitney [qq.v.], the other two geologists on survey; thereafter, resumed work in Boston laboratory and prepared reports for various mining interests; 1873, committed to hospital for insane. MEMBERSHIPS: Am. Ac. Arts Scis.; Bos. Soc. Nat. Hist.; and other societies. SCIENTIFIC CONTRIBUTIONS: Studied geology and mineralogy of Nova Scotia with F. Alger. Boston laboratory became a center for scientific interests and was one of first such laboratories in U.S. to take in students. First state geologist for Maine, Rhode Island, and New Hampshire, but in work for these states no significant new geological problems were generated; chief interest was in mineralogical—rather than stratigraphical—geology, and he was concerned with economic impact of his work. Engaged in series of priority controversies over discovery of guncotton, electric telegraph, and use of ether as anesthesia, all problems with which he had at one time or another been concerned.
WORKS: Incomplete bibliography in *American Geologist* 20 (1897): 87-110. MSS: No major collection has been found; *DSB* gives references to locations of several small bodies of manuscripts. WORKS ABOUT: *DAB,* 9: 536-38 (George P. Merrill and John F. Fulton); *DSB,* 7: 44-46 (George E. Gifford, Jr.).

JACKSON, JOHN BARNARD SWETT (June 5, 1806, Boston, Mass.—January 6, 1879, Boston, Mass.). *Pathological anatomy.* Son of Capt. Henry and Hannah (Swett) Jackson. Married Emily J. Andrews, 1853; survived by two children. EDUCATION: 1825, A.B., Harvard College; 1829, M.D., Harvard Univ.; 1829-1831, studied surgery and medicine in Paris, London, and Edinburgh. CAREER: 1827, appointed house apothecary for one year, Massachusetts General Hospital; 1831, began medical practice, Boston; at Massachusetts General Hospital, house physician and surgeon (generally without salary) (1835-1839) and physician "in full standing" (1839-ca. 1864); 1847-1879, professor of pathological anatomy and curator of Warren Anatomical Museum, Harvard Medical School; 1853-1855, dean, Harvard Medical School. MEMBERSHIPS: Am. Ac. Arts Scis.; Bos. Soc. Nat. Hist. SCIENTIFIC CONTRIBUTIONS: Judged to have been better prepared than any Boston physician before him. Dissected and studied several large animals, including Galápagos tortoise, anteater, male and female dromedaries and spermaceti whale; studied fossil bones and dentition of mastodon; and published on these subjects in *Boston Journal of Natural History* and in *Proceedings of Bos. Soc. Nat. Hist.;* also studied Eskimo skulls. Chief work done in study of organic effects of disease, building important collections as curator of museum of Boston Society for Medical Improvement and Warren Museum; prepared descriptive catalogues for both museums, and catalogue of latter (1847) was considered pioneer American contribution to pathological anatomy; museum collections were supported by careful records of medical case histories to which specimens related.

WORKS: *Descriptive Catalogue of the Anatomical Museum of the Boston Society for Medical Improvement* (Boston, 1847), 352 pp.; *A Descriptive Catalogue of the Warren Anatomical Museum* (Boston, 1870), 759 pp. See Meisel, *Bibliog. Amer. Nat. Hist.* MSS: Harv. U.-MS (*NUCMC* 62-3639). WORKS ABOUT: [Oliver W. Holmes], memoir, *Proceedings of Am. Ac. Arts Scis.* 14 (1878-1879): 344-52; Thomas T. Bouvé, "Historical Sketch of the Boston Society Natural History," Bos. Soc. Nat. Hist., *Anniversary Memoirs* (Boston, 1880), pp. 199-201.

JACOBS, MICHAEL (January 18, 1808, near Waynesboro, Franklin County, Pa.—July 22, 1871, Gettysburg, Pa.). *Meteorology.* Son of Henry, farmer, and Anna Maria (Miller) Jacobs. Married Julianna M. Eyster, 1833; at least one child. EDUCATION: 1828, A.B., Jefferson College (Canonsburg, Pa.); 1832, licensed by West Pennsylvania Synod (Lutheran), having pursued private study of theology. HONORS: Referred to as D.D. CAREER: 1828-1829, teacher, boarding school at Belair, Md.; 1829-1832, teacher, Gettysburg Gymnasium; 1832-1866, professor of mathematics and natural sciences, Pennsylvania (later, Gettysburg) College, successor to gymnasium. MEMBERSHIPS: Am. Assoc. Advt. Sci. SCIENTIFIC CONTRIBUTIONS: Working without benefit of association with other scientists, prepared own scientific apparatus and carried out original experiments in addition to classroom demonstrations. Reputation as scientist based chiefly on meteorological studies; participated in weather observations coordinated in Pennsylvania by Franklin Institute and on national basis by Smithson. Instn.; in 1841 began what is said to have been second course of lectures on meteorology offered in an American college, and continued study of weather throughout life. About 1845, devised means of preservation of fruit by canning, although process had been known in France for some years.

WORKS: No scientific works listed in *Natl. Un. Cat. Pre-1956 Imp.* or *Roy. Soc. Cat.* MSS: Lutheran Theological Seminary Library, Philadelphia (*NUCMC* 61-2514); Am. Phil. Soc. (Hamer); meteorological records, belonging to Smithson. Instn. and now only on microfilm, available in Natl. Arch. (Ludlum [below]). WORKS ABOUT: *DAB* 9: 567-68 (George H. Genzmer); David M. Ludlum, "The Weather at Gettysburg [battle]," *Weatherwise* 13 (1960): 101-5, 130.

JAMES, EDWIN (August 27, 1797, Weybridge, Addison County, Vt.—October 28, 1861, Rock Spring, Iowa). *Botany; Natural history exploration.* Son of Deacon Daniel and Mary (Emmes) James. Married Clarissa Rogers, 1827; one child. EDUCATION: 1816, A.B., Middlebury College; ca. 1816-ca. 1820, studied botany and geology with J. Torrey and A. Eaton [qq.v.] and medicine with brother, at Albany. CAREER: 1820, botanist, geologist, and surgeon with Maj. Stephen H. Long's expedition to explore country between Mississippi and Rocky Mountains; following return to East, spent some time preparing report on expedition (published in 1823); 1823, appointed to accompany second Long Expedition, but failed to receive notice of appointment in time; 1823-1833, assistant surgeon, U.S. Army, stationed at various frontier locations; 1833-ca. 1836, involved in activities of New York State Temperance Society, including editorial work with *Temperance Herald and Journal* (Albany); ca. 1836-1861, resided on farm at Rock Spring, Iowa; 1837-1838, subagent for Potawatamie Indians at Old Council Bluffs, Neb. MEMBERSHIPS: American Geological Society (founded at Yale College, 1819). SCIENTIFIC CONTRIBUTIONS: Known chiefly for work done on Long Expedition, during which he and two others were first whites to ascend Pike's Peak; as member of expedition, explored botany in West, was first in North America to collect alpine flora above timberline, and sent specimens to Torrey, who published descriptions in series of papers. With access to own notes and those of Long and other members of expedition, wrote chief account [see Works], in which West was characterized as "Great American desert." Also interested in geology; during later years as army surgeon, studied and published on Indian languages.

WORKS: *Account of an Expedition from Pittsburgh to the Rocky Mountains, Performed in the Years 1819 and 1820* . . . (Philadelphia, 1823, in 2 vols., and London, 1823, in 3 vols., the two editions having variations); "Catalogue of Plants Collected During a Journey to and From the Rocky Mountains During the Summer of 1820," *Transactions of Am. Phil. Soc.* n.s. 2 (1825): 172-90. See also Meisel, *Bibliog. Amer. Nat. Hist.* MSS: See references in Benson [below]; also Northwestern Univ. Medical School Library (Hamer). WORKS ABOUT: *DAB*, 9: 576 (Frank E. Ross); Joseph Ewan, *Rocky Mountain Naturalists* ([Denver, Colo.], 1950), pp. 13-20; Maxine F. Benson, "Edwin James: Scientist, Linguist, Humanitarian," *Dissertation Abstracts*, sec. A, vol. 29 (1968): 1485-A.

JAMES, THOMAS POTTS (September 1, 1803, Radnor, Pa.—February 22, 1882, Cambridge, Mass.). *Botany.* Son of Dr. Isaac and Henrietta (Potts) James. Married Isabella Batchelder, 1851; four children. EDUCATION: Studied pharmacy and medicine. CAREER: 1831-1866, with older brother engaged in wholesale drug business, Philadelphia; for number of years, professor and examiner, Philadelphia College of Pharmacy; 1866-1882, lived in Cambridge, Mass., devoting full efforts to study of mosses. MEMBERSHIPS: Am. Phil. Soc.; Am. Ac. Arts Scis.; and other societies. SCIENTIFIC CONTRIBUTIONS: Botanical interests apparently developed from study of materia medica, and he came to specialize as student of mosses and liverworts. In 1850s, published on local botany in Pennsylvania, and later prepared catalogues of mosses collected by explorers in Alaska (1867) and in the West, including C. R. King [q.v.], U.S. Geological Survey of Fortieth Parallel (1871), and Lt. G. M. Wheeler [q.v.], Surveys West of 100th Meridian (1878). After death of W. S. Sullivant [q.v.] in 1873, James and L. Lesquereux [q.v.] were chief authorities on American mosses and executed Sullivant's planned systematic work on mosses of North America [see Works], James nearly completing his portion of project before death.

WORKS: *Manual of the Mosses of North America* (Boston, 1884), with L. Lesquereux; chief works listed in *DAB*. MSS: Hist. Soc. Penn. (*NUCMC* 60-2802); Harv. U.-FL (Hamer). WORKS ABOUT: *DAB,* 9: 589-90 (Hugh M. Raup); [Asa Gray], "Memoir," *Proceedings of Am. Ac. Arts Scis.* 17 (1881-1882): 405-6.

JAMES, WILLIAM (1842-1910). *Psychology; Philosophy.* Professor of psychology and philosophy, Harvard Univ.

MSS: Harv. U.-HL (Hamer). WORKS ABOUT: See *AMS; DAB; DSB.*

JAYNE, HORACE FORT (1859-1913). *Zoology; Anatomy.* Professor of vertebrate morphology and dean, Univ. of Pennsylvania, and director and professor of zoology in its Wistar Institute.

WORKS ABOUT: See *AMS; DAB.*

JEFFERSON, THOMAS (April 13, 1743, Goochland [now Albemarle] County, Va.—July 4, 1826 "Monticello," Albemarle County, Va.). *Natural history.* Son of Peter, surveyor and land developer, and Jane (Randolph) Jefferson. Married Martha Wayles Skelton, 1772; six children. EDUCATION: 1762, graduated, College of William and Mary; 1767, admitted to bar, having studied law for five years with George Wythe. CAREER: Through father and father-in-law inherited large amount of land; 1767-1774, law practice, beginning at Williamsburg, Va.; 1769-1775, member, House of Burgesses; 1770, appointed county lieutenant, Albemarle County, Va.; 1773, appointed county surveyor; 1775-1776, delegate to Continental Congress; 1776-1779, member of Virginia House of Delegates; 1779-1781, governor of Virginia; 1781-1783, retired to Monticello; 1783-1784, served in Continental Congress; 1784-1785, appointed to assist B. Franklin [q.v.] and John Adams in negotiation of commerce treaties, Paris; 1785-1789, minister to France; 1790-1793, secretary of state; 1793-1797, again retired to Monticello; 1797-1801, vice-president of United States; 1801-1809, president of United States; 1809, retired to Monticello; 1819, aided in founding Univ. of Virginia and served as rector. MEMBERSHIPS: President, Am. Phil. Soc., 1797-1815. SCIENTIFIC CONTRIBUTIONS: Has been recognized as contributor to number of fields of science and technology, including agriculture, botany, cartography, ethnology, meteorology, paleontology, and surveying. During first period of retirement, 1781-1783, produced *Notes* for publication [see Works], based on information about Virginia gathered over number of years, and touched on geography, climate, flora, fauna, topography, ethnology, and civil aspects of that state; *Notes* was widely admired as pioneer study of multiple aspects of a part of United States. Interested in paleontology and published on fossilized bones of ground sloth [see Works]. Introduced several European agricultural products to United States and perfected moldboard for plow. As secretary of state was interested in problems of standardization of weights and measures and patent rights. As president supported surveys of Louisiana Territory and was involved in detailed planning for Lewis and Clark Expedition; also promoted survey of U.S. coast.

WORKS: *Notes on the State of Virginia . . .* (Paris 1782; printed in 1784-1785; 1st American ed., Philadelphia, 1788); "A Memoir of the Discovery of Certain Bones of an Unknown Quadruped . . .," *Transactions of Am. Phil. Soc.* 4 (1799): 246-60. See also J. P. Boyd and L. C. Butterfield, eds., *Papers of Thomas Jefferson,* 18 vols. (Princeton, 1950-). MSS: Chief collection in Lib. Cong. WORKS ABOUT: *DAB,* 10:17-35 (Dumas Malone); *DSB,* 7:88-90 (Silvio A. Bedini). Also see Edwin T. Martin, *Thomas Jefferson, Scientist* (New York, 1952).

JEFFRIES, JOHN (February 5, 1744/1745, Boston, Mass.—September 6, 1819, Boston, Mass.). *Aeronautics; Meteorology.* Son of David, businessman and Boston town treasurer, and Sarah (Jaffrey) Jeffries. Married Sarah Roads, 1773; Hannah Hunt, 1787; three children by first marriage, eleven by second. EDUCATION: 1763, A.B., Harvard College; studied medicine in colonies and, 1768-1769, in England and

Scotland; 1769, M.D., Marischal College (Aberdeen). HONORS: M.D., Harvard Univ., 1819. CAREER: 1766, began medical practice in Massachusetts; 1769-1771, returned from studies abroad and engaged in successful medical practice in Boston; 1771-1774, surgeon at marine shore hospital; 1774, appointed physician and surgeon to Boston Almshouse; 1776, left Boston for Halifax; 1776-1779, chief of military surgical staff of Nova Scotia; 1779, went to England and there received appointment as apothecary for forces at Nova Scotia, surgeon to General Hospital at New York, and surgeon major of forces in America; 1780, went to Savannah, Ga., and Charleston, S.C., and later, at New York, sold commission and returned to England; 1780-1789, medical practice, London; 1789-1819, medical practice, Boston, Mass. SCIENTIFIC CONTRIBUTIONS: Remembered for two early balloon ascensions, in 1784 over London and in 1785 from Dover across English Channel, both with Frenchman Jean Pierre Blanchard; Jeffries financed flights and arranged for scientific work that accompanied them; after channel flight, the two men were loudly received in Paris. Flights were provided with quality scientific instruments and resulted in collection of first scientific data on free air; gathered air samples at various heights, later analyzed by Cavendish, and also observed temperature, pressure, humidity as high as 9,309 feet; Jeffries's account of flight and scientific observations was read to Roy. Soc. and Am. Ac. Arts Scis., and newspaper excerpts made him widely known. After return to U.S. in 1789, began first Boston lecture course in anatomy, promptly terminated when cadaver was taken by mob.

WORKS: *A Narrative of the Two Aerial Voyages of Dr. Jeffries with Mons. Blanchard; with Meteorological Observations and Remarks* (London, 1786; reprinted, New York, 1971), 60 pp. MSS: Harv. U.-HL (Hamer); Mass. Hist. Soc. (Hamer). WORKS ABOUT: *DAB*, 10: 38-39 (Alexander McAdie); Shipton, *Biog. Sketches, Harvard*, 15 (1970): 418-27.

JENKINS, EDWARD HOPKINS (1850-1931). *Agricultural chemistry.* Director, Connecticut Agricultural Experiment Station.

WORKS ABOUT: See *AMS; DAB.*

JENKS, JOHN WHIPPLE POTTER (May 1, 1819, West Boylston, Mass.—September 26, 1895, Providence, R.I.). *Zoology.* Son of Nicholas, physician and teacher, and Betsey (Potter) Jenks. Married Sarah Tucker, 1842; four children. EDUCATION: Attended Peirce Academy (Middleboro, Mass.); 1838, A.B., Brown Univ.; ca. 1840, studied medicine at Washington, Ga. CAREER: 1838-1840, conducted school and preached, at Americus, Ga.; 1840-1842, assistant pastor at Baptist Church, Washington, Ga., and thereafter took charge of school in Taliaferro County, Ga.; 1842-1871, principal, Peirce Academy; 1871, established museum and served as curator without salary, Brown Univ.; 1872-1895, director and (after 1874) curator of Museum of Natural History, and lecturer on special branches of agriculture (in 1875 title changed to professor of agricultural zoology), Brown Univ. MEMBERSHIPS: Am. Assoc. Advt. Sci. SCIENTIFIC CONTRIBUTIONS: While principal at Peirce Academy, pursued study of ornithology and taxidermy and gathered sizable collection, which was given to academy. Created museum at Brown Univ., where he prepared majority of zoological specimens and also worked on collections for botany, anthropology, geology, and mineralogy. Said to have assisted friend J. L. R. Agassiz [q.v.] with work on embryology of turtle; was asked to assist S. F. Baird [q.v.] at time of establishment of U.S. Commission on Fish (1871), but declined because of duties at Brown Univ.; however, served commission as time allowed, and received duplicates for museum at Brown. In 1874, explored swamps and everglades of southern Florida; subsequently traveled throughout North America, Hawaii, and Europe, and during all of travels gathered specimens for Brown. Chief work as scientist was as collector; disliked public appearance as author and in consequence published little.

WORKS: *Fourteen Weeks in Zoology* (New York and Chicago, 1877) appeared without his name as part of series by J. D. Steele [q.v.] and was published with name as *Popular Zoology* (New York and Chicago, 1887; 3d ed., 1895). No works in *Roy. Soc. Cat.* WORKS ABOUT: *NCAB*, 10 (1909 [copyright 1900]): 22; Reuben A. Guild, *Memorial Address on the Late John Whipple Potter Jenks . . . Delivered Before the Faculty and Students of Brown University . . .* (Providence, R.I., 1895).

JOHNSON, SAMUEL WILLIAM (1830-1909). *Agricultural chemistry.* Professor of theoretical and agricultural chemistry, Sheffield Scientific School, Yale; director, Connecticut Agricultural Experiment Station.

MSS: Connecticut Historical Society, Hartford (Hamer); Connecticut Agricultural Experiment Station—Biochemistry Department (New Haven). WORKS ABOUT: See *AMS*; *DAB.*

JOHNSON, WALTER ROGERS (June 21, 1794, Leominster, Mass.—April 26, 1852, Washington, D.C.). *Physics; Geology; Chemistry.* Son of Luke Johnson, farmer. Married; no information on children. EDUCATION: Attended Groton Academy; 1819, A.B., Harvard College. CAREER: 1819-1821, taught in Framingham and Salem, Mass.; 1821-

1826, principal, Germantown (Pa.) Academy; 1826-ca. 1836, principal of school begun by Franklin Institute and after 1829 continued independently by Johnson; 1828-1837, lecturer and then professor of mechanics and natural philosophy, Franklin Institute; 1836, appointed to Wilkes Expedition scientific staff, but did not accompany expedition; 1839-1843, professor of chemistry and natural philosophy, Pennsylvania Medical College, Philadelphia (a branch of Pennsylvania College at Gettysburg); acted as scientific consultant, including work for federal government, investigating coal and iron locations and various physical questions, and during later years maintained no formal position, but continued variety of activities including geological surveys; 1848, moved to Washington, D.C. MEMBERSHIPS: Ac. Nat. Scis. Philad. and other societies; first secretary, Am. Assoc. Advt. Sci. SCIENTIFIC CONTRIBUTIONS: Early interests were in problems of education, while connections with Franklin Institute later directed attention to scientific concerns, especially to matters having practical significance. First important work done as chairman of Franklin Institute boiler explosion committee's subcommittee on strength of materials, carrying out experiments from 1832 to 1837 that were reported in institute's journal and as separate publication. In 1836, began geological studies (especially in Pennsylvania), which led to important investigations of uses of coal, after 1841 conducted partly for U.S. Navy. Other interests and activities included early concern with daguerreotype and report in 1845 on Boston water supply.

WORKS: *A Report to the Navy Department of the United States, on American Coals Applicable to Steam Navigation, and to Other Purposes* (Washington, D.C., 1844); *The Coal Trade of British America* (Washington, D.C., 1850). See bibliography of works in Pettengill [below], pp. 102-13. WORKS ABOUT: *NCAB*, 12 (1904): 260-61; George E. Pettengill, "Walter Rogers Johnson," *Journal of Franklin Institute* whole ser. 250 (1950): 93-113.

JOHNSON, WILLIAM WOOLSEY (1841-1927). *Mathematics*. Professor of mathematics, U.S. Naval Academy, Annapolis.

WORKS ABOUT: See *AMS; DAB*.

JOHNSTON, JOHN (August 23, 1806, Bristol, Maine—December 2, 1879, Clifton, Staten Island, N.Y.). *Chemistry; Physics*. No information on parentage. Married N. Maria Hamilton, 1835; five children. EDUCATION: 1832, A.B., Bowdoin College. HONORS: LL.D., McKendree College (Lebanon, Ill.), 1850. CAREER: teacher (1832-1834) and principal (1835), Oneida Conference Seminary, Cazenovia, N.Y.; at Wesleyan Univ. (Connecticut), assistant professor of mathematics and lecturer on natural science (1835-1839) and professor of natural sciences (1839-1873). MEMBERSHIPS: Am. Assoc. Advt. Sci.; Am. Phil. Soc. SCIENTIFIC CONTRIBUTIONS: Enjoyed reputation for skill in design and construction of apparatus used in laboratory. Wrote several articles on mineralogical, geological, and physical subjects, including papers on apparatus for solidification of carbonic acid and on electrical properties of pyroxiline paper and guncotton, published in *Am. Journ. Sci.* and *Proceedings of Am. Assoc. Advt. Sci.* Best known for textbooks on chemistry and natural philosophy, which had wide circulation and went to several editions; *Manual of Chemistry* [see Works] had distinction of being one of earliest U.S. textbooks to include basic principles of analytical chemistry.

WORKS: *A Manual of Chemistry, on the Basis of Dr. Turner's Elements of Chemistry . . .* (Philadelphia, 1842); *A Manual of Natural Philosophy . . . for Use as a Textbook in High Schools and Academies* (Philadelphia, 1846); *Elements of Chemistry for the Use of Schools* (Philadelphia, 1850). Bibliography in *Alumni Record of Wesleyan University*, 3d ed. (Middletown, Conn., 1881-1883), pp. 591-92. MSS: Small collection in Wesleyan Univ. Library. WORKS ABOUT: *Appleton's CAB*, vol. 3: 458; Guralnick, *Science and American College*, pp. 196-97; clippings and letter from Wesleyan Univ. Archives.

JONES, HUGH (ca. 1692, native of Herefordshire, England—September 8, 1760, Cecil County, Md.). *Mathematics*. No information on parentage. Married ca. 1721; no information on children. EDUCATION: A.B. in 1712 and M.A. in 1716, Jesus College, Oxford Univ. CAREER: 1716, received master's degree, took priest's orders, and went to Virginia; 1716-1722, professor of mathematics, College of William and Mary; 1718, became chaplain for Virginia House of Burgesses and assistant at Bruton Church, Williamsburg, and also served as minister at Jamestown; 1721-1724, returned to England; ca. 1724-1726, engaged in ministerial duties, St. Stephen's Parish, King and Queen County, Va.; 1726-1731, was minister, William and Mary Parish, Md., and also taught school; 1731-1760, rector, St. Stephen's Church, North Sassafras Parish, Cecil County, Md. SCIENTIFIC CONTRIBUTIONS: Gave much time to mathematical and astronomical studies, in which he became proficient, and wrote on calendar, on computation of time, and the like [see Works]; reportedly made recommendations for reformed calendar of 1752, and is said to have published an "Accidence to the Mathematicks," though no printed copy is known. One of more significant activities was as chief mathematician for conferences on Maryland-Pennsylvania boundary line, later determined by Mason-Dixon survey. Wrote early history of Virginia, which was useful to later historians.

WORKS: "Essays on the British Computation of Time, Coins, Weights and Measures," *Gentleman's Magazine* 15 (1745): 377-79; notice of "Panchronometer, or Universal Georgian Calendar," ibid. 23 (1753): 394. WORKS ABOUT: *DAB*, 10: 175 (Armistead C. Gordon, Jr.); Richard M. Gummere, *Seven Wise Men of Colonial America* (Cambridge, Mass., 1967), pp. 1-11.

JONES, WILLIAM LOUIS (March 27, 1827, Liberty County, Ga.—August 22, 1914, Atlanta, Ga.). *Chemistry; Geology; Agriculture.* Son of William, planter and botanist, and Mary Jane (Robarts) Jones; cousin to John and Joseph Le Conte [qq.v.]. Married Mary Williams, 1852; five children living in 1896. EDUCATION: 1845, A.B., Univ. of Georgia; attended lectures at Jefferson Medical College (Philadelphia); 1848, M.D., College of Physicians and Surgeons, New York; 1851, S.B., Lawrence Scientific School, Harvard. CAREER: For a time practiced medicine; at Univ. of Georgia, professor of chemistry and geology (1851-1852 and 1861-1872) and professor of agriculture (1883-1892); 1864-1865, chemist to Confederate Powder Mills, Augusta, Ga., when Univ. of Georgia suspended teaching due to war; for long period edited *Southern Cultivator;* at one time, head of Georgia Agricultural Experiment Station. MEMBERSHIPS: Am. Assoc. Advt. Sci.; Ac. Nat. Scis. Philad.; and other societies. SCIENTIFIC CONTRIBUTIONS: Said to have given major portion of life to teaching and to scientific research, and made significant contributions through editorship of *Southern Cultivator.* As professor of agriculture, taught botany, zoology, and geology, and seemingly preferred work in purer aspects of science to practical agriculture.

WORKS: *Roy. Soc. Cat.* lists one work, "Description of a New Species of Woodpecker," in *Annals of N.Y. Lyc. Nat. Hist.*, 4 (1847): 489-90. No works in *Natl. Un. Cat. Pre-1956 Imp.* WORKS ABOUT: *NCAB*, 9 (1907 [copyright 1899 and 1907]): 184; obituary notices and other items, especially manuscript "History of the University of Georgia" by Thomas Reed, from Univ. of Georgia Archives.

JORDAN, DAVID STARR (1851-1931). *Ichthyology.* Professor of zoology and president, Indiana Univ.; first president, Leland Stanford Univ.

MSS: Stanford Univ. Archives (*NUCMC* 72-1692); see additional references in Hamer. WORKS ABOUT: See *AMS; DAB; DSB.*

JOSLIN, BENJAMIN FRANKLIN (November 25, 1796, Exeter, R.I.—December 31, 1861, New York, N.Y.). *Physics; Meteorology; Medicine.* Son of Potter, farmer, and Anna Wightman Joslin. Married Phoebe Titus, ca. 1827 [?]; at least two children. EDUCATION: 1821, A.B., Union College; 1826, M.D., College of Physicians and Surgeons, New York. HONORS: LL.D., Union College, 1853. CAREER: 1821-1822, principal, Schenectady (N.Y.) Academy; 1822-1824, tutor, Union College; ca. 1826-1827, professor of chemistry and natural sciences, Polytechny, Chittenango, N.Y., and also practiced medicine in that town; 1827-1837, professor of mathematics and natural philosophy, Union College, and also practiced medicine; 1837, established medical practice, New York City; 1838-1844, professor of mathematics and natural philosophy, Univ. of City of New York (New York Univ.); 1844-1861, devoted full attention to medical practice. MEMBERSHIPS: Several scientific societies. SCIENTIFIC CONTRIBUTIONS: Combined active medical practice with teaching and study in physics, meteorology, and mechanics. Writings on scientific topics appeared especially during 1830s and included papers on radiation of cold, vision, geomagnetism and electromagnetic apparatus, irradiation, observations of comets, crystallization of snow and its relation to atmospheric origin of aurora borealis; also wrote papers entitled "Physiology of Respiration and Chemistry of the Blood Applied to Epidemic Cholera" and "Physiological Explanation of the Beauty of Form"; for some thirty years maintained daily meteorological record. Also wrote on medical topics; after ca. 1842 was particularly interested in homoeopathic medicine, and at time of death was president of Homoeopathic Medical Society of New York. Interested in mechanics and is said to have made several useful inventions.

WORKS: *Meteorological Observations and Essays, Furnished for the Appendix to the Report of the Regents of the Univ. of the State of New York . . .* (Albany, 1836), 69 pp. See *Natl. Un. Cat. Pre-1956 Imp.* and *Roy. Soc. Cat.*, vols. 3, 6 (additions). WORKS ABOUT: Benjamin F. Bowers, *Address on the Life and Character of the Late Benjamin F. Joslin . . . Delivered Before the Homeopathic Medical Societies of New York and Brooklyn* (New York, 1862); Andrew V. A. Raymond, *Union University . . .* (New York, 1907), vol. 3, app., pp. 42 and 140.

JOSSELYN, JOHN (ca. 1608, England, probably Essex—1675, England). *Natural history.* Son of Sir Thomas Josselyn of Essex, knight and councillor of province of Maine, and Theodora (Cooke) Bere Josselyn. Never married. EDUCATION: No certain information on education; apparently trained in medicine and surgery. CAREER: 1638-1639, with father visited brother at Black Point (Scarborough), Maine, and also visited in Boston; 1639, returned to England; 1663-1671 (second visit to New England), was with brother in Maine, and apparently practiced medicine in that province while engaged in observation and study later reported in publications; 1671, returned to England and may have received royal pension. SCIENTIFIC CONTRIBUTIONS: Reputation, and most biographical information, based on two works resulting from travels and observations in New England [see Works]. Chief scientific interest was botany and did best work in that field, presenting fullest account of region's flora to be prepared by any writer before M. Cutler [q.v.]. *Voyages* was most ambitious work, and it and earlier *Rarities* gave vast amount of observation and information on natural history of region, especially upper New England; work on birds incomplete and characterized by lack of technical description and terminology and by neglect of life histories, anatomy, etc.; notices of "beasts," numbering seventeen in all, included their relations to fur trade and medical uses; included twenty fishes, including several mollusks, but made little of the "serpents" and insects; made significant reports on folk medicine, both English and Indian, and this concern dominated botanical interests; also included information on mineralogy, geology, etc. In general, works were characterized by lack of scientific description or concern and fell outside newer approaches to natural history as developed by John Ray and others, and yet for decades, they ranked as most complete natural description of region. Apparently did not collect seeds or herbarium for shipment abroad.

WORKS: *New England's Rarities Discovered: In Birds, Beasts, Fishes, Serpents, and Plants of That Country. Together With the Physical and Chyrurgical Remedies Wherewith the Natives Constantly Use to Cure Their Distempers, Wounds and Sores* . . . (London, 1672) ; *An Account of Two Voyages to New England* (London, 1674; 2d ed., 1675). For information on reprints see both *DAB* and Stearns [below]. WORKS ABOUT: *DAB,* 10: 219-20 (Fulmer Mood) ; Stearns, *Sci. in Brit. Colonies,* esp. pp. 139-50.

JUDD, ORANGE (July 26, 1822, near Niagara Falls, N.Y.—December 27, 1892, Evanston, Ill.). *Agriculture.* Son of Ozias, probably farmer, and Rheuama (Wright) Judd. Married Sarah L. Ford, 1847; Harriet Stewart, 1855; at least two children. EDUCATION: 1847, graduated, Wesleyan Univ. (Connecticut) ; 1850-1853, studied agricultural chemistry, Yale Univ. CAREER: Ca. 1847, became teacher at Wilbraham Academy, Mass.; 1850-1853, student at Yale; in 1853 became joint editor and in 1856 owner and publisher of *American Agriculturist,* New York; 1855-1863, agricultural editor, *New York Times;* formed Orange Judd Co. for publication of *Agriculturist* and other agricultural and scientific publications, as well as *Hearth and Home;* during Civil War, served with United Christian Commission and later with U.S. Sanitary Commission; president of railroad connecting Flushing, Long Island, and New York City; 1883, suffered business losses, retired from *Agriculturist,* and moved to Chicago; 1884-1888, edited *Prairie Farmer* at Chicago; 1888-1892, owned and edited *Orange Judd Farmer* at Chicago. MEMBERSHIPS: Am. Assoc. Advt. Sci. SCIENTIFIC CONTRIBUTIONS: While adding nothing of note to scientific knowledge, played role in promotion of scientific agriculture; as agricultural writer combined experience in practical farming and laboratory sciences, and by 1864 had increased circulation of *American Agriculturist* to more than 100,000. Helped to found sorghum industry in U.S. through first importation and distribution of seeds, and was originator of crop reporting percentage system, adopted by governments throughout the world. Accumulated wealth, and in 1871 presented Orange Judd Hall of Natural Science to Wesleyan Univ.; after period of disillusionment with scientific agriculture, lent personal aid and support to establishment of first state agricultural experiment station, at Wesleyan in 1875.

WORKS: Particularly noted for editorial labors with *American Agriculturist* and other farming publications. MSS: Numerous letters in W. O. Atwater Papers at Wesleyan Univ. Library. WORKS ABOUT: *DAB,* 10: 231-32 (William B. Shaw) ; Rossiter, *Emergence of Agricultural Science,* pp. 113, 165-69, passim.

JULIEN, ALEXIS ANASTAY (1840-1919). *Geology; Petrography.* Curator of geology, Columbia Univ.

WORKS ABOUT: See *AMS*; *NCAB,* 18 (1922) : 40.

K

KEATING, WILLIAM HYPOLITUS (August 11, 1799, Wilmington, Del.—May 17, 1840, London, England, while on business trip). *Mineralogy.* Son of Baron John, former colonel in Irish Brigade of French army who settled in Delaware and later in Philadelphia, and Eulalia (Deschapelles) Keating. Married Elizabeth Bollman; no information on children. EDUCATION: 1816, A.B., Univ. of Pennsylvania; ca. 1816-1820, studied geology, mineralogy, chemistry, and mining engineering in Great Britain and Europe; 1834, admitted to Philadelphia bar following period of legal studies. CAREER: 1820-1822, engaged in mineralogical studies and writing; 1822-1828, professor of mineralogy and chemistry, Univ. of Pennsylvania; 1823, geologist and historiographer, expedition of Maj. Stephen H. Long to Minnesota country; served as professor of chemistry, Franklin Institute; for four years, supervised silver mine at Temascalapa, Mexico; 1832-1834, member of Pennsylvania legislature; 1834, began successful law practice; served as president of Little Schuylkill Navigation, Railroad and Coal Co. and later as a manager of Reading Railroad. MEMBERSHIPS: Am. Phil. Soc.; Wernerian Society (Europe); and other societies. SCIENTIFIC CONTRIBUTIONS: Upon completion of studies in Europe was one of best prepared American scientists of his time and may have been country's only professional mining engineer; his *Considerations Upon Mining* [see Works] was reputedly first scientific work on mining by American author. With own notes and those of others, prepared *Narrative* of Long Expedition [see Works], including analysis of minerals collected and valuable information on Indians; report on geology in the *Narrative*, while criticized by some as inadequate, remained standard source for many years. An early practitioner of laboratory method in teaching. From investigations near Franklin, N.J., contributed to discovery of several new minerals.

WORKS: *Considerations Upon the Art of Mining: To Which Are Added, Reflections on Its Actual State in Europe, and the Advantages Which Would Result from an Introduction of This Art into the U.S.* (Philadelphia, 1821), 87 pp.; *Narrative of an Expedition to the Source of St. Peter's River, Lake Winnepeek, Lake of the Woods, etc.*, 2 vols. (Philadelphia, 1824). References to scientific papers and to locations of manuscripts appear in footnotes to Miles [below]. WORKS ABOUT: *DAB,* 10: 276-77 (John H. Frederick); Wyndham D. Miles, "A Versatile Explorer: A Sketch of William H. Keating, *Minnesota History* 36 (1959): 294-99.

KEDZIE, ROBERT CLARK (January 28, 1823, Delhi, Delaware County, N.Y.—November 7, 1902, East Lansing, Mich.). *Chemistry.* Son of William Kedzie, probably farmer, who settled in Michigan when Robert was quite young. Married Harriet Fairchild, 1850; at least six children. EDUCATION: 1847, A.B., Oberlin College; 1851, M.D., Univ. of Michigan. HONORS: D.Sc., Michigan Agricultural College, 1898; LL.D., Univ. of Michigan, 1901. CAREER: 1847-1849, in charge of Rochester Academy (Michigan); ca. 1851, medical practice, Kalamazoo, Mich.; ca. 1852-ca. 1862, medical practice, Vermontville, Mich.; ca 1862-1863, was surgeon, Twelfth Michigan Infantry, and taken prisoner at Shiloh; ca. 1863, entered medical practice, Lansing, Mich.; 1863-1902, professor of chemistry, Michigan Agricultural College (Michigan State Univ.); 1877, became president, Michigan State Board of Health, and served on board for number of years; 1888-1902, chemist, Agricultural Experiment Station of Michigan. MEMBERSHIPS: President, Association of Official Agricultural Chemists, 1898. SCIENTIFIC CONTRIBUTIONS: Chief contributions made in establishment and work of Michigan Agricultural College. Wrote on variety of topics, including sanitation and public health; played significant role in establishment of quarantine at New Orleans, first such action anywhere in the world; interested in accident prevention, investigated explosive point of kerosene, and produced oil tester for that purpose; made earliest studies of dangers of arsenical wallpapers. Largely responsible for establishment of commercial fertilizer inspection in Michigan and was credited with establishment of beet-sugar industry in that state. Also interested in railway car ventilation, dangers of deforestation, climate and meteorology of Michigan, and other subjects; interests were practical and educational, and studies led to governmental regulations.

WORKS: See *Natl. Un. Cat. Pre-1956 Imp.*, which lists various pamphlets, articles in Agricultural Experiment Station's *Bulletin* and State Board of Health's *Annual Report*, and the like; also *Roy. Soc. Cat.*, vols. 12, 16, which lists papers appearing chiefly in *Proceedings of Society for Promotion of Agricultural Science*. WORKS ABOUT: *DAB*, 10: 277 (Edward Preble); L. S. Munson, "Memoir," *Bulletin of U.S. Department of Agriculture, Bureau of Chemistry*, no. 73 (1903), pp. 85-86.

KEELER, JAMES EDWARD (September 10, 1857, La Salle, Ill.—August 12, 1900, San Francisco, Calif.). *Astronomy*. Son of William F., U.S. Navy paymaster during Civil War, and Anna (Dutton) Keeler. Married Cora S. Matthews, 1891; no information on children. EDUCATION: 1869, family moved to Mayport, Fla., and secondary-level education was acquired through study at home; 1881, B.A., Johns Hopkins Univ.; 1883-1884, studied at Heidelberg and Berlin, Germany. HONORS: Sc.D., Univ. of California, 1893; recipient of Rumford Medal, Am. Ac. Arts Scis., 1899; also other honors. CAREER: 1878, while student at Johns Hopkins, went as assistant to C. S. Hastings [q.v.] to observe solar eclipse in Colorado; 1881-1883 and 1884-1886, assistant to S. P. Langley [q.v.], Allegheny Observatory, Pittsburgh; 1886, began work at Lick Observatory, California, then under construction, and was charged with establishing time service; 1888-1891, held title of astronomer, Lick Observatory; 1891-1898, director, Allegheny Observatory; 1895-1900, coeditor, *Astrophysical Journal;* 1898-1900, director, Lick Observatory. MEMBERSHIPS: Royal Astronomical Society (London); Natl. Ac. Scis.; and other societies. SCIENTIFIC CONTRIBUTIONS: Before entering college, built astronomical observatory at Mayport, Fla. As new member of Allegheny staff, in 1881 accompanied expedition to Mt. Whitney, Calif., where S. P. Langley's bolometer was used to measure sun's infrared radiation; at Lick Observatory (1890), used thirty-six-inch telescope and a spectroscope to measure wavelengths of bright lines in nebular spectra, demonstrating that nebulae, like stars, have discernible motion; as director at Allegheny, devised instrument (spectrograph) to photograph spectral lines, and with this equipment was able to confirm James Clerk Maxwell's theory that rings of Saturn are actually congregations of meteoritic bodies in independent orbits; after returning to Lick, worked on Crossley telescope, previously troublesome instrument, and produced photographs revealing large number of spiral nebulae. Earned reputation as leading American astrophysicist of his time.

WORKS: Major works listed in *DSB;* full bibliography, with 126 entries, in *Biog. Mems. of Natl. Ac. Scis.* 5 (1905): 241-46. MSS: Univ. of Pittsburgh—Archives of Industrial Society (Allegheny Observatory records); Univ. of California at Santa Cruz Library (Lick Observatory archives). WORKS ABOUT: *DAB*, 10: 278-79 (Raymond S. Dugan); *DSB*, 7: 270-71 (Sally H. Dieke).

KELLERMAN, WILLIAM ASHBROOK (1850-1908). *Botany*. Professor of botany, Ohio State Univ.

WORKS ABOUT: See *AMS; NCAB*, 26 (1937): 159-60.

KELLOGG, ALBERT (December 6, 1813, New Hartford, Conn.—March 31, 1887, Alameda, Calif.). *Botany*. Son of Isaac, successful farmer, and Aurilla (Barney) Kellogg. Never married. EDUCATION: At early age, began study of medicine with doctor in Middleton, Conn., and later studied in Charleston, S.C.; M.D., Transylvania Univ. (Lexington, Ky.). CAREER: Practiced medicine in Kentucky, Georgia, and Alabama; 1845, in San Antonio, Tex., and soon thereafter returned to Connecticut; 1849, arrived at Sacramento, Calif., after sailing around Cape Horn; before 1853, moved to San Francisco to take up medical practice; 1867, surgeon and botanist, U.S. Coast and Geodetic Survey expedition to Alaska. MEMBERSHIPS: A founder, Calif. Ac. Scis. (1853). SCIENTIFIC CONTRIBUTIONS: Before going to West Coast, traveled extensively in South and Mississippi Valley, studying natural history. Noted as first botanist to take up residence in California, where his botanical work was closely associated with Calif. Ac. Scis.; collected extensively in California; published numerous botanical papers in *Proceedings* and *Bulletin* of academy, especially on flowering herbs and shrubs, and described 215 botanical species, of which about 50 are now acknowledged. While not first to publish on Sequoia tree, accounts were of special importance as early firsthand observations; *Forest Trees* [see Works] summarized years of effort and ranks as earliest general work on California silva. Collected extensively during 1867 Alaska trip, and sent specimens to Smithson. Instn., Ac. Nat. Scis. Philad., and Calif. Ac. Scis. Numerous botanical writings said to contain elements of poetry and theology in addition to botany.

WORKS: *Forest Trees of California* (Sacramento, Calif., 1882), 148 pp., also printed as appendix to

Second Report of California State Mineralogist, 1880-1882. See works in *Roy. Soc. Cat.*, vols. 8, 10, 16. WORKS ABOUT: *DAB*, 10: 300-301 (Willis L. Jepson); *DSB*, 7: 285 (Joseph Ewan).

KEMP, JAMES FURMAN (1859-1926). *Geology*. Professor of geology, Columbia Univ.
WORKS ABOUT: See *AMS; DAB*.

KENDALL, EZRA OTIS (May 17, 1818, Wilmington, Mass.—January 5, 1899, Philadelphia, Pa.). *Astronomy; Mathematics*. Son of Ezra and Susanna (Cook) Walker Kendall; half-brother of S. C. Walker [q.v.]. Married Emma Lavinia Dick, 1844; at least one child. EDUCATION: Attended academy at Woburn, Mass.; 1835, began study of mathematics under half-brother, S. C. Walker, at Philadelphia. HONORS: LL.D., Univ. of Pennsylvania, 1888. CAREER: 1838, became professor of theoretical mathematics and astronomy, Central High School, Philadelphia; at Univ. of Pennsylvania, professor of mathematics (1855-1899), vice-provost (1883-1894), Flower professor of astronomy (1892-1894), at times dean of faculty, and honorary dean of faculty (1894-1899). MEMBERSHIPS: Vice-president of Am. Phil. Soc. at time of death; member of Am. Math. Soc. and other societies. SCIENTIFIC CONTRIBUTIONS: At Central High School, assisted S. C. Walker in establishment of observatory and devoted considerable time to observations and preparation of results for publication. For several years, engaged in series of observations for determination of longitudes for U.S. Coast Survey. 1851-1882, responsible for computation of ephemerides of Neptune and of Jupiter and its satellites, for publication in *American Ephemeris and Nautical Almanac*.

WORKS: *Uranography; or, A Description of the Heavens; Designed for Academics and Schools; Accompanied by an Atlas of the Heavens . . .* (Philadelphia, 1844). See also *Roy. Soc. Cat.*, vol. 3. WORKS ABOUT: *NCAB*, 2 (1899 [copyright 1891 and 1899]): 415; *Lamb's Biographical Dictionary of the U.S.* (Boston, 1901).

KENNICOTT, ROBERT (November 13, 1835, New Orleans, La.—May 13, 1866, Fort Nulato, Alaska). *Zoology; Natural history collecting*. Son of John Albert, physician and horticulturist, and Mary Shutts (Ransom) Kennicott. No information on marital status. EDUCATION: Raised in Illinois; poor health as child made attendance at school impossible; studied natural history with J. P. Kirtland [q.v.] (1852-1853), with Philo R. Hoy (1854) at Racine, Wis., and with S. F. Baird [q.v.] and others; 1855-1856 and 1856-1857 (winters), attended medical lectures in Chicago. CAREER: 1855, conducted natural history survey of southern Illinois for Illinois Central Railroad Co.; 1857-1858, worked to bring together museum for Northwestern Univ., collecting in field as far as St. Paul, and spent winter of 1857-1858 at Smithson. Instn., selecting duplicates and working up his collections; 1858 (summer), made expedition to northern Wisconsin; 1858-1859 (winter), at Smithsonian, where he worked on zoological collections gathered by Lt. W. P. Trowbridge [q.v.] in California, part of which went to Univ. of Michigan; 1859-1862, conducted explorations in British and Arctic America, sponsored by Smithsonian, Audubon Club of Chicago, and other sources; 1862-1863 (winter), worked on Arctic-American collections at Smithsonian; ca. 1864 appointed curator (later director), Chicago Academy of Sciences, and spent 1864 moving Arctic-American collections from Washington to Chicago; 1865, headed a Western Union Telegraph Co. party to Alaska and Yukon River, where died. MEMBERSHIPS: A founder, Chicago Academy of Sciences (1856). SCIENTIFIC CONTRIBUTIONS: Early reputation based on report on mammals for Patent Office [see Works], published at age twenty-one. Remembered chiefly for natural history explorations and collecting. Played important role in early history of Chicago Academy. Explorations in British and Arctic America resulted in discovery of number of new animals. Published relatively little, mainly on vertebrates; in addition to descriptions, works included useful observations on animal relations, habits, and distribution.

WORKS: "The Quadrupeds of Illinois, Injurious or Beneficial to the Farmer," in U.S. Patent Office, *Agricultural Report for 1856*, pp. 52-110, and *Agricultural Report for 1857*, pp. 72-107. Bibliography (twelve entries) in *Transactions of Chicago Academy of Sciences* [below], pp. 225-26. MSS: Minnesota Historical Society (*NUCMC* 60-655); Smithson. Instn. Library (Hamer). WORKS ABOUT: *DAB*, 10: 338-39 (Frank C. Baker); biographical sketch, *Transactions of Chicago Academy of Sciences* 1, pt. 2 (1869), pp. 113-227, with extensive extracts from journal of 1859-1862 Arctic-American expedition.

KERR, WASHINGTON CARUTHERS (May 24, 1827, Guilford County, N.C.—August 9, 1885, Asheville, N.C.). *Geology*. Son of William M., farmer, and Euphence (Doak) Kerr. Married Emma Hall, 1853; no information on children. EDUCATION: 1850, A.B., Univ. of North Carolina; 1853-1858, student in Lawrence Scientific School, Harvard. CAREER: ca. 1850-1852, taught school in Williamston, N.C., and then became professor at Marshall Univ. (Texas); 1852-1857, computer in *Nautical Almanac* office, Cambridge, Mass.; 1856-1865, professor of chemistry and geology, Davidson College (North Carolina); 1862-1864, on leave from Davidson, employed as chemist and superintendent, Mecklenburg Salt

Works, Mt. Pleasant, S.C.; 1864, appointed North Carolina geologist, without pay, while directing attentions to state's chemical and mineral productions that were vital to civilian and war needs; 1866-1882, reappointed state geologist of North Carolina; 1882-1883, associated with U.S. Geol. Surv., with headquarters at Washington, D.C. MEMBERSHIPS: Am. Phil. Soc. and other societies. SCIENTIFIC CONTRIBUTIONS: Work for *Nautical Almanac* was mainly preparation of astronomical computations. Chiefly known for work done in connection with North Carolina survey and all scientific publications apparently related to this survey, major outcome of which was a detailed and accurate map intended as base for future delineation of geological features of state. Continued work of earlier state geologist, E. Emmons [q.v.], and undertook efforts in topography, climatology, mineralogy, water power, and lithology; main work was done in physical geology; wrote on the influence of earth's rotation on stream flow and effects of frost on inclined surfaces ("soil creep"). Brief work for U.S. Geol. Surv. related to planned systematic and detailed survey of Appalachian region.

WORKS: *Report of the Geological Survey of North Carolina,* vol. 1, *Physical Geography, Resumé, Economic Geology* (Raleigh, 1875); *Map of North Carolina* (New York, 1882). See bibliography in Holmes [below], pp. 21-24. WORKS ABOUT: *DAB,* 10: 358-59 (George P. Merrill); J. A. Holmes, "A Sketch of Professor Washington Caruthers Kerr, M.A., D.Ph.," *Journal of Elisha Mitchell Scientific Society* 4 pt. 2 (1887), pp. 1-24.

KEYT, ALONZO THRASHER (January 10, 1827, Higginsport, Ohio—November 9, 1885, Cincinnati, Ohio). *Physiology.* Son of Nathan and Mary (Thrasher) Keyt. Married Susannah D. Hamlin, 1848; seven children. EDUCATION: Attended Parker's Academy, Felicity, Ohio; 1848, M.D., Medical College of Ohio (Cincinnati). CAREER: 1848, intern at Commercial Hospital (now Cincinnati General Hospital); 1849-1850, medical practice, Moscow, Ohio; 1850-1885, medical practice, Walnut Hills, near Cincinnati, Ohio. SCIENTIFIC CONTRIBUTIONS: Chief medical interests related to circulatory diseases. In 1873, began studies of circulation using sphygmographs as devised in Europe; later constructed much improved instrument called multigraph sphygmometer and cardiograph, or compound sphygmograph, which made possible graphic register of activity of artery and heart, or of two arteries, at one time, in association with chronograph; with this instrument Keyt studied circulatory phenomena of both diseased and normal persons. First studies were conducted to determine length of time of systole and diastole of heart, based on large number of cases; made first accurate determinations of velocity of pulse wave and showed its relation to various states and conditions of arteries and of heart valves. Reached number of general conclusions regarding heart and circulation, as well as perfecting methods of medical diagnosis of abnormalities; published series of papers in American medical journals, but was better known in France and England; papers collected and republished after death, as *Sphygmography* [see Works], which established reputation as important early American researcher in physiology and medicine.

WORKS: *Sphygmography and Cardiography, Physiological and Clinical* (New York, 1887). WORKS ABOUT: *DAB,* 10: 366-67 (Albert P. Mathews); David A. Tucker, Jr., "Alonzo Thrasher Keyt, Cardiologist," *Bull. Hist. Med.* 21 (1947): 753-57.

KING, CLARENCE RIVERS (January 6, 1842, Newport, R.I.—December 24, 1901, Phoenix, Ariz.). *Geology.* Son of James Rivers, businessman in China trade, and Caroline Florence (Little) King. Married Ada Todd, 1888; five children. EDUCATION: 1862, Ph.B., Sheffield Scientific School, Yale. CAREER: 1863-1866, assistant, California Geological Survey; 1865-1866 (winter), studied geology of Arizona as scientific assistant with military road survey; 1866, returned to East; 1867-1878, head, U.S. Geological Survey of Fortieth Parallel, by 1873 completing most of necessary field work; 1879-1881, first director, U.S. Geol. Surv.; 1879-1882, conducted mining surveys for Tenth U.S. Census; 1882, established practice as mining geologist. MEMBERSHIPS: Natl. Ac. Scis.; Geological Society of London; and other societies. SCIENTIFIC CONTRIBUTIONS: Notable for his administrative as well as scientific talents; geological work characterized by practical, descriptive, and theoretical aspects. Work of Fortieth Parallel Survey involved investigation of topography, petrology, and geological history along Union Pacific and Central Pacific railway routes from eastern Colorado to California border; survey's reports established new standards for official geological publications and introduced new techniques to American geology, including microscopic study of thin sections of rocks (prepared by Ferdinand Zirkel) and use of contour mapping in topography. King made early use of laboratory studies in dealing with geophysical questions, including problem of age of earth [see Works].

WORKS: *Report of the Geological Exploration of the Fortieth Parallel,* 7 vols. (Washington, D.C., 1870-1880) of which vol. 1, *Systematic Geology,* was written by King; "Catastrophism and Evolution," *American Naturalist* 11 (1877): 449-70; "The Age of the Earth" *Am. Journ. Sci.* 3d ser. 45 (1893): 1-20. Not a prolific author; see bibliography of technical and popular works in Thurman Wilkins, *Clar-*

ence King (New York, 1958), which also gives locations of manuscripts. WORKS ABOUT: *DAB,* 10: 384-86 (George P. Merrill); *DSB,* 7: 370-71 (Michele L. Aldrich).

KINGSLEY, JOHN STERLING (1854-1929). *Zoology; Comparative anatomy.* Professor of zoology, Tufts Univ.; professor of zoology, Univ. of Illinois.

WORKS ABOUT: See *AMS; NCAB,* 12 (1904): 119; *World Who's Who in Science.*

KINNERSLEY, EBENEZER (November 30, 1711, Gloucester, England [brought to America 1714]—March 1778, Pennepack, Pa.). *Electricity.* Son of William Kinnersley, Baptist minister. Married Sarah Duffield, 1739; survived by two children. EDUCATION: Studied with father at home. HONORS: M.A., College of Philadelphia, 1757. CAREER: At young age, established school in Philadelphia; 1743, ordained as Baptist minister, but never received pastoral appointment; 1749-1753, as friend and collaborator of B. Franklin [q.v.], toured colonies and lectured on electricity; 1753-1772, professor of English and oratory, College of Philadelphia (later, Univ. of Pennsylvania). MEMBERSHIPS: Am. Phil. Soc. SCIENTIFIC CONTRIBUTIONS: Became chief collaborator with Franklin in electrical studies; ability as public speaker and as demonstrator of scientific experiments made him effective public lecturer, spreading word of Philadelphia experiments on electricity and lightning in that city and elsewhere, including Boston, Newport (R.I.), and New York. While in Boston in 1752, rediscovered Dufay's two electricities ("of glass and sulphur"), of which Franklin was not aware; continued work in electricity after beginning association with College of Philadelphia, displayed production of heat from electricity, and suggested theory to account for repulsion of negatively charged bodies, ascribing repulsion to attractions of surrounding air. Best-known contribution to early study of electricity in America was invention of electrical air thermometer, devised to measure rise in pressure as result of movement of spark through confined volume of air.

WORKS: "New Experiments in Electricity," *Phil. Trans. of Roy. Soc.* 53 (1763): 84-97, includes description of air thermometer and theory of repulsion. Several works referred to in *DAB* and in *DSB.* Full bibliography in Joseph A. L. Lemay, *Ebenezer Kinnersley* (Philadelphia, 1964), pp. 123-24, with references to manuscripts on pp. 125-26 of that work. WORKS ABOUT: *DAB,* 10: 416-17 (Frank A. Taylor); *DSB,* 7: 371-72 (John L. Heilbron).

KINNICUTT, LEONARD PARKER (1854-1911). *Sanitary chemistry.* Professor of chemistry, Worcester Polytechnic Institute.

WORKS ABOUT: See *AMS; DAB.*

KIRKWOOD, DANIEL (September 27, 1814, Harford County, Md.—June 11, 1895, Riverside, Calif.). *Astronomy.* Son of John, farmer, and Agnes (Hope) Kirkwood. Married Sarah A. McNair, 1845; no information on children. EDUCATION: 1834-1838, attended York County (Pa.) Academy. HONORS: M.A., Washington College (later Washington and Jefferson College, Pennsylvania), 1848; LL.D., Univ. of Pennsylvania, 1852. CAREER: 1833-1834, teacher, Hopewell, Pa.; 1838, appointed first assistant and instructor in mathematics, York County Academy; 1843, appointed principal, Lancaster High School; 1849, became principal, Pottsville Academy; at Delaware College (Univ. of Delaware), professor of mathematics (1851-1856) and president (1854-1856); 1856-1865 and 1867-1886, professor of mathematics, Indiana Univ.; 1865-1867, professor of mathematics and astronomy, Washington and Jefferson College; ca. 1886, moved to California; 1891, made nonresident lecturer in astronomy, Stanford Univ. MEMBERSHIPS: Am. Phil. Soc.; Am. Assoc. Advt. Sci.; and other societies. SCIENTIFIC CONTRIBUTIONS: Worked chiefly in mathematical astronomy and particularly on character and history of solar system; produced detailed critique of nebular hypothesis of Kant and Laplace during period of great interest in that theory. As early as 1861, produced first persuasive presentation of case for connection between meteors and comets, an idea later confirmed, and in years 1866-1867 took up question of relations of comets and asteroids and of shower meteors and stony meteorites. Principal contribution to astronomy was discovery of "gaps" in orbital distribution of asteroids, explained in terms of gravitational influences of Jupiter; first noted this phenomenon about 1857 when only about fifty asteroids were known, and later explained gaps in rings of Saturn as result of similar relations.

WORKS: *Meteoric Astronomy* (Philadelphia, 1867); *Comets and Meteors* (Philadelphia, 1873); "On the Theory of Meteors," *Proceedings of Am. Assoc. Advt. Sci.* 15 (1866): 8-14, is initial presentation regarding gaps in asteroid distribution and in rings of Saturn. Extensive bibliography of chief works in *DSB.* MSS: Papers were destroyed in fire at Indiana Univ. WORKS ABOUT: *DAB,* 10: 436 (Raymond S. Dugan); *DSB,* 7: 384-87 (Brian G. Marsden).

KIRTLAND, JARED POTTER (November 10, 1793, Wallingford, Conn.—December 10, 1877, Rockport [near Cleveland], Ohio). *Zoology.* Son of Turhand, stockholder and general agent of Connecticut Land Co., and Mary (Potter) Kirtland; raised by maternal grandfather, a physician. Married Caroline Atwater, 1815; Hannah Fitch Tousey, 1825; survived by one child from first marriage. EDUCA-

TION: 1811, began private study of medicine; 1813, entered medical school, Yale Univ.; 1814, studied in medical school at Univ. of Pennsylvania; 1815, M.D., Yale Univ. HONORS: LL.D., Williams College, 1861. CAREER: 1810-1811, teacher, Poland, Ohio; 1815-1818, medical practice, Wallingford, Conn.; 1818-1823, medical practice, Durham, Conn.; 1823, established medical practice in Poland, Ohio; 1828-1834, served in Ohio legislature; 1837, moved to farm near Cleveland, Ohio, and that year was appointed zoologist on Ohio Geological Survey; 1837-1842, professor of theory and practice of medicine, Medical College of Ohio (Cincinnati); 1843-1864, a founder and professor of theory and practice of medicine, Cleveland Medical College of Western Reserve College. MEMBERSHIPS: Natl. Ac. Scis and other societies; founder, Cleveland Academy of Natural Science (1845), later named Kirtland Society of Natural Science. SCIENTIFIC CONTRIBUTIONS: At age fifteen, made first discovery of parthenogenesis in silkworm moth; discovered bisexual character of bivalve freshwater mollusks. Made early studies of natural history of Ohio and played important role in promoting scientific interests in that state; in *Second Annual Report* of Ohio survey (1838), published checklist and notes on 585 animals of the state, encompassing large number of the mammals, birds, reptiles, fishes, and mollusks; collected and wrote on birds, published on Ohio's climate, insects, and especially fishes of the state, and pursued experimental work in floriculture and horticulture.

WORKS: "Observations on the Sexual Characters of the Animals Belonging to Lamarck's Family of Naiades," *Am. Journ. Sci.* 26 (1834): 117-20; work on fishes of Ohio in *Boston Journal of Natural History*, 3-5 (1839, 1842, 1845). See Meisel, *Bibliog. Amer. Nat. Hist.* MSS: In several collections at Western Reserve Historical Society (Cleveland). WORKS ABOUT: *DAB*, 10: 438-39 (Frederick C. Waite); J. S. Newberry, "Memoir," *Biog. Mems. of Natl. Ac. Scis.* 2 (1886): 129-38.

KNEELAND, SAMUEL (August 1, 1821, Boston, Mass.—September 27, 1888, Hamburg, Germany). *Zoology.* Son of Samuel, merchant, and Nancy (Johnson) Kneeland. Married Eliza Maria Curtis, 1849; two children. EDUCATION: A.B. in 1840 and M.D. in 1843, Harvard Univ.; 1843-1845, studied medicine and surgery, Paris. CAREER: 1845, established medical practice, Boston; 1845-1847, physician to Boston Dispensary; 1851-1853, demonstrator in anatomy, Harvard Medical School; 1856-1857, physician and surgeon to copper mining companies, Lake Superior region; 1862-1866, as army surgeon, achieved brevet rank of lieutenant colonel of volunteers; at Massachusetts Institute of Technology, secretary (1865-1878) and professor of zoology and physiology (1869-1878). MEMBERSHIPS: Secretary, Am. Ac. Arts Scis. and Bos. Soc. Nat. Hist.; member of other societies. SCIENTIFIC CONTRIBUTIONS: Medical thesis, "On the Contagiousness of Puerperal Fever," prepared under influence of Oliver Wendell Holmes, won Boylston Prize (1843). Extensive travels included expeditions to Brazil, Lake Superior copper region, Iceland, Hawaii, and Philippine Islands; on these trips collected data and specimens of scientific interest; several trips were made to observe active volcanoes. Prepared popular articles and books, and gave public lectures on zoology; wrote over 1,000 medical and zoological articles for Appleton's *American Cyclopaedia* (1873-1876) and from 1866 to 1871 served as editor of *Annual of Scientific Discovery*; also published articles on zoological and other topics in scientific journals, including papers on skeletal and other anatomical features, with several items on crania. During later years was chiefly interested in ethnological matters.

WORKS: Several works mentioned in *DAB*. See bibliography in *Roy. Soc. Cat.*, vols. 3, 8, 10, 16. WORKS ABOUT: *DAB*, 10: 459 (John F. Fulton); memoir, *Proceedings of Am. Ac. Arts Scis.* 24 (1888-1889): 438-41.

KNIGHT, WILBUR CLINTON (December 13, 1858, Rochelle, Ill.—July 28, 1903, Laramie, Wyo.). *Geology.* Son of David A. Knight, farmer. Married E. Emma Howell, 1889; four children. EDUCATION: B.Sc. in 1886, M.A. in 1893, and Ph.D. in 1901, Univ. of Nebraska; studied at Univ. of Chicago. CAREER: 1886-1887, assistant territorial geologist, Wyoming; 1887-1888, assayer, Cheyenne, Wyo.; 1888-1893, superintendent of mines in Colorado and Wyoming; 1893-1903, professor of geology and mining engineering and principal of School of Mines, Univ. of Wyoming; 1898-1899, Wyoming state geologist; ca. 1901-1902, on leave from Univ. of Wyoming to serve as consultant to Belgian-American Oil Co.; served for number of years as consultant to Union Pacific Railroad Co., was associated with U.S. geological and hydrographic survey, and served as geologist of Wyoming experiment station and curator of Wyoming state museum. MEMBERSHIPS: Geol. Soc. Am.; Am. Inst. Min. Engrs.; and other societies. SCIENTIFIC CONTRIBUTIONS: Because of industrial and developmental concerns of state of Wyoming, most publications related to economic geology. Nevertheless, his chief interest was in vertebrate paleontology and stratigraphy, and published on these subjects as well; described several new fossil forms, and shortly before death published list of Wyoming birds. In 1899, organized Fossil Field Scientific Expedition in Wyoming, in which over 300 colleges and scientific societies were invited to participate, and conducted annual collecting expeditions for Univ. of Wyoming.

WORKS: Bibliographies of works appended to all three publications listed below. MSS: Small collection with geology department records in Univ. of Wyoming Library—Rare Books and Special Collections. WORKS ABOUT: Aven Nelson, memoir, *Science* n.s. 18 (1903): 406-9; S. W. Williston, "Wilbur Clinton Knight," *American Geologist* 33 (1904): 1-5; Erwin H. Barbour, "Memoir," *Bulletin of Geol. Soc. Am.* 15 (1904): 544-49.

KNOWLTON, FRANK HALL (1860-1926). *Paleobotany*; *Ornithology*. Paleontologist and geologist, U.S. Geol. Surv.

WORKS ABOUT: See *AMS*; *DAB*.

KOENIG, GEORGE AUGUSTUS (1844-1913). *Chemistry*; *Mineralogy*. Professor of mineralogy, Univ. of Pennsylvania; professor of chemistry, Michigan College of Mines.

WORKS ABOUT: See *AMS*; *DAB*.

KUHN, ADAM (November 17, 1741, Germantown, Pa.—July 5, 1817, Philadelphia, Pa.). *Botany*; *Medicine*. Son of Adam Simon, physician and magistrate, and Anna Maria Sabina Schrack Kuhn. Married Elizabeth (Hartman) Markhoe, ca. 1780; two children. EDUCATION: Studied medicine with father; 1761-1764, studied medicine, Univ. of Upsala (Sweden); 1764-1765, studied in London; 1767, M.D., Univ. of Edinburgh. CAREER: 1768-1789, professor of materia medica and botany, College of Philadelphia (after 1791, part of Univ. of Pennsylvania); 1774-1781 and 1782-1798, member of medical staff, Pennsylvania Hospital; 1786, became a consulting physician to Philadelphia Dispensary; 1789, appointed professor of theory and practice of medicine, Univ. of State of Pennsylvania (after 1791, part of Univ. of Pennsylvania); 1792-1797, professor of physic, Univ. of Pennsylvania; until ca. 1814 [age seventy-three], engaged in extensive private medical practice. MEMBERSHIPS: Am. Phil. Soc.; president, College of Physicians of Philadelphia, 1808. SCIENTIFIC CONTRIBUTIONS: At Upsala, studied botany with Linnaeus; with appointment at College of Philadelphia in 1768, became first professor of botany in what is now United States; while instruction in materia medica continued until 1789, botanical lectures seem to have lasted only two years. Other than taking living specimen of North American plant, a new genus, to Linnaeus, made no contribution to botanical knowledge; medical and materia medica teaching was sound, but unoriginal, following closely the work of European teachers, especially William Cullen.

WORKS: Except for medical thesis on cold baths, dedicated to Linnaeus, apparently published nothing. References to manuscript notes of lectures in footnotes to Brown [below], and also in *Natl. Un. Cat. Pre-1956 Imp.* WORKS ABOUT: *DAB*, 10: 510-11 (Willis P. Jepson); Marion E. Brown, "Adam Kuhn: Eighteenth Century Physician and Teacher," *Journal of the History of Medicine* 5 (1950): 163-77.

L

LADD, GEORGE TRUMBULL (1842-1921). *Psychology*; *Philosophy*. Professor of philosophy, Yale Univ.

MSS: See Mills [below]. WORKS ABOUT: See *AMS*; *DAB*; Eugene S. Mills, *George Trumbull Ladd: Pioneer American Psychologist* (Cleveland, 1969).

LANE, JONATHAN HOMER (August 9, 1819, Genesee, N.Y.—May 3, 1880, Washington, D.C.). *Physics*. Son of Mark, farmer, and Henrietta (Tenny) Lane. Never married. EDUCATION: 1839, entered Phillips Academy (Exeter, N.H.); 1846, A.B., Yale College. CAREER: For short time, teacher in Vermont; 1847-1848, attached to U.S. Coast Survey; 1848-1857, examiner, U.S. Patent Office; 1857-ca. 1860, may have worked as patent agent in Washington, D.C.; failed in attempt to find financial backing for development of low-temperature apparatus, and in 1860 went to live with brother in Franklin, Pa.; 1866, returned to Washington, D.C.; 1869, accompanied U.S. Coast Survey solar eclipse expedition to Des Moines, Iowa, and in 1870 was sent to Spain for similar purpose; 1869-1880, worked for U.S. Office of Weights and Measures. MEMBERSHIPS: Natl. Ac. Scis. SCIENTIFIC CONTRIBUTIONS: About 1840, became interested in problem of determining absolute zero experimentally, a concern that continued until about 1870, though results were never published. During period 1846-1851, published four papers on electricity that attempted to deal with electrical phenomena mathematically. Paper on temperature of sun [see Works] attempted to examine mathematically various theories of heat and was particularly significant for its careful presentation of relationships between mass and heat within sun; that paper also implied law of heat loss and gain in contracting gaseous body, a conclusion that Lane was said to have presented verbally, in more explicit form, to S. Newcomb [q.v.] before 1869. Details of life and work not entirely clear, and evaluations of scientific efforts are uncertain, although J. Henry [q.v.] considered Lane an early mathematical physicist.

WORKS: "On the Theoretical Temperature of the Sun, Under the Hypothesis of a Gaseous Mass Maintaining Its Volume by Its Internal Heat and Depending on the Laws of Gases as Known to Terrestrial Experiments," *Am. Journ. Sci.* 2d ser. 50 (1870): 57-74. Bibliography in Abbe [below], pp. 263-64. MSS: Natl. Arch. (see this and other references in *DSB*). WORKS ABOUT: *DSB*, 8: 1-3 (Nathan Reingold); Cleveland Abbe, "Memoir," *Biog. Mems. of Natl. Ac. Scis.* 3 (1895): 253-64 (see *DSB* for evaluation of usefulness of this memoir).

LANGLEY, SAMUEL PIERPONT (1834-1906). *Astrophysics*. Professor of physics and astronomy and director of Allegheny Observatory, Western Univ. of Pennsylvania (Pittsburgh); third secretary, Smithson. Instn.

MSS: Smithson. Instn. Archives (*NUCMC* 72-1242); Univ. of Michigan Library (Hamer). WORKS ABOUT: See *AMS*; *DAB*; *DSB*.

LAPHAM, INCREASE ALLEN (March 7, 1811, Palmyra, N.Y.—September 14, 1875, Oconomowoc Lake, near Milwaukee, Wis.). *Natural history*; *Meteorology*. Son of Seneca, canal contractor and engineer, and Rachel (Allen) Lapham. Married Ann M. Alcott, 1838; five children. EDUCATION: At age thirteen, began work on Erie Canal; ca. 1827, attended school at Shippingsport, Ky., for several months. HONORS: LL.D., Amherst College, 1860. CAREER: Ca. 1824, employed on Erie Canal, and then worked on Welland Canal in Canada and Miami Canal in Ohio; 1827, began work on Portland and Louisville Canal (Kentucky); 1830-1833, employed on Ohio canal, at Portsmouth; 1833-1836, secretary, Ohio Board of Canal Commissioners; 1836, moved permanently to Milwaukee, Wis., and became involved in various surveying and development projects as assistant to Byron Kilbourn, a founder of Milwaukee; 1871-1872, connected with U.S. Weather Bureau headquarters at Chicago; 1873-1875, Wisconsin state geologist; active in civic affairs of Milwaukee and held number of public offices. MEMBERSHIPS: An original member, Wisconsin Academy of Sciences, Arts, and Letters. SCIENTIFIC CONTRIBUTIONS: Noted as first scientist in Wisconsin and made contributions in botany, geology, conchology, map making, meteorology, archeology, and agriculture. Scientific activities began while in Kentucky, where prepared first published paper, on geology in vicinity of Louisville and Shippingsport Canal. In Wisconsin, began scientific exploration and writing almost immediately; *Catalogue* (1836) [see Works] was first Wisconsin imprint; 1844 work entitled *Wisconsin* [see Works] was based on extensive travels and included information on natural history, climate, and civic affairs, while 1846 edition included first map of the state; in 1849 induced Am. Ant. Soc. to sponsor his studies of Indian mounds, and resulting "Antiquities of Wisconsin (*Smithson. Contr. Knowl.*, 1855) was one of his chief publications; also investigated grasses of Wisconsin region, and in 1867 headed state commission to study forests. In 1869, prepared resolution leading to establishment of U.S. Weather Bureau. Began first successful State Geological Survey of Wisconsin; active as lecturer and organizer of science in Wisconsin.

WORKS: *Catalogue of Plants and Shells found in the Vicinity of Milwaukee . . .* (Milwaukee, Wis., 1836), 12 pp.; *Wisconsin: Its Geography and Topography, History, Geology, and Mineralogy* (Milwaukee, Wis., 1st ed., 1844; 2d ed., 1846). Bibliography of works, in *American Geologist* 13 (1894): 31-38. MSS: St. Hist. Soc. Wis. (*NUCMC* 62-2911); Ohio Historical Society (*NUCMC* 75-1071). WORKS ABOUT: *DAB*, 10: 611-12 (Louise P. Kellogg); James I. Clark, *Increase A. Lapham: Scientist and Scholar* (Madison, Wis., 1957), 21 pp. Also see Graham H. Hawks, "Increase A. Lapham, Wisconsin's First Scientist," *Dissertation Abstracts* 21, no. 6 (1960), p. 1542.

LATTIMORE, SAMUEL ALLAN (1828-1913). *Chemistry.* Professor of chemistry, Univ. of Rochester.

MSS: Univ. of Rochester Library (Hamer). WORKS ABOUT: See *AMS*; *NCAB*, 12 (1904): 244; *Who Was Who in America,* vol. 1.

LAWRENCE, GEORGE NEWBOLD (October 20, 1806, New York, N.Y.—January 17, 1895, New York, N.Y.). *Ornithology.* Son of John Burling, businessman dealing in wholesale drugs and chemicals, and Hannah (Newbold) Lawrence. Married Mary Ann Newbold, 1834; had children. EDUCATION: No information on education. CAREER: Ca. 1822 [age sixteen], began work in father's wholesale drug and chemicals business, became partner at age twenty, and in 1835, assumed charge of company; ca. 1862 [age fifty-six], retired and devoted efforts to study of ornithology. MEMBERSHIPS: Founding member of Am. Ornith. Un.; honorary member of Zoological Society of London; member of other societies. SCIENTIFIC CONTRIBUTIONS: Began ornithological collecting at age fourteen, but most specimens later described were collected by other persons, including explorers employed by himself or by Smithson. Instn.; publications were chiefly lists of collections and descriptions of new species and dealt initially with birds of U.S.; in early 1840s became friend of S. F. Baird [q.v.] and later collaborated with him on ornithological account of Pacific Railroad Surveys, contributing to report on water birds of North America (1858). In late 1850s began study of birds of America south of the U.S., and it is in this area that most work was done, especially birds of Mexico, Central America, and West Indies; all but 17 of more than 300 new species described were from area south of U.S. Collection of some 8,000 bird skins is in Am. Mus. Nat. Hist. (New York).

WORKS: "Catalogue of Birds Observed in New York, Long and Staten Islands and the Adjacent Parts of New Jersey," *Annals of N.Y. Lyc. Nat. Hist.* 8 (1866): 279-300; "A Catalogue of the Birds Found in Costa Rica," ibid. 9 (1870): 86-149; "The Birds of Western and Northwestern Mexico," *Memoirs of Bos. Soc. Nat. Hist.* 2 (1874): 265-319; "Birds of Southwestern Mexico," *Bulletin of U.S. Natl. Mus.*, no. 4 (1876), pp. 1-56. Elaborate bibliography, with lists of species, in Foster [below]. MSS: Am. Mus. Nat. Hist. (Hamer). WORKS ABOUT: *DAB*, 11: 49 (Witmer Stone); L. S. Foster, "The Published Writings of . . . Lawrence," *Bulletin of U.S. Natl. Mus.*, no. 40 (1892); W. L. McAtee, "George Newbold Lawrence," *Nature Magazine* 47 (1954): 320.

LAWSON, JOHN (died 1711, North Carolina). *Travels*; *Natural history.* No information on parentage. Married; at least one child. EDUCATION: Nothing definite known of educational background; may have served apprenticeship at London Society of Apothecaries. CAREER: Surveyor; 1700, arrived in Charleston, S.C., and later that year began explorations in North Carolina; later, secured land in North Carolina; 1705, a founder of Bath, N.C.; 1708-1710, in England; 1709, commissioned surveyor general of North Carolina and appointed by Lords Proprietors as a commissioner to determine boundary between North Carolina and Virginia; 1710, having met Christopher de Graffenried in London, became involved in plans to establish German and Swiss Palatine colony in North Carolina and returned to

America with Palatine refugees; 1711, executed by Indians. SCIENTIFIC CONTRIBUTIONS: Remembered primarily for *New Voyage to Carolina* [see Works], especially useful as source on ways of Indians and on early civil history of area, but also valuable for natural history; book was essentially travel account, dealt with topography, fauna, and flora, and showed author to be effective observer of natural phenomena, although operating in older tradition, not being aware of advances in methods of study and presentation in natural history. Corresponded with James Petiver, English naturalist, and sent him zoological and botanical specimens; lists and descriptions of specimens generally better than those that had appeared before him. *Natural History of Virginia; or, The Newly Discovered Eden*, credited to William Byrd II, largely borrowed verbatim from Lawson's *Voyage*. Zirkle [below] poses unanswered question—whether Lawson was first consciously to describe a hybrid plant.

WORKS: *A New Voyage to Carolina, Containing the Exact Description and Natural History of that Country; Together with the Present State thereof and A Journal of a Thousand Miles, traveled thro' several Nations of Indians, Giving a particular Account of their Customs, Manners, etc.* (London, 1709), republished in 1714 and 1718 under title *The History of Carolina*, and modern edition issued with introductory sketch, notes, etc. (Chapel Hill, N.C., 1967). WORKS ABOUT: *DAB*, 11: 57-58 (William K. Boyd); Stearns, *Sci. in Brit. Colonies*, pp. 305-15; Conway Zirkle, "The First Recognized Plant Hybrid?" *Journal of Heredity* 49 (1958): 137-38.

LEA, ISAAC (March 4, 1792, Wilmington, Del.—December 8, 1886, Philadelphia, Pa.). *Malacology.* Son of James, wholesale merchant, and Elizabeth (Gibson) Lea. Married Frances Anne Carey, 1821; four children, including Mathew C. Lea [q.v.]. EDUCATION: Attended academy at Wilmington, Del. HONORS: LL.D., Harvard Univ., 1852. CAREER: Ca. 1807 [age fifteen], began work in brother's wholesale and importing firm, Philadelphia, and later became partner; 1821-1851, from time of marriage, associated with M. Carey & Sons, publishers, and eventually became president; 1851, retired, a man of wealth. MEMBERSHIPS: President of Am. Assoc. Advt. Sci., 1860, and of Ac. Nat. Scis. Philad., 1858-1863; member of other societies. SCIENTIFIC CONTRIBUTIONS: Earliest article, on minerals in vicinity of Philadelphia, appeared in 1818, and other early papers dealt with variety of topics, including impressions in sandstone, hibernation, earthquakes, Northwest Passage, halos with parhelia; after 1827, came to specialize in study of freshwater mollusks and eventually described 1,842 molluscan species, both living and fossil, terrestrial and freshwater; best-known work done with pearly freshwater mussels; works on mollusks dealt not only with shells, but also touched on study of soft parts, embryology, and other aspects; also worked to clear up accumulated confusions in names and classification of Unionidae (family of mussels). In 1833 published *Contributions to Geology*, which related mainly to specimens from Tertiary formations of Claiborne, Ala. In later life became interested in study of crystals and was American pioneer in use of microscope in mineralogy.

WORKS: Over period of years 1827-1874, collected papers and reissued them in thirteen volumes, with title *Observations on the Genus Unio; Synopsis of the Family of Naiades* (Philadelphia and London, 1836; 4th ed., 1870). See Newton P. Scudder, "Published Writings of Isaac Lea, LL.D.," *Bulletin of U.S. Natl. Mus.*, no. 23 (1885), with analysis of species described. MSS. Ac. Nat. Scis. Philad. (*NUCMC* 66-55); Smithson. Instn. Archives (*NUCMC* 72-1243); Am. Phil. Soc. (*DSB*). WORKS ABOUT: *DAB*, 11: 70-71 (William B. Marshall); *DSB*, 8: 103-4 (Whitfield J. Bell, Jr.).

LEA, MATHEW CAREY (August 18, 1823, Philadelphia, Pa.—March 15, 1897, Philadelphia, Pa.). *Chemist.* Son of Isaac [q.v.]—merchant, publisher, and malacologist—and Frances Anne (Carey) Lea. Married Elizabeth Lea Jaudon, 1852; Eva Lovering, sometime after 1881; one child, by first marriage. EDUCATION: Taught by private tutors at home; studied chemistry in Philadelphia laboratory of J. C. Booth [q.v.]; ca. 1847, admitted to bar, having studied law in office of Philadelphia lawyer. CAREER: Practiced law for short time, but abandoned it because of ill health; a man of wealth, established private chemical laboratory in home. MEMBERSHIPS: Franklin Institute; Natl. Ac. Scis. SCIENTIFIC CONTRIBUTIONS: First paper, 1841, dealt with Pennsylvania coal fields, but interests soon turned to more strictly chemical subjects; worked in both organic and inorganic chemistry, touching on wide variety of topics in research carried out alone in private laboratory; chemical studies encompassed subjects such as platinum metals, nature of solutions, affinities of acids, picric acid and its compounds, ethyl and methyl bases, and others. Particularly interested in chemical theory, pursuing speculative work on theory of atoms, including numerical relations between them. Chief work was in photochemistry, and Lea was recognized by Europeans as international authority in that field; investigated particularly the chemical and physical properties of silver haloid salts, studying these substances under various combinations and conditions. In addition to work in chemistry, was also interested in physics.

WORKS: "Identity of the Photosalts of Silver with the Material of the Latent Photographic Image," *Am. Journ. Sci.* 3d ser. 33 (1887) : 480-88. Published several hundred papers, especially in *Am. Journ. Sci.*, and made large number of contributions to photographic journals. Bibliography of important papers in Barker [below], pp. 204-8. WORKS ABOUT: *DAB*, 11: 71 (Edgar F. Smith) ; George F Barker, "Memoir," *Biog. Mems. of Natl. Ac. Scis.* 5 (1905) : 155-208.

LeCONTE, JOHN (December 4, 1818, Liberty County, Ga.—April 29, 1891, Berkeley, Calif.). *Physics*; *Natural history*. Son of Louis, plantation owner and amateur botanist, and Ann (Quarterman) LeConte; brother of Joseph LeConte [q.v.] and cousin of John L. LeConte and W. L. Jones [qq.v.]. Married Eleanor Josephine Graham, 1841; three children. EDUCATION: 1838, A.B., Franklin College (Univ. of Georgia) ; 1841, M.D., College of Physicians and Surgeons, New York. CAREER: Ca. 1841-1846, medical practice, Savannah, Ga.; 1846-1855, professor of physics and chemistry, Franklin College; 1855, professor of chemistry, College of Physicians and Surgeons; 1856-1869, professor of physics, South Carolina College (Univ. of South Carolina) ; during Civil War, superintendent of Confederate niter works located near college; at Univ. of California, Berkeley, professor of physics (1869-1891) and president (1869-1870 and 1876-1881). MEMBERSHIPS: Natl. Ac. Scis. SCIENTIFIC CONTRIBUTIONS: Published on wide range of scientific topics, especially medicine and physics, including work in physiology, botany, astronomy, and geophysics. Not a willing experimentalist, he tended to rely on published works of other investigators, and contributions were characterized by variety rather than by fundamental importance in any field. A notable early paper was on nervous system of alligator [see Works]; chief work in physics dealt with effects of sound on flame [see Works], providing visual means of studying effects of sound; wrote on Laplace's theory regarding velocity of sound (1864), which helped to clear up certain misunderstandings about the theory; also wrote notable papers on sound shadows in water and on color of lakes, which he explained on basis of selective reflection.

WORKS: "Experiments Illustrating the Seat of Volition in the Alligator . . . ," *New York Journal of Medicine and the Collateral Sciences* 5 (1845) : 335-47; "On the Influence of Musical Sounds on the Flame of a Jet of Coal Gas," *Am. Journ. Sci.*, 2d ser. 23 (1858) : 62-67. Chief works mentioned in *DAB* and *DSB*. Full bibliography in *Popular Science Monthly* 36 (1889), 112-13. MSS: Univ. Calif.-Ban. (see *DSB*, which also lists other locations). WORKS ABOUT: *DAB*, 11: 88-89 (W. W. Kemp) ; *DSB*, 8: 121-22 (Gisela Kutzbach). Also see John S. Lupold, "From Physician to Physicist: The Scientific Career of John LeConte," *Dissertation Abstracts International* 31 (1971) : 5327-A.

LeCONTE, JOHN EATTON, Jr. (February 22, 1784, near Shrewsbury, N.J.—November 21, 1860, Philadelphia, Pa.). *Natural history*, especially *Entomology*. Son of John Eatton, land and real estate owner and apparently man of some means, and Jane (Sloane) LeConte. Married Mary Anne H. Lawrence, ca. 1821; three children, of whom the only one to live to maturity was John L. LeConte [q.v.]. EDUCATION: 1803, A.B., Columbia College. CAREER: Inherited landed property from father; 1818-1831, topographical engineer, U.S. Army, and in 1828 attained rank of brevet major; 1831, retired in ill health and, until 1852, lived in New York City; 1852-1860, resided in Philadelphia. MEMBERSHIPS: A founder of N.Y. Lyc. Nat. Hist.; member of Am. Phil. Soc. and other societies. SCIENTIFIC CONTRIBUTIONS: Pursued studies in several areas of natural history. In 1811 published paper in *American Medical and Philosophical Register* that listed plants of New York City; while botanical interests continued, in later years zoological interests came to predominate; during career published twelve papers on plants, eight on mammals, eleven on reptiles and amphibians, five on insects, two on crustaceans, one each on birds and on medicine. During part of army service, situated in South and studied specimens from that region, and also wrote on mammals of West Africa. Best-known publication is that on North American Lepidoptera, prepared with French entomologist J. A. Boisduval [see Works] and illustrated primarily by J. Abbot [q.v.]; this was only volume of projected series that was published. In later years, gave increasing attention to study of Coleoptera, a subject pursued by son, John L. LeConte.

WORKS: *Histoire Générale et Iconographie des Lépidoptères et des Chenilles de l'Amérique Septentrionale* (Paris, 1833), with J. A. Boisduval. Early papers appeared especially in *Annals of N.Y. Lyc. Nat. Hist.* and, after 1853, in *Proceedings of Ac. Nat. Scis. Philad.* See Meisel, *Bibliog. Amer. Nat. Hist.* Watercolor and other illustrations of insects in Am. Phil. Soc. (see Rehn [below]). WORKS ABOUT: Mary Graham, "Reminiscences of Major John E. LeConte," *Pittonia* 1 (1887-1889) : 303-11; James A. G. Rehn, "The John Eatton LeConte Collection of Paintings of Insects, Arachnids, and Myriopods," *Proceedings of Am. Phil. Soc.* 98 (1954) : 442-48.

LeCONTE, JOHN LAWRENCE (May 13, 1825, New York, N.Y.—November 15, 1883, Philadelphia, Pa.). *Entomology*. Son of John Eatton [q.v.], army topographical engineer and naturalist, and Mary

Anne H. (Lawrence) LeConte; cousin of John and Joseph LeConte [qq.v.]. Married Helen S. Grier, 1861; survived by two children. EDUCATION: 1842, A.B., Mount St. Mary's College (Maryland); 1846, M.D., College of Physicians and Surgeons, New York. CAREER: Received medical degree, but inherited income made practice unnecessary; as early as 1843, began extensive travels, including California and other parts of West, Florida, Lake Superior region, Honduras and Panama, Europe, and elsewhere, studying fauna and collecting specimens; ca. 1861-1865, surgeon of volunteers, medical corps, and attained rank of lieutenant colonel and medical inspector; 1867, geologist for railroad survey, Kansas and New Mexico; 1869-1872, was in Europe and also traveled in Algeria and Egypt; 1878-1883, assistant director, U.S. Mint, Philadelphia. MEMBERSHIPS: President, Am. Assoc. Advt. Sci., 1874; member of Natl. Ac. Scis. and other societies. SCIENTIFIC CONTRIBUTIONS: Has been called most important American student of Coleoptera; first paper appeared at age nineteen; described or named nearly 5,000 beetles and did pioneer work on geographic distribution of species, including European and North American Coleoptera. Wrote on faunal and ecological relations of insects and of other animals, and produced first map of faunal regions of western United States, beginning (1851) with study of distribution of California beetles; chief contributions made to classification of Coleoptera; dealt with vast number of species and made fundamental revisions in relationship of groups. Also published on mineralogy, recent fossil mammals, ethnology, and other subjects.

WORKS: "The Rhynchophora of America, North of Mexico," *Proceedings of Am. Phil. Soc.* 15 (1876): 1-455, and "Classification of the Coleoptera of North America," *Smithson. Misc. Coll.* 26 (1883), both works with G. H. Horn [q.v.]. Reference to bibliography of works in *DAB*; index to Coleoptera described by LeConte in *Transactions of American Entomological Society* 9 (1881-1882): 197-272. MSS: Ac. Nat. Scis. Philad. (Hamer); Am. Phil. Soc. (Hamer). WORKS ABOUT: *DAB*, 11: 89-90 (Leland O. Howard); Samuel H. Scudder, "Memoir," *Biog. Mems. of Natl. Ac. Scis.* 2 (1886): 261-93; Mallis, *Amer. Entomologists*, pp. 242-48.

LeCONTE, JOSEPH (February 26, 1823, Liberty County, Ga.—July 6, 1901, Yosemite Valley, Calif.). *Natural history*; *Physiology*; *Geology*. Son of Louis, plantation owner and amateur botanist, and Ann (Quarterman) LeConte; brother of John LeConte [q.v.], and cousin to John L. LeConte and W. L. Jones [qq.v.]. Married Caroline Elizabeth Nisbet, 1847: four children. EDUCATION: 1841, A.B., Franklin College (Univ. of Georgia); 1845, M.D., College of Physicians and Surgeons, New York; 1851, S.B., Lawrence Scientific School, Harvard. HONORS: LL.D., Princeton Univ., 1896. CAREER: 1847-1850, medical practice, Macon, Ga.; ca. 1852/1853, professor of science, Oglethorpe Univ. (Atlanta, Ga.); 1853-1856, professor of natural history, Franklin College; 1857-1862 and 1866-1869, professor of chemistry and geology, College of South Carolina (Univ. of South Carolina); during Civil War, served Confederacy as government arbitrator and as chemist for Niter and Mining Bureau; 1869-1901, professor of geology and natural history, Univ. of California. MEMBERSHIPS: Am. Phil. Soc.; Natl. Ac. Scis.; and other societies. SCIENTIFIC CONTRIBUTIONS: At Harvard studied with J. L. R. Agassiz [q.v.] and with him made scientific excursions to fossil regions of New York, shores of Massachusetts, and Florida; in 1856 presented notable paper on formation of Florida peninsula [see Works]. Chief scientific work done after going to California, and covered range of subjects that included binocular vision, glycogenic functions of liver, and other physiological topics and various geological subjects, especially history and structure of mountain ranges, genesis of metals, critical periods in earth's history, lava flood of Northwest, and Ozarkian (Sierran) period. Also wrote philosophical works on evolution.

WORKS: "On the Agency of the Gulf Stream in the Formation of the Peninsula and Keys of Florida," *Am. Journ. Sci.* 2d ser. 23 (1857): 46-60; "On the Critical Periods in the History of the Earth and Their Relation to Evolution; and on the Quaternary as Such a Period," ibid. 3d ser. 14 (1877): 99-114; *Elements of Geology* (New York, 1875); *Outlines of the Comparative Physiology and Morphology of Animals* (New York, 1900). Chief articles listed in *DAB*; books listed in *DSB*. Bibliography of works in *Biog. Mems. of Natl. Ac. Scis.* 6 (1909): 212-18. MSS: Univ. N.C.-SHC (*NUCMC* 64-543); Univ. S.C.-SCL (*NUCMC* 66-567): see additional references in *DSB*. WORKS ABOUT: *DAB*, 11: 90-91 (George P. Merrill); *DSB*, 8: 122-23 (H. Lewis McKinney).

LEEDS, JOHN (May 18, 1705, Bay Hundred, Talbot County, Md.—March 1790, Wade's Point, Md.). *Mathematics*; *Astronomy*. Son of Edward and Ruth (Ball) Leeds. Married Rachel Harrison, 1726; three children. EDUCATION: Probably self-educated. CAREER: Apparently resided in Talbot County throughout life; 1734, became county commissioner and justice of peace; 1738-1777, clerk of county court; in 1760 appointed assistant or surveyor and in 1762 received full appointment, as commissioner for Maryland, in settlement of Maryland-Pennsylvania boundary question; 1766, treasurer of Eastern Shore; 1766-ca. 1777, justice of Provincial Court and naval officer of Pocomoke; ca. 1767-ca. 1777, surveyor general of Maryland; after Revolutionary War, reappointed surveyor general and served until death. SCIENTIFIC CONTRIBUTIONS: Reputation for ability in science and mathematics earned him appointment to joint commission to settle Maryland-Pennsylvania boundary, and Gov. Horatio Sharpe of Maryland considered him one of leading mathematicians in colonies; but various problems of health, lack of equipment, and Indian threat led boundary commissioners in 1763 to hire professional surveyors, Mason and Dixon, to carry out work. With watch and reflecting telescope, observed transit of Venus in 1769 and results were published by Roy. Soc. [see Works]; report was written as though figures were accurate, but these observations were of little value, since Leeds did not know exact time of event or longitude of location near Annapolis. Appointment as Maryland surveyor general was recognition of abilities in science.

WORKS: "Observation of the transit of Venus taken in Maryland by Mr. John Leeds," *Phil. Trans. of Roy. Soc.* 59 (1769): 444-45. WORKS ABOUT: *DAB*, 11: 136-37 (Mary E. Fittro); Oswald Tilghman, *History of Talbot County, Md., 1661-1861* (Baltimore, 1915); Brooke Hindle, *Pursuit of Science in Revolutionary America, 1735-1789* (Chapel Hill, N.C., 1956), pp. 156-74, passim.

LEFFMANN, HENRY (1847-1930). *Chemistry.* Professor of chemistry, Wagner Free Institute of Science and Woman's Medical College of Pennsylvania.

WORKS ABOUT: See *AMS*; *DAB*.

LEIDY, JOSEPH (September 9, 1823, Philadelphia, Pa.—April 30, 1891, Philadelphia, Pa.). *Zoology; Paleontology.* Son of Philip, successful hatmaker, and Catherine (Mellick) Leidy. Married Anna Harden, 1864; one adopted child. EDUCATION: 1844, M.D., Univ. of Pennsylvania. HONORS: Recipient of medals from Geological Society of London and Institute of France; also other honors. CAREER: 1844-1846, tried with little success to establish medical practice; 1846, demonstrator of anatomy, Franklin Medical College; 1848 and 1850, visited Europe; 1849, began course of lectures on physiology, Medical Institute of Philadelphia; at Univ. of Pennsylvania, became prosector to W. E. Horner [q.v.] (1845), lecturer on anatomy during Horner's illness (1852-1853), and professor of anatomy (1853-1891); during Civil War, surgeon, Satterlee Army Hospital, Philadelphia; 1870-1885, professor of natural history, Swarthmore College. MEMBERSHIPS: President, Ac. Nat. Scis. Philad., 1881-1891; member of Natl. Ac. Scis. and other societies. SCIENTIFIC CONTRIBUTIONS: Interests ranged over study of anatomy, vertebrate paleontology, parasitology, and protozoology, and Leidy made important contributions in all these fields. Work began in early 1840s with publications on fossil shells and anatomy of snail; in 1847, published important paper on fossil horse, and in years before 1855 studied most of important collections of vertebrate paleontology specimens from American West, making significant contributions in this field before O. C. Marsh and E. D. Cope [qq.v.] began their work; in 1869 produced major work on extinct mammalia of Dakota and Nebraska [see Works]. Interest in parasitology began in 1840s, and in 1848 began publishing studies of flora and fauna in the intestines of healthy animals; in mid-1870s, devoted attentions to investigations of freshwater protozoa. All work was characterized by great care and attention to detail and by preference for fact over theory or speculation.

WORKS: "A Flora and Fauna Within Living Animals," *Smithson. Contr. Knowl.* 5 (1853); "On the Extinct Mammalia of Dakota and Nebraska," *Journal of Ac. Nat. Scis. Philad.*, n.s. 7 (1869): 23-362; *Fresh-Water Rhizopods of North America*, Report of U.S. Geological Survey of the Territories (Washington, D.C., 1879). Other works, and bibliographies of works, in *DAB* and *DSB*; bibliography in *Biog. Mems. of Natl. Ac. Scis.* 7 (1913): 370-94. MSS: College of Physicians of Philadelphia (*NUCMC* 66-1024); Ac. Nat. Scis. Philad. (*NUCMC* 66-33). WORKS ABOUT: *DAB*, 11: 150-52 (George P. Merrill); *DSB*, 8: 169-70 (Philip C. Ritterbush).

LEMMON, JOHN GILL (January 2, 1832, Lima, Mich.—November 24, 1908, Oakland, Calif.). *Botany.* Son of William, probably farmer, and Amila (Hudson) Lemmon. Married Sara Allen Plummer, 1880; no information on children. EDUCATION: Educated in common schools; is said to have attended Michigan State Normal School and Univ. of Michigan, but name not in catalogue of graduates and nongraduates at Univ. of Michigan. CAREER: For a time, was schoolteacher, apparently in Michigan; 1862, enlisted in Union army, and during 1864-1865 was imprisoned at Andersonville Prison; 1866, in ill health, emigrated to Sierra Valley, Calif.; sent frequent communications to California newspapers recounting Civil War memories, explored botany of Southwest, and conducted small private school; 1880, moved to Oakland, Calif., and with wife devoted efforts to botanical collecting and sales and to reform activities; 1884, superintendent, California exhibit on forestry and botany, Cotton Centennial Exposition in New Orleans; 1888-1892, botanist to California State Board of Forestry; 1900-1902, councilman, Oakland, Calif. SCIENTIFIC CONTRIBUTIONS: In California, hoping to recover health, became interested in local flora and sent specimens to A. Gray [q.v.] that included two new species, one of which was named for Lemmon; pursued botanical explorations and collecting in California, Nevada, and Arizona. His enthusiastic newspaper articles played important part in spreading botanical knowledge and interest in California; promoted establishment of California State Board of Forestry and as its botanist prepared two contributions for board's biennial report, which made notable addition to literature on forestry and forest management [see Works]; later prepared popular booklets and pamphlets on trees. Widely known as botanical collector, and corresponded with leading American and foreign botanists.

WORKS: "Pines of the Pacific Slope . . . ," *Report of California State Board of Forestry* 2 (1888): 69-120, plus plate; "Cone-bearers of California. Pt. I-IV," ibid. 3 (1890): 79-212, plus plates. Also see Alfred Rehder, *The Bradley Bibliography: A Guide to the Literature of the Woody Plants of the World* . . . 5 vols. (Cambridge, Mass., 1911-1918); *Roy. Soc. Cat.*, vols. 12, 16. MSS: *DAB* account used manuscript sources (but location not given); Herbarium at Univ. of California, Berkeley (Reifschneider [below]). WORKS ABOUT: *DAB*, 11: 162-63 (Willis L. Jepson); Olga Reifschneider, *Biographies of Nevada Botanists, 1844-1963* (Reno, Nev., 1964), pp. 39-42; *Who Was Who in America*, vol. 1.

LESLEY, J. PETER (September 17, 1819, Philadelphia, Pa.—June 1, 1903, Milton, Mass.). *Geology*. Son of Peter, cabinetmaker, and Elizabeth Oswald (Allen) Lesley. Married Susan Inches Lyman, 1849; survived by two children. EDUCATION: 1838, A.B., Univ. of Pennsylvania; 1841-1844, attended theological seminary, Princeton Univ.; 1844-1845 studied several months at Halle, Germany. CAREER: 1838-1841, employed on Pennsylvania Geological Survey and, after suspension of survey in 1841, continued to assist H. D. Rogers [q.v.] with reports during vacations from studies at Princeton seminary; 1844-1845, traveled and studied in Europe; 1845-1846, preacher and distributor, American Tract Society; 1846, again became associated with H. D. Rogers, now in Boston, in preparation of maps and sections for geological reports; 1847, appointed minister, Milton, Mass.; 1852, gave up ministry and became geologist, being employed in completion of Pennsylvania survey and thereafter conducting coal and iron surveys in various parts of country for private interests; 1856-1864, secretary, American Iron Association; at Univ. of Pennsylvania, professor of mining (1859-1883) and appointed dean of Science Department (1872) and dean of Towne Scientific School (1875); librarian (1859-1885) and secretary (1859-1887), Am. Phil. Soc.; 1869-1873, editor, *U.S. Railroad and Mining Register*; 1873-1887, director, Second Geological Survey of Pennsylvania; employed as consultant by various companies. MEMBERSHIPS: Member Natl. Ac. Scis. and other societies. SCIENTIFIC CONTRIBUTIONS: Began work in geology for reasons of health, but showed great ability. *Manual of Coal* [see Works] was important for its generalized view of Appalachian coal beds and their correlation with similar regions of Europe and elsewhere; *Iron . . . Guide* [see Works] was likewise a significant publication in relation to ironworks and deposits in Pennsylvania and elsewhere. As director, paid close attention to all details of second Pennsylvania survey, and its numerous reports set new standards; but breakdown in health prevented his completion of final report.

WORKS: *A Manual of Coal and Its Topography* (Philadelphia, 1856); *The Iron Manufacturers' Guide to Furnaces, Forges and Rolling Mills of the U.S.* (New York, 1859); also, reports of Second Geological Survey of Pennsylvania. Bibliography in Nickles, *Geol. Lit. on N. Am.* MSS: Am. Phil. Soc. (*NUCMC* 60-2793); Ac. Nat. Scis. Philad. (Hamer). WORKS ABOUT: *DAB*, 11: 183-84 (George P. Merrill); *DSB*, 8: 260-61 (Martha B. Kendall). Also see Mary Lesley Ames, ed., *Life and Letters of Peter and Susan Lesley* (New York, 1909).

LESQUEREUX, LEO (November 18, 1806, Fleurier, Neuchâtel, Switzerland—October 25, 1889, Columbus, Ohio). *Botany; Paleontology*. Son of V. Aimé, manufacturer of watch springs, and Marie Anne Lesquereux. Married Sophia von Wolffskel von Reichenberg, 1830; five children. EDUCATION: Prepared at Neuchâtel, but lack of funds prevented planned enrollment in university. CAREER: Ca. 1825-ca. 1832, teacher of French at Eisenach, Saxony; ca. 1832, gave up teaching because of increasing deafness; worked as engraver of watches and joined father in manufacture of watch springs; 1844, won government-sponsored prize for essay on peat bogs; later employed by Swiss government to prepare textbook on this subject, appointed director of peat bogs, and also engaged to prepare reports on peat bogs of Germany, Sweden, Denmark, and France; 1848, political disturbances terminated government appointments, so emigrated to Boston, where employed by J. L. R. Agassiz [q.v.] in classification of plants collected around Lake Superior; 1848, moved to Columbus, Ohio; joined W. S. Sullivant [q.v.] in bryological studies and operated jewelry establishment; 1867-1872, employed in organization of fossils at Museum of Comparative Zoology, Harvard Univ.; contributed paleobotanical reports for various geological surveys. MEMBERSHIPS: Natl. Ac. Scis. and other societies. SCIENTIFIC CONTRIBUTIONS: An authority on mosses before coming to U.S. In this country became associated with W. S. Sullivant, chief American bryologist, and assisted latter with *Musci Boreali-Americani* (1856) and *Icones Muscorum* (1864); began *Manual* [see Works] with Sullivant, but completed that work with T. P. James [q.v.]. Was first in U.S. to gain eminence as student of fossil plants, and became chief authority on Appalachian coal fields; earliest work on paleobotany appeared 1854; worked on fossil plants for surveys of Kentucky (1860), Illinois (1866, 1870), Indiana (1876, 1884), and Mississippi, as well as survey by F. V. Hayden [q.v.] of Dakotas. Chief paleontological work was *Coal Flora* [see Works], prepared for Pennsylvania survey.

WORKS: *Manual of the Mosses of North America* (Boston, 1884), with T. P. James; *Description of the Coal Flora of the Carboniferous Formation in Pennsylvania and Throughout the U.S.*, Second Geological Survey of Pennsylvania: Report of Progress, P, 3 vols. in 2 (Harrisburg, Pa., 1879-1884). References to bibliographies and to manuscripts in *DSB*. MSS: See additional references in Hamer and *NUCMC*. WORKS ABOUT: *DAB*, 11: 188-89 (George P. Merrill) ; *DSB*, 8: 263-65 (Joseph Ewan).

LESUEUR, CHARLES ALEXANDRE (January 1, 1778, Le Havre, France—December 12, 1846, Le Havre, France). *Zoology*. Son of Jean-Baptiste Denis, officer of admiralty, and Charlotte Geneviève Thieullent Lesueur. No information on marital status. EDUCATION: 1787-1796, attended Royal Military School, Beaumont-en-Auge. CAREER: 1797-1799, served with national guard at Le Havre; 1800-1804, accompanied expedition sent by first consul to explore coast of Australia and Tasmania, beginning as apprentice helmsman, but talents as artist soon won position on expedition's scientific staff; through loss of scientific corps during voyage, Lesueur, naturalist François Péron, and Louis de Freycinet prepared final report of expedition; 1815-1817, as traveling companion and naturalist to W. Maclure [q.v.], explored West Indies and traveled through northeastern U.S.; 1817-1825, lived in Philadelphia, engaged in engraving and printing his own plates, taught drawing and painting, and served as curator of Ac. Nat. Scis.; 1819, employed on mapping of northeast boundary between U.S. and Canada; 1826-1837, engraver and teacher of drawing, New Harmony, Ind., and spent much time in travel; 1837-1845, resided in Paris; 1845-1846, director, Museum of Natural History, Le Havre. MEMBERSHIPS: Am. Phil. Soc.; Ac. Nat. Scis. Philad. SCIENTIFIC CONTRIBUTIONS: Australian expedition of 1800-1804 collected more than 100,000 specimens (including 2,500 new species) for Museum of Natural History at Paris, and during this time Lesueur prepared some 1,500 drawings, which are now at Le Havre. In travels with Maclure, sketched and collected fossils, fishes, mollusks, and insects and prepared notes for projected work on North American fishes, which was later begun at New Harmony, but abandoned after publication of five plates. First student of fishes of interior of North America, published numerous papers in *Journal of Ac. Nat. Scis. Philad.* on that subject, as well as on reptiles, crustaceans, and other zoological topics.

WORKS: "A New Genus of Fishes, of the Order Abdominales Proposed Under the Name of Catostomus . . . ," *Journal of Ac. Nat. Scis. Philad.* 1 (1817): 88-96, 102-11. See references to bibliographies of works in *DAB* and *DSB*. MSS: Univ. of Illinois—Illinois Historical Survey (*NUCMC* 61-1743) ; Ac. Nat. Scis. Philad. (*NUCMC* 66-56) ; also Muséums d'Histoire Naturelle, Le Havre and Paris (*DAB* and *DSB*). WORKS ABOUT: *DAB*, 11: 190-91 (David S. Jordan) ; *DSB*, 8: 266-67 (John W. Wells).

LEWIS, ENOCH (January 29, 1776, Radnor, Pa.—July 14, 1856, Philadelphia, Pa.). *Mathematics*. Son of Evan, farmer, and Jane (Meredith) Lewis. Married Alice Jackson, 1799; Lydia Jackson, 1815; fifteen children. EDUCATION: Attended school at Radnor; 1793, studied for short time in mathematical department, Friends' Academy, Philadelphia; thereafter, pursued studies independently. CAREER: Ca. 1791-1793 and 1793-1794, teacher at Radnor, Pa.; 1793, 1794, teacher in Philadelphia; 1795, assistant to A. Ellicott [q.v.] in surveys in western Pennsylvania; 1795-1799, mathematics teacher, Friends' Academy, Philadelphia; 1799-1808, mathematics teacher, Friends' Westtown Boarding School; 1808-1825, operated boarding school for mathematics, New Garden, Pa., and also engaged in farming; 1825-1827, moved boarding school to Wilmington, Del.; 1827, editor, *African Observer*, Philadelphia; ca. 1828, appointed Philadelphia city regulator and began employment as surveyor and engineer; 1834, again became teacher, Westtown Boarding School; 1836, moved to New Garden, continued connection for a time with Westtown school, and engaged in farming; 1847-1856, founder and editor, *Friends' Review*. SCIENTIFIC CONTRIBUTIONS: Mathematical activity mainly as teacher; prepared revised editions of school texts, originally written by other authors, on trigonometry and algebra; planned series of elementary mathematical texts, and published textbook on arithmetic that had several editions; also issued text on algebra and, later, one on plane and spherical trigonometry. Left unfinished a translation of Biot's treatise on physics. In later life became more and more involved in activities and concerns of Society of Friends.

WORKS: See *Natl. Un. Cat. Pre-1956 Imp.* No works listed in *Roy. Soc. Cat.* WORKS ABOUT: *DAB*, 11: 211-12 (Arthur E. Case) ; Joseph J. Lewis, *A Memoir of Enoch Lewis* (West Chester, Pa., 1882).

LILLEY, GEORGE (February 9, probably 1851, Kewanee, Ill.—June 8, 1904, Eugene, Ore.). *Mathematics.* Son of William and Harriet (Huntley) Lilley. Married Sophia Adelaide Munn, 1879; no children. EDUCATION: 1869-1873, attended Knox College (Illinois); 1872-1874, enrolled in Univ. of Michigan; 1883, Ph.D. in mathematics (nonresident program), Illinois Wesleyan Univ. HONORS: M.A. [hon.?], Washington and Jefferson College, 1878; M.A., Knox College, 1886; LL.D., Craddock College (now part of Illinois Wesleyan Univ.), 1887. CAREER: Ca. 1874-1878, apparently teacher of mathematics; ca. 1878-1880, engaged in grain and implement business, Corning, Iowa; 1880-1884, no certain information on activities during this period—he may have been at Corning, Iowa; was president from 1884 to 1886 and served until 1890 as professor of mathematics, South Dakota State College; 1891-1892, president, director of experiment station, and teacher of mathematics and elementary physics, Washington Agricultural College (now Washington State Univ.); 1894-ca. 1897, principal, Park School, Portland, Ore.; 1897-1902, professor of mathematics, Univ. of Oregon; 1902-1904, activities not known. SCIENTIFIC CONTRIBUTIONS: Much uncertainty surrounds life and career. Wrote several textbooks on mathematics, and these and other publications are said to have earned him wide reputation as mathematician.

WORKS: *Higher Algebra* (Boston and New York, 1894), 504 pp., of which first 400 pages are same as his *Elements of Algebra,* published in 1892. No works in *Roy. Soc. Cat.* MSS: Doctoral dissertation in Illinois Wesleyan Univ. Library. WORKS ABOUT: *NCAB* 25 (1936): 374 [some inaccuracies, especially in dates]; William H. Wilder, ed., *An Historical Sketch of the Illinois Wesleyan University Together With a Record of the Alumni, 1857-1895* (Bloomington, Ill., 1895), p. 143; unpublished draft biographical sketch by Roland B. Botting, Washington State Univ. Library (Archives); correspondence with several educational institutions with which Lilley was connected.

LINCECUM, GIDEON (April 22, 1793, Warren County, Ga.—November 28, 1874, near Long Point, Washington County, Tex.). *Natural history.* Son of Hezekiah, pioneer and farmer, and Sally (Hickman) Lincecum. Married Sarah Bryan, 1814; thirteen children. EDUCATION: At age fourteen spent five months in country school; studied medicine privately. CAREER: Ca. 1808-ca. 1812, clerk and tax collector, Eatonton, Ga.; served briefly in War of 1812 and then joined father in farming; 1818, resided at Tuscaloosa, Ala., where practiced medicine and engaged in sawing business and other activities; 1818, moved to location near Columbus, Miss.; 1821, appointed commissioner by Mississippi legislature to organize Monroe County; served in various civic capacities at Columbus, and until about 1830 conducted mercantile business, and engaged in Indian trade; medical practice at Cotton Gin Port (1830-1841) and at Columbus, Miss. (1841-1848); 1835, made exploring trip to Texas; 1848, moved to Long Point, Tex., continued medical practice, and acquired land; after Civil War produced and sold medicines made from plants; 1868-1873, lived at Tuxpan, Mexico; 1873, returned to Texas. MEMBERSHIPS: Corresponding member, Ac. Nat. Scis. Philad. SCIENTIFIC CONTRIBUTIONS: Interest in study of nature began in boyhood and encompassed wide range of natural phenomena as both an observer and collector of data and specimens; largely self-taught and isolated for time even from basic books on natural history. In Texas, collected fossils, shells, rocks, plants, insects, and other objects and sent specimens to scientists in North who became aware of his interests and activities in early 1860s; studied native Texas grasses, submitted weather data to Smithson. Instn. Most notable contribution to natural history was study of activities of ants, especially agricultural ants, which according to his reports engage in seed planting, an observation not yet disproved. Natural history publications related especially to insects, but also to other subjects.

WORKS: "Agricultural Ants of Texas," *Journal of Linnaean Society of London* 6 (1862): 29-31, originally read to society by Charles Darwin. Bibliography of works in Burkhalter [below], pp. 328-29. MSS: Univ. Tex.-TA (*NUCMC* 70-866). WORKS ABOUT: *DAB,* 11: 241-42 (Samuel W. Geiser); Lois W. Burkhalter, *Gideon Lincecum* (Austin, Tex., 1965).

LINDENKOHL, ADOLPH (March 6, 1833, Niederkaufungen, Hesse Cassel, Germany—June 22, 1904, Washington, D.C.). *Cartography; Oceanography.* Son of George C. F. and Anna Elizabeth (Krug) Lindenkohl. Married Pauline Praeger, 1872; six children. EDUCATION: 1844-1849, attended *realschule* at Cassel; 1852, graduated Polytechnische Schule, Cassel. CAREER: 1852-1854, came to U.S. and taught in York, Pa., and in Washington, D.C.; 1854-1904, cartographic draftsman, U.S. Coast and Geodetic Survey, achieving rank of senior draftsman; 1862-1864, did topographic and cartographic work for army. MEMBERSHIPS: Am. Assoc. Advt. Sci.; Natl. Geog. Soc.; and other societies. SCIENTIFIC CONTRIBUTIONS: Work for Coast Survey was chiefly in preparation of hydrographic charts; an authority on mathematical principles of projections and in this line made several notable contributions to cartography. On own initiative, pursued broader questions relating especially to physical geography, oceanography, and temperature, density, and currents of oceans; studied submarine channel of Hudson River and accounted for it in terms of glacial action; wrote on specific gravity of Gulf of Mexico and Gulf Stream and on salinity and temperature of Pacific Ocean.

WORKS: "Geology of the Sea-Bottom in the Approaches to the New York Bay," *Am. Journ. Sci.* 3d ser. 29 (1885) : 475-80; "Resultate der Temperatur- und Dichtigkeitsbeobachtungen in den Gewässern des Golfstroms und des Golfs von Mexico durch das Bureau des U.S. Coast and Geodetic Survey," *Petermanns Geographische Mitteilungen* 42 (1896) : 25-29. Wrote comparatively few papers, published chiefly in *Annual Reports of Coast Survey, Am. Journ. Sci.,* and *Petermanns Mitteilungen.* See *Roy. Soc. Cat.,* vols. 10, 16. WORKS ABOUT: *DAB,* 11: 272-73 (H. A. Marmer) ; Herbert G. Ogden, obituary, *Bulletin of Phil. Soc. Wash.* 14 (1900-1904) : 296-99.

LINDGREN, WALDEMAR (1860-1939). *Geology.* Chief geologist, U.S. Geol. Surv.; professor and head of Geology Department, Massachusetts Institute of Technology.

MSS: M.I.T.-Arch. (DSB). WORKS ABOUT: See *AMS; DAB,* supp. 2; *DSB.*

LINDHEIMER, FERDINAND JACOB (May 21, 1801, Frankfurt am Main, Germany—December 2, 1879, New Braunfels, Comal County, Tex.). *Botany.* Son of Johann H. Lindheimer, successful merchant. Married Eleanore Reinarz, 1846; at least one child. EDUCATION: Probably educated at gymnasium in Frankfurt; 1825-1827, attended Univ. of Bonn. CAREER: 1827-1833, teacher, Georg Bunsen's preparatory school, Frankfurt; 1834, arrived with Bunsen at Belleville, Ill.; 1834-ca. 1836 [sixteen months], in Mexico, where managed distillery and oversaw banana and pineapple plantation; 1836, joined Texans in fight for independence; activities immediately after Texas war uncertain; in 1839, on trip to New Orleans and St. Louis collected specimens at San Felipe for G. Engelmann [q.v.], friend from university days, and winter of 1839-1840 was spent with Engelmann in St. Louis; 1840-1842, made unsuccessful try at farming near Houston; 1842-1851, collected botanical specimens in Texas for Engelmann and A. Gray [q.v.]; ca. 1845, established home in newly founded colony of New Braunfels; 1852-ca. 1870, editor, *Neu Braunfelser Zeitung;* conducted free private school and served as county superintendent of education and justice of peace. SCIENTIFIC CONTRIBUTIONS: Began natural history collecting while in Mexico (1834-ca. 1836) and there gathered numerous specimens of insects and plants. Under terms of collecting arrangements in force during period 1842-1851, gathered plants in sets, which were named and mounted by Gray and Engelmann; proceeds from sales went to Lindheimer; specimens so gathered were of great scientific interest, and twenty plants honor Lindheimer's name. Published essays of scientific, philosophical, and historical interest in *Neu Braunfelser Zeitung* and *New Yorker Staats-Zeitung*; former student Gustav Passavant gathered selections of writings for *Aufsätze* [see Works], which included detailed descriptions of Mexican plants and animals based on observations and collecting in mid-1830s.

WORKS: "Plantae Lindheimerianae," *Boston Journal of Natural History* 5 (1845): 210-64, and 6 (1850): 141-240, by A. Gray and G. Engelmann; *Aufsätze und Abhandlungen von Ferdinand Lindheimer in Texas* (Frankfurt, 1879). MSS: Mo. Bot. Gard. and Harv. U.-GH (Geiser, *Naturalists* [below]). WORKS ABOUT: *DAB,* 11: 273-74 (Samuel W. Geiser) ; Geiser, *Naturalists of the Frontier,* 2d ed. ([Dallas, Tex.], 1948), pp. 132-47.

LINING, JOHN (April, 1708, Walston, Scotland—September 21, 1760, [Charleston?], S.C.). *Physiology; Meteorology; Electricity.* Son of Thomas, Presbyterian minister, and Ann (Hamilton) Lining. Married Sarah Hill, 1739; eleven children. EDUCATION: Probably received medical training through apprenticeship (did not attend Edinburgh Univ.). CAREER: 1730, emigrated to America and settled at Charleston, S.C.; began as pharmacist and soon turned to medical practice; ca. 1737-1740, held appointment as parish doctor and affiliated with St. Philip's Hospital; 1739, married into notable South Carolina family, thereafter lived in Charleston and on Hill plantation at Hillsborough; mid-1750s, left medical practice and retired to Hillsborough. SCIENTIFIC CONTRIBUTIONS: Interested in effects of weather on bodily functions as means of explaining origin of epidemic diseases; began weather records as early as 1737, and about 1740 began sending reports to Roy. Soc. and continued until 1753; these observations were first continuing record of American weather. For one year, kept record of own weight, pulse, food and water, and weight of excretions, as well as daily record of weather, using recently introduced Fahrenheit thermometer; also recorded humidity, rainfall, wind, and clouds; metabolic and meteorological record that resulted from this study was published by Roy. Soc. in early 1740s [see Works] and was first published record of colonial weather. In 1756, published earliest description of yellow fever in North America, in the Edinburgh Philosophical Society's *Essays and Observations, Physical and Literary.* Also interested in electricity and pursued ideas of B. Franklin [q.v.], but made no new contributions to subject.

WORKS: ". . . Account of Statical Experiments Made Several Times a Day Upon Himself, for One Whole Year, Accompanied with Meteorological Observations; To Which are Subjoined Six General Tables,

Deduced from the Whole Year's Course," *Phil. Trans. of Roy. Soc.* 42 (1742-1743) : 491-509, and 43 (1744-1745) : 318-30. Bibliography in Mendelsohn [below], p. 292, and also references in *DAB*. WORKS ABOUT: *DAB,* 11: 280-81 (Franklin C. Bing) ; Everett Mendelsohn, "John Lining and His Contribution to Early American Science," *Isis* 51 (1960) : 278-92.

LINTNER, JOSEPH ALBERT (February 8, 1822, Schoharie, N.Y.—May 5, 1898, Rome, Italy). *Entomology.* Son of Rev. George Ames and Maria (Wagner) Lintner. Married Frances C. Hutchinson, 1856; no information on children. EDUCATION: 1837, graduated, Schoharie Academy. HONORS: Ph.D., State Univ. of New York, 1884. CAREER: Ca. 1837, began business career in New York City; 1848-1860, engaged in business at Schoharie; 1860-1868, woolens manufacturer, Utica, N.Y.; in 1868 became assistant in zoology and in 1874 took charge of entomological department, New York State Museum; 1880-1898, New York state entomologist. MEMBERSHIPS: President, Association of Economic Entomologists, 1892; member of other societies. SCIENTIFIC CONTRIBUTIONS: Began writings while in New York City, preparing occasional items for *New York Tribune.* Undertook collecting of insects in 1853 as aid to A. Fitch [q.v.] in entomological survey for New York state and in 1862 published first paper on subject. Writings were of practical nature and appeared in newspapers as well as in official publications; these works were based on some personal observations in field, but depended more generally on compilation and correspondence; yet they were carefully done and were well thought of. For twenty-five years, served as entomological editor for *Country Gentleman.* Chief contributions appeared in thirteen annual reports on injurious and other insects, issued as entomologist of state of New York.

WORKS: Bibliography in E. P. Felt, "Memorial of Life and Entomologic Work of Joseph A. Lintner," *Bulletin of New York State Museum* 5 (1899) : 303-611. WORKS ABOUT: *DAB,* 11: 283-84 (Leland O. Howard) ; Mallis, *Amer. Entomologists,* pp. 52-54.

LIPPINCOTT, JAMES STARR (April 12, 1819, Philadelphia, Pa.—March 17, 1885, Greenwich, Cumberland County, N.J.). *Horticulture; Meteorology.* Son of John and Sarah West (Starr) Lippincott. Married Susan Haworth Ecroyd, 1857; Anne E. Sheppard, 1861; no children. EDUCATION: 1834-1835, attended Haverford College. CAREER: Worked first as teacher, but later took up farming at Cole's Landing near Haddonfield, N.J.; became interested in agricultural science; 1868, moved to Haddonfield and pursued study and writing on subjects relating to agriculture and other matters. SCIENTIFIC CONTRIBUTIONS: As farmer, devised so-called vapor index for measurement of humidity. At Haddonfield maintained meteorological records; for Smithson. Instn., tabulated and reduced meteorological observations made during period 1856-1861 by Benjamin Sheppard at Greenwich, N.J., and served as Smithsonian meteorological observer at Cole's Landing (1864-1866) and at Haddonfield (1869-1870). Contributed articles on climate and grapevines, geography of plants, atmospheric humidity, fruit regions of northern U.S. and their local climates, and other subjects for reports of U.S. commissioner of agriculture (1862-1866) ; published other items in various horticultural and agricultural journals, and wrote articles on botany and geology for *Friend* (Philadelphia). Also engaged in other literary activities, including genealogical research and work on Lippincott's *Universal Pronouncing Dictionary of Biography and Mythology* (1870).

WORKS: References in *DAB* to articles appearing in commissioner of agriculture's reports. WORKS ABOUT: *DAB,* 11: 286-87 (Warren B. Mack).

LLOYD, JOHN URI (1849-1936). *Chemistry; Pharmacy; Materia medica.* Professor of chemistry and president, Eclectic Medical Institute (Cincinnati) ; president, Lloyd Brothers, Pharmacists, Inc.

MSS: Lloyd Library and Museum, Cincinnati (Hamer) ; also see *DSB.* WORKS ABOUT: See *AMS; DAB,* supp. 2; *DSB.*

LOCKE, JOHN (February 19, 1792, Lempster, N.H.—July 10, 1856, Cincinnati, Ohio). *Botany; Geology; Physics.* Son of Samuel Barron, farmer and millwright, and Hannah (Russell) Locke. Married Mary Morris, 1825; ten children. EDUCATION: Ca. 1815-1816, attended chemical lectures and acted as laboratory assistant to B. Silliman [q.v.], Yale College; 1816-1818, studied medicine privately in Keene, N.H.; 1819, M.D., Yale. CAREER: 1816-1818, delivered botanical lectures at various New England schools and colleges; ca. 1817, curator at Harvard Botanical Garden; 1818, appointed assistant surgeon, U.S. Navy, but planned exploration of Columbia River was canceled after ship was wrecked; became associated with girls' academy, Windsor, Vt.; 1821, established academy at Lexington, Ky; 1823-1835, conducted Locke's Female Academy, Cincinnati, Ohio; 1835-1837, in Europe and England; 1835-1853, professor of chemistry and pharmacy, Ohio Medical College (Cincinnati) ; after 1853, conducted preparatory school, Lebanon, Ohio; connected with Ohio Geological Survey and also associated with geological activities of U.S. government. MEMBERSHIPS: Association of American Geologists; Am. Assn. Advt. Sci. SCIENTIFIC CONTRIBUTIONS: Earliest interests were in botany, and *Outlines of Botany* (1819) included more than 200 illustrations drawn and engraved by author. During period 1835-1840, conducted investigations in geology and paleontology, including work for Ohio state survey, and study of mineral regions of Iowa, Illinois, and Wisconsin was published in 1840 by U.S. Congress; about 1838, began studies in terrestrial magnetism, making observations at various locations in northern U.S., and also made geological observations at those places; later years devoted to these geophysical studies and to electricity. Talented inventor and instrument maker and as such most notable contribution was electro-chronograph; this device was adopted by U.S. Coast Survey in 1848 and greatly improved accuracy of longitude determinations.

WORKS: "Observations Made in the Years 1838-1843, to Determine the Magnetical Dip and the Intensity of Magnetical Force, in Several Parts of the U.S.," *Transactions of Am. Phil. Soc.* n.s. 9 (1846): 283-328; "On the Electro-chronograph," *Am Journ. Sci.* 2d ser. 8 (1849): 231-52. Bibliography in *American Geologist* 14 (1894): 354-56. WORKS ABOUT: *DAB*, 11: 337-38 (Carl C. Mitman); Adolph E. Waller, "Dr. John Locke, Early Ohio Scientist," *Ohio Archaeological and Historical Quarterly* 55 (1946): 346-73.

LOGAN, JAMES (October 20, 1674, Lurgan, Ireland—October 31, 1751, Germantown, Pa.). *Botany; Optics.* Son of Patrick, clergyman and schoolmaster, and Isabel (Hume) Logan. Married Sarah Read, 1714; five children. EDUCATION: In early years, taught by father; largely self-educated. CAREER: Ca. 1694 [age nineteen], put in charge of school at Bristol, England, when father, school's master, returned to Ireland; ca. 1697-1699, involved in shipping trade between Dublin and Bristol; 1699, became secretary to William Penn and went to Pennsylvania; 1701, given charge of Penn family interests when William returned to England; 1701-1717, secretary of province and clerk of Provincial Council of Pennsylvania; 1702-1747, member of council, and became its president; in 1717 elected to board of alderman and during years 1722-1723 served as mayor, Philadelphia; 1731-1739, chief justice, Supreme Court of Pennsylvania; 1736-1737, acting governor of Pennsylvania; acquired wealth from investment in land and in Indian trade. SCIENTIFIC CONTRIBUTIONS: In America built up large library, acquired notable level of knowledge in mathematics, natural history, and astronomy, and owned and used telescope; with such background and facilities, aided other colonial scientists, especially T. Godfrey, J. Bartram, B. Franklin, and C. Colden [qq.v.]. Contributions to *Phil. Trans. of Roy. Soc.* included letters on T. Godfrey's quadrant, paper on apparent increase in size of sun and moon near horizon, and paper on angular appearance of lightning. Most notable scientific work related to pollination of plants, beginning experiments on Indian corn in ca. 1727, with preliminary report to Roy. Soc. in 1735; work on plants was received with considerable favor among botanists on continent of Europe. Also published work on optics.

WORKS: *Experiments and Considerations on the Generation of Plants* (London, 1747), translation of Latin edition, which was published in Leyden, 1739. Works published in *Phil. Trans. of Roy. Soc.* are listed in *DAB;* other works mentioned in *DSB*. MSS: Hist. Soc. Penn. (Hamer); extensive library in Lib. Co. Philad. WORKS ABOUT: *DAB*, 11: 360-62 (Marion P. Smith); *DSB*, 8: 459-61 (Edwin Wolf II). Also see Frederick B. Tolles, *James Logan and the Culture of Provincial America* (Boston, 1957).

LOOMIS, ELIAS (August 7, 1811, Willington, Conn.—August 15, 1889, New Haven, Conn.). *Meteorology; Mathematics; Astronomy.* Son of Hubbel, Baptist clergyman, and Jerusha (Burt) Loomis. Married Julia Elmore Upson, 1840; two children. EDUCATION: 1830, A.B., Yale College; 1831-1833, attended Andover Theological Seminary; 1836-1837, studied science in Paris. CAREER: 1830-1831, teacher of mathematics, academy near Baltimore; 1833-1836, tutor, Yale College; 1836-1844, professor of mathematics and natural philosophy, Western Reserve College; 1844-1847 and 1849-1860, professor of mathematics and natural philosophy, Univ. of City of New York (New York Univ.); 1848, at Princeton Univ.; 1860-1889, professor of natural philosophy and astronomy, Yale Univ. MEMBERSHIPS: Natl. Ac. Scis. and other societies. SCIENTIFIC CONTRIBUTIONS: Active in several branches of science. As tutor at Yale did work in astronomy (including rediscovery and computation of orbit of Halley's comet) with Prof. D. Olmsted [q.v.]. Exerted chief influence among contemporaries through series of textbooks on natural philosophy, astronomy, meteorology, analytic geometry, calculus, and other topics, which in translation helped make Western mathematics better known in China and regions of Middle East. During years 1859-1861, published series of papers on aurora borealis in *Am. Journ. Sci.*, and produced first map of frequency distribution of auroras; work on auroras was related to lifelong interest in geomagnetism that began as early as 1833; investigations in astronomy related mainly to meteors and longitude and latitude determinations. Most notable studies were in meteorology; with first "synoptic" map, published in 1846, he introduced new way of representing weather data, which had considerable influence on development of means of weather prediction and in work on theories of storms; after 1871 and the availability of daily weather maps, undertook detailed statistical studies of cyclones and anticyclones.

WORKS: "Contributions to Meteorology," twenty-three papers in *Am. Journ. Sci.* (1874-1889). Bibliography in *Biog. Mems of Natl. Ac. Scis.* 3 (1895): 241-52. MSS: Yale Univ.—Beinecke Lib. (*DSB*). WORKS ABOUT: *DAB*, 11: 398-99 (David E. Smith); *DSB*, 8: 487 (Gisela Kutzbach).

LOOMIS, MAHLON (July 21, 1826, Oppenheim, N.Y.—October 13, 1886, Terre Alta, W.Va.). *Electricity.* Son of Prof. Nathan—astronomer, educator, civil engineer, and public official—and Waitie Jenks (Barber) Loomis. Married Achsah Ashley, 1856; one child. EDUCATION: Attended district school near Springvale, Va.; 1848-ca. 1851, studied with dentist, Cleveland, Ohio; largely self-taught. CAREER: Ca. 1843-ca 1846, teacher, Springvale, Va.; 1848-ca. 1851, studied dentistry and did some teaching, Cleveland; ca. 1851, began dental practice, Earlville, N.Y.; for a time, practiced dentistry in Cambridgeport, Mass.; ca. 1852, apparently moved to West Springfield, Mass., and later practiced dentistry in Philadelphia; 1860, established dental practice at Washington, D.C., and about that time began experimental work in electricity; 1873, Loomis Aerial Telegraph Co. chartered by U.S. Congress, but financial support never materialized; ca. 1877, began several years' employment as mineralogist for Great Magnetic Iron Ore Co. of Mt. Athos, at Lynchburg, Va.; during last years, practiced dentistry at Chicago for more than a year, went to Parkersburg, W. Va., and in 1884 moved with brother to Terra Alta, W. Va. SCIENTIFIC CONTRIBUTIONS: Interest in electricity began in early 1850s, and among early work was burial of metal plates attached to batteries as experiment in forcing growth of plants. Now acknowledged as discoverer and inventor of wireless telegraphy, although series of financial circumstances prevented commercial development of idea. In ca. 1860, turned attention to study of electrical charges obtainable from upper atmosphere through use of kites with metal wires, using this electrical source as substitute for battery. In 1864, produced drawing of system of wireless communication, and in 1866 experimentally verified two-way wireless communication over distance of eighteen miles, between two mountain peaks in Virginia; in 1868 demonstrated experiment for members of Congress and leading scientists, and in 1872 received first U.S. wireless patent, entitled "Improvement in Telegraphing," a general patent for system of aerial communication rather than for specific apparatus.

WORKS: Appleby [below], p. 70, lists patents granted and also reproduces address delivered at Franklin Institute, 1881, though apparently written in 1872. MSS: Lib. Cong. (*NUCMC* 62-4526); extant apparatus at Smithson. Instn. (Appleby [below]). WORKS ABOUT: *DAB*, 11: 399-400 (George H. Clark); Thomas Appleby, *Mahlon Loomis, Inventor of Radio* (Washington, D.C., 1967); Otis B. Young, "Mahlon Loomis, the Discoverer and Inventor of Radio," committee report, *Transactions of Illinois State Academy of Science* 60 (1967): 3-8.

LOVERING, JOSEPH (December 25, 1813, Charlestown, Mass.—January 18, 1892, Cambridge, Mass.). *Physics; Mathematics.* Son of Robert, town official, and Elizabeth (Simonds) Lovering. Married Sarah Gray Hawes, 1844; four children. EDUCATION: 1833, A.B., Harvard College; 1834-1836, attended Harvard Divinity School. HONORS: LL.D., Harvard Univ., 1879. CAREER: 1833-1834, teacher in private school, Charlestown; at Harvard Univ., instructor in mathematics (1834-1838), tutor and lecturer in natural philosophy (1836-1838), Hollis professor of mathematics and natural philosophy (1838-1888), and regent (1857-1869); beginning in 1840, gave nine separate courses of lectures for Lowell Institute, and also lectured in Washington, Baltimore, and other locations; 1867-1876, associated with U.S. Coast Survey, engaged in work on computations for longitude. MEMBERSHIPS: President (1873) and permanent secretary (1854-1873), Am Assoc. Advt. Sci.; member of Natl. Ac. Scis. and other societies. SCIENTIFIC CONTRIBUTIONS: Noted chiefly as educator and as public lecturer on science; edited several publications, including number of volumes of proceedings and memoirs of Am. Assoc. Advt. Sci. and Am. Ac. Arts Scis. Did little work in experimental physics, but was active in observation and correlation of data and was much involved in study of terrestrial magnetism as coordinated by Roy. Soc. Chief scientific work was on periodicity of aurora, which involved study, correlation, and discussion of numerous observations [see Works].

WORKS: "On the Periodicity of the Aurora Borealis," *Memoirs of Am. Ac. Arts Scis.*, n.s. 10, pt. 1 (1868), pp. 9-351. Bibliography in Peirce [below], pp. 339-44. MSS: Miscellaneous items in several collections in Harv. U.-HL and Harv. U.-UA. WORKS ABOUT: *DAB*, 11: 442-43 (Edwin H. Hall); B. Osgood Peirce, "Memoir," *Biog. Mems. of Natl. Ac. Scis.* 6 (1909): 327-44.

LYMAN, BENJAMIN SMITH (1835-1920). *Geology; Surveying; Mining engineering.* Geologist and mining engineer to Japanese government; assistant geologist, Pennsylvania Geological Survey; engaged in private geological work.

MSS: Hist. Soc. Penn. (*NUCMC* 61-129); Am.

Phil. Soc. (*NUCMC* 61-1124); Forbes Library, Northampton, Mass. (*NUCMC* 62-3746). WORKS ABOUT: See *AMS; DAB; DSB.*

LYMAN, CHESTER SMITH (January 13, 1814, Manchester, Conn.—January 29, 1890, New Haven, Conn.). *Astronomy; Geology.* Son of Chester, farmer and/or miller, and Mary (Smith) Lyman. Married Delia Williams Wood, 1850; no information on children. EDUCATION: 1837, A.B., Yale College; 1839-1840, attended Union Theological Seminary; 1842, graduated, Yale Divinity School. HONORS: M.A., Beloit College, 1864. CAREER: 1830-1832, school teacher, Manchester; 1837-1839, superintendent, Ellington Academy; 1843-1845, Congregational minister, New Britain, Conn.; left ministry for reasons of health, sailed around Cape Horn to Hawaii, and there worked as missionary, teacher, and surveyor; 1847, went to California, where worked as surveyor and gold digger; 1850, returned to Connecticut; engaged for time in preparation of scientific terms for Webster's *Dictionary;* at Sheffield Scientific School, Yale, professor of industrial mechanics and physics (1859-1872), professor of physics and astronomy (1872-1884), and professor of astronomy (1884-1889). MEMBERSHIPS: Vice-President, Am. Assoc. Advt. Sci., 1874; president, Connecticut Academy of Arts and Sciences, 1859-1877. SCIENTIFIC CONTRIBUTIONS: During years in Hawaii and California, wrote letters to *Am. Journ. Sci.,* including what is thought to be first informed report to reach East of discovery of gold in California. Invented combination transit and zenith instrument for latitude determination, and also devised other apparatus. Through observations made at Yale Observatory in 1866 and 1874, was first astronomer to present convincing evidence of atmosphere around planet Venus.

WORKS: Observations on Venus reported in *Am. Journ. Sci.,* 2d ser. 43 (1867): 129-30, and 3d ser. 9 (1875): 47-48. See *Roy. Soc. Cat.,* vols. 4, 8, 10. WORKS ABOUT: *DAB,* 11: 515-16 (Frank A. Taylor); *DSB,* 8: 577-78 (Sally H. Dieke).

LYMAN, THEODORE (August 23, 1833, Waltham, Mass.—September 9, 1897, Nahant, Mass.). *Zoology.* Son of Theodore—man of wealth, author, and Boston mayor—and Mary Elizabeth (Henderson) Lyman. Married Elizabeth Russell, 1856; at least one child. EDUCATION: 1855, A.B., Harvard College; 1858, B.S., Lawrence Scientific School, Harvard. HONORS: LL.D., Harvard Univ., 1891. CAREER: While in Lawrence Scientific School, studied with J. L. R. Agassiz [q.v.] and accompanied research expedition to Florida; 1859, elected trustee, Museum of Comparative Zoology, Harvard; 1861-1863, engaged in scientific activities in Europe, including collection of specimens for museum; 1863-1865, aide-de-camp to Gen. George Meade; at Museum of Comparative Zoology, assistant in zoology (1863-1887) and treasurer (1865-1872 and 1874-1876); 1866-ca. 1883, chairman, Massachusetts Commission of Inland Fisheries; 1868-1880 and 1881-1888, an overseer, Harvard Univ.; 1882, elected to single term in U.S. Congress; last years spent as invalid. MEMBERSHIPS: Natl. Ac. Scis.; president, American Fish Cultural Association, 1884. SCIENTIFIC CONTRIBUTIONS: Had early interest in ornithology, and wrote papers on coral. Through extensive studies and descriptions of new species of Ophiuridae (serpent stars), including preparation of reports on specimens collected by several deep-sea exploring expeditions, became authority on that group of echinoderms.

WORKS: See bibliography in Bowditch [below], pp. 151-53. MSS: Harv. U.-MCZ; Mass. Hist. Soc. WORKS ABOUT: *DAB,* 11: 519 (H. M. Varrell); H. P. Bowditch, "Memoir," *Biog. Mems. of Natl. Ac. Scis.* 5 (1905): 141-53.

M

MABERY, CHARLES FREDERIC (1850-1927). *Chemistry.* Professor of chemistry, Case School of Applied Science (Case Institute of Technology).

WORKS ABOUT: See *AMS; DAB.*

McARTHUR, WILLIAM POPE (April 2, 1814, St. Genevieve, Mo.—December 23, 1850, Panama). *Hydrography.* Son of John and Mary (Linn) McArthur. Married Mary Stone Young, 1838; at least three children. EDUCATION: 1832, appointed midshipman, U.S. Navy; 1837, attended naval school at Norfolk. HONORS: U.S. government built a schooner named *McArthur* to be used in Pacific Coast Survey, 1876. CAREER: 1832, entered U.S. Navy as midshipman; spent several years in South Pacific Station; 1837-1838, commanded a ship in Everglades expedition during Seminole War and was wounded; 1838-1840, served on several ships; 1840, assigned to service with Coast Survey and took part in surveys of Gulf Coast and other areas; 1841, promoted to lieutenant; 1848, given command of hydrographic survey of Pacific Coast; 1849, arrived in San Francisco by way of Panama; 1849-1850, with vessel *Ewing,* traveled to Hawaii; 1850, conducted survey of Pacific Coast northward from San Francisco, and same year died at Panama on return trip to Washington. SCIENTIFIC CONTRIBUTIONS: In California, chose Mare Island as navy yard. Is remembered for first reconnaissance of Pacific Coast, from Monterey to Columbia River, and for initial survey of entrance to that river; chief result of Pacific Coast work was reconnaissance chart [see Works]. At time of death was on return eastward to examine ship contracted for by Coast Survey and to plan for further survey work on Pacific Coast; assisted on Pacific Coast by Lt. Washington A. Bartlett.

WORKS: *Notices of the Western Coast of the United States, U.S. Coast Survey* (Washington, D.C., 1851), 24 pp. (McArthur [below], pp. 263-72, includes synopses or reprints of the eight notices); chart (three sheets) of Pacific Coast reconnaissance accompanied *Annual Report of U.S. Coast Survey,* 1851. WORKS ABOUT: *DAB,* 11: 552-53 (Frank E. Ross); Lewis A. McArthur, "The Pacific Coast Survey of 1849 and 1850," *Oregon Historical Society Quarterly* 16 (1915): 246-74.

MacBRIDE, THOMAS HUSTON (1848-1934). *Botany; Geology.* Professor of botany and president, State Univ. of Iowa.

WORKS ABOUT: See *AMS; NCAB,* 11 (1909 [copyright 1901 and 1909]): 473; *Who Was Who in America,* vol. 1.

McCALLEY, HENRY (February 11, 1852, Huntsville, Ala.—November 21, 1904, Huntsville, Ala.). *Geology.* Son of Thomas Sanford, farmer [?], and Caroline (Landford) McCalley. Never married. EDUCATION: 1875, received B.S. degree and graduated in civil and mining engineering, Univ. of Virginia. HONORS: A.M., Univ. of Alabama, 1878. CAREER: 1876, after college, returned home and engaged in farming; ca. 1877-1878, teacher, Demopolis, Ala.; 1878-1883, assistant in chemistry, Univ. of Alabama, and volunteer assistant on state's geological survey; 1879 (summer), directed work on levelings and soundings as part of U.S. War Office survey of Warrior River; 1883-1904, assistant state geologist, Alabama. MEMBERSHIPS: Am. Assoc. Advt. Sci.; Geol. Soc. Am.; and other societies. SCIENTIFIC CONTRIBUTIONS: Work on Alabama survey related especially to coal and iron locations; in connection with that work in 1886 published report on Warrior basin that gave first systematic account of geology of coal in that area. Studied Tennessee, Coosa, and other valleys where limestones, iron ores, and bauxites are found, and later again conducted investigations in Warrior region and elsewhere.

WORKS: Writings appeared especially in Geological Survey of Alabama reports, local newspapers, and mining publications. See bibliography in Smith [below], pp. 200-201. WORKS ABOUT: Eugene A.

Smith, "Biographical Sketch," *American Geologist* 35 (1905): 197-201; obituary, *Science,* n.s. 20 (1904): 773-74.

McCAY, CHARLES FRANCIS (March 8, 1810, Danville, Pa.—March 13, 1889, Baltimore, Md.). *Mathematics; Actuarial science.* Son of Robert and Sarah (Read) McCay. Married Narcissa Harvey Williams, 1840; had children. EDUCATION: 1829, graduated, Jefferson College. HONORS: Said to have held LL.D. degree. CAREER: 1832-1833, taught mathematics, natural philosophy, and astronomy, Lafayette College; 1833-1853, professor of mathematics, Univ. of Georgia; 1846-1853, agent for Mutual Life Insurance Co. of New York; 1848-1855, actuary to life department, Southern Mutual Insurance Co., Georgia; at South Carolina College (Univ. of South Carolina), professor of mathematics (1853-1857) and president (1855-1857); subsequently entered into insurance business at Augusta, Ga., where served as president of bank and made small fortune; 1869, moved to Baltimore; during period 1848-1889, served at times as actuarial consultant to number of life insurance companies, as actuary to several organizations, and from 1871 to 1889 as actuary for Maryland Insurance Department. MEMBERSHIPS: Am. Assoc. Advt. Sci. SCIENTIFIC CONTRIBUTIONS: Prepared textbook on differential and integral calculus (1839). Did early work in preparation of mortality tables, and ca. 1859 his Southern Mutual Mortality Table was adopted for purposes of valuation by state of Georgia, reportedly the first such table adopted by any state; drawing on background of several insurance companies, in 1887 produced first "select and ultimate" life insurance mortality table in U.S.

WORKS: "On the Laws of Human Mortality," *Proceedings of Am. Assoc. Advt. Sci.* 9 (1856): 21-29; two papers in *Journal of Institute of Actuaries* (London), vols. 16 (1872) and 22 (1881), on tables of mortality (see *Roy. Soc. Cat.,* vol. 16); also see *Natl. Un. Cat. Pre-1956 Imp.* WORKS ABOUT: *DAB,* 11: 577-78 (Edwin W. Kopf).

McCLINTOCK, EMORY (1840-1916). *Mathematics; Actuarial science.* Actuary to Northwestern Life Insurance Co. (Milwaukee) and Mutual Life Insurance Co. (New York).

WORKS ABOUT: See *AMS; DAB.*

McCULLOH, RICHARD SEARS (March 18, 1818, Baltimore, Md.—September 15, 1894, Oldfields, Glencoe, Md.). *Chemistry; Physics.* Son of James William, banker and farmer, and Abigail Hall (Sears) McCulloch. Married Mary Stewart Vowell, 1845; one child. EDUCATION: 1836, A.B., Princeton Univ.; 1838-1839, studied chemistry in laboratory of J. C. Booth [q.v.], Philadelphia. HONORS: LL.D., Washington and Lee Univ., 1878. CAREER: For period of time before 1840, served as observer in Magnetic Observatory, Girard College (Philadelphia); 1841-1843, professor of natural philosophy, mathematics, and chemistry, Jefferson College (Pennsylvania); 1846-1849, melter and refiner, U.S. Mint, Philadelphia; 1849-1854, professor of natural philosophy, Princeton Univ.; at Columbia College, professor of natural and experimental philosophy and chemistry (1854-1857) and professor of physics (1857-1863); 1863, joined Confederacy and became consulting chemist, Confederate Nitre and Mining Bureau; 1866-1877, professor of natural philosophy, Washington and Lee Univ.; 1877-1888, professor of general and agricultural chemistry and of physics, Louisiana State Univ. MEMBERSHIPS: Am. Phil. Soc. SCIENTIFIC CONTRIBUTIONS: At U.S. Mint, devised improved method for refining California gold, through use of zinc, and published several pamphlets attempting to substantiate related patent claims. In 1844, published *Plan of Organization for the Naval Observatory,* prepared at request of secretary of navy. Chief scientific work was analysis of sugars, molasses, and the like, done for secretary of treasury, partly performed in Cuba and Louisiana; resulting papers, collected and published in substantial volume [see Works], constituted significant summary of knowledge regarding sugar; later (1879-1880), again took up experimental work on sugar, but no report was prepared. Assisted J. C. Booth in compiling *Encyclopedia of Chemistry* (1850). During Civil War, prepared combustible intended for destruction of U.S. military property.

WORKS: . . . *Report of Chemical Analyses of Sugars, Molasses, Etc., and of Researches on Hydrometers, Made Under the Superintendence of Prof. A. D. Bache* [q.v.], Senate Executive Document no. 50, 30th Congress, 1st session [Washington, 1848], 653 pp.; *Treatise on the Mechanical Theory of Heat and Its Applications to the Steam Engine* (New York, 1876). See references to other works in Thomas [below]. WORKS ABOUT: Milton H. Thomas, "Professor McCulloh . . . ," *Princeton Univ. Library Chronicle* 9 (1947): 17-29; Guralnick, *Science and American College,* pp. 199-200.

MacFARLANE, JOHN MUIRHEAD (1855-1943). *Botany.* Professor of botany and director of department, director of botanical garden, Univ. of Pennsylvania.

WORKS ABOUT: See *AMS; Who Was Who in America,* vol. 4, *1961-1968*; *Science,* n.s. 98 (1943): 487-88.

McGEE, WILLIAM JOHN (1853-1912). *Geology; Hydrology; Anthropology.* Geologist, U.S. Geol.

Surv.; chief ethnologist, Bureau of American Ethnology; member of Inland Waterways Commission; studied water resources for U.S. Department of Agriculture.

MSS: Lib. Cong. (*NUCMC* 62-4622). WORKS ABOUT: See *AMS; DAB.*

MACLEAN, JOHN (March 1, 1771, Glasgow, Scotland—February 17, 1814, Princeton, N.J.). *Chemistry.* Son of John, surgeon, and Agnes (Lang) Maclean. Married Phebe Bainbridge, 1798; six children. EDUCATION: Before age thirteen, entered Univ. of Glasgow; ca. 1787, left Glasgow to study in Edinburgh, London, and Paris; 1790, resumed studies at Glasgow; 1791, received diploma from Faculty of Physicians and Surgeons of City of Glasgow, authorizing practice of surgery and pharmacy. HONORS: M.D. [honorary?], Univ. of Aberdeen, 1797. CAREER: 1791-1814, member of Faculty of Physicians and Surgeons, Glasgow; ca. 1791, began practice of medicine and surgery; 1795, came to America; 1795-1797, practiced medicine and surgery, Princeton, N.J.; at College of New Jersey (Princeton), professor of chemistry and natural history (1795-1797) and professor of mathematics and natural philosophy, including chemistry and natural history (1797-1812); 1812-1813, professor of natural philosophy and chemistry, College of William and Mary; 1813, returned to Princeton, N.J. MEMBERSHIPS: Am. Phil. Soc. SCIENTIFIC CONTRIBUTIONS: During student days at Glasgow read several papers before Chemical Society (synopses of three of these papers appear in Maclean, *Memoir* [below], pp. 9-10). With appointment at Princeton, became first professor of chemistry in U.S. not affiliated with medical school. While student in Paris during late 1780s, came to accept Lavoisier's antiphlogistic theory of chemistry, and in U.S. helped introduce Lavoisier's "new chemistry"; of special importance in this respect was *Two Lectures* [see Works], which led to debate between Maclean, J. Priestley, J. Woodhouse, and S. L. Mitchill [qq.v.] in latter's New York *Medical Repository.*

WORKS: *Two Lectures on Combustion: Supplementary to a Course of Lectures on Chemistry Read at Nassau Hall; Containing an Examination of Dr. Priestley's Considerations on the Doctrine of Phlogiston, and the Decomposition of Water* (Philadelphia, 1797). MSS: Small collection in Princeton Univ. Library (*DAB*). WORKS ABOUT: *DAB,* 12: 126-27 (William Foster); *DSB,* 8: 612-13 [contains little additional information]; John Maclean [son], *A Memoir of John Maclean, M.D.* (Princeton, 1876).

MACLURE, WILLIAM (October 27, 1763, Ayr, Scotland—March 23, 1840, St. Angelo, Mexico). *Geology.* Son of David and Ann (Kennedy) McClure. Never married. EDUCATION: Instructed by private tutors. CAREER: 1782, visited U.S. for business reasons; after return to England, became partner in mercantile firm of Miller, Hart & Co., London, and made fortune; 1796, returned to U.S., traveled about country studying geological features; made U.S. citizen; 1803, began several years' involvement as member of commission to deal with spoliation claims between U.S. and France; 1815, while visiting in France, engaged C. A. Lesueur [q.v.] as cartographer-naturalist; 1815-1816 (winter), explored West Indies with Lesueur; explored and collected in region of Allegheny Mountains; 1817-1818, conducted expedition to Georgia and Spanish Florida; 1820-1824, lived in Spain; 1824, returned to U.S. and became associated with Robert Owen in experimental community at New Harmony, Ind.; 1827, made visit to Mexico, and subsequently passed most of remainder of life there. MEMBERSHIPS: President of Ac. Nat. Scis. Philad., 1817-1840, and of American Geological Society, 1819-1826. SCIENTIFIC CONTRIBUTIONS: Fortune acquired in business made possible a life devoted to geological studies and philanthrophy. In business travels, and during years as American claims commissioner in France, studied geology and natural history of Europe. In 1809, published geological map of United States, based on personal observations throughout region east of Mississippi River; best-known work is "Observations" [see Works], a revision and expansion of 1809 study, and together these works constitute first systematic study of geology of U.S. in English language; wrote on geology of West Indies [see Works], and published paper presenting terminology and theory regarding origin of rocks. Was generous patron of science, and took Lesueur, T. Say, and G. Troost [qq.v.] to New Harmony settlement; also engaged in educational projects.

WORKS: "Observations on the Geology of the U.S. . . . ," *Transactions of Am. Phil. Soc.,* n.s. 1 (1818), pp. 1-92 and colored map, also published separately (Philadelphia, 1817) and reprinted (New York, 1962); "Observations on the Geology of the West India Islands . . . ," *Journal of Ac. Nat. Scis. Philad.* 1 (1817): 134-49; "Essay on the Formation of Rocks . . . ," ibid., pp. 261-76, 285-310, 327-45 (also published separately). See Meisel, *Bibliog. Amer. Nat. Hist.* MSS: Univ. of Illinois—Illinois Historical Survey (*NUCMC* 69-1605); New Harmony (Ind.) Workingmen's Institute (Hamer). WORKS ABOUT: *DAB,* 12: 135-37 (George P. Merrill); *DSB,* 8: 615-17 (George W. White).

McMURRICH, JAMES PLAYFAIR (1859-1939). *Animal morphology.* Professor of anatomy, Univ. of Michigan; professor of anatomy and dean, Univ. of Toronto.

WORKS ABOUT: See *AMS; DAB,* supp. 2.

McMURTRIE, WILLIAM (1851-1913). *Chemistry.* Chemist and special agent in agricultural technology, U.S. Department of Agriculture; chemist and vice-president, Royal Baking Powder Co.
WORKS ABOUT: See *AMS; DAB.*

MADISON, JAMES (August 27, 1749, near Staunton, Va.—March 6, 1812, Williamsburg, Va.). *Astronomy; Physics; Geology.* Son of John, clerk of large district of Virginia, and Agatha (Strother) Madison. Married Sarah Tate, 1779; survived by two children. EDUCATION: 1771, A.B., College of William and Mary; studied law with George Wythe and admitted to bar; 1775, went to England for additional study. HONORS: D.D., Univ. of Pennsylvania, 1785. CAREER: 1773-1784, professor of natural philosophy and mathematics, William and Mary; 1775, in England, ordained in Church of England; at William and Mary, president (1777-1812) and professor of natural and moral philosophy (1784-1812); 1779, appointed commissioner for Virginia to determine Virginia-Pennsylvania boundary; 1790, chosen first bishop of Episcopal church in Virginia, and that year was consecrated in ceremonies in England. MEMBERSHIPS: Am. Phil. Soc. SCIENTIFIC CONTRIBUTIONS: Interested in wide range of scientific subjects, he corresponded extensively with T. Jefferson [q.v.] and with other American and European scientists on electricity, geology, Indian mounds, fossil bones, scientific instruments, meteorology, astronomy, and other subjects, and his several published papers are nearly as diverse. Served as chief commissioner for Virginia in extension of Mason-Dixon line between that state and Pennsylvania, performing associated astronomical determinations with great precision; conducted surveys of Virginia later used as basis for so-called Madison Map [see Works].
WORKS: "A Letter from J. Madison, Esq. to D. Rittenhouse, Esq." [meteorological observations for 1777-1778], *Transactions of Am. Phil. Soc.* 2 (1786): 141-58; "Observations of a Lunar Eclipse, Nov. 2d, 1789 . . . ," ibid. 3 (1793): 150-54; "Experiments Upon Magnetism . . . ," ibid. 4 (1799): 323-28; "The Supposed Fortifications of the Western Country," ibid. 6 (1809): 132-42; *A Map of Virginia Formed from Actual Surveys . . .* (Richmond, Va., 1807). See several other works in *Roy. Soc. Cat.*, vol. 4. David L. Holmes is preparing an edition of the papers of Bishop Madison, to be published by the College of William and Mary. WORKS ABOUT: *DAB*, 12: 182-84 (G. MacLaren Brydon); Charles Crowe, "Bishop James Madison and the Republic of Virtue," *Journal of Southern History* 30 (1964): 58-70; private communication from David L. Holmes [see above].

MALLET, JOHN WILLIAM (1832-1912). *Chemistry.* Superintendent, Ordnance Laboratories, Confederate States of America; professor of chemistry, Univ. of Virginia.
WORKS ABOUT: See *AMS; DAB.*

MALTBY, MARGARET ELIZA (1860-1944). *Physics; Physical chemistry.* Associate professor of physics and head of department, Barnard College, Columbia Univ.
WORKS ABOUT: See *AMS; Nota. Am. Wom.*

MANSFIELD, JARED (May 23, 1759, New Haven, Conn.—February 3, 1830, New Haven, Conn.). *Mathematics; Physics.* Son of Stephen, sea captain, and Hannah (Beach) Mansfield. Married Elizabeth Phipps, 1800; four children. EDUCATION: 1777, during senior year expelled from Yale College for misconduct; 1787, A.M., Yale College, and listed with class of 1777. HONORS: LL.D., Yale, 1825. CAREER: 1786-1795, rector, Hopkins Grammar School, New Haven; ca. 1795, taught several months in Friends' Academy, Philadelphia; ca. 1796-1802, conducted private school, New Haven; 1802, appointed captain of engineers, U.S. army; 1802-1803, acting professor of mathematics, U.S. Military Academy, West Point; 1803-1812, U.S. surveyor general, with rank of lieutenant colonel, to survey Ohio and Northwest Territory; 1812-1828, professor of natural and experimental philosophy, West Point, but assigned during war years, 1812-1814, to supervise fortifications at New London and Stonington, Conn.; 1828, took up residence in Cincinnati. MEMBERSHIPS: Am. Phil. Soc. SCIENTIFIC CONTRIBUTIONS: *Essays* [see Works] touched on algebra, geometry, calculus, astronomy in relation to navigation, and ballistics; it is considered first work of original mathematical research written by native American and established author's scientific reputation; section of *Essays* dealing with gunnery (ballistics) presented pioneering work on significance of air resistance, both as retarding force and in effects on projectile. Subsequently published papers on orbit of comet of 1807, figure of earth, duplication of cube and trisection of angle, and vanishing fractions.
WORKS: *Essays, Mathematical and Physical* (New Haven, 1801), 274 pp.; other papers listed in *DAB*. MSS: Ohio Historical Society (*NUCMC* 63-262); U.S. Mil. Ac. Library (*NUCMC* 70-1346). WORKS ABOUT: *DAB*, 12: 256-57 (Alois Kovarik); Franklin B. Dexter, *Biographical Sketches of the Graduates of Yale College . . .* (New York, 1903), 3: 691-94.

MAPES, CHARLES VICTOR (July 4, 1836, New York, N.Y.—January 23, 1916, New York, N.Y.). *Agricultural chemistry.* Son of James J. [q.v.]—agricultural chemist, farmer, and editor—and Sophia

(Furman) Mapes. Married Martha Meeker Halsted, 1863; five children. EDUCATION: 1857, A.B., Harvard College. CAREER: 1858, entered employment in counting room of New York wholesale grocers; 1858, became assistant editor and, later, editor of father's publication, *Working Farmer*; 1859, in partnership with B. M. Whitlock (an owner of the wholesale grocery), began manufacture and sale of fertilizers and agricultural implements, with factory near Newark, N.J., but business failed during Civil War; after about 1862, chiefly involved in manufacture and sale of chemical fertilizers; 1877-1916, vice-president and general manager, and then president, Mapes Formula and Peruvian Guano Co.; first president, New York Chemical and Fertilizer Exchange. MEMBERSHIPS: Am. Assoc. Advt. Sci.; Am. Chem. Soc. SCIENTIFIC CONTRIBUTIONS: Especially noted for development of fertilizers designed to meet specific needs of different crops and soils; in 1874 produced fertilizer for use on potatoes, the first such specialized preparation to be produced in U.S. Published articles in various agricultural journals and reports. Took leading part in early development of Peruvian guano beds. For a time, engaged in soil test work with W. O. Atwater [q.v.], of U.S. Department of Agriculture.

WORKS: "Some Rambling Notes on Agriculture and Manures," in New Jersey State Board of Agriculture, *Sixth Annual Report*, 1878 (Trenton, N.J., 1878), and "The Effects of Fertilizers on Different Soils," in New Jersey State Board of Agriculture, *Seventh Annual Report*, 1879-1880 (Trenton, N.J., 1880). MSS: Letters in W. O. Atwater Papers at Wesleyan Univ. Library. WORKS ABOUT: *DAB*, 12: 263-64 (Lee Garby); Harvard College Class of 1857, *Twenty-Fifth Anniversary Report* (Cambridge, Mass., 1882), pp. 85-87.

MAPES, JAMES JAY (May 29, 1806, Maspeth, Long Island, N.Y.—January 10, 1866, New York, N.Y.). *Chemistry*; *Agriculture*. Son of Jonas—importer, merchant tailor, War of 1812 army officer—and Elizabeth (Tylee) Mapes. Married Sophia Furman, 1827; survived by four children, including Charles V. Mapes [q.v.]. EDUCATION: Attended classical school at Hempstead, Long Island; largely self-educated. CAREER: Served as clerk in father's mercantile establishment; 1827-1834, operated his own mercantile business; 1834-1847, worked as consulting and analytical chemist; 1835-1838, professor of chemistry and natural philosophy of colors, National Academy of Design, New York; 1840-1842, editor, *American Repertory of Arts, Sciences, and Manufactures*; 1842-1843, associate editor, *Journal of Franklin Institute*; 1845, became president, Mechanics' Institute of City of New York; 1847, acquired farm near Newark, N.J., restored it to model of productivity, there taught principles of scientific agriculture, and became consulting agricultural chemist; 1849-1863, founder and editor of *Working Farmer*; manufacturer of fertilizers and several other chemical products. MEMBERSHIPS: N.Y. Lyc. Nat. Hist. and other societies. SCIENTIFIC CONTRIBUTIONS: Activities as chemist included contributions to development of processes and apparatus for sugar refining, color making and dyeing, tempering of steel, and the like. Conducted analyses of beer and wine for New York senate and other groups; chief source of income was work as expert witness in chemical patent cases. Increasingly turned attentions to agriculture, actively promoted improved methods of agricultural practice, including use of chemical fertilizers, and was one of first American manufacturers of superphosphates; in 1852 began selling that product, and in 1859 received patent for formula; quality of fertilizer product was strongly criticized by agricultural chemists such as S. W. Johnson [q.v.].

WORKS: See above for journals edited; published results of own agricultural and chemical experiments in his *Working Farmer* and elsewhere. WORKS ABOUT: *DAB*, 12: 264 (Lee Garby); Williams Haynes, *Chemical Pioneers: The Founders of the American Chemical Industry* (New York, 1939), pp. 74-87; Margaret Rossiter, *Emergence of Agricultural Science*, especially pp. 150-65.

MARCOU, JULES (April 20, 1824, Salins, France—April 17, 1898, Cambridge, Mass.). *Geology*; *Paleontology*. No information on parentage. Married Jane Belknap, 1850; two children. EDUCATION: Educated in school at Salins and in lycée at Besançon; 1842-1844, attended College of St. Louis (Paris). HONORS: Received Grand Cross of Legion of Honor, 1867. CAREER: 1846, appointed professor of mineralogy, Sorbonne; in 1847 became curator of fossil conchology in museum and in 1848 received traveling fellowship, Jardin des Plantes (Paris); 1848, joined J. L. R. Agassiz [q.v.] in U.S., accompanied him on Lake Superior expedition, and collected geological specimens for Paris Museum; 1850, through marriage attained independent means; that year returned to Europe and terminated association with Paris Museum; 1853, appointed geologist with U.S. Army surveys for Pacific railroad; 1854, again returned to Europe; 1856-1859, professor of paleontology, Ecole Polytechnique, Zurich; 1859, returned to U.S., established residence in Cambridge, Mass., and aided Agassiz in work with new Museum of Comparative Zoology in Cambridge; after 1864 returned to France several times, but maintained permanent residence in Cambridge; conducted geological field work, and made last expedition in 1875 with U.S. Survey West of 100th Meridian (under G. M. Wheeler [q.v.]). MEMBERSHIPS: Am. Ac. Arts Scis.; Geological Society of Paris; and other societies. SCIENTIFIC CONTRIBUTIONS: While student in Paris, published several papers on mathematics. In 1845, published first geological paper, and soon became authority on fossils; chief work done in stratigraphy and in geological mapping, and map on world geology [see Works, 1875] stood as landmark in that subject; became involved in series of controversies with American geologists, producing map in 1853 of U.S. geological strata based on limited observations. Wrote over 180 works, most published in French.

WORKS: *A Geological Map of the United States and the British Provinces of North America* (Boston, 1853, with later editions in 1855 and 1858); *Geology of North America* (Zurich, 1858); *Carte Géologique de la Terre* (Zurich, 1875). Major works listed in *DSB*. Also see bibliography in *Bulletin of U.S. Natl. Mus.*, no. 30 (1885-1886), pp. 241-44. WORKS ABOUT: *DAB*, 12: 272-73 (Hubert L. Clark); *DSB*, 9: 99-101 (Edward Lurie).

MARCY, OLIVER (February 13, 1820, Coleraine, Franklin County, Mass.—March 19, 1899, Evanston, Ill.). *Geology*. Son of Thomas and Anna (Henry) Marcy. Married Elizabeth E. Smith, 1847; four children. EDUCATION: 1846, graduated, Wesleyan Univ. HONORS: LL.D., Chicago Univ., 1873. CAREER: 1846-1862, teacher of mathematics and science, Wilbraham Academy (Mass.); during 1851, teacher of natural sciences, Amenia Seminary (N.Y..); at Northwestern University, professor of physics and natural history (1862-1869), professor of natural history (1870-1881), acting president (1876-1881 and 1890), professor of geology and curator of museum (1882-1899), dean of College of Liberal Arts; 1866, geologist with survey of U.S. government road from Lewiston, Idaho, to Virginia City, Mont. MEMBERSHIPS: Geol. Soc. Am.; Am. Assoc. Advt. Sci.; and other societies. SCIENTIFIC CONTRIBUTIONS: Interest in geology grew out of assignment to teach that subject at Wilbraham Academy, but duties there, and at Northwestern Univ. (where he taught nearly all science subjects at one time or another), diverted attention from original investigations. Wrote very few scientific papers, and "Enumeration of Fossils" [see Works] was his only geological monograph. Accumulated museum for Northwestern, including the classification and labeling of over 70,000 specimens in zoology, botany, archeology, and geology; published number of newspaper articles on science education, and on metaphysical, geological and travel topics. Chief contribution was as teacher of science.

WORKS: "Enumeration of Fossils Collected in the Niagara Limestone at Chicago, Ill., with Description of Several New Species," *Memoirs of Bos. Soc. Nat. Hist.* 1 (1865): 81-113, with A. Winchell [q.v.]. WORKS ABOUT: *NCAB*, 13 (1906): 536-37; Alja R. Crook, "Memoir," *Proceedings of Geol. Soc. Am.* 11 (1900): 537-42.

MARK, EDWARD LAURENS (1847-1946). *Zoology*. Professor of anatomy, director of Zoological Laboratory, Harvard Univ.; director of Bermuda Biological Station for Research.

MSS: Harv. U.-UA (*NUCMC* 69-230). WORKS ABOUT: See *AMS*; *NCAB*, 9 (1907 [copyright 1899 and 1907]): 271-72; *New York Times*, December 17, 1946, p. 38.

MARSH, GEORGE PERKINS (March 15, 1801, Woodstock, Vt.—July 23, 1882, Vallombrosa [near Florence], Italy). *Physical geography*; *Humanity and environment*. Son of Charles, lawyer, and Susan (Perkins) Arnold Marsh. Married Harriet Buell, 1828; Caroline Crane, 1839; two children by first marriage. EDUCATION: 1820, A.B., Dartmouth College; ca. 1821-1825, studied law while recovering from serious problems with eyes; 1825, admitted to Vermont bar. CAREER: 1820-ca. 1821, professor of Greek and Latin languages, Norwich (Vt.) Academy; 1825, established law practice, Burlington, Vt., and there engaged with little success in various business ventures; ca. 1833, inherited property from father-in-law; 1835, member of Vermont Supreme Legislative Council; 1842, ceased active law practice; 1843, elected to first of four consecutive terms in U.S. Congress; 1849-1854, U.S. minister to Turkey; 1854, returned to Vermont, attempted to salvage business interests, designed new statehouse for Vermont, served as railroad commissioner; 1856, lectured on English language, Columbia Univ.; 1857, appointed Vermont state fish commissioner; ca. 1860, rendered nearly bankrupt through railroad failure; 1860-1861, gave Lowell Institute lectures (Boston) on English language; 1861-1882, U.S. minister to Kingdom of Italy. SCIENTIFIC CONTRIBUTIONS: As a member of Congress, played central role in formulating program for Smithson. Instn.; in 1847 was made a regent, and as minister in Turkey collected botanical and zoological specimens for the institution. Chief contribution to scientific thought was *Man and Nature* [see Works], based on personal observations during travels and on reading in wide range of literature; the work had international influence in drawing attention to ways in which humans affect their environment and became major source for conservation movement; in 1873, book instigated drive to establish U.S. Forestry Commission and to set aside forest reserves. Scholar of great depth and range of interests, especially noted for work in linguistics. Wrote work on camel and its importation into U.S.

WORKS: *Man and Nature; or, Physical Geography as Modified by Human Action* (New York and London, 1864; reprinted Cambridge, Mass., 1965). Bibliography of publications and references to manuscripts in Lowenthal [below]. MSS: Chief collection in Univ. of Vermont (Lowenthal). WORKS ABOUT: *DAB*, 12: 297-98 (Walter L. Wright, Jr.); David Lowenthal, *George Perkins Marsh: Versatile Vermonter* (New York, 1958).

MARSH, OTHNIEL CHARLES (October 29, 1831, Lockport, N.Y.—March 18, 1899, New Haven, Conn.). *Paleontology*. Son of Caleb, farmer and shoe manufacturer, and Mary Gaines (Peabody) Marsh. Never married. EDUCATION: 1860, A.B., Yale College; 1861-1862, attended Sheffield Scientific School, Yale; for three years, studied in Germany. HONORS: Bigsby Medal of Geological Society of London, 1877; Cuvier Prize from French Academy, 1897. CAREER: 1866-1899, professor of paleontology, Yale Univ.; 1882-1892, vertebrate paleontologist, U.S. Geol. Surv.; through aid of uncle, George Peabody, had independent means and was able personally to finance his early exploring expeditions. MEMBERSHIPS: President, Natl. Ac. Scis., 1883-1895. SCIENTIFIC CONTRIBUTIONS: Made several exploring excursions into West, including leadership during 1870-1873 of four Yale scientific expeditions. Gathered enormous amount of material, depending on hired collectors during later years, and built up large vertebrate fossil collection at Peabody Museum (Yale) along with an accumulation of osteological, mineralogical, archeological, ethnological, and invertebrate fossil specimens. Paleontological accomplishments included description of earliest mammals known at that time, work on evolution of horse in North America, first descriptions of fossil serpents and flying reptiles from American West, proof of reptilian origin of birds, demonstration of existence of early primates in North America, and other contributions; engaged in bitter rivalry with E. D. Cope [q.v.] while the two men established study of vertebrate paleontology in America.

WORKS: "Introduction and Succession of Vertebrate Life in America," *Proceedings of Am. Assoc. Advt. Sci.* 26 (1877): 211-58, also issued separately; *Odontornithes: A Monograph on the Extinct Toothed Birds of North America*, U.S. Geological Exploration of Fortieth Parallel, vol. 7 (Washington, D.C., 1880); *Dinocerata: A Monograph of an Extinct Order of Gigantic Mammals* (Washington D.C., 1886). Bibliography in Charles Schuchert and Clara M. LeVene, *O. C. Marsh* (New Haven, 1940), pp. 503-26. MSS: Yale Univ. Library (*NUCMC* 65-1076); also, Yale—Peabody Museum of Natural History (Hamer); Ac. Nat. Scis. Philad. (Hamer). WORKS ABOUT: *DAB*, 12: 302-3 (George P. Merrill); *DSB*, 9: 134 (Elizabeth N. Shor).

MARSHALL, HUMPHRY (October 10, 1722, Chester County, Pa.—November 5, 1801, Marshallton, Chester County, Pa.). *Botany*. Son of Abraham—stone cutter, farmer, and man of wealth—and Mary (Hunt) Marshall; cousin of John Bartram [q.v.]. Married Sarah Pennock, 1748; Margaret Minshall, 1788; no children. EDUCATION: No formal schooling beyond age twelve; apprenticed to stonemason; self-educated. CAREER: For few years, worked as stonemason; after 1748, managed father's farm; operated grist mill, and served as treasurer of county; 1767, inherited fortune from father; 1774, moved to house at Marshallton and there established botanic garden. MEMBERSHIPS: Am. Phil. Soc. SCIENTIFIC CONTRIBUTIONS: About 1767, began collecting plants, birds' nests, eggs, and other zoological specimens for John Fothergill in London; through extensive reading and study, became knowledgeable in all fields of natural history and astronomy, and in 1772 submitted paper on sunspot observations to Am. Phil. Soc. Botanic garden at Marshallton came to include impressive selection of American and foreign specimens. Chiefly remembered for *Arbustum Americanum*, begun about 1780, the first work published in America on trees and shrubs native to U.S.; *Arbustum* brought together work on American botany from number of authors, included Marshall's own observations, and employed the Linnaean system, although genera were arranged alphabetically as aid to use by gardeners; the publication can be viewed as promotion of sales of seeds and plants by its author; work was considered useful by Europeans and within three years appeared in translation in Leipzig and Paris.

WORKS: *Arbustrum* [i.e., *Arbustum*] *Americanum: The American Grove; or, An Alphabetical Catalogue of Forest Trees and Shrubs, Natives of the American United States* (Philadelphia, 1785; reprinted, New York, 1967, with introduction by Joseph Ewan). MSS: Hist. Soc. Penn. (Hamer). WORKS ABOUT: *DAB*, 12: 311-12 (Marion P. Smith); Ewan, Introduction to *Arbustrum* [above].

MARTIN, HENRY NEWELL (July 1, 1848, Newry, County Down, Ireland—October 27, 1896, Burley-in-Wharfedale, Yorkshire, England). *Physiology*. Son of Congregational minister and schoolmaster. Married Hetty (Cary) Pegram 1878; no children. EDUCATION: There is some confusion about Martin's various degrees, of which the following have been attributed to him (only the Cambridge A.B. and A.M. have been confirmed by reference to lists of graduates published by the universities): M.B. in 1871 and D.Sc. in 1872, Univ. of London;

A.B. in 1874, D.Sc. in 1875, and A.M. in 1877, Cambridge Univ. CAREER: 1870, became demonstrator to instructor of physiology, Michael Foster, in Trinity College, Cambridge; 1874, assistant to T. H. Huxley in biology course at Royal College of Science (South Kensington); 1874, chosen fellow, Trinity College, Cambridge; 1876-1893, professor of biology, Johns Hopkins Univ.; 1893, returned to England in ill health. MEMBERSHIPS: Roy. Soc.; a founder of American Physiological Society. SCIENTIFIC CONTRIBUTIONS: Said to have received first doctorate in physiology from Cambridge Univ., and in 1875, with guidance from Huxley, published *A Course of Practical Instruction in Elementary Biology*. Played significant part in development of American interest in experimental biology and in establishment of physiology as field separate from medicine; research work related chiefly to cardiac physiology, and Martin succeeded in developing technique for isolating mammalian heart in order to study effects on it of variations in factors such as temperature, alcohol, arterial and venous pressure; in 1883, delivered Royal Society's Croonian lectures, discussing results of study of temperature changes on heartbeat of dog. Founder and editor of *Studies from the Biological Laboratory of the Johns Hopkins University*, and wrote several textbooks.

WORKS: Articles reprinted as *Physiological Papers*, Memoirs from the Biological Laboratory of the Johns Hopkins University, vol. 3 (Baltimore, 1895). MSS: Numerous letters in Daniel C. Gilman Papers at Johns Hopkins Univ. Library (*DSB*). WORKS ABOUT: *DAB*, 12: 337-38 (Russell H. Chittenden); *DSB*, 9: 142-43 (Charles E. Rosenberg).

MARX, GEORGE (June 22, 1838, Laubach, Hesse, Germany—January 3, 1895, Washington, D.C.). *Entomology*. Son of George Marx, clergyman. Married Minnie Maurer, 1869; no information on children. EDUCATION: Ca. 1852 [age fourteen], entered gymnasium at Darmstadt in preparation for ministry; subsequently, abandoned plans for career as clergyman in favor of pharmacy, and completed pharmaceutical studies at Giessen; 1885, M.D., Columbian Univ. (George Washington Univ.). CAREER: 1860, came to U.S.; 1861, enlisted in New York Volunteers and during years 1861-1862 served as assistant surgeon in medical corps; 1862, was honorably discharged from army, went to New York, and began work as pharmacist; 1865-1878, engaged in business, Philadelphia; 1878-1889, natural history draftsman, U.S. Department of Agriculture, during earlier years attached to Division of Entomology; 1889-1894, chief, Agriculture Department's Division of Illustrations. MEMBERSHIPS: President, Entomological Society of Washington. SCIENTIFIC CONTRIBUTIONS: As student at gymnasium in Germany, became noted for botanical and artistic talents and prepared illustrations for flora of Gross-Gerau. After settling in Philadelphia in 1865, began collecting and studying spiders and other arachnids and became a leading arachnologist; during later years, began study of ticks, but illness prevented publication of projected monograph; completed E. Keyserling's *Die Spinnen [Spiders] Amerikas* (Nuremberg, 1891) after death of originating author. Earliest paper in bibliography has 1881 date. Well known as scientific illustrator, preparing figures for own works, and during period 1878-1883 did many illustrations for Division of Entomology.

WORKS: "On the Morphology of Scorpionidae," *Proceedings of Entomological Society of Washington* 1 (1888): 108-12; "On the Effect of the Poison of Latrodectus mactans Walck. Upon Warm-Blooded Animals," ibid. 2 (1890-1892): 85-86; "On the Morphology of the Ticks," ibid, pp. 271-87; "Catalogue of the Described Araneae of Temperate North America," *Proceedings of U.S. Natl. Mus.* 12 (1889-1890): 497-594. Bibliography in Riley [below], pp. 199-201. WORKS ABOUT: C. V. Riley and others, memoir, *Proceedings of Entomological Society of Washington* 3 (1893-1896): 195-201; Mallis, *Amer. Entomologists*, pp. 408-10 [gives little additional information].

MASCHKE, HEINRICH (1853-1908) *Mathematics*. Professor of mathematics, Univ. of Chicago.

WORKS ABOUT: See *AMS; DAB*.

MASON, JAMES WEIR (April 22, 1836, New York, N.Y.—January 10, 1905, Easton, Pa.). *Mathematics*. Son of Ebenezer, Presbyterian minister, and Sarah Locke (Weir) Mason. Never married. EDUCATION: 1855, graduated, Free Academy of New York (later, City College of New York). CAREER: 1855-1858, teacher at Anthon School, founded by George C. Anthon; 1858-1862, owner and operator of school in New York City; 1862-1863, principal, Yonkers High School; 1863-1868, principal, Albany Academy; agent in Albany (1868-1869) and actuary in Springfield (1869-1872) for Massachusetts Mutual Life Insurance Co.; 1872-1879, actuary, Penn Mutual Life Insurance Co., Philadelphia; 1879-1903, professor of pure mathematics, City College of New York. MEMBERSHIPS: Actuarial Society of America. SCIENTIFIC CONTRIBUTIONS: Said to have done original and imaginative work as actuary and to have been an American leader in that profession, especially after move to Philadelphia. Served as head of department of mathematics at City College.

WORKS: No works listed in *Roy. Soc. Cat.* or *Natl. Un. Cat. Pre-1956 Imp.* MSS: A few letters in archives of City College of New York. WORKS

ABOUT: *Who Was Who in America,* vol. 1; Russell Sturgis, "The Life of Prof. James Weir Mason," *College Mercury* [City College of New York], commencement number (June 1903), pp. 200-202; death notice, *Science,* n.s. 21 (1905) : 117.

MASTERMAN, STILLMAN (January 28, 1831, Weld, Franklin County, Maine—July 19, 1863, Weld, Franklin County, Maine). *Meteorology; Astronomy.* No information on parentage or marital status; father may have been farmer. EDUCATION: Attended district school, entering at somewhat later age than normal; largely self-educated. CAREER: Ca. 1851, spent more than a year in Minnesota Territory, then returned to Weld in ill health; 1861-1863, worked on reduction of observations accumulated for some fifteen years by Naval Observatory (Washington, D.C.). SCIENTIFIC CONTRIBUTIONS: Self-taught in physical sciences, Latin, mathematics, and other subjects; constructed apparatus, ground and polished small lenses, and made prisms for studies of fluid media. First scientific paper gave results of observations of sixteen thunderstorms, 1850-1854, at Weld, Maine, and Stillwater, Minn. While lacking instruments, made and published number of astronomical observations visible to naked eye, including study of maxima and minima of variable stars, zodiacal light, auroras, meteors, and planets; these studies were done with great accuracy. Full realization of astronomical talents thwarted by isolation and early death.

WORKS: "Observations on Thunder and Lightning," *Annual Report of Smithson. Instn.* (1856) : 265-82. Other papers appeared in B. A. Gould's *Astronomical Journal* (1858-1868), *Am. Journ. Sci.* (1860-1863), *Annual Report of Smithson. Instn.* (1858), F. F. E. Brunnow's *Astronomical Notices* (1861), and *Astronomische Nachrichten* (1863). See *Roy. Soc. Cat.* vols. 4, 8, 10, which lists seventeen papers. WORKS ABOUT: [Benjamin A. Gould], obituary, *Am. Journ. Sci.* 2d ser. 36 (1863) : 448-50.

MATHER, COTTON (February 12, 1662/1663, Boston, Mass.—February 13, 1727/1728, Boston). *Medicine*; *Science.* Son of Increase [q.v.], clergyman and president of Harvard College, and Maria (Cotton) Mather. Married Abigail Phillips, 1686; Elizabeth (Clark) Hubbard, 1703; Lydia (Lee) George, 1715; fifteen children. EDUCATION: 1678, A.B., Harvard College; studied medicine. CAREER: Ca. 1680, began preaching; soon thereafter, undertook to assist father at Second Church, Boston; 1685, ordained at Second Church and from 1685 to 1727/1728, minister to that congregation, where father also served until 1723; 1690-1703, fellow, Harvard College. MEMBERSHIP: Roy. Soc. SCIENTIFIC CONTRIBUTIONS: Known chiefly as clergyman and scholar, Mather was a leading American intellectual of his day, most important single work being *Magnalia Christi Americana* (1702). Showed interest in science at early age; while never neglecting theological perspective, nor wholly separating physical and supernatural worlds, he appreciated importance of observation and explanation of phenomena in terms of physical or scientific causes; this perspective was evident in his *Christian Philosopher* [see Works], one of the chief repositories of his scientific outlook. Beginning 1712, sent long series of letters to Roy. Soc. (and elsewhere) that were given collective title of *Curiosa Americana,* but only a few were published in *Phil. Trans. of Roy. Soc.; Curiosa* dealt with subjects now classifiable as anthropology, astronomy, geology, mathematics, meteorology, zoology, and especially medicine, botany, psychology, and philosophy. In 1721, played major part in introduction of smallpox inoculation into Boston, having wide familiarity with medical as well as general scientific literature of the day; advanced a theory of animalculae, found with miscroscope in smallpox pustules, as cause of disease. Did work on plant pollination and much interested in study of mental problems and treatment of mentally ill. Promoted science in colonies, sent specimens to Roy. Soc., and was first person born in British mainland colonies of North America who contributed original hypotheses and ideas to science in addition to collecting data and specimens for European scientists.

WORKS: *The Christian Philosopher: A Collection of the Best Discoveries in Nature, with Religious Improvements* (London, 1721). See bibliographical references in *DAB* and Stearns [below]. MSS: Harv. U.-HL, Roy. Soc., Mass. Hist. Soc., Am. Ant. Soc. (Stearns [below]). WORKS ABOUT: *DAB,* 12: 386-89 (Kenneth B. Murdock); Stearns, *Sci. in Brit. Colonies,* pp. 403-26. Also see Otho T. Beall, Jr., and Richard H. Shryock, *Cotton Mather: First Significant Figure in American Medicine* (Baltimore, 1954).

MATHER, INCREASE (June 21, 1639, Dorchester, Mass.—August 23, 1723, Boston, Mass.). *Astronomy; Promotion of science.* Son of Richard, clergyman, and Katherine (Holt) Mather. Married Maria Cotton, 1662; Ann (Lake) Cotton, 1715; ten children by first marriage, including Cotton Mather [q.v.]. EDUCATION: 1656, A.B., Harvard College, having studied under Rev. John Norton; 1658, M.A., Trinity College, Dublin. CAREER: 1657, went to Dublin to study; 1658-1659, preached at Great Torrington, Devonshire, England; 1659, served as chaplain to English garrison at Guernsey and later that year went to Gloucester; 1660, returned to Guernsey; 1661, preached at Weymouth and Dorchester, England; 1661, returned to Massachusetts; preached at Dorchester and Boston; 1664, ordained at Second Church,

Boston and, 1664-1723, served that congregation as minister; at Harvard College, fellow (1675-1685) and president (1685-1701); 1688-1692, in England on mission to arrange restoration of Massachusetts's charter; 1692, returned to Boston with new charter and royal governor whom he had nominated. SCIENTIFIC CONTRIBUTIONS: Preached and wrote on comets, and viewed comet of 1682 with telescope; in 1681 published sermon on comets under title *Heaven's Alarm to the World,* dealing with comets as signs; was familiar with latest scientific writing in *Phil. Trans. of Roy. Soc.* and elsewhere, and came to conceive of comets as natural phenomena; *Kometographia* [see Works] presented comets as natural heavenly bodies moving according to fixed laws and yet having at same time theological significance as signs of warning. Apparently was chief stimulus to formation of Philosophical Society, which met in 1683-1688 in Boston; *Illustrious Providences* [see Works], while set in tradition of God's intervention in human affairs, emphasized importance of observation—as opposed to written or traditional authority—and is seen as one of first American works written from scientific perspective. As president of Harvard, promoted study of science.

WORKS: *Kometographia; or, A Discourse Concerning Comets* (Boston, 1683); *Essays for the Recording of Illustrious Providences* (Boston, 1684). See references to printed and manuscript sources in *DAB* and in Kenneth B. Murdock, *Increase Mather* (Cambridge, Mass., 1925). WORKS ABOUT: *DAB,* 12: 390-94 (Kenneth Murdock); Stearns, *Sci. in Brit. Colonies,* pp. 154-59.

MATHER, WILLIAM WILLIAMS (May 24, 1804, Brooklyn, Conn.—February 25, 1859, Columbus, Ohio). *Geology.* Son of Eleazar, hatter and hotel owner, and Fanny (Williams) Mather. Married Emily Maria Baker, ca. 1830; Mrs. Mary (Harry) Curtis; six children by first marriage, one by second. EDUCATION: 1828, graduated, U.S. Military Academy, West Point. CAREER: 1828, brevetted second lieutenant, U.S. Army, and served for short time in Louisiana; 1829-1835, acting professor of chemistry and mineralogy, U.S. Military Academy; 1833, professor of chemistry, mineralogy, geology, Wesleyan Univ.; 1834, promoted to first lieutenant; 1835, topographical engineer with federal geological survey in Wisconsin and Minnesota; 1836, frontier duty, Indian Territory, and that year resigned from army; for brief period, professor of chemistry, Univ. of Louisiana; 1836-1844, geologist of first district, New York Natural History [Geological] Survey; 1837-1838, director, Geological Survey of Ohio; 1838-1839, Kentucky state geologist; 1842-1845 and 1847-1850, professor of natural science and, at times, vice-president and acting president, Ohio Univ. (Athens); 1846, acting professor of chemistry, mineralogy, and geology, Marietta College; 1850-1854, secretary and chemist, Ohio State Board of Agriculture; engaged as consultant for various mineral and coal interests. MEMBERSHIPS: Numerous scientific societies. SCIENTIFIC CONTRIBUTIONS: Chief geological work done for New York geological survey, the final report of which [see Works] contained great amount of detail and was largest of survey's reports, although not its most important. Work carried out at same time with Ohio and Kentucky surveys was mainly administrative. Recognized as expert on geology of coal and was associated with development of coal areas of Ohio.

WORKS: *Geology of New York: Part I, Comprising the Geology of the First Geological District* (Albany, 1843). Bibliography of works in *American Geologist* 19 (1897): 7-15. MSS: No collection is known. WORKS ABOUT: *DAB,* 12: 399-400 (George P. Merrill); *DSB,* 9: 172-73 (John W. Wells).

MAURY, MATTHEW FONTAINE (January 14, 1806, near Fredericksburg, Va.—February 1, 1873, Lexington, Va.). *Oceanography; Meteorology.* Son of Richard, small planter, and Diana (Minor) Maury. Married Ann Hull Herndon, 1834; eight children. EDUCATION: 1825, graduated, Harpeth Academy (near Franklin, Tenn.). HONORS: LL.D., Cambridge Univ., 1868; honors from various foreign governments. CAREER: 1825, made midshipman, U.S. Navy; 1825-1834, had three long periods of sea duty; 1834, received leave of absence and worked on book on navigation; 1836, promoted to lieutenant; 1837, appointed astronomer to South Seas exploring expedition (Wilkes Expedition), but soon resigned; assigned harbor survey duty in southeastern states; 1839, suffered leg injury that made duty at sea impossible; 1842, appointed superintendent of Depot of Charts and Instruments and head of new Naval Observatory; 1858, after forced leave of absence (1855-1858), reinstated and promoted to commander; 1861, resigned directorship of U.S. Naval Observatory, received commission as commander in Confederate navy, and was assigned to harbor defense; 1862-1865, agent for Confederate government in England; 1865-1866, in service to Emperor of Mexico as commissioner of immigration; 1866-1868, in England, engaged in writing textbooks; 1868-1873, professor of physics, Virginia Military Institute (Lexington). MEMBERSHIPS: Am. Assoc. Advt. Sci. and other societies. SCIENTIFIC CONTRIBUTIONS: In 1836, published work on navigation that, together with efforts at naval reform, helped secure position as head of Naval Observatory and Hydrographical Office (so renamed in 1854). Neglected astronomical work of Naval Observatory in favor of hydrographic and meteorological interests, concentrating on matters related to navigation. Motivated by desire to aid oceanic trade; chief work was attempt to chart global movements of wind and ocean currents based on ships' logs; solicited cooperation of mariners throughout world, and organized international congress in 1853 at Brussels. Widely recognized for organizational abilities in data collection and systematization, but was less successful in scientific interpretation of data; came to be largely ostracized by American scientific community, based on its assessment of his scientific work and his aggressive attempts to control areas of science outside his area of knowledge.

WORKS: *Wind and Current Chart of the North Atlantic* (Washington, D.C., 1847), with later charts for other regions, together with explanatory texts; *Physical Geography of the Sea* (New York, 1855). See bibliography in Frances L. Williams, *Matthew Fontaine Maury* (New Brunswick, N.J., 1963), pp. 693-710. MSS: Lib. Cong. (*NUCMC* 63-378); see other references in *NUCMC* and Hamer. WORKS ABOUT: *DAB,* 12: 428-31 (H. A. Marmer); *DSB,* 9: 195-97 (Harold L. Burstyn).

MAYER, ALFRED MARSHALL (November 13, 1836, Baltimore, Md.—July 13, 1897, Hoboken, N.J.). *Physics.* Son of Charles F., lawyer, and Eliza (Blackwell) Mayer. Married Katherine Duckett Goldsborough, 1865; Louisa Snowden, 1869; one child by first marriage, two by second. EDUCATION: Attended St. Mary's College (Baltimore); 1863-1865, studied physics, mathematics, and physiology, Univ. of Paris. HONORS: Ph.D., Pennsylvania College, 1866. CAREER: Ca. 1852 [age sixteen], began work as machinist for Baltimore engineer; ca. 1854, began practice as analytical chemist; ca. 1857, appointed assistant professor of physics and chemistry, Univ. of Maryland; 1859, appointed professor of physical science, Westminister College (Fulton, Mo.); 1865-1867, professor of physical sciences, Pennsylvania College (Gettysburg); 1867, appointed professor of physics and astronomy, Lehigh Univ.; 1869, accompanied *U.S. Nautical Almanac* solar eclipse expedition; 1871-1897, associated with Stevens Institute of Technology, he organized and directed its department of physics. MEMBERSHIPS: Natl. Ac. Scis.; Am. Phil. Soc.; and other societies. SCIENTIFIC CONTRIBUTIONS: At age nineteen (1855) produced first scientific paper, on apparatus for determination of carbonic acid. Work with 1869 solar eclipse expedition resulted in some forty photographs of especially high quality for that period; published works on observations of Jupiter and on electricity, heat, gravity, and magnetism; at Stevens Institute, conducted experiments in acoustics and became most important American investigator in that field, while also pursuing research in other areas of physics. Most significant single contribution, published in 1878, was series of experiments in which small magnetized needles placed in corks were floated on water, under electromagnet, an arrangement used by later investigators in study of atomic structure. Designed number of instruments for scientific measurement. During 1880s, active as writer and participant in amateur sports, especially fishing.

WORKS: Publications appeared especially in *Am. Journ. Sci.*, including papers on floating magnets in 3d ser. vols. 15 and 16 (1878). See bibliography in *Biog Mems. of Natl. Ac. Scis.* 8 (1916): 266-72. MSS: Syracuse Univ. Library (*NUCMC* 68-1740); N.Y. Pub. Lib. (*NUCMC* 72-1059); Princeton Univ. Library (Hamer). WORKS ABOUT: *DAB,* 12: 448 (Franklin De R. Furman); *DSB,* 9: 230-31 (I. B. Cohen).

MEADE, WILLIAM ([?], Ireland—August 29, 1833, Newburgh, N.Y.). *Mineralogy.* No information on parentage. Apparently married (one source refers to a Mrs. Catherine Meade, who conducted sale of mineral collection after William's death). EDUCATION: M.D. (probably Edinburgh Univ., 1790). CAREER: Last twenty-five years of life spent in U.S.; may have engaged in medical practice; during later years, lived at Newburgh. SCIENTIFIC CONTRIBUTIONS: Visited number of the chief mineral locations in northern U.S., and through personal collecting and exchanges, accumulated significant cabinet of some 1,200 foreign and 2,000 American mineral and fossil specimens. Published articles appeared especially in *American Mineralogical Journal* and *Am. Journ. Sci.*, including papers on lead ore, elastic marble from Massachusetts, anthracite, relations of European and American minerals to their geology, analyses of coal and mineral springs, and related subjects. Probably was author of medical dissertation at Univ. of Edinburgh (1790) on mineral waters [see *Natl. Un. Cat. Pre-1956 Imp.*]; in 1817, published book on New York mineral waters [see Works], and in later years was particularly interested in production of artificial mineral water; in 1832 published paper on this subject in *Am. Journ. Sci.;* after death patent rights to mineral concoctions were offered for sale.

WORKS: *An Experimental Inquiry into the Chemical Properties and Medicinal Qualities of the Principal Mineral Waters of Ballston and Saratoga in the State of New York . . . [also] a Chemical Analysis of the Waters of the Lebanon Spring [N.Y.]* (Philadelphia, 1817). See also Meisel, *Bibliog. Amer. Nat. Hist.* WORKS ABOUT: Obituary, *Am. Journ. Sci.* 25 (1834): 215-16; notice concerning mineralogical cabinet, in ibid. 26 (1834): 209-10.

MEASE, JAMES (August 11, 1771, Philadelphia, Pa.—May 14, 1846, Philadelphia, Pa.). *Medicine;*

General science. Son of John, well-to-do shipping merchant, and Esther (Miller) Mease. Married Sarah Butler, 1800; five children. EDUCATION: A.B. in 1787 and M.D. in 1792, Univ. of Pennsylvania. CAREER: Medical practice, Philadelphia; 1794-1798, resident physician, Health Office for Port of Philadelphia, charged with prevention of recurrence of yellow fever epidemic; 1814-1815, hospital surgeon in War of 1812; 1816 and 1817, gave private lectures on pharmacy, held in building of Univ. of Pennsylvania. MEMBERSHIPS: Am. Phil. Soc.; Pennsylvania Horticultural Society. SCIENTIFIC CONTRIBUTIONS: Medical dissertation on hydrophobia was first important American work on that subject. Wrote about and promoted agriculture, serving as a manager of "Company for the Improvement of the Vine" and establishing a vineyard; in 1807, in association with several other persons, formed Philadelphia Mineral Water Association, a business venture that eventually proved unsuccessful. In 1813, presented series of lectures, "Upon Comparative Anatomy and Diseases of Animals," published the following year; *Geological Account* [see Works], while largely a compilation, was an early and useful work in physical and commercial geography; also wrote on silk culture, and published articles on American alligator, the streaked basse, Texas, large tree in Mexico, plants used in making rope and thread, dry rot in ships, spontaneous combustion, and bills of mortality. In addition to medical and scientific interests, also wrote on prison discipline, public finance, and other subjects.

WORKS: *Geological Account of the United States* (Philadelphia, 1807); *Thermometrical Observations as Connected with Navigation* (Philadelphia, 1841), 24 pp. See additional works mentioned in *DAB* and *Roy. Soc. Cat.*, vol. 4. MSS: Hist. Soc. Penn.; Am. Phil. Soc.; Lib. Co. Philad.; Philadelphia College of Physicians (Bell, *Early Am. Sci.*). WORKS ABOUT: *DAB*, 12: 486 (Joseph Jackson); William S. Miller, "James Mease," *Ann. Med. Hist.* 7 (1925): 6-30.

MEEHAN, THOMAS (March 21, 1826, near London, England—November 19, 1901, Germantown, Pa.). *Botany; Horticulture.* Son of Edward, gardener, and Sarah (Denham) Meehan. Married Catherine Colflesh, 1852; at least six children. EDUCATION: At age twelve, began learning gardening and botany under father; 1845, began training in horticulture, Royal Botanic Gardens at Kew, and later received certificate; self-educated, learning Greek, Latin, French. CAREER: 1848, came to America and worked in Buist's Nursery, Philadelphia; ca. 1849, made superintendent, Bartram Garden, then owned by Andrew Eastwick; 1852, given direction of grounds and conservatories of Caleb Cope near Holmesburg, Pa.; 1853, established own nurseries at Ambler and Germantown, Pa., which became highly successful business; 1859-1887, editor, *Gardener's Monthly;* 1877-1901, botanist, Pennsylvania State Board of Agriculture; 1882-1901, elected member of Philadelphia Common Council; 1891-1901, editor, *Meehan's Monthly,* continued by sons; for number of years, agricultural editor, *Forney's Weekly Press.* MEMBERSHIPS: Ac. Nat. Scis. Philad.; Am. Phil. Soc.; and other societies. SCIENTIFIC CONTRIBUTIONS: Publications began in teens, including early paper on sensitive stamens of portulaca, recently introduced into England from Mexico. One of first to produce hybrid of fuchsia; in last half of nineteenth century, became a leading American authority on horticulture; nursery business specialized in trees and shrubs native to America, and *Handbook* [see Works] grew out of Meehan's earlier catalogue of trees in Bartram Garden; wrote extensively on botany, and chief publication was *Flowers and Ferns* [see Works]. As member of Philadelphia's city council was instrumental in development of park system, including incorporation of Bartram Garden.

WORKS: *The American Handbook of Ornamental Trees* (Philadelphia, 1853); *The Native Flowers and Ferns of the United States,* 4 vols. (Boston, 1878-1880). See *Roy. Soc. Cat.*, vols. 4, 8, 10, 17, which lists over 200 papers. MSS: Ac. Nat. Scis. Philad. (*NUCMC* 66-62). WORKS ABOUT: *DAB*, 12: 492-93 (John W. Harshberger); Frederick McGourty, Jr., "Thomas Meehan, Nineteenth Century Plantsman," *Plants and Gardens* 23 (Autumn 1967): 81, 85.

MEEK, FIELDING BRADFORD (December 10, 1817, Madison, Ind.—December 21, 1876, Washington, D.C.). *Paleontology; Geology.* Son of lawyer. Never married. EDUCATION: Attended public schools in Madison. CAREER: Attempted unsuccessfully to establish mercantile business both at Madison and at Owensboro, Ky.; for several years thereafter, found intermittent employment, including portrait painting, while apparently residing chiefly at Owensboro and pursuing study of fossils; 1848-1849, an assistant to D. D. Owen [q.v.] on U.S. geological survey of Iowa, Wisconsin, Minnesota; 1852-1858, assistant to J. Hall [q.v.], New York state paleontologist, at Albany; 1853 (summer), with F. V. Hayden [q.v.], Badlands of South Dakota region; 1854 and 1855 (summers), worked for Geological Survey of Missouri; 1858, took up residence as paleontologist at Smithson. Instn., where worked thereafter, especially on fossils collected in West by F. V. Hayden, and lived frugally on small income. MEMBERSHIPS: Natl. Ac. Scis. SCIENTIFIC CONTRIBUTIONS: In poor health and of retiring disposition, nevertheless did enormous amount of work and became a leading American invertebrate paleontologist; worked especially with F. V. Hayden, with A. H. Worthen [q.v.] of Illinois Geological Survey, and also for Ohio and California surveys; in 1855, with J. Hall, produced first of over 100 publications, and in 1857 was first to recognize Permian fossils in North America; noted for paleontological work for U.S. Geological and Geographical Survey of Territories (Hayden survey) although never officially employed by that organization. Work in invertebrate paleontology ranged widely over geological time and geographical area, and his interpretation of specimens helped explain geological character of western half of U.S.

WORKS: "Palaeontology of the Upper Missouri," *Smithson. Contr. Knowl.* 14 (1865), with F. V. Hayden; *Report on the Invertebrate Cretaceous and Tertiary Fossils of the Upper Missouri Country,* Report of U.S. Geological Survey of the Territories, vol. 9 (Washington, D.C., 1876), 629 pp. plus 45 plates. Bibliography in *Biog. Mems. of Natl. Ac. Scis.* 4 (1902) : 80-91. Other references to bibliographies in *DAB* and *DSB*. MSS: Smithson. Instn. Archives (*NUCMC* 72-1244). WORKS ABOUT: *DAB,* 12: 493-94 (George P. Merrill) ; *DSB,* 9: 255-56 (Ellis L. Yochelson).

MELLUS, EDWARD LINDON (1848-1922). *Anatomy; Physiology.* Had medical degree; resided part of life in Baltimore; spent time in Obersteiner's laboratory in Vienna, and with Sir Victor Horsley in London; retired at young age, and pursued research in neurological physiology.

WORKS ABOUT: See *AMS;* death notice, *Journal of Am. Med. Assoc.* 80 (1923) : 126.

MELSHEIMER, FRIEDRICH VALENTIN (September 25, 1749, Negenborn, Duchy of Brunswick, Germany—June 30, 1814, Hanover, Pa.). *Entomology.* Son of Joachim Sebastian, superintendent of ducal forests, and Clara M. (Reitemeyer) Melsheimer. Married Mary Agnes Man, 1779; eleven children. EDUCATION: 1769-1772, attended Univ. of Helmstedt. CAREER: 1772-1776, private tutor; 1776, commissioned chaplain to Dragoon Regiment of Brunswick Auxiliaries employed by British to fight in America; 1777, wounded at Battle of Bennington and taken prisoner; 1779, released from prison, resigned commission, took charge of several Lutheran congregations in Lancaster County, Pa., and preached also in Lebanon and Dauphin counties, Pa.; 1784-1787, minister, apparently preached chiefly at Manheim and New Holland, Pa.; at Franklin College (Lancaster, Pa.), professor of Greek, Latin, and German (1787-1789) and president (ca. 1788-1789); 1789-1814, minister at Hanover, Pa. MEMBERSHIPS: Am. Phil. Soc. SCIENTIFIC CONTRIBUTIONS: The pioneer student of insects in America; *Catalogue* [see Works], now rare, was first separate publication in American entomology and maintained that status for some twenty-five years; *Catalogue* dealt only with beetles, listing 1,363 species of which more than 400 are now identified, and included notes on habits of the insects; also gathered first important collection of American insects and planned second volume of *Catalogue* to encompass insects other than beetles, but this project never was completed. Apparently also interested in other aspects of natural history and also in mineralogy and astronomy. Sons Johan Friedrich and Friedrich Ernst Melsheimer inherited father's collections and interest in entomology, and latter was first president of Entomological Society of Pennsylvania (1842) and published *Catalogue of the Described Coleoptera of the U.S.,* revised by S. S. Haldeman and J. L. LeConte [qq.v.] (Washington, D.C., 1853).

WORKS: *A Catalogue of Insects of Pennsylvania* (Hanover, Pa., 1806), 60 pp. No other works listed in Meisel, *Bibliog. Amer. Nat. Hist.;* no works in *Roy. Soc. Cat.; DAB* includes references to several publications in German. WORKS ABOUT: *DAB,* 12: 518-19 (George H. Genzmer) ; M. Luther Heisey, "Frederick Valentine Melsheimer," *Papers of Lancaster County Historical Society* 41 (1937) : 103-11, 162-63; Mallis, *Amer. Entomologists,* pp. 9-13.

MENDENHALL, THOMAS CORWIN (1841-1924). *Physics.* Superintendent, U.S. Coast and Geodetic Survey; president, Worcester Polytechnic Institute.

MSS: American Institute of Physics—Center for History of Physics, New York (*NUCMC* 65-1719) ; presidential papers in Worcester Polytechnic Institute Archives. WORKS ABOUT: See *AMS*; *DAB*.

MERRILL, GEORGE PERKINS (1854-1929). *Geology.* Curator and head curator of geology, U.S. Natl. Mus.

MSS: Univ. of Maine—Folger Library, Orono (*NUCMC* 74-422). WORKS ABOUT: See *AMS*; *DAB*.

METCALFE, SAMUEL LYTLER (September 21, 1798, near Winchester, Va.—July 17, 1856, Cape May, N.J.). *Chemistry.* Son of Joseph, probably farmer, and Rebecca (Littler or Sittler) Metcalfe. Married first before 1831 [name of wife not known]; Ellen Blondel, 1846; one child, by second marriage. EDUCATION: 1823, M.D., Transylvania Univ. (Lexington, Ky.). CAREER: 1823-1831, medical practice, first at New Albany, Ind., and later in Mississippi, residing chiefly at Natchez; 1831, went to England to pursue studies in chemistry and geology; upon return to U.S., lived in New York City, where engaged in writing on chemical and other scientific topics; 1835-1845, returned to London and there pursued scientific research and writing; 1845, returned to U.S. SCIENTIFIC CONTRIBUTIONS: While in Mississippi, conducted walking tour through eastern Tennessee and North Carolina, and wrote articles on chemistry, geology, botany, and zoology of that area. After return from first visit to England, engaged in scientific writing, several papers appearing in *Knickerbocker Magazine* under initial "M." Became intensely interested in nature of heat and returned to England 1835 to work in libraries and to elaborate ideas finally published as *Caloric* [see Works], an expansion of ideas presented earlier in *New Theory* [see Works]. During period in England after 1835, was proposed as candidate for Gregorian professorship at Edinburgh Univ., but declined in order to continue research and writing.

WORKS: *A New Theory of Terrestrial Magnetism* (New York, 1833); *Caloric: Its Mechanical, Chemical, and Vital Agencies in the Phenomena of Nature*, 2 vols. (London, 1843; 2d ed., Philadelphia, 1859). WORKS ABOUT: *DAB*, 12: 584 (Lyman C. Newell); E.M., "Dr. Metcalfe's Life," in Metcalfe, *Caloric*, 2d ed., 1: xv-xviii.

MICHAEL, ARTHUR (1853-1942). *Chemistry.* Professor of chemistry, Tufts Univ.; professor of organic chemistry, Harvard Univ.

WORKS ABOUT: See *AMS*; *DAB*, supp. 3; *DSB*.

MICHAUX, ANDRÉ (March 7, 1746, Satory, near Versailles, France—November 1802, Madagascar). *Botany.* Father was manager of royal farm at Satory. Married Cécile Claye, 1769; one child, François A. Michaux [q.v.]. EDUCATION: For four years, attended boarding school; later studied botany with Bernard de Jussieu and others and at Jardin des Plantes. CAREER: Succeeded father as manager of royal farm at Satory; after death of wife (the daughter of well-to-do Beauce farmer) in 1770, began study of botany; 1779-ca. 1781, moved closer to Jardin des Plantes and conducted botanical searches in England, Auvergne, and the Pyrenees; ca. 1781, appointed secretary to French consul in Persia; 1782-1785, traveled throughout Persia, and there collected plants and seeds; 1785, sent by French government to study and collect forest trees of North America; 1785-1787, investigated local flora and established garden near Hackensack, N.J.; 1787, removed to Charleston, S.C., where bought plantation and set up second garden; 1787-ca. 1789, explored and collected in southern Appalachians, Spanish Florida, Bahamas, and mountains of North Carolina; 1789, with the Revolution, support from French government terminated; 1792, conducted botanical explorations in Canada to Hudson Bay; 1793, began explorations in Kentucky, southern Appalachians (1794), and Illinois; 1796, left Charleston for France; 1800, departed from France as naturalist with expedition of Nicolas Baudin to Australia and in 1802, while engaged in plans to establish garden on Madagascar, died there of fever. SCIENTIFIC CONTRIBUTIONS: Collected and sent thousands of American plants and seeds to France, in addition to similar collections from other geographical areas. *Flora* [see Works] was largely based on personal collections and notes and was first such publication relating to eastern America.

WORKS: *Histoire de Chênes* [*Oaks*] *de l'Amérique* (Paris, 1801); *Flora Boreali-Americana* (Paris, 1803; reprinted, New York, 1973, with introduction by Joseph Ewan). Another work, and notes on authorship of *Flora*, in *DSB*. MSS: Am. Phil. Soc. (Hamer). WORKS ABOUT: *DAB*, 12: 591-92 (Horace B. Barker); *DSB*, 9: 365-66 (Joseph Ewan).

MICHAUX, FRANÇOIS ANDRÉ (August 16, 1770, Satory, near Versailles, France—October 23, 1855, France). *Botany.* Son of André [q.v.], botanist and botanical collector, and Cécile (Claye) Michaux. Married late in life; no children. EDUCATION: After return to France in 1790, studied medicine. HONORS: Chevalier of Legion of Honor. CAREER: 1785, accompanied father to U.S., helped establish botanical nursery at Hackensack, N.J., and there grew trees for shipment to France; 1787, assumed primary responsibility for management of second nursery near Charleston, S.C.; 1787, explored in Florida, Bahamas, and Tennessee Valley; 1790, returned to France and took up study of medicine, while participating in French Revolution; beginning in 1796, assisted father in cultivation of plants brought back from America; 1801, under commission from French government, went to U.S. to dismantle and sell gardens at Hackensack and Charleston; 1802, traveled from Philadelphia to Kentucky and Tennessee; 1803, returned to France; 1806, returned again to U.S. to collect trees for France; 1806-ca. 1809, explored eastern U.S. from Maine to Georgia, Great Lakes, and elsewhere; ca. 1809, returned to France and engaged in preparation of *Sylva* [see Works]; thereafter, spent remainder of life in growing trees, as administrator of experimental estate belonging to Société Centrale de l'Agriculture, and in related work elsewhere in France. MEMBERSHIPS: Am. Phil. Soc. SCIENTIFIC CONTRIBUTIONS: Botanical work characterized by attention to properties and uses of wood and was less scientific and more practically oriented than work of father. *Voyage* [see Works] included descriptions of plants as well as soil, agriculture, and other such matters. Chief work was *Sylva*, to which T. Nuttall [q.v.] later added supplement on trees of lands farther west.

WORKS: *Voyage à l'Ouest des Monts Alléghanys* (Paris, 1804), with English and German translations (1805); *The North American Sylva; or, A Descrip-*

tion of the Forest Trees . . . Considered Particularly with Respect to Their Use in the Arts and Their Introduction into Commerce, 3 vols. (Paris, 1818-1819), which is translation from original French edition (Paris, 1810-1813). MSS: Am. Phil. Soc. (*NUCMC* 60-2773). WORKS ABOUT: *DAB,* 12: 592-93 (Horace B. Baker); Rodney H. True, "François André Michaux, the Botanist and Explorer," *Proceedings of Am. Phil. Soc.* 78 (1937-1938): 313-27.

MICHELSON, ALBERT ABRAHAM (1852-1931). *Physics.* Professor of physics, head of department, Univ. of Chicago.

MSS: Michelson Museum, U.S. Department of Navy, China Lake, Calif. (*NUCMC* 71-1184); also see *DSB.* WORKS ABOUT: See *AMS*; *DAB*; *DSB.*

MICHENER, EZRA (November 24, 1794, London Grove Township, Chester County, Pa.—June 24, 1887, New Garden, Pa.). *Natural history.* Son of Mordecai, probably farmer, and Alice (Dunn) Michener. Married Sarah Spencer, 1819; Mary S. Walton, 1844; seven children. EDUCATION: 1818, M.D., Univ. of Pennsylvania. CAREER: Until age twenty-two, worked as teacher, surveyor, and farm worker; ca. 1816-1818, associated with Philadelphia Dispensary and was student at Univ. of Pennsylvania; 1818, began medical practice, Chester County, Pa., and continued there for remainder of life, after 1829 at New Garden township; on occasion, lectured on anatomy, physiology, diseases of women, experimental philosophy. MEMBERSHIPS: Corresponding member of Ac. Nat. Scis. Philad. and of West Chester Microscopical Society. SCIENTIFIC CONTRIBUTIONS: Activities as collector and student of natural history related especially to Chester County; began botanical studies in youth and continued in adult years, accumulating large collection of flowering and other plants, and also collected zoological specimens; assisted W. Darlington [q.v.] with *Florula Cestrica,* and contributed zoological and botanical catalogues to G. Cope and J. S. Futhey's *History of Chester County, Pennsylvania* (1881). *Autobiographical Notes* [see Works About] included an essay on electricity as aid to understanding lightning rod and an essay on tornadoes in which author attempted to analyze and deal with forces of tornado in terms of established physical laws of matter and motion; wrote extensively on medicine, agriculture, science, religion, and other topics, and corresponded with number of scientists.

WORKS: *Manual of Weeds* (Philadelphia, 1872); *Conchologia Cestrica* (Philadelphia, 1874), with W. D. Hartman. WORKS ABOUT: *DAB,* 12: 596-97 (E. Estelle Wells); *Autobiographical Notes from the Life and Letters of Erza Michener, M.D.* (Philadelphia, 1893).

MICHIE, PETER SMITH (March 24, 1839, Brechin, County Forfar, Scotland [1843, came to U.S. with parents, settled in Cincinnati]—February 16, 1901, West Point, N.Y.). *Physics*; *Engineering.* Son of William and Ann D. (Smith) Michie. Married Marie Louise Roberts, 1863; three children. EDUCATION: 1863, graduated, U.S. Military Academy, West Point. HONORS: Ph.D., College of New Jersey (Princeton), 1871; M.A., Dartmouth College, 1873. CAREER: 1863, made first lieutenant, Corps of Engineers; 1863-1864, assistant engineer in military action against Charleston, S.C.; 1864, chief engineer, district of Florida; 1864-1865, assistant engineer, then chief engineer, Army of the James; 1865, brevetted brigadier general of volunteers and lieutenant colonel; 1865, promoted to captain, Corps of Engineers; 1865-1866, engaged in survey of Richmond, Va., area; 1866-1867, leave of absence; at West Point, instructor (1867-1871) and professor of natural and experimental philosophy (1871-1901); 1870, member of commission to Europe to gather information on iron used in coastal defense; 1871-1901, member of Board of Overseers, Thayer School of Civil Engineering, Dartmouth College. SCIENTIFIC CONTRIBUTIONS: During Civil War, at a young age perfomed important military engineering work. At U.S. Military Academy, wrote several textbooks on mechanics, physics, and astronomy.

WORKS: *Elements of Wave Motion Relating to Sound and Light* (New York, 1882); *Elements of Analytical Mechanics* (West Point, 1886); *Hydromechanics* ([West Point], 1888); *Practical Astronomy* ([West Point], 1891), with F. S. Harlow. Several other works listed in *DAB.* No works listed in *Roy. Soc. Cat.* WORKS ABOUT: *DAB,* 12: 597-98 (Gustav J. Fiebeger); Cullum, *Biographical Register, West Point* 2: 866-67.

MILES, MANLY (July 20, 1826, Homer, Cortland County, N.Y.—February 15, 1898, Lansing, Mich.). *Agriculture*; *Zoology.* Son of Manly, probably farmer, and Mary (Cushman) Miles. Married Mary E. Dodge, 1851; no information on children. EDUCATION: 1850, M.D., Rush Medical School, Chicago. HONORS: D.V.S., Columbia Veterinary College (N.Y.), 1880. CAREER: Ca. 1850, began medical practice, Flint, Mich.; 1859-1861, zoologist, Michigan State Geological Survey; at Michigan State Agricultural College (Michigan State Univ.), professor of zoology and animal physiology (1861-1865), professor of animal physiology and practical agriculture (1865-1869), professor of agriculture (1869-1875), and superintendent of farm (1865-1875); 1874, on leave of absence, passed part of time with field crop experimenters, Lawes and Gilbert, in England; 1875, appointed professor of agriculture, Univ. of Illinois; 1878, became experimentalist, Houghton Farm, Mountainville, N.Y.; 1883, made professor of agriculture, Massachusetts Agricultural College (Univ. of Massachusetts); 1886-1898, returned to Lansing, Mich., established office and laboratory, and engaged in scientific and agricultural research and writing. MEMBERSHIPS: An original member of Michigan Academy of Science; member of Am. Assoc. Advt. Sci. and other societies. SCIENTIFIC CONTRIBUTIONS: On his own and in connection with Michigan survey, gathered large collection of fauna of the state, and as state zoologist prepared and published report [see Works]; particularly interested in mollusks, and also collected and studied birds, mammals, reptiles, and fishes; corresponded with and sent specimens to numerous other naturalists. With appointment at Michigan State Univ., 1865, became first U.S. professor of practical agriculture, and later published books on stock breeding, Indian corn, silos, land drainage; also wrote papers on various agricultural and related topics.

WORKS: "Zoology," in *First Biennial Report of the Progress of the Geological Survey of Michigan* (Lansing, Mich., 1861), pp. 211-41. See *Roy. Soc. Cat.*, vol. 17. Chief agricultural works listed in *DAB*. WORKS ABOUT: *DAB*, 12: 613-14 (Rufus P. Hibbard); "Sketch," Michigan Academy of Science, *Second Report . . . for the Year Ending June 30, 1900* (Lansing, Mich. 1901), pp. 101-7 [appeared in earlier form in *Popular Science Monthly*, April 1899].

MILLSPAUGH, CHARLES FREDERICK (1854-1923). *Botany.* Curator of botany, Field Museum (Chicago).

WORKS ABOUT: See *AMS*; *DAB*.

MINOT, CHARLES SEDGWICK (1852-1914). *Embryology*; *Biology.* Professor of histology and embryology, Harvard Univ. Medical School.

MSS: Harv. U.-MS (*NUCMC* 62-3755). WORKS ABOUT: See *AMS*; *DAB*; *DSB*.

MINTO, WALTER (December 6, 1753, near Cowdenknows, County Merse, Scotland—October 31, 1796, Princeton, N.J.). *Mathematics.* No information on parentage, except that they were poor. Married Mary Skelton (of Princeton); no children. EDUCATION: At age fifteen, was in attendance at Univ. of Edinburgh; studied theology, residing at that time in Perthshire; while in Italy (beginning 1776), studied mathematics under direction of Prof. Slop. HONORS: LL.D., Univ. of Aberdeen. CAREER: 1776, went to Italy as tutor to sons of George Johnstone, and took up residence at Pisa with Giuseppe Slop, astronomer; 1779, returned to Edinburgh, and soon thereafter became teacher of mathematics, Univ. of Edinburg; 1786, went to U.S. and was made principal at Erasmus Hall, Flatbush, Long Island; at College of New Jersey (Princeton), professor of mathematics and natural philosophy (1787-1796) and treasurer (1795-1796). MEMBERSHIPS: Am. Phil. Soc. SCIENTIFIC CONTRIBUTIONS: In 1781, William Herschel discovered Uranus, and shortly thereafter Prof. Slop made observations of planet at Pisa; Slop communicated results to Minto, and the two men worked out mathematics of new planet's orbit; in 1783, Minto published *Researches* [see Works], which included mathematical formulae, as well as data on observations on Uranus made by various persons. Won reputation for mathematical efforts, corresponded with English philosophers, prepared several astronomical papers, and worked with David Steuart Erskine, Earl of Buchan, on life of John Napier (published in Edinburgh, 1787), which credited Napier with invention of logarithms. At Princeton, is said to have prepared manuscript of mathematics textbook, which was not published before early death and apparently was subsequently lost.

WORKS: *Researches into Some Parts of the Theory of the Planets in Which Is Solved the Problem, To Determine the Circular Orbit of a Planet by Two Observations; Exemplified in the New Planet* (London, 1783); *An Inaugural Oration, on the Progress and Importance of the Mathematical Sciences* (Trenton, N.J., 1788). MSS: Princeton Univ. Library and letters to Earl of Buchan at Univ. of Edinburgh (Eisenhart, "Minto" [below]). WORKS ABOUT: *DAB*, 13: 32 (Luther P. Eisenhart); Eisenhart, "Walter Minto and the Earl of Buchan," *Proceedings of Am. Phil. Soc.* 94 (1950): 282-94.

MITCHEL, ORMSBY MacKNIGHT (July 28, 1809, Morganfield, Union County, Ky.—October 30, 1862, Beaufort, S.C.). *Astronomy.* Son of John, surveyor and planter, and Elizabeth (MacAlister) Mitchel. Married Louisa (Clark) Trask, 1831; six children. EDUCATION: 1829, graduated, U.S. Military Academy, West Point; studied law and admitted to bar. HONORS: M.A., Harvard Univ., 1851. CAREER: 1829-1831, assistant professor of mathematics, West Point; 1831-1832, stationed at Fort Marion, St. Augustine, Fla.; 1832, resigned from army and went to Cincinnati; ca. 1833, began law practice at Cincinnati; 1836-ca. 1845, professor of mathematics, philosophy, and astronomy, Cincinnati College; 1836-1837, chief engineer, Little Miami Railroad; public lecturer on astronomy in New York, Boston, and elsewhere; 1842, sent to Europe by Cincinnati Astronomical Society to purchase telescope; 1845-1861, director, Cincinnati Observatory; 1846-1848, founder and editor, *Sidereal Messenger*; 1852, became consulting engineer, Ohio and Mississippi

Railroad, and in 1853 went to Europe to sell bonds for that railroad; 1859-1861, director, Dudley Observatory, Albany; 1861-1862, general in Union army. MEMBERSHIPS: Am. Phil. Soc.; Royal Astronomical Society of England. SCIENTIFIC CONTRIBUTIONS: Considerable success as lecturer stimulated public interest in astronomy, and he was largely responsible for establishment of Cincinnati Observatory, which was begun in 1842 and completed in 1845 with world's second largest telescope; studies at observatory included investigations of double and multiple stars, which made important additions to similar work done by Europeans. Devised chronograph for use in meridian observations. *Sidereal Messenger* was first journal of popular astronomy. Offered, but declined professorships at Harvard (1846) and Univ. of Albany (1851).

WORKS: *Planetary and Stellar Worlds* (New York, 1848); *Popular Astronomy* (New York, 1860). See several works listed in *Roy. Soc. Cat.*, vols. 4, 6 (additions). MSS: Cincinnati Historical Society (*NUCMC* 71-1549). WORKS ABOUT: *DAB*, 13: 38-39 (Jermain G. Porter); F. A. Mitchel (son), *Ormsby MacKnight Mitchel* (Boston and New York, 1887); Stephen Goldfarb, "Science and Democracy: A History of the Cincinnati Observatory, 1842-1872," *Ohio History* 78 (1969): 172-78.

MITCHELL, ELISHA (August 19, 1793, Washington, Conn.—June 27, 1857, on Mt. Mitchell, N.C.). *Natural history.* Son of Abner, farmer, and Phoebe (Eliot) Mitchell. Married Maria Sybil North, 1819; seven children. EDUCATION: 1813, A.B., Yale College; for brief period, attended Andover Theological Seminary; 1817, licensed to preach. HONORS: D.D., Univ. of Alabama, 1838. CAREER: Ca. 1813-1816, teacher, Long Island, N.Y., and New London, Conn.; 1816-1817, tutor, Yale College; at Univ. of North Carolina (Chapel Hill), professor of mathematics and natural philosophy (1818-1825), professor of chemistry, geology, and mineralogy (1825-1857); 1821, ordained Presbyterian minister, Hillsborough, N.C.; 1826, North Carolina state geologist. MEMBERSHIPS: Am. Assoc. Advt. Sci. SCIENTIFIC CONTRIBUTIONS: Traveled extensively in North Carolina, studying geology, topography, botany, and zoology; these observations were published especially in *Am. Journ. Sci.*; also wrote on meteorology and agriculture and produced manuals for students in natural history, botany, chemistry, geology, and mineralogy. *Elements* [see Works] was accompanied by first geological map of North Carolina. First measured height of Mt. Mitchell, highest peak in U.S. east of Rockies, beginning exploration in that area in 1835; on fifth excursion to the mountain in 1857 died in an accident.

WORKS: *Elements of Geology With an Outline of the Geology of North Carolina* ([n.p.], 1842). List of works in *DSB*. MSS: Univ. N.C.-SHC (*NUCMC* 64-580). WORKS ABOUT: *DAB*, 13: 45-46 (Collier Cobb); *DSB*, 9: 420-21 (Ellen J. Moore).

MITCHELL, HENRY (September 16, 1830, Nantucket, Mass.—December 1, 1902, New York, N.Y.). *Hydrography.* Son of William [q.v.]—teacher, banker, and astronomer—and Lydia (Coleman) Mitchell; younger brother of Maria Mitchell [q.v.]. Married Mary Dawes, date uncertain; Margaret Hayward, 1873; Mary Hayward, 1877; one child, by second marriage. EDUCATION: Educated in private schools and at home. HONORS: M.A., Harvard Univ., 1867. CAREER: 1849-1888, connected with U.S. Coast Survey, associated initially with triangulation work, but later joined hydrographic branch; 1869-1876, professor of physical hydrography, Massachusetts Institute of Technology; 1874, appointed member of commission on construction of interoceanic canal and of board of engineers to survey mouth of Mississippi River; 1879, appointed member of Mississippi River Commission; also served in several other advisory roles on matters related to harbors. MEMBERSHIPS: Natl. Ac. Scis. SCIENTIFIC CONTRIBUTIONS: Investigations with Coast Survey were particularly significant in demonstrating dynamics of tides and currents and their geological effects as determinants of shore line and channel configurations. Came to hold rank as leading American hydrographer. Until 1866, survey work related chiefly to northeast coast, especially New York harbor; in 1866, began work on southeastern coast of U.S., including hydrographic work related to proposed cable between U.S. and Cuba and study of Gulf Stream; in 1868-1869, sent to Europe to study latest hydrographic and engineering practices; between 1869 and 1874 again took up study of New England and New York harbors, and in 1874, as member of canal commission, visited Central America; later involved in work on Mississippi and Delaware rivers and elsewhere. In addition to other contributions, devised several instruments that were used in hydrographic studies.

WORKS: Chief publications, which appeared mainly in *Annual Reports of U.S. Coast Survey*, dealt with "Reclamation of Tide Lands and Its Relation to Navigation" (1869), Delaware River (1878), Gulf of Maine (1879), New York harbor (1886); *Tides and Tidal Phenomena* (Washington, D.C., 1868), 56 pp. Chief works listed in *DAB*; full bibliography in *Biog. Mems.* [below], pp. 148-50. WORKS ABOUT: *DAB*, 13: 47-48 (H. A. Marmer); Marmer, "Memoir," *Biog. Mems. of Natl. Ac. Scis.* 20 (1939): 141-50.

MITCHELL, JAMES ALFRED (November 21, 1852, Ireland—October 18, 1902, Emmitsburg, Md.). *Geology.* Son of M. and Jane (Petrie) Mitchell. Married Margaret J. Willson, 1889; at least one child. EDUCATION: Received diploma in science, Royal School of Mines, England; studied astronomy and meteorology, Lord Ross's Observatory, Birr Castle, Ireland; 1895-1897, student in geology, Graduate School of Philosophy, Johns Hopkins Univ. HONORS: A.M., Mount St. Mary's College, 1888; Ph.D. [hon.?], Niagara Univ., 1894. CAREER: Taught at College of St. Stanislaus and at Clongowes Wood College, Ireland, and also is said to have taught at Stonyhurst; 1888-1902, professor of geology, mathematics, natural science, and mechanical drawing, Mount St. Mary's College (Emmitsburg, Md.); also lecturer in natural science, St. Joseph's Academy (Emmitsburg). MEMBERSHIPS: Washington Geological Society, Natl. Geog. Soc. (Washington, D.C.). SCIENTIFIC CONTRIBUTIONS: Chief work was in geology, especially discovery of fossil footprints in Maryland [see Works]. Information on life and career sparse and uncertain. Said to have worked on reports for state [?] geological survey and U.S. Weather Bureau.

WORKS: "The Discovery of Fossil Tracks in the Newark System (Jura-Trias) of Frederick County, Maryland," *Johns Hopkins Univ. Circulars* 15 (1895): 15-16. WORKS ABOUT: *World Who's Who in Science;* information provided by archivist of Mount St. Mary's College.

MITCHELL, JOHN (April 3, 1711, probably White Chapel Parish, Lancaster County, Va.—February 29, 1768, England). *Natural history; Medicine; Chemistry; Cartography.* Son of Robert—merchant, farmer, and tobacco receiver—and Mary (Chilton) Sharpe Mitchell. Married Helen [?]; no information on children. EDUCATION: Attended Univ. of Edinburgh and very likely received A.M. in 1729; studied medicine at Univ. of Edinburgh; ca. 1731, may have received M.D. degree from European university. CAREER: Ca. 1732, returned to Virginia after some years of study abroad; ca. 1732-1734, apparently practiced medicine in Lancaster County, Va.; 1734-ca. 1745, lived and practiced medicine in and about Urbanna, Middlesex County, Va.; 1735, became physician to the poor, Christ Church Parish, Middlesex County; 1746, gave up medical practice because of health, moved to England, and there apparently lived at least partly on investments; 1750, began service with Lords Commissioners of Trade and Plantations; 1759-1761, aided in establishment of Kew Gardens. MEMBERSHIPS: Am. Phil. Soc.; Roy. Soc. SCIENTIFIC CONTRIBUTIONS: Actively interested in wide range of subjects. Collected botanical specimens and sent them to England; carried out study of pines in Virginia, including consideration of ecological and physiological matters and their medical and commercial uses, as well as classification and description (this work never published); wrote early account of what was thought to be yellow fever (published 1804); investigated preparation and uses of potash. "Dissertatio Brevis" [see Works] was first work on principles of taxonomy written by North American. Made early study of anatomy and reproduction of opossum; wrote on electricity and on pigmentation in humans. Best known for map [see Works], which played important historical role in boundary questions for many years thereafter. In 1756, was serious candidate for librarian of British Museum.

WORKS: "Dissertatio Brevis de Principiis Botanicorum et Zoologicorum . . . ," *Acta Physico-Medica Academae Caesarae . . . Ephemerides* 8 (1748): 188-224; *A Map of the British and French Dominions in North America . . .*, under auspices of Lords Commissioners for Trade and Plantations (London, 1755), 8 sheets. List of publications in Berkeley and Berkeley [below], p. 274, with references to manuscripts in footnotes and on p. 269. WORKS ABOUT: *DAB*, 13: 50-51 (Lawrence Martin); Edmund Berkeley and Dorothy S. Berkeley, *Dr. John Mitchell* (Chapel Hill, N.C., 1974).

MITCHELL, JOHN KEARSLEY (May 12, 1793, Shepherdstown, Va. [now W.Va.]—April 4, 1858, Philadelphia, Pa.). *Chemistry; Physiology.* Son of Alexander, physician and farmer, and Elizabeth (Kearsley) Mitchell. Married Sarah Matilda Henry, 1822; nine children, including S. Weir Mitchell [q.v.]. EDUCATION: Ca. 1805, both parents having died, sent to Scotland to be educated, with relatives at Ayr and in academic department of Univ. of Ediburgh; 1819, M.D., Univ. of Pennsylvania. CAREER: 1814, returned to Virginia after several years in Scotland; 1816, began private study of medicine; 1817-1818, made first of four voyages to Asia as ship's surgeon; 1822, established medical practice, Philadelphia, and there became a leading physician; 1824, lecturer, institutes of medicine and physiology, Philadelphia Medical Institute; 1833-1838, lecturer on chemistry, Franklin Institute; 1841-1858, professor of theory and practice of medicine, Jefferson Medical College (Philadelphia); served as visiting physician, Pennsylvania Hospital and city hospital. MEMBERSHIPS: Am. Phil. Soc. and other societies. SCIENTIFIC CONTRIBUTIONS: While engaged in teaching chemistry, also conducted several investigations in that subject; an early writer on subject of osmosis and on liquefaction of carbonic acid gas [see Works]; studied arsenic and its effects. Published lectures presenting view that certain diseases had parasitic origins [see Works]; also wrote other works on medicine and on mesmerism. Published book of poetry.

WORKS: "On the Penetrativeness of Fluids," *American Journal of Medical Sciences* 7 (1830) : 36-67; "On the Penetration of Gases," ibid. 13 (1833) : 100-112; "On the Liquefaction and Solidifaction of Carbonic Acid," *Journal of Franklin Institute* 2d ser. 22 (1838) : 289-95; *On the Cryptogamous Origin of Malarious and Epidemical Fevers* (Philadelphia, 1849). Also see *Natl. U. Cat. Pre-1956 Imp.* MSS: Duke Hospital Library, Durham, N.C. (Hamer). WORKS ABOUT: *DAB,* 13: 54-55 (Charles W. Burr) ; Anna R. Burr, *Weir Mitchell: His Life and Letters* (New York, 1929), especially p. 3-23.

MITCHELL, MARIA (August 1, 1818, Nantucket, Mass.—June 28, 1889, Lynn, Mass.). *Astronomy.* Daughter of William [q.v.]—teacher, banker, and astronomer—and Lydia (Coleman) Mitchell; sister of Henry Mitchell [q.v.]. Never married. EDUCATION: Educated in Nantucket schools and by father. HONORS: LL.D., Columbia College, 1887; honorary degrees from other institutions. CAREER: 1836, appointed librarian, Nantucket Atheneum, and maintained that position for some twenty years: 1849-1868, computer for *U.S. Nautical Almanac* office; 1857-1858, went abroad to inspect observatories and to meet European scientists; 1861, with father, moved to Lynn, Mass.; 1865-1889, professor of astronomy, director of observatory, Vassar College. MEMBERSHIPS: Am. Ac. Arts Scis. (first female member) ; Am. Phil. Soc. SCIENTIFIC CONTRIBUTIONS: Ranks as first woman in America to be recognized for work in astronomy. Began at young age to assist father with observations used to rate chronometers of Nantucket whaling ships, and in 1831 helped him in work used to determine longitude of Nantucket; carried out telescopic survey of heavens, including observations and calculations of orbits of several comets, and in 1847 discovered new comet, an achievement that brought wide recognition, including gold medal from king of Denmark; telescope from group of American women made possible further astronomical work. Interested in promotion of cause of women, and at Vassar College her wide reputation for work in astronomy helped establish educational status of that new institution. Work for *Nautical Almanac* involved computations of ephemerides of planet Venus.

WORKS: Published only a few, brief papers. See *Roy. Soc. Cat.,* vols. 4, 8, 10, 17. MSS: Vassar College Library (*NUCMC* 66-1997) ; Nantucket Maria Mitchell Association (*Nota. Am. Wom.,* vol. 2). WORKS ABOUT: *DAB,* 13: 57-58 (Caroline Furness) ; *DSB,* 9: 421-22 (Dorrit Hoffleit). Also see Helen Wright, *Sweeper in the Sky: The Life of Maria Mitchell* (New York, 1950).

MITCHELL, SILAS WEIR (1829-1914). *Medicine; Neurology; Physiology.* Physician; associated with Philadelphia Orthopaedic Hospital and Infirmary for Nervous Diseases; novelist.

MSS: See references to several repositories in Hamer. WORKS ABOUT: See *AMS; DAB; DSB.*

MITCHELL, WILLIAM (December 20, 1791, Nantucket, Mass.—April 19, 1869, Poughkeepsie, N.Y.). *Astronomy.* Son of Peleg, shipowner and businessman engaged in whaling and oil business, and Lydia (Cartwright) Mitchell. Married Lydia Coleman, 1812; ten children, including Maria and Henry Mitchell [qq.v.]. EDUCATION: Attended school in Nantucket; at age fifteen, began study of cooper's trade. HONORS: M.A., Brown Univ. (1848) and Harvard Univ. (1860). CAREER: Ca. 1809 [age eighteen], became school principal; soon thereafter, assisted father in oil factory and cooperage; during War of 1812, lived on farm and resumed teaching; after war, returned to oil and cooperage business; 1820, delegate to Massachusetts Constitutional Convention; 1822, opened private school; 1827, became master of first free school of Nantucket, but soon thereafter reestablished private school; 1833, became secretary, Phoenix Bank; 1837-1861, cashier, Pacific Bank; 1845, Massachusetts state senator; 1848 and 1849, member of Governor's Council; 1857-1865, member of Board of Overseers, Harvard Univ.; 1861, retired, left Nantucket, and went to reside at Lynn, Mass.; 1865, removed to Poughkeepsie, N.Y., where daughter Maria was appointed professor of astronomy at Vassar College. MEMBERSHIPS: Am. Ac. Arts Scis.; Am. Assoc. Advt. Sci. SCIENTIFIC CONTRIBUTIONS: Interested in astronomy from early age. Was called on to check chronometers, using sextant, of all returning whaling ships. Acquired telescopes, and later received loan of instruments from Military Academy at West Point and telescope from Coast Survey and carried out observations of positions of stars for Coast Survey; wrote on tails of comets and on aurora of May 29, 1840, and published account of discovery by daughter Maria of comet of October 1, 1847; for number of years, maintained meteorological record. Served as member of visiting committee to Harvard College Observatory, 1848-1865, and in 1855 became chairman.

WORKS: Wrote several papers, published chiefly in *Am. Journ. Sci.* See *Roy. Soc. Cat.,* vol. 4. MSS: Nantucket Maria Mitchell Association (*DAB*). WORKS ABOUT: *DAB,* 13: 66-67 (Margaret Harwood) ; Helen Wright, "William Mitchell of Nantucket," *Proceedings of Nantucket Historical Association,* 1949, pp. 33-46.

MITCHILL, SAMUEL LATHAM (August 20, 1764, North Hempstead, Long Island, N.Y.—September 7, 1831, Brooklyn, N.Y.). *Chemistry*; *Natural history*; *Geology.* Son of Robert, farmer and "pounder" and overseer of highways, and Mary (Latham) Mitchill. Married Catherine (Akerly) Cock, 1799; adopted two children. EDUCATION: Ca. 1780-1783, private study of medicine in New York; 1786, M.D., Univ. of Edinburgh; returned to New York and pursued study of law. CAREER: 1787, returned to New York, licensed to practice medicine; 1788, appointed a commissioner to negotiate land purchase from Six Nations; 1791, 1798, and 1810, member of New York state legislature; 1792-1801, professor of natural history, chemistry, and agriculture (1793-1795, also professor of botany), Columbia Univ.; 1797-1824, a founder and chief editor, *Medical Repository*; 1799, through marriage acquired financial means; 1801-1804 and 1809-1813, member, U.S. Congress; 1804-1809, U.S. senator; at College of Physicians and Surgeons, New York, professor of chemistry (1807-1808), professor of natural history (1808-1820), and professor of botany and materia medica (1820-1826); 1818, surgeon general, state militia; 1826-ca. 1830, a founder and vice-president of short-lived Rutgers Medical College; for some twenty years, physician to New York Hospital. MEMBERSHIPS: Chief founder, N.Y. Lyc. Nat. Hist.; member of many other societies. SCIENTIFIC CONTRIBUTIONS: Added to number of scientific fields and promoted development of science and medicine in America, through *Medical Repository* and other activities. One of first Americans to teach new chemistry of Lavoisier. Published number of chemical papers, and produced theory of disease based on so-called Septon, chemically based concept that (although incorrect) promoted interest in problems of sanitation. Studied geology of New York state, and produced early descriptive work on American geology; of writings on zoology, that on New York fishes was particularly significant; did useful work in bibliography of American botany. Chief contribution was probably as organizer and promoter of science.

WORKS: "Sketch of the Mineralogical and Geological History of the State of New York," series of articles in *Medical Repository* (1798, 1800, 1801); *Explanation of the Synopsis of Chemical Nomenclature and Arrangement* (New York, 1801); "Fishes of New York," *Transactions of New York Literary and Philosophical Society* 1 (1815): 355-92. Bibliography in Hall [below], pp. 141-50. MSS: See Hall, which refers to collections in N.Y. Hist. Soc. and East Hampton (N.Y.) Public Library. WORKS ABOUT: *DAB*, 13: 69-71 (Lyman C. Newell); Courtney R. Hall, *A Scientist in the Early Republic: Samuel Latham Mitchill* (New York, 1934; reprinted, 1967).

MIXTER, WILLIAM GILBERT (1846-1936). *Chemistry*. Professor of chemistry, Sheffield Scientific School, Yale.

WORKS ABOUT: See *AMS*; *Who Was Who in America*, vol. 1.

MOHR, CHARLES THEODORE (December 28, 1824, Esslingen, Württemberg, Germany—July 17, 1901, Asheville, N.C.). *Botany*. Son of Louis M. Mohr, associated with manufacture of chemicals. Married Sophia Roemer, 1852; five children. EDUCATION: 1845, graduated, Polytechnic School, Stuttgart. CAREER: 1845-1846, accompanied August Kappler as botanical collector on excursion to Dutch Guiana; 1846, returned to Germany because of illness; 1847-1848, chemist with Hochstetter & Schickard Co., Brunn, Austria; 1848, removed to Cincinnati, Ohio, and there was employed in chemical manufacturing establishment; 1849-1850, joined California gold rush; for a time, became involved in farming venture with brothers in Clarke County, Ind.; removed to Louisville, Ky., and entered pharmaceutical business; 1857, settled in Mobile, Ala., where engaged in pharmaceutical business; 1892, gave pharmaceutical business over to son and turned full attentions to botany, including work for Division of Forestry, U.S. Department of Agriculture; 1900, moved to Asheville, N.C., in order to work in Biltmore Herbarium. SCIENTIFIC CONTRIBUTIONS: Botanical interests began in polytechnic school, and in U.S. he undertook extensive botanical study and collecting. Chiefly remembered for work on botany of Alabama; investigated mosses and ferns, prepared work on Alabama grasses and forage plants for U.S. Department of Agriculture (1878, 1879), and investigated forests of Gulf states for Tenth U.S. Census (1880). In 1896, Division of Forestry published his study of southern timber pines, and other works also were prepared for division; collected and studied southern botany for number of organizations, and assembled collections for several expositions. Chief work was *Plant Life* [see Works], first published by Department of Agriculture (1901); at death, had in preparation important study, "Economic Botany of Alabama," representing an interest characteristic of his botanical studies. Also interested in mineral resources.

WORKS: *Plant Life of Alabama*, Alabama edition, Geological Survey of Ala. (Montgomery, Ala., 1901); also *Roy. Soc. Cat.*, vols. 4 (Carl Mohr), 10, 12, 17. WORKS ABOUT: *DAB*, 13: 77-78 (Samuel W. Geiser); Eugene A. Smith, biographical sketch, in Mohr's *Plant Life*, pp. v-xii.

MORE, THOMAS (fl. 1670-1724). *Natural history*. No information on date or place of birth and death, parentage, marital status, education, or occupation. CAREER: Ca. 1670-ca. 1724, mention of More appears periodically in connection with scientific personalities in England; 1690s, by this time had become associated with Jacob Bobart, Edward Lhwyd, and

Oxford naturalists, probably in subordinate role; 1704, engaged by Lhwyd in collecting specimens and selling subscriptions to Lhwyd's work, in southwestern England, and that year visited Roy. Soc. in London; referred to as "Pilgrim Botanist" and employed by several English naturalists from time to time as collector, apparently mainly in British Isles; 1722-1724, visited New England, sponsored chiefly by William Sherard, who at about same time also promoted explorations by M. Catesby [q.v.] in southern colonies; while in colonies, More desired land for botanical experiments, made claim to discovery of tin mine, and sought appointment as surveyor general of woods; 1723-1724 (winter), returned to England apparently to press these claims at court; did not return to New England, and no evidence exists as to career after that time. SCIENTIFIC CONTRIBUTIONS: Natural history collector for several English naturalists. In 1704, presented before Roy. Soc. a scheme of "Tables for Reducing Nature under several heads," apparently not taken very seriously by other scientists. In New England, became involved in political and personal difficulties and accomplished very little as collector there; wrote two letters to Sherard from New England and sent some common specimens, not of much interest to English scientists, and descriptions reveal lack of familiarity with botanical knowledge of the time; two trees, grown from seeds gathered in New England, represent sole contribution of any lasting importance from More's sojourn in colonies.

WORKS: Frick and Stearns [below] reproduce More's two letters to Sherard, written from New England, October 27, 1722, and December 12, 1723 (pp. 120-27); this source also refers to other manuscripts. WORKS ABOUT: George F. Frick and Raymond P. Stearns, "Thomas More and His Expedition to New England," appendix to their *Mark Catesby: The Colonial Audubon* (Urbana, Ill., 1961), pp. 114-27; see summary in Stearns, *Sci. in British Colonies*, pp. 473-77.

MOREY, SAMUEL (October 23, 1762, Hebron, Conn.—April 17, 1843, Fairlee, Vt.). *Invention*; *Physics*; *Chemistry*. Son of Israel, farmer and tavern owner, and Martha (Palmer) Morey. Married Hannah Avery, ca. 1793; one child. EDUCATION: Attended public school at Orford, N.H. CAREER: Established successful lumbering business at Orford, N.H., where in 1766 parents had relocated; associated with construction of locks on Connecticut River, serving as chief engineer at Bellows Falls, Vt.; devoted much time to experimental and inventive work, and received several patents; during later years, moved to Fairlee, Vt., across river from Orford. SCIENTIFIC CONTRIBUTIONS: In 1793, received first patent, for steam motor used to turn roasting spit; in 1795 was granted patent for rotary steam engine designed for propelling boat, but failed to gain financial support for development, and later made claim that ideas had been appropriated by Robert Fulton; during period 1795-1818 received several additional patents for wind, water, and tide mills, for additional improvements in steam engine and in boilers, for fireplace, and especially, in 1815, for revolving steam engine, sale of which brought some financial return. Publications demonstrated familiarity with up-to-date developments in chemistry and physics. From work on engines and boilers grew an interest in problems of fuels and combustion, and in *Am. Journ. Sci.* published results of experiments on steam directed upon burning combustible substances such as oil, tar, and turpentine; related these experiments to invention of device for producing heat and light. Published brief paper on bubbles in molten rosin and also on experiments on carbonated waters. Interest in reactions of turpentine vapor and air led to 1826 patent for internal combustion engine, probably Morey's most important invention.

WORKS: "On Heat and Light," *Am. Journ. Sci.* 2 (1820): 118-32; "Bubbles Blown in Melted Rosin," ibid., p. 179; "An Account of a New Explosive Engine" [turpentine engine], ibid. 11 (1826): 104-10. List of works and patents in Getman [below]. MSS: In private hands as of 1936 (Getman [below]); small collection (chiefly patents) in Dartmouth College Library. WORKS ABOUT: *DAB*, 13: 161-62 (Carl W. Mitman); Frederick H. Getman, "Samuel Morey," *Osiris* 1 (1936): 278-302. Also see George C. Carter, *Samuel Morey: The Edison of His Day* (Concord, N.H., 1945).

MORFIT, CAMPBELL (November 19, 1820, Herculaneum, Mo.—December 8, 1897, South Hampstead, England). *Chemistry*. Son of Henry Mason, lawyer and public official, and Catherine (Campbell) Morfit. Married Maria Clapier Chancellor, 1855; one child. EDUCATION: Attended Columbian College (George Washington Univ.); studied in private chemical laboratory of J. C. Booth [q.v.], Philadelphia. HONORS: M.D., Univ. of Maryland, 1853. CAREER: After leaving Booth laboratory, worked in another Philadelphia laboratory devoted to industrial chemistry and eventually became its owner; 1854-1858, professor of applied chemistry, Univ. of Maryland; 1858-1861, engaged in analytical and industrial work in New York; 1861-1897, lived in England, and there pursued work related to refinement of chemical manufacturing processes; apparently a person of means. MEMBERSHIPS: Chemical Society of London. SCIENTIFIC CONTRIBUTIONS: While in U.S., carried out research in industrial chemistry related to topics such as glycerine, guano, gums, salt, coal, sugar, and animal oils, some of which work was done with J. H. Alexander [q.v.]. In early 1850s, with J. C. Booth, conducted research on gunmetal for U.S. Ordnance Department, Morfit carrying out analytical work at his own laboratory at Pikesville (Md.) Arsenal; also worked with Booth on other projects, including *Encyclopedia of Chemistry* (1850). Published books on manufacture of soap and candles, on chemical and pharmaceutical manipulations, and on fertilizers. In England, concerned chiefly with improvement of technical processes, such as condensed food rations, paper, soap, and candle manufacture, and especially problems in refining cotton and linseed oils and use of cottonseed oil as food substance. In 1851, offered to provide funds for establishment of school of applied chemistry, connected with medical school at Univ. of Maryland, but offer was turned down as being inappropriate.

WORKS: Major books listed in *DAB*; also see *Roy. Soc. Cat.*, vols. 4, 8. MSS: Family papers in Md. Hist. Soc. (*NUCMC* 67-1614). WORKS ABOUT: *DAB*, 13: 162 (Lyman C. Newell); obituary, *Chemical News* (London) 76 (1897): 301-2.

MORGAN, ANDREW PRICE (1836-1906). *Botany*; *Mycology*. Resident of Ohio.

WORKS ABOUT: See *AMS*; *Ohio Naturalist* 1 (1901): 37.

MORLEY, EDWARD WILLIAMS (1838-1923). *Chemistry*; *Physics*. Professor of natural history and chemistry, Western Reserve Univ.

MSS: Lib. Cong. (*NUCMC* 59-213); Case Western Reserve Univ. Archives (*NUCMC* 74-71); also see *DSB*. WORKS ABOUT: See *AMS*; *DAB*; *DSB*.

MORRIS, JOHN GOTTLIEB (November 14, 1803, York, Pa.—October 10, 1895, Lutherville, Md.). *Entomology*. Son of John, physician, and Barbara (Myers) Morris. Married Eliza Hay, 1827; several children. EDUCATION: 1820-ca. 1822, attended College of New Jersey (Princeton); 1823, graduated, Dickinson College; 1823-1824, studied theology privately; 1825-1826, attended Princeton Theological Seminary; 1826-1827, attended Gettysburg Theological Seminary; 1827, ordained at Frederick, Md. CAREER: 1827-1860, minister, First English Lutheran Church, Baltimore; 1831-1833, founded and conducted *Lutheran Observer*; 1851, cofounder, Lutherville, Md.; 1860-1865, librarian, Peabody Institute, Baltimore; 1864-1873, minister, Third Church, Baltimore; ca. 1874, became minister, Lutherville, Md.; served as president of Maryland Synod and in 1843 and 1883 as president of General Synod; for many years, a director of Gettysburg Seminary and a trustee of Pennsylvania (later, Gettysburg) College; 1869-1894, delivered course of lectures at those institutions, including teaching of natural history at Pennsylvania College; lectured number of times at Smithson. Instn. MEMBERSHIPS: Am. Assoc. Advt. Sci. and other societies. SCIENTIFIC CONTRIBUTIONS: Ranks as one of earliest of American students of Lepidoptera and teachers of entomology; among publications were papers on entomology and other zoological topics that were contributed to *Literary Record*, the organ of Linnaean Association of Pennsylvania College; also served as that association's president (organized 1844). Chief works in entomology were "Catalogue" and "Synopsis" [see Works], the latter a work of marked importance and usefulness to contemporaries. Also wrote on theological and other topics.

WORKS: "Catalogue of the Described Lepidoptera of North America," *Smithson. Misc. Coll.* 3 (1860); "Synopsis of the Described Lepidoptera of North America; Part 1: Diurnal and Crepuscular Lepidoptera," ibid. 4 (1862). See Meisel, *Bibliog. Amer. Nat. Hist.*; also see bibliographic references in *DAB*. MSS: Ac. Nat. Scis. Philad. (Hamer). WORKS ABOUT: *DAB*, 13: 212-13 (George H. Genzmer); Mallis, *Amer. Entomologists*, pp. 284-85.

MORRIS, MARGARETTA HARE (December 3, 1797, apparently Philadelphia, Pa.—May 29, 1867, Germantown, Pa.). *Entomology*. Daughter of Luke, apparently man of some means, and Ann (Willing) Morris. Never married. EDUCATION: No information. CAREER: Evidently lived in same house in Germantown most of life, residing with mother, who died in 1853, and with unmarried sister. MEMBERSHIPS: First female member, Ac. Nat. Scis. Philad. SCIENTIFIC CONTRIBUTIONS: For number of years, studied habits of so-called Hessian fly and seventeen-year locust, results of which had particular significance for agriculture; as result of study of Hessian fly, based on observations of insects raised for that purpose, delineated life history and concluded that eggs were laid in grain rather than stalk. Also said to have investigated fungi as enemy of plants. Prepared illustrations for botanical paper by W. Gambel [q.v.] (1848), and with mother attended special lectures on science in Germantown; an acquaintance of T. Nuttall [q.v.] and other scientists. Sister, Elizabeth Carrington Morris (1795-1865), was interested in study of botany and reportedly corresponded with A. Gray [q.v.].

WORKS: "On the Cecidomyia destructor, or Hessian Fly," *Transactions of Am. Phil. Soc.* n.s. 8 (1843): 49-52; "Observations on the Development of the Hessian Fly," *Proceedings of Ac. Nat. Scis. Philad.* 1 (1841-1843): 66-68; "On the Discovery of the Larvae of the Cicada septemdecim," ibid. 3 (1846-1847): 132-34; "On the Cecidomyia culmicola," ibid. 4 (1848-1849): 194; "On the Seventeen

Year Locusts," *Proceedings of Bos. Soc. Nat. Hist.* 4 (1851) : 110. Also see Meisel, *Bibliog. Amer. Nat. Hist.* WORKS ABOUT: Robert C. Moon, *The Morris Family of Philadelphia* (Philadelphia, 1898), 2: 399-404, 581-83; Nicholas B. Wainwright, ed., *A Philadelphia Perspective: The Diary of Sidney George Fisher Covering the Years 1834-1871* (Philadelphia, 1967), pp. 107-8, 286, 328; Jeannette E. Graustein, *Thomas Nuttall* (Cambridge, Mass., 1967), p. 374.

MORSE, EDWARD SYLVESTER (1838-1925). *Zoology*; *Ethnology*. Director, Peabody Museum (Salem, Mass.) ; curator of Japanese pottery, Boston Museum of Fine Arts.

WORKS ABOUT: See *AMS*; *DAB*; *DSB*.

MORSE, HARMON NORTHROP (1848-1920). *Chemistry*. Professor of analytical chemistry, Johns Hopkins Univ.

WORKS ABOUT: See *AMS*; *DAB*.

MORTON, HENRY (December 11, 1836, New York, N.Y.—May 9, 1902, New York, N.Y.). *Chemistry*; *Physics*. Son of Henry J., Episcopal minister, and Helen (MacFarlan) Morton. Married Clara Whiting Dodge; two children. EDUCATION: 1857, A.B., Univ. of Pennsylvania; 1859, studied law. CAREER: Lectured on chemistry and physics, Episcopal Academy of Philadelphia; 1863, appointed professor of chemistry, Philadelphia Dental College; at Franklin Institute, was resident secretary (1864-1870) and editor of *Journal* (1867-1871), and also gave public lectures for benefit of institute; 1868, became professor of chemistry, Univ. of Pennsylvania, serving initial year as acting professor; 1869, conducted expedition to Iowa to photograph eclipse under auspices of U.S. *Nautical Almanac* office; 1870-1902, first president, Stevens Institute of Technology (Hoboken, N.J.) ; after assuming Stevens presidency, often served as paid consultant in patent disputes and became leading expert in such cases. MEMBERSHIPS: Natl. Ac. Scis. SCIENTIFIC CONTRIBUTIONS: As undergraduate, prepared translation of hieroglyphic inscriptions of Rosetta stone. Was great success as public lecturer on science, and in association with this work produced several papers in *Journal of Franklin Institute* on various aspects of projection and production of suitable lantern. In conjunction with 1869 eclipse expedition, helped differentiate optical and photographic phenomena in astronomical work. At Stevens Institute, helped establish mechanical engineering as field of study, while continuing his own laboratory studies; developed plastic materials for filling teeth, investigated fluorescence, and also published on variety of other topics in physics and chemistry; in later years more frequently contributed useful accounts of work by other researchers. Also known for work in Biblical criticism.

WORKS: Papers appeared especially in *Journal of Franklin Institute* and in *Engineering* (London). See abbreviated footnote references to chief works in Nichols [below]; also see *Roy. Soc. Cat.*, vols. 8, 10, 12, 17. MSS: Presidential appointment books in Stevens Institute of Technology Library. WORKS ABOUT: *DAB*, 13: 254-55 (Franklin De R. Furman) ; Edward L. Nichols, "Memoir," *Biog. Mems. of Natl. Ac. Scis.* 8 (1919) : 143-51.

MORTON, SAMUEL GEORGE (January 26, 1799, Philadelphia, Pa.—May 15, 1851, Philadelphia, Pa.). *Natural history*; *Anthropology*. Son of George and Jane (Cummings) Morton. Married Rebecca Grellet Pearsall, 1827; eight children. EDUCATION: Attended Quaker schools at Westtown, Pa., and Burlington, N.J.; 1820, M.D., Univ. of Pennsylvania; 1823, M.D., Univ. of Edinburgh. CAREER: 1824, began medical practice, Philadelphia; 1829, physician to Almhouse; 1830, began lecturing at Philadelphia Association for Medical Instruction, associated with Dr. Joseph Parrish; 1839-1843, professor of anatomy, medical department of Pennsylvania (Gettysburg) College, Philadelphia. MEMBERSHIPS: President, Ac. Nat. Scis. Philad. SCIENTIFIC CONTRIBUTIONS: Wrote on variety of subjects, including geology, paleontology, zoology, medicine, craniology. In 1830 and 1833, published papers on vertebrate fossils, in *Am. Journ. Sci.* [see Works]; [See Works], *Snyopsis*, published 1834, described fossils gathered by Lewis and Clark Expedition and helped establish serious study of invertebrate paleontology in U.S.; medical writings included work on prescription of fresh air as aid in pulmonary consumption (1834), and in 1849 published book, *Human Anatomy*. In 1830 began collecting human crania and most influential work was done in measurement and comparison of skulls from various locations; based on these studies, concluded that races were differentiated not only by skin color, but also by nature of skull; most controversial conclusion argued that races were not derived from common origin.

WORKS: "Synopsis of the Organic Remains of the Ferruginous Sand Formation of the U.S.," *Am. Journ. Sci.* 17-18 (1830) and 23-24 (1833) ; *Synopsis of the Organic Remains of the Cretaceous Group of the U.S.* (Philadelphia, 1834) ; *Crania Americana* (Philadelphia, 1839) ; *Crania Aegyptiaca* (Philadelphia and London, 1844) ; "Hybridity in Animals, Considered in Reference to the Question of the Unity of the Human Species," *Am. Journ. Sci.* 2d ser. 3 (1847) : 39-50, 203-12. See Meisel, *Bibliog. Amer. Nat. Hist.*; also additional bibliographic reference in *DSB*. MSS: Am. Phil. Soc. (*NUCMC* 62-4974) ; Lib. Co. Philad. (Hamer). WORKS ABOUT: *DAB*,

13: 265-66 (Daniel M. Fisk); *DSB*, 9: 540-41 (Whitfield J. Bell, Jr.).

MOTT, HENRY AUGUSTUS, Jr. (October 22, 1852, Clifton, Staten Island, N.Y.—November 8, 1896, New York, N.Y.). *Chemistry.* Son of Henry Augustus Mott. No information on marital status. EDUCATION: Engineer of Mines degree and Ph.B. in metallurgy in 1873 and Ph.D. in 1875, Columbia College. HONORS: LL.D., Univ. of Florida, 1886. CAREER: 1874-1881, chemist to Havemeyer & Elder, sugar refining company; 1881-1886, professor of chemistry, New York Medical College and Hospital for Women; for three years, food inspector, U.S. Bureau of Indian Affairs; 1890-1896, chemist to New York Medico-Legal Society; ca. 1890-1896, delivered annual course of lectures on chemistry, sponsored by New York City Board of Education; served as consultant to several components of food industry, was called upon as expert witness in court cases, and conducted investigations for private individuals. MEMBERSHIPS: Am. Chem. Soc. and other societies. SCIENTIFIC CONTRIBUTIONS: At early stage of industry, became interested in problems of manufacture of artificial butter and invented process for preventing its crystallization that made commercial production feasible; also carried out other studies of food products and uncovered adulteration of baking powders with alum. Was noted for work in toxicology, and wrote on air and ventilation. Also interested in philosophical questions; is said to have been first to doubt validity of wave theory of sound, and wrote on that topic. Later studies dealt with question of nature of force; Mott argued for entitative character of force, assigning it an objective, though not a material, existence.

WORKS: *Chemists' Manual* (New York, 1877); *The Fallacy of the Present Theory of Sound* (New York, 1885); *The Philosophy of Substantialism* (New York, 1895), 46 pp. Abbreviated references to writings included in works cited in Works About. See also *Natl. Un. Cat. Pre-1956 Imp.* and *Roy. Soc. Cat.*, vols. 10, 12, 17. WORKS ABOUT: *NCAB*, 3 (1893 [copyright 1891]): 171-72; William McMurtrie, "Obituary," *Journal of Am. Chem. Soc.* 19 (1897): 90-92; Rossiter Johnson, ed., *Twentieth Century Biographical Dictionary of Notable Americans* (Boston, 1904), vol. 7.

MÜHLENBERG, GOTTHILF HENRY ERNEST (November 17, 1753, Trappe, Montgomery County, Pa.—May 23, 1815, Lancaster, Pa.). *Botany.* Son of Henry Melchior, Lutheran clergyman, and Anna Maria (Weiser) Mühlenberg. Married Mary Catharine Hall, 1774; eight children. EDUCATION: 1763, sent to be educated at Halle, Germany; 1769, matriculated at Univ. of Halle. CAREER: 1770, ordained to Lutheran ministry, Reading, Pa.; during early 1770s, assisted father with pastoral duties at Philadelphia and in Raritan Valley, N.J.; 1774, named third pastor at Philadelphia; 1777-1778, during occupation of Philadelphia by British, resided with father at Trappe, Pa.; 1778-1779, returned to Philadelphia; 1780-1815, minister, Holy Trinity, Lancaster, Pa.; 1787, first president, Franklin College. MEMBERSHIPS: Honorary member of several learned societies; member of Am. Phil. Soc. SCIENTIFIC CONTRIBUTIONS: Active interest in botany began in 1778, and later corresponded and exchanged specimens with botanists in England and Europe; earliest botanical contributions apparently were in bryology and were made through contacts with foreign naturalists; conducted intensive field studies in local area, and before 1791 compiled list of 1,100 botanical species within three-mile radius of Lancaster. Became noted for careful and accurate work in description and classification; *Catalogus* [see Works] listed names of plants, with notes, and introduced number of new names to American botanical studies; at death, left uncompleted manuscript, later published as *Descriptio Uberior* [see Works], giving fuller descriptions of species. Interested in medical and economic uses of plants. Benefited especially by contributions from W. Baldwin and S. Elliott [qq.v.].

WORKS: *Catalogus Plantarum Americae Septentrionalis . . .* (Lancaster, Pa., 1813); *Descriptio Uberior Graminum et Plantarum Calamariarum Americae Septentrionalis Indigenarum et Circurum* (Philadelphia, 1817). See other works listed in *DAB*; also Meisel, *Bibliog. Amer. Nat. Hist.* MSS: Hist. Soc. Penn. (*NUCMC* 60-2676); Am. Phil. Soc. (*NUCMC* 61-1093); Harv. U.-GH (Hamer); Ac. Nat. Scis. Philad. (Hamer). WORKS ABOUT: *DAB*, 13: 308-9 (George H. Genzmer); C. Earle Smith, Jr. "Henry Mühlenberg: Botanical Pioneer," *Proceedings of Am. Phil. Soc.* 106 (1962): 443-60. Also see Paul A. W. Wallace, *Muhlenbergs of Pennsylvania* (Philadelphia, 1950).

MUIR, JOHN (1838-1914). *Natural history; Geology.* Traveler and explorer; promoted establishment of national parks, especially Yosemite.

MSS: American Academy of Arts and Letters, New York (*NUCMC* 61-2403); Hunting. Lib. (*NUCMC* 62-397); Univ. Calif.-Ban. (*NUCMC* 71-770); Univ. of Pacific Library, Stockton, Calif. (*NUCMC* 72-779); also see references in Hamer. WORKS ABOUT: See *AMS*; *DAB*.

MUNROE, CHARLES EDWARD (1849-1938). *Chemistry.* Professor of chemistry, U.S. Naval Academy; professor of chemistry, George Washington Univ.

WORKS ABOUT: See *AMS*; *DAB*, supp. 2.

N

NASON, HENRY BRADFORD (June 22, 1831, Foxboro, Mass.—January 18, 1895, Troy, N.Y.). *Chemistry.* Son of Elias and Susanna (Keith) Nason. Married Frances Kellogg Townsend, 1864; two children. EDUCATION: 1851, graduated, Williston Seminary, Easthampton, Mass.; 1855, A.B., Amherst College; 1857, Ph.D., Univ. of Göttingen (Germany); also studied at Heidelberg and Freiberg. HONORS: M.D., Union Univ., 1880; LL.D., Beloit College, 1880. CAREER: 1857-1858, teacher, Raymond Collegiate Institute, Carmel, N.Y.; 1858-1866, professor of natural history at Rensselaer Polytechnic Institute and also at Beloit College; 1866-1895, professor of chemistry and natural science, Rensselaer; 1880-1890, adviser to Standard Oil Co. MEMBERSHIPS: Am. Chem. Soc.; London Chemical Society; and other societies. SCIENTIFIC CONTRIBUTIONS: As adviser to Standard Oil, produced several technical advances relating to processing of crude oil, and in 1881 was appointed inspector by New York state board of health to deal with problems resulting from use of petroleum. During extensive travels in U.S. and Europe, studied local geological and mineralogical features including volcanoes, fjords, glaciers, and mining regions; chemical researches limited to improvements in techniques of analysis, especially in relation to geology and mineralogy. Chief works prepared as compiler, translator, or editor; played important role in organization and activities of scientific societies, particularly Geol. Soc. Am. and Am. Chem. Soc. Prepared *Biographical Record . . . of the Rensselaer Polytechnic Institute* (1887).

WORKS: *Table of Reactions for Qualitative Analysis* (Troy, N.Y., [1865]). See additional works listed in *DAB*. WORKS ABOUT: *DAB*, 13: 390 (Raymond P. Baker); W. P. Mason, obituary, *Journal of Am. Chem. Soc.* 17 (1895): 339-41.

NEWBERRY, JOHN STRONG (December 22, 1822, Windsor, Conn.—December 7, 1892, New Haven, Conn.). *Paleontology*; *Geology*. Son of Henry, businessman and founder of Cuyahoga Falls, Ohio, and Elizabeth (Strong) Newberry. Married Sarah Brownell Gaylord, 1848; six children. EDUCATION: 1846, A.B., Western Reserve College; 1848, M.D., Cleveland Medical School; 1849-1851, studied medicine and science in Paris. HONORS: Murchison Medal, Geological Society of London, 1888; president of International Geological Congress, 1891. CAREER: 1851-1855, medical practice, Cleveland, Ohio; 1855-1856, physician-naturalist, Pacific Railroad Survey party under Lt. R. S. Williamson; 1856, while in Washington working on survey report, became associated with Smithson. Instn.; 1856-1857, professor of chemistry and natural history, Columbian (now George Washington) Univ.; 1857-1858, physician-naturalist, exploration of Colorado River under Lt. J. C. Ives; 1859, attached to survey of Santa Fe region, under Capt. J. N. Macomb; 1861-1865, physician-administrator with U.S. Sanitary Commission; 1866-1892, professor of geology and paleontology, Columbia Univ. School of Mines; 1869-1874, director, Ohio Geological Survey; during 1870s, paleobotanist for geological surveys led by F. V. Hayden and J. W. Powell [qq.v.]. MEMBERSHIPS: President, Am. Assoc. Advt. Sci., 1867; member of Natl. Ac. Scis. and other societies. SCIENTIFIC CONTRIBUTIONS: Field geologist with interest in fossil fishes, but more especially in paleobotany; contributed to knowledge of relations of strata and to study of fossil plants in coal areas of U.S. Made significant contributions through Ohio survey and in other publications listed below; in 1861, as part of Ives's report, gave notable account of Grand Canyon in terms of large-scale erosion. Played important role in establishing reputation of Columbia School of Mines.

WORKS: "Geological Report,' in Joseph C. Ives, *Report Upon the Colorado River . . .* (Washington, D.C., 1861); *Report of Geological Survey of Ohio*, vols. 1 (Columbus, Ohio, 1873) and 2 (Columbus, Ohio, 1875); *Fossil Fishes and Fossil Plants of the Triassic Rocks of New Jersey and the Connecticut*

Valley, U.S. Geol. Surv. Monograph 14 (Washington, D.C., 1888); *The Paleozoic Fishes of North America*, U.S. Geol. Surv. Monograph 16 (Washington, D.C., 1889). Bibliography in *Biog. Mems. of Natl. Ac. Scis.* 6 (1909): 15-24. Other bibliographic references in *DAB* and *DSB*. MSS: Letters in Columbia College Papers, Colum. U. Lib. WORKS ABOUT: *DAB*, 13: 445-46 (George P. Merrill); *DSB*, 10: 32-33 (Michele L. Aldrich).

NEWCOMB, SIMON (1835-1909). *Astronomy*. Professor of mathematics, U.S. Navy, stationed at Naval Observatory; superintendent, *American Ephemeris and Nautical Almanac*; professor of mathematics and astronomy, Johns Hopkins Univ.

MSS: Lib. Cong. (*NUCMC* 63-384). WORKS ABOUT: See *AMS*; *DAB*; *DSB*.

NEWTON, HENRY (August 12, 1845, New York, N.Y.—August 5, 1877, Deadwood, S. Dak.). *Geology*; *Mining engineering*. Son of Isaac, steamboat designer-owner, and Hannah (Cauldwell) Newton. No information on marital status. EDUCATION: 1866, A.B., College of City of New York (City College); 1869, E.M., School of Mines, Columbia Univ.; 1877, Ph.D., Columbia Univ. CAREER: 1870 (summer), began several years of association with Ohio Geological Survey, as assistant to J. S. Newberry [q.v.]; 1870-1875, assistant in mineralogy and geology, School of Mines, Columbia; 1875, appointed assistant geologist by secretary of interior, to explore gold deposits of Black Hills, S. Dak.; 1877, returned to Black Hills exploration and died there; 1877, appointed professor of mining and metallurgy by Ohio State Univ., but died before assuming office. SCIENTIFIC CONTRIBUTIONS: Interested chiefly in metallurgy of iron and steel, in 1870 he visited chief locations of iron and steel production in Great Britain and Europe, and through several years of study became an American expert on subject; during association with Ohio survey, investigated resources for development of iron industry in that state. In exploration of Black Hills, was chiefly responsible for geological work; as result of that expedition, report was issued that accurately delineated the geology and topography of the region while correcting earlier inaccurate accounts.

WORKS: "A Sketch of the Present State of the Steel Industry," in Ohio Geological Survey, *Report of Progress for 1870* (Columbus, Ohio, 1871), pt. 9, pp. 527-55; "The Ores of Iron, Their Geographical Distribution and Relation to the Great Centers of the World's Iron Industries," *Transactions of Am. Inst. Min. Engrs.* 3 (1874-1875): 360-91; *Report on the Geology and Resources of the Black Hills of Dakota* (U.S. Geological and Geographical Survey of the Rocky Mountain Region, under J. W. Powell) (Washington, D.C., 1880), with Walter P. Jenney. WORKS ABOUT: J. S. Newberry, "Biographic Notice," in *Report on the Black Hills* [above], pp. ix-xii; *NCAB*, 4 (1897 [copyright 1891 and 1902]): 190-91.

NEWTON, HUBERT ANSON March 19, 1830, Sherburne, N.Y.—August 12, 1896, New Haven, Conn.). *Astronomy*; *Mathematics*. Son of William, involved in canal and railroad construction, and Lois (Butler) Newton. Married Anna C. Stiles, 1859; survived by two children. EDUCATION: 1850, A.B., Yale College; 1850-1853, studied mathematics at Sherburne and New Haven; ca. 1855-1856 [one year], studied mathematics at Sorbonne, Paris. HONORS: LL.D., Univ. of Michigan, 1868; J. Lawrence Smith Medal for work on meteoroids, Natl. Ac. Scis., 1888. CAREER: At Yale Univ., appointed tutor and became head of mathematics department (1853), then became professor of mathematics (1855-1896); served single term as alderman, New Haven. MEMBERSHIPS: President, Am. Assoc. Advt. Sci., 1885; member, Natl. Ac. Scis. and other societies. SCIENTIFIC CONTRIBUTIONS: While teaching mathematics, chief work was done in astronomy, especially in study of meteors; in early 1860s began publishing on that subject, especially on orbits and velocities of fireballs; in 1864, published study of meteor shower of November 1861, giving account of its periodic reappearance and calculation of its orbit; work on meteors helped identify similarities in movements of meteoroids and comets. During 1870s, compared statistical distribution of known orbits of comets with theoretical distributions according to Kant's and Laplace's theories on origin of solar system, and while inconclusive, his results tended to favor ideas of Laplace; in 1891, published important paper on effects of planets on distribution of orbits of comets. Continued comet and meteor studies until death, at which time he probably ranked as leading American investigator of meteors. Promoted adoption of metric weights and measures and was a founder of American Metrological Society.

WORKS: "On November Star-Showers," *Am. Journ. Sci.* 2d ser. 37 (1864): 377-89; "On the Origin of Comets," ibid. 3d ser. 16 (1878): 165-79; "On Shooting Stars," *Memoirs of Natl. Ac. Scis.* 1 (1866): 291-312; "On the Capture of Comets by Planets, Especially . . . by Jupiter," ibid. 6 (1891): 7-23; *The Metric System of Weights and Measures* (Washington, D.C., 1868). Chief works listed in *DAB* and *DSB*; bibliography in *Biog. Mems. of Natl. Ac. Scis.* 4 (1902): 120-24. WORKS ABOUT: *DAB*, 13: 470-71 (David E. Smith); *DSB*, 10: 41-42 (Richard Berendzen).

NICHOLS, EDWARD LEAMINGTON (1854-1937). *Experimental physics*. Professor of physics

and head of department, Cornell Univ.
WORKS ABOUT: See *AMS*; *DAB*, supp. 2.

NICHOLS, JAMES ROBINSON (July 18, 1819, West Amesbury [Merrimac], Mass.—January 2, 1888, Haverhill, Mass.). *Chemistry.* Son of Stephen, apparently farmer, and Ruth (Sargent) Nichols. Married Harriet Porter, 1844; Margaret Gale, 1851; survived by two children. EDUCATION: Attended course of medical lectures in 1841-1842 and received M.D. in 1867, Dartmouth College. HONORS: M.A., Dartmouth. CAREER: Ca. 1837 [age eighteen], joined uncle in drug business at Haverhill, Mass.; 1843, after attendance at medical school, returned to drug business at Haverhill, giving chief attention to working in laboratory, writing, and lecturing; 1857-1872, conducted J. R. Nichols & Co., in Boston, devoted to production of quality chemicals and medical preparations; 1863, established experimental farm at Haverhill; 1866-1888, founder and editor, *Boston Journal of Chemistry and Pharmacy*, later known as *Popular Science News and Boston Journal of Chemistry*; 1873-1878, president, Vermont & Canada Railroad; 1873-1888, a director, Boston & Maine Railroad; 1878, became member of Massachusetts State Board of Agriculture. MEMBERSHIPS: Am. Assoc. Advt. Sci. SCIENTIFIC CONTRIBUTIONS: During years as head of J. R. Nichols & Co. in Boston, made several voyages to Europe in attempt to discover methods for production of fine chemicals used in medicine, dyeing, photography, painting, and so on, and previously imported from Europe. Introduced new chemical and pharmaceutical substances and methods for their efficient production; among inventions were soda-water apparatus, carbonic-acid fire extinguisher, and improved hot air furnace that enjoyed widespread use. Also contributed to establishment of leatherboard industry. *Boston Journal* had distinction of being first popular but reliable journal devoted to chemistry, and later encompassed a wider range of scientific concerns; Nichols was author of numerous articles and editorials and several popular books on science. On farm at Haverhill, carried out experiments with chemical fertilizers.

WORKS: Among books were *Chemistry of the Farm and the Sea* (Boston, 1867), and *Whence, What, Where? A View of the Origin, Nature and Destiny of Man* (Boston, 1882). Also see *DAB* and *Natl. Un. Cat. Pre-1956 Imp.* MSS: Letters to E. N. Horsford [q.v.] at Rensselaer Polytechnic Institute. WORKS ABOUT: *DAB*, 13: 494-95 (Lyman C. Newell); Kelly and Burrage, *Med. Biog.* (1928).

NICHOLSON, HENRY HUDSON (1850-1940). *Chemistry*; *Mineralogy.* Professor of chemistry and director of chemical laboratories, Univ. of Nebraska; consulting mining and industrial engineer.
WORKS ABOUT: See *AMS*; *NCAB*, 48 (1965): 279-80.

NICOLLET, JOSEPH NICOLAS (July 29, 1786, Cluses, Savoy [now in France]—September 11, 1843, Washington, D.C.). *Astronomy*; *Mathematics*; *Exploration.* Son of poor parents, Joseph spent early years as herdsman. No information on marital status. EDUCATION: Studied at college at Cluses; said to have been student in Collège Louis-le-Grand. CAREER: Ca. 1805 [age nineteen], teaching at Chambéry; later, went to Paris, was naturalized, and in 1817 appointed secretary and librarian at the observatory, working with Laplace; 1822, became astronomical assistant, Bureau of Longitudes, under Royal Observatory; served as professor of mathematics, Collège Louis-le-Grand; Revolution of 1830 resulted in financial loss and apparent disgrace through speculations on stock market; 1832, arrived in U.S.; 1832-1835, traveled through South; 1835, went to St. Louis; 1836, conducted exploration to sources of Mississippi River; 1838, under sponsorship of U.S. War Department, explored upper Missouri River; 1839, made second journey up Missouri River and reached points in North Dakota; ca. 1840-1843, resided in Baltimore and Washington, engaged in preparation of report and map on region northwest of Mississippi. MEMBERSHIPS: Am. Phil. Soc.; Ac. Nat. Scis. Philad. SCIENTIFIC CONTRIBUTIONS: Association with French Bureau of Longitudes involved work on map of France; there published on cartography and astronomical surveying and also was known for discovery of comets. In U.S., explorations resulted in first map of a large portion of this country carried out with highest standards of accuracy, involving barometric and astronomical observations; also made geological and magnetic observations, and these were published by Am. Phil. Soc. and Ac. Nat. Scis. During 1838 expedition up Missouri, was accompanied by Lt. John C. Frémont, who later conducted explorations in Rocky Mountain region.

WORKS: "Map of the Hydrographical Basin of the Upper Mississippi River" and accompanying report (both published as U.S. Senate Documents) (Washington, D.C., 1843). See also *Roy. Soc. Cat.*, vol. 4. Journals of American exploration have been issued by Minnesota Historical Society (St. Paul, Minn., 1970 and 1976), edited by Martha C. Bray and Edmund C. Bray [not seen]. MSS: Lib. Cong. (Hamer). WORKS ABOUT: *DAB*, 13: 514 (Louise P. Kellogg); Martha C. Bray, "Joseph N. Nicollet, Geographer," in John F. McDermott, ed., *Frenchmen and French Ways in the Mississippi Valley* (Urbana, Ill., 1969), pp. 29-42.

NILES, WILLIAM HARMON (1838-1910). *Geol-*

ogy; *Geography*. Professor of geology and geography, Massachusetts Institute of Technology; professor of geology, Boston Univ. and Wellesley College.
WORKS ABOUT: See *AMS*; *NCAB*, 12 (1904): 481-82; *Who Was Who in America*, vol. 1.

NIPHER, FRANCIS EUGENE (1847-1926). *Physics*. Professor of physics, Washington Univ. (St. Louis).
WORKS ABOUT: See *AMS*; *DAB*.

NORTON, JOHN PITKIN (July 19, 1822, Albany, N.Y.—September 5, 1852, Farmington, Conn.). *Agricultural chemistry*. Son of John Treadwell, well-to-do-farmer, and Mary Hubbard (Pitkin) Norton. Married Elizabeth P. Marvin, 1847; two children. EDUCATION: 1840-1842, attended scientific and other lectures at Yale College; studied experimental chemistry and mineralogy in private laboratory of B. Silliman, Jr. [q.v.]; 1842-1843, attended lectures, Harvard Law School; 1844-1846, studied agricultural chemistry with James F. W. Johnston, Edinburgh; 1846-1847, studied with Gerardus Mulder at Utrecht. HONORS: M.A., Yale Univ., 1846; prize for research on oats, from Highland Agricultural Society of Scotland, ca. 1845. CAREER: 1846-1852, was professor of agricultural chemistry, Yale Univ., and helped to establish scientific department there; 1851-1852, involved in establishment of university at Albany. MEMBERSHIPS: Am. Assoc. Advt. Sci. SCIENTIFIC CONTRIBUTIONS: Demonstrated talents as chemical analyst and investigator while still a student. Became leading American proponent for application of scientific methods to agriculture, and remained interested in practical problems of farmers while holding what probably was first U.S. professorship devoted to agricultural chemistry; interested especially in soil analysis as means of improving agricultural practice; wrote popular articles for *Cultivator* and *American Agriculturist*; on basis of work done while in Europe, published papers on analysis of oats, protein bodies of peas and almonds, and potato disease, and later wrote paper on value of soil analysis. Chief publication was *Elements* [see Works], a school textbook. His Yale Analytical Laboratory was beginning of Sheffield Scientific School.
WORKS: *Elements of Scientific Agriculture* (Albany, 1850); chief works listed in *DSB* and *DAB*. MSS: Yale Univ. Library (*NUCMC* 71-2029). WORKS ABOUT: *DAB*, 13: 574-75 (Edward M. Bailey); *DSB*, 10: 150-51 (Louis I. Kuslan); Rossiter, *Emergence of Agricultural Science*, pp. 89-124.

NORTON, WILLIAM AUGUSTUS (October 25, 1810, East Bloomfield, N.Y.—September 21, 1883, New Haven, Conn.). *Astronomy*; *Physics*. Son of Herman and Julia (Strong) Norton. Married Elizabeth Emery Stevens, 1839; no information on children. EDUCATION: 1831, graduated, U.S. Military Academy, West Point. CAREER: 1831-1833, assistant professor of natural and experimental philosophy, West Point; 1832, took part in expedition to West during Black Hawk War; 1833, resigned from army; 1833-1839, professor of natural philosophy and astronomy, Univ. of City of New York (New York Univ.); at Delaware College (Univ. of Delaware), professor of mathematics and philosophy (1839-1850) and president (1850); 1850-1852, professor of natural philosophy and civil engineering, Brown Univ.; 1852-1883, professor of civil engineering, Sheffield Scientific School, Yale. MEMBERSHIPS: Natl. Ac. Scis. SCIENTIFIC CONTRIBUTIONS: Involved in several areas of physical sciences, earliest publication being textbook on astronomy (1839). Wrote several papers on terrestrial magnetism and on comets and other astronomical phenomena; also wrote papers of engineering interest; work entitled "Ericsson's . . . Caloric Engine" [see Works] helped delineate limits of hot air engines. Chief interest was in study of molecular physics, but extended work on that subject was left uncompleted at death.
WORKS: "An Inquiry into the Constitution and Mode of Formation of the Tails of Comets," *Am. Journ. Sci.* 46 (1844): 104-29; "On Ericsson's Hot Air, or Caloric Engine," ibid. 2d ser. 15 (1853): 393-413; "Laws of the Deflection of Beams Exposed to a Transverse Strain. Tested by Experiment," *Proceedings of Am. Assoc. Advt. Sci.* 18 (1869): 47-63. Chief works listed as abbreviated references in Trowbridge [below], p. 194; see also *Roy. Soc. Cat.*, vols. 4, 8, 10. MSS: Some items in E. N. Horsford [q.v.]. Papers at Rens. Poly. Inst.-Arch. WORKS ABOUT: *NCAB*, 9 (1907 [copyright 1899 and 1907]): 187; W. P. Trowbridge, "Memoir," *Biog. Mems. of Natl. Ac. Scis.* 2 (1886): 189-99.

NUTTALL, THOMAS (January 5, 1786, Long Preston, near Settle, Yorkshire, England—September 10, 1859, Nut Grove Hall, near St. Helens, Lancashire, England). *Botany*; *Ornithology*. Son of James and Mary (Hardacre) Nuttall. Never married. EDUCATION: 1800 [age fourteen], apprenticed to uncle in printing trade. CAREER: 1808, emigrated to Philadelphia, and soon thereafter became plant collector for new friend, B. S. Barton [q.v.]; 1809, made two collecting trips for Barton; 1810-1811, set out on botanical exploration to Canada for Barton, but joined Pacific Fur Co. expedition up Missouri River; 1811-1815, in England; 1818-1820, engaged in explorations along Arkansas and Red rivers, in Arkansas, Louisiana, and Indian Territory; 1822-1834, curator of botanic garden and lecturer on natural history, Harvard College; during this period also went on several collecting trips; 1834-1836, crossed continent to Oregon, with Nathaniel Jarvis Wyeth expedition, and spent time also in Hawaii; 1836, returned to Boston by way of Cape Horn; 1836-1841, most of time spent in Philadelphia, working on specimens collected in West; 1842-1859, lived in England because of conditions of inheritance from uncle; 1847-1848, visited U.S. MEMBERSHIPS: Linnaean Society of London; Am. Phil. Soc.; and other societies. SCIENTIFIC CONTRIBUTIONS: Noted as first important collector and student of plants in American West; his chief contributions were based on abilities as field worker and resulting familiarity with plants in natural setting; used Linnaean classification, although also aware of natural relations of genera; produced three-volume supplement to F. A. Michaux [q.v.], *North American Sylva,* including numerous references to western trees. Active interest in ornithology began while in Cambridge, but work in that subject was not extensive; *Manual* [see Works] was inexpensive guide and drew on knowledge of birds observed in field. Also wrote on shells and on mineralogy and made early attempt to correlate American and European geological strata.

WORKS: *The Genera of North American Plants, and a Catalogue of the Species, to the Year 1817* (Philadelphia, 1818); *A Manual of the Ornithology of the U.S. and Canada: The Land Birds* (Cambridge, Mass., 1832); *The Water Birds* (Boston, 1834). Chief works listed in *DAB* and *DSB*. MSS: See references in Jeannette E. Graustein, *Thomas Nuttall* (Cambridge, Mass., 1967). WORKS ABOUT: *DAB*, 13: 596-97 (Witmer Stone); *DSB*, 10: 163-65 (Phillip D. Thomas).

NUTTING, CHARLES CLEVELAND (1858-1927). *Marine zoology*; *Ornithology*. Professor of zoology, State Univ. of Iowa.

WORKS ABOUT: See *AMS*; *DAB*.

O

OBER, FREDERICK ALBION (February 13, 1849, Beverly, Mass.—May 31, 1913, Hackensack, N.J.). *Ornithology.* Son of Andrew Kimball, manufacturer and merchant, and Sarah Abigail (Hadlock) Ober. Married Lucy Curtis, 1870; Jean MacCloud, early 1890s; Nellie F. McCarthy [or McCartny], 1895; two children by third marriage. EDUCATION: Educated in Beverly public schools; for one semester attended Massachusetts Agricultural College (Univ. of Massachusetts). CAREER: 1862-1866, worked as shoemaker; 1867-1869, employed in drugstore; 1870, returned to shoemaking trade; 1872-1874, collected birds in Florida, including Everglades and Lake Okeechobee, sponsored by Smithson. Instn. and *Field and Stream* magazine; 1876-1878, collected birds in Lesser Antilles (West Indies), sponsored by Smithsonian; 1880, returned to Lesser Antilles; 1881, 1883, and 1885, visited Mexico and there conducted archaeological explorations; 1888, went to Spain in pursuit of interest in European conquest of America and also visited North Africa; made trip to South America and again to West Indies; 1891, appointed U.S. special commissioner for Columbian Exposition, Chicago, in charge of ornithological exhibits, and explored routes of Columbus and early Spanish settlements, collecting objects related to those events; 1908, entered real estate business, Hackensack, N.J. MEMBERSHIPS: A founder of Explorers Club, New York; member of N.Y. Ac. Scis. SCIENTIFIC CONTRIBUTIONS: At early age stuffed and mounted collection of local birds, which was purchased in 1872 by Harvard Univ. museum. Under pseudonym of Fred Beverly, wrote articles on Florida explorations for *Field and Stream*; in collecting in Lesser Antilles, discovered twenty-two new ornithological species, and is remembered especially for these zoological explorations in West Indies. Wrote number of books based on travels and also adventure and historical tales, most intended for young people.

WORKS: "Ornithological Exploration of the Caribee Islands," *Annual Report of Smithson. Instn.*, 1878 (Washington, D.C., 1879), pp. 446-51; *Guide to the West Indies and Bermudas* (London and New York, 1908). See list of works in *NCAB* [below]. WORKS ABOUT: *DAB*, 13: 606-7 (Herbert Friedmann); *NCAB*, 54 (1973): 322.

OLIVER, ANDREW (November 13, 1731, Boston, Mass.—December 4, 1799, Salem, Mass.). *Meteorology*; *Astronomy.* Son of Andrew, secretary and lieutenant governor of Massachusetts, and Mary (Fitch) Oliver. Married Mary Lynde, 1752; seven children. EDUCATION: 1749, A.B., Harvard College. CAREER: Inherited wealth; after marriage in 1752, lived in Boston and held several civic appointments in that city; 1759, elected Boston selectman; 1760, moved to Salem; 1761-ca. 1775, judge of Court of Common Pleas; 1762-1767, served in Massachusetts House of Representatives; ca. 1783-1799, confined to home by illness. MEMBERSHIPS: A founder, Am. Ac. Arts Scis.; member, Am. Phil. Soc. SCIENTIFIC CONTRIBUTIONS: Attributed interest in science to Prof. J. Winthrop [q.v.] at Harvard, to whom most significant work, that on comets [see Works], was dedicated; among striking ideas in this essay was suggestion that because of their tails, comets may be inhabited. Also wrote on thunder and lightning and on waterspouts, the latter argued on analogy with movement of liquid in quill. Said to have been able mathematician.

WORKS: *An Essay on Comets* (Salem, Mass., 1772), reprinted in French (Amsterdam, 1777) and with *Two Lectures on Comets*, by J. Winthrop (Boston, 1811); "A Theory of Lightening and Thunder Storms," *Transactions of Am. Phil. Soc.* 2 (1786): 74-101; "Theory of Water Spouts," ibid., pp. 101-19. MSS: See references in footnotes to Shipton [below]. WORKS ABOUT: *DAB*, 14: 15-16 (Edward E. Curtis); Shipton, *Biog. Sketches, Harvard*, 12 (1962): 455-61.

OLIVER, JAMES EDWARD (July 27, 1829, Portland, Maine—March 27, 1895, Ithaca, N.Y.). *Mathematics.* Son of James, bank cashier, and Olivia (Cobb) Oliver. Married Sara T. Van Petten, 1888; apparently no children. EDUCATION: 1849, A.B., Harvard College. CAREER: 1850-ca. 1868, assistant in U.S. *Nautical Almanac* office, Cambridge, Mass.; 1868-1869, went to New York, and while there apparently studied at School of Mines, Columbia Univ.; 1869-1870, in Philadelphia, worked as private teacher; 1870, returned to Lynn, Mass., where had spent his childhood; 1871, lectured on thermodynamics, Harvard Univ.; at Cornell Univ., assistant professor (1871-1873) and professor of mathematics (1873-1895); 1889, visited Europe and there studied methods for teaching mathematics. MEMBERSHIPS: Am. Phil. Soc.; Natl. Ac. Scis.; and other societies. SCIENTIFIC CONTRIBUTIONS: At Harvard, came under influence of Prof. B. Peirce [q.v.], and worked with Peirce at *Nautical Almanac.* During early period of career was particularly interested in modern algebra; worked on application of mathematics to questions in economics, physiology, and psychology. Had tendency to abandon projects before completion, and heavy teaching load at Cornell prevented full realization of mathematical potential. Made important contribution to development of Cornell mathematical program, and coauthored textbooks on algebra and trigonometry; said to have published few mathematical studies, although bibliography includes some twenty entries for years 1858-1892.

WORKS: See bibliography, Hill [below], p. 74. WORKS ABOUT: *NCAB,* 14 (1910): 342; G. W. Hill, "Memoir," *Biog. Mems. of Natl. Ac. Scis.* 4 (1902): 57-74.

OLMSTED, DENISON (June 18, 1791, East Hartford, Conn.—May 13, 1859, New Haven, Conn.). *Geology; Astronomy; Physics.* Son of Nathaniel, farmer, and Eunice (Kingsbury) Olmsted. Married Eliza Allyn, 1818; Julia Mason, 1831; seven children by first marriage, three by second. EDUCATION: 1813, A.B., Yale College; later studied theology with Yale president Timothy Dwight; 1817-1818, studied science with B. Silliman [q.v.], Yale. HONORS: LL.D., New York Univ., 1845. CAREER: 1813-1815, teacher, Union School, New London, Conn.; 1815, appointed tutor, Yale College; 1818-1825, professor of chemistry, Univ. of North Carolina; 1823, appointed North Carolina state geologist and mineralogist; at Yale College, professor of mathematics and natural philosophy (1825-1836) and professor of natural philosophy and astronomy (1836-1859). MEMBERSHIPS: Am. Assoc. Advt. Sci. SCIENTIFIC CONTRIBUTIONS: Particularly noted as teacher and as writer of textbooks on astronomy and natural philosophy for colleges, schools, and general public; in teaching, used experiments to accompany lectures and also involved students in laboratory exercises. Research efforts were mainly in astronomy, with papers on meteoric showers of November 1833 and subsequent years; also wrote on hailstorms, zodiacal light, aurora borealis, and other astronomical phenomena, contributing to theoretical consideration of these subjects. Promoted Geological Survey of North Carolina, which was supported by legislative appropriation and resulted in published report [see Works] stressing economic and agricultural concerns; also published other geological papers, especially on mineralogical resources. Procured several patents, including use of cottonseed in illumination.

WORKS: *Report on the Geology of North Carolina, Conducted Under the Auspices of the Board of Agriculture,* 2 pts. (Raleigh, 1824-1825); "Of the Phenomena and Causes of Hailstorms," *Am. Journ. Sci.* 18 (1830): 1-11; papers on meteoric shower of November 13, 1833, in ibid. 25 (1834): 363-411, 26 (1834): 132-74, and 29 (1836): 376-83. Full bibliography in Franklin B. Dexter, *Biographical Sketches of the Graduates of Yale College* (New Haven, 1912), 6: 594-600. MSS: Yale Univ. Library (*NUCMC* 74-1197). WORKS ABOUT: *DAB,* 14: 23-24 (Alois F. Kovarik); Theodore R. Treadwell, "Denison Olmsted, an Early American Astronomer," *Popular Astronomy* 54 (1946): 237-41.

ORD, GEORGE (March 4, 1781, probably Philadelphia, Pa.—January 24, 1866, Philadelphia, Pa.). *Zoology.* Son of George—sea captain, ship chandler, and ropemaker—and Rebecca (Lindemeyer) Ord. Married 1815; two children. EDUCATION: No information on education. CAREER: 1800, joined father in ship chandler business; after father's death in 1806, continued business with mother; 1818, accompanied T. Say, T. R. Peale, W. Maclure [qq.v.] on expedition to Georgia and Florida; ca. 1829, retired and thereafter lived as person of wealth and leisure. MEMBERSHIPS: President, Ac. Nat. Scis. Philad., 1851-1858; member of Am. Phil. Soc. and other societies. SCIENTIFIC CONTRIBUTIONS: Closely associated with A. Wilson [q.v.], whom he accompanied on outings in vicinity of Philadelphia; made contributions to Wilson's *American Ornithology,* and after Wilson's early death (1813) completed volume 8, wrote all of text for volume 9, and later (1824-1825) published another edition of the *Ornithology,* to which new information was added; attempted to carry out these labors as anonymously as possible, and engaged in controversy on Wilson's behalf; wrote life of Wilson, one of few such publications that appeared under own name. In 1815, prepared section on zoology of North America for second American edition of William Guthrie's *New Geographical, Historical, and Commercial Grammar*

(Philadelphia); while *Grammar* is now rare, Ord's contribution was reprinted in 1894 and ranks as one of earliest attempts by an American author to give systematic account of North American zoology; included descriptions of specimens gathered by Lewis and Clark; also wrote papers on various zoological topics, published in proceedings of scientific societies. For some forty years, pursued studies in philology.

WORKS: See Meisel, *Bibliog. Amer. Nat. Hist.* MSS: Am. Phil. Soc. (*NUCMC* 61-769). WORKS ABOUT: *DAB*, 14: 49-50 (Witmer Stone); Samuel N. Rhoads, "George Ord," *Cassinia: Proceedings of Delaware Valley Ornithological Club*, 12 (1908), pp. 1-8.

ORTON, EDWARD FRANCIS BAXTER (March 9, 1829, Deposit, Delaware County, N.Y.—October 16, 1899, Columbus, Ohio). *Geology.* Son of Samuel Gibbs, Presbyterian minister, and Clara (Gregory) Orton. Married Mary M. Jennings, 1855; Anna Davenport Torrey, 1875; four children by first marriage, two by second. EDUCATION: 1848, graduated, Hamilton College; 1849-1850, attended Lane Theological Seminary, Cincinnati; 1852-1853, attended Lawrence Scientific School, Harvard; ca. 1854, began studies at Andover (Mass.) Theological Seminary, but did not graduate. CAREER: Ca. 1848-1849, assistant principal, academy at Erie, Pa.; 1851, appointed teacher, Delaware Literary Institute, Franklin, N.Y.; 1855-1856, minister, Downsville, N.Y.; 1856-1859, professor of natural sciences, New York State Normal School at Albany; 1859-1865, principal, academy at Chester, Orange County, N.Y.; at Antioch College, appointed principal of preparatory department (1865), served as professor of natural history (1866-1872), and president (1872-1873); 1869, appointed assistant in Ohio Geological Survey; at Ohio College of Agriculture and Mechanics (Ohio State Univ.), president (1873-1881) and professor of geology (1873-1899); 1882-1899, Ohio state geologist. MEMBERSHIPS: President of Geol. Soc. Am., 1897, and of Am. Assoc. Advt. Sci., 1899. SCIENTIFIC CONTRIBUTIONS: As assistant in Ohio survey, worked on southwestern portion of state and produced number of county reports; as head of survey after 1882, completed work begun under J. S. Newberry [q.v.], while giving greater emphasis to economic concerns and yet also arguing against unbridled exploitation of energy sources. Study of Carboniferous strata of Ohio and correlation of the series with neighboring regions established basis for future studies of geology of eastern Ohio and adjacent areas. A pioneer in relating geological structure and rock characteristics to presence of oil and gas; noted as interpreter, rather than innovator, in geological studies.

WORKS: Contributions appeared especially in reports of Ohio Geological Survey, and as director he completed volumes 5-7 of final report. See Nickles, *Geol. Lit. on N. Am.* WORKS ABOUT: *DAB*, 14:62-63 (George P. Merrill); A. C. Swinnerton, "Edward Orton, Geologist," *Science* n.s. 89 (1939): 373-78.

ORTON, JAMES (April 21, 1830, Seneca Falls, N.Y.—September 24, 1877, Lake Titicaca, Bolivia). *Natural history; Exploration.* Son of Azariah Giles, minister, and Minerva (Squire) Orton. Married Ellen Foote, 1859; at least one child. EDUCATION: 1855, A.B., Williams College; 1858, graduated, Andover Theological Seminary. CAREER: 1860, ordained to Presbyterian ministry; ca. 1860-ca. 1866, held three pastorates, in New York state and in Maine; 1866, appointed acting instructor in natural history, Univ. of Rochester, while friend H. A. Ward [q.v.] was on leave; 1867, led expedition to South America sponsored by Williams College and with cooperation of Smithson. Instn., and explored region of Andes and Amazon; 1869-1877, professor of natural history, Vassar College; 1873, headed second Andean expedition; 1876, third Andean exploration, with financial backing from E. D. Cope [q.v.]. MEMBERSHIPS: Am. Assoc. Advt. Sci. SCIENTIFIC CONTRIBUTIONS: Before graduating from college, published *Miner's Guide and Metallurgist's Directory* (1849), and as undergraduate participated in two scientific excursions to Nova Scotia and Newfoundland. Chief scientific work related to three South American expeditions, with initial exploration (1867) undertaken to study deposits of upper Amazon Valley, variously attributed to marine and glacial origins; on first and second trips, procured natural history specimens for wide range of fields of study and benefited collections of several museums; died while on third trip, after having been abandoned by porters and escort, and notes and specimens were lost in shipment to New York. *Andes and Amazon* [see Works] brought together wide variety of information on geology, climate, animal and vegetable life, civil conditions, and resources of region, which he also presented in numerous brief communications to various journals. *Comparative Zoology* [see Work], a successful textbook, demonstrated influence of J. L. R. Agassiz [q.v.] in featuring importance of function as well as structure in zoological studies. Interested in education of women.

WORKS: *Andes and the Amazon* (New York, 1870; 2d ed., 1876); *Comparative Zoology, Structural and Systematic* (New York, 1876). See *DSB* for list of other works and for references to miscellaneous manuscripts in Smithson. Instn. and De Golyer Library at Univ. of Oklahoma. WORKS ABOUT: *DAB*, 14: 64-65 (Aaron L. Treadwell); *DSB*, 10: 240-41 (Joseph Ewan).

OSBORN, HENRY FAIRFIELD (1857-1935). *Vertebrate paleontology*; *Comparative anatomy*; *Biology*. Professor of zoology, Columbia Univ.; curator of vertebrate paleontology and president, Am. Mus. Nat. Hist.; geologist and paleontologist, U.S. Geol. Surv.

MSS: Ac. Nat. Scis. Philad. (Hamer); also see *DAB*. WORKS ABOUT: See *AMS*; *DAB*, supp. 1; *DSB*.

OSTEN SACKEN, CARL ROBERT ROMANOVICH VON DER (August 21, 1828, St. Petersburg, Russia—May 20, 1906, Heidelberg, Germany). *Entomology*. No information on parentage or marital status; family presumably part of Russian nobility. EDUCATION: Educated in St. Petersburg. CAREER: Russian baron; 1849, entered Russian diplomatic service; 1856-1862, secretary to Russian Legation, Washington, D.C.; 1862-1871, consul general for Russia, in New York City; 1871-1873, several trips to Europe; 1873-1877, in U.S. as private citizen; 1873-1875, associated with Museum of Comparative Zoology, Harvard; 1875-1876, traveled and collected entomological specimens in California, Sierra Nevadas, and Rocky Mountains; 1877, settled at Heidelberg and there continued entomological work. SCIENTIFIC CONTRIBUTIONS: Interest in entomology began 1839 (age eleven), while visiting at Baden-Baden, and in 1854 published work on insects of St. Petersburg; chief work during mature years was done on Diptera (flies, mosquitoes, and the like) of North America, but collected in all orders with exception of Lepidoptera (butterflies and moths). During years in U.S., was engaged in study of Diptera north of Isthmus of Panama, which involved collaboration with Viennese entomologist Dr. Hermann Loew; Osten Sacken sent number of specimens to Loew and also worked on translating and editing latter's manuscripts; in 1877, collection was purchased from Loew by Museum of Comparative Zoology. Published papers on various groups of dipterous and hymenopterous insects, and in 1858 and 1878 Smithson. Instn. published catalogues of Diptera, that of 1878 [see Works] a critical work of considerable significance. At Heidelberg, published number of short papers, studying Diptera of world outside Europe.

WORKS: "Catalogue of the Described Diptera of North America," 2d ed., *Smithson. Misc. Coll.* 16 (1878); "Monographs of the Diptera of North America," 4 vols., *Smithson. Misc. Coll.* 6, 11, 8 (1862-1875), with H. Loew. Bibliography of works in Osten Sacken, *Record of My Life in Entomology*, pt. 3 (Heidelberg, 1904). MSS: Harv. U.-MCZ. WORKS ABOUT: *DAB*, 14: 87-88 (Leland O. Howard); Mallis, *Amer. Entomologists*, pp. 381-84.

OWEN, DAVID DALE (June 24, 1807, New Lanark, Scotland—November 13, 1860, New Harmony, Ind.). *Geology*. Son of Robert, mill owner and utopian philanthropist, and Anne Caroline (Dale) Owen. Married Caroline C. Neef, 1837; four children. EDUCATION: 1824-1827, attended progressive school of Philipp Emanuel von Fellenberg near Berne, Switzerland; for one year, studied chemistry in Glasgow; 1831-ca. 1832, studied chemistry and geology, London Univ.; 1837, M.D., Ohio Medical College, Cincinnati. CAREER: 1828, arrived at New Harmony, Ind., and maintained residence there for remainder of life; 1831-1833, went to England, accompanied by H. D. Rogers [q.v.]; 1836 (summer), volunteer assistant, Tennessee Geological Survey; 1837-1838, Indiana state geologist; 1839-1840, director of survey of Dubuque and Mineral Point districts in Wisconsin and Iowa, for U.S. Land Office; 1847-1852, U.S. geologist for survey of mineral region of Chippewa Land District, eventually covering Wisconsin, Iowa, and Minnesota; 1854-1860, Kentucky state geologist; 1857-1860, Arkansas state geologist; 1859-1860, Indiana state geologist; never practiced medicine. MEMBERSHIPS: Am. Assoc. Advt. Sci. SCIENTIFIC CONTRIBUTIONS: While studying medicine, passed summers working on fossil collections gathered by W. Maclure [q.v.]. Worked alone on first Indiana survey, and special attention given to practical considerations on that survey also characterized later geological labors as well; Dubuque and Mineral Point survey involved mobilization of 139 assistants to study major U.S. lead region, while Chippewa survey reached into South Dakota Badlands and resulted in publication of great importance [see Works], illustrated with maps and plates based on Owen's own sketches; in work for second Indiana survey, brother Richard Owen conducted field work under David's direction. In general, geological work was of preliminary nature, though he was a skilled and careful field worker capable of effective analysis and correlations, and was also well grounded in chemistry. Attracted and trained number of young geologists.

WORKS: *Report of a Geological Survey of Wisconsin, Iowa, and Minnesota; and Incidentally of a Portion of Nebraska Territory* (Philadelphia, 1852). Major works listed in *DSB*. Bibliography and references to manuscripts in W. B. Hendrickson, *David Dale Owen* (Indianapolis, 1943). WORKS ABOUT: *DAB*, 14: 116-17 (George P. Merrill); *DSB*, 10: 257-59 (George W. White).

P

PACKARD, ALPHEUS SPRING, Jr. (February 19, 1839, Brunswick, Maine—February 14, 1905, Providence, R.I.). *Entomology.* Son of Alpheus Spring, professor of Greek and Latin at Bowdoin College, and Frances Elizabeth (Appleton) Packard. Married Elizabeth Derby Walcott, 1867; four children. EDUCATION: 1861, A.B., Bowdoin College; 1864, S.B., Lawrence Scientific School, Harvard; 1864, M.D., Maine Medical School, Bowdoin. HONORS: Ph.D. in 1879 and LL.D. 1901, Bowdoin College. CAREER: 1860 and 1864, accompanied expeditions to Labrador; 1861-1862, assistant, Maine Geological Survey; 1863-1864, assistant, Museum of Comparative Zoology, Harvard; 1864-1865, assistant surgeon, Maine Veteran Volunteers; 1865-1866, librarian and custodian, Bos. Soc. Nat. Hist.; 1866, curator, Essex Institute; at Peabody Academy of Science (Salem, Mass.), appointed curator (1867) and then director (1877-1878); 1867-1887, a founder and editor-in-chief, *American Naturalist;* 1870, lecturer on economic entomology, Maine College of Agriculture and Mechanics (Univ. of Maine); 1870-1878, lecturer on economic entomology, Massachusetts Agricultural College (Univ. of Massachusetts); 1871-1874, lecturer on entomology and comparative anatomy, Bowdoin College; 1873, teacher, Anderson School of Natural History, Penikese Island; 1874, associated with Kentucky Geological Survey; 1875-1876, connected with F. V. Hayden [q.v.] and Geological Survey of Territories; 1877-1882, member of U.S. Entomological Commission; 1878-1905, professor of zoology and geology, Brown Univ. MEMBERSHIPS: Natl. Ac. Scis. and other societies. SCIENTIFIC CONTRIBUTIONS: While noted for entomological contributions, also worked in other areas of zoology and in geology. Promoted study of insects with *Guide to the Study of Insects* (1869) and other such works; is remembered especially for work in taxonomy, described some 50 genera and 580 species of invertebrate animals; also did notable work on life histories of insects, as well as embryological investigations that included insects, crustaceans, and other invertebrates. A founder of neo-Lamarckianism.

WORKS: *Insects Injurious to Forest and Shade Trees* (Washington, D.C., 1881); *The Cave Fauna of North America* (Washington, D.C. 1888); *Monograph of the Bombycine Moths of North America,* 3 vols. (Washington, D.C., 1895-1914). Bibliography in *Biog. Mems. of Natl. Ac. Scis.* 9 (1920): 207-36. WORKS ABOUT: *DAB,* 14: 126-27 (Leland O. Howard); *DSB,* 10: 272-74 (Calvert E. Norland).

PAGE, CHARLES GRAFTON (January 25, 1812, Salem, Mass.—May 5, 1868, Washington, D.C.). *Physics.* Son of Jeremiah Lee, sea captain, and Lucy (Lang) Page. Married Priscilla Sewall Webster, 1844; six children. EDUCATION: A.B. in 1832 and M.D. in 1836, Harvard Univ. CAREER: 1836, began medical practice, Salem, Mass.; 1838 moved to Fairfax County, Va.; 1839-ca. 1842 practiced medicine in that location; 1842-1852 and 1861-1868, patent examiner, U.S. Patent Office; 1844-1849, professor of chemistry and pharmacy, medical department of Columbian College (George Washington Univ.); 1852-1861, engaged in business as patent agent (inventor's advocate); 1853-1854, a founder and editor, *American Polytechnic Journal.* MEMBERSHIPS: Am. Assoc. Advt. Sci.; Am. Ac. Arts Scis. SCIENTIFIC CONTRIBUTIONS: During period 1836-1839 carried out significant experiments that won him rank with J. Henry [q.v.] as early American researcher in electromagnetism. One noteworthy result of this early experimental work was design and use of what was, in effect, an induction coil; in 1867 published *History* [see Works] in attempt to insure claim to discovery of coil, and in 1868 U.S. Congress granted essentially honorary patent; later exploitation of this patent by heirs damaged Page's reputation. After early interest in experimental physics, turned attentions especially to applied science; in 1849, received $20,000 grant from Congress to construct and demonstrate battery-powered electric motor as substitute for steam engine in locomotive. The proven impracticability of this project in a test in 1851 and the support given by Page to inventors' lobby in its attempt to lower standards of originality as criterion for patent grants helped obscure his earlier contributions to study of electromagnetism; by mid-1850s Page was alienated from ranks of leadership of American scientific community. Interested in floriculture during later years.

WORKS: *History of Induction: The American Claim to the Induction Coil and Its Electrostatic Developments* (Washington, D.C., 1867); published numerous papers, especially in *Am. Journ. Sci.* between 1834 and 1852. See *Roy. Soc. Cat.*, vol. 4, and Post [below], pp. 207-13. MSS: No substantial collection of Page papers is known, but see footnote references and pp. 206-7 in Post [below]. WORKS ABOUT: *DAB*, 14: 135-36 (Carl W. Mitman); Robert C. Post, *Physics, Patents and Politics: A Biography of Charles Grafton Page* (New York, 1976).

PAINE, ROBERT TREAT (October 12, 1803, Boston, Mass.—June 3, 1885, Brookline, Mass.). *Astronomy.* Son of Robert Treat—publisher, lawyer, writer, and supporter of theater—and Eliza (Baker) Paine. Married Anne Stevens, 1843; no children. EDUCATION: 1822, A.B., Harvard College; thereafter, studied law and was admitted to bar. CAREER: Practiced law; 1828, 1833, and 1834, member Boston Common Council; 1830-1842, prepared astronomical content for *American Alamac,* founded with Jared Sparks and Joseph E. Worcester; 1840, appointed chief engineer, Trigonometrical Survey of State of Massachusetts; 1875, moved to Brookline; a person of wealth, devoted great part of life to scientific and benevolent activities. MEMBERSHIPS: Am. Phil. Soc.; Am. Ac. Arts Scis. SCIENTIFIC CONTRIBUTIONS: Shortly after leaving college, began meteorological record maintained for almost sixty years, and for number of years published results of these observations in Boston *Daily Traveler.* Held office as chief engineer on Massachusetts survey for only short period, but completed all of astronomical work, which proved to have been accurately executed. Chief astronomical interest was motion of moon, and he computed some 2,000 occultations; also observed three transits of Mercury and one of Venus; conducted number of observations of solar eclipses at various places, including South Carolina, Iowa, Colorado, California (1880), and Cuba (1846). At death, left fortune to Harvard College for support of observatory and for establishment of professorship in practical astronomy.

WORKS: See *Roy. Soc. Cat.*, vols. 4, 8, 10. MSS: Notebooks in Mass. Hist. Soc. WORKS ABOUT: Memoir, *Proceedings of Am. Ac. Arts Scis.* 21 (1885-1886): 532-35; newspaper clippings, Harvard Univ. Archives; Mary C. Crawford, *Famous Families of Massachusetts* (Boston, 1930), 2: 3-23.

PALMER, ARTHUR WILLIAM (February 17, 1861, London, England—February 3, 1904, Urbana, Ill.). *Chemistry.* Son of William and Harriet (Fairchild) Palmer. Married Anna Shattuck, 1893; one child. EDUCATION: 1883, B.S. in chemistry, Univ. of Illinois; 1886, S.D., Harvard Univ.; 1888-1889, studied at Berlin and Göttingen. CAREER: At Univ. of Illinois, chief assistant in chemistry (1883-1884 and 1886-1888), assistant professor (1889-1890), and professor of chemistry (1890-1904), and head of department; member, Illinois State Board of Health. MEMBERSHIPS: Deutsche Chemische Gesellschaft; Am. Chem. Soc. SCIENTIFIC CONTRIBUTIONS: Worked chiefly in organic chemistry when in Germany, and subsequent studies were done mainly in that area; in 1892, discovered arsine series, the outcome of research begun while at Berlin. 1895, put in charge of chemical survey of waters of Illinois, an extensive undertaking resulting in significant reports; 1899, at request of trustees of Sanitary District of Chicago, undertook work on examination of waters between Lake Michigan at Chicago and Mississippi River at St. Louis.

WORKS: *Chemical Survey of the Water Supplies of Illinois: Preliminary Report* ([Urbana, Ill.?], 1897); *Chemical Survey of the Waters of Illinois: Report for the Years 1897-1902* ([Urbana, Ill.?], 1903). See several papers in *Roy. Soc. Cat.*, vol. 17. MSS: Univ. of Illinois Archives (*NUCMC* 71-1163). WORKS ABOUT: *Who Was Who in America,* vol. 1; Franklin W. Scott, ed., *Alumni Record of University of Illinois* (Lakeside Press, 1918), p. 35; obituary, Urbana *Herald,* February 10, 1904 (clipping in Harvard Univ. Archives); death notice, *Science,* n.s. 19 (1904): 276.

PARISH, SAMUEL BONSALL (1838-1928). *Botany; Phytogeography.* Farmer; explored flora of southern California; honorary curator of herbarium, Univ. of California; lecturer in botany, Stanford Univ.

WORKS ABOUT: See *AMS; Publication of Botanical Society of America* 99 (1929): 26-28.

PARRY, CHARLES CHRISTOPHER (August 28, 1823, Admington, Gloucestershire, England [came to U.S. with parents, 1832]—February 20, 1890, Davenport, Iowa). *Botany.* Son of Joseph, clergyman in Church of England and farmer, and Eliza (Elliott) Parry. Married Sarah M. Dalzell, 1853; Emily R. Preston, 1859; one child by first marriage, who died at young age. EDUCATION: 1842, A.B., Union College; 1846, M.D., Columbia College. CAREER: 1846, began medical practice, Davenport; 1848, surgeon-naturalist under D. D. Owen [q.v.] on geological survey of Wisconsin, Iowa, and Minnesota; 1849-1852, botanist, U.S. and Mexican boundary survey; 1861, made first collecting excursion in Colorado and for some years thereafter, on nearly annual basis, returned to Rocky Mountain region; after 1877 conducted botanical studies especially in California; 1869-1871, first botanist to U.S. Department of Agriculture, at Washington; 1884-1885, visited England to study specimens at Royal Botanic Gardens at Kew. MEMBERSHIPS: Active in Davenport (Iowa) Academy of Natural Sciences. SCIENTIFIC CONTRIBUTIONS: Began botanical studies as early as 1842, collecting initially in northeastern New York state and later in Iowa and surrounding region; carried out pioneering botanical studies in Colorado, Wyoming, Utah, and California, during which time his traveling companions included naturalists such as Elihu Hall, J. P. Harbour, Dr. J. W. Velie, J. Duncan Putnam, and others; at times Parry's botanical collecting was done on an individual basis, while on other occasions it was carried out in conjunction with surveying expeditions or other enterprises; corresponded with and sent specimens to A. Gray, J. Torrey [qq.v.], and others; discovered hundreds of new botanical forms. In addition to more specialized papers contributed number of pieces on natural history and other features of West to various newspapers. Did significant work in investigating plants both in natural setting and in herbaria. Has been criticized for lack of care in record keeping in his collecting activities.

WORKS: Published number of short botanical papers and contributed to publications of other persons. Bibliography in *Proceedings of Davenport Academy of Natural Sciences* 6 (1897) : 46-52. References to several works given in footnotes to Ewan [below] and in *DAB*. MSS: Iowa State Department of History and Archives (*NUCMC* 62-2804) ; Harv. U.-GH (Hamer) ; also see references in Ewan and in *DAB*. WORKS ABOUT: *DAB*, 14: 261-62 (Willis L. Jepson) ; Joseph A. Ewan, *Rocky Mountain Naturalists* ([Denver, Colo.], 1950), pp. 34-44.

PATTERSON, ROBERT (May 30, 1743, Hillsborough, County Down, Ireland—July 22, 1824, Philadelphia, Pa.). *Mathematics*. Son of Robert and Jane (Walker) Patterson. Married Amy Hunter Ewing, 1774; eight children, including Robert M. Patterson [q.v.]. EDUCATION: Completed schooling before coming to America. HONORS: LL.D., Univ. of Pennsylvania, 1816. CAREER: Ca. 1759, joined militia company at time of French landing on coast of Ireland; 1768, arrived in Philadelphia virtually without funds; ca. 1768, became schoolmaster, Buckingham, Bucks County, Pa.; later, returned to Philadelphia and there taught method of navigation employing calculation of longitude based on observations of moon; 1772, opened country store in New Jersey; ca. 1774, appointed principal, academy at Wilmington, Del.; at outbreak of Revolution, given charge of training three companies of militia; later, entered army with rank of brigade major; at Univ. of Pennsylvania, professor of mathematics and natural philosophy (1779-1813) and vice-provost (1810-1813) ; member of Select Council of Philadelphia and in 1799 elected council president; 1805-1824, director, U.S. Mint. MEMBERSHIPS: Elected president, Am. Phil. Soc., 1819. SCIENTIFIC CONTRIBUTIONS: At early age, demonstrated ability in mathematics; later developed particular interest in application of mathematics to practical problems and is said to have been sought after for advice by accomplished mechanicians. Published editions of foreign works on mechanics and astronomy, as well as papers on physical and mathematical subjects in *Transactions of Am. Phil. Soc.;* contributed problems and solutions to mathematical journals. *Treatise* [see Works] based on lectures at Univ. of Pennsylvania, where son Robert succeeded him in professorship.

WORKS: *Newtonian System of Philosophy* (Philadelphia, 1808) ; *A Treatise of Practical Arithmetic* (Philadelphia, 1818). Contributions to *Transactions of Am. Phil. Soc.* listed in *Roy. Soc. Cat.*, vol. 4. MSS: Am. Phil. Soc. (Hamer). WORKS ABOUT: *DAB*, 14: 305-6 (John R. Kline) ; memoir, *Transactions of Am. Phil. Soc.*, n.s. 2 (1825) : ix-xii.

PATTERSON, ROBERT MASKELL (March 23, 1787, Philadelphia, Pa.—September 5, 1854, Philadelphia, Pa.). *Chemistry; Natural philosophy*. Son of Robert [q.v.], mathematician and educator, and Amy Hunter (Ewing) Patterson. Married Helen Hamilton Leiper, 1814; six children. EDUCATION: A.B. in 1804 and M.D. in 1808, Univ. of Pennsylvania; 1809-1812, studied medicine and science in Paris for two years and chemistry with Humphry Davy in London. CAREER: 1809, acting American consul, Paris; 1812, returned to Philadelphia from Europe; at Univ. of Pennsylvania, elected professor of natural philosophy in medical department and professor of natural history (1813), then professor of mathematics and natural philosophy (1814-1828), and vice-provost; during War of 1812, had charge of topographical defense of Philadelphia; 1828-1835, professor of natural sciences, Univ. of Virginia; 1835-1851, director, U.S. Mint, Philadelphia; president of life insurance company. MEMBERSHIPS: President, Am. Phil. Soc., 1845-1846 and 1849-1853; a founder, Franklin Institute. SCIENTIFIC CONTRIBUTIONS: Succeeded father in professorship in Univ. of Pennsylvania, and there taught mathematics, physics, chemistry, and astronomy. When applying (unsuccessfully) for professorship in Medical School of Univ. of Pennsylvania in 1818, made probable claim to have been first person to teach atomic theory in U.S., as part of 1812 chemistry lectures to seniors; probably learned about John Dalton's theory in contacts with Humphry Davy. Only one scientific paper is known [see Works], but *Proceedings of Am. Phil. Soc.* during period 1837-1854 include abstracts of oral contributions to society's meetings.

WORKS: "Ueber die Beschaffenheit und das Vorkommen des Geldes, Platin, und der Diamanten in der Vereinigten Staaten," *Zeitschrift Deutsche Geologische Gesellschaft* 11 (1850): 60-64. MSS: Am. Phil. Soc. (Hamer); also see references to manuscripts in Miles [below]. WORKS ABOUT: *NCAB,* 26 (1937): 59-60; Wyndham D. Miles, "Robert Maskell Patterson: First to Teach Atomic Theory in America?" *Journ. Chem. Educ.* 38 (1961): 561-63.

PEALE, CHARLES WILLSON (April 15, 1742, St. Paul's Parish, Queen Anne County, Md.—February 22, 1827, Philadelphia, Pa.). *Natural history; Museum direction.* Son of Charles, London postal clerk and educator, and Margaret (Triggs) Peale. Married Rachel Brewer, 1762; Elizabeth De Peyster, 1791; Hannah Moore, 1805; six children who survived infancy by first marriage and six children by second marriage, including Titian R. Peale [q.v.]. EDUCATION: Ca. 1755-ca. 1762, served apprenticeship to saddle maker; sought out instruction in painting from John Hesselius and John Singleton Copley; 1767-1769, studied under Benjamin West in England. CAREER: 1764, gave up trade as saddle maker and thereafter turned increasingly to portrait painting; 1765, traveled to New England, where painted several portraits and also met Copley; 1766, returned to Annapolis after year in Virginia; 1767-1769, in England; 1776, took up residence in Philadelphia; 1776-1778, officer in militia; 1779, elected Philadelphia representative in Pennsylvania General Assembly; 1786, established natural history museum at Philadelphia and subsequently came to devote chief efforts to care of museum rather than to painting; 1810, retired to country home. MEMBERSHIPS: Am. Phil. Soc. SCIENTIFIC CONTRIBUTIONS: Museum was conceived as public institution and was first in U.S. devoted to natural history. Interest in establishment of museum grew out of commission to prepare drawings of prehistoric bones taken from banks of Ohio River; while begun in Peale's home, in 1794 the museum was moved to headquarters of Am. Phil. Soc. and in 1802 to Pennsylvania Statehouse (Independence Hall); gifts or deposits constituted bulk of collections, including Lewis and Clark Expedition specimens deposited by Pres. T. Jefferson [q.v.]; also contained items procured by Peale himself, including mastodon bones from Orange County, N.Y.; arrangement of exhibits based on modified Linnaean system, placed in natural postures with scenes painted by Peale. Sons Titian, Rubens, and Franklin also did work in science and related areas.

WORKS: *Scientific and Descriptive Catalogue of Peale's Museum* (Philadelphia, 1796), with A. M. F. J. Palisot de Beauvois; *Guide to the Philadelphia Museum* (Philadelphia, 1804); *Introduction to a Course of Lectures on Natural History* (Philadelphia, 1800). Other works listed in *DAB.* MSS: Fordham Univ. Library (Hamer); Am. Phil. Soc. (Hamer); Hist. Soc. Penn. (Hamer). WORKS ABOUT: *DAB,* 14: 344-47 (Horace W. Sellers); *DSB,* 10: 438-39 (Whitfield J. Bell, Jr.). Also see Charles C. Sellers, *Charles Willson Peale* (New York, 1969).

PEALE, TITIAN RAMSAY (November 2, 1799, Philadelphia, Pa.—March 13, 1885, Philadelphia, Pa.). *Natural history.* Son of Charles Willson [q.v], artist and museum director, and Elizabeth (De Peyster) Peale. Married Eliza Cecilia Laforgue, 1822; Lucy Mullen; six children by first marriage. EDUCATION: At age thirteen, formal schooling terminated; apprenticed to spinning machine manufacturer; ca. 1816 [age seventeen], began work with brother Rubens at father's Philadelphia Museum of natural history; attended lectures in anatomy, Univ. of Pennsylvania. CAREER: 1817-1818, natural history collecting expedition to Georgia and Florida with T. Say, G. Ord, W. Maclure [qq.v.]; 1819-1820, assistant naturalist and artist, Stephen Long's expedition to Rocky Mountains; at Philadelphia Museum, served as assistant manager (1821-1833) and appointed manager (1833); 1824-1825, collected birds and prepared plates for C. L. Bonaparte [q.v.]; 1829, trip to Maine; 1830-1832, collecting trip to Colombia; 1838-1842, a naturalist with U.S. Exploring Expedition to the Pacific (Wilkes); 1849-1872, examiner, U.S. Patent Office. MEMBERSHIPS: Ac. Nat. Scis. Philad.; Am. Phil. Soc. SCIENTIFIC CONTRIBUTIONS: While in teens, worked on sketches for T. Say's *American Entomology*; in 1833, issued prospectus for work on American lepidoptera, never completed; prepared drawings for number of reports of U.S. Exploring Expedition to the Pacific, but Peale's own report on birds and mammals was suppressed by C. Wilkes [q.v.], on basis of criticism of nomenclature, and report was completed later (1858) by J. Cassin [q.v.]. Peale was notable collector and field observer, though less talented in taxonomy. Later years spent working with photography, painting, and work on manuscript of butterflies of North America.

WORKS: *Lepidoptera Americana: Prospectus* (Philadelphia, 1833); *Mammalia and Ornithology,* vol. 8, Reports of U.S. Exploring Expedition (Philadelphia, 1848). Both these works are exceedingly rare. MSS: See references in *DSB*; manuscript for unpublished "Butterflies of North America" at Am. Mus. Nat. Hist. WORKS ABOUT: *DAB*, 14: 351-52 (Horace W. Sellers); *DSB*, 10: 439-40 (Jessie Poesch). Also see Poesch, *Titian Ramsay Peale . . . and His Journals of the Wilkes Expedition*, Am. Phil. Soc. Memoirs 52 (Philadelphia, 1961).

PECK, CHARLES HORTON (1833-1917). *Botany*; *Mycology*. Botanist, New York State Museum.

WORKS ABOUT: See *AMS*; *DAB*.

PECK, WILLIAM DANDRIDGE (May 8, 1763, Boston, Mass.—October 3, 1822, Cambridge, Mass.). *Natural history*. Son of John, naval architect, and Hannah (Jackson) Peck. Married Harriet Hilliard, 1810; one child. EDUCATION: 1782, A.B., Harvard College. CAREER: Ca. 1782, commenced employment in accounting house; thereafter, for some twenty years, lived on farm at Kittery, Maine; 1805-1822, professor of natural history and curator of botanical garden, Harvard Univ.; 1805-1808, traveled and studied in Europe, in preparation for assumption of duties at Harvard. MEMBERSHIPS: Am. Ac. Arts Scis.; Am. Phil. Soc. SCIENTIFIC CONTRIBUTIONS: While living in Maine, pursued studies in natural history, is said to have constructed a microscope, and collected specimens of insects, aquatic plants, and fishes; began in 1795 to write on insects, and following year won prize for work on cankerworm [see Works]; and paper on slug-worm [see Works] was awarded similar recognition. A pioneer among American investigators of the life history and control of insect pests and probably first teacher of entomology in the United States. Had ability at drawing and used this talent in illustrating publications; also wrote on fishes, and helped organize the botanic garden at Harvard.

WORKS: "The Description and History of the Canker-Worm," *Massachusetts Magazine; or Monthly Museum* 7 (1795): 323-27, 415-16; *Natural History of the Slug-Worm* (Boston, 1799), 14 pp. and 1 plate; "Description of Four Remarkable Fishes Taken Near the Piscataqua in New Hampshire" (1794), *Memoirs of Am. Ac. Arts Scis.* 2, pt. 2 (1804): 46-57; *A Catalogue of American and Foreign Plants Cultivated in the Botanic Garden, Cambridge, Mass.* (Cambridge, Mass., 1818). See Meisel, *Bibliog. Amer. Nat. Hist.* Several additional entomological works listed in *DAB*. MSS: Harv. U.-UA (*NUCMC* 65-1260). WORKS ABOUT: *DAB*, 14: 383-84 (Leland O. Howard); Mallis, *Amer. Entomologists*, pp. 13-16.

PECK, WILLIAM GUY (October 16, 1820, Litchfield, Conn.—February 7, 1892, Greenwich, Conn.). *Mathematics*. No information on parentage. Married Miss Davies; no information on children. EDUCATION: 1844, graduated, U.S. Military Academy, West Point. HONORS: LL.D., Trinity College, 1863; Ph.D., Columbia Univ., 1877. CAREER: 1844, attached to U.S. Topographical Engineers, and assigned to military survey of Portsmouth, N.H., harbor; 1845, accompanied John C. Frémont's third expedition through Rocky Mountains; 1846-1847, attached to Army of the West during war with Mexico; 1847-1855, assistant professor of mathematics, West Point; 1855, resigned from army with rank of first lieutenant of topographical engineers; 1855-1857, professor of physics and civil engineering, Univ. of Michigan; at Columbia Univ., adjunct professor of mathematics (1857-1859), professor of mathematics (1859-1861), professor of mathematics and astronomy (1861-1892), and professor of mechanics in School of Mines (1864-1892); 1868, member, Board of Visitors, U.S. Military Academy. MEMBERSHIPS: Am. Assoc. Advt. Sci. SCIENTIFIC CONTRIBUTIONS: Particularly noted as teacher and author of textbooks; during years 1859-1878 produced series of mathematical texts, encompassing arithmetic, algebra, geometry, trigonometry, analytical geometry, and calculus; also prepared textbooks on mechanics, physics, and astronomy. With father-in-law, Charles Davies, edited *Dictionary* [see Works].

WORKS: *Mathematical Dictionary and Cyclopedia of Mathematics* (New York, 1855), with Charles Davies. No works listed in *Roy. Soc. Cat.* WORKS ABOUT: *NCAB*, 5 (1907 [copyright 1891 and 1907]): 520; J. H. Van Amringe, memoir, *Columbia [University] Spectator* 30 (February 18, 1892): 4-5; Cullum, *Biographical Register, West Point*, 2 (1891): 192.

PEIRCE, BENJAMIN (April 4, 1809, Salem, Mass. —October 6, 1880, Cambridge, Mass.). *Mathematics*; *Astronomy*. Son of Benjamin, Massachusetts legislator and Harvard College librarian, and Lydia Ropes (Nichols) Peirce. Married Sarah Hunt Mills, 1833; five children, including Charles S. and James M. Peirce [qq.v.]. EDUCATION: 1829, A.B., Harvard College. CAREER: 1829-1831, teacher, George Bancroft's Round Hill School, Northampton, Mass.; at Harvard Univ., tutor in mathematics (1831-1833), professor of mathematics and natural philosophy (1833-1842), and professor of astronomy and mathematics (1842-1880); 1849-1867, consulting astronomer, *American Ephemeris and Nautical Almanac*; 1847, appointed to committee of Am. Ac. Arts Scis. to prepare program of organization for Smithson. Instn.; with U.S. Coast Survey, director of longitude determinations (1852-1867), superintendent (1867-1874), and consulting geometer (1874-1880); 1855-1858, served with A. D. Bache and J. Henry [qq.v.]

to organize Dudley Observatory (Albany). MEMBERSHIPS: President, Am. Assoc. Advt. Sci., 1853; member of Natl. Ac. Scis and other societies. SCIENTIFIC CONTRIBUTIONS: Greatly influenced by early association with N. Bowditch [q.v.] in native Salem; achieved rank as leading American mathematician of the time; during 1830s and early 1840s, prepared several textbooks. Took part in establishment of Harvard College Observatory, and did notable work in computing general perturbations of Uranus and the recently discovered planet Neptune, an interest associated with his efforts at revision of theory of planets; also carried out mathematical studies of rings of Saturn. In work for Coast Survey, proposed criterion for discriminating between normal and abnormal observations (1852), a useful contribution, though it was eventually proved to be fallacious; as head of Coast Survey, promoted its expansion into a nationwide geodetic enterprise. Approximately one-quarter of works were in pure mathematics, and others related especially to questions in astronomy, geodesy, and mechanics; most important and original mathematical work was *Linear Associative Algebra* [see Works], first published in 1870.

WORKS: *A System of Analytic Mechanics* (Boston, 1855); *Linear Associative Algebra*, new ed. (New York, 1882), edited by C. S. Peirce. See list of works in *DSB* and bibliography in Raymond C. Archibald, *Benjamin Peirce* (Oberlin, Ohio, 1925). MSS: Harv. U.-HL; also Natl. Arch. (*DSB*). WORKS ABOUT: *DAB*, 14: 393-97 (Raymond C. Archibald); *DSB*, 10: 478-81 (Carolyn Eisele).

PEIRCE, BENJAMIN OSGOOD (1854-1914). *Mathematical physics.* Professor of mathematics and physics, Harvard Univ. MSS: Harv. U.-UA (*NUCMC* 76-1993). WORKS ABOUT: See *AMS*; *DAB*; *DSB*.

PEIRCE, CHARLES SANDERS (1839-1914). *Logic*; *Physics.* Associated with U.S. Coast and Geodetic Survey; founder of pragmatism; in later life lived near Milford, Penn.

MSS: Harv. U.-HL (Hamer); also see *DSB*. WORKS ABOUT: See *AMS*; *DAB*; *DSB*.

PEIRCE, JAMES MILLS (1834-1906). *Mathematics.* Professor of mathematics and dean, Harvard Univ.

WORKS ABOUT: See *AMS*; *DAB*.

PENFIELD, SAMUEL LEWIS (1856-1906). *Mineralogy.* Professor of mineralogy, Yale Univ.

WORKS ABOUT: *See AMS*; *Who Was Who in America*, vol. 1.

PERCIVAL, JAMES GATES (September 15, 1795, Kensington, Conn.—May 2, 1856, Hazel Green, Wis.). *Geology.* Son of James, physician, and Elizabeth (Hart) Percival. Never married. EDUCATION: 1815, A.B., Yale College; studied medicine at Yale and at Univ. of Pennsylvania; 1820, M.D., Yale Univ. CAREER: 1820, made unsuccessful attempt to establish medical practice in Connecticut, published early poems that were well received; 1821-1822, attempted without success to establish medical practice, Charleston, S.C.; 1822, returned to New Haven, Conn.; 1823, editor, *Connecticut Herald*, and then moved to Berlin, Conn., where engaged in editorial and translating work for *Am. Journ. Sci.*; 1824, for brief periods, professor of chemistry, U.S. Military Academy at West Point, and assistant surgeon, army recruiting office, Boston; engaged in series of editorial projects, but alienated publishers and public through complaining and petulant attitude; 1827-1828, employed on revisions and proofreading for Noah Webster's *American Dictionary of the English Language*; 1835, appointed Connecticut state geologist; took up residence in State Hospital at New Haven, continued writing poetry, and performed translations and other editorial and scholarly labors; 1848, surveyed railroad route from Fishkill, N.Y., to New Haven; 1851, surveyed iron mines in Nova Scotia, coal areas of Albert County, New Brunswick, and geology of Bass Islands of Lake Erie; 1853-1854, examined lead mining areas of Illinois, Wisconsin, and Iowa for American Mining Co.; 1854, appointed Wisconsin state geologist. MEMBERSHIPS: Connecticut Academy of Arts and Sciences. SCIENTIFIC CONTRIBUTIONS: During 1820s, recognized as leading American poet. Final report on Connecticut survey [see Works] contained extensive, though undigested, lithological detail; but map was still of fundamental value for state's geology a century later. Earned reputation for astuteness and ability among mining interests, giving correct and profitable advice regarding the iron and coal potential in Canada and the existence of deep veins of lead in Wisconsin region.

WORKS: *Report on the Geology of the State of Connecticut* (New Haven, 1842), 495 pp. and map; first and second *Annual Report* of Wisconsin Geological Survey (Madison, Wis., 1855 and 1856), the second published posthumously. See Meisel, *Bibliog. Amer. Nat. Hist.* MSS: Yale Univ.—Collection of American Literature (Hamer). WORKS ABOUT: *DAB*, 14: 460-61 (Harry R. Warfel); Warfel, Introduction, *Uncollected Letters of James Gates Percival* . . . (Gainesville, Fla., 1959), pp. xiii-xx.

PERRINE, HENRY (April 5, 1797, Cranbury, N.J. —August 7, 1840, Indian Key, Fla.). *Botanical exploration.* Son of Peter and Sarah (Rozengrant) Perrine. Married Ann Fuller Townsend, 1822; three children. EDUCATION: Studied medicine. CA-

REER: Taught school at Rockyhill, N.J.; 1819-ca. 1824, medical practice, Ripley, Ill.; ca. 1824-1827, medical practice, Natchez, Miss.; 1827-1837, U.S. consul, Campeche, Yucatan Peninsula, Mexico; 1837, returned to New Orleans, La., then went to Havana, Florida, and New York, and in Washington, D.C., appealed to Congress for land grant in Florida; 1838, with two associates, received grant of township on Biscayne Bay, Fla.; 1838-1840, resided at Indian Key, Fla., awaiting end of Seminole War and opportunity to settle at land grant on Florida mainland. SCIENTIFIC CONTRIBUTIONS: While residing in Mexico, collected plants that later were found in herbarium of N.Y. Bot. Gard.; responded to general appeal by treasury secretary (1827), which called for cooperation in finding and introducing into U.S. foreign plants of potential usefulness; sent detailed reports on plants to several offices at Washington, demonstrating particular concern for plants having potentially useful fiber. As early as 1832, proposed establishment of station in southern Florida for introduction of tropical plants into U.S., on land to be granted by Congress; in 1833 began nursery on Indian Key that contained over 200 species of tropical plants when land grant was received five years later; that grant may have been earliest such means of promoting agriculture by U.S. Congress. In 1840, killed by Indians at Indian Key before taking up residence at nursery planned for land granted by Congress. Of plants introduced to Florida, sisal and henequen, both fiber plants, were most significant.

WORKS: "Random Records and Recollections Respecting the Establishment of the Tropical Plant Co., Indian Key, Florida," *Magazine of Horticulture and Botany* 6 (1840): 161-70, 321-33. Also see two items in *Natl. Un. Cat. Pre-1956 Imp.* WORKS ABOUT: *DAB*, 14: 480-81 (William R. Maxon); Frances F. C. Preston, "One of the Heroes of Horticulture," *Bulletin of Garden Club of America* (November, 1931), pp. 2-8.

PETER, ROBERT (January 21, 1805, Launceston, Cornwall, England—April 26, 1894, near Lexington, Ky.). *Chemistry.* Son of Robert and Johanna (Dawe) Peter. Married Frances Paca Dallam, 1835; eleven children. EDUCATION: Brought by parents to U.S. in 1817 and at young age, began employment in wholesale drug store, Pittsburgh, Penn.; ca. 1828, attended Rensselaer School (later, Rensselaer Polytechnic Institute); 1834, M.D., Transylvania Univ. (Lexington, Ky.). CAREER: 1830-1831, lecturer in chemistry, Western Univ. of Pennsylvania (Univ. of Pittsburgh); 1833, appointed professor of chemistry, Morrison College of Transylvania Univ.; for a time, engaged in medical practice; at Transylvania Univ., professor of chemistry and pharmacy (1838-1857) and dean of Medical College (1847-1857); 1850-1853, professor of chemistry and toxicology, Kentucky School of Medicine (Louisville); 1854-1860, chemist to Geological Survey of Kentucky; 1859-1860, performed chemical work for Arkansas and Indiana geological surveys; 1861, given charge of military hospitals, Lexington; 1865-1878, professor of chemistry and experimental philosophy, Kentucky Univ. (incorporating Transylvania Univ.); 1878-1887, professor of chemistry, Agricultural and Mechanical College (formerly part of Kentucky Univ., and now Univ. of Kentucky). MEMBERSHIPS: Am. Assoc. Advt. Sci. SCIENTIFIC CONTRIBUTIONS: Interest in chemistry grew out of early employment in drug firm. During 1830s contributed papers to *Transsylvania Journal of Medicine and Associate Sciences* on chemistry, medicine, and meteorological observations; in 1839, went to Europe to purchase books and equipment for Transylvania medical department, and upon return carried out chemical studies in relation to medicine, including analyses of urinary calculi and investigations of guncotton. As early as 1849, made pioneer contributions in pointing out that high phosphate content of lower Silurian limestones was in part responsible for fertility of bluegrass soils; instigated formation of Kentucky Geological Survey (1854); over long period of years performed thousands of soil and other chemical analyses for that and later geological surveys, although some of his analytical work was criticized by other scientists, including S. W. Johnson [q.v.]. Also interested in mineralogy and zoology and collected an herbarium.

WORKS: Several publications listed in *DAB*. See also Nickles, *Geol. Lit. on N. Am.*, including especially references to reports of Kentucky Geological Survey. MSS: Univ. of Kentucky Library (*NUCMC* 61-961); Transylvania College Library, Lexington (Hamer). WORKS ABOUT: *DAB*, 14: 499-500 (George Roberts); J. S. McHargue, "Dr. Robert Peter," *Journ. Chem. Educ.* 5 (1928): 151-56; John D. Wright; "Robert Peter and the First Kentucky Geological Survey," *Register of Kentucky Historical Society* 52 (1954): 201-12.

PETERS, CHRISTIAN HENRY FREDERICK (September 19, 1813, Coldenbüttel, Schleswig, Denmark [now Germany]—July 18, 1890, Clinton, N.Y.). *Astronomy.* Son of Hartwig Peters, minister. Never married. EDUCATION: 1825-1832, attended gymnasium at Flensburg; 1836, Ph.D., Univ. of Berlin, studying with Encke; later studied under Gauss at Göttingen. HONORS: French Legion of Honor, 1887. CAREER: 1838-1843, member of privately sponsored survey of Mt. Etna; ca. 1838-1848, director, government trigonometric survey of Sicily; 1848, was asked to leave Sicily because of revolutionary sympathies, but later returned and served as captain of engineers and then major; 1849-1854, at

Constantinople; 1854, came to U.S.; joined staff of Coast Survey, stationed at Cambridge, Mass., and at Dudley Observatory, Albany, N.Y.; 1858-1890, professor of astronomy and director of Litchfield Observatory, Hamilton College (Clinton, N.Y.). MEMBERSHIPS: Natl. Ac. Scis.; Royal Astronomical Society; and other societies. SCIENTIFIC CONTRIBUTIONS: As early as 1845, began studies of sun, including movement of sunspots. Particular interest was in observations of positions of stars; in 1860 began star chart without aid of photography, a project that eventually involved some 100,000 observations, intended as source for comparison of similar observations by future astronomers; 1876, began work on perfected edition of star catalogue of Ptolemy's *Almagest* [see Works], and in so doing examined manuscripts in Greek, Latin, and Arabic languages in various cities of Europe. Discovered two comets (1846, 1857), and forty-eight asteroids; in 1869, headed expedition to observe total solar eclipse, at Des Moines, and in 1874 had charge of one of eight government parties to observe transit of Venus, in New Zealand; attended 1887 International Astrophotographic Congress (Paris).

WORKS: *Celestial Charts Made at the Litchfield Observatory* (Clinton, N.Y., 1882); *Heliographic Positions of Sun-spots Observed at Hamilton College from 1860-1870*, Carnegie Institution of Washington, Publication no. 43 (Washington, D.C., 1907), edited by E. B. Frost; *Ptolemy's Catalogue of Stars; A Revision of the Almagest*, Carnegie Institution of Washington, Publication no. 86 (Washington, D.C., 1915), with E. B. Knobel. See also additional works listed in *DAB* and *DSB*; also *Roy. Soc. Cat.*, vols. 4, 8, 10, 17. MSS: Hamilton College Library. WORKS ABOUT: *DAB*, 14: 502-4 (Raymond S. Dugan); *DSB*, 10: 543 (Deborah J. Warner).

PHELPS, ALMIRA HART LINCOLN (July 15, 1793, Berlin, Conn.—July 15, 1884, Baltimore, Md.). *Author of science textbooks.* Daughter of Capt. Samuel, perhaps farmer, and Lydia (Hinsdale) Hart. Married Simeon Lincoln, 1817; John Phelps, 1831; three children by first marriage, two by second, EDUCATION: Early education with parents; ca. 1804, began formal education under sister, Emma (Hart) Willard, at Berlin district school; attended Berlin Academy; ca. 1810-ca. 1811, lived and studied with sister Emma at Middlebury, Vt.; 1812, attended Female Academy conducted by cousin Nancy Hinsdale, at Pittsfield, Mass. CAREER: 1809, taught at district school near Hartford, Conn.; 1813, engaged as teacher at Berlin Academy and (winter) at district school; ca. 1814, opened boarding school for young ladies, Berlin; 1816, took charge of academy at Sandy Hill, N.Y.; 1823-1831, following death of first husband, taught at sister Emma's Troy (N.Y.) Female Seminary; 1831, married second time, moved to Guilford, Vt., and later to Brattleboro, Vt.; 1838-1839, principal, West Chester (Pa.) Young Ladies' Seminary; 1839, became head, Rahway (N.J.) Female Institute; 1841-1856, principal, Patapsco Female Institute (Ellicott's Mills, Md.), where husband served as businesss manager until his death (1849); 1856, retired to Baltimore and continued to write and speak on various subjects. MEMBERSHIPS: Second female member (after M. Mitchell [q.v.]), Am. Assoc. Advt. Sci.; member, Maryland Academy of Science. SCIENTIFIC CONTRIBUTIONS: Especially important as educator. While at Troy, N.Y., was assisted by A. Eaton [q.v.] in pursuing her interest in science, and wrote series of popular school textbooks on subjects that included chemistry, geology, and natural philosophy; first text, on botany (1829) [see Works]), was particularly notable and appeared in numerous subsequent editions. As educator, helped to make more widespread the study of science by girls, while texts promoted acceptance of science as part of American school curriculum.

WORKS: *Familiar Lectures on Botany* (Hartford, Conn., 1829). References to manuscripts, and bibliography of writings, in Emma L. Bolzau, *Almira Hart Lincoln Phelps* (Philadelphia, 1936), pp. 482-89, 499-503. WORKS ABOUT: *DAB*, 14: 524-25 (Thomas Woody); *Nota. Am. Wom.*, 3: 58-60 (Frederick Rudolph).

PHILLIPS, FRANCIS CLIFFORD (1850-1920). *Chemistry.* Professor of chemistry, Western Univ. of Pennsylvania (Univ. of Pittsburgh).

WORKS ABOUT: See *AMS*; *DAB*.

PICKERING, CHARLES (November 10, 1805, near Starucca, Susquehanna County, Pa.—March 17, 1878, Boston, Mass.). *Natural history.* Son of Timothy (Jr.), at one time naval midshipman, and Lurena (Cole) Pickering; raised by mother and paternal grandfather, Col. Timothy Pickering, Salem, Mass. Married Sarah Stoddard Hammond, 1851; not survived by any children. EDUCATION: Attended Harvard College, received A.B. 1823 (1849); 1826, M.D., Harvard Univ. CAREER: 1827, began medical practice, Philadelphia, and became active member of Ac. Nat. Scis.; 1838-1842, chief zoologist, U.S. Exploring Expedition to the Pacific (Wilkes); 1843, visited the Middle East, India, eastern Africa; ca. 1844, settled in Boston, and there worked on Wilkes Expedition reports and other publications. MEMBERSHIPS: Ac. Nat. Scis. Philad.; Am. Phil. Soc.; and other societies. SCIENTIFIC CONTRIBUTIONS: Active on committees and as librarian (1828-1833) and curator (1833-1837) of Ac. Nat. Scis.; among other activities before 1838 prepared catalogue of academy's American plants; as early as 1827, read paper to Am. Phil. Soc. on geographical distribution of plants. On Wilkes Expedition, was particularly interested in anthropology and in geographical distribution of plants and animals, and these subjects constituted chief interest for remainder of life; travels of 1843 were undertaken in pursuit of investigations begun on Wilkes Expedition. Last sixteen years of life devoted to research and writing of *Chronological History* [see Works], completed by wife. Despite wide range of interests and studies, works are noted for accuracy.

WORKS: *Races of Man and Their Geographical Distribution*, Reports of U.S. Exploring Expedition, vol. 9 (Boston, 1848); *The Geographical Distribution of Animals and Plants*, Reports of U.S. Exploring Expedition, vol. 19 (Boston, 1854), with supplement, *Plants in Their Wild State* (Salem, Mass., 1876); *The Chronological History of Plants: Man's Record of His Own Existence Illustrated through Their Names, Uses, and Companionship* (Boston, 1879). See also references in Ruschenberger [below]. MSS: Ac. Nat. Scis. Philad. (Hamer); Mass. Hist. Soc. WORKS ABOUT: *DAB*, 14: 562 (F. Estelle Wells); W. S. W. Ruschenberger, memoir, *Proceedings of Ac. Nat. Scis. Philad., 1878* (1879): 166-70.

PICKERING, EDWARD CHARLES (1846-1919). *Astronomy.* Professor of astronomy and director of observatory, Harvard Univ.

MSS: Harv. U.-UA (*NUCMC* 65-1263; *DSB*). WORKS ABOUT: See *AMS*; *DAB*; *DSB*.

PICKERING, WILLIAM HENRY (1858-1938). *Astronomy.* Assistant professor of astronomy, Harvard College Observatory; established Harvard Observatory station at Mandeville, Jamaica, which later became his private facility.

MSS: Harv. U-UA (*NUCMC* 65-1265). WORKS ABOUT: See *AMS*; *DSB*.

PIKE, NICHOLAS (October 6, 1743, Somersworth, N.H.—December 9, 1819, Newburyport, Mass.). *Arithmetic.* Son of Rev. James and Sarah (Gilman) Pike. Married Hannah Bowman, ca. 1767; Eunice Smith, 1779; five children by first marriage, one by second. EDUCATION: 1766, A.B., Harvard College. CAREER: For a time, taught school at York, Mass. (now Maine); for nearly two decades, operated town school at Newburyport, Mass., and during much of same period there also conducted private school for boys interested in navigation and surveying; for time, also ran school for girls; 1776-1780, Newburyport town clerk; 1776, appointed justice of peace; 1782-1783, Newburyport selectman; 1784, appointed to so-called Quorum, after unsuccessful attempt to secure appointment as clerk of Supreme Court; 1791, "laid down" the public school [Shipton]; 1792, elected to Newburyport town school committee and on several occasions thereafter served as moderator; a projector and shareholder in Newburyport Turnpike; 1805, entered upon single term in Massachusetts House of Representatives; 1814, an incorporator, Merrimack Insurance Co. MEMBERSHIPS: Am. Ac. Arts Scis. SCIENTIFIC CONTRIBUTIONS: Began work on *Arithmetick* [see Works] as early as 1786, and during half-century after its publication in 1788 the book went through numerous editions; place in history of mathematics in America rests chiefly on this work, which won recommendations from presidents of Harvard and Yale and from others, including George Washington; in 1793, produced abridgement of *Arithmetick* that was especially useful in elementary schools, and also published edition of Daniel Fenning's *Ready Reckoner; or, The Trader's Useful Assistant* (1794). Ranks as first American to win wide popular reputation on basis of preparation of arithmetic texts for use in schools.

WORKS: *A New and Complete System of Arithmetick, Composed for the Citizens of the U.S. . . .* (Newburyport, Mass., 1788). MSS: Boston Public Library (*NUCMC* 73-28); Mass. Hist. Soc. (Shipton [below]). WORKS ABOUT: *DAB*, 14: 597-98 (Louis C. Karpinski); Shipton, *Biog. Sketches, Harvard*, 16 (1972): 406-9.

PIRSSON, LOUIS VALENTINE (1860-1919). *Geology.* Professor of physical geology, Yale Univ.

MSS: See *DAB*. WORKS ABOUT: See *AMS*; *DAB*.

PITCHER, ZINA (April 12, 1797, near Fort Edward, Washington County, N.Y.—April 5, 1872, Detroit, Mich.). *Natural history.* Son of Nathaniel, probably farmer, and Margaret Stevenson Pitcher. Married Anne Sheldon, 1824; Emily L. (Montgomery) Backus, 1867; two children by first marriage, one by second. EDUCATION: At about age twenty-one, began medical apprenticeship; attended medical school at Castleton, Vt.; 1822, M.D., Middlebury College. HONORS: 1851, given title of emeritus professor of medicine, Univ. of Michigan. CAREER: 1822, appointed assistant surgeon, U.S. Army; until ca. 1830, stationed at Detroit and at several other locations in Michigan; ca. 1830-1836, stationed at posts in Indian Territory and finally at Fortress Monroe, Va.; 1836, resigned from army and established medical practice in Detroit; 1837-1852, member of Michigan State Board of Regents; 1840, 1841, and 1843, elected mayor of Detroit; 1848-1867, chief physician, St. Mary's Hospital; a founder (1853) and coeditor (1855-1858) *Peninsular Journal of Medicine*; 1857-1858 (summers), carried out teaching duties at Detroit hospitals as clinical instructor in medicine, Univ. of Michigan; 1859, appointed examiner, U.S. Mint; for a time, surgeon with U.S. Marine Hospital at Detroit and associate editor of *Richmond and Louisville Medical Journal.* MEMBERSHIPS: Ac. Nat. Scis. Philad. and other societies; president, Am. Med. Assoc., 1856. SCIENTIFIC CONTRIBUTIONS: During army medical career, stationed chiefly at frontier posts, and there pursued interests in botany, geology, and meteorology; maintained contacts with leading American botanists and geologists, and name was given to several newly discovered plants and fossils; contributed specimens to A. Gray and J. Torrey [qq.v.] for *Flora of North America.* Also interested in Indian studies, and prepared section on Indian medicine for H. R. Schoolcraft [q.v.], *Information Respecting . . . Indian Tribes of the U.S.*, vol. 4 (1854). Publications mainly on medical topics, including numerous papers on clinical cases, epidemics, and medical education. Called founder of medical school at Univ. of Michigan, and helped organize university's natural history collections.

WORKS: "Collection of Reptiles and Other Geological Objects Collected at Fort Brady on the Northwestern Frontier," *Am. Journ. Sci.* 16 (1829) : 356; "Collection of Mammalia . . . in Vicinity of Fort Gratiot," ibid. 18 (1830) : 194. Also see bibliography (chiefly medical) in Novy [below], pp. 62-64. MSS: Univ. Mich.-MHC. WORKS ABOUT: *DAB*, 14 636-37 (James M. Phalen); Frederick G. Novy, "Memoir," *Physician and Surgeon* 30 (1908) : 49-64.

PLATT, FRANKLIN (November 19, 1844, Philadelphia, Pa.—July 24, 1900, Cape May, N.J.). *Geology.* Son of Franklin and Clara A. Greenough Platt. Never married. EDUCATION: Attended Univ. of Pennsylvania; after Civil War, studied geology with B. S. Lyman [q.v.]. CAREER: 1863, served in Pennsylvania reserve regiment; 1864, appointed aid in U.S. Coast Survey, and was first assigned to North Atlantic squadron; later, served under Gen. Orlando M. Poe, chief engineer of military division of the Mississippi, and in that capacity is said to have accompanied Sherman's "march to the sea"; 1870, associated for one year with P. Frazer [q.v.] as "reporting geologist"; 1874-1881, assistant geologist, Second Pennsylvania Geological Survey; later established practice as consulting geologist with brother, William Greenough Platt; served as consulting engineer and for a time as president, Rochester and Pittsburgh Coal and Iron Co.; ca. 1891. retired and during later years was invalid. MEMBERSHIPS: Geol. Soc. Am.; Ac. Nat. Scis. Philad. SCIENTIFIC CONTRIBUTIONS: Chief geological work done with Pennsylvania survey, engaged particularly in study of coal areas in western part of that state. Contributed seven volumes and several other reports to survey's publication series; two volumes were prepared with brother, W. G. Platt.

WORKS: See Frazer [below], and also Nickles, *Geol. Lit. on N. Am.* Works listed in these sources, with one exception, were published by Pennsylvania survey. WORKS ABOUT: *Appleton's CAB*, 5: 38; *Who Was Who in America*, vol. 1; Persifor Frazer, "Memoir," *Bulletin of Geol. Soc. Am.* 12 (1900-1901) : 454-55.

PLUMMER, JOHN THOMAS (March 12, 1807, Montgomery County, Md.—April 10, 1865, Richmond, Ind.). *Natural history*; *Chemistry.* Son of Joseph Pemberton, merchant, and Susanna (Husband) Plummer. Married Hannah Wright, 1833; Sarah O. Pierce, 1837; one child by first marriage, four by second. EDUCATION: 1828, M.D., Yale Univ. CAREER: Until shortly before death, practiced medicine in Richmond, Ind., having moved there in 1823 with father; 1839, editor, *Schoolmaster*; for short period, stockholder and first president, Richmond Gas Light and Coke Co. SCIENTIFIC CONTRIBUTIONS: Actively engaged in study of natural history, and gathered and organized collection of specimens particularly related to region about Richmond, Ind. Also interested in pharmacy and chemistry, and in latter field was considered particularly accomplished. During 1840s contributed papers to *Am. Journ. Sci.* on topics such as insects, experiments in horticulture, dry rot, suburban geology about Richmond, quadrupeds, mollusks, and (with O. P. Hubbard [q.v.]) on removal of carbonic acid gas from wells and on wood ashes; during 1850s, published several papers in *Am. Journ. Pharm.* on analysis of concretion of hairs in esophagus of ox, decolorizing property of essential oils, experiments with sulphindigotic acid and ozonous atmospheres, spontaneous generation of prussic acid, red sandalwood, and xanthosantalic acid.

WORKS: See above; also *Roy. Soc. Cat.*, vol. 4. WORKS ABOUT: Andrew W. Young, *History of Wayne County, Indiana* (Cincinnati, 1872), pp. 427-30; obituary notice, *Am. Journ. Sci.* 2d ser. 40 (1865) : 396.

PORCHER, FRANCIS PEYRE (December 14, 1825, St. John's, Berkeley Parish, S.C.—November 19, 1895, Charleston, S.C.). *Botany.* Son of Dr. William, apparently plantation owner, and Isabella Sarah (Peyre) Porcher. Married Virginia Leigh; Margaret Ward; had children. EDUCATION: 1844, A.B., South Carolina College (Univ. of South Carolina); 1847, M.D., Medical College of South Carolina; for two years, studied medicine in Paris, and several months in Italy. HONORS: LL.D., Univ. of South Carolina, 1891. CAREER: Practiced medicine,

Charleston, S.C.; during five years in 1850s and from 1873 to 1876, edited *Charleston Medical Journal and Review*; 1852, helped found Charleston Preparatory Medical School, and served there as principal and professor of materia medica and therapeutics; 1855, established hospital for Blacks; for some years before and after Civil War, headed Charleston Marine Hospital; during Civil War, surgeon to Holcombe Legion and at several hospitals; at Medical College of South Carolina, professor of clinical medicine (1872-1874) and professor of materia medica and therapeutics (1874-1891); for number of years, member of Charleston Board of Health. MEMBERSHIPS: Elliott Society of Natural History. SCIENTIFIC CONTRIBUTIONS: Medical thesis published as *Medico-Botanical Catalogue of the Plants and Ferns of St. John's, Berkeley*, the first of several books on southern medical botany. During Civil War, engaged by surgeon general of Confederacy to prepare work useful to people of South in taking advantage of native plants that have medicinal value; result was *Resources* [see Works], a publication of great value to Confederacy, in preparation of which Porcher was assisted by mother (a botanist), and wife, Virginia Leigh; in revised form, book was considered as basis for international reputation. Also interested in geology. Wrote on clinical studies using microscope; considered an expert in diseases of heart and chest; in 1890, one of ten American delegates to International Medical Congress, Berlin.

WORKS: *The Resources of the Southern Fields and Forests* (Charleston, 1863; rev. ed., 1869). Additional works mentioned in *DAB* and in Townsend [below]. MSS: Medical College of South Carolina, Charleston (*NUCMC* 70-1621). WORKS ABOUT: *DAB*, 15: 79-80 (Arney R. Childs); Jonathan M. Townsend, "Francis Peyre Porcher," *Ann. Med. Hist.*, 3d ser. 1 (1939): 177-88.

PORTER, JOHN ADDISON (March 15, 1822, Catskill, N.Y.—August 25, 1866, New Haven, Conn.). *Chemistry*. Son of Addison, Presbyterian minister, and Ann (Hogeboom) Porter. Married Josephine Sheffield, 1855; two children. EDUCATION: 1842, A.B., Yale College; ca. 1842-1844, studied at Philadelphia and New Haven; 1847-1850, studied in Germany, especially agricultural chemistry with Justus Liebig at Giessen; 1855, M.D. [honorary?], Yale Univ. CAREER: 1844-1847, professor of rhetoric, Delaware College (Univ. of Delaware); 1850 (several months), assistant to E. N. Horsford [q.v.], Harvard Univ.; 1850-1852, professor of chemistry applied to arts, Brown Univ.; at Yale's scientific school (after 1861, known as Sheffield Scientific School of Yale Univ.), professor of analytical and agricultural chemistry (1852-1856), professor of organic chemistry (1856-1864), and dean. MEMBERSHIPS: Connecticut State Agricultural Society; Am. Assoc. Advt. Sci. SCIENTIFIC CONTRIBUTIONS: Although perhaps with less enthusiasm than J. P. Norton [q.v.], carried out public lectures to farmers and other such labors initiated by Norton, his predecessor at Yale; in 1860, took leading role in organizing course of lectures on agricultural subjects, held at New Haven and attended by several hundred people, that is said to have influenced subsequent passage of Morrill Land-Grant Act (1862). Father-in-law, Joseph E. Sheffield, made large gifts to Yale's scientific school, which was renamed for him and of which Porter was first dean. Published several early papers in Liebig's *Annalen der Chemie und Pharmacie* and in the Albany *Cultivator*, and also wrote several textbooks on chemistry and agricultural chemistry.

WORKS: See works listed in *DAB* and in *Amercan Chemist* 5 (December 1874): 196. MSS: Some items in E. N. Horsford Papers at Rens. Poly. Inst.-Arch. WORKS ABOUT: *DAB*, 15: 96-97 (Lyman C. Newell); Rossiter, *Emergence of Agricultural Science*, pp. 136-39.

PORTER, THOMAS CONRAD (January 22, 1822, Alexandria, Pa.—April 27, 1901, Easton, Pa.). *Botany*. Son of John and Maria (Bucher) Porter. Married Susan Kunkel, 1850; no information on children. EDUCATION: 1840, graduated, Lafayette College; 1843, graduated, Princeton Theological Seminary. CAREER: 1846-1847, headed Presbyterian mission, Monticello, Ga.; 1848, ordained to ministry at Lebanon, Pa., and became pastor, Second German Church, Reading, Pa.; 1849-1866, professor of chemistry, zoology, and botany, Marshall College (later, Franklin and Marshall College); 1866-1897, professor of botany, zoology, and geology, Lafayette College; 1869-1874, carried out botanical work in Rocky Mountains under F. V. Hayden [q.v.] on U.S. Geological Survey of Territories; 1877-1884, minister, church at Easton, Pa. MEMBERSHIPS: Am. Phil. Soc.; Torrey Bot. Club; and other societies. SCIENTIFIC CONTRIBUTIONS: Began botanical collecting in 1840 in vicinity of Alexandria, Pa.; in 1846, while residing at Monticello, accompanied Joseph LeConte [q.v.] in explorations of northern Georgia, gathered number of previously unknown botanical species, and began characteristically generous sharing of notes and specimens with A. Gray [q.v.] and others, a practice that tended to detract from credit he received for botanical efforts; became noted for herbarium on Pennsylvania species. During years 1869-1874, conducted pioneering studies of flora of Rocky Mountains with Hayden survey, working closely with J. M. Coulter [q.v.]; Porter carried out some explorations, but *Synopsis* [see Works] relied especially on field work done by Coulter and others, while Porter's chief contribution was to assemble notes and descriptions, make efforts at identification, and give the work its final form. Notable works on Pennsylvania botany appeared posthumously [see Works].

WORKS: *Synopsis of the Flora of Colorado* (Washington, D.C., 1874), with J. M. Coulter; *The Flora of Pennsylvania* (Boston, 1903) and *Catalogue of the Bryophyta and Pteridophyta Found in Pennsylvania* (Boston, 1904), both edited by John K. Small. See other works mentioned in *DAB* and in Ewan [below]; also see *Roy. Soc. Cat.*, vols. 8, 12, 17. MSS: Ac. Nat. Scis. Philad. (*NUCMC* 66-353). WORKS ABOUT: *DAB*, 15: 104-5 (Beverly W. Kunkel); Joseph A. Ewan, *Rocky Mountain Naturalists* ([Denver, Colo.], 1950), 67-72.

POURTALÈS, LOUIS FRANÇOIS De (March 4, 1823/1824, Neuchâtel, Switzerland—July 17/18, 1880, Beverly Farms, Mass.). *Oceanography.* Born into titled and wealthy family. Married Elise Bachmann; one child. EDUCATION: Ca. 1838 [about age fifteen], became student of J. L. R. Agassiz [q.v.] at Neuchâtel; trained as engineer. CAREER: 1847-1848, resided with Agassiz at Cambridge, Mass.; 1848-1873, attached to U.S. Coast Survey, after 1854 as head of tidal division; ca. 1870, on death of father, inherited title of count and thereby achieved financial independence; 1870, began association with Museum of Comparative Zoology, Harvard, and from 1873 to 1880, served as "keeper" of museum; 1871-1872, accompanied Agassiz on *Hassler* voyage around Cape Horn. MEMBERSHIPS: Am. Ac. Arts Scis.; Natl. Ac. Scis. SCIENTIFIC CONTRIBUTIONS: At young age, came under Agassiz's influence, and in 1840 accompanied him on early glacial studies in Alps. Mature work confined to oceanography, in which he pioneered in collection and study of both life and geology of the sea; continued work of J. W. Bailey [q.v.] in microscopic study of ocean sediments, collected by U.S. Coast Survey, and in 1870 published colored chart showing distribution of such sediments on ocean floor, from Cape Cod to Florida. Devised improved means of dredging at greater depths, and chief work was done on animals collected at deeper level than hitherto possible; during period 1867-1869, carried out dredging operations for Coast Survey and discovered so-called Pourtalès Plateau off southeast Florida, an area particularly abundant in sea life. Chief publication was *Corals* [see Works], and also did work on other special groups of coelenterates and echinoderms. Handled dredging work on *Hassler* expedition, and afterward, at Museum of Comparative Zoology, took charge of administrative matters and worked on deep-sea collections.

WORKS: "Der Boden des Golfstroms und der Atlantischen Kuste Nord Amerikas," *Petermanns Geographische Mitteilungen* 16 (1870): 393-98; *Deep-Sea Corals* (Cambridge, Mass., 1871). Bibliography of major writings in *Biog. Mems. of Natl. Ac. Scis.* 5 (1905): 87-89. MSS: Natl. Arch.—Coast Survey records and in Harv. U.-MCZ (*DSB*). WORKS ABOUT: *DAB*, 15: 141-42 (Hubert L. Clark); *DSB*, 11: 113-14 (Harold L. Burstyn).

POWELL, JOHN WESLEY (March 24, 1834, Mount Morris, N.Y.—September 23, 1902, Haven, Maine). *Geology*; *Ethnology.* Son of Joseph, Methodist preacher, and Mary (Dean) Powell. Married Emma Dean, 1862; one child. EDUCATION: Studied at Illinois Institute (now Wheaton College), Illinois College, and Oberlin College, but did not receive degree. HONORS: Ph.D., Univ. of Heidelberg, 1886; LL.D., Harvard Univ., 1886. CAREER: Worked on family farm, taught school, and went on long solitary excursions on Ohio and Mississippi rivers; 1861-1865, in U.S. Army, achieved rank of major; 1865, appointed professor of geology, Illinois Wesleyan College; 1866-1872, lecturer and curator of museum, Illinois State Normal Univ.; 1867 and 1868, led expeditions of students and amateur naturalists to Rocky Mountains of Colorado; 1869 and 1871-1872, led pioneer explorations down Green and Colorado rivers; 1870-1879, engaged in western explorations under sponsorship of U.S. Congress, as division of U.S. Geological and Geographical Survey of Territories (1875-1877) and as Survey of Rocky Mountain Region (1877-1879); 1879-1902, first director of Smithsonian Institution's Bureau of American Ethnology; 1881-1894, director, U.S. Geol. Surv. MEMBERSHIPS: Became secretary, Illinois Society of Natural History, 1858; member of Natl. Ac. Scis. SCIENTIFIC CONTRIBUTIONS: Boat trip in 1869 through unexplored canyons of Green and Colorado rivers, including Grand Canyon, closed an era in western exploration and brought Powell to national attention; in subsequent geological work, demonstrated abilities as administrator, bringing together geologists of talent; in report on explorations of Colorado [see Works] introduced idea that Uinta canyons had been formed by action of rivers on rock in process of gradual elevation. Made unsuccessful attempt at land reform through report on arid lands [see Works], a publication significant for its realization of need for modification of laws to suit western terrain and climate. During years as director of U.S. Geol. Surv. and of Bureau of Ethnology was one of most influential scientists in U.S. Was a student of Indian languages, and as head of Bureau of Ethnology helped establish systematic study of Indian tribes.

WORKS: *Report on the Exploration of the Colorado River of the West and Its Tributaries* (Washington, D.C., 1875); *Report on the Lands of the Arid Region of the U.S.* (Washington, D.C., 1878); *Introduction to the Study of the Indian Languages* (Washington, D.C., 1877). Other works and references to bibliographies in *DSB*. MSS: Natl. Arch. (*DSB*). WORKS ABOUT: *DAB*, 15: 146-68 (George P. Merrill); *DSB*, 11: 118-20 (Wallace Stegner). Also see William C. Darrah, *Powell of the Colorado* (Princeton, 1951).

PRESCOTT, ALBERT BENJAMIN (December 12, 1832, Hastings, N.Y.—February 25, 1905, Ann Arbor, Mich.). *Chemistry.* Son of Benjamin, apparently farmer, and Experience (Huntley) Prescott. Married Abigail Freeburn, 1866; adopted one child. EDUCATION: 1864, M.D., Univ. of Michigan. HONORS: Ph.D. in 1886 and LL.D. in 1896, Univ. of Michigan; LL.D., Northwestern Univ., 1902. CAREER: Wrote for *Liberator* (antislavery publication); 1853, went to work as reporter for *New York Tribune*; ca. 1854, began preparation for medical career, did some teaching in neighborhood school, and for three years served in physician's office in Brewerton, N.Y.; 1860, entered Univ. of Michigan as medical student and, 1863-1864, served there as assistant in chemistry; 1864-1865, assistant surgeon, U.S. Army; at Univ. of Michigan, assistant professor of chemistry and lecturer on organic chemistry and metallurgy (1865-1870), professor of organic chemistry and applied chemistry and pharmacy (1870-1889), dean of School of Pharmacy (1878-1905), director of Chemical Laboratory (1884-1905), professor of organic chemistry and pharmacy (1889-1890), and professor of organic chemistry (1890-1905). MEMBERSHIPS: President of Am. Chem. Soc., 1886, and of Am. Assoc. Advt. Sci., 1891. SCIENTIFIC CONTRIBUTIONS: Through pioneering efforts at Michigan, beginning in 1868, played major part in replacing apprenticeship system with sound academic education in pharmacy. Author of several textbooks and of some 200 papers, including publications relating to pharmacy, toxicology, analytical and organic chemistry; interested in sanitary and food chemistry, worked to improve drinking water in Michigan, and promoted legislation related to purity of food products and drugs; also contributed techniques for detection of additives in food, and has been given credit for development of first U.S. laboratory for work in public health. Performed important service on U.S. *Pharmacopoeia*. Through writings, teaching, and other efforts helped promote greater interest and high standards of careful investigation in chemistry.

WORKS: Among earliest publications were *Qualitative Chemical Analysis . . .* (Ann Arbor, Mich., 1874), with S. H. Douglas [q.v.], and *Outlines of Proximate Organic Analysis* (New York, 1875). See *DAB* for references to bibliographies of other works. MSS: Small collection in Univ. Mich.-MHC. WORKS ABOUT: *DAB*, 15: 192-93 (Lyman C. Newell); Henri R. Manasse, Jr., "Albert B. Prescott's Legacy to Pharmaceutical Education in America," *Pharmacy in History* 15 (1973): 22-28.

PRIESTLEY, JOSEPH (March 13, 1733, Birstal Fieldhead, Yorkshire, England—February 6, 1804, Northumberland, Pa.). *Chemistry; Electricity; Natural philosophy.* Son of Jonas, cloth dresser, and Mary (Swift) Priestley. Married Mary Wilkinson, 1762; four children. EDUCATION: Attended parish schools and studied privately; 1752-1755, attended dissenting academy at Daventry. HONORS: LL.D., Univ. of Edinburgh, 1764. CAREER: 1755, became preacher, Needham Market, Suffolk; later, preached at Nantwich, Cheshire, and there also conducted private school; 1761-1767, tutor in languages and belles lettres in dissenting academy at Warrington; 1762, ordained at Warrington; 1767-1772, minister at Leeds; 1773-1780, librarian to earl of Shelburne; 1780-1791, minister at Birmingham; 1791, driven from Birmingham by destructive mob opposed to his liberal religious views and his sympathy with French Revolution; 1791, went to London; for short time, taught and preached at Hackney; 1794, emigrated to U.S. and established residence at Northumberland, declining offer of professorship in chemistry at Univ. of Pennsylvania. MEMBERSHIPS: Roy. Soc.; Am. Phil. Soc.; and other societies. SCIENTIFIC CONTRIBUTIONS: Noted for advocacy of political and theological liberalism. Work in science initially an outgrowth of educational concerns, and he envisioned *The History of Electricity* [see Works] and the later *History of Optics* as parts of uncompleted series of similar works on all experimental sciences; experimental work in electricity related especially to conductivity of different materials, and *History* included deductive discovery of inverse-square law as related to force between electrical charges. Later studies related mainly to gases, and is best known for discovery of oxygen (called "dephlogisticated air" by Priestley); developed methods for study in "pneumatic" chemistry, and work helped undermine chemical theory of the day, but Priestley himself continued to adhere to older terminology and concepts and never accepted new chemistry of Antoine Lavoisier; in U.S., carried on debates with disciples of Lavoisier in New York *Medical Repository* and elsewhere.

WORKS: *The History and Present State of Electricity, with Original Experiments* (London, 1767); *Experiments and Observations on Different Kinds of Air, and Other Branches of Natural Philosophy* (Birmingham, England, 1790). Other works and references to bibliographies and manuscripts in

DSB. WORKS ABOUT: *DAB*, 15: 223-26 (John F. Fulton); *DSB*, 11: 139-47 (Robert E. Schofield). Also see Frederick W. Gibbs, *Joseph Priestley, Adventurer in Science and Champion of Truth* (London, 1965).

PRIME, TEMPLE (September 14, 1832, New York, N.Y.—February 25, 1903, Huntington, Long Island, N.Y.). *Conchology*. Son of Rufus Prime, successful financier, and his wife, probably Augusta T. Temple. Never married. EDUCATION: Early instruction in Europe and with private tutors; 1850-1852, attended Lawrence Scientific School, Harvard; 1853, LL.B., Harvard Univ.; apparently also LL.B., 1861, Columbia Univ. CAREER: Never practiced law; ca. 1860, secretary to American legation at The Hague; after father's death in 1885, made permanent home at paternal estate at Huntington, residing there with unmarried sister; throughout life, active in local charitable, intellectual, and political affairs, serving as president of Citizens' League for Good Government, at Huntington, and president of several business organizations. SCIENTIFIC CONTRIBUTIONS: Studied with J. L. R. Agassiz [q.v.]. In 1851 (age nineteen), published first of some 167 descriptions of species of mollusks, although in later years he designated more than 60 of his names as synonyms; conchological works appeared chiefly through the late 1870s, especially in serial publications of Bost. Soc. Nat. Hist., Ac. Nat. Scis. Philad., and N.Y. Lyc. Nat. Hist., and are noted for their accuracy and authority; in 1895, deposited collection with Museum of Comparative Zoology, Harvard. Also interested in genealogy and history, and at time of death is said to have been working on French history.

WORKS: "Monograph of American Corbiculadae (recent and fossil)," *Smithson. Misc. Coll.* 7 (1865); *Catalogue of the Species of Corbiculadae in the Collection of Temple Prime, Now Forming Part of the Collection of the Museum of Comparative Zoology at Cambridge, Mass.* (New York, 1895). Bibliography of works on recent Mollusca in Johnson [below], pp. 436-40. WORKS ABOUT: Richard I. Johnson, "The Types of Corbiculidae and Sphaeriidae (Mollusca: Pelecypoda) in the Museum of Comparative Zoology, and a Bio-bibliographic Sketch of Temple Prime, an Early Specialist of the Group," *Bulletin of Museum of Comparative Zoology, Harvard* 120 (1959): 431-79; clippings on Prime in Harvard Univ. Archives.

PRITCHETT, HENRY SMITH (1857-1939). *Astronomy*; *Geodesy*. President, Massachusetts Institute of Technology; first president, Carnegie Foundation for Advancement of Teaching.

MSS: Lib. Cong. (*NUCMC* 66-1449). WORKS ABOUT: *AMS*; *DAB*, supp. 2.

PROCTER, JOHN ROBERT (March 16, 1844, Mason County, Ky.—December 12, 1903, Washington, D.C.). *Geology*. Son of George Morton and Anna Maria (Young) Procter. Married Julia Leslie Dobyns, 1869; three children. EDUCATION: 1863-1864, completed freshman year of scientific study, Univ. of Pennsylvania; ca. 1875, studied geology, Harvard Univ. CAREER: 1864-1865, served in Confederate army and achieved rank of lieutenant of artillery; 1865-1873, engaged in farming, Kentucky; for Kentucky Geological Survey, assistant (1873-1880), then director (1880-1893), and during latter period also Kentucky state commissioner of immigration; 1875, and for several years thereafter, assisted N. S. Shaler [q.v.] with Harvard geology summer camp associated with Kentucky survey; 1893-1903, member and, for most of period, president of U.S. Civil Service Commission. MEMBERSHIPS: Am. Assoc. Advt. Sci.; Geol. Soc. Am. SCIENTIFIC CONTRIBUTIONS: Interest in geology grew out of contacts with N. S. Shaler, director of Kentucky Geological Survey (1874-1880); Robinson article [below] says Procter was Shaler's "office assistant" and claims he was not a geologist. Succeeded Shaler as head of Kentucky survey, but division of attention between geological work and responsibilities to immigration bureau prevented accomplishment of significant work in either, despite employment of able geological assistants; chief accomplishment of survey during Procter's directorship was series of county surveys, issued as pamphlets; refusal to appoint governor's son to survey resulted in its termination, but Procter's courage displayed in this instance won appointment to U.S. Civil Service Commission.

WORKS: Geological work issued mainly as reports of Kentucky Geological Survey during period 1880-1892. See Nickles, *Geol. Lit. on N. Am.* MSS: Lib. Cong. (Hamer). WORKS ABOUT: *DAB*, 15: 241-42 (E. Merton Coulter); George P. Merrill, *Contributions to a History of American State Geological and Natural History Surveys*, U.S. Natl. Mus. Bulletin no. 109 (Washington, D.C., 1920), pp. 118-23; L. C. Robinson, "The Kentucky Geological Survey," *Register of Kentucky State Historical Society* 25 (1927): 87.

PROSSER, CHARLES SMITH (1860-1916). *Geology*; *Paleontology*. Professor of geology, Ohio State Univ.

WORKS ABOUT: See *AMS*; *DAB*.

PUGH, EVAN (February 29, 1828, Jordan Bank, East Nottingham, Pa.—April 29, 1864, Bellefonte, Pa.). *Chemistry*. Son of Lewis, farmer and blacksmith, and Mary (Hutton) Pugh. Married Rebecca Valentine, 1864; no children. EDUCATION: Ca.

1844-ca. 1847 [age sixteen to nineteen], apprenticeship to blacksmith; for two years, studied at Manual Labor Academy, Whitestown (Whitesboro), N.Y.; 1853, entered Univ. of Leipzig; 1856, Ph.D. in chemistry, Univ. of Göttingen, where studied under Friedrich Wöhler. CAREER: Taught in district school and managed family farm; for two years, operated private school for boys at Oxford, Pa.; 1853, went to Europe; ca. 1856, for six months, engaged in research in laboratories of Heidelberg Univ.; ca. 1857-1859, carried out researches at laboratories of Sir John Bennett Lawes and Sir Joseph Henry Gilbert, Rothamsted, England; 1859-1864, first president and professor of chemistry, scientific and practical agriculture, and mineralogy, Agricultural College of Pennsylvania (Pennsylvania State Univ.). MEMBERSHIPS: Chemical Society of London; Am. Phil. Soc. SCIENTIFIC CONTRIBUTIONS: At Göttingen, produced doctoral thesis entitled *Miscellaneous Chemical Analyses* (published in 1856), and at Heidelberg began researches on possibility that plants make use of free nitrogen in the air; in 1857, with Gilbert and Lawes, showed that nitrogen did not directly enter plants, and these studies at Rothamsted, especially as presented in 1862 paper [see Works], made a contribution of considerable significance to knowledge of plant growth. Last years of life devoted to task of firmly establishing the Agricultural College of Pennsylvania; during this same period also did some research work; multitudinous duties contributed to death at age thirty-six.

WORKS: "On a New Method for the Quantitative Estimation of Nitric Acid," *Quarterly Journal of Chemical Society of London* 12 (1860): 35-42; "On the Sources of the Nitrogen of Vegetation," *Phil. Trans. of Roy. Soc.* 151 (1862): 431-577, with J. B. Lawes and J. H. Gilbert. MSS: Pennsylvania State Univ. Library (Hamer). WORKS ABOUT: *DAB*, 15: 257-58 (Fred L. Pattee); Margaret T. Riley, "Evan Pugh of Pennsylvania State University and the Morrill Land-Grant Act," *Pennsylvania History* 27 (1960): 339-60.

PUMPELLY, RAPHAEL (1837-1923). *Geology.* Conducted geological studies in Japan and China; associated with New England division, U.S. Geol. Surv.; state geologist of Michigan and Missouri; professor of mining geology, Harvard Univ.

MSS: Hunting. Lib. (*NUCMC* 71-1083). WORKS ABOUT: See *AMS*; *DAB*; *DSB*.

PURSH, FREDERICK (February 4, 1774, Grossenhain, Saxony, Germany—July 11, 1820, Montreal, Canada). *Botany.* No information on parentage. Said to have married, though details are not known. EDUCATION: Attended public schools at Grossenhain; studied horticulture under court gardener, Johann Heinrich Seidel, apparently at Dresden. CAREER: Until ca. 1799, member of staff, Royal Botanic Gardens, Dresden; 1799, came to U.S. and entered employment in botanic garden near Baltimore; 1803, took charge of gardens at "Woodlands," estate of William Hamilton near Philadelphia; 1806, under sponsorship of B. S. Barton [q.v.], carried out botanical explorations southward to North Carolina border; 1807, also for Barton, botanized in Pennsylvania, New York, and Vermont; after return, engaged in work on specimens collected by Lewis and Clark Expedition; 1809, entered employ of David Hosack in development of Elgin Botanic Garden in New York City; 1810-1811, traveled in West Indies; 1811, returned to New York City by way of Wiscasset, Maine; ca. 1811, went to England, won patronage of Aylmer B. Lambert, and published *Flora* (1814); worked on the publication of catalogues of gardens at Cambridge, England, and St. Petersburg, Russia; 1816-1820, in Montreal, engaged in some botanic work and received financial aid from friends; 1818, botanized on Anticosti Island. SCIENTIFIC CONTRIBUTIONS: From early years at Baltimore, gathered specimens of flora from all parts of North America; travels for B. S. Barton were first extensive botanical explorations carried out under American sponsorship. When going to England, took notes, drawings, and parts of collection; *Flora* [see Works] produced there drew upon these and British collections and included first publication of Lewis and Clark specimens from Pacific Coast; *Flora* was his chief work and the first such publication aiming to cover all of North America above Mexico and was of great instructional and inspirational value to a generation of botanists. At Montreal, worked on flora of Canada, but collection, and probably notes, perished in fire.

WORKS: *Flora Americae Septentrionalis*, 2 vols. (London, 1814), of which facsimile with notes is planned (see *DSB*); *Journal of a Botanical Excursion in the Northeastern Parts of the States of Pennsylvania and New York During the year 1807* (Philadelphia, 1869; reprinted, New York, 1969). WORKS ABOUT: *DAB*, 15: 271 (William R. Maxon); *DSB*, 11: 217-19 (Joseph Ewan).

PURYEAR, BENNET (July 23, 1826, Mecklenburg County, Va.—March 30, 1914, Madison County, Va.). *Chemistry*; *Agriculture.* Son of Thomas, apparently plantation owner, and Elizabeth (Marshall) Puryear. Married Virginia Catherine Ragland, 1858; Ella Marian Wyles, 1871; five children by first marriage, six by second. EDUCATION: 1847, A.B., Randolph-Macon College; 1848-1849, studied medicine, Univ. of Virginia. HONORS: LL.D., Georgetown College (Ky.) and Howard College (Ala.), 1878. CAREER:

1847-1848, teacher, Monroe County, Ala.; at Richmond College (Univ. of Richmond), tutor in chemistry (1849-1850), professor of natural science (1850-1858 and 1866-1873), professor of chemistry (1873-1895), and chairman of faculty (1869-1885 and 1889-1895); 1858-1863, professor of chemistry and geology, Randolph-Macon College; ca. 1863-1865, operated Classical School on Randolph-Macon campus; ca. 1895-1914, retired to home at "Edgewood," Madison County, Va. SCIENTIFIC CONTRIBUTIONS: Added to science chiefly through teaching. Interested in agriculture and lectured on agricultural chemistry; had important influence with farmers of Virginia; actively engaged in gardening and had productive plot of land near Richmond College; during period 1866-1867, published in *Farmer* a series of articles, "The Theory of Vegetable Growth." Widely known for writings on political and educational questions, publishing widely debated items in opposition to public support of education, during Reconstruction period.

WORKS: See reference to *Farmer*, above; general reference to political and educational writings in *DAB*. No works in *Roy. Soc. Cat.* WORKS ABOUT: *DAB*, 15: 272-73 (Samuel C. Mitchell); Woodford B. Hackley, *Faces on the Wall: Brief Sketches of the Men and Women Whose Portraits and Busts Were on the Campus of the Univ. of Richmond in 1955* ([Richmond, Va., 1972]), pp. 83-84; manuscript sketch in Univ. of Richmond Archives.

PUTNAM, FREDERIC WARD (1839-1915). *Anthropology*; *Natural history*. Professor of American archaeology and ethnology, curator of Peabody Museum, Harvard Univ.

MSS: Harv. U.-UA (*NUCMC* 65-1267); Lib. Cong. (Hamer). WORKS ABOUT: See *AMS*; *DAB*; *DSB*.

R

RAFINESQUE, CONSTANTINE SAMUEL (October 22, 1783, Galata, near Constantinople, Turkey —September 18, 1840, Philadelphia, Pa.). *Natural history.* Son of Georges F., successful French merchant, and Madeleine (Schmaltz) Rafinesque. Married Josephine Vaccaro, 1809; two children. EDUCATION: Instructed under private tutors, at Leghorn, Italy, and at Marseilles; 1800, apprenticed to merchant. CAREER: 1802-1804, was in U.S., where was employed for time in Philadelphia countinghouse, met scientists of that city, and conducted botanical explorations in New Jersey and Dismal Swamp in Virginia; 1804, returned to Leghorn with botanical specimens; 1805-1815, at Palermo, Sicily, first employed as secretary and chancellor to American consul, and after 1808 engaged in business of exporting squills and medicinal plants; 1815, returned to U.S., arriving at New York, where served for time as tutor, and met naturalists in New York and Philadelphia; 1818, went to Lexington, Ky.; 1819-1826, professor of botany, natural history, and modern languages, Transylvania Univ. (Lexington); 1826-1840, lived in Philadelphia, continued travels, lectured at Franklin Institute (1826-1827), sold vegetable concoction for treatment of tuberculosis, founded savings bank still in existence at death, but died in poverty. SCIENTIFIC CONTRIBUTIONS: While in Sicily studied natural history, including fishes. During return to U.S. in 1815 lost manuscripts and sketches, as well as other possessions, in shipwreck. In U.S., traveled more extensively than any other naturalist and met scientists throughout country. Having begun publishing while in Sicily, eventually issued over 900 works, many now rare; particularly interested in botany and ichthyology; also wrote on financial topics, and suggested that ideographs of Mayan writing were in part syllabic; helped introduce Jussieu's natural plant classification into U.S., replacing artificial scheme of Linnaeus. Descriptions in natural history said frequently to be inadequate or incorrect, and had tendency to elevate varieties to species, though this practice grew out of ideas on development of new forms, which later were acknowledged by Charles Darwin. Perhaps because of immensity and range of output, writings were largely ignored by contemporaries, but have won attention of later researchers.

WORKS: T. J. Fitzpatrick, *Rafinesque: A Sketch of His Life With Bibliography* (Des Moines, Iowa, 1911), lists 938 publications. MSS: Am. Phil. Soc. (*NUCMC* 61-766); Ac. Nat. Scis. Philad. (*NUCMC* 66-354); also see Fitzpatrick [above] and Hamer. WORKS ABOUT: *DAB*, 15: 322-24 (George H. Genzmer); *DSB*, 11: 262-64 (Joseph Ewan).

RAMSAY, ALEXANDER (probably 1754, at or near Edinburgh, Scotland—November 24, 1824, Parsonsfield, Maine). *Anatomy.* Born into family of some means. No information on marital status. EDUCATION: Thought to have attended Aberdeen Univ.; studied medicine with noted teachers in London, Dublin, and Edinburgh. HONORS: M.D., Univ. of St. Andrews, 1805. CAREER: Ca. 1790, established successful school of anatomy at Edinburgh, adding museum and other facilities; 1801, came to U.S., finally settled at Fryeburg, Maine, and there established school of anatomy; also occasionally engaged in medical practice, and lectured in various locations in northeastern states; spent some time in New York during yellow fever epidemic, and in conjunction with several other persons made unsuccessful attempt to establish new medical school there; 1808, lectured at Dartmouth Medical School; 1810-1816, in England, where lectured and attempted to raise funds for anatomical institute at Fryeburg; 1816, returned to U.S., traveled through northeast states, eastern Canada, and Charleston, S.C., lecturing on anatomy and natural philosophy. SCIENTIFIC CONTRIBUTIONS: Biographical accounts stress great abilities as anatomist, but lack of fulfillment is ascribed to querulous temperament displayed both in England and U.S.; this and other factors led to failure of lifelong scheme to establish anatomical museum of importance in America. While in England, 1810-1816, published first of projected five volumes on system of anatomy, with illustrations drawn and engraved by author at Fryeburg; this was work of great beauty and accuracy, but only portion on heart and brain ever appeared [see Works]; did significant work in 1803 New York yellow fever epidemic; lectures ordinarily were designated "The Animal and Intellectual Economy of Human Nature as Founded on Comparative Anatomy," and "Dissection as a Basis of Physiology, Anatomy, Surgery and Medicine." Private anatomical museum, at one time estimated at value of $14,000, was dispersed after death.

WORKS: *Anatomy of the Heart, Cranium, and Brain*, 2d ed. (Edinburgh, 1813). MSS: Maine Historical Society (Hamer). WORKS ABOUT: *DAB*, 15: 337-38 (Henry R. Viets); Kelly and Burrage, *Med. Biog.* (1928).

RATHBUN, RICHARD (1852-1918). *Marine invertebrates.* Associated with U.S. Commission on Fish; assistant secretary, Smithson. Instn., in charge of U.S. Natl. Mus.

MSS: Smithson. Instn. Arch. (*NUCMC* 72-1252). WORKS ABOUT: See *AMS*; *DAB*.

RAVENEL, EDMUND (December 8, 1797, Charleston, S.C.—July 27, 1871, Charleston, S.C.). *Conchology.* Son of Daniel, apparently planter, and Catherine (Prioleau) Ravenel. Married Charlotte Ford; Louisa C. Ford; one child by first marriage, eight by second. EDUCATION: 1819, M.D., Univ. of Pennsylvania. CAREER: Ca. 1819, began medical practice in Charleston; 1824-1835, professor of chemistry, Medical College of South Carolina; ca. 1835, purchased plantation on Cooper River, S.C., and resided there until after Civil War; on several occasions, headed Fort Moultrie Hospital; ca. 1866, returned to Charleston, nearly blind. MEMBERSHIPS: Corresponding member, Ac. Nat. Scis. Philad.; vice-president, Elliott Society of Natural History (Charleston), 1853-1871. SCIENTIFIC CONTRIBUTIONS: Noted for pioneer work in conchology, and in 1834 produced catalogue of 735 species [see Works], a publication referred to as first of its kind in U.S. Entertained visiting naturalists on plantation; there and on Sullivan's Island gathered conchological specimens, and eventually accumulated collection of 3,500 recent and fossil shells, including land, freshwater, and marine forms from all areas of the globe. Also conducted experiments with fruit growing.

WORKS: *Catalogue of the Recent Shells in the Cabinet of Edmund Ravenel* (Charleston, 1834), 20 pp.; *Echinidae, Recent and Fossil of South Carolina* (Charleston, 1848), 4 pp.; three papers published in *Proceedings of Elliott Society of Natural History* (vols. 1, 1858, and 2, 1860) listed in *DAB*. See also Meisel, *Bibliog. Amer. Nat. Hist.* MSS: Charleston Museum (Hamer). WORKS ABOUT: *DAB*, 15: 394-95 (Anne K. Gregorie); Kelly and Burrage, *Med. Biog.* (1928).

RAVENEL, HENRY WILLIAM (May 19, 1814, "Pooshee" plantation, St. John's, Berkeley Parish, S.C.—July 17, 1887, Aiken, S.C.). *Botany.* Son of Henry, physician and planter, and Catherine (Stevens) Ravenel. Married Elizabeth Gilliard Snowden, 1835; Mary Huger Dawson, 1858; survived by nine children. EDUCATION: 1832, A.B., South Carolina College (Univ. of South Carolina). CAREER: After college, became established as planter at St. John's; 1853-1887, resided at Aiken, S.C.; during Civil War, lost fortune, and thereafter conducted nursery and seed business, published newspaper, and wrote for agricultural journals; 1869, went as agent of U.S. government to Texas to study outbreak of cattle disease; collected and classified botanical specimens for other scientists and for societies; 1882-1887, agricultural editor, *Weekly News and Courier.* MEMBERSHIPS: Royal Zoological and Botanical Society of Vienna and other societies. SCIENTIFIC CONTRIBUTIONS: Chief contributions resulted from field work as collector and disseminator of botanical specimens, especially cryptogamic, in study of which Ravenel was a pioneer and leading U.S. authority at the time. Early work with flowering plants culminated in paper read in 1850 to Am. Assoc. Advt. Sci. [see Works]; this paper also included cryptogamic plants, an interest that originated about 1846; during years 1852-1860 issued sets of fungi [see Works], and during period 1878-1882 issued second series [see Works], which contained specimens collected by Ravenel and named by English botanist M. C. Cooke; in 1893 cryptogamic specimens in herbarium were sold to British Museum.

WORKS: "A Catalogue of the Natural Orders of Plants Inhabiting the Vicinity of the Santee Canal, S.C., as Represented by Genera and Species; with Observations on the Meteorological and Topographical Conditions of that Section of Country," *Proceedings of Am. Assoc. Advt. Sci.* 3 (1850): 2-17; *Fungi Caroliniani Exsiccati*, 5 vols., each with 100 specimens (Charleston, 1852-1860); *Fungi Americani Exsiccati*, 8 sets of 100 specimens (London, 1878-1882), with M. C. Cooke; *Private Journal . . . 1859-1887* (Columbia, S.C., 1947), 428 pp. MSS: Clemson College Library (*NUCMC* 62-2034); Univ. S.C.-SCL (*NUCMC* 74-997); Charleston Museum (Hamer). WORKS ABOUT: *DAB*, 15: 396-97 (Arney R. Childs); Neil E. Stevens, "The Mycological Work of Henry W. Ravenel," *Isis* 18 (1932-1933): 133-49.

RAVENEL, ST. JULIEN (December 19, 1819, Charleston, S.C.—March 17, 1882, Charleston, S.C.). *Agricultural chemistry.* Son of John, merchant and ship owner, and Anna Elizabeth (Ford) Ravenel. Married Harriott Horry Rutledge, 1851; survived by nine children. EDUCATION: Studied medicine with Charleston physician; 1840, M.D., Charleston Medical College; thereafter, studied medicine in Philadelphia and in Paris. CAREER: Upon completion of medical studies, for brief period served as demonstrator of anatomy at Charleston Medical College; until 1851, practiced medicine, but found it disagreeable and thereafter turned attentions to scientific studies; 1857, established first South Carolina stone lime works, at plantation on Cooper River, Stoney Landing; during Civil War, served first with Phoenix Rifles, then as chief surgeon in Confederate hospital at Columbia, and lastly took charge of Confederate medical and pharmaceutical laboratory at Columbia; 1868, Wando Fertilizer Co. began production, utilizing process developed by Ravenel; during remainder of lifetime was involved with various chemical-agricultural firms in South Carolina. MEMBERSHIPS: Am. Assoc. Advt. Sci. and other societies. SCIENTIFIC CONTRIBUTIONS: Interested in natural history and physiology, but particularly in chemical research as applied to agriculture. Following Civil War, conducted experiments on phosphate deposits along Ashley and Cooper rivers, with intent to develop that resource for commercial purposes; devised manufacturing process for ammoniated fertilizer, utilized by Wando Co.; subsequently, also developed phosphate fertilizer not involving ammonia, which he later neutralized by addition of marl, further increasing its fertilizing value. Discovered soil-enriching value of planting and plowing legumes. Among other interests, helped promote artesian wells for Charleston.

WORKS: Apparently no publications of his own. Arthur R. Guerard, *An Essay on the Application of Finely-Ground Mineral Phosphate of Lime and 'Ash' Element As Manure: Being an Exposition of the Views and Experiments of the Late Dr. St. Julien Ravenel* (Charleston, 1882), 20 pp. WORKS ABOUT: *DAB*, 15: 397 (Arney R. Childs); *NCAB*, 10 (1909 [copyright 1900]): 272-73.

REDFIELD, WILLIAM C. (March 26, 1789, Middleton, Conn.—February 12, 1857, New York, N.Y.). *Meteorology; Paleontology.* Son of Peleg, seafarer, and Elizabeth (Pratt) Redfield. Married Abigail Wilcox, 1814; Lucy Wilcox, 1820; Jane Wallace, 1828; three children by first marriage, one child by second. EDUCATION: 1803-1810, served apprenticeship to saddle and harness maker; largely self-educated in science. HONORS: A.M., Yale Univ., 1839. CAREER: 1810-1811, trip to Ohio, where mother and stepfather were living; 1811-ca. 1821, worked as saddle and harness maker and operated store, Cromwell, Conn.; 1822, entered on career in transportation and naval engineering with steamboat on Connecticut River; 1824-1857, lived in New York City, engaged in steam transportation business, devised so-called safety barges used between Albany and New York (first for passengers and later for freight), and served as superintendent of Steam Navigation Co.; actively engaged in railroad promotion, as early as 1829 he published suggestion for railroad from Hudson to Mississippi and promoted Harlem and Hartford/New Haven railroads; on board of directors, Hudson Railroad. MEMBERSHIPS: A founder and first president, Am. Assoc. Advt. Sci., 1848. SCIENTIFIC CONTRIBUTIONS: In 1821, observed pattern of felled trees after hurricane in western Massachusetts; from this and other evidence later presented his conclusion that such a storm was a "progressive whirlwind" whose winds blew counterclockwise about center moving in direction of prevailing winds; with this insight won international reputation; continued to write on subject for many years with aid of data from seamen, a debt repaid with information that helped sailors cope with such storms. Carried on heated debate with J. P. Espy [q.v.] about nature of storms. Beginning in late 1830s, also became interested in paleontological studies and achieved place as American pioneer in study of fossil fishes, to which he made contributions of permanent usefulness.

WORKS: "Remarks on the Prevailing Storms of the Atlantic Coast of the North American States," *Am. Journ. Sci.* 20 (1831): 17-51; paleontological papers (1838-1856) in *Am. Journ. Sci.* and *Proceedings of Am. Assoc. Advt. Sci.* See bibliography in *Am. Journ. Sci.*, 2d ser. 24 (1857): 370-73. MSS: Yale U.-HMC (Hamer); Lib. Cong. (Hamer). WORKS ABOUT: *DAB*, 15: 441-42 (William J. Humphreys); *DSB*, 11: 340-41 (Harold L. Burstyn).

REES, JOHN KROM (1851-1907). *Astronomy; Geodesy.* Professor of geodesy and astronomy and director of observatory, Columbia Univ.

WORKS ABOUT: See *AMS*; *DAB*.

REMSEN, IRA (1846-1927). *Chemistry.* Professor of chemistry and president, Johns Hopkins Univ.

MSS: Johns Hopkins Univ. Library (*NUCMC* 64-1382); also see *DSB*. WORKS ABOUT: See *AMS*; *DAB*; *DSB*.

RENOUF, EDWARD (1848-1934). *Chemistry.* Collegiate professor of chemistry, Johns Hopkins Univ.

WORKS ABOUT: See *AMS*; *Who Was Who in America*, vol. 1.

RENWICK, JAMES, Sr. (May 30, 1792, Liverpool, England—January 12, 1863, New York, N.Y.). *Engineering*; *Physics*. Son of William, merchant, and Jane (Jeffrey) Renwick; brother-in-law of C. Wilkes [q.v.]. Married Margaret Anne Brevoort, 1816; four children. EDUCATION: 1807, A.B., Columbia Univ. HONORS: LL.D., Columbia Univ., 1829. CAREER: After graduation from Columbia, traveled in Europe with Washington Irving; 1812, lectured on natural philosophy, Columbia Univ.; 1814, engaged as topographical engineer in employ of government; for a time, conducted father's mercantile business; 1817, commissioned colonel of engineers in state militia; at Columbia Univ., trustee (1817-1820) and professor of natural philosophy and experimental chemistry (1820-1853); 1838, appointed by President Van Buren to commission for testing devices for prevention of steam boiler explosions; 1840, appointed to U.S. commission for survey of Maine-Canada boundary; 1853, retired and thereafter lived active life as person of wealth. MEMBERSHIPS: Am. Ac. Arts Scis. SCIENTIFIC CONTRIBUTIONS: During lifetime, recognized as leading American engineer and was frequently consulted on engineering projects; maintained special interest in canal and railroad problems and in questions related to steam boilers; in 1826, won Franklin Institute Medal for proposals regarding use of inclined planes or railways for transporting canal boats in cradles over elevations, used in Morris Canal connecting Delaware and Hudson rivers. 1827, published report on weights and measures, prepared for New York commission on revision of state laws. *Outlines* [see Works], based on lectures at Columbia, achieved distinction as first comprehensive work on physics by American author; publications also included textbook on chemistry (1840) and book on steam engine [see Works], later translated into French and German; extensive writings also included works on artillery and on applications of mechanics.

WORKS: *Outlines of Natural Philosophy*, 2 vols. (New York, 1822-1823); *Treatise on the Steam Engine* (New York, 1830). See *Natl. Un. Cat. Pre-1956 Imp.* and *Roy. Soc. Cat.*, vol. 5. MSS: Items in various collections (especially Renwick Family Papers) at Colum. U. Lib. WORKS ABOUT: *DAB*, 15: 506-7 (James P. C. Southall); Guralnick, *Science and American College*, pp. 205-6.

RICE, CHARLES (October 4, 1841, Munich, Germany—May 13, 1901, New York, N.Y.). *Pharmacology*; *Chemistry*. Son of Austrian parents. Never married. EDUCATION: Early education at Munich, then Passau and Vienna; for a time, attended Jesuit college in Paris; received sound training in both natural sciences and classics. HONORS: Ph.D., Univ. of City of New York (New York Univ.), 1879; honorary masterate, Philadelphia College of Pharmacy and Science. CAREER: 1862, came to U.S.; ca. 1862-1865, surgeon's steward, U.S. Navy; ca. 1865, appointed to subordinate position, drug department, Bellevue Hospital (New York); soon thereafter, became chief chemist, Department of Public Charities and Corrections, City of New York; later appointed, and served for life, as superintendent and pharmacist, general drug department of Bellevue Hospital; 1870, became trustee and librarian, College of Pharmacy of City of New York; 1880-1901, chairman of committee on revision, U.S. *Pharmacopoeia*; worked on proofreading of *The Index Catalogue of the Library of the Surgeon General's Office*. MEMBERSHIPS: Am. Chem. Soc.; N.Y. Ac. Scis.; and other societies. SCIENTIFIC CONTRIBUTIONS: Professional activities at Bellevue led to involvement in chemical and pharmaceutical studies and concerns; during years 1871-1901 published numerous papers in pharmaceutical journals; perhaps best known for work on revision of U.S. *Pharmacopoeia*, his abilities and high standards helping to make it the equal of pharmacopoeias published anywhere in world. Also noted as preeminent philologist and Sanskrit scholar.

WORKS: See *Natl. Un. Cat. Pre-1956 Imp.* and *Roy. Soc. Cat.*, vols. 11, 18; *Charles Rice* (Philadelphia, 1904, printed for private circulation, by J. B. Lippincott Co.) is said to contain bibliography of writings. WORKS ABOUT: *DAB*, 15: 535 (Virgil Coblentz); Frederick J. Wulling, *Samuel W. Melendy Memorial Lectures: I-IV* (La Crosse, Wis., 1946), pp. 59-63.

RICE, WILLIAM NORTH (1845-1928). *Geology*. Professor of geology and natural history, Wesleyan Univ. (Connecticut).

WORKS ABOUT: See *AMS*; *DAB*.

RICHARDS, ELLEN HENRIETTA SWALLOW (1842-1911). *Sanitary chemistry*; *Hygiene*. Instructor of sanitary chemistry, Massachusetts Institute of Technology.

MSS: Smith College Library—Sophia Smith Collection (see this and additional references in *Nota. Am. Wom.*). WORKS ABOUT: See *AMS* (Mrs. Robert H.); *DAB*; *Nota. Am. Wom.*

RIDDELL, JOHN LEONARD (February 20, 1807, Leyden, Mass.—October 7, 1865, New Orleans, La.). *Botany*; *Microscopy*; *Invention*. Son of John, small farmer, and Lephe (Gates) Riddell. Married Mary Knocke, 1836; Anne Hennifin; Angelica Brown, ca. 1845/1850; one child by first marriage, two children by second, three by third. EDUCATION: 1829, A.B., Rensselaer School (Rensselaer Polytechnic Institute); 1836, received M.D. degree, Cincinnati Medical College. CAREER: 1830, began as traveling lecturer on chemistry and related subjects; 1832-1834, lecturer, Ohio Reformed Medical College (Worthington, Ohio); ca. 1834, appointed adjunct professor of chemistry and professor of botany, Cincinnati Medical College; 1836 (summer), conducted investigations for Ohio Geological Survey; 1836, married and moved to Louisiana, where wife had property, and established medical practice at New Orleans; 1836-1865, professor of chemistry, Medical College of Louisiana (later Tulane Univ.); 1838, connected with government-sponsored scientific explorations in Texas; ca. 1838-1849, melter and refiner, U.S. Mint at New Orleans; short time before Civil War, appointed postmaster at New Orleans, and by circumstance continued under Confederacy to end of war. MEMBERSHIPS: A founding member, New Orleans Academy of Science. SCIENTIFIC CONTRIBUTIONS: Best remembered for *Synopsis* [see Works], significant early catalogue of western flora; later prepared catalogue of Louisiana plants, original manuscript of which apparently was lost, though abridgement appeared in 1852 in *New Orleans Medical and Surgical Journal.* Early 1850s, devised binocular microscope involving prisms dividing light from single objective, although difficulties in use prevented serious application; carried out microscopic studies of Mississippi River and of blood; *Miasm and Contagion* (1836) was early work suggesting organisms as cause of fevers. Also interested in aerial navigation; wrote on mercurial barometer, preservation of natural history specimens, geology, use of logarithms, and other topics.

WORKS: *Synopsis of the Flora of the Western States* (Cincinnati, 1835); "On the Binocular Microscope," *Proceedings of Am. Assoc. Advt. Sci.* 7 (1853): 16-22. See *Roy. Soc. Cat.*, vol. 5. MSS: Tulane Univ. Library (Hamer). WORKS ABOUT: *DAB*, 15: 589-90 (Rudolph Matas and Virginia Gray); Adolph Waller, "The Vaulting Imagination of John L. Riddell," *Ohio Archaeological and Historical Quarterly* 54 (1945): 331-60.

RIDGWAY, ROBERT (1850-1929). *Ornithology.* Curator of birds, U.S. Natl. Mus.

MSS: Ac. Nat. Scis. Philad. (*NUCMC* 66-78); Am. Mus. Nat. Hist. (Hamer). WORKS ABOUT: See *AMS*; *DAB*; *DSB*.

RILEY, CHARLES VALENTINE (September 18, 1843, Chelsea, London, England—September 14, 1895, Washington, D.C.). *Entomology.* Son of Charles, clergyman in Church of England, and Mary Valentine (Cannon) Riley. Married Emilie Conzelman, 1878; survived by five children. EDUCATION: Until ca. 1860 [age seventeen], attended school in England and boarding schools in Dieppe, France, and in Bonn, Germany. HONORS: Many honors, including Ph.D., Univ. of Missouri, 1873. CAREER: Ca. 1860 [age seventeen], came to U.S.; found employment on farm in Illinois; ca. 1864-1868, excepting six months' service with Illinois Volunteers, lived in Chicago and worked at various jobs, finally as reporter, artist, and then entomological editor of *Prairie Farmer*; 1868-1877, Missouri state entomologist, and during this period lectured at Univ. of Missouri and elsewhere; 1877, appointed chairman, U.S. Entomological Commission; 1878-1879 and 1881-1894, entomologist, U.S. Department of Agriculture, from 1881 as head of new Division of Entomology. MEMBERSHIPS: Honorary member, Entomological Society of London; a founder, American Association of Economic Entomologists. SCIENTIFIC CONTRIBUTIONS: While in school had interest in nature studies and showed ability as artist; while on Illinois farm, attention turned to study of insect damage to vegetation and began contributions to *Prairie Farmer.* Position as Missouri entomologist secured in part through association with B. D. Walsh [q.v.]; nine Missouri reports included notable work on life histories of insects, illustrations (by Riley himself), and superior typography and format that set new standard for American publications on economic entomology; largely responsible for establishment in 1877 of U.S. Entomological Commission, instigated by grasshopper problem in western states, and commission prepared five volumes and seven bulletins on grasshoppers and on other insect pests. In Department of Agriculture, helped establish study of insects as important government activity; helped control cottony-cushion scale in California citrus industry by importation of predator; investigated relationship of Yucca moth to fertilization of Yucca flower, and carried out other studies related to biology and economic significance of insects. During 1889-1894, published journal, *Insect Life*; credited in *DAB* with well over 2,000 published items. Donated large entomology collection to U.S. Natl. Mus.

WORKS: Missouri and Entomological Commission reports; annual reports as chief of entomology in Department of Agriculture. See also *Roy. Soc. Cat.*, vols. 8, 11, 12, 18; also references to other bibliographies in Max Arnim, *Internationale Personalbibliographie, 1800-1943* (Stuttgart, 1952). MSS: Smithson. Instn. Archives (*NUCMC* 74-978); Ac. Nat. Scis. Philad. (Hamer). WORKS ABOUT: *DAB*, 15: 609-10 (Leland O. Howard); Mallis, *Amer. Entomologists*, pp. 69-79.

RISING, WILLARD BRADLEY (1839-1910). *Chemistry.* Professor of chemistry, Univ. of California, Berkeley.

WORKS ABOUT: See *AMS; NCAB*, 25 (1936): 38.

RITTENHOUSE, DAVID (April 8, 1732, Paper Mill Run, near Germantown, Pa.—June 26, 1796, Philadelphia, Pa.). *Astronomy*; *Natural philosophy*; *Scientific instruments*. Son of Matthias, farmer, and Elizabeth (Williams) Rittenhouse; uncle of Benjamin S. Barton [q.v.]. Married Eleanor Coulston, 1766; Hannah Jacobs, 1772; two children by first marriage, one by second. EDUCATION: Little formal education; largely self-educated. CAREER: Ca. 1751 [age nineteen], established instrument shop, especially for clock making, on father's farm at Norriton, Pa.; 1763-1764, engaged in surveying of Pennsylvania-Maryland boundary, and in subsequent years served on number of boundary surveys and commissions involving majority of American colonies and also engaged in surveys of rivers and canals; 1770, moved permanently to Philadelphia, where continued trade as clock and instrument maker and over years held numerous civic and scientific positions; 1775, became engineer, Committee of Safety; 1776, made vice-president, Committee of Safety, and served as member of Pennsylvania General Assembly and state constitutional convention; 1777, became president, Council of Safety; 1777-1789, Pennsylvania state treasurer; 1792-1795, first director, U.S. Mint. MEMBERSHIPS: Roy. Soc.; president, Am. Phil. Soc., 1791-1796. SCIENTIFIC CONTRIBUTIONS: Reputation first established with design in 1767 of orrery; wide range of scientific instruments were of higher quality than those theretofore produced in America, and scientific work largely grew out of interest in instruments. Constructed observatory for viewing transit of Venus in 1769, and his observations were highly accurate; later continued astronomical studies at Philadelphia; did work in mathematics, especially as related to astronomy. Engaged in experimentation on compensation pendulums and expansion of steel and of wood and on magnetism and electricity; made plane transmission gratings used in diffraction studies; in optics studied and explained causes of illusion of reversal of raised and depressed surfaces. Also published observations on lightning and on meteorology, geology, and natural history; scientific work largely experimental and observational and related to number of separate problems or concerns.

WORKS: Nearly all papers appeared in *Transactions of Am. Phil. Soc.*, vols. 1-4 (1771-1799), of which vol. 1 contains accounts of orrery and transit of Venus observations; "To Determine the True Place of a Planet," ibid. 4 (1799): 21-26. See publications listed in both *DAB* and *DSB*. MSS: For information on scattered locations of manuscripts, see Brooke Hindle, *David Rittenhouse* (Princeton, 1964), pp. 367-70. WORKS ABOUT: *DAB*, 15: 630-32 (W. Carl Rufus); *DSB*, 11: 471-72 (Brooke Hindle).

ROBIE, THOMAS (March 20, 1688/1689, Boston, Mass.—August 28, 1729, Salem, Mass.). *Meteorology*; *Astronomy*; *Medicine*. Son of William and Elizabeth (Greenough) Robie. Married Mehitable Sewall, 1723; four children. EDUCATION: 1708, A.B., Harvard College; apparently self-taught in medicine. CAREER: 1708, taught school, Watertown, Mass.; 1708-1720, published almanac, Boston; at Harvard College, librarian (1712-1713) and tutor (1713-1723); 1722-1723, fellow of Harvard Corporation; while in Cambridge, Mass., preached and engaged in medical practice; 1723-1729, medical practice, Salem, Mass. MEMBERSHIPS: Fellow, Roy. Soc. SCIENTIFIC CONTRIBUTIONS: Almanacs included essays on scientific topics for benefit of general reader; they showed author's familiarity with both American and English works and helped promote scientific outlook and Newtonianism in colonies, almanacs for 1716 and 1720 being particularly notable in this regard. Kept weather record for years 1715-1722 and used Harvard instruments for astronomical observing; corresponded with scientists in England, especially Rev. William Derham, and much of this correspondence was presented by Derham to Roy. Soc.; these communications included meteorological and astronomical observations begun as early as 1714 and related to satellites of Jupiter (1714), eclipses (1717 and 1722), and aurora borealis (1719), the latter explained by Robie in wholly natural terms; work on aurora was published at Boston in pamphlet form; observed transit of Mercury in 1723, results of which were communicated to Roy. Soc. by C. Mather [q.v.]. Also wrote to Derham concerning old oak tree chemically analyzed after burning; as published in *Phil. Trans. of Roy. Soc.* (1720), this ranks as first detailed chemical analysis published by colonial American. Chief contribution was promotion of acceptance of scientific outlook and method in Boston community.

WORKS: See above; see also "Concerning the Effects of Inoculation," "The Eclipse of the Sun in Nov. 1722," and "The Venom of Spiders" [extracts from a letter], *Phil. Trans. of Roy. Soc.* 33 (1724): 67-71; meteorological record, as published by W. Derham, in ibid. 37 (1732): 261-73. See bibliography in Shipton, *Biog. Sketches, Harvard*, 5 (1937): 454-55, including reference to manuscripts at Mass. Hist. Soc. and Am. Ant. Soc. WORKS ABOUT: Stearns, *Sci. in Brit. Colonies*, pp. 426-35; Frederick G. Kilgour, "Thomas Robie," *Isis* 30 (1939): 473-90.

ROGERS, HENRY DARWIN (August 1, 1808, Philadelphia, Pa.—May 29, 1866, Shawlands, near Glasgow, Scotland). *Geology*. Son of Patrick Kerr, physician and professor of physical sciences, and Hannah (Blythe) Rogers; brother of James B.,

Robert E., and William B. Rogers [qq.v.]. Married Eliza S. Lincoln, 1854; one child. EDUCATION: Educated mainly by father; 1832-1833, studied science in England. CAREER: For short period, taught school at Windsor, Md.; 1828, appointed lecturer in chemistry, Maryland Institute (Baltimore); 1829, appointed professor of chemistry and natural philosophy, Dickinson College; 1832-1833, accompanied social reformer Robert Dale Owen to London; on return to U.S., lectured on geology, Franklin Institute; 1835-1845, professor of geology and mineralogy, Univ. of Pennsylvania; 1835-1837, director, New Jersey State Geological Survey; 1836-1842, director, Pennsylvania State Geological Survey; 1845, moved to Boston, worked on completion of report on geology of Pennsylvania; 1855, went to Edinburgh; 1857-1866, Regius professor of natural history, Univ. of Glasgow. MEMBERSHIPS: Chairman, Association American Geologists, 1843; first American fellow, Geological Society of London, 1833. SCIENTIFIC CONTRIBUTIONS: During trip to England in 1832-1833, met number of scientists and became greatly interested in geology. Chief geological work done on Pennsylvania survey; when funding stopped in 1842, continued work on report at personal expense; later received legislative assistance, and final report of 1858 [see Works] ranked as one of leading contributions to American geology up to that time, emphasizing physical rather than paleontological aspects. Worked closely with brother, W. B. Rogers, and produced in 1843 joint paper on mountain elevation for which Henry probably was chiefly responsible; this paper was important early theoretical contribution to science in America, although the suggested dynamical cause was not generally accepted.

WORKS: "On the Physical Structure of the Appalachian Chain, as Exemplifying the Laws Which Have Regulated the Elevation of Great Mountain Chains Generally," *Reports of Meetings of Association of American Geologists and Naturalists* (Boston, 1843), pp. 474-531, with W. B. Rogers; *Geology of Pennsylvania* [final report], 2 vols. (Edinburgh and Philadelphia, 1858). See Meisel, *Bibliog. Amer. Nat. Hist.* MSS: Various collections at Am. Phil. Soc.; small number of letters in W. B. Rogers Papers at M.I.T.-Arch. WORKS ABOUT: *DAB*, 16: 94-95 (George P. Merrill); *DSB*, 11: 504-6 (John Rodgers); Patsy Gerstner, "A Dynamic Theory of Mountain Building: H. D. Rogers, 1842," *Isis* 66 (1975): 26-37.

ROGERS, JAMES BLYTHE (February 11, 1802, Philadelphia, Pa.—June 15, 1852, Philadelphia, Pa.). *Chemistry.* Son of Patrick Kerr, physician and professor of physical sciences, and Hannah (Blythe) Rogers; brother of Henry D., Robert E., and William B. Rogers [qq.v.]. Married Rachel Smith, 1830; survived by three children. EDUCATION: 1820-1821, attended College of William and Mary; 1822, M.D., Univ. of Maryland. CAREER: For several years, engaged in medical practice; 1827, became superintendent, Tyson and Ellicott chemical works, Baltimore; for short time, also professor of chemistry, Washington Medical College (Baltimore); ca. 1829, appointed lecturer, Maryland Institute (Baltimore); 1835-1839, professor of chemistry, medical department, Cincinnati College; 1837, began work with brother, W. B. Rogers, on Virginia State Geological Survey; 1840, settled in Philadelphia, where worked with brother, H. D. Rogers, on Pennsylvania Geological Survey; 1841, appointed professor of chemistry, Medical Institute of Philadelphia; 1844, appointed professor of chemistry, Franklin Institute; 1847, became professor of chemistry, Univ. of Pennsylvania. MEMBERSHIPS: Am. Phil. Soc.; Ac. Nat. Scis. Philad. SCIENTIFIC CONTRIBUTIONS: In work for firm of Tyson and Ellicott, helped improve manufacturing procedures and products; also in Baltimore, investigated arsenic in soap, and carried out studies on voltaic battery [see Works]. Chief work was in chemical analysis, and in this area made significant contributions to brothers' geological reports on Virginia and Pennsylvania.

WORKS: "Minutes of an Analysis of Soap Containing Arsenic," *Journal of Philadelphia College of Pharmacy* 6 (1834): 94, with George W. Andrews and William R. Fisher; "Experiments with the Elementary Voltaic Battery," *Am. Journ. Sci.* 28 (1835): 33-42, with James Green; *Elements of Chemistry* (Philadelphia, 1846) and "On the Alleged Insolubility of Copper in Hydrochloric Acid . . . ," *Am. Journ. Sci.*, 2d ser. 6 (1848): 395-96, both with R. E. Rogers. MSS: Some items in W. B. Rogers Papers at M.I.T.-Arch. and Va. St. Lib. Arch. Div. WORKS ABOUT: *DAB*, 16: 99-100 (Harris E. Starr); Edgar F. Smith, *James Blythe Rogers . . . Chemist* (Philadelphia, 1927): 14 pp.

ROGERS, ROBERT EMPIE (March 29, 1813, Baltimore, Md.—September 6, 1884, Philadelphia, Pa.). *Chemistry.* Son of Patrick Kerr, physician and professor of physical sciences, and Hannah (Blythe) Rogers; brother of Henry D., James B., and William B. Rogers [qq.v]. Married Fanny Montgomery, 1843: Delia Saunders, 1866; no information on children. EDUCATION: Early education under father, and later with brothers; 1836, M.D., Univ. of Pennsylvania. HONORS: LL.D., Dickinson College, 1883. CAREER: 1831-1832, participated in railway surveys in New England; 1836-1842, chemist, Pennsylvania Geological Survey; 1842, appointed professor of general and applied chemistry, Univ. of Virginia; at Medical School, Univ. of Pennsylvania, professor of chemistry (1852-1877) and dean (1856-1877);

during Civil War, assistant surgeon, West Philadelphia military hospital; 1872-1884, chemist, gas trust of Philadelphia; 1874-1879, member of U.S. Assay Commission; 1877-1884, professor of medical chemistry and toxicology, Jefferson Medical College (Philadelphia). MEMBERSHIPS: Am. Phil. Soc.; Natl. Ac. Scis.; and other societies. SCIENTIFIC CONTRIBUTIONS: Medical thesis, *Experiments upon the Blood*, based on original work and demonstrated skill as investigator. Worked with brother Henry on Pennsylvania survey, including analysis of limestone done with M. H. Boyé [q.v.]. While at Univ. of Virginia, was interested especially in pure chemistry, and during years 1846-1854 published series of papers with brother William (also at Virginia), including articles on procedures for obtaining chlorine gas, formic acid, and preparation of aldehyde and acetic acid, as well as papers on determination of carbon in graphite, volatility of potassium and sodium carbonates, decomposition of rocks by meteoric water, and absorption of carbon dioxide by liquids, published in *Am. Journ. Sci.* and *Proceedings of Am. Assoc. Advt. Sci.* Later involvement in applied chemistry included work in petroleum and on wastage of silver in Philadelphia mint (1872); in 1875 designed refinery equipment for San Francisco mint and also conducted mining studies. Also produced number of inventions, including steam boiler and work on electrical apparatus.

WORKS: *Elements of Chemistry* (Philadelphia, 1846), with J. B. Rogers. See bibliography in Smith [below], p. 309. MSS: Some items in W. B. Rogers Papers at M.I.T.-Arch. and Va. St. Lib. Arch. Div. WORKS ABOUT: *DAB*, 16: 109-10 (Lyman C. Newell); Edgar F. Smith, "Memoir," *Biog. Mems. of Natl. Ac. Scis.* 5 (1905): 291-309.

ROGERS, WILLIAM AUGUSTUS (November 13, 1832, Waterford, Conn.—March 1, 1898, Waterville, Maine). *Astronomy*; *Physics*. Son of David Potter, master of fishing vessel and farmer, and Mary Anna (Rogers) Rogers. Married Rebecca Jane Titsworth, 1857; three children. EDUCATION: Attended De Ruyter and Alfred academies (New York); 1857, B.A., Brown Univ.; 1860-1861, studied astronomy at Harvard College Observatory; for one year, studied mechanics at Sheffield Scientific School, Yale. HONORS: Received several honorary degrees. CAREER: Before entering college in 1854, taught school for short time at New Market, N.J.; 1857, made tutor and instructor in mathematics, Alfred Academy; at Alfred Univ., appointed professor of mathematics (1859) and became professor of industrial mechanics (1860); 1864-1865, naval duty during war, on leave from Alfred; at Harvard Univ., assistant in observatory (1870-1877) and assistant professor of astronomy (1877-1886); 1886-1898, professor of physics and astronomy, Colby College. MEMBERSHIPS: Honorary fellow, Roy. Soc.; member of Natl. Ac. Scis. and other societies. SCIENTIFIC CONTRIBUTIONS: Early work was in astronomy; erected observatory at Alfred Univ., and while there carried out computations of orbits of asteroids and investigated effects of fatigue and hunger on observations; at Harvard Observatory worked with eight-inch meridian circle, contributing to general star catalogue as envisioned by Astronomische Gesellschaft, and continued to direct preparation of catalogues after going to Colby; interest in precision apparatus, which grew out of work with meridian circle, led to work on line etching on glass, using hydrofluoric acid. In 1879, sent by Am. Ac. Arts Scis. to Europe to procure copies of imperial yard and French meter, and comparisons of measures led to work in thermometry in attempt to detect effects of temperature change; work in metrology, including association with A. A. Michelson and E. W. Morley [qq.v.] on application of optical methods in detecting changes in length, was of first-rank importance.

WORKS: Astronomical observations appeared especially in *Annals of Harvard College Observatory*. See bibliography in Searle [below], pp. 113-17. WORKS ABOUT: *DAB*, 16: 114-15 (Leon Campbell); Edward W. Morley, "Memoir," *Biog. Mems. of Natl. Ac. Scis.* 4 (1902): 185-99; Arthur Searle, "Memoir. Part II," ibid. 6 (1909): 109-17.

ROGERS, WILLIAM BARTON (December 7, 1804, Philadelphia, Pa.—May 30, 1882, Boston, Mass.). *Geology*; *Science education.* Son of Patrick Kerr, physician and professor of physical sciences, and Hannah (Blythe) Rogers; brother of Henry D., James B., and Robert E. Rogers [qq.v.]. Married Emma Savage, 1849; no children. EDUCATION: 1820-1821, attended College of William and Mary. CAREER: For time, operated school at Windsor, Md.; 1827, appointed lecturer, Maryland Institute (Baltimore); 1828, succeeded father as professor of chemistry and physics, College of William and Mary; 1835-1853, professor of natural philosophy and geology, Univ. of Virginia; 1835-1842, Virginia state geologist; 1853, moved to Boston; 1861, appointed Massachusetts state inspector of gas meters; at Massachusetts Institute of Technology, first president (1862-1870 and 1878-1881) and professor of physics and geology (1865-1870). MEMBERSHIPS: Chairman, Association of American Geologists and Naturalists, 1845 and 1847; president, Natl. Ac. Scis., 1878-1882; member, other societies. SCIENTIFIC CONTRIBUTIONS: Had early interest in chemistry with special concern for its applications, an outlook that continued throughout career; produced series of chemical studies with brother Robert in late 1840s and early 1850s. As geologist of Virginia, issued annual reports, but no final report was called for; chief geological contribution was work done with brother Henry on structure of Appalachian chain, especially 1843 paper [see sketch of H. D. Rogers]; work on Virginia survey included aspects of chemical geology, and also demonstrated relation of coal beds to enclosing strata. In Boston, turned to physical problems, including variations of so-called ozone in atmosphere, binocular vision, smoke rings, and rotating rings in liquids. Made significant contribution to science in America through leadership in establishment and early shaping of Massachusetts Institute of Technology.

WORKS: See Works under H. D. Rogers for joint paper (1843); *A Reprint of Annual Reports and Other Papers on the Geology of the Virginias* (New York, 1884); bibliography in *Life and Letters of William Barton Rogers*, 2 vols. (Boston, 1896). MSS: Va. St. Lib. Arch. Div. (*NUCMC* 61-2418); M.I.T.-Arch. (Hamer). WORKS ABOUT: *DAB*, 16: 115 (George P. Merrill); *DSB*, 11: 504-6 (John Rodgers); Francis A. Walker, "Memoir," *Biog. Mems. of Natl. Ac. Scis.* 3 (1895): 1-13.

ROMANS, BERNARD (ca. 1720, Netherlands—ca. 1784). *Civil engineering*; *Natural history*; *Cartography.* No information on parentage. Married Elizabeth Whiting, 1779; at least one child. EDUCATION: Learned engineering in England. CAREER: Ca. 1757, sent by British government to North America to carry out engineering work; 1766, appointed deputy surveyor of Georgia, but went instead to East Florida to survey estates of Lord Egmont; 1767-1769, acquired land in Georgia and Florida; 1769-1770, appointed chief deputy surveyor for Southern District, by W. G. De Brahm [q.v.]; at personal expense, completed exploration of Florida and Bahama banks and of western coast to Pensacola, and there engaged by Florida governor and superintendent of Indian affairs to take part in survey of West Florida; also engaged as king's botanist in Florida; 1773, sailed from New Orleans to Charleston, S.C.; 1773-1774, visited in New York, Philadelphia, and Boston, and there solicited subscribers for projected book on Floridas; ca. 1774, settled at Hartford, Conn.; 1775, as a member of Connecticut committee to capture Fort Ticonderoga, personally took over the abandoned Fort George; 1775-1776, engaged in construction of fortifications for New York Committee of Safety; 1776-1778, captain in Pennsylvania artillery; 1778, in Wethersfield, Conn.; 1780, en route to joining southern army in South Carolina, captured by British, taken to Jamaica, and is said to have died at sea while being returned to a U.S. port after war. MEMBERSHIPS: Am. Phil. Soc. SCIENTIFIC CONTRIBUTIONS: Man of wide talents in botany, engineering, mathematics, artistry, engraving, cartography, and other fields. To Am. Phil. Soc., presented account of Florida plants, improvements in mariner's compass, and navigation chart for Florida waters; wrote articles on indigo and madder; produced number of printed maps, especially on Florida. Best remembered for *Natural History* [see Works], a publication perhaps most noteworthy for its sections on sailing directions.

WORKS: *A Concise Natural History of East and West Florida* (New York, 1775). See chief works listed in *DAB*; annotated bibliography in Phillips [below], pp. 74-99. WORKS ABOUT: *DAB*, 16: 126-27 (Wilbur H. Siebert); P. Lee Phillips, *Notes on the Life and Works of Bernard Romans* (Deland, Fla., 1924).

ROOD, OGDEN NICHOLAS (February 3, 1831, Danbury, Conn.—November 12, 1902, New York, N.Y.). *Physics.* Son of Anson, Congregational clergyman, and Alida Gouverneur (Ogden) Rood. Married Mathilde Prunner, 1858; five children. EDUCATION: Attended Yale College; 1852, A.B., College of New Jersey (Princeton); 1852-1854, postgraduate study, Sheffield Scientific School, Yale; 1854-1858, studied physics and chemistry at Berlin and Munich. HONORS: LL.D., Yale Univ., 1901. CAREER: 1858, appointed professor of chemistry, Univ. of Troy, a short-lived denominational college at Troy, N.Y.; 1863-1902, professor of physics, Columbia Univ. MEMBERSHIPS: Natl. Ac. Scis.; vice-president, Am. Assoc. Advt. Sci., 1869. SCIENTIFIC CONTRIBUTIONS: While largely uninterested in mathematical physics, demonstrated great skill as experimentalist and devised means for accurate measurement or observation of phenomena. While at Troy, published number of papers in *Am. Journ. Sci.*, especially in optics, touching on the polarization of light passing through glass block strained during annealing process, stereoscopic studies, and particularly the continuation of investigations on theory of luster begun while in Germany; concern with optics, and especially physiological aspects, was persistent interest throughout life, and while at Troy also did work on relation of perception of distance and color, and afterimage produced by observation of bright source through gaps in revolving disk. Did pioneering work in union of photography with microscope, and studied duration of electric spark; greatly increased means for more precise measurement of time and of vacuums; during career investigated nearly all aspects of physics of the day, including later work on X rays and on high electrical resistances; of particular importance was development of means for comparing brightness of light of different colors by means of flicker photometer. An accomplished painter, and in *Modern Chromatics* [see Works] produced work of interest to physicists and artists;

book is said to have greatly influenced impressionist painters.

WORKS: *Modern Chromatics with Applications to Art and Industry* (New York, 1879). Bibliography of works in *Biog. Mems. of Natl. Ac. Scis.* 6 (1909): 469-72. MSS: Colum. U. Lib. (*NUCMC* 66-1601). WORKS ABOUT: *DAB*, 16: 131-32 (James P. C. Southall); *DSB*, 11: 531-32 (Daniel J. Kevles).

ROWLAND, HENRY AUGUSTUS (November 27, 1848, Honesdale, Pa. —April 16, 1901, Baltimore, Md.). *Physics*. Son of Henry Augustus, Protestant clergyman, and Harriette (Heyer) Rowland. Married Henrietta Troup Harrison, 1890; survived by three children. EDUCATION: Attended Phillips Academy, Andover; 1870, degree of civil engineer, Rensselaer Polytechnic Institute. HONORS: Among other recognitions, received Rumford and Draper medals from Natl. Ac. Scis. CAREER: 1870-1872, railroad surveyor, and then teacher of natural science at Univ. of Wooster (Ohio); at Rensselaer, instructor (1872-1874) and assistant professor of physics (1874-1875); 1875-1876, was in Europe to examine laboratories and purchase apparatus for Johns Hopkins Univ. and conducted research in laboratory of von Helmhotz at Berlin; 1876-1901, professor of physics, Johns Hopkins Univ.; U.S. representative to international bodies concerned with determination of electrical units. MEMBERSHIPS: Roy. Soc.; Natl. Ac. Scis.; and other societies. SCIENTIFIC CONTRIBUTIONS: Worked in both mathematical and experimental physics, carrying out laboratory work related to theoretical problems, while also showing marked mechanical and engineering ability; had significant success in precise measurements of physical constants. While at Troy, undertook study of magnetic permeability of iron, steel, and nickel; started with Faraday's idea of magnetic lines of force, translated these ideas into mathematical terms, and subjected ideas to experimental tests. While in von Helmhotz's laboratory in 1876, carried out experiments to test ideas originated as undergraduate at Rensselaer, again drawing upon Faraday, and proved for first time that movement of charged body brought about magnetic effects, an experimental result of importance in modern electron theory. Did notable work in arriving at more precise values for ohm, the mechanical equivalent of heat, and other physical phenomena. Best remembered for invention and ruling of concave spectral grating, an instrument of unprecedented precision and ease of use, and at Paris Exposition (1890) won gold medal and grand prize for spectral grating and revision of solar spectrum.

WORKS: Publications collected in *The Physical Papers of Henry Augustus Rowland* (Baltimore, 1902). MSS: Johns Hopkins Univ. Library (*NUCMC* 60-1646); also see other references to manuscript locations in *DSB*. WORKS ABOUT: *DAB*, 16: 198-99 (Joseph S. Ames); *DSB*, 11: 577-79 (Daniel J. Kevles).

RUFFIN, EDMUND (January 5, 1794, Prince George County, Va.—June 18, 1865, "Redmoor," Amelia County, Va.). *Agricultural chemistry*. Son of George, plantation owner, and Jane (Lucas) Ruffin. Married Susan Travis, 1813; eleven children. EDUCATION: Early education at home; 1811-1812, attended College of William and Mary. CAREER: 1812-1813, served as private during War of 1812; 1813, upon death of father, took possession of farm at Coggin's Point, Va.; 1823-1826, served in Virginia state senate; 1833-1842, published *Farmer's Register*; 1841, elected corresponding secretary, Virginia State Board of Agriculture; 1842, appointed South Carolina agricultural surveyor; 1843-1855, resided at "Marlbourne," estate in Hanover County, Va.; 1855, retired from farming, and thereafter devoted efforts particularly to cause of southern secession. MEMBERSHIPS: President, Virginia State Agricultural Society, 1852. SCIENTIFIC CONTRIBUTIONS: From suggestion in work of Humphry Davy, undertook investigation in soil chemistry; in 1818 began use of marl as soil additive and achieved good results; these studies were published in 1821 in *American Farmer* and later in separate volume [see Works, *Manures*]. Wrote number of articles in *Farmer's Register*, and as editor of this monthly journal established it as leading American farming periodical of its time; *Report of . . . Agricultural Survey of South Carolina* (1843) said to have been important event in agricultural history of that state; campaigned for establishment of experimental farms and for agricultural education; in 1855, collected number of previous writings in *Essays and Notes on Agriculture*. While he made no fundamentally new contributions to soil chemistry, performed important work in publicizing successful practices of scientific agriculture, and promoted interest in scientific study of agricultural problems.

WORKS: *Essays on Calcareous Manures*, 5th ed. (Richmond, Va., 1852). See also other titles mentioned in *DAB*; bibliography of works in *Bulletin of Virginia State Library* 11 (1919): 36-114. MSS: Univ. N.C.-SHC (*NUCMC* 64-1098); Lib. Cong. (*NUCMC* 68-2067); Virginia Historical Society (*NUCMC* 69-549). WORKS ABOUT: *DAB*, 16: 214-16 (Avery O. Craven); Aaron J. Ihde, "Edmund Ruffin, Soil Chemist of the Old South," *Journ. Chem. Educ.* 29 (1952): 407-14. Also see A. O. Craven, *Edmund Ruffin, Southerner: A Study in Secession* (New York and London, 1932).

RUMFORD, COUNT. See THOMPSON, BENJAMIN.

RUNKLE, JOHN DANIEL (October 11, 1822, Root, Montgomery County, N.Y.—July 8, 1902, Southwest Harbor, Maine). *Mathematics; Engineering education.* Son of Daniel, apparently farmer, and Sarah (Gordon) Runkle. Married Catharine Robbins Bird, 1862; four children. EDUCATION: 1851, S.B. and M.A., Lawrence Scientific School, Harvard. HONORS: Ph.D., Hamilton College, 1869; LL.D., Wesleyan Univ., 1871. CAREER: Worked on farm, and taught in district school; 1844-1847, teacher, academy at Onandaga, N.Y.; 1849-1884, associated with *Nautical Almanac*; 1858-1861, founder and editor of *Mathematical Monthly*; at Massachusetts Institute of Technology, appointed secretary (1862), professor of mathematics (1865-1868 and 1880-1902), acting president (1868-1870), and president (1870-1878); 1878-1880, visited Europe and there studied technical education. MEMBERSHIPS: Am. Ac. Arts Scis. SCIENTIFIC CONTRIBUTIONS: Did not enter Lawrence Scientific School until age twenty-five, having pursued self-study in mathematics before that time. In 1852, published two papers on elements of Thetis and Psyche in *Astronomical Journal*; four years later, published tables relating to motion of planets [see Works], and number of years later published textbook [see *Elements*, Works]. Was associated with planning of M.I.T. from beginning; as president organized field excursions and summer schools in engineering, which began in 1871 with engineering school that examined mining operations in American West; in 1872 established Lowell School of Practical Design, and in 1876 helped develop School of Mechanic Arts for manual training at secondary-school level, both associated with M.I.T. Wrote on technical and manual education, including reports on two years abroad (1878-1880).

WORKS: "New Tables for Determining the Values of the Coefficients in the Perturbative Function of Planetary Motion, Which Depend Upon the Ratio of the Mean Distances," and "Asteroid Supplement," *Smithson. Contr. Knowl.* 9 (1856, 1857); *Elements of Plane Analytic Geometry* (Boston, 1888). See works on technical education listed in *DAB*; also see *Roy. Soc. Cat.*, vol. 5 (no works after 1860 listed in that source). MSS: Small collection in M.I.T.-Arch. WORKS ABOUT: *DAB*, 16: 225 (David E. Smith); William T. Bawden, "Some Leaders in Industrial Education: John Daniel Runkle," *Industrial Arts and Vocational Education* 37 (1948): 191-93, 221-24.

RUSH, BENJAMIN (January 4, 1746, Byberry, Pa.—April 19, 1813, Philadelphia, Pa.). *Chemistry; Medicine; Psychiatry.* Son of John, gunsmith and farmer, and Susanna (Hall) Harvey Rush. Married Julia Stockton, 1776; thirteen children, including James Rush [q.v.]. EDUCATION: 1760, A.B., College of New Jersey (Princeton); 1761-1766, studied medicine under Dr. John Redman, Philadelphia, and attended medical lectures at College of Philadelphia; 1768, M.D., Univ. of Edinburgh. CAREER: 1769, began medical practice at Philadelphia; at College of Philadelphia, professor of chemistry (1769-1789) and professor of theory and practice of medicine (1789-1791); at Univ. of Pennsylvania (incorporating College of Philadelphia), appointed professor of institutes of medicine (physiology) and clinical practice (1791) and also appointed professor of theory and practice of medicine (1796); 1776, member of Continental Congress and signer of Declaration of Independence; 1777, served as surgeon general to armies of Middle Department; 1783-1813, staff member, Pennsylvania Hospital; 1787, member of Pennsylvania convention to ratify U.S. Constitution; 1789, took part in movement resulting in new Pennsylvania state constitution; 1797-1813, treasurer, U.S. Mint. MEMBERSHIPS: Am. Phil. Soc. SCIENTIFIC CONTRIBUTIONS: Professorship in chemistry at Philadelphia was first such appointment in America, Rush having prepared through chemical studies at Univ. of Edinburgh, where he attended lectures by Joseph Black and submitted doctoral dissertation on digestive processes involving self-experimentation; *Syllabus* [see Works], prepared in conjunction with professorship, was first chemistry text prepared in colonies, but Rush did next to no original work in that subject. At Edinburgh, took up medical system of William Cullen, and eventually reduced causes of disease essentially to excess of tension in arterial system, calling for bloodletting; while medical theory was criticized by some, Rush became leading American physician and teacher of medicine. Promoted use of hypotheses in scientific and medical work, while neglecting need for experimental testing; but despite failings is to be credited with role in promotion of science in America. A pioneer in dealing with insane, and *Medical Inquiries* [see Works] ranks as earliest work on psychiatry by native American.

WORKS: *Syllabus of a Course of Lectures on Chemistry* (Philadelphia, 1770; reprinted, Philadelphia, 1954); *Medical Inquiries and Observations Upon the Diseases of the Mind* (Philadelphia, 1812; reprinted, New York, 1962). Other works mentioned in *DAB* and *DSB*. MSS: Lib. Co. Philad. (*NUCMC* 61-2703); see also DAB and Hamer. WORKS ABOUT: *DAB*, 16: 227-31 (Richard H. Shryock); *DSB*, 11: 616-18 (Eric T. Carlson). Also see Carl Binger, *Revolutionary Doctor: Benjamin Rush* (New York, 1966).

RUSH, JAMES (March 15, 1786, Philadelphia, Pa.—May 26, 1869, Philadelphia, Pa.). *Psychology.* Son of Benjamin [q.v.], physician and medical educator, and Julia (Stockton) Rush. Married Phoebe Anne

Ridgway, 1819; no children. EDUCATION: 1805, A.B., College of New Jersey (Princeton); 1809, M.D., Univ. of Pennsylvania; 1809-1811, studied medicine at Univ. of Edinburgh. CAREER: 1811, began medical practice in Philadelphia, and in 1813 took over father's practice; gave private lectures; 1819, married woman of wealth; ca. 1828-1869, retired from active medical practice, and lived as recluse devoted to studies. MEMBERSHIPS: Am. Phil. Soc.; French Academy of Sciences. SCIENTIFIC CONTRIBUTIONS: From early years of medical career, became increasingly interested in questions related to human thought processes, a subject that also had come to interest father in his work with insane in Pennsylvania Hospital; in 1815, delivered lectures on human intellect. Most influential scientific study was *Human Voice* [see Works], published in 1827 as first volume of projected study on processes of human thought; with work on voice, including important idea of thought as subvocal speech, achieved rank as a pioneer of American psychology, and during lifetime work was praised by students of oratory and rhetoric. Devoted large part of life to what was conceived as major work, published in 1865 as *Human Intellect* [see Works]; based on self-analysis and library research, book presented essentially mechanistic view of processes of intellect; while lacking appreciation of need for empirical data, did view mind as capable of objective study and suggested fruitful areas for future investigation; although it never achieved wide notice, *Human Intellect* was final outcome of first extended psychological study by an American author.

WORKS: *The Philosophy of the Human Voice* (Philadelphia, 1827); *A Brief Outline of an Analysis of the Human Intellect*, 2 vols. (Philadelphia, 1865); *Collected Works*, 4 vols. (Weston, Mass., 1974). MSS: Lib. Co. Philad. (*NUCMC* 61-2704). WORKS ABOUT: *DAB*, 16: 231 (Richard H. Shryock); Stephen G. Kurtz, "James Rush, Pioneer in American Psychology," *Bull. Hist. Med.* 28 (1954): 50-59.

RUSSELL, ISRAEL COOK (1852-1906). *Geology*; *Physical geography*. Geologist. U.S. Geol. Surv.; professor of geology, Univ. of Michigan.

WORKS ABOUT: See *AMS*; *DAB*.

RUTHERFURD, LEWIS MORRIS (November 25, 1816, New York, N.Y.—May 30, 1892, Tranquility, N.J.). *Astrophysics*. Son of Robert Walter and Sabina (Morris) Rutherfurd, family of financial means and social prominence. Married Margaret Stuyvesant Chanler, 1841; seven children. EDUCATION: 1834, graduated, Williams College; studied law under William H. Seward; 1837, admitted to bar. CAREER: Ca. 1837-1849, practiced law, and in 1841 increased personal wealth through marriage; 1849-1856, lived in France, Germany, and Italy, partly on account of wife's health; 1856, returned to New York, and there built private observatory; for over twenty-five years, a trustee of Columbia Univ.; 1881, helped establish department of geodesy and practical astronomy at Columbia Univ. and in 1883, donated personal observatory equipment to that university. MEMBERSHIPS: Natl. Ac. Scis. SCIENTIFIC CONTRIBUTIONS: As student at Williams, aided professor in preparation of physical and chemical experiments as accompaniment to lectures, and later pursued private studies in chemistry, mechanics, and astronomy; while in Florence, Italy, became friend of G. B. Amici, who was interested in astronomy and microscopy. On return to U.S., constructed private observatory and workshop and in 1858 produced his first photograph of moon; showed great skill in designing equipment for work in astronomical photography and spectroscopy. By 1864 had in use eleven-and-one-half inch objective lens employable only for photographic purposes, thus eliminating visual component; in following year commenced work on mapping heavens, producing photographs of star clusters, and devised and built instrument for measuring stellar plates; after 1868, had available thirteen-inch refractor telescope convertible for either visual or photographic work; in early 1860s, became interested in spectroscopy, and in 1863 published early classification of star spectra; perfected means of constructing spectroscope of hollow prisms filled with liquid of consistent density, and also constructed apparatus for preparing interference gratings of unprecedented precision. In 1878 carried out final observations, and in 1890 donated photographic plates and measurements to Columbia Univ.; these later were measured and reduced by other scientists.

WORKS: Papers appeared in *Am. Journ. Sci.*, 1848 and 1863-1876. See bibliography in *DSB*. MSS: Collection of papers not located; see *DSB* for references to manuscripts in other collections. WORKS ABOUT: *DAB*, 16: 256-57 (Raymond S. Dugan); *DSB*, 12: 36-37 (Nathan Reingold).

RYDBERG, PER AXEL (1860-1931). *Botany*. Curator, N.Y. Bot. Gard.

WORKS ABOUT: See *AMS*; *DAB*.

S

SADTLER, SAMUEL PHILIP (1847-1923). *Chemistry.* Professor of organic and industrial chemistry, Univ. of Pennsylvania; professor of chemistry, Philadelphia College of Pharmacy; consulting chemical expert, Philadelphia.

WORKS ABOUT: See *AMS*; *DAB*.

SAFFORD, JAMES MERRILL (1822-1907). *Geology; Natural history; Chemistry.* Professor of geology, Vanderbilt Univ.; Tennessee state geologist.

WORKS ABOUT: See *AMS*; *DAB*.

SAFFORD, TRUMAN HENRY (January 6, 1836, Royalton, Vt.—June 13, 1901, Newark, N.J.). *Astronomy*; *Mathematics.* Son of Truman Hopson, farmer, and Louisa (Parker) Safford. Married Elizabeth Marshall Bradbury, 1860; six children. EDUCATION: 1854, A.B., Harvard College. CAREER: For a time, worked in Cambridge office of *Nautical Almanac*; at Harvard College Observatory, assistant (1854-1865) and acting director (1865-1866); 1866-ca. 1871, professor of astronomy, (old) Univ. of Chicago, and director of Dearborn Observatory; ca. 1871-1876, carried out geodetic work for U.S. geographical surveys; 1876-1901, Field memorial professor of astronomy, Williams College. MEMBERSHIPS: Am. Ac. Arts Scis. SCIENTIFIC CONTRIBUTIONS: At early age manifested exceptional skill in mathematical manipulation and prepared computations for 1846 almanac for Bradford, Vt.; for following year prepared almanacs for Bradford, Cincinnati, Philadelphia, and Boston. Scientific work related especially to positional astronomy; at Harvard Observatory, assisted in observing and reducing zones of faint stars near equator; studied orbit of star Sirius, and successfully indicated relative position of companion to Sirius at about same time that A. G. Clark [q.v.] first observed it; also at Harvard, worked with G P. Bond [q.v.] on study of position of stars in Great Orion nebula, and after Bond's death prepared observations for publication; later continued work on correct determination of position of fixed stars, and at Chicago was assigned star zone as part of system of observations coordinated by Astronomische Gesellschaft. While teaching and other duties at Williams (including service as librarian) took much time, he continued astronomical work and produced *The Willams College Catalogue of North Polar Stars* (1888).

WORKS: See above; also several works referred to in *DAB* and in *Roy. Soc. Cat.*, vols. 5, 8, 11, 12, 18. WORKS ABOUT: *DAB*, 16: 287-88 (Joseph M. Poor); Arthur Searle, memoir, *Proceedings of Am. Ac. Arts Scis.* 37 (1901-1902) : 654-56.

SALISBURY, JAMES HENRY (October 13, 1823, Scott, N.Y.—August 23, 1905, Dobbs Ferry, N.Y.). *Microscopy*; *Medicine.* Son of Nathan and Lucretia (Babcock) Salisbury. Married Clara Brasee, 1860; two children. EDUCATION: 1846, Bachelor of natural science, Rensselaer Polytechnic Institute; 1850, M.D., Albany Medical College. HONORS: A.M., Union College, 1852. CAREER: For New York Natural History Survey, Albany, assistant chemist (1846-1849) and chief chemist (1849-1852); 1851-1852, lecturer on elementary and applied chemistry, New York State Normal School; later, established medical practice at Newark, Ohio; a founder, and professor of physiology, histology, and pathology (1864-1866), Charity Hospital Medical School, Cleveland; after ca. 1880, lived in New York City. MEMBERSHIPS: Am. Assoc. Advt. Sci. SCIENTIFIC CONTRIBUTIONS: Won prize in 1848 for essay entitled "Anatomy and Histology of Plants," and in 1849 also won prize for paper, *History and Chemical Investigation of Maize*, originally published in New York State Agricultural Society's *Transactions*; subsequently published analyses of number of vegetable products, appearing especially in New York Agricultural *Transactions*. Said to have begun studies in microscopic medicine as early as 1849, and in early 1860s not only produced works on plant pathology, but also wrote on theory of human disease, ascribed to fungus and later to what he called "algoid vegetations"; has been credited with early glimpse of germ theory of disease. Also in 1860s, studied origin and function of blood and traced it to spleen and to mesenteric and lymphatic glands; in later years, turned attentions to study of relation of food and drink to the occurrence of disease. Was considered accomplished student of microscopy, and research efforts suggested fruitful areas for future investigation.

WORKS: See titles in *DAB*, of which several originally were journal articles rather than separate publications. See also *Roy. Soc. Cat.*, vols. 5, 8, 11, 12, and *Natl. Un. Cat. Pre-1956 Imp.* WORKS ABOUT: *DAB*, 16: 309 (James M. Phalen); *NCAB*, 8 (1900 [copyright 1898]): 469-70.

SALISBURY, ROLLIN D. (1858-1922). *Geology.* Professor of geographic geology and dean of school of science, Univ. of Chicago.

WORKS ABOUT: See *AMS*; *DAB*; *DSB*.

SAMUELS, EDWARD AUGUSTUS (July 4, 1836, Boston, Mass.—May 27, 1908, Fitchburg, Mass.). *Ornithology*; *Nature study.* Son of Emanuel and Abigail Samuels. Married Sarah B. Caldwell; at least one child. EDUCATION: Attended Boston public schools. CAREER: Lived in Boston most of life; at early age, began writing for publication; 1860-1880, assistant secretary, Massachusetts State Board of Agriculture, and curator of state natural history collections; 1870-ca. 1890, engaged in publication of music; 1885-ca. 1892, president, Massachusetts Fish and Game Protective Association; after ca. 1890, suffered from failing health and eventual blindness. SCIENTIFIC CONTRIBUTIONS: A lover of nature who was concerned with study and preservation of wildlife (especially birds). While working for Massachusetts Board of Agriculture, began publication of papers in its *Reports* intended to meet need for popular work on Massachusetts birds; this interest culminated in *Birds* [see Works], first published in 1867 as *Ornithology and Oology of New England*; while largely dependent on works of S. F. Baird [q.v.] and other naturalists, this book proved extremely popular and had important influence in developing widespread interest in study of birds. *With Fly-Rod and Camera* (1890) was significant in promoting use of photography rather than gun in capturing animals; also wrote for sport and natural history publications, especially *Field and Stream* magazine. As president of Massachusetts Fish and Game Protective Association, did important work in promoting its interests; considered by contemporaries as authority on natural history of New England.

WORKS: *The Birds of New England* (Boston, 1870); *The Living World*, 2 vols. (Boston, 1868), with Augustus C. L. Arnold; *With Fly-Rod and Camera* (New York, 1890). WORKS ABOUT: *DAB*, 16: 323-24 (Hubert L. Clark); obituary notice, *Auk* 25 (1908): 341.

SARGENT, CHARLES SPRAGUE (1841-1927). *Botany*; *Dendrology.* Professor of arboriculture and director of arboretum, Harvard Univ.

MSS: Univ. N.C.-SHC (*NUCMC* 64-635); Harvard Univ.—Arnold Arboretum (see this and other references in Sutton [below]. WORKS ABOUT: See *AMS*; *DAB*; Stephanne B. Sutton, *Charles Sprague Sargent and the Arnold Arboretum* (Cambridge, Mass., 1970).

SARTWELL, HENRY PARKER (April 18, 1792, Pittsfield, Mass.—November 15, 1867, Penn Yan, N.Y.). *Botany.* Son of Levi, mechanic, and Eleanor (Crofut) Sartwell. Reportedly married four times. EDUCATION: At early age, commenced study of medicine with physician at Utica, N.Y.; 1811, received medical license, Oneida County Medical Society. HONORS: Ph.D., Hamilton College, 1864. CAREER: Ca. 1811, began medical practice at New Hartford, N.Y.; during War of 1812, surgeon, U.S. Army; ca. 1815, took up residence at Springville, N.Y.; later, transferred medical practice to Bethel (Gorham), N.Y.; 1832-1867, medical practice at Penn Yan, N.Y. SCIENTIFIC CONTRIBUTIONS: Interested in general field of natural history, and while at Bethel came to devote attentions to botany, especially carices (sedges); during lifetime, carried out extensive exploration of flora of western New York and distributed specimens widely to other botanists; herbarium of some 8,000 specimens, many acquired by exchange with botanists throughout world, deposited with Hamilton College; interested especially in genus Carex, and distributed sets of these plants in *Exsiccatae* in 1848 and 1850 [see Works], but planned third volume was never completed. Published little, but generously contributed to works of others, including John Alsop Paine's *Catalogue of Plants Found in Oneida County and Vicinity* (1865). Also collected insects, meteorological data, and minerals, and was interested in geology and horticulture.

WORKS: "Catalogue of Plants Growing Without Cultivation in the Vicinity of Seneca and Crooked Lakes, in Western New York," *Annual Report of Regents of Univ. of State of New York* 58 (1845): 273-90; *Carices Americae Septentrionalis Exsiccatae*, 2 vols. (Penn Yan, N.Y., 1848-1850). WORKS ABOUT: *DAB*, 16: 374 (Charles W. Dodge); Kelly and Burrage, *Med. Biog.* (1928).

SAUGRAIN de VIGNI, ANTOINE FRANÇOIS (February 17, 1763, Paris, France—May 18/19,

1820, apparently St. Louis, Mo.). *Natural history; Physical sciences; Instrument making.* Son of Antoine and Marie (Brunet) Saugrain; ancestors were involved with libraries and book trade. Married Genevieve Rosalie Michau, 1793; six children. EDUCATION: Apparently enjoyed sound education in physics, chemistry, and mineralogy; no information on medical training. CAREER: As youth, entered service of Spanish king; 1785 and 1786, in Mexico, where engaged in mining and mineralogical studies; 1787, traveled to U.S. from France and stayed for time at Philadelphia; 1788, went West to find location for French settlement and reached Louisville, Ky., after encounter with Indians; ca. 1789, returned to France; 1790, came again to U.S. and, in employ of Scioto Company, established residence at Gallipolis, Ohio, where rendered medical aid and apparently operated inn; later, removed to Lexington, Ky., and to Portage des Sioux, Mo.; 1800, settled at St. Louis; appointed post surgeon by Spanish lieutenant governor, at St. Louis; 1805-1811, army surgeon, appointed by Pres. T. Jefferson [q.v.]; until 1820, continued medical practice at St. Louis, and on death left large landed estate. SCIENTIFIC CONTRIBUTIONS: Has been labeled "First Scientist of the Mississippi Valley." At various locations engaged in chemical investigation and production, had electric battery, and carried out studies in electricity; made and sold ink, thermometers, barometers, phosphorus matches, the latter thought to have been produced by him in advance of European inventors. Mineralogical knowledge led to involvement in advising on mining developments in Ohio Valley and elsewhere. In 1809, introduced smallpox vaccination to St. Louis.

WORKS: See *Natl. Un. Cat. Pre-1956 Imp.*, which includes chiefly references to posthumous publications based on his notebooks. MSS: Missouri Historical Society (Hamer). WORKS ABOUT: *DAB*, 16: 377-78 (Stella M. Drumm); N. P. Dandridge, "Antoine François Saugrain (de Vigni)," *Ohio Archaeological and Historical Quarterly* 15 (1906): 192-206. Also see H. Fouré Selter, *L'Odyssée américaine d'une famille française: Le Docteur Antoine Saugrain* (Baltimore, 1936).

SAVAGE, THOMAS STAUGHTON (June 7, 1804, upper Middletown [now Cromwell], Conn.—December 29, 1880, Rhinecliff, N.Y.). *Zoology.* Son of Josiah, wealthy shipowner engaged especially in West Indies trade, and Mary (Roberts) Savage. Married Susan Metcalf, 1838; Maria Chapin, 1842; Elizabeth Rutherford, 1844; survived by four children. EDUCATION: 1825, A.B., Yale College; 1833, M.D., Yale Medical School; 1836, ordained to ministry, following graduation from Theological Seminary in Virginia (Alexandria). HONORS: D.D., Delaware College, 1876. CAREER: Ca. 1833, traveled to Cincinnati, New Orleans, and north along Atlantic Coast; 1836-1838, 1839-1843, and 1844-1847, in Liberia as first Protestant Episcopal missionary to Africa; 1838 and 1843-1844, returned to U.S. for health reasons; ca. 1847-1851, Episcopal priest in parishes in Mississippi and Alabama; 1851-1861 and 1865-1867, resided at Pass Christian, Miss.; 1868, left the South; received appointment as associate secretary, Episcopal Board of Missions; 1869-1880, priest, Church of the Ascension, Rhinecliff, N.Y. SCIENTIFIC CONTRIBUTIONS: While in Africa, conducted studies in natural history, and as early as 1841 published paper on African insects; later also wrote on ants and termites of Africa and on habits of certain reptilian forms. Published paper of some significance on chimpanzee [see Works], which made early contribution to knowledge of behavior of those animals; best remembered for 1847 paper [see Works], in which appeared first description of the previously unknown gorilla, Savage having procured several skulls and bones as well as descriptions and impressions of the animal from African natives. Does not appear to have actively pursued natural history interests after return to U.S.

WORKS: "Observations on the External Characters, Habits and Organisation of the Troglodytes niger, Geof. [chimpanzee]," *Boston Journal of Natural History* 4 (1843-1844): 362-86, and "Notice of the External Characters, Habits, and Osteology of Troglodytes gorilla . . . ," ibid. 5 (1845-1847): 417-42, both with J. Wyman [q.v.], and latter paper reprinted in John C. Burnham, ed., *Science in America* (New York, 1971), pp. 115-27. See several other works listed in *DAB*; also *Roy. Soc. Cat.*, vols. 5, 6 (additions). WORKS ABOUT: *DAB*, 16: 391-92 (Remington Kellogg); *Obituary Record of Graduates of Yale University* (1880-1890), pp. 15-16.

SAXTON, JOSEPH (March 22, 1799, Huntington, Pa.—October 26, 1873, Washington, D.C.). *Instrument making.* Son of James, banker and nail manufacturer, and Hannah (Ashbaugh) Saxton. Married Mary H. Abercrombie, 1850; survived by one child. EDUCATION: At age twelve, left school and began work in father's nail factory; for two years, apprenticed to watchmaker. HONORS: Among awards, won gold medal for precision balances at Great Exhibition (London), 1851. CAREER: Ca. 1818-1829, worked as watchmaker in Philadelphia, and after a time became associated with well-known machinist, Isaiah Lukens; 1829-1837, in England, constructed exhibits for Adelaide Gallery of Practical Science (London), produced number of patentable inventions, and built experimental apparatus; 1837-1844, built and maintained balance scales, Philadelphia Mint; 1844-1873, director, U.S. Coast Survey Office of Weights and Measures (later, National Bureau of Standards).

MEMBERSHIPS: Am. Phil. Soc.; Natl. Ac. Scis. SCIENTIFIC CONTRIBUTIONS: Showed great ability in design and construction of apparatus useful in practical and scientific work, although never well grounded in scientific knowledge himself. During early years in Philadelphia, produced clock with temperature-compensating pendulum and special escapement, and with Lukens constructed clock for belfry of Independence Hall. Among achievements in England was preparation of scientific apparatus used by Charles Wheatstone in electrical studies; acquaintances in England also included Michael Faraday; in 1833, demonstrated innovative magneto-electric machine to British Association for Advancement of Science, but failed to publish account of it, though he acquired number of other British patents. After return to Philadelphia, did early work with daguerreotype photography (1839). At Coast Survey Office, produced standard balances, weights, and measures distributed to the several states, and constructed number of other instruments relating to survey's work.

WORKS: "Notice of Electro-Magnetic Experiments," *Journal of Franklin Institute*, 2d ser. 10 [whole ser. 14] (1832): 66-72; "Description of a Revolving Keeper Magnet for Producing Electric Currents," ibid., 2d ser. 13 [whole ser. 17] (1834): 155-56; "On the Application of the Rotating Mirror to the Aneroid Barometer," *Proceedings of Am. Assoc. Advt. Sci.* 12 (1858): 40-42. See *DSB* for references to patents. MSS: See references in *DSB*, including Natl. Arch. WORKS ABOUT: *DAB*, 16: 400 (Carl W. Mitman); *DSB*, 12: 131-32 (Arthur H. Frazier).

SAY, THOMAS (June 27, 1787, Philadelphia, Pa.—October 10, 1834, New Harmony, Ind.). *Entomology*; *Conchology*. Son of Benjamin—physician-apothecary, legislator, and man of wealth—and Ann (Bonsall) Say. Married Lucy Way Sistaire, 1827; no children. EDUCATION: Until ca. 1802 [age fifteen], attended Quaker school at Westtown, Pa.; studied pharmacy with father; largely self-taught in natural history. CAREER: For a time, partner in pharmaceutical business, which soon met with failure; 1812, took part in establishing Ac. Nat. Scis. Philad. and thereafter devoted efforts to natural history and lived for a period in academy's quarters; 1814, served with volunteers; 1818, with G. Ord, W. Maclure, T. R. Peale [qq.v.], made excursion to Georgia and Florida; 1819, appointed zoologist on Maj. Stephen H. Long's expedition to Rocky Mountains; 1821-1827, curator, Am. Phil. Soc.; 1822-1828, professor of natural history, Univ. of Pennsylvania; 1823, zoologist, Maj. Long's exploration of sources of Minnesota River; 1825-1834, residence at New Harmony, having gone there with W. Maclure and others to establish community; 1827-1828, accompanied Maclure to New Orleans and Mexico. MEMBERSHIPS: Linnaean Society of London and other societies. SCIENTIFIC CONTRIBUTIONS: In 1816 was first American to publish paper on native shells, and in that year commenced work on a major effort, the *American Entomology* [see Works]. Writings dealt almost wholly with descriptive and taxonomic matters, contributing little of note to study of life histories and habits of animals; publications included descriptions for more than 1,000 previously unknown insect species, and established study of descriptive entomology in U.S. Supervised publication of C. L. Bonaparte [q.v.], *American Ornithology*, vol. 1 (1825). Also wrote on paleontological and other topics.

WORKS: *American Entomology; or, Descriptions of the Insects of North America*, 3 vols. (Philadelphia, 1817-1828); *American Conchology*, 6 vols. (New Harmony, Ind., 1830-1834); John L. LeConte, ed., *The Complete Writings of Thomas Say on the Entomology of North America*, 2 vols. (New York, 1859). Also see Meisel, *Bibliog. Amer. Nat. Hist.* MSS: Am. Phil. Soc. (*NUCMC* 61-779); Ac. Nat. Scis. Philad. (*NUCMC* 66-81); New Harmony (Ind.) Workingmen's Institute (Hamer). WORKS ABOUT: *DAB*, 16: 401-2 (Leland O. Howard); *DSB*, 12: 132-34 (Elizabeth N. Shor). Also see Harry B. Weiss and Grace M. Ziegler, *Thomas Say, Early American Naturalist* (Springfield, Ill., 1931).

SCHAEBERLE, JOHN MARTIN (1853-1924). *Astronomy*. Instructor in astronomy and assistant in observatory, Univ. of Michigan; member of original staff, Lick Observatory, Univ. of California; latter part of life spent in retirement at Ann Arbor, Mich.

WORKS ABOUT: See *AMS*; *DAB*; *DSB*.

SCHOOLCRAFT, HENRY ROWE (March 28, 1793, Albany County [now Guilderland], N.Y.—December 10, 1864, Washington, D.C.). *Geology*; *Ethnology*. Son of Lawrence, glass maker, and Margaret Anne Barbara (Rowe) Schoolcraft. Married Jane Johnston (granddaughter of Ojibwa chief), 1823; Mary Howard, 1847; three children by first marriage, after second marriage adopted a child. EDUCATION: Attended Union College and for time studied informally at Middlebury College. HONORS: LL.D., Univ. of Geneva, 1846. CAREER: 1809-ca. 1816, engaged in glass-making business with father in New York and New England; 1817-1818, explored mineral regions in southern Missouri and Arkansas; 1820, appointed naturalist to Lewis Cass's abbreviated expedition to upper Mississippi River and Lake Superior copper country; 1822-1841, Indian agent for tribes about Lake Superior, and acting superintendent of Indian affairs for northwest (1836-1841); 1832, led expedition to sources of Mississippi River; 1841, went to New York and there oversaw his many literary productions; 1847, returned to Office of Indian Affairs, Washington, D.C., and engaged in editing comprehensive data on Indian tribes. MEMBERSHIPS: N.Y. Lyc. Nat. Hist.; a founder, American Ethnological Society (1842). SCIENTIFIC CONTRIBUTIONS: Following early southern expedition, published work on Missouri lead mines (1819) [see Works]; during next few years, published several papers of geological and mining interest, including accounts of native copper and silver in Michigan region; these and other writings rank among first such productions relating to American mineral wealth. Known chiefly for studies of American Indians, especially tribes about Great Lakes; wrote large number of works on habits, customs, and lore of Indians, and was one of earliest Americans to undertake collection and publication of such ethnological data; after 1847, devoted much attention to editing six-volume work of Office of Indian Affairs, *Historical and Statistical Information Respecting the History, Condition and Prospects of the Indian Tribes of the United States* (1851-1857).

WORKS: See above; also *A View of the Lead Mines of Missouri* (New York, 1819). See other works listed in *DAB* and *DSB*, including early travel narratives. Bibliography in Charles S. Osborn and Stellanova Osborn, *Schoolcraft-Longfellow-Hiawatha* (Lancaster, Pa., 1942), pp. 631-45. MSS: Description of those in Lib. Cong. and elsewhere, in Osborn and Osborn [above], pp. 645-53. WORKS ABOUT: *DAB*, 16: 456-57 (Walter Hough); *DSB*, 12: 203-5 (Elizabeth N. Shor).

SCHÖPF, JOHANN DAVID (March 8, 1752, Wunsiedel, Germany—September 10, 1800, Germany). *Natural history*; *Travels.* Son of wealthy merchant. No information on marital status. EDUCATION: 1767-1770, attended gymnasium at Hof; 1770-1773, studied medicine and natural sciences, Univ. of Erlangen; 1773, at Berlin for study in forestry, and thereafter studied in Prague and Vienna; 1776, M.D., Univ. of Erlangen. CAREER: 1773-ca. 1776, traveled through Bohemia, Austria, and northern Italy; ca. 1776, established medical practice at Ansbach; 1777-1783, chief surgeon to Ansbach regiment in service of British army, stationed at New York, Long Island, Rhode Island, and Philadelphia; 1783-1784, traveled through U.S., from New York to western Pennsylvania to Charleston, S.C., and sailed from latter place to East Florida and then to Bahamas; 1784, returned to Germany, and there apparently engaged in medical practice, traveled in Italy and Holland, and held number of positions of civic responsibility; 1795-1800, president, Ansbach medical college, which, after 1797, was united with medical college of Bayreuth. SCIENTIFIC CONTRIBUTIONS: Chief German student of American natural history during his lifetime; in science writings and travel accounts touched on wide range of aspects of U.S. culture and environment; published in German. Produced first work on American geology (1787), and also published pioneering works on American frogs, fish, and turtles; also wrote on climate and disease in America, and in 1787 published first American materia medica. Chief work dealing with years in America, the *Reise* [see Works], combined traveler's impressions with objective reporting of facts, including description of Pennsylvania mines and topography of Alleghenies, botany (encompassing its economic considerations), and other aspects of the civil and natural conditions in America during that time.

WORKS: *Reise durch einige der mittlern und südlichen vereinigten nordamerikanischen Staaten nach Ost-Florida und den Bahama-Inseln*, 2 vols. (Erlangen, 1788), translated as *Travels in the Confederation* (Philadelphia, Pa., 1911). See several other works listed in *DAB* and in Morrison [below], pp. 256-57. WORKS ABOUT: *DAB*, 16: 457-58 (George H. Genzmer); Alfred J. Morrison, "Dr. Johann David Schoepf," *German American Annals*, n.s. 8 (1910): 255-64.

SCHOTT, ARTHUR CARL VICTOR (February 27, 1814, Stuttgart, Germany—July 26, 1875, apparently Washington, D.C.). *Geology*; *Natural history collector.* No information on parentage or marital status. EDUCATION: At age fifteen, completed program in gymnasium and technical school at Stuttgart; for one year, served apprenticeship in Royal Gardens of Stuttgart; attended Institute of Agriculture at Hohenheim. CAREER: Manager of several rural estates in Germany; for ten years, directed mining property in Hungary; 1848, traveled in southern Europe, Turkey, and Arabia; 1850, came to U.S. and was employed by U.S. Topographical Engineers office, Washington; ca. 1853-ca. 1857, assistant surveyor and collector, U.S.-Mexico boundary survey under W. H. Emory [q.v.]; 1857-1858, member of party to survey interoceanic canal near Isthmus of Darien (Central America); 1864-1866, engaged in geological survey of Yucatan under commission from governor; thereafter, served in office of U.S. Topographical Bureau and finally, until death, worked in office of Coast Survey. SCIENTIFIC CONTRIBUTIONS: Remembered especially for work on Mexican boundary survey, which he served as surveyor, collector, artist, and geologist. Collected insects described by J. L. LeConte [q.v.], botanical specimens worked up by friend J. Torrey [q.v.], and zoological specimens described by S. F. Baird and C. F. Girard [qq.v.]; also collected fossils and minerals. Prepared reports and papers on geology of Mexican border region, and also prepared topographical and ethnographical illustrations for survey report. On survey for canal near Isthmus of Darien, contributed accounts of geology, botany, and zoology to report and collected botanical specimens for Torrey.

WORKS: For contributions to Mexican boundary survey reports (1857-1858) and to report on interoceanic canal survey (1861), see Meisel, *Bibliog. Amer. Nat. Hist.*; also see *Roy. Soc. Cat.*, vols 5, 8, 12. WORKS ABOUT: Charles S. Sargent, *Silva of North America* (Boston, 1896), 10: 18; Goetzmann, *Army Exploration*, pp. 182-205, passim; Andrew D. Rodgers III, *John Torrey* (Princeton, 1942).

SCHOTT, CHARLES ANTHONY (August 7, 1826, Mannheim, Germany—July 31, 1901, Washington, D.C.). *Geophysics.* Son of Anton Carl, merchant, and Anna Maria (Hoffman) Schott. Married Teresa Gildermeister, 1854; Bertha Gildermeister, 1863; one child by first marriage, four by second. EDUCATION: 1847, graduated as civil engineer, *technische hochschule* in Karlsruhe. HONORS: Received Wilde Prize, French Academy of Sciences, 1898. CAREER: 1848, for brief period took part in revolution of that year and then emigrated to U.S.; 1848-1855, computer in Washington office, U.S. Coast Survey, with one year as hydrographic draftsman on survey ship; 1855-1899, head of computing division, Coast Survey. MEMBERSHIPS: Natl. Ac. Scis. and other societies. SCIENTIFIC CONTRIBUTIONS: As head of section charged with responsibility for calculating, correlating, and interpreting data gathered in field, helped further high standards of scientific work within Coast Survey. Interests necessarily touched on several fields, including questions of instrumentation and observation, and also on astronomy, geodesy, magnetism, tides, and map making; most significant labors were in geodesy and terrestrial magnetism, and included a study of influence of aurora and a correlation of sunspots and magnetic storms; also did work in climatology, especially study of Arctic region, and played important part in carrying out triangulation across North American continent, a work pursued into retirement. Wilde Prize (1898), given in recognition of a life of work of highest quality and precision, referred particularly to contributions in terrestrial magnetism. In 1869 took part in government-sponsored solar eclipse observation in Illinois, and next year (1870) observed eclipse in Sicily.

WORKS: Large number of works appeared in Coast Survey reports; see bibliography in *Biog. Mems. of Natl. Ac. Scis.* 8 (1915): 116-33: MSS: Natl. Arch.—Coast Survey records (*DSB*). WORKS ABOUT: *DAB*, 16: 458-59 (H. A. Marmer); *DSB*, 12: 209-10 (Nathan Reingold).

SCHUBERT, ERNST (October 20, 1813, Gleinitz bei Gross-Glogau, Silesia—January 9, 1873, Frankfurt am Main, Germany). *Astronomy.* No information on parentage, marital status, or education. CAREER: Ca. 1840, lived in village in Silesia, and there engaged in independent study in mathematics and astronomy; 1843-ca. 1848, assistant at observatory, Univ. of Breslau; ca. 1848-1849, assistant, observatory at Berlin; 1849-ca. 1859, employed by U.S. *Nautical Almanac*, part of time in office at Cambridge, Mass.; 1853, left Cambridge for Berlin because of ill health; by about 1858, apparently again in U.S. (Cambridge), and by about 1859, back in Berlin; ca. 1862, in Ann Arbor and Washington; thereafter, astronomer or astronomical computer in Germany, especially at Berlin and also at Jena, Eisenach, Cassel, Dessau, and Stuttgart. MEMBERSHIPS: Astronomische Gesellschaft. SCIENTIFIC CONTRIBUTIONS: While at Breslau, engaged in work on computations for Uranus and carried out computative studies on comets; thereafter, became known for work on elements and ephemerides of various asteroids, carried out for *Berliner astronomisches Jahrbuch* and *American Ephemeris and Nautical Almanac*; calculated general perturbations and prepared tables for several of the small planets. Was brought to U.S. in 1849 by M. F. Maury [q.v.], with recommendation from Johann F. Encke, under whom he had worked at Berlin; played significant role in introducing European knowledge and techniques at time of foundation and early production of U.S. *Nautical Almanac.* Association with *Nautical Almanac* continued after return to Germany, number of his later publications being communicated to its superintendents, and several works in late 1860s and early 1870s were computed for, and published by, *Nautical Almanac.*

WORKS: *Roy. Soc. Cat.*, vols. 5 and 8, lists total of sixty-nine works, appearing especially in Gould's *Astronomical Journal* and later in *Astronomische Nachrichten*; see also Poggendorff [below]. MSS: Several letters in Natl. Arch.—Naval Observatory records are cited by Rothenberg [below]. WORKS ABOUT: Poggendorff, *Biographisch-literarisches Handwörterbuch*, vols. 2 and 3; Breslau Universität, Sternwarte, *Mittheilungen . . .* (Breslau, 1879), by J. G. Galle, p. 122; incidental biographical information with articles and indexes to *Astronomische Nachrichten*; Marc Rothenberg, "The Educational and Intellectual Background of American Astronomers, 1825-1875" (Ph.D. diss., Bryn Mawr, 1974), pp. 149-51.

SCHWARZ, EUGENE AMANDUS (April 21,

1844, Liegnitz, Silesia—October 15, 1928, Washington, D.C.). *Entomology.* Son of Amandus, cloth merchant and member of Liegnitz Common Council, and Luise (Harnwolf) Schwarz. Never married. EDUCATION: Attended Liegnitz Gymnasium; before and after Franco-Prussian War, attended Univ. of Breslau; 1872, enrolled at Univ. of Leipzig. HONORS: D.S., Univ. of Maryland, 1923. CAREER: 1870, near Paris with German army, serving in medical corps; 1872, emigrated to U.S. and found employment in entomology at Museum of Comparative Zoology, Harvard; 1874, founded Detroit Scientific Association in that city, with H. G. Hubbard [q.v.], and began formation of collections; 1874-1875 (winter), and 1876, carried out entomological studies in Florida, first year with Hubbard; 1876 and 1877, with Hubbard, made two entomological excursions to Lake Superior region; 1878, entomological collecting trip to Colorado for J. L. LeConte [q.v.]; 1878-1879, employed by U.S. Department of Agriculture, studied cotton worm in South; 1879, became attached to U.S. Entomological Commission; 1881-1926, member of staff, U.S. Bureau of Entomology; 1898-1928, custodian of Coleoptera (beetles), U.S. Natl. Mus. MEMBERSHIPS: A founder, Entomological Society of Washington; honorary fellow, Entomological Society of America. SCIENTIFIC CONTRIBUTIONS: In 1869, published first work on insects, while in Europe; early entomological labor in America done with H. G. Hubbard, with whom he published an important work on Coleoptera of Michigan [see Hubbard]. At Bureau of Entomology was one of most important members of staff, worked in Washington, and also took significant role in bureau's field work; ranked as most knowledgable American coleopterist and one of the best in world, concentrating taxonomic work on this order, while also working on other insect groups as well; collected insects throughout U.S. and in Cuba and Central America; through many publications and through personal contacts, especially in Washington, influenced number of other entomologists. Later years devoted especially to museum work.

WORKS: Credited with some 400 publications; see bibliography in *Proceedings of Entomological Society of Washington* 30 (1928): 166-83. MSS: Smithson. Instn. Archives (*NUCMC* 74-981). WORKS ABOUT: *DAB*, 16: 480-81 (Leland O. Howard); Mallis, *Amer. Entomologists*, pp. 252-57.

SCHWEINITZ, EMIL ALEXANDER de (January 18, 1864[?], Salem, N.C.—February 15, 1904, Washington, D.C.). *Biochemistry.* Son of Bishop Emil Adolphus, executive of southern province of Moravian church and head of Salem Academy, and Sophia A. (Herman) de Schweinitz; grandson of Lewis D. von Schweinitz [q.v.]. Never married. EDUCATION: Attended Nazareth Hall High School and Moravian College of Bethlehem, Pa.; A.B. in 1882 and Ph.D. in 1885, Univ. of North Carolina; studied at Univ. of Berlin; 1886, Ph.D. in chemistry, Univ. of Göttingen. HONORS: M.D., Columbian Univ. (now George Washington), 1895. CAREER: Ca. 1886-1888, taught chemistry, first at Tufts College and then at Agricultural and Mechanical College of Kentucky (Univ. of Kentucky, Lexington); 1888-1890, assistant in Division of Chemistry, U.S. Department of Agriculture; 1890-1904, in charge of biochemical research, Agriculture Department's Bureau of Animal Industry, after 1894 as head of separate Biochemic Division; 1894, became professor of chemistry and toxicology and dean of medical faculty, Columbian Univ.; 1900, U.S. representative to international congress on medicine and hygiene, Paris. MEMBERSHIPS: American Public Health Association. SCIENTIFIC CONTRIBUTIONS: At Bureau of Animal Industry, research work related primarily to study of disease-causing bacteria, encompassing investigation of their metabolic products, their chemical composition, and mechanisms for immunity from the bacteria; in keeping with the bureau's mission, directed attention especially to diseases of domestic animals, including tuberculosis, hog cholera, swine plague, glanders. Most significant contributions dealt with questions of immunity from tuberculosis, and made first reported use of attenuated human tubercle bacilli to produce immunity in cattle; during later years, engaged in studies of immunization for hog cholera, and pursued questions related to transmission of tuberculosis of humans and bovines.

WORKS: Publications include number of items published by Department of Agriculture. "The Comparative Virulence of Human and Bovine Tubercle Bacilli for Some Large Animals," in U.S. Department of Agriculture, Bureau of Animal Industry, *Experiments Concerning Tuberculosis* (Washington, D.C., 1905), pp. 31-100, with Marion Dorset and E. C. Schroeder. See *Natl. Un. Cat. Pre-1956 Imp.* [vol. 140, "De Schweinitz"], and *Roy. Soc. Cat.*, vol. 18. WORKS ABOUT: *DAB*, 16: 483 (Albert G. Rau); Marion Dorset, "Emil Alexander de Schweinitz," *Proceedings of Washington Academy of Sciences* 10 (1908): 208-10.

SCHWEINITZ, LEWIS DAVID von (February 13, 1780, Bethlehem, Pa.—February 8, 1834, Bethlehem, Pa.). *Botany,* especially *Mycology.* Son of Baron Hans Christian Alexander, head of secular affairs of Moravian church in America at time of son's birth, and Anna Dorothea Elizabeth (de Watteville) von Schweinitz. Married Louisa Amelia Ledoux, ca. 1812; survived by four children; grandfather of Emil A. de Schweinitz [q.v.]. EDUCATION: For ten years, studied at Nazareth Hall, near Bethlehem;

1798, went with family to Germany and enrolled in Moravian Theological Seminary at Niesky, Silesia (Lusatia). HONORS: Ph.D., Univ. of Kiel, ca. 1812. CAREER: Until 1807, teacher at Niesky; 1807, became pastor to Moravian Brethren at Gnadenburg, Silesia, and later at Gnadau in Saxony; 1812-1821, administrator of Moravian church in North Carolina and head of Salem (N.C.) Academy; 1817-1818 and ca. 1825-1826, in Germany; 1821-1834, administrator of northern province of Moravian church, at Bethlehem, Pa. MEMBERSHIPS: Ac. Nat. Scis. Philad.; Am. Phil. Soc.; and other societies. SCIENTIFIC CONTRIBUTIONS: Interested in all aspects of botany, but concentrated particularly on fungi and is known as founder of mycological studies in America. First botanical work, on fungi of Lusatia, appeared at Leipzig in 1805 and was coauthored with Prof. Johannes B. von Albertini; continued botanical labor in North Carolina and there collected and described large number of specimens, including some 300 new species of North Carolina fungi that were discussed in 1818 work on that subject; in 1821, published pamphlet on hepatic mosses that suggested plans to prepare comprehensive work on cryptogamic plants of North America; also wrote on genus Viola (violet). At Bethlehem, completed descriptions by T. Nuttall [q.v.] of plants gathered by T. Say [q.v.] on Long Expedition to Northwest; also wrote on genus Carex (sedges) and rare plants in vicinity of Easton, Pa. Chief work was "Synopsis" [see Works], encompassing over 3,000 species of fungi, of which more than 1,200 were either entirely new or previously published by Schweinitz. Of equal importance with publications was large herbarium, now in Ac. Nat. Scis. Philad.

WORKS: "Synopsis Fungorum in America Boreali degentium," *Transactions of Am. Phil. Soc.* n.s. 4 (1834): 141-316. MSS: Ac. Nat. Scis. Philad. (*NUCMC* 66-82). WORKS ABOUT: *DAB*, 16: 483-84 (Albert G. Rau); several articles about life and work, in *Bartonia: Proceedings of the Philadelphia Botanical Club*, no. 16 (1934).

SCOTT, CHARLOTTE ANGAS (1858-1931). *Mathematics.* Professor of mathematics, Bryn Mawr College.

WORKS ABOUT: See *AMS*; *Nota. Am. Wom.*

SCUDDER, SAMUEL HUBBARD (1837-1911). *Entomology.* Custodian, Bost. Soc. Nat. Hist.; assistant librarian, Harvard Univ.; paleontologist, U.S. Geol. Surv.; engaged in private study at Cambridge, Mass.

MSS: Bos. Soc. Nat. Hist.—Boston Museum of Science (Hamer); Ac. Nat. Scis. Philad. (Hamer); Harv. U.-MCZ. WORKS ABOUT: See *AMS*; *DAB*; *DSB*.

SEAMAN, WILLIAM HENRY (1837-1910). *Chemistry.* Professor of chemistry, Medical Department, Howard Univ.; examiner, U.S. Patent Office.

WORKS ABOUT: See *AMS*; *NCAB*, 14 (1910): 231; *Who Was Who in America*, vol. 1.

SEARLE, ARTHUR (1837-1920). *Astronomy.* Professor of astronomy, Harvard Univ.

MSS: Harv. U.-UA (*NUCMC* 65-1274). WORKS ABOUT: See *AMS*; *DAB*.

SENNETT, GEORGE BURRITT (July 28, 1840, Sinclairville, Chautauqua County, N.Y.—March 18, 1900, Youngstown, Ohio). *Ornithology.* Son of Pardon, successful owner of blast furnaces, and Mary (Burritt) Sennett. Married Sarah Essex; one child. EDUCATION: Graduated from Erie Academy; for four years, studied at preparatory school, Delaware County, N.Y., but ill-health prevented expected entrance to Yale College. CAREER: Ca. 1861-1865, traveled in Europe, including Germany, Austria, and France; 1865, established business as manufacturer of oil-well machinery, Meadville, Pa.; 1876, made ornithological expedition to western Minnesota; 1877, 1878, and 1882, ornithological expeditions to Rio Grande; 1877-1881, mayor of Meadville; early 1880s-ca. 1896, spent winters in New York City; 1896, moved manufacturing facilities to Youngstown. MEMBERSHIPS: Am. Ornith. Un.; president, Linnaean Society of New York, 1887-1889. SCIENTIFIC CONTRIBUTIONS: Active interest in ornithology began about 1874, and first expedition was to Minnesota (1876), though outcome of that trip was never published. Remembered chiefly for work on birds of Texas, collected during excursions to lower Rio Grande; reports on first two Texas trips [see Works] included nearly 200 species; never published results of third Texas expedition, but continued to gather birds from that region, through collectors, for work on Rio Grande birds that was never completed; reports on Texas birds edited by E. Coues [q.v.], who helped mainly with matters of nomenclature and synonymy, Sennet having good command of technical aspects of subject at that early date; work on Texas birds also gave evidence of ability as field worker. In 1883, gave collection to Am. Mus. Nat. Hist., and during winters in New York continued ornithological labors; chief activity was as businessman, and in 1892 last publication appeared. While in Texas, also collected mammals, reptiles, fishes, and insects.

WORKS: "Notes on the Ornithology of the Lower Rio Grande, Texas . . . during the season of 1877 . . . ," *Bulletin of U.S. Geological and Geographical Survey of the Territories* [Hayden] 4 (1878): 1-66, and "Further Notes on the Ornithology of the Lower Rio Grande of Texas . . . during the spring of 1878," ibid. 5 (1879) : 371-440, both edited by E. Coues. See bibliography in Allen [below], pp. 20-23. MSS: Univ. Tex.-TA (*NUCMC* 70-1969). WORKS ABOUT: *DAB*, 16: 585-86 (Herbert Friedmann) ; J. A. Allen, "In Memoriam," *Auk* 18 (1901) : 11-23.

SESTINI, BENEDICT (March 20, 1816, Florence, Italy—January 17, 1890, Frederick, Md.). *Astronomy*; *Mathematics*. No information on parentage. EDUCATION: Early education at Scuola Pia, near Florence; 1836, entered Society of Jesus at Rome; 1839, began studies in theology and philosophy, Roman College; 1844, ordained as Jesuit priest. CAREER: Ca. 1834 [age eighteen], appointed assistant to director, Osservatorio Ximeniano, Florence; 1836, entered Jesuit order; until 1848, assistant director, Roman Observatory; ca. 1848-1869, taught mathematics and natural sciences, Georgetown Univ. (Washington, D.C.), and during this period spent one year at Frederick, Md., three years at Boston College, and two years at Gonzaga College (Washington, D.C.) ; 1866-1885, founder and editor, *Messenger of the Sacred Heart*; 1869-1885, taught mathematics and natural sciences, Jesuit Seminary, Woodstock, Md.; 1885-1890, retired to Jesuit novitiate, Frederick, Md. SCIENTIFIC CONTRIBUTIONS: At early age, showed ability in mathematical computation, and while at Roman Observatory, in 1845 and 1847 published two studies of star colors. After settling at Georgetown, continued astronomical work, most important contribution being study of sun's surface during period of more than one month in 1850; prepared daily drawings of changes in sunspots, which were subsequently engraved and published by U.S. Naval Observatory [see Works]; in period before general use of photography, these drawings ranked as one of best available records of changes in surface of sun. In 1878, performed last work in astronomy in organizing and leading party to Denver, Colo., for observation of total solar eclipse. Published number of well-received textbooks on mathematics, of which the first, *Analytical Geometry* (1852), has been thought the best; also prepared texts on algebra (1855 and 1857), geometry and trigometry (1856), and geometrical and infinitesimal analysis (1871) ; at Woodstock, prepared works on mechanics (1873), animal physics (1874), and cosmography (1878) for private use of students.

WORKS: Engravings of sunspot observations in U.S. Naval Observatory, *Astronomical Observations for 1847*, vol. 3 (Washington, D.C., 1853), app. A. *DAB* lists two works on star colors done while in Italy and also titles of textbooks. WORKS ABOUT: *DAB*, 16: 594-95 (Francis A. Tondorf) ; biographical sketch in *Woodstock Letters* 19 (1890) : 259-63.

SEYBERT, HENRY (December 23, 1801, Philadelphia, Pa.—March 3, 1883, Philadelphia, Pa.). *Mineralogy*. Son of Adam—physician, apothecary-chemist, and congressman—and Maria Sarah (Pepper) Seybert. Never married. EDUCATION: Early education carried out under father's guidance; attended Ecole des Mines, Paris. CAREER: 1825, on death of father, inherited large fortune and thereafter spent periods of time in residence in U.S. and in Europe; engaged in philanthropic endeavors. MEMBERSHIPS: Am. Phil. Soc. SCIENTIFIC CONTRIBUTIONS: Father (Adam Seybert), one of first Americans to attend Paris School of Mines, became noted for abilities in chemistry and in mineral analysis and, during late 1790s, read papers to Am. Phil. Soc., reporting on investigations of marsh air and land and sea air. Son's active interest in mineralogy lasted for brief period in early 1820s, before father's death, and except for single paper that appeared in 1830, apparently published no works on mineralogy after 1825. Analyzed numerous American minerals, especially from Pennsylvania, New York, Connecticut, and New Jersey, including tourmalines, manganesian garnets, glassy actynolite, crysoberyls, pyroxene, tabular spar, chromite, colophonite, fluosilicate of magnesia, and bog iron ore; made independent discovery of fluorine in Maclurite or chondrodite from New Jersey [see Works]; in 1822 published analysis of hydraulic lime used in construction of Erie Canal and in 1830 published similar work on Tennessee meteorite [see Works]. During later years, became interested in spiritualism, and bequest to Univ. of Pennsylvania for professorship in philosophy also called for inquiry into spiritualism, carried out after his death by commission appointed for the purpose.

WORKS: "On the Discovery of Fluoric Acid in the Chondrodite," *Am. Journ. Sci.* 6 (1823) : 356-61; "Tennessee Meteorite," ibid. 17 (1830) : 326-28. See also *Roy. Soc. Cat.*, vol. 5. WORKS ABOUT: *DAB*, 17: 3 (Courtney R. Hall) ; Moncure Robinson, "Obituary Notice," *Proceedings of Am. Phil. Soc.* 21 (1883-1884) : 241-63.

SHALER, NATHANIEL SOUTHGATE (1841-1906). *Geology*. Professor of paleontology and geology, dean of Lawrence Scientific School, Harvard Univ.; geologist, U.S. Geol. Survey.

WORKS ABOUT: See *AMS*; *DAB*; *DSB*.

SHARPLES, STEPHEN PASCHALL (1842-1923). *Chemistry*. Analytical and consulting chemist; professor of chemistry, Boston Dental College; Massachusetts state assayer of ores and metals, and inspector and assayer of liquors.

WORKS ABOUT: See *AMS*; *Who Was Who in America*, vol. 1.

SHATTUCK, LYDIA WHITE (June 10, 1822, East Landaff [later Easton], N.H.—November 2, 1889, South Hadley, Mass.). *Botany.* Daughter of Timothy, farmer, and Betsey (Fletcher) Shattuck. Never married. EDUCATION: Ca. 1838, one term in academy at Haverhill, N.H.; ca. 1845, attended select school at Centre Harbor, N.H.; for time, studied at Newbury, Vt.; 1851, graduated, Mt. Holyoke Seminary; 1873 (summer), student in Anderson School of Natural History on Penikese Island, conducted by J. L. R. Agassiz [q.v.]. HONORS: Memorialized by science building at Mt. Holyoke College. CAREER: Ca. 1837-ca. 1847, taught in district schools in several locations; 1851-1889, teacher of botany and other sciences, Mt. Holyoke Seminary (College); 1869, traveled to Europe; 1886-1887, went to Hawaii; 1889, elected corporate member, Marine Biological Laboratory at Woods Hole, Mass. MEMBERSHIPS: President, Connecticut Valley Botanical Association. SCIENTIFIC CONTRIBUTIONS: Interested in native New Hampshire flora from early age, and as student at Mt. Holyoke began study of botany in its technical aspects. Known chiefly for work in science education; at Mt. Holyoke taught especially botany, but also very interested in chemistry and at times taught algebra, geometry, physiology, physics, and astronomy; helped to further tradition of science education established by Mary Lyon, founder of Mt. Holyoke. Developed botanical interests through travel, including trip to Hawaii; work in botany related especially to classification of plants and to establishment of herbarium and botanical garden at Mt. Holyoke. A correspondent of A. Gray [q.v.] and J. L. R. Agassiz, and at time of death was considered one of most knowledgeable American women working in botanical and natural history studies.

WORKS: No scientific works are known. MSS: Mt. Holyoke College Library. WORKS ABOUT: *Nota. Am. Wom.*, 3: 273-74 (Charlotte Haywood); *Memorial of Lydia W. Shattuck* (Boston, 1890), 46 pp.

SHECUT, JOHN LINNAEUS EDWARD WHITRIDGE (December 4, 1770, Beaufort, S.C.—June 1, 1836, Charleston, S.C.). *Botany.* Son of Abraham and Marie (Barbary) Shecut. Married Sarah Cannon, 1792; Susanna Ballard, 1805; four children by first marriage, five by second. EDUCATION: 1786, began private study of medicine with Dr. David Ramsay; thereafter, said to have pursued medical studies in Philadelphia. CAREER: Ca. 1791-1836, medical practice, Charleston; 1808-ca. 1812, associated with South Carolina Homespun Co. MEMBERSHIPS: Founder, Antiquarian Society of South Carolina (1813), incorporated following year as Literary and Philosophical Society of South Carolina. SCIENTIFIC CONTRIBUTIONS: In addition to large medical practice, devoted much time to botanical studies, and in 1806 published *Flora* [see Works] by subscription, which at time ranked as most complete work on South Carolina botany ever compiled; *Flora* was attempt to promote popular interest in subject through simplification of Linnaean system. Interested also in employment of electricity in treatment of disease and ideas on relation of yellow fever to absence of atmospheric electricity were supported by extensive meteorological and thermometrical observations; in 1806, publicly displayed invention of electrical machine used in medical practice.

WORKS: *Flora-Carolinaeensis* (Charleston, 1806), of which only first of planned two volumes appeared; *Sketches of the Elements of Natural Philosophy . . . with . . . a New Theory of the Earth* (Charleston, 1826); *Medical and Philosophical Essays* (Charleston, 1819), including *An Inquiry into the Properties and Powers of the Electric Fluid, and, Its Artificial Application to Medical Uses.* See list of works in Gee [below], p. 32. WORKS ABOUT: *DAB*, 17: 53 (Arney R. Childs); Wilson Gee, "South Carolina Botanists," *Bulletin of Univ. of South Carolina*, no. 72 (September 1918), pp. 29-32.

SHEPARD, CHARLES UPHAM (June 29, 1804, Little Compton, R.I.—May 1, 1886, Charleston, S.C.). *Mineralogy.* Son of Mase, Congregational minister, and Deborah (Haskins) Shepard. Married Harriet Taylor, 1831; three children. EDUCATION: 1820-1821, attended Brown Univ.; 1824, graduated, Amherst College; for period after Amherst graduation, studied botany and mineralogy with T. Nuttall [q.v.], Cambridge, Mass. HONORS: M.D., Dartmouth College, 1836; LL.D., Amherst College, 1857. CAREER: Ca. 1824, became teacher of natural science in Boston schools; 1827-1831, assistant to B. Silliman, Sr. [q.v.], New Haven; at Yale Univ., lecturer in botany (1830-1831) and lecturer in natural history (1833-1847); ca. 1831-ca. 1833, director, Brewster Scientific Institute (New Haven); 1834-1861 and 1865-1869, professor of chemistry, South Carolina Medical College; 1835, assistant, Connecticut Geological Survey; 1839, made first of several European trips for exchange and purchase of minerals; at Amherst College, professor of chemistry and natural history (1844-1852) and professor of natural history (1852-1877). MEMBERSHIPS: Societies of natural science at St. Petersburg and Vienna; Royal Society of Göttingen; and other societies. SCIENTIFIC CONTRIBUTIONS: Began mineral collection at age fifteen; subsequently collected minerals throughout U.S. east of Mississippi and discovered number of new species; discoveries of pink and green tourmalines at Paris, Maine, and rutiles in Georgia made possible creation of large and important mineral collection that was greatly expanded through European exchanges; in 1882, when his collection at Amherst partly burned, it was one of the best in world. Also collected fossils, and compiled largest U.S. collection of meteorites; also engaged in study of mines. 1832-1833, while working for Silliman, engaged in study of sugar industry in South. Discovered deposits of phosphate of lime near Charleston, S.C., and son, Charles U. Shepard, Jr., who succeeded him as professor at South Carolina Medical College, later helped develop phosphate and other chemical industries there.

WORKS: *Treatise on Mineralogy*, 2 pts. (New Haven, 1832, 1835). Wrote large number of papers, published especially in *Am. Journ. Sci.* See *Roy. Soc. Cat.*, vols. 5, 8, 11, 18. MSS: Amherst College Library (*NUCMC* 75-20). WORKS ABOUT: *DAB*, 17: 71-72 (Frederic B. Loomis); Youmans, *Pioneers of Sci.*, pp. 419-27.

SHEPARD, JAMES HENRY (April 14, 1850, Lyons, Mich.—February 21, 1918, St. Petersburg, Fla.). *Chemistry.* Son of Daniel Ensign, farmer, and Lydia Maria (Pendell) Shepard. Married Clara R. Durand, 1888; survived by three children. EDUCATION: For two or three years, attended Albion College; B.S. in 1875 and until 1881 did occasional postgraduate study in chemistry, Univ. of Michigan. CAREER: For one year, principal of schools, Athens, Mich.; 1875-1880, superintendent of public schools, Holly, Marquette, and Saline, Mich.; 1882-1888, teacher of science, Ypsilanti, Mich.; at South Dakota State College of Agriculture and Mechanic Arts, professor of chemistry (1888-1918) and vice-president (1890-1900); for South Dakota Agricultural Experiment Station, chemist (1888-1918) and director (1895-1901); 1901-1918, chemist, South Dakota pure food commission; for two years, director of South Dakota farmers' institutes; 1901, appointed state engineer of irrigation. MEMBERSHIPS: Am. Assoc. Advt. Sci.; Am. Chem. Soc.; and other societies. SCIENTIFIC CONTRIBUTIONS: Work dealt especially with questions relating to agriculture and food products. Textbook on *Elements of Inorganic Chemistry* (1885) was widely used and helped promote laboratory work by students. Investigations were related to surface waters of South Dakota, chemical study of native grasses and forage plants in that state, nitrogen control of cereals, and milling and analysis of macaroni wheat then being introduced into the state; also developed sugar beets of high sugar content capable of being produced in South Dakota; as chemist for state pure food commission, conducted experiments related to effects of preservatives and coal tar dyes on digestion; carried out studies of nitrogen peroxide, used in bleaching flour, and as government witness gave expert testimony in U.S. and England in cases relating to prohibition of bleaching; performed first effective analysis of whiskey and later took part in legal actions against whiskey trust. For Louisiana Purchase Exposition at St. Louis, prepared influential exhibit of adulterated foods.

WORKS: Published numerous bulletins and pamphlets on subjects such as those mentioned above; several specific works mentioned in *DAB* and *NCAB.*; also see *Natl. Un. Cat. Pre-1956 Imp.* WORKS ABOUT: *DAB*, 17: 73-74 (B. A. Dunbar); *NCAB*, 17 (1920 [copyright 1921]): 218-20.

SHERMAN, JOHN (December 26, 1613, Dedham, England—August 8, 1685, Watertown, Mass.). *Mathematics.* Son of Edmund and John (Makin) Sherman. Married Mary [?] (died 1644); Mary Launce; reportedly six children by first marriage, twenty by second. EDUCATION: 1631, matriculated sizar in Cambridge Univ. from St. Catherine's College, but because of refusal to subscribe to Thirty-Nine Articles of English Church, did not take degree. CAREER: 1634, came to Massachusetts Bay, and for a time served as assistant to minister at Watertown, Mass.; 1635, moved to Wethersfield, Conn., and in 1636 helped organize church there; 1639, listed among "Free Planters" at Milford, Conn.; 1643, sent as deputy from Milford to General Court of New Haven Colony; 1644-1647, preacher and teacher, Branford, Conn., and other locations in colony; 1647-1685, minister, Watertown, Mass.; became an overseer (1647), then a fellow (1678-1685) at Harvard College; for some thirty years, delivered fortnightly lectures at Watertown. SCIENTIFIC CONTRIBUTIONS: Said to have occupied first rank among colonial contemporaries in study of mathematics and astronomy, and reportedly left number of astronomical calculations in manuscript; published several almanacs. Fortnightly lectures were attended by students from Harvard College as well as by other persons in vicinity, and may have dealt with mathematics and astronomy.

WORKS: *Almanack of Coelestial Motions* (Cambridge, Mass.); issues for 1674, 1676, and 1677 are known. WORKS ABOUT: *DAB*, 17: 83-84 (Edward H. Dewey); William B. Sprague, *Annals of the American Pulpit* (New York, 1857), 1: 44-46.

SHORT, CHARLES WILKINS (October 6, 1794, "Greenfield," Woodford County, Ky.—March 7, 1863, estate near Louisville, Ky.). *Botany.* Son of Peyton, owner of large farm, and Mary (Symmes) Short. Married Mary Henry Churchill, 1815; survived by six children. EDUCATION: 1810, graduated, Transylvania Univ. (Lexington, Ky.); studied medicine privately, and 1813 became student of C. Wistar [q.v.]; 1815, M.D., Univ. of Pennsylvania. CAREER: For some years, practiced medicine at Hopkinsville,

Ky.; at Transylvania Univ., professor of materia medica and medical botany (1825-1838) and dean of faculty (1827-1837); 1828, with John Esten Cooke, established *Transylvania Journal of Medicine and the Associate Sciences*; 1838-1849, professor of materia medica and medical botany, Medical Institute of Louisville; 1849, retired to "Hayfield," estate near Louisville, having increased personal fortune through inheritance from uncle. MEMBERSHIPS: Am. Phil. Soc. SCIENTIFIC CONTRIBUTIONS: While at Hopkinsville, carried out botanical studies in addition to medical practice, exploring flora of surrounding region; particularly noted for extensive botanical collecting activity west of Alleghenies, and was one of first Americans to distribute large numbers of dried specimens of high quality to other researchers and collectors; during a five-year period prior to 1836, in association with colleagues, distributed some 25,000 botanical specimens in both U.S. and Europe. Lent financial and other assistance to several botanical expeditions, and among other collecting activities purchased important collection of flora of Texas and Mexico; collection of some 15,000 species finally deposited with Ac. Nat. Scis. Philad. Publications, medical and botanical, appeared especially in *Transylvania Journal*, but never yielded to urgings of associates to prepare book on botany.

WORKS: "Notices of Western Botany and Conchology," *Transylvania Journal of Medicine* . . . 4 (1831): 69-82, with H. Halbert Eaton; "A Catalogue of the Native Phaenogamous Plants and Ferns of Kentucky," ibid. 6 (1833): 490-501, 7 (1834): 598-600, 8 (1835): 575-82, and 10 (1837): 435-40, with R. Peter and H. A. Griswold. See Meisel, *Bibliog. Amer. Nat. Hist.* MSS: See several locations in *NUCMC* and Hamer. WORKS ABOUT: *DAB*, 17: 127-28 (Mary L. Didlake); Howard A. Kelly, *Some American Medical Botanists* (New York and London, 1929), pp. 129-35.

SHUFELDT, ROBERT WILSON (1850-1934). *Zoology*; *Biology*; *Ornithology*. Surgeon; officer, Medical Department, U.S. Army.

MSS: Ac. Nat. Scis. Philad. (*NUCMC* 66-83). WORKS ABOUT: See *AMS*; *NCAB*, 6 (1929 [copyright 1892 and 1929]): 242-44; *Who Was Who in America*, vol. 1.

SHUMARD, BENJAMIN FRANKLIN (November 24, 1820, Lancaster, Pa.—April 14, 1869, St. Louis, Mo.). *Geology*; *Paleontology*. Son of John, businessman, and Ann Catherine (Getz) Shumard. Married Elizabeth Maria Allen, 1852; survived by two children. EDUCATION: Ca. 1835-ca. 1838 [three years], attended Miami Univ. (Oxford, Ohio); ca. 1840 [one year], attended Jefferson Medical College, Philadelphia; 1842, M.D., Medical Institute of Louisville, Ky. CAREER: Ca. 1842, for less than one year, practiced medicine at Hodgenville, Ky.; thereafter, returned to Louisville, apparently practiced medicine, and with a Prof. Cobb carried out geological explorations and collection of fossils in Louisville vicinity; 1846-1850, assistant to D. D. Owen [q.v.] in U.S. survey of northwestern territories (Iowa, Wisconsin, and Minnesota); 1851, with Dr. J. Evans [q.v.], undertook geological survey of Oregon Territory; 1852, returned to Louisville; 1852-1853, worked on paleontological specimens collected by brother (Dr. George G. Shumard) and Capt. R. B. Marcy in Red River exploration; 1853-1858, assistant geologist and paleontologist, Missouri Geological Survey; 1858-1860, conducted State Geological Survey of Texas; 1861-1869, medical practice, St. Louis; 1866, appointed professor of obstetrics, Univ. of Missouri. MEMBERSHIPS: Geological Society of London and other societies; president, St. Louis Academy of Science. SCIENTIFIC CONTRIBUTIONS: During 1846 visit of M. Edouard de Verneuil to Louisville, Shumard's collection of Kentucky fossils aided in demonstrating relation of European and North American paleozoic formations. Made significant contributions to published reports on geology of Iowa, Wisconsin, and Minnesota under D. D. Owen survey; prepared paleontological report on explorations with J. Evans in Oregon, but it was not published, although an article on the Oregon fossils did appear in *Transactions of St. Louis Academy of Science* (1858); in Texas, preliminary reconnaissance revealed geological deposits to be most complete of those known on North American continent, but political patronage and Civil War brought survey to an end. Ranked after D. D. Owen as major contributor to early knowledge of geology west of Mississippi, and was responsible for discovery of number of new fossil species. After 1861, conducted geological and mineralogical surveys for private commercial interests.

WORKS: *Contributions to the Geology of Kentucky* (Louisville, Ky., 1847), 36 pp., with L. P. Yandell [q.v.]; reports on work in Texas, and other labors, appeared especially in *Transactions of St. Louis Academy of Science*. See Meisel, *Bibliog. Amer. Nat. Hist.* MSS: No collection is known. WORKS ABOUT: *NCAB*, 8 (1900 [copyright 1898]): 256-57; "Benjamin Franklin Shumard," *American Geologist* 4 (1889): 1-6.

SILLIMAN, BENJAMIN (August 8, 1779, North Stratford [now Trumbull], Conn.—November 24, 1864, New Haven, Conn.). *Chemistry*; *Mineralogy*; *Geology*. Son of Gold Selleck, lawyer, and Mary (Fish) Silliman. Married Harriet Trumbull, 1809; Sarah Isabella (McClellan) Webb, 1851; nine children by first marriage, including Benjamin Silliman,

Jr. [q.v.]; father-in-law to James D. Dana and Oliver P. Hubbard [qq.v.]. EDUCATION: 1796, A.B., Yale College; ca. 1798, began study of law at New Haven; 1802, admitted to bar; 1802-1804, pursued study of science, attended lectures at medical school in Philadelphia, worked in chemical laboratory of R. Hare [q.v.] in that city, and visited J. Maclean [q.v.] at Princeton; 1805-1806, went to England to purchase books and apparatus for Yale and pursued science studies, especially in London and Edinburgh. CAREER: 1796-ca. 1798, part of this time, taught in private school, Wethersfield, Conn.; at Yale College, tutor (1799-1802) and professor of chemistry and natural history (1802-1853); 1813, was a founder and appointed professor of chemistry and pharmacy, Yale Medical School; 1808, introduced public lectures in New Haven, later extended to other cities and eventually reaching throughout U.S.; 1818, established *Am. Journ. Sci.*; 1839-1840 (winter), gave first series of Lowell Institute lectures, Boston; engaged as chemical and geological consultant. MEMBERSHIPS: Natl. Ac. Scis. and other societies. SCIENTIFIC CONTRIBUTIONS: Conducted scientific research chiefly during early years of career; description and chemical analysis of Weston (Conn.) meteor of 1807 and fusion experiments with oxy-hydrogen blowpipe and deflagrator (inventions of R. Hare) were particularly significant. During course of long career, achieved rank as most influential American scientist of generation; is best remembered for central role in encouraging science studies, through editorship of *Am. Journ. Sci.*, promotion of scientific education (especially at Yale, which became chief American center for study of his subjects), and popularization of interest in science through public lectures, of which those before Lowell Institute were climax; also took central part in scientific organizations, and edited and wrote textbooks.

WORKS: *Elements of Chemistry*, 2 vols. (New Haven, 1830-1831). See *Roy. Soc. Cat.*, vol. 5; also, Franklin B. Dexter, *Biographical Sketches of the Graduates of Yale College* (New York, 1911), 5: 224-27. MSS: Yale Univ. Library (*NUCMC* 74-1201); see additional references in *NUCMC*, Hamer, and *DSB*. WORKS ABOUT: *DAB*, 17: 160-63 (Charles H. Warren); *DSB*, 12: 432-34 (John C. Greene). Also see John F. Fulton and Elizabeth H. Thomson, *Benjamin Silliman, Pathfinder in American Science* (New York, 1947).

SILLIMAN, BENJAMIN, Jr. (December 4, 1816, New Haven, Conn.—January 14, 1885, New Haven, Conn.). *Chemistry*; *Geology*. Son of Benjamin [q.v.], professor of chemistry and natural history, and Harriet (Trumbull) Silliman; brother-in-law to James D. Dana and Oliver P. Hubbard [qq.v.]. Married Susan Huldah Forbes, 1840; seven children. EDUCATION: 1837, A.B., Yale College. HONORS: M.D., Medical College of South Carolina, 1849; LL.D., Jefferson Medical College, 1884. CAREER: 1837-1846, assistant in chemistry, Yale College; 1838-1885, involved in editorial work on *Am. Journ. Sci.* (founded by father); late 1830s, began to assist father on lecture tours and with mining surveys; 1846, was a founder and was appointed professor of practical chemistry in what was beginnings of Yale's Sheffield Scientific School (so named in 1861), and until 1869 maintained association with that school; 1849-1854, professor of chemistry, medical department, Univ. of Louisville; 1853-1870, professor of chemistry, Yale College; 1853-1885, professor of chemistry, Yale Medical School; for number of years, a director of New Haven Gas Works; engaged in chemical and geological consulting work. MEMBERSHIPS: Natl. Ac. Scis. and other societies. SCIENTIFIC CONTRIBUTIONS: Research in chemistry was varied, though related especially to minerals, petroleum, coal, rare metals, and use of gas in illumination. Wrote textbooks on chemistry (1847) and physics (1859) that were widely popular. Most important publication grew out of consulting work for company interested in Pennsylvania oil; in this *Report on Rock Oil* [see Works], Silliman employed little-used technique of fractional distillation to isolate various petroleum products and demonstrated their uses, thus outlining methods and products that were basis of petroleum industry for fifty years; in successful consulting work conducted examinations of mineral and petroleum regions, but in 1860s his enthusiastic reports on oil in southern California contradicted reports of California Geological Survey and led to controversy; in this instance, J. D. Whitney [q.v.] and other scientists accused him of misleading public for personal gain and tried unsuccessfully to have him stripped of scientific positions and associations; but subsequent developments proved Silliman's predictions as correct.

WORKS: *Report on the Rock Oil, or Petroleum, from Venango Co., Pennsylvania* (New Haven, 1855), now rare. See chief works listed in *DSB*. MSS: Yale Univ. Library (*NUCMC* 74-1201); see additional references in *DSB*. WORKS ABOUT: *DAB*, 17: 163-64 (Harry W. Foote); *DSB*, 12: 434-37 (Elizabeth H. Thomson).

SKINNER, AARON NICHOLS (1845-1918). *Astronomy*. Professor of mathematics, U.S. Navy; associated with Naval Observatory.

WORKS ABOUT: See *AMS; DAB*.

SMITH, EDGAR FAHS (1854-1928). *Chemistry*.

Professor of chemistry and provost, Univ. of Pennsylvania.

WORKS ABOUT: See *AMS; DAB; DSB.*

SMITH, ERMINNIE ADELE PLATT (April 26, 1836, Marcellus, N.Y.—June 9, 1886, Jersey City, N.J.). *Geology; Ethnology.* Daughter of Joseph, successful farmer and Presbyterian deacon, and Ermina (Dodge) Platt. Married Simeon H. Smith, 1855; four children. EDUCATION: 1853, graduated, Troy (N.Y.) Female Seminary; later, accompanied four young sons to Germany and studied crystallography at Strasbourg and German language and literature at Heidelberg; graduated from two-year course in mineralogy, School of Mines, Freiberg, Germany. HONORS: Name given to annual award in geology or mineralogy, Vassar College, 1888. CAREER: Husband was a man of wealth; during early years of marriage, she devoted attentions to raising four sons; 1866, with husband, moved from Chicago to Jersey City, N.J.; after return to Jersey City from Germany, began lecturing on geology and other subjects; 1876-1886, founder and president, Aesthetic Society, to which much effort was devoted; 1880-1885, with financial assistance from Smithson. Instn.'s Bureau of American Ethnology, conducted field studies among Iroquois of New York and Canada. MEMBERSHIPS: First female fellow, N.Y. Ac. Scis.; member, Am. Assoc. Advt. Sci. SCIENTIFIC CONTRIBUTIONS: Early interest in geology and botany encouraged by father, himself a rock collector; while in Chicago participated in classifying and labeling of mineral specimens intended for European museums; accumulated large collection of minerals, and after return from Germany, home in New Jersey took on appearance of mineralogical museum. Achieved success as popular lecturer, and was leading spirit in Aesthetic Society, organization of women who read and discussed papers on science, literature, and art. In 1879, read paper on amber to Am. Assoc. Advt. Sci. [see Works]. Remembered especially for work in anthropology; first woman to do field work in ethnography, concentrating studies on Six Nations of Iroquois federation; work on legends of these tribes was published in 1883 [see Works], and she compiled Iroquois-English dictionary, not published. At time of death, secretary of anthropology section, Am. Assoc. Advt. Sci.

WORKS: "Concerning Amber," *American Naturalist* 14 (1880): 179-90; "Myths of the Iroquois," *Second Annual Report of Bureau of American Ethnology* (Washington, D.C., 1883), pp. 47-116. References to bibliography of ethnological works in *Nota. Am. Wom.* [below]. MSS: Smithson. Instn.—Bureau of American Ethnology; see additional references in *Nota. Am. Wom.* WORKS ABOUT: *DAB,* 17: 262 (Walter Hough); *Nota. Am. Wom.,* 3: 312-13 (Nancy O. Lurie).

SMITH, EUGENE ALLEN (1841-1927). *Geology.* Professor of mineralogy and geology, Univ. of Alabama; state geologist of Alabama.

WORKS ABOUT: See *AMS; DAB.*

SMITH, HAMILTON LANPHERE (November 5, 1818, New London, Conn.—August 1, 1903, New London, Conn.). *Astronomy; Microscopy.* Son of Anson and Amy C. (Beckwith) Smith. Married Susan Beecher, 1841; Julia Buttles, 1847; one child by first marriage, two by second. EDUCATION: 1839, A.B., Yale College. HONORS: LL.D., Trinity College, 1871; D.S., Hobart College, 1900. CAREER: For some years, flour merchant in Cleveland, Ohio, and during this time also carried on scientific studies; 1852-1854, editor, *Annals of Science;* 1854-1868, professor of astronomy, natural philosophy, and chemistry, Kenyon College (Ohio); at Hobart College (Geneva, N.Y.), professor of astronomy and natural philosophy (1868-1900) and acting president (1883-1884). MEMBERSHIPS: President, American Microscopical Society, 1880 and 1885; fellow, Royal Microscopical Society of London; member, other societies. SCIENTIFIC CONTRIBUTIONS: While student, constructed largest telescope then in U.S., and at about that time commenced extensive series of observations of nebulae with Ebenezer P. Mason, a Yale classmate (reported by Mason in *Transactions of Am. Phil. Soc.* n.s. 7 [1841]: 165-214). In addition to holding editorship of *Annals of Science* (Cleveland), wrote textbooks on astronomy and natural philosophy. Did early work in U.S. on daguerreotype photography; in 1856 applied for patent on tintype and is credited with inventing that process. Accomplished in microscopy, and chief scientific publications dealt with Diatomaceae and marine Algae; these studies, which gave Smith an international reputation, appeared in *Am. Journ. Sci., American Monthly Microscopical Journal, Proceedings of American Microscopical Society,* and elsewhere.

WORKS: See *Roy. Soc. Cat.,* vols. 5, 8, 11, 12, 18, which lists over thirty papers, especially on microscopical studies. WORKS ABOUT: *NCAB,* 12 (1904): 466; *Obituary Record of Graduates of Yale University* (1900-1910), pp. 296-97.

SMITH, JOHN LAWRENCE (December 17, 1818, near Charleston, S.C.—October 12, 1883, Louisville, Ky.). *Chemistry; Mineralogy.* Son of Benjamin Smith, merchant. Married Sarah Julia Guthrie, 1852; no children. EDUCATION: Attended College of Charleston; 1835-1837, studied physical sciences and engineering, Univ. of Virginia; 1840, M.D.,

Medical College of South Carolina; 1840-1844, studied medicine and science in Europe and was first American to study chemistry under Justus Liebig at Giessen. HONORS: Decorated by governments of France, Turkey, Russia. CAREER: Ca. 1837 [one year], assistant engineer, Charleston and Cincinnati Railroad; 1844, began medical practice and also lectured, Charleston; 1846, cofounder, *Southern Journal of Medicine and Pharmacy;* 1847-1850, went to Turkey as adviser on cotton culture to Turkish government, and there was appointed to conduct survey of mineral resources; 1850-1852, resided in New Orleans, lectured, and held professorship of chemistry in projected Univ. of Louisiana; 1852, through marriage was assured of financial independence; 1852-1853, professor of chemistry, Univ. of Virginia; 1854-1866, professor of medical chemistry and toxicology, Univ. of Louisville; 1857-1858, associated with E. R. Squibb [q.v.] in commercial pharmaceutical laboratory; 1866-1883, traveled extensively, but maintained home in Louisville; president, Louisville Gas Works. MEMBERSHIPS: President, Am. Assoc. Advt. Sci., 1872; member, Natl. Ac. Scis. and other societies. SCIENTIFIC CONTRIBUTIONS: In 1842, while in Paris, published analysis of spermaceti [see Works], probably first extensive work in organic chemistry by an American. Later analyses of soils and related investigations probably led to Turkish appointment, where chief outcome was discovery of deposits of emery and coal; reports of his experience in Turkey later helped uncover emery deposits in U.S. While in Paris on return trip to U.S. from Turkey (1850), had idea for inverted microscope that he later developed and published. With G. J. Brush [q.v.] published important series of articles in *Am. Journ. Sci.* (1853-1855), "Reexamination of American Minerals." Did notable work in analytical chemistry; his method for separation of alkali metals from silicates was long a standard procedure that bears his name. During later years mainly interested in meteorites, having published first paper on that subject in 1854; notable collection of meteorites later purchased by Harvard Univ.

WORKS: "The Composition and Products of Distillation of Spermaceti . . . ," *Am. Journ. Sci.* 43 (1842): 301-21. J. B. Marvin, ed., *Original Researches in Mineralogy and Chemistry by Prof. J. Lawrence Smith* (Louisville, Ky., 1884) reprints number of papers and also includes bibliography and biography. List of 145 works also in Silliman [below], pp. 239-48. WORKS ABOUT: *DAB,* 17: 304-5 (T. Cary Johnson, Jr.); Benjamin Silliman, Jr. "Memoir," *Biog. Mems. of Natl. Ac. Scis.* 2 (1886): 219-48; John R. Sampey, "J. Lawrence Smith," *Journ. Chem. Educ.* 5 (1928): 123-28.

SMITH, SIDNEY IRVING (1843-1926). *Biology.* Professor of comparative anatomy, Yale Univ.

MSS: Yale Univ. Library (*NUCMC* 74-1202). WORKS ABOUT: See *AMS; DSB.*

SMITH, WILLIAM BENJAMIN (1850-1934). *Mathematics; New Testament criticism.* Professor of mathematics and professor of philosophy, Tulane Univ.

MSS: Univ. of Missouri—Western Historical Manuscripts Collection (*NUCMC* 60-2641). WORKS ABOUT: See *AMS; NCAB,* 35 (1949): 223-24; Warren Browne, *Titan vs. Taboo: The Life of William Benjamin Smith* (Tucson, Ariz., 1961).

SMYTH, WILLIAM (February 2, 1797, Pittston, Maine—April 4, 1868, Brunswick, Maine). *Mathematics.* Son of Caleb, ship carpenter and teacher of music, and Abia (Colburn) Smyth. Married Harriet Porter Coffin, 1827; six children. EDUCATION: College preparation largely through self-study; 1822, A.B., Bowdoin College; 1822-1823, studied at Andover Theological Seminary. CAREER: During War of 1812, entered army and served as secretary to an officer; thereafter, became clerk at Wiscasset, Maine, and then opened private school there, having care of his orphaned brother and sister; 1817-1819, assistant at Gorham (Maine) Academy; at Bowdoin College, appointed proctor and instructor in Greek (1823), then tutor (1824), adjunct professor (1825), and professor of mathematics and natural philosophy (1828-1868); 1825, licensed by Cumberland Association and for several years preached at locations around Brunswick, Maine; for seventeen years, member and usually chairman, Board of Agents, concerned with supervision of public schools. SCIENTIFIC CONTRIBUTIONS: Said to have been first American college teacher to make use of blackboard in classroom. Best remembered for series of mathematical textbooks that began to appear in 1830; these texts met with unprecedented success and went through numerous editions; first work on algebra was attempt to introduce best French mathematical work to American students, while enlarged work on trigonometry included application of that subject to surveying and navigation; work on calculus won praise from A. D. Bache [q.v.] for its originality; later version of algebra (1852) most extensively used textbook on subject in America at time. Much interested in public education and in antislavery movement.

WORKS: *Elements of Analytic Geometry* (Brunswick, Maine, 1830); *Elements of Plane Trigonometry, with Its Application to Mensuration of Heights and Distances, Surveying and Navigation* (Portland, Maine, 1852); *Treatise on Algebra* (Portland,

Maine, 1852); *Elements of Differential and Integral Calculus* (Portland, Maine, 1854). WORKS ABOUT: *DAB*, 17: 378-79 (Kenneth C. M. Sills); Alpheus S. Packard, *Address on the Life and Character of William Smyth* (Brunswick, Maine, 1868); Guralnick, *Science and American College*, pp. 212-13.

SNELL, EBENEZER STRONG (October 7, 1801, North Brookfield, Mass.—September 18, 1876, Amherst, Mass.). *Mathematics; Physics.* Son of Rev. Dr. Thomas, Congregational minister, and Tirzah (Strong) Snell. Married ca. 1829; apparently had children. EDUCATION: 1819-1821, attended Williams College; 1822, A.B. (conferred in 1825), Amherst College. HONORS: LL.D., Amherst College in 1860 and Western Reserve College in 1865. CAREER: 1822 (senior year in college), taught in North Brookfield and at Amherst Academy; 1822-1825, teacher, and then principal, Amherst Academy; at Amherst College (upon organization of faculty under new charter), became tutor (1825), then appointed instructor (1827) and adjunct professor of mathematics and natural philosophy (1829-1833), and finally became professor of mathematics and natural philosophy (1833-1876). MEMBERSHIPS: Am. Assoc. Advt. Sci. SCIENTIFIC CONTRIBUTIONS: Noted chiefly as teacher of science, for which purpose he constructed numerous pieces of scientific apparatus, and among many detailed geometrical diagrams prepared for students were examples illustrating modern French analysis; through critical review and correction of textbooks of D. Olmsted [q.v.], was able to make these works more appropriate for use in colleges; with Snell's revisions, and added diagrams, new edition of Olmsted's *Introduction to Natural Philosophy* was used for number of years (first edition by Snell, 1860); also published revision of Olmsted's *Compendium of Natural Philosophy, adapted to the Use of Schools and Academies* (first edition by Snell, 1863). In *Am. Journ. Sci.* and elsewhere published several scientific papers that related to descriptions of instruments for demonstrating physical (including optical) phenomena and to observations on temperature at Amherst (1839), parhelion and theory of ordinary halos, cones and spheres, rainbow caused by light reflected from water, and vibration of fall over dam.

WORKS: See above; also *Roy. Soc. Cat.*, vol. 5. MSS: Amherst College Library (Guralnick [below]). WORKS ABOUT: *NCAB*, 5 (1907 [copyright 1891 and 1907]): 11; *Addresses at the Annual Meeting of the Amherst College Alumni Commemorative of the Late Prof. Snell* (Springfield, Mass., 1877); Guralnick, *Science and American College*, pp. 213-14.

SNOW, FRANCIS HUNTINGTON (1840-1908). *Entomology.* Professor of mathematics, natural history, entomology, and meteorology, and chancellor, Univ. of Kansas.

MSS: Univ. of Kansas Libraries (Hamer). WORKS ABOUT: See *AMS; DAB.*

SPALDING, VOLNEY MORGAN (1849-1918). *Botany.* Professor of botany, Univ. of Michigan; member of staff, Carnegie Institution of Washington's Desert Botanical Laboratory (Tucson, Ariz.).

WORKS ABOUT: See *AMS; DAB.*

SQUIBB, EDWARD ROBINSON (July 4, 1819, Wilmington, Del.—October 25, 1900, Brooklyn, N.Y.). *Chemistry; Pharmacy.* Son of James R. and Catherine H. (Bonsall) Squibb. Married Caroline L. Cook, 1852; four children. EDUCATION: 1837-1842, apprenticeship under two Philadelphia druggists; 1845, M.D., Jefferson Medical College (Philadephia). CAREER: 1845-1847, medical practice, Philadelphia; 1847, commissioned assistant surgeon, U.S. Navy; 1847-1851, medical officer at sea; 1852, began assignment as house physician, Brooklyn Naval Hospital, and also served as manufacturing chemist; 1853-1857, director, Naval Laboratory (Brooklyn); 1857, resigned from navy; 1857-1858, partner with Thomas E. Jenkins and J. L. Smith [q.v.] in Louisville (Ky.) Chemical Works and also supervisor; 1858, established Squibb laboratory, Brooklyn, for manufacture of pharmaceuticals and chemicals, and in 1892, changed name to E. R. Squibb and Sons; 1869-1872, lecturer, College of Pharmacy of City and County of New York; 1871, became investor and supervisor and, later, full owner of National Chemical Wood Treatment Co.; 1882-1900, published *An Ephemeris of Materia Medica, Pharmacy, Therapeutics, and Collateral Information;* 1885-1886, 1890-1891, and 1897, in Europe; ca. 1893, retired from business. MEMBERSHIPS: A number of medical and pharmaceutical associations. SCIENTIFIC CONTRIBUTIONS: Throughout career, motivated by desire to assure purity and uniformity of drugs; at Naval Laboratory invented means of producing pure ether of standard quality, in 1854 perfecting use of steam in ether still as substitute for less safe open-flame method; also developed means of producing chloroform. One of chief contributions to science of pharmacy was in study and improvement of process of percolation. Throughout career, maintained perspective of physician and always was actively interested in actual effects of products such as anesthetics. Published large number of papers, especially in *Am. Journ. Pharm.*, and dominated *Ephemeris* [see above], which promoted desire for pure and effective drugs and resulted in number of controversies; delegate to 1860 and 1870 pharmacopoeial conventions; in 1880 served on committee on revision of U.S. *Pharmacopoeia*, a subject on which he was a recognized authority.

WORKS: See reference to bibliography of most significant papers, in *DAB;* also see *Roy. Soc. Cat.*, vols. 5, 8, 11, 12, 18. MSS: Some papers apparently with family; also Squibb archives at New Brunswick, N.J. (Blochman [below]). WORKS ABOUT: *DAB*, 17: 487-88 (Andrew G. Du Mez); Lawrence G. Blochman, *Dr. Squibb* (New York, 1958).

STALLO, JOHANN BERNHARD (March 16, 1823, Sierhausen, Oldenburg, Germany—January 6, 1900, Florence, Italy). *Philosophy of science.* Son of Johann Heinrich, teacher, and Maria Adelheid (Moormann) Stallo. Married Helene Zimmermann, 1855; seven children. EDUCATION: Ca. 1836 [age thirteen], began study in normal school at Vechta and also attended the gymnasium; 1841-1844, student and teacher, St. Xavier's College (Cincinnati); 1847-1849, studied law in Cincinnati; 1849, admitted to bar. CAREER: 1839, emigrated to U.S., settled in Cincinnati, and began teaching in Catholic parish school; 1841-1844, taught language and mathematics, St. Xavier's College; 1844-1847, professor of physics, chemistry, and mathematics, St. John's College (now Fordham Univ.), New York; 1849-1885, law practice, Cincinnati; 1852-1855, judge in Hamilton County (Ohio) Court of Common Pleas; 1884-1889, U.S. ambassador to Italy; 1889-1900, lived in Florence. SCIENTIFIC CONTRIBUTIONS: Wrote *General Principles* (1848) [see Works] during early years as devotee of *Naturphilosophie*, but later repudiated it; nonetheless, that work favorably impressed Ralph Waldo Emerson, who later saw in it suggestions of evolutionary thought. Chief contribution to philosophy of science was *Concepts* (1881) [see Works], subsequently translated into several European languages; *Concepts* was attempt to reveal metaphysical assumptions of science of the day and foreshadowed some of chief concerns of later philosophical thought as developed by Ernst Mach and others; from perspective of epistemology, criticized what he called atomo-mechanical view of matter; questioned absolutism of Newtonian science, and promoted idea that all physical properties are dependent on relations between various parts of the world.

WORKS: *The General Principles of the Philosophy of Nature . . . ; Embracing the Philosophical Systems of Schelling and Hegel and Oken's System of Nature* (Boston, 1848); *The Concepts and Theories of Modern Physics* (New York, 1882; reprinted, Cambridge, Mass., 1960). Also see references to several other works in *DAB* and *DSB*. WORKS ABOUT: *DAB*, 17: 496-97 (Adolf E. Zucker); *DSB*, 12: 606-10 (M. Čapek).

STEARNS, ROBERT EDWARDS CARTER (1827-1909). *Zoology; Agriculture; Botany.* Paleontologist, U.S. Geol. Surv.; assistant curator of mollusks and honorary associate in zoology, U.S. Natl. Mus.

MSS: Smithson. Instn. Archives (*NUCMC* 72-1257). WORKS ABOUT: See *AMS; DAB.*

STEELE, JOEL DORMAN (May 14, 1836, Lima, N.Y.—May 25, 1886, Elmira, N.Y.). *Science textbook writer.* Son of Allen, Methodist itinerant minister and farm owner, and Sabra (Dorman) Steele. Married Esther Baker, 1859; apparently no children, adopted one child. EDUCATION: Studied under Charles Anthon at Boys' Classical Institute, Albany; attended Boys' Academy, Troy, N.Y.; 1858, graduated, Genesee College (later part of Syracuse Univ.). HONORS: Ph.D., Regents of Univ. of State of New York, 1870. CAREER: 1853 (summer), taught at country district school; 1854, bookkeeper and clerk, New York City; at Mexico (N.Y.) Academy, instructor (1858-1859) and principal (1859-1861); 1861-1862, captain with Eighty-first New York Volunteers, until battle wound ended military service; 1862-1866, principal, high school in Newark, N.Y.; 1866-1872, principal and science teacher, Elmira (N.Y.) Free Academy; 1870-1886, a trustee, Syracuse Univ.; 1872, gave up teaching to devote full efforts to textbook writing, but continued to live in Elmira; beginning ca. 1871, made four trips to Europe to gather material for history texts; ordained preacher (exhorter), Methodist Episcopal Church. MEMBERSHIPS: Am. Assoc. Advt. Sci. SCIENTIFIC CONTRIBUTIONS: As teacher, experienced the need for textbooks in sciences that were suited especially to high schools and academies; in 1867 published first work, on chemistry, the beginning of so-called Fourteen Weeks series of well-written and widely popular elementary works on sciences; eventually prepared texts on botany (with assistance of Alphonso Wood), geology, physics, physiology, zoology (with J. W. P. Jenks [q.v.]), and astronomy; other persons also aided or advised on content of various volumes; also prepared, with collaboration of wife, widely used texts on history. Books were criticized by some as superficial, and yet aim was to prepare works for beginners; they were used for many years and played important part in popularizing study of science at high school level, while introducing students to simple laboratory procedures and study of specimens.

WORKS: See *Natl. Un. Cat. Pre-1956 Imp.* WORKS ABOUT: *DAB*, 17: 556-57 (Ethel W. Faulkner); Anna C. Palmer, *Joel Dorman Steele: Teacher and Author* (New York, 1900).

STEVENS, WALTER LeCONTE (1847-1927). *Physics.* Professor of physics, Washington and Lee Univ.

WORKS ABOUT: See *AMS; Who Was Who in America,* vol. 1.

STEVENSON, JOHN JAMES (1841-1924). *Geology.* Professor of geology, New York Univ.

WORKS ABOUT: See *AMS; DAB.*

STIMPSON, WILLIAM (February 14, 1832, Roxbury, Mass.—May 26, 1872, Ilchester, Md.). *Marine zoology.* Son of Herbert Hawthorne, successful stove merchant, and Mary Ann Devereau (Brewer) Stimpson. Married Annie Gordon, 1864; survived by at least one child. EDUCATION: 1848, graduated, Cambridge (Mass.) High School; for one year, attended Cambridge Latin School; 1850, became special student in laboratory of J. L. R. Agassiz [q.v.], Harvard Univ. HONORS: M.D., Columbia Univ., 1860. CAREER: Ca. 1848, for brief period, associated with civil engineering establishment; ca. 1849, accompanied fishing vessel to Grand Manan, New Brunswick, and there collected marine animals; 1850-1853, curator of mollusks, Bos. Soc. Nat. Hist.; 1853-1856, naturalist, U.S. North Pacific Surveying and Exploring Expedition; 1856-1865, as head of invertebrate department at Smithson. Instn., worked on specimens collected during North Pacific Expedition; 1865, appointed director, Chicago Academy of Sciences, which in 1871 was destroyed in great Chicago fire. MEMBERSHIPS: Natl. Ac. Scis. and other societies. SCIENTIFIC CONTRIBUTIONS: Persisted in natural history interests despite early paternal opposition. During North Pacific expedition, collected more than 5,000 specimens, especially invertebrates, and also compiled extensive notes; in years subsequent to expedition, devoted efforts to identification and description of crustaceans and other invertebrates, while A. A. Gould [q.v.] worked on mollusks; went to Europe to compare specimens. Took to Chicago the North Pacific invertebrates (minus mollusks), plus his personal collection of shells, the deep-sea crustaceans and mollusks of L. F. de Pourtalès [q.v.], and other collections, an unrivaled accumulation, all destroyed in fire. Published number of works on Mollusca and Crustacea, contributing descriptions for 948 new marine invertebrate species. Credited as pioneer in methodical dredging of Atlantic Coast.

WORKS: "Synopsis of the Marine Invertebrata of Grand Manan . . . ," *Smithson. Contr. Knowl.* 6, no. 4 (1854); "Researches Upon Hydrobiinae and Allied Forms," *Smithson. Misc. Coll.* 7 (1865); "Report on the Crustacea (Brachyura and Anomura, collected by the North Pacific Exploring Expedition, 1853-1856)," ibid. 49 (1907), edited by Mary Rathbun. Bibliography in A. G. Mayer, "Memoir," *Biog. Mems. of Natl. Ac. Scis.* 8 (1918): 429-33. MSS: Smithson. Instn. Library (Hamer); Ac. Nat. Scis. Philad. (Hamer); manuscript notes and drawings lost in Chicago fire. WORKS ABOUT: *DAB,* 18: 31-32 (Frank C. Baker); *DSB,* 13: 65-66 (Richard I. Johnson) [in cases of conflicting information between these sources, the latter has been favored].

STOCKWELL, JOHN NELSON (1832-1920). *Astronomy.* Professor of mathematics and astronomy, Case School of Applied Science (Case Institute of Technology); engaged in private research.

WORKS ABOUT: See *AMS; DAB.*

STODDARD, JOHN FAIR (July 20, 1825, farm in Greenfield, N.Y.—August 6, 1873, Kearny, N.J.). *Arithmetic.* Son of Phineas, perhaps farmer, and Marilda (Fair) Stoddard. Married Eliza Ann Platt, 1865; one child. EDUCATION: Attended Montgomery Academy, Orange County, N.Y., and Nine Partners' School, Dutchess County, N.Y.; 1847, graduated, State Normal College, Albany, N.Y.; 1853, A.M., Univ. of City of New York (New York Univ.). CAREER: At about age sixteen, began teaching district school; 1847-1851, head, Liberty Normal Institute (Liberty, N.Y.?); 1851-1854, president, Univ. of Northern Pennsylvania; 1854, became county superintendent [of schools?], Wayne County, Pa.; 1855-1857, principal, Lancaster County (Pa.) Normal School; 1857, bought property of defunct Univ. of Northern Pennsylvania and there established teachers' college, which burned shortly after opening; 1857, elected president, Pennsylvania State Teachers' Association; 1857-1859, established and conducted Susquehanna County Normal School, Montrose, Pa.; 1859, moved to New York City and served there for several years as principal of a public grammar school; 1864, in ill health, retired to Greenfield, N.Y.; 1867, moved to New Jersey. SCIENTIFIC CONTRIBUTIONS: Primarily an educator, in science remembered for series of widely used school textbooks on arithmetic; as young teacher, chiefly interested in mathematics, and in 1849 published *American Intellectural Arithmetic,* followed by others, including works on algebra (1857, 1859) written with W. D. Henkle; school texts proved so popular that before 1860 some 1.5 million copies were issued, and textbooks continued to be issued, in revised form, by publishers (Sheldon and Company of New York) long after author's death. Approached teaching of mathematics from both practical and "discipline" points of view, having as objectives teaching of arithmetic and also development of powers of reasoning. Move to New York City partly motivated by desire to continue study in advanced mathematics.

WORKS: See *Natl. Un. Cat. Pre-1956 Imp.* WORKS ABOUT: *DAB*, 18: 54-55 (Ethel W. Faulkner); brief biographical sketch in *American Journal of Education* 15 (1865): 677-78.

STODDARD, JOHN TAPPAN (1852-1919). *Chemistry.* Professor of physics and chemistry, Smith College.
WORKS ABOUT: See *AMS; DAB.*

STONE, ORMOND (1847-1933). *Astronomy.* Professor of practical astronomy and director of Leander McCormick Observatory, Univ. of Virginia.
MSS: Cincinnati Observatory (*NUCMC* 75-1938). WORKS ABOUT: See *AMS; DAB*, supp. 1.

STORER, DAVID HUMPHREYS (March 26, 1804, Portland, Maine—September 10, 1891, Boston, Mass.). *Zoology*, especially *Ichthyology.* Son of Woodbury, chief justice of common pleas, and Margaret (Boyd) Storer. Married Abby Jane Brewer, 1829; five children, including Francis H. Storer [q.v.]. EDUCATION: 1822, graduated, Bowdoin College; 1825, M.D., Harvard Univ. Medical School. HONORS: LL.D., Bowdoin College, 1876. CAREER: Ca. 1825, began medical practice, Boston, and came to specialize in obstetrics; 1837, appointed to Massachusetts Zoological and Botanical Survey; 1839, with several other Boston practitioners, established Tremont Street Medical School; 1849-1858, member of staff, Massachusetts General Hospital; 1854-1868, member of staff, Lying-in Hospital; at Harvard Medical School, professor of obstetrics and medical jurisprudence (1854-1868) and dean (1854-1864); for some thirty-five years, insurance company medical examiner. MEMBERSHIPS: Vice-president, Bos. Soc. Nat. Hist., 1843-1860; member, Am. Ac. Arts Scis. SCIENTIFIC CONTRIBUTIONS: Early interest in natural history centered on mollusks, but assignment to report on fishes of Massachusetts for state Zoological and Botanical Survey redirected attentions, and chief scientific contributions were made in ichthyology; report (1839) [see Works], despite lack of prior experience in study of fishes, was well done. For more than twenty-five years thereafter, devoted scientific efforts to perfecting report hurriedly done for survey, and revision appeared serially, 1853-1867 [see Works], and later as a separate volume; final work on Massachusetts fishes described all species then known, their natural history, and their economics, and was model for study of fish fauna of other regions of U.S.; in studies, solicited aid of fishermen and fish merchants; "Synopsis" [see Works] was essentially compilation. Also a leading American obstetrician. Sons Francis H. and Horatio Robinson Storer also known for work in science.

WORKS: "Reports on the Ichthyology and Herpetology of Massachusetts," in Storer and William B. O. Peabody, *Report on Fishes, Reptiles and Birds of Massachusetts* (Boston, 1839), pp. 1-253, published by Commissioners on Zoological and Botanical Survey; "A History of the Fishes of Massachusetts," *Memoirs of Am. Ac. Arts Scis.*, n.s. 5, 6, 8, 9 (1853-1867), with plates; "Synopsis of the Fishes of North America," ibid., n.s. 2 (1846): 253-550. See *Roy. Soc. Cat.*, vols 5, 8. MSS: Harv. U.-MS; Boston Museum of Science. WORKS ABOUT: *DAB*, 18: 93-94 (Henry R. Viets); memoir, *Proceedings of Am. Ac. Arts Scis.* 27 (1891-1892): 388-91.

STORER, FRANCIS HUMPHREYS (1832-1914). *Agricultural chemistry.* Professor of agricultural chemistry and dean of Bussey Institution, Harvard Univ.
MSS: See *DAB.* WORKS ABOUT: See *AMS; DAB.*

STORY, WILLIAM EDWARD (1850-1930). *Mathematics.* Professor of mathematics, Clark Univ.
WORKS ABOUT: See *AMS*; *DAB.*

STRECKER, HERMAN [Ferdinand Heinrich Herman] (March 24, 1836, Philadelphia, Pa.—November 30, 1901, Reading, Pa.). *Entomology.* Son of Ferdinand H., sculptor and dealer in marble, and Ann (Kern) Strecker. Married; survived by two children. EDUCATION: Ca. 1845, settled with family at Reading, Pa., and attended schools there; ca. 1847 [age eleven], became apprenticed to father as sculptor; apparently self-taught in science; drew on libraries of New York and Philadelphia. HONORS: Ph.D., Franklin and Marshall College, 1890. CAREER: 1856, on father's death, took over marble and sculpture business at Reading, Pa.; also did tombstone lettering; referred to as architect, designer, and sculptor; traveled extensively and, 1855-1856, visited West Indies, Mexico, and Central America to study Aztec monuments and to collect butterflies. SCIENTIFIC CONTRIBUTIONS: During early years, spent free days in study of natural history in library of Ac. Nat. Scis. Philad., finally settling on study of Lepidoptera (moths and butterflies); extensive correspondence and collection done in evenings and on holidays, and over forty-year period accumulated some 200,000 butterflies and moths from all parts of world, which ranked as most extensive and best collection of these insects, public or private, in U.S. Best-known work, *Lepidoptera* [see Works], was privately printed in limited edition, with fifteen plates of illustrations drawn on stone and colored by author; also published *Butterflies* [see Works], which included instructions for collecting, breeding, and classifying the butterflies, as well as other such material of instructive or reference use, and this work achieved considerable sales. Maintained high standards of work and was at times critical of efforts of others. In addition to entomological collecting, sold moths and butterflies.

WORKS: *Lepidoptera, Rhopaloceres and Heteroceres, Indigenous and Exotic, with Descriptions and Colored Illustrations,* 15 pts. (Reading, Pa., 1872-1878, with supplements issued in 1898, 1899, 1900); *Butterflies and Moths of North America* (Reading, Pa., 1878). See *Roy. Soc. Cat.*, vols. 11, 18. MSS: Ac. Nat. Scis. Philad. (*NUCMC* 66-86); Chicago Natural History Museum (*NUCMC* 66-764). WORKS ABOUT: *NCAB,* 10 (1909 [copyright 1900]): 317-18; obituary, *Entomological News* 13 (1902): 1-4; Mallis, *Amer. Entomologists*, pp. 294-96.

STRINGHAM, WASHINGTON IRVING (1847-1909). *Mathematics.* Professor of mathematics, Univ. of California (Berkeley).

WORKS ABOUT: See *AMS; DAB.*

STRONG, THEODORE (July 26, 1790, South Hadley, Mass.—February 1, 1869, New Brunswick, N. J.). *Mathematics.* Son of Joseph, Congregational minister, and Sophia (Woodbridge) Strong. Married Lucy Dix, 1818; seven children. EDUCATION: 1812, A.B., Yale College. HONORS: LL.D., Rutgers College, 1835. CAREER: At Hamilton College, tutor in mathematics (1812-1816) and professor of mathematics and natural philosophy (1816-1827); at Rutgers College, professor of mathematics and natural philosophy (1827-1861) and vice-president (1839-1863). MEMBERSHIPS: Am. Ac. Arts Scis.; Natl. Ac. Scis.; and other societies. SCIENTIFIC CONTRIBUTIONS: Published first mathematical paper while still undergraduate, on Stewart's properties of the circle *(Memoirs of Connecticut Academy of Arts and Sciences* 1 [1810]). During subsequent career achieved place as one of leading American mathematicians of time, although not considered in class with N. Bowditch [q.v.], whose interests he shared, or with B. Peirce [q.v.], and he made no great contribution to mathematical knowledge; mastered geometry of Newton and higher analytical works of French mathematicians such as Lagrange and Laplace, and ranked as one of first Americans to achieve truly intimate acquaintance with European work in mathematical analysis; published number of short papers in American mathematical and scientific journals. His own natural grasp of mathematical thought often made his teaching difficult for students lacking commensurate abilities, but in spite of this weakness as teacher, his services as professor were sought by several universities; while textbooks on *Algebra* and *Calculus* [see Works] included original matter, organization was such as to make them unusable by most students. Nevertheless, Strong played central role in introduction of study of higher mathematics into U.S.

WORKS: *A Treatise on Elementary and Higher Algebra* (New York, 1859); *A Treatise on the Differential and Integral Calculus* (New York, 1869). See references to bibliographies in *DAB;* Bradley [below], especially pp. 21-23, gives brief summaries of contents of chief papers. WORKS ABOUT: *DAB,* 18: 152 (Raymond C. Archibald); Joseph P. Bradley, "Memoir," *Biog. Mems. of Natl. Ac. Scis.* 2 (1886): 1-28.

STURTEVANT, EDWARD LEWIS (January 23, 1842, Boston, Mass.—July 30, 1898, South Framingham, Mass.). *Botany; Agriculture.* Son of Lewis W. and Mary Haight (Leggett) Sturtevant. Married Mary Elizabeth Mann, 1864; Hattie Mann, 1883; four children by first marriage, one by second. EDUCATION: 1859-1861, attended Bowdoin College and in 1863 received A.B. from that institution; 1866, M.D., Harvard Medical School. CAREER: 1861-1863, officer in Twenty-fourth Maine Volunteers; 1867, with brothers, began establishment of Waushakum Farm near South Farmingham, Mass.; 1876-1879, an editor, *Scientific Farmer;* 1882-1887, first director, New York Agricultural Experiment Station, Geneva; 1887, retired to Waushakum Farm and pursued interest in history of plants; 1893-ca. 1896 [three winters], in California for reasons of health; though holding M.D. degree, never practiced medicine. MEMBERSHIPS: Am. Assoc. Advt. Sci.; a founder of Society for Promotion of Agricultural Science. SCIENTIFIC CONTRIBUTIONS: Purchased Waushakum with brothers Joseph N. and Thomas L. Sturtevant and there bred Ayrshire dairy cattle, on which subject Edward and Joseph published important works. Studied milk; constructed first lysimeter in U.S. for testing percolation in soil, and during period 1875-1880 kept records of instrument's use; also had special interest in Indian corn; for some years carried out studies of that plant, published number of items on its culture and classification, and developed productive new variety called Waushakum. A leader in promotion of idea of agricultural experiment stations. Developed remarkable botanical library, especially of pre-Linnaean works; investigated history of edible plants and published various articles on the subject; summation of this research was published as posthumous *Notes* [see Works]; left uncompleted manuscript, "Encyclopedia of Agriculture."

WORKS: *Varieties of Corn,* U.S. Dept. of Agriculture, Office of Experiment Stations Bulletin no. 57 (Washington, D.C., 1899); *Sturtevant's Notes on Edible Plants* (Albany, 1919), with biographical sketch by U. P. Hedrick and bibliography of works, pp. 13-16. MSS: Corn. U-CRH (*NUCMC* 70-1130); Mo. Bot. Gard. (which also received library and herbarium) (*NUCMC* 72-1577). WORKS ABOUT: *DAB,* 18: 185-86 (Everett E. Edwards); see also Hedrick [above].

SULLIVANT, WILLIAM STARLING (January 15, 1803, Franklinton [now part of Columbus], Ohio—April 30, 1873, Columbus, Ohio). *Botany,* especially *Bryology.* Son of Lucas, surveyor and proprietor of Franklinton, and Sarah (Starling) Sullivant. Married Jane Marshall, 1824; Eliza Griscom Wheeler, 1834; Caroline Eudora Sutton, 1851; thirteen children. EDUCATION: Attended Ohio University (Athens); 1823, A.B., Yale College. HONORS: Sullivant Moss Society named for him. CAREER: 1823, following father's death in that year, took over family business interests, including gristmills, and began study of surveying; later, expanded business interests to include stone quarries and stage coaches; also was bank president and owner of farm lands; after 1856, essentially free from active business concerns. MEMBERSHIPS: Ac. Nat. Scis. Philad.; Natl. Ac. Scis.; and other societies. SCIENTIFIC CONTRIBUTIONS: First interested in ornithology and then conchology. During early 1830s, attention turned to botany, an interest shared with wife Eliza; outcome of early interest in flowering plants was *Catalogue of Plants . . . in the Vicinity of Columbus, Ohio* (1840). About that time, chief botanical concerns settled on mosses, and first important work on that group was issue of dried specimens of mosses of Alleghenies, done with assistance of wife; contributed section on mosses and hepatics to *Manual* of A. Gray [q.v.] and as reprinted separately (1856) it was contribution of fundamental importance; received mosses and hepatics from Wilkes Expedition and various explorations in West and elsewhere and employed L. Lesquereux, C. Wright [qq.v.], and Augustus Fendler as collectors; aided number of correspondents with identification of species, and with Lesquereux, issued dried specimens under title *Musci Boreali-Americani* (1856). Climax to career was *Icones* [see Works], and remainder of life was devoted to preparation of *Supplement.* Also achieved reputation as microscopist.

WORKS: See above; also *Musci Alleghanienses,* 2 vols. (Columbus, Ohio, 1845-1846); *Icones Muscorum; or, Figures and Descriptions of Most of Those Mosses Peculiar to Eastern North America Which Have Not Been Heretofore Figured* (Cambridge, Mass., 1864), with 129 copperplates; *Supplement* to latter (1874) prepared after death by L. Lesquereux. Bibliography in Rodgers [below], pp. 321-24. MSS: Harv. U.-GH (Hamer); Harv. U.-FL (Hamer); Ohio Historical Society (Hamer). WORKS ABOUT: *DAB,* 18: 201 (William R. Maxon); Andrew D. Rodgers III, *"Noble Fellow": William Starling Sullivant* (New York, 1940).

SWALLOW, GEORGE CLINTON (November 17, 1817, Buckfield, Oxford County, Maine—April 20, 1899, Evanston, Ill.). *Geology.* Son of Larned, farmer and manufacturer of implements, and Olive Fletcher (Proctor) Swallow. Married Martha Ann Hill, 1844; two children. EDUCATION: 1843, graduated, Bowdoin College; 1867, M.D., Missouri Medical College (St. Louis). CAREER: 1843-1849, principal, Brunswick (Maine) Female Seminary; 1845-1850, lectured at teachers' institutes at various locations in Maine; 1849, appointed principal, Hampden (Maine) Academy and also became member of state board of education; 1851-1853 and 1857-1858, professor of chemistry and natural sciences, Univ. of Missouri; 1853-1861, Missouri state geologist; with Kansas State Geological Survey, paleontologist (1864-1865) and director (1865-1866); 1867-1870, superintendent of mining operations, Highland, Mont.; at Agricultural and Mechanical College, Univ. of Missouri, appointed professor of agriculture (1870), later professor of natural sciences, dean (1872-1882); 1872, also made professor of botany, comparative anatomy, and physiology, medical school, Univ. of Missouri; 1882-1890, lived at Helena, Mont., and there edited *Daily Independent*; 1888-1890, Montana state inspector of mines. MEMBERSHIPS: Am. Assoc. Advt. Sci.; Geol. Soc. Am.; and other societies. SCIENTIFIC CONTRIBUTIONS: Chief geological work done in relation to Missouri survey, in course of which notes and observations on soil, plants, and animals also were taken, though never published; geological work in Missouri was quickly—though carefully—done, in keeping with highest contemporary standards, and resulted in admirable classification of rocks and general outline of distribution of formations; second Missouri report (1854), which included work of assistants, was most important of the reports of that survey, but added nothing that was strikingly new to geological science. Work on Kansas survey marred by need to work rapidly under public pressure. In 1858, first to announce Permian rocks on American continent, though based on initial judgment of F. B. Meek [q.v.]. Also wrote on grape culture and on Indian mounds.

WORKS: Issued five reports for Missouri survey and preliminary report for Kansas. Bibliography in Broadhead [below], pp. 4-5. MSS: Small collection in Ellis Library, Univ. of Missouri (Columbia). WORKS ABOUT: *DAB,* 18: 232-33 (George P. Merrill); G. C. Broadhead, "Biographical Sketch," *American Geologist* 24 (1899): 1-6; information from Univ. of Missouri Archives.

SWIFT, LEWIS (1820-1913). *Astronomy.* Director of Warner Observatory, Rochester, N.Y., and of Lowe Observatory, Echo Mountain, Calif.

WORKS ABOUT: See *AMS; DAB.*

SYLVESTER, JAMES JOSEPH [originally surnamed Joseph, later adopted name Sylvester] (September 3, 1814, London, England—March 15, 1897, London, England). *Mathematics.* Son of Abraham Joseph. Never married. EDUCATION: For short period, attended Univ. of London (later University College) ; 1829, entered school at Royal Institution, Liverpool; 1831-1833 and 1836-1837, attended St. John's College, Cambridge; unable to receive Cambridge degree because of adherence to Jewish faith, but in 1872 (following passage of Test Act) received B.A. and M.A. degrees; 1841, B.A. and M.A., Trinity College, Dublin; 1846, entered Inner Temple; 1850, admitted to bar. HONORS: Among many honors was Royal Society's creation of Sylvester Medal for pure mathematics research. CAREER: 1838-1841, professor of natural philosophy, University College, London; 1841-1842, professor of mathematics, Univ. of Virginia; 1843, returned to England; 1844-1855, engaged in actuarial work; 1855-1870, professor of mathematics, Royal Military Academy, Woolwich; 1855, became editor, *Quarterly Journal of Pure and Applied Mathematics;* 1876-1883, professor of mathematics, Johns Hopkins Univ.; 1878-1884, founder and editor, *American Journal of Mathematics;* 1883-1897, Savilian professor of geometry, Oxford Univ. MEMBERSHIPS: Roy. Soc. and other societies; president, London Mathematical Society, 1866-1868. SCIENTIFIC CONTRIBUTIONS: Most significant mathematical work done in algebra, and is especially remembered for that done with Arthur Cayley in establishing theory of algebraic invariants; with Cayley contributed to development of theory of determinants and their use in relation to nonalgebraic concerns; interested in theory of numbers, and made particularly important contributions to question of partition of numbers; among other areas in which important advances were made was study of number of imaginary roots in an algebraic equation. Also wrote poetry and published *Laws of Verse* (1870). Although years at Johns Hopkins were relatively few, through reputation as one of world's foremost mathematicians and by dedication to research, made significant contribution to promotion of teaching and research in higher mathematics in U.S.; editorship of *American Journal of Mathematics* gave that journal an international standing.

WORKS: *The Collected Mathematical Papers of James Joseph Sylvester,* 4 vols. (Cambridge, 1904-1912), edited by H. F. Baker. MSS: Brown Univ. Library (*NUCMC* 60-2152). WORKS ABOUT: *DAB,* 18: 256-57 (David E. Smith) ; *DSB,* 13: 216 22 (J. D. North).

T

TAKAMINE, JOKICHI (1854-1922). *Chemistry.* Chemist, Japanese department of agriculture and commerce; in U.S., acted as consulting chemist to Parke, Davis & Co.; established private laboratory, Clifton, N. J.

WORKS ABOUT: See *AMS,* 2d ed. (1910) ; *DAB.*

TAYLOR, CHARLOTTE De BERNIER SCARBROUGH (1806, Savannah, Ga.—November 26, 1861, Isle of Man, Great Britain). *Entomology.* Daughter of William, planter and merchant, and Julia (Bernard) Scarbrough. Married James Taylor, 1829; three children. EDUCATION: Graduated, Madam Binze's School, New York City; thereafter, made tour of Europe; probably chiefly self-taught in science. CAREER: After marriage, apparently continued to live in Savannah, a person of wealth; devoted attentions to raising of children, social affairs, and scientific studies and writing; ca. 1861, left Savannah and went to England. SCIENTIFIC CONTRIBUTIONS: During 1830s began serious study of insects, but appearance of scientific writings in general literary magazines rather than scientific journals probably minimized influence on other students of subject; carried out close study of insects associated with cotton culture for more than fifteen years before attempting publication, which began in 1850s in American magazines, especially *Harper's New Monthly;* in addition to study of cotton insects, also investigated insects associated with wheat. Entomological work involved use of powerful magnifying glasses, though probably not compound microscope, and articles were illustrated with drawings in preparation of which daughters lent assistance. For that time, showed sensitive awareness of relations of entomology and agriculture, and promoted necessity for informed control of insect pests; published study of habits and anatomy of silkworm and also investigations of natural history and anatomy of spiders. While crossing Atlantic (ca. 1861), conducted microscopic studies of soundings from ocean. On Isle of Man began, but did not complete, work on plantation life.

WORKS: Nineteen works on entomological and related topics listed in index to *Harper's New Monthly Magazine,* vols. 18 (1859)—29 (1864); "Insects Destructive to Wheat," ibid. 20 (December 1859) : 38-52; "The Silk-Worm," ibid. 20 (May 1860) : 753-64; "Insects Belonging to the Cotton Plant," ibid. 21 (June 1860) : 37-51; "Spiders: Their Structure and Habits," ibid. 21 (September 1860) : 461-77; "Soundings," ibid. 29 (July 1864) : 179-86. WORKS ABOUT: *DAB,* 18: 319-20 (Richard H. Shryock) ; *NCAB,* 2 (1899 [copyright 1891 and 1899]) : 164.

TAYLOR, FRANK BURSLEY (1860-1938). *Geology.* Field assistant, U.S. Geol. Surv.; investigated glacial history of Great Lakes.

WORKS ABOUT: See *AMS; DAB,* supp. 2; *DSB.*

TAYLOR, RICHARD COWLING (January 18, 1789, Hinton, Suffolk, or Banham, Norfolk, England—October 27, 1851, Philadelphia, Pa.). *Geology.* Son of Samuel Taylor, well-to-do farmer. Married Emily Errington, 1820; four children. EDUCATION: Early instruction at Halesworth, Suffolk, including sound training in higher mathematics and in mapping; 1805-ca. 1811, apprenticed to land surveyor in Gloucestershire. HONORS: Awarded Isis Medal by Society of Arts for what reportedly was first plaster model of mines, prepared for British Iron Co. CAREER: After completion of apprenticeship, engaged in surveying work in various locations in England, and was associated with William Smith (1769-1839), so-called father of British geology; for time, headed a Department of Ordnance [topographical] Survey for Buckingham and Bedford; 1813, settled at Norwich as land surveyor; 1826, moved to London; while in England, became involved in studies of various mining properties, including report for British Iron Co. of South Wales; 1830, moved to U.S., took up residence at Phillipsburg, Pa., and undertook survey of Blossburg coal region in that state; ca. 1834, removed to Philadelphia; directed coal and iron explorations in Dauphin County (Pa.) for Dauphin and Susquehanna Coal Co.; subsequently, carried out number of mineral explorations, including Cuban copper mines, Panamanian gold fields, and asphaltum of New Brunswick. MEMBERSHIPS: Geological Society of London; Am. Phil. Soc.; and other societies. SCIENTIFIC CONTRIBUTIONS: Mainly involved in economic geology, especially surveys of mineral, coal, and metal regions, but also published numerous papers on general geology, paleontology, and ancient ruins of England; knowledgeable in theoretical geology, and in survey work in Pennsylvania produced first correlation of old red sandstone with corresponding strata of European rocks. Chief work was *Statistics of Coal* [see Works], which covered all parts of world and was warmly praised by contemporary scientists.

WORKS: *Statistics of Coal* (Philadelphia, 1848), 754 pp. Several works done in England listed with biographical account in *Dictionary of National Biography;* see also *Roy. Soc. Cat.*, vol. 5. MSS: Ac. Nat. Scis. Philad. WORKS ABOUT: *DAB*, 18: 341 (Hazel S. Garrison); Isaac Lea, memoir, *Proceedings of Ac. Nat. Scis. Philad.* 5 (1850-1851): 290-96.

TESCHEMACHER, JAMES ENGELBERT (June 11, 1790, Nottingham, England—November 9, 1853, near Boston, Mass.). *Geology; Mineralogy; Botany.* No information on parentage, marital status, or education. CAREER: 1804 [age fourteen], began business career with foreign mercantile establishment in London; 1830, went to Cuba, having received offer of business partnership in Havana, but soon returned to England; 1832, arrived in New York; ca. 1832-1853, engaged in commercial activities in Boston, Mass. MEMBERSHIPS: Bos. Soc. Nat. Hist. SCIENTIFIC CONTRIBUTIONS: Pursued scientific work in moments when free from business engagements. Wrote papers on botany, mineralogy, and geology, including well-respected studies of carboniferous formations; toward end of life engaged in study of structure of coal and associated fossil forms; noted for careful and accurate work, including microscopic studies, contributing significantly to knowledge of American minerals. For time, held office as curator of botany in Bos. Soc. Nat. Hist. Apparently published no scientific works before coming to U.S.

WORKS: Papers appeared mainly in *Am. Journ. Sci., Boston Journal of Natural History,* and *Proceedings of Bos. Soc. Nat. Hist.* See *Roy. Soc. Cat.*, vol. 5, which lists first paper as in 1837. WORKS ABOUT: *Appleton's CAB*, vol. 6: 67; T. T. Bouvé, "Historical Sketch," in Bos. Soc. Nat. Hist. *Anniversary Memoirs* (Boston, 1880), pp. 59-60.

THOMAS, BENJAMIN FRANKLIN (1850-1911). *Physics.* Professor of physics, Ohio State Univ.

WORKS ABOUT: See *AMS; Who Was Who in America*, vol. 1.

THOMAS, CYRUS (1825-1910). *Ethnology; Entomology.* Lawyer; archaeologist, Bureau of American Ethnology.

MSS: U.S. Bureau of American Ethnology Archives, Washington, D.C. (Hamer). WORKS ABOUT: See *AMS; DAB.*

THOMPSON, ALMON HARRIS (September 24, 1839, Stoddard, N.H.—July 31, 1906, Washington, D.C.). *Topographical engineering.* Son of Lucas and Mary (Sawyer) Thompson. Married Ellen L. Powell, sister of John W. Powell [q.v.], 1862; no information on children. EDUCATION: 1848-1856, attended schools at Southboro, Mass.; 1857-1861, took scientific course at Wheaton College (Illinois). CAREER: During Civil War, served as first lieutenant, 139th Illinois Volunteer Infantry; 1865-1867, superintendent of schools, Lacon, Ill., and during period 1867-1868 held similar position at Bloomington, Ill.; 1869-1870, acting curator, Illinois Natural History Society; 1870-1878, topographical engineer for J. W. Powell on surveys of Colorado River and Rocky Mountain regions; 1882-1906, as geographer with U.S. Geol. Surv., headed geographical work west of Mississippi (1884-1895), engaged in field and office work (1896-1903), and was in charge of Geol. Surv. exhibits at Louisiana Purchase Exposition (1904). MEMBERSHIPS: A founder, Natl. Geog. Soc.; member, other societies. SCIENTIFIC CONTRIBUTIONS: Scientific career largely determined by friendship with brother-in-law, Powell. Responsible for exploration and mapping operations, and related astronomical determinations, with Powell surveys; continued similar work as chief geographer with U.S. Geol. Surv. until death.

WORKS: See *Roy. Soc. Cat.*, vol. 19. MSS: N.Y. Pub. Lib. (Hamer). WORKS ABOUT: *Who Was Who in America*, vol. 1; "Diary of Almon Harris Thompson" [relating to work with Powell surveys], *Utah Historical Quarterly* 7 (1939): 1-140, with biographical introduction.

THOMPSON, BENJAMIN [COUNT RUMFORD] (March 26, 1753, Woburn, Mass.—August 21, 1814, Auteuil, France). *Physics.* Son of Benjamin, small farmer, and Ruth (Simonds) Thompson. Married Sarah (Walker) Rolfe, 1772 (separated in 1775); married widow of Antoine Lavoisier, 1805 (separated in 1809); one child by first marriage. EDUCATION: Enjoyed relatively little formal instruction; in efforts at self-education, had assistance of friends and local clergy; 1766-ca. 1769, apprenticed to merchant; 1771, began private study of medicine; attended lectures by J. Winthrop [q.v.] at Harvard. CAREER: For short period, taught school in various places, finally at Concord, N.H., where married wealthy widow; ca. 1772, commissioned major in New Hampshire regiment and took up farming; 1776, went to England as loyalist; entered British Colonial Office; 1780, became under-secretary of state for Northern Department; ca. 1781-1783, in America as officer in British army; 1784, knighted by British king, and entered into military and civil service to elector of Bavaria; 1788, became head of Bavarian war department; 1791, made count of Holy Roman Empire; 1795-1796 and 1798-1802, chiefly in England; thereafter, traveled in Europe; ca. 1804, settled in France. MEMBERSHIPS: Roy. Soc.; Am. Ac. Arts Scis.; and other societies. SCIENTIFIC CONTRIBUTIONS: While in U.S., engaged in scientific study and experimentation, especially on gunpowder, an interest later pursued in England and in Europe; investigations included study of firing vents in cannons and relation of velocity of shot to makeup of gunpowder. Cannon studies led to interest in nature of heat, and through experiments promoted vibratory theory of heat as opposed to theory of caloric fluid; made number of technical innovations, including studies of insulation, devised and installed improvements in heating and cooking, and invented more efficient fireplace; devised calorimeter to gauge heats of combustion of different fuels. Also invented shadow photometer, introduced idea of standard candle, engaged in studies of relation of heat and light. Studied nutrition, an interest that grew out of military duties in Bavaria. Founder of Royal Institution of Great Britain; in 1796 donated money to Roy. Soc. and to Am. Ac. Arts Scis. for prizes in research on heat and light; founded professorship on application of science to useful arts at Harvard Univ.

WORKS: Modern edition of papers in 5 vols., edited by Sanborn C. Brown (Cambridge, Mass., 1968-1970); see *DSB*. MSS: New Hampshire Historical Society (*NUCMC* 73-189); Am. Ac. Arts Scis. (Hamer); Harv. U.-HL (Hamer). WORKS ABOUT: *DAB*, 18: 449-52 (Tenney L. Davis); *DSB*, 13: 350-52 (Sanborn C. Brown). Also see George E. Ellis, *Memoir of Benjamin Thompson, Count Rumford . . .* (Boston, 1871; reprinted, Boston, 1972).

THOMPSON, ZADOCK (May 23, 1796, Bridgewater, Vt.—January 19, 1856, Burlington, Vt.). *Natural history*; *Mathematics*. Son of Capt. Barnabas, farmer, and Sarah (Fuller) Thompson. Married Phebe Boyce, 1824; two children. EDUCATION: 1823 (age twenty-seven), graduated, Univ. of Vermont; later, studied theology. CAREER: Did farm work, taught school, and wrote and sold almanacs; 1824, published gazeteer of Vermont; 1825, appointed tutor, Univ. of Vermont; until 1833, taught and wrote in Burlington and published several textbooks; 1828, edited *Iris* and *Green Mountain Repository;* mid-1830s, taught in Canada, at Hatley and in Sherbrooke; 1835, ordained as deacon in Episcopal church, and thereafter frequently preached in and around Burlington; taught in Episcopal boys' school near Burlington and continued literary work; 1845, appointed assistant geologist on Vermont state survey under C. B. Adams [q.v.]; 1851-1856, professor of natural history and chemistry, Univ. of Vermont; 1853-1856, state naturalist, Vermont. MEMBERSHIPS: Bos. Soc. Nat. Hist. SCIENTIFIC CONTRIBUTIONS: Carried out astronomical computations for own almanacs and for *Walton's Vermont Register and Farmer's Almanack*; during 1820s, published school textbooks on arithmetic and, in 1835, *Geography and History of Lower Canada*. Best-known work was *History* [see Works], planned for number of years; its initial section was devoted to natural history, including list of rocks, fossils, and minerals, descriptions of birds, fishes, amphibians, reptiles, and mammals, and pages on botany by William Oakes; an 1853 appendix, bound with third edition of *History*, dealt mainly with natural history. Had notable natural history cabinet, and descriptions of specimens were based on personal observation; as Vermont state naturalist after 1853, planned volumes on geology, botany, and zoology, but ill health prevented realization of scheme.

WORKS: *History of Vermont, Natural, Civil, and Statistical* (Burlington, Vt., 1842), of which first part was republished as *Natural History of Vermont* (Rutland, Vt., 1972), with introduction by T. D. Seymour Bassett. See bibliography in Perkins, *American Geologist* [below], pp. 70-71; also see Meisel, *Bibliog. Amer. Nat. Hist.* MSS: Vermont Historical Society (*NUCMC* 62-1752). WORKS ABOUT: *DAB*, 18: 480-81 (Henry F. Perkins); George H. Perkins, "Sketch of Life," *American Geologist* 29 (1902): 65-71.

THURBER, GEORGE (September 2, 1821, Providence, R.I.—April 2, 1890, Passaic, N.J.). *Botany*. Son of Jacob, businessman, and Alice Ann (Martin) Thurber. Never married. EDUCATION: Attended Union Classical and Engineering School, Providence; served apprenticeship with Providence pharmacist. HONORS: M.D., New York Medical College, 1859. CAREER: Entered pharmaceutical partnership in Providence; for a time, honorary lecturer on chemistry, Franklin Society of Providence; 1850-ca. 1853, botanist, quartermaster, and commissary, U.S.-Mexico boundary survey; 1853-1856, employed in U.S. Assay Office, New York; lecturer on botany, Cooper Union; 1856-1861 and 1865-1866, lecturer on botany and materia medica, College of Pharmacy, New York; 1859-1863, professor of botany and horticulture, Michigan State Agricultural College (Michigan State Univ.); 1863-1885, editor, *American Agriculturist.* MEMBERSHIPS: President, Torrey Bot. Club, 1873-1880; member, N.Y. Ac. Scis. and other societies. SCIENTIFIC CONTRIBUTIONS: Chief outcome of collecting activities with Mexican boundary survey was paper published by A. Gray [q.v.] [see Works]. As editor of *American Agriculturist,* drew heavily on work done in experimental garden at the Pines, his farm at Passaic, and for number of years prepared "Notes from the Pines" and also contributed elementary science series called "The Doctor's Talks"; wrote relatively few papers under own name; prepared revision of *Agricultural Botany* by W. Darlington [q.v.], published as *American Weeds and Useful Plants* (1859); prepared botanical entries for Appleton's *American Cyclopedia* (1873-1876) and wrote section on grasses for California Geological Survey's *Botany* (1880). Noted as expert on grasses, but never completed projected monograph on that subject; supervised preparation of many agricultural books published by O. Judd [q.v.]. Influential in interesting farmers in study of botany.

WORKS: See above; also Asa Gray, "Plantae Novae Thurberianae" [Mexican boundary survey collection], *Memoirs of Am. Ac. Arts Scis.,* n.s. 5 (1855): 297-328. MSS: Harv. U.-GH (Hamer). WORKS ABOUT: *DAB,* 18: 514-15 (Carl R. Woodward); H. H. Rusby, "Biographical Sketch," *Bulletin of Torrey Bot. Club* 17 (1890): 204-10.

TILGHMAN, RICHARD ALBERT (May 24, 1824, Philadelphia, Pa.—March 24, 1899, Philadelphia, Pa.). *Chemistry.* Son of Benjamin and Anna Maria (McMurtrie) Tilghman. Married Susan Price Toland, 1860; survived by five children. EDUCATION: 1841, B.A., Univ. of Pennsylvania; studied for one year in laboratory of J. C. Booth [q.v.], Philadelphia. CAREER: Engaged in investigations in chemistry, and promoted application of innovative processes in chemical manufacturing; spent some years in Scotland, where industrial opportunities in relation to chemical work were more advanced, including two years at Tennant chemical works, Glasgow; 1852-1855, in Philadelphia, employed in chemical work; 1859, returned to U.S. from England, and thereafter spent some ten years in litigation over disputed discovery of chemical process for decomposition of fats; for two years, worked with brother on problems related to preparation of wood pulp; after 1870, worked with brother on development of technique for sandblasting; served as a director of George Richards and Company, Ltd., machine tool manufacturers, and of Tilghman Sand Blast Co. near Manchester, England; illness necessitated abandonment of work some years before death. MEMBERSHIPS: Am. Phil. Soc. SCIENTIFIC CONTRIBUTIONS: Scientific work and career activities closely interrelated. Earliest studies dealt with technique for production of potassium dichromate, which was taken over by a Baltimore manufacturing concern; discovered action of high temperature steam in decomposing number of inorganic salts; 1847 paper on that subject, read at Am. Phil. Soc. (*Trans.* [1853], pp. 173-76) [see Works], ranked as pioneering investigation of hydration; in Great Britain, promoted method for production of caustic soda based on these studies in hydrolysis; later, developed means for decomposition of fats into fat acids and glycerine, rights in Great Britain being sold to Price Patent Candle Co. Other studies and projects related to manufacture of gas from coal, use of gas in evaporation prccesses, and sulphurous acid decomposition of wood pulp for use in making paper.

WORKS: "On the Decomposing Power of Water at High Temperatures," *Transactions of Am. Phil Soc.,* n.s. 10 (1853): 173-76; "On the Decomposition of the Alkaline Sulphates by Hydrochloric Acid and Chlorine," ibid., pp. 359-62. WORKS ABOUT: *DAB,* 18: 544 (Owen L. Shinn); Isaac J. Wistar, "Memoir," Am. Phil. Soc., *Proceedings: Memorial Volume* 1 (1900): 189-95.

TODD, JAMES EDWARD (1846-1922). *Geology.* Professor of natural science, Tabor College (Iowa); professor of geology and mineralogy, Univ. of South Dakota, and state geologist of South Dakota; assistant professor of geology, Univ. of Kansas.

WORKS ABOUT: See *AMS*; *NCAB,* 10 (1909 [copyright 1900]): 117; *Who Was Who in America,* vol. 1.

TORREY, BRADFORD (October 9, 1843, Weymouth, Mass.—October 7, 1912, Santa Barbara, Calif.). *Ornithology.* Son of Samuel and Sophronia (Dyer) Torrey. Never married. EDUCATION: Ca. 1861 [age eighteen], graduated from high school, Weymouth. CAREER: For short period, employed in local shoe factory; for few years, taught school; thereafter, held positions with two or three business houses, Boston; for sixteen years, worked in office of treasurer, American Board of Commissioners for Foreign Missions, Boston; 1886-1901, an editor, *Youth's Companion,* charged with selection and preparation of miscellany section; 1901, resigned to devote efforts to writing; after 1907, situated in Santa Barbara, Calif. MEMBERSHIPS: Nuttall Ornithological Club; Am. Ornith. Un. SCIENTIFIC CON-

TRIBUTIONS: Became interested in study of birds after moving to Boston, and first article related to birds on Boston Common (*Atlantic Monthy*, 1883); in 1885 published first book, *Birds in the Bush*; spent time in New Hampshire and other parts of New England, and after about 1894 traveled in South and to Arizona and California, and carried out observational studies of bird life. Wrote number of books on birds and on general landscape, based on travels; while written especially for bird lovers, these books combined scientific interest with an aesthetic appreciation of natural world and presented both in style of an accomplished essayist; writings were marked by accuracy, based on close and careful observation, but Torrey was not recognized as academic ornithologist. Also wrote essays for *Atlantic*, *Boston Transcript*, and elsewhere and contributed article entitled "The Booming of the Bittern" to *Auk* (January 1889) and series of notes to that journal (1886-1905); also published two articles in *Bird-Lore*, and during years 1907-1910 published notes in *Condor*. Achieved notable reputation as leading student of hummingbirds. Edited fourteen volumes of journal of Henry David Thoreau (1906).

WORKS: See above; titles of books in *DAB*; also see *Roy. Soc. Cat.*, vol. 19. WORKS ABOUT: *DAB*, 18: 594-95 (Henry S. Chapman); F. H. Allen, obituary, *Auk* 30 (1913): 157-59.

TORREY, JOHN (August 15, 1796, New York City, N.Y.—March 10, 1873, New York City, N.Y.). *Botany*; *Chemistry*. Son of Capt. William, merchant and fiscal agent of state prison, and Margaret (Nichols) Torrey. Married Eliza Shaw, 1824; survived by four children. EDUCATION: 1818, M.D., College of Physicians and Surgeons, New York. HONORS: Name given to Torrey Bot. Club (organized in 1867). CAREER: Ca. 1818, began medical practice, New York City; 1824-1827, professor of chemistry, mineralogy, and geology, U.S. Military Academy, West Point; 1827-1855, professor of chemistry and botany, College of Physicians and Surgeons; 1830-1854, professor of chemistry and natural history, College of New Jersey (Princeton); 1836, appointed New York state botanist; 1853-1873, U.S. assayer, Assay Office in New York; 1865 and 1872, visited California. MEMBERSHIPS: President, Am. Assoc. Advt. Sci., 1855; member, Natl. Ac. Scis. and other societies. SCIENTIFIC CONTRIBUTIONS: Throughout life, chief employment was in chemistry and early work was in mineralogy, but major research and publishing were in botany, an interest aroused in 1810 through contact with A. Eaton [q.v.], who was then in the prison in which Torrey's father was officer. Among earliest botanical works was *Catalogue of Plants . . . Within Thirty Miles of the City of New York* (1819), prepared for committee of N.Y. Lyc. Nat. Hist.; in 1824, published first (and only) volume of *Flora of the Northern and Middle Sections of the U.S.*, which was beginning of project to gather in single work all known information on American botany. Early became dissatisfied with Linnaean system, and was leader in introducing natural classification into U.S. Planned comprehensive work, *Flora of North America* [see Works], to be based on natural system; in 1836, A. Gray [q.v.] became associated with project; between 1838 and 1843 issued nearly all of two volumes, but botanical work for New York state survey and receipt of increasing numbers of specimens collected by various government surveys finally brought work on *Flora* to halt. Through work for government surveys, issued numerous reports on western botany, and gathered herbarium of prime importance, later the foundation of N.Y. Bot. Gard. Had wide influence on American botany and botanists, and relations with Gray were of particular importance.

WORKS: See above; also *A Flora of North America . . . Arranged According to the Natural System*, 2 vols. (New York, 1838-1843), with A. Gray; *Flora of the State of New York*, 2 vols. (Albany, 1843), for state survey. Other major works mentioned in *DAB*; list of works in Andrew D. Rodgers III, *John Torrey* (Princeton, 1942), pp. 316-23. MSS: N.Y. Bot. Gard. (*DSB*); also see additional references in *NUCMC* and Hamer. WORKS ABOUT: *DAB*, 18: 596-98 (John H. Barnhart); *DSB*, 13: 432-33 (A. Hunter Dupree).

TOWNSEND, JOHN KIRK (August 10, 1809, Philadelphia, Pa.—February 6, 1851, Washington, D.C.). *Ornithology*. Son of Charles, watchmaker, and Priscilla (Kirk) Townsend. Married Charlotte Holmes; one child. EDUCATION: Attended Quaker boarding school at Westtown, Pa.; ca. 1845, studied dentistry in Philadelphia. CAREER: Much of early life passed at West Chester, Pa., where engaged in ornithological collection and preparation; 1834, accompanied T. Nuttall [q.v.] on Nathaniel J. Wyeth's overland expedition to Oregon; 1835, visited Hawaiian Islands; 1835-1836, post surgeon at Fort Vancouver; 1836-1837, returned to Philadelphia by way of Cape Horn, again visiting Hawaii, and also stopping at Tahiti and Valparaiso, Chile; 1837-1842, apparently in Philadelphia; 1839-1840 and 1845-1846, a curator, Ac. Nat. Scis. Philad.; 1842, was in Washington and there engaged in curatorial work on bird collections of National Institute, housed in Patent Office; 1845, again in Philadelphia, where pursued study of dentistry, though apparently never practiced; 1851, planned voyage as naturalist on navy ship around Cape of Good Hope, but died before departure. MEMBERSHIPS: Ac. Nat. Scis. Philad. SCIENTIFIC CONTRIBUTIONS: At early age, became accomplished collector and taxidermist; first significant ornithological find, a unique specimen, occurred 1833 in Chester County, and was named Townsend's bunting by J. J. Audubon [q.v.]. Chief contributions grew out of Oregon expedition, during which number of new ornithological specimens were collected, both in western U.S. and in Hawaii. Undertook ambitious project entitled *Ornithology of the United States of North America,* with illustrations, of which a single part (twelve pages and four plates) appeared in 1839; natural history specimens sold to Audubon and others, and chief outcome of ornithological work appeared in Audubon's *Birds of America,* while mammals collected in West appeared in Audubon and J. Bachman [q.v.], *Viviparous Quadrupeds of North America*; considered one of America's best ornithologists, but throughout career was overshadowed by Audubon. Best-known work is *Narrative* [see Works], including catalogue of birds and mammals.

WORKS: See above. Descriptions and list of western birds in *Journal of Ac. Nat. Scis. Philad.* 7 (1837) : 187-93, 8 (1842) : 148-59; *Narrative of a Journey Across the Rocky Mountains to the Columbia River* (Philadelphia, 1839). MSS: Ac. Nat. Scis. Philad. (Hamer). WORKS ABOUT: *DAB,* 18: 617-18 (Witmer Stone) ; Richard G. Beidelman, "John K. Townsend on the Oregon Trail," *Audubon Magazine* 59 (March-April 1957) : 64.

TRACY, CLARISSA TUCKER (November 12, 1818, Jackson, Susquehanna County, Pa.—November 13, 1905, Ripon, Wis.). *Botany.* Daughter of Stephen, pioneer and apparently farmer, and Lucy (Harris) Tucker. Married Horace Hyde Tracy, 1844; two children. EDUCATION: At age three, began attending local schools; 1835-1840, student and teacher, Franklin Academy (Harford, Pa.) ; 1840, became assistant at school in Honesdale and there continued studies; 1844, one term at Troy (N.Y.) Seminary. CAREER: Ca. 1832 [age fourteen], began period of years of teaching while continuing studies; 1835-1840, teacher and student, Franklin Academy; at Ladies' Seminary, Honesdale, Wayne County, Pa., assistant (1840-1842) and head (1842-ca. 1846) ; 1848, widowed; ca. 1849-ca. 1851, operated private school at Honesdale; ca. 1851-1856, associated with academy at Honesdale; 1856-1859, operated private school, Neenah, Wis.; 1859, appointed matron in charge of domestic operations, head of ladies' department, and teacher, Ripon College (Wisconsin) ; 1893, purchased home off campus, continued to engage in tutoring, and maintained affiliation with college until death. SCIENTIFIC CONTRIBUTIONS: Academic duties at Ripon College included teaching of number of subjects, encompassing at times Latin, algebra, arithmetic, English literature, and composition; especially noted for botanical teaching. For nearly thirty years studied flora of region about Ripon, and in 1889 published *Catalogue* [see Works], based on personal collecting or that of students, all specimens having been seen in fresh condition; while the largest list of flora of region up to that time, no claims were made for completeness, and had plans for preparation of supplements.

WORKS: *Catalogue of Plants Growing Without Cultivation in Ripon and the Near Vicinity* ([Ripon, Wis., 1889]), 26 pp. WORKS ABOUT: Victoria Brown, *Uncommon Lives of Common Women* (Madison, Wis., 1975), pp. 22-23; Ada C. Merrell, *Life and Poems of Clarissa Tucker Tracy* (Chicago, 1908).

TRASK, JOHN BOARDMAN (1824, Roxbury, Mass.—July 3, 1879, San Francisco, Calif.). *Geology.* No information on parentage or marital status. EDUCATION: Passed examinations in science and medical subjects and was licensed to practice medicine by Yale Univ. HONORS: M.D., Yale Univ., 1859; also said to have received honorary degrees from Italian and German universities. CAREER: 1850, went to California; said to have been associated with U.S.-Mexico boundary survey; 1853, appointed to prepare geological report for state of California; 1858, cofounder and, for some years thereafter, editor, *Pacific Medical and Surgical Journal*; during Civil War, served as assistant surgeon of volunteers; after 1866, devoted efforts mainly to medical practice. MEMBERSHIPS: A founder, Calif. Ac. Scis. (1853). SCIENTIFIC CONTRIBUTIONS: As first state geologist of California, prepared five reports (three of which appeared as state documents) between 1853 and 1856; first report was based on personal observations made during years 1850-1852; this and subsequent reports were concerned with geology and with agricultural and mineral resources. Read number of papers before Calif. Ac. Scis., published in its *Proceedings.* Was early student of recent and fossil shells of California, wrote on earthquakes of that state and also on microscopic organisms; during early years in West, studied medical flora of Pacific Coast, and introduced several medicinal plants; also introduced to medical practitioners of West a number of chemical and other medical preparations. Foreign honorary degrees reportedly granted for investigations in organic chemistry, mineralogy, microscopy, and medical botany.

WORKS: See Vogdes [below], pp. 28-30. WORKS ABOUT: Anthony W. Vogdes, "A Bibliographical Sketch of Dr. John B. Trask," *Transactions of San Diego Society of Natural History* 1 (1907) : 27-30; Robert E. C. Stearns, "Dr. John B. Trask . . . ," *Science,* n.s. 28 (1908) : 240-43.

TRAUTWINE, JOHN CRESSON (March 30, 1810, Philadelphia, Pa.—September 14, 1883, Philadelphia, Pa.). *Engineering.* Son of William and Sarah (Wilkinson) Trautwine. Married Eliza Ritter, 1838; three children. EDUCATION: Studied with J. P. Espy [q.v.]; 1828, became student in Philadelphia office of William Strickland, civil engineer and architect. CAREER: While associated with Strickland, worked on construction of several public buildings and on Delaware breakwater; 1831, employed on Philadelphia section of Columbia Railroad, and until 1836 engaged in various other projects; 1836-1843, chief engineer, Hiwassee Railroad, from Tennessee to Georgia; 1838-1843, resided in Knoxville; 1843, returned to Philadelphia; 1844-1849, engaged in canal construction, Colombia, South America; 1849, returned to Philadelphia; 1849-1851, involved in railroad survey in Panama; 1851, again in Philadelphia; 1852-ca. 1854, involved in survey for interoceanic canal, Panama; in subsequent years, engaged in various surveying and engineering projects, including survey (1857) for route of interoceanic railroad in Honduras, planning (1858) for dock system for Montreal, and planning (1864) for harbor in Nova Scotia; after 1864, engaged in occasional consulting work only. MEMBERSHIPS: Franklin Institute; Am. Phil. Soc.; and other societies. SCIENTIFIC CONTRIBUTIONS: Noted for work in explorations in Panama, partly with George M. Totten, and prepared first comprehensive and reliable reports on the isthmus, but expressed doubt at feasibility of ever constructing ship canal in that area. Wrote important standard works on engineering, and *Pocket Book* [see Works] earned author international reputation. Also did work as architect, of which Old Dorm at Gettysburg College is extant example. Assembled impressive collection of minerals.

WORKS: *Rough Notes of an Exploration for an Inter-Oceanic Canal Route by Way of the Rivers Atrato and San Juan, in New Granada, South America* (Philadelphia, 1854), 96 pp., with plates and maps; *The Civil Engineers' Pocket Book . . .* (Philadelphia, 1872), with many subsequent editions. Also see *Roy. Soc. Cat.*, vols. 6, 8, 11, for references to several articles appearing in *Journal of Franklin Institute.* MSS: Corn. U.-CRH (Hamer). WORKS ABOUT: *DAB*, 18: 628-29 (F. Lynwood Garrison); obituary, *Journal of Franklin Institute*, whole ser. 116 (1883): 390-96; for works on building designs, see *Journal of Society of Architectural Historians* 17 (1958): 32-33, 18 (1959): 161-63.

TREADWELL, DANIEL (October 10, 1791, Ipswich, Mass—February 27, 1872, Cambridge, Mass.). *Invention*; *Physics*. Son of Capt. Jabez, apparently farmer, and Elizabeth (Dodge) Treadwell. Married Adeline Lincoln, 1831; no children. EDUCATION: Until age fourteen, attended school; ca. 1805, entered into apprenticeship in silversmithing, first with brother and then in Boston; later, for year and a half, studied medicine with Dr. John Ware. HONORS: A.M., Harvard College, 1829. CAREER: Until War of 1812, worked as silversmith; for time, operated screw-making business at Saugus, Mass., employing machine of own invention; thereafter, devised machine for making nails, but soon turned attentions to study of medicine; 1819-1820, made unsuccessful bid to introduce improved hand printing press into England; thereafter, returned to Boston and began manufacture of press; 1823-1826, an editor, *Boston Journal of Philosophy and the Arts*; 1825 and 1837, appointed to study feasibility of water supply for Boston; 1826, patented "Treadwell's Power (Steam) Press," and in 1829 gave up press business, having made sizable profit; 1827, elected vice-president, and later president, Boston Mechanics' Institute, and gave lectures for working classes; 1828-ca. 1836, devoted efforts to development of machinery for spinning hemp, and in 1833 rope-making company merged with larger concern; 1834-1845, Rumford professor on application of science to useful arts, Harvard Univ.; 1835-1836, served on commission to examine state weights and measures; 1842, formed Steel Cannon Co., but met with little success, and last years were clouded by patent controversy concerning cannon invention. MEMBERSHIPS: Am. Ac. Arts Scis. SCIENTIFIC CONTRIBUTIONS: Known especially as inventor; in addition to contributions mentioned above, other inventions included means for operating single-track railroad with system of turnouts; steam press and hemp-spinning machine were widely used. During period of medical study, became acquainted with scientific community in Boston, and edited *Boston Journal* with Dr. Ware and J. W. Webster [qq.v.]; was active in American Academy and a founder of Cambridge Scientific Club (1842). Model for theologian in Longfellow's *Tales of a Wayside Inn.*

WORKS: See *Roy. Soc. Cat.*, vol. 6. MSS: Harv. U.-UA (*NUCMC* 65-1285). WORKS ABOUT: *DAB*, 18: 631-32 (Carl W. Mitman); John van Schaick, "Characters in the Tales of a Wayside Inn," *Christian Leader* 121 (February 25, 1939): 181-83. Also see Morrill Wyman, "Memoir," *Memoirs of Am. Ac. Arts Scis.*, 11 (1882-1888): 325-524, which includes list of his papers and patents (p. 496) and extracts from a number of his letters and journal.

TRELEASE, WILLIAM (1857-1945). *Botany*. Professor of botany, Washington Univ. (St. Louis); director, Mo. Bot. Gard.; professor of botany, Univ. of Illinois.

MSS: Univ. of Illinois Archives (NUCMC 65-1937); Corn. U.-CRH (*NUCMC* 70-1135); Mo. Bot. Gard. (DSB). WORKS ABOUT: See *AMS*; *DAB*, supp. 3; *DSB*.

TROOST, GERARD (March 15, 1776, 's Hertogenbosch, Netherlands—August 14, 1850, Nashville, Tenn.). *Geology; Mineralogy.* Son of Everhard Joseph and Anna Cornelia (van Haeck) Troost. Married Margaret Tage, 1811; a Mrs. O'Reilly, some time after 1819; two children by first marriage. EDUCATION: M.D., Univ. of Leyden; 1801, master of pharmacy, Amsterdam Athenaeum; 1807, sent to Paris by King Louis Napoleon of Holland for further study in science, especially mineralogy. CAREER: For short time, worked as pharmacist, Amsterdam and The Hague; served in army; 1807-1809, collected minerals throughout Europe for king of Holland; studied with Abraham G. Werner; 1809, member Dutch scientific expedition to Java, terminated by action of French pirates; 1810, attempted to reach Java again, by way of U.S.; decided to remain in U.S. and established chemical and pharmaceutical laboratory, Philadelphia; 1811, initiated establishment of first facility in U.S. for producing alum, Cape Sable, Md.; 1821, made professor of mineralogy, Philadelphia Museum; 1821-1822, professor of pharmaceutical and general chemistry, Philadelphia College of Pharmacy; 1821-1825, conducted geological survey of Philadelphia region for Philadelphia Society for Promoting Agriculture; 1825, went with W. Maclure [q.v.] and others to New Harmony, Ind.; 1827, moved to Nashville, Tenn.; 1828-1850, was professor of geology and mineralogy and also taught chemistry and natural philosophy, Univ. of Nashville (George Peabody College for Teachers); 1831-1839, Tennessee state geologist. MEMBERSHIPS: First president, Ac. Nat. Scis. Philad., 1812-1817; member of number of other societies. SCIENTIFIC CONTRIBUTIONS: In 1826, published results of geological studies of area around Philadelphia; while in that city explored areas of New Jersey and New York; and from New Harmony explored Missouri lead mines. Carried number of geological and natural history specimens, including some 400 species of Javanese birds, to Nashville, where collection (with later additions) was opened as public museum. Made early stratigraphic studies in U.S., investigated Tennessee mineral resources and prepared first geological map of state; shortly before death, completed work on crinoid fossils of Tennessee, not published until 1909. Interested in nearly all aspects of science, including Indian archaeology, but especially proficient in chemical and mineralogical studies.

WORKS: Bibliography of works in *American Geologist* 35 (1905): 90-94. WORKS ABOUT: *DAB*, 18: 647-48 (Leonidas C. Green); *DSB*, 13: 466-67 (Ellen J. Moore).

TROWBRIDGE, JOHN (1843-1923). *Physics.* Professor of physics and director of physical laboratory, Harvard Univ.

MSS: See *DSB*. WORKS ABOUT: See *AMS*; *DAB*; *DSB*.

TROWBRIDGE, WILLIAM PETIT (May 25, 1828, Troy, Mich.—August 12, 1892, New Haven, Conn.). *Engineering; Geophysics; Instrumentation.* Son of Stephen Van Rensselaer, state legislator and farmer, and Elizabeth (Conkling) Trowbridge. Married Lucy Parkman, 1857; survived by six children. EDUCATION: 1848, graduated, U.S. Military Academy, West Point. HONORS: Several honorary degrees, including Ph.D., Princeton Univ., 1880. CAREER: During last student year at West Point, taught chemistry; 1848-1850, assistant in observatory, West Point; 1851-1856, assigned to duty with Coast Survey, first in Maine, later in Virginia, and, 1853-1856, on Pacific Coast survey; 1856, resigned from army; 1856-1857, professor of mathematics, Univ. of Michigan; 1857, appointed assistant superintendent, U.S. Coast Survey; 1862-1865, headed army engineer agency, New York City, and also supervised construction or repairs to fortifications in New York harbor; 1865-1871, vice-president, Novelty Iron Works, New York City; 1871-1877, professor of dynamic engineering, Sheffield Scientific School, Yale; 1872-1876, adjutant general of Connecticut; 1877-1892, professor of engineering, School of Mines, Columbia Univ.; at New Haven and New York, served on several engineering commissions. MEMBERSHIPS: N.Y. Ac. Scis.; Natl. Ac. Scis.; and other societies. SCIENTIFIC CONTRIBUTIONS: Duties with Coast Survey included variety of assignments, encompassing observations of Pacific Coast winds, work on arrangement of observations of Gulf Stream, work related to deep-sea soundings and collection of specimens from ocean bottom (including design of apparatus for that purpose); also was charged with installation of self-registering instruments at magnetic station, Key West. During early period of Civil War, performed service related to military requirements of navy, including hydrographic survey of Narragansett Bay. In later career, made significant contributions to engineering education and to development of engineering as applied science; published papers on various subjects in *Proceedings of N.Y. Ac. Scis.* and elsewhere.

WORKS: *Heat as a Source of Power: ... An Introduction to the Study of Heat Engines* (New York, 1874); *Turbine Wheels* (New York, 1879); work for Coast Survey mentioned in annual *Report of Superintendent*, 1857-1861. Also see *Roy. Soc. Cat.*, vols. 6, 11, 19. WORKS ABOUT: *DAB*, 18: 656-57 (George W. Littlehales); C. B. Comstock, "Memoir," *Biog. Mems. of Natl. Ac. Scis.* 3 (1895): 365-67.

TRUE, FREDERICK WILLIAM (1858-1914). *Zoology.* Head curator, Department of Biology, U.S.

Natl. Mus.; assistant secretary in charge of library and exchanges, Smithson. Instn.

MSS: Ac. Nat. Scis. Philad. (Hamer); Smithson. Instn. Archives. WORKS ABOUT: See *AMS*; *DAB*.

TRUMBULL, GURDON (May 5, 1841, Stonington, Conn.—December 28, 1903, Hartford, Conn.). *Ornithology.* Son of Gurdon and Sarah Ann (Swan) Trumbull. Married. EDUCATION: Studied art in Hartford and with James M. Hart in New York. CAREER: Known as artist; particularly noted for paintings of fishes. MEMBERSHIPS: Am. Ornith. Un. SCIENTIFIC CONTRIBUTIONS: Chief paintings of fishes entitled *Over the Fall*, *A Plunge for Life*, and *A Critical Moment*, which were copied and widely sold; a notable smaller painting was that of common sunfish. A lover of nature and devoted sportsman, also much attracted to cause of well-being of lower animals and is said to have written rather extensively on matters of humane interest. During later years, particularly interested in ornithological studies, and wrote several works on subject [see Works], that on woodcock giving first account of a bird's ability to curve its upper mandible; articles on scoters presented results of careful and minute study from recent specimens.

WORKS: *Names and Portraits of Birds Which Interest Gunners, with Descriptions in Languages Understanded of the People* (New York, 1888); "American Woodcock," *Forest and Stream*, December 11, 1890; "Our Scoters," *Auk* 9 (1892): 153-60, 10 (1893): 165-76. WORKS ABOUT: Obituary, *Auk* 21 (1904): 310-11; *Appleton's CAB*, 6: 166-67 (also includes accounts of brothers James Hammond and Henry Clay Trumbull); *Who Was Who in America*, vol. 1; obituary, *New York Times*, December 29, 1903, p. 9.

TRYON, GEORGE WASHINGTON (May 20, 1838, Philadelphia, Pa.—February 5, 1888, Philadelphia, Pa.). *Conchology.* Son of Edward K., manufacturer and dealer of firearms and sportsman's appliances, and Adeline (Savidt) Tryon. Never married. EDUCATION: Attended private schools; 1850-1853, enrolled in Friends' Central School, Philadelphia; for time thereafter, studied languages and music with private tutors. CAREER: 1857, received share in father's business and in 1864 took over its management; 1865-1872, owner and editor, *American Journal of Conchology*; 1868, retired from business and thereafter devoted efforts to scientific and literary interests; at Ac. Nat. Scis. Philad., a curator (1869-1876) and conservator of conchology (1875-1888); edited and published number of librettos from standard and popular operas. MEMBERSHIPS: Ac. Nat. Scis. Philad.; Bos. Soc. Nat. Hist.; and other societies. SCIENTIFIC CONTRIBUTIONS: Began collecting as child, and by 1867 was able to deposit with Ac. Nat. Scis. Philad. more than 10,000 species of shells and in excess of 100 jars of specimens in alcohol, chiefly naked mollusks. First paper on conchology appeared in 1861, and eventually published more than seventy papers on mollusks from land, freshwater, and sea; *Manual* [see Works] was chief publication, an ambitious project intended to encompass descriptions and illustrations of all living mollusks; resulted in nine volumes on marine shells and three on land shells published during his lifetime; this work was continued by Henry A. Pilsbry after Tryon's death. Also had secondary interest in botany. Took leading role in erection of new building for Ac. Nat. Scis. Philad.

WORKS: "Monograph of the Terrestrial Mollusca of the U.S.," *American Journal of Conchology* 2 (1866): 218-77, 306-27, and 4 (1869): 5-22; "Strepomatidae," pt. 4 of W. G. Binney [q.v.], "Land and Fresh-water Shells of North America," *Smithson. Misc. Coll.* 16 (1873); *Structural and Systematic Conchology*, 3 vols. (Philadelphia, 1882-1884); *Manual of Conchology, Structural and Systematic, with Illustrations of the Species* (Philadelphia, 1879-, vol. 1-). See bibliography in Ruschenberger [below], pp. 413-18. MSS: Ac. Nat. Scis. Philad. (*NUCMC* 66-359). WORKS ABOUT: *DAB*, 19: 24-25 (Frank C. Baker); W. S. W. Ruschenberger, "Biographical Notice," *Proceedings of Ac. Nat. Scis. Philad.*, 39 (1888): 399-418.

TUCKERMAN, ALFRED (1848-1925). *Chemistry*; *Bibliography.* Assistant for four years, Astor Library (New York); engaged in work on science bibliographies.

WORKS ABOUT: See *AMS*; *Who Was Who in America*, vol. 1.

TUCKERMAN, EDWARD (December 7, 1817, Boston, Mass.—March 15, 1886, Amherst, Mass.). *Botany.* Son of Edward, merchant, and Sophia (May) Tuckerman. Married Sarah Eliza Sigourney Cushing, 1854; no children. EDUCATION: 1837, A.B., Union College; later earned M.A. there; 1839, LL.B., Harvard Law School; 1841-1842, traveled and studied in Europe, including period at Upsala, Sweden, under lichenologist Elias Fries; 1847, A.B., Harvard College; 1852, completed course of study at Harvard Divinity School. HONORS: LL.D., Amherst College, 1875; name given to Tuckerman Ravine, Mt. Washington, N.H. CAREER: At Amherst College became lecturer (1854), professor of history (1855-1858), and professor of botany (1858-1886). MEMBERSHIPS: Natl. Ac. Scis. and other societies. SCIENTIFIC CONTRIBUTIONS: One of first American botanists to specialize in study of lichens, he became noted authority on that subject; explored extensively in mountains of New England, and in late 1830s, began publishing on lichens of that region; from association with Elias Fries in Sweden, was given introduction to scientific study of lichens and thereafter followed Fries's ideas on classification. In "Synopsis" (1848) [see Works] presented first attempt at systematization of lichens found in temperate areas of American continent; later, extended geographical view, encompassing lichens from California and other parts of American West, the South, Cuba, and foreign areas, including specimens collected by Wilkes Expedition; chief botanical work was *Genera Lichenum* (1872) [see Works]. Also published work on other plant groups, including *Enumeratio Methodica Caricum Quarundam* (1843), on sedges, an early demonstration of abilities in systematic botany; also wrote work on pondweeds (1849) and *Catalogue of Plants Growing Without Cultivation Within Thirty Miles of Amherst College* (1875). Wrote extensively on historical and theological subjects.

WORKS: "A Synopsis of the Lichenes of the Northern United States and British America," *Proceedings of Am. Ac. Arts Scis.* 1 (1846-1848) : 195-285; *Genera Lichenum: An Arrangement of North American Lichens* (Amherst, Mass., 1872). Bibliography in Farlow [below], pp. 26-28. MSS: Brown Univ. Library (*NUCMC* 60-2207); Am. Ant. Soc. (*NUCMC* 62-3178); Harv. U.-GH (Hamer). WORKS ABOUT: *DAB*, 19: 42-44 (William R. Maxon); W. G. Farlow, "Memoir," *Biog. Mems. of Natl. Ac. Scis.* 3 (1895) : 17-28.

TUOMEY, MICHAEL (September 29, 1805, Cork, Ireland—March 30, 1857, Tuscaloosa, Ala.). *Geology.* Son of Thomas, man of mechanical ability, and Nora (Foley) Tuomey, who was descended from noble family. Married Sarah E. Handy, 1837; two children. EDUCATION: Educated at home, and with grandmother while in Ireland; 1835, graduated, Rensselaer Polytechnic Institute. CAREER: Ca. 1822 [age seventeen], began teaching at Yorkshire, England; shortly thereafter, came to Philadelphia; took up farming in Pennsylvania; afterwards, taught at country school in Virginia; for several years, was employed as private tutor in family of John H. Dennis, Maryland; after graduating in 1835, worked as engineer for railroad in North Carolina; ca. 1838, became teacher of mathematics and natural sciences, Miss Mercer's School, Loudoun County, Va.; ca. 1840, with wife, established seminary at Petersburg, Va.; 1844, became state geologist of South Carolina; 1847, appointed professor of geology, mineralogy, and agricultural chemistry, Univ. of Alabama; 1848, appointed state geologist of Alabama, under auspices of university; 1854, assumed duties of Alabama state geologist with legislative appropriation, and for a time resigned duties in university. MEMBERSHIPS: Am. Assoc. Advt. Sci.; Bos. Soc. Nat. Hist. SCIENTIFIC CONTRIBUTIONS: While at Petersburg, Va., collected in geology, mineralogy, and paleontology, entertained Sir Charles Lyell, and corresponded with noted American geologists; in 1843, published discovery of "infusorial stratum" near Petersburg. During geological work in South Carolina, became associated with F. S. Holmes [q.v.] in important study of fossils of that state; published two official reports as South Carolina geologist. Published first report on Alabama geology in 1850 and issued geological map of the state in 1853; second Alabama report appeared in 1858 under editorship of J. W. Mallet [q.v.].

WORKS: *Report on the Geology of South Carolina* (Columbia, S.C., 1848); *Pleiocene Fossils of South Carolina* (Charleston, 1857), with F. S. Holmes. See annotated bibliography in Smith [below], pp. 210-12. WORKS ABOUT: *NCAB*, 13 (1906) : 95; Eugene A. Smith, "Sketch of Life . . . ," *American Geologist* 20 (1897) : 205-12; Merrill, *Hundred Years of Geology*, pp. 239, 264-69.

TUTTLE, CHARLES WESLEY (November 1, 1829, Newfield, Maine—July 17, 1881, Boston, Mass.). *Astronomy.* Son of Moses and Mary (Merrow) Tuttle. Married Mary Louisa Park, 1872; apparently no children. EDUCATION: Attended schools at Newfield and at Dover, N.H.; apprenticed to carpenter; largely self-taught in advanced studies; 1854-1855, studied at Harvard Law School. HONORS: Ph.D., Dartmouth College, 1880. CAREER: 1849, began carpenter's work at Cambridge, Mass.; 1850-1854, assistant, Harvard College Observatory; 1855, went to England on chronometric expedition for U.S. Coast Survey; 1856-1857 and ca. 1859-1881, law practice, Boston; 1857-ca. 1859, law practice, Newburyport, Mass.; 1858, admitted to practice in U.S. Circuit Courts, and in 1861 admitted to U.S. Supreme Court; 1860, appointed a U.S. commissioner; 1874, appointed to take testimony for U.S. Court of *Alabama* Claims. MEMBERSHIPS: Bos. Soc. Nat. Hist. SCIENTIFIC CONTRIBUTIONS: While still in teens, constructed telescope; entered Harvard Observatory as paid student, but soon became second assistant. With G. P. Bond [q.v.] worked on zone observations of fixed stars; also participated in observations of Saturn, and in 1850 made most significant contribution to astronomy by offering explanation of "dusky" ring of Saturn (W. C. Bond, "Observations on . . . Saturn," *Annals of Harvard College Observatory* 2, pt. 1, [1857], especially p. 48); in 1852, undertook comparison of brightness of stars and planets at sea level (Isles of Shoals, N.H.) and on Mt. Washington, but clouds prevented latter observation; in March 1853, discovered telescopic comet, and computed orbit and ephemerides of this and other comets. In 1854, gave up work at observatory because of problems with eyes, but thereafter continued some observational and computing work in astronomy, as well as occasionally lecturing on subject. Also known for later historical writings. Brother, Horace Parnell Tuttle (1839-1893), at U.S. Naval Observatory, discovered "Tuttle's comet" and other comets and asteroids.

WORKS: "Zone Catalogue of 5500 stars . . . Observed During the Years 1852-1853," *Annals of Harvard College Observatory* 1, pt. 2 (1855), with W. C. Bond and G. P. Bond; announcement of new comet in *Astronomical Journal* 3 (1854) : 47-48. See *Roy. Soc. Cat.*, vol. 6. WORKS ABOUT: *DAB*, 19: 74 (John M. Poor) ; John W. Dean, "Memoir," in Charles W. Tuttle, *Capt. Francis Champernowne . . . and Other Historical Papers* (Boston, 1889), pp. 1-59.

TWINING, ALEXANDER CATLIN (July 5, 1801, New Haven, Conn.—November 22, 1884, New Haven, Conn.). *Invention*; *Engineering*; *Astronomy*. Son of Stephen, financial officer of Yale College, and Almira (Catlin) Twining. Married Harriet Amelia Kinsley, 1829; seven children. EDUCATION: 1820, A.B., Yale College; subsequently pursued advanced studies there; for one year, attended Andover Theological Seminary; later, enrolled in civil engineering course, U.S. Military Academy, West Point. CAREER: 1823-1825, tutor, Yale College; at one time, employed on state works of Pennsylvania; 1835-1837, directed survey for Hartford and New Haven Railroad; subsequently, acted as chief or consulting engineer for nearly all railroads out of New Haven and engaged in similar work in Vermont, western New York and Ohio, Illinois, and Michigan; 1839-1848, professor of mathematics and natural philosophy, Middlebury College (Vermont) ; thereafter, resumed engineering practice at New Haven, and devoted particular efforts to inventions, especially development of apparatus for manufacture of ice. MEMBERSHIPS: Am. Assoc. Advt. Sci. SCIENTIFIC CONTRIBUTIONS: As early as 1826, began writing on mathematical and astronomical questions, including papers on meteors, aurora borealis, telescopes, astronomical and optical phenomena, and the like, as well as several mathematical papers. While at West Point, observed meteoric showers of November 1833 [see Works], and independently of D. Olmsted [q.v.] suggested theory of meteors as originating in outer space and thus a part of astronomy rather than terrestrial physics. In 1848, began experiments leading to most significant invention, an ether refrigeration machine; in 1849 filed original caveat with U.S. Patent Office for his ice-making process (patented 1853, patent no. 10,221), and in 1864 and 1871 received extensions on patent; about 1850, built at Cleveland earliest-known facility for large-scale production of ice by vapor refrigeration, but never achieved commercial success.

WORKS: "Investigations Respecting the Meteors of 13 November 1833," *Am. Journ. Sci.* 26 (1834) : 320-52; a brochure entitled *Manufacture of Ice By Mechanical Means on a Commercial Scale (By Steam and Water Power)* (New Haven, 1851). See also *Roy. Soc. Cat.*, vols. 6, 8. WORKS ABOUT: *DAB*, 19: 83-84 (Carl W. Mitman) ; Willis R. Woolrich, *Men Who Created Cold* (New York, 1967), pp. 117-18.

TYSON, PHILIP THOMAS (June 23, 1799, Baltimore, Md.—December 16, 1877, Baltimore, Md.). *Geology*; *Chemistry*. Grandson of Elisha Tyson, manufacturing chemist and philanthropist. Married Rebecca Webster, 1824; no children. EDUCATION: Attended Baltimore public schools. CAREER: No definite information on early career; possibly associated with family chemical and copper mining business; 1849, went to California at outbreak of gold rush; ca. 1858, appointed agricultural chemist for state of Maryland. MEMBERSHIPS: First president, Maryland Academy of Sciences. SCIENTIFIC CONTRIBUTIONS: During period in California, carried out geological investigations and published *Geology and Industrial Resources* [see Works], first issued as U.S. Senate Document (1850) ; one of early geological reports on California after discovery of gold, it took conservative view regarding extensiveness of gold reserves. Two reports as Maryland state agricultural chemist [see Works] related chiefly to agricultural interests, but amounted also to geological survey of state; of particular significance was earliest comprehensive geological map of Maryland, published with first report (1860) ; second report, in 1862, dealt very largely with topics of economic interest; discoverer of fourteen-foot vein of bituminous coal in area of western Maryland, the largest known at that time. Among early works were several papers on Maryland geology and mineralogy, in *Am. Journ. Sci.* (1830) and in *Transactions of Maryland Academy of Sciences.*

WORKS: *Geology and Industrial Resources of California* (Baltimore, 1851) ; Maryland Agricultural Chemist, *Report*, 2 vols. (Annapolis, Md. 1860, 1862). See also Meisel, *Bibliog. Amer. Nat. Hist.* WORKS ABOUT: *NCAB*, 13 (1906) : 543; Merrill, *Hundred Years of Geology*, pp. 290-92.

U

UHLER, PHILIP REESE (1835-1913). *Entomology; Geology.* Librarian, Peabody Institute (Baltimore).
WORKS ABOUT: See *AMS*; *DAB*.

UNDERWOOD, LUCIEN MARCUS (1853-1907). *Botany.* Professor of botany, Columbia Univ.
WORKS ABOUT: See *AMS*; *DAB*.

UPHAM, WARREN (1850-1934). *Geology*; *Archaeology.* Associated with geological surveys of New Hampshire and Minnesota and with U.S. Geol. Surv.; secretary, librarian, and archaeologist, Minnesota Historical Society.
WORKS ABOUT: See *AMS*; *DAB*.

V

VAN HISE, CHARLES RICHARD (1857-1918). *Geology.* Professor of metallurgy and geology and president, Univ. of Wisconsin; associated with U.S. Geol. Surv.

MSS: St. Hist. Soc. Wis. (*NUCMC* 62-2894). WORKS ABOUT: See *AMS*; *DAB*; *DSB.*

VAN RENSSELAER, JEREMIAH (August 4, 1793, Greenbush, Rensselaer County, N.Y.—March 7, 1871, New York, N.Y.). *Natural history,* especially *Geology.* Son of Gen. John Jeremiah and Catherine (Glen) Van Rensselaer. Married Charlotte Willis Foster, 1822; at least one child. EDUCATION: 1813, A.B., Yale College; studied medicine with uncle, Archibald Bruce [q.v.], New York City; 1819, said to have received M.D. degree; for three years, studied medicine in Edinburgh, London, and Paris. HONORS: M.D., Vermont Academy of Medicine, 1823. CAREER: After completion of studies, established large medical practice, New York City; 1825, presented course of lectures on geology, New York Athenaeum; 1840-ca. 1843, visited Europe; 1852, retired from medical practice and other activities, and turned to management of his estates at Greenbush; 1867-1870, again visited Europe. MEMBERSHIPS: N.Y. Lyc. Nat. Hist. and other societies. SCIENTIFIC CONTRIBUTIONS: Active member of N.Y. Lyc. Nat. Hist. during its early days, and participated in digging up bones of mastodon near Monmouth, N.J., subsequently presented to lyceum. One of earliest Americans to reach summit of Mont Blanc, and in 1820 published account of climb in *Am. Journ. Sci.*; wrote several papers on fossils; also wrote on natural history of the ocean and of Orange County (N.Y.), lightning rods, and other topics.

WORKS: *An Essay on Salt* (New York, 1823); *Lectures on Geology . . . Delivered in the New York Athenaeum* (New York, 1825). See Meisel, *Bibliog. Amer. Nat. Hist.* WORKS ABOUT: *NCAB*, 7 (1897 [copyright 1892]): 525; Kelly and Burrage, *Med. Biog.* (1920); *Obituary Record of Graduates of Yale University* (1870-1880), p. 8.

VANUXEM, LARDNER (July 23, 1792, Philadelphia, Pa.—January 25, 1848, Bristol, Pa.). *Geology.* Son of James, successful shipping merchant, and Rebecca (Clarke) Vanuxem. Married Elizabeth Newbold, ca. 1830; had children. EDUCATION: 1819, after three years' study, graduated from Ecole des Mines, Paris. CAREER: Until ca. 1816 [age twenty-four], employed in father's mercantile business; 1821-1827, professor of chemistry, mineralogy, and geology, South Carolina College (Univ. of South Carolina); 1827, entered upon career as geologist, first engaging in examination of mines in Mexico, and later carrying out geological survey work in New Jersey, New York, Ohio, Kentucky, Tennessee, and Virginia; 1830, acquired farm at Bristol, Pa., and made home there for remainder of life; 1836, appointed geologist with New York Natural History Survey, first associated with Fourth District and soon thereafter becoming head of Third District; ca. 1842, worked for short time with J. Hall [q.v.], arranging New York state geological collections at Albany; thereafter, retired from public duty; apparently had independent financial resources. MEMBERSHIPS: Ac. Nat. Scis. Philad. and other societies. SCIENTIFIC CONTRIBUTIONS: While in South Carolina, engaged in geological survey work in that state and in North Carolina and published results in newspapers and in Robert Mills, *Statistics of South Carolina* (1826); as early as 1829, argued in favor of fossils as primary aids in geological classification, relegating mineral content and rock position to secondary consideration; in study of geology along Atlantic Coast, by 1828 was able to suggest correlation of strata in that region with Cretaceous of Europe, based on fossil evidence. Most important geological work was on New York state survey, assigned to district of central New York; in addition to high quality of work in general, made significant contributions in relation to classification of New York strata, which later became standard for eastern U.S. Assembled most extensive and important private collection of mineral and geological specimens in country at the time.

WORKS: *Geology of New York; Part 3: Comprising the Survey of the Third Geological District* (New York, 1842). See also Meisel, *Bibliog. Amer. Nat. Hist.* MSS: Ac. Nat. Scis. Philad. (Hamer). WORKS ABOUT: *DAB*, 19: 218 (F. Lynwood Garrison); *DSB*, 13: 581 (John W. Wells).

VASEY, GEORGE (February 28, 1822, near Scarborough, England—March 4, 1893, Washington, D.C.). *Botany.* No information on parentage; family one of limited means. Married Miss Scott, 1846; Mrs. John W. Cameron, 1867; survived by six children. EDUCATION: As infant, settled with parents at Oriskany, Oneida County, N.Y.; at age twelve, left school to work in village store; later, graduated from Oneida Institute; 1846, M.D., Berkshire Medical Institute (Pittsfield, Mass.). HONORS: M.A. [honorary?], Illinois Wesleyan Univ. CAREER: Ca. 1846, began medical practice, Dexter, N.Y.; 1848-1866, medical practice at Elgin and at Ringwood, in northern Illinois; 1866, moved to southern Illinois because of wife's health; 1868, went as botanist with J. W. Powell [q.v.] expedition to Colorado; ca. 1869, became curator, natural history museum, State Normal Univ. of Illinois; 1870, worked with C. V. Riley [q.v.] as editor of *American Entomologist and Botanist*; 1872-1893, botanist, U.S. Department of Agriculture, and curator of U.S. National Herbarium; 1892, represented Department of Agriculture and Smithson. Instn., and served as one of vice-presidents, at Botanical Congress at Genoa. MEMBERSHIPS: Am. Assoc. Advt. Sci.; Am. Ac. Arts Scis.; and other societies. SCIENTIFIC CONTRIBUTIONS: At early age, became actively interested in botany, began correspondence with leading American botanists, and collected extensively in New York state and later in Illinois; during early years had special interest in genus Carex (sedges), and botanical contacts included C. Dewey [q.v.], leading American caricographer. One of chief labors became enlargement and organization of National Herbarium; in that capacity developed familiarity with botanical specimens gathered by number of expeditions in American West and elsewhere. Chief publications, not begun until after going to Washington (1872), related to grasses and forage plants, and also promoted research regarding suitable grasses for arid West.

WORKS: *Agricultural Grasses of the U.S.* (Washington, D.C., 1884); *Illustrations of North American Grasses*, vol. 1, *Grasses of the Southwest*, and vol. 2, *Grasses of the Pacific Slope* (Washington, D.C., 1891-1893); *Monograph of the Grasses of the U.S. and British America* (Washington, D.C., 1892). See also several other works mentioned in *DAB*; also bibliography in Canby and Rose [below], pp. 176-83. MSS: A few items in Smithson. Instn. Arch. WORKS ABOUT: DAB, 19: 229-30 (William R. Maxon); William M. Canby and J. N. Rose, "Biographical Sketch," *Botanical Gazette* 18 (May, 1893): 170-83.

VAUGHAN, DANIEL (ca. 1818, Glenomara, near Killaloe, County Clare, Ireland—April 6, 1879, Cincinnati, Ohio). *Astronomy*; *Mathematics*; *Physiology.* Son of John Vaughan [or Vaughn], reportedly man of wealth. Never married. EDUCATION: Early instruction by private tutor; later attended Killaloe Classical Academy. CAREER: 1840, came to U.S. and traveled in Virginia and other parts of South; 1842, became tutor in Bourbon County, Ky., and soon thereafter expanded activities to become teacher of neighborhood school; several years later, appointed professor of Greek in college at Bardstown, Ky.; 1850, moved to Cincinnati, and in that city lectured on chemistry at Eclectic Medical Institute; for some years, lectured on astronomy and other scientific subjects at teachers' institutes, schools, academies, and colleges in Cincinnati region; 1860-1872, professor of chemistry, Cincinnati College of Medicine and Surgery; for time after 1872, continued to lecture in Kentucky; later returned to Cincinnati and died forgotten and poverty-stricken. MEMBERSHIPS: Am. Assoc. Advt. Sci. SCIENTIFIC CONTRIBUTIONS: Originally intended for priesthood, came to U.S. in search of opportunities involving study in higher mathematics; eventually published works on wide range of scientific and mathematical topics, especially on astronomy and related subjects, but also on earthquakes, volcanoes, various aspects of geology, meteorology, physics of the earth, electricity and magnetism, and other subjects; in 1850, published on experiments in physiology [see Works]. Also in early 1850s, studied problem of rings of Saturn [See Works, "Stability," 1852]; published several volumes of lectures on astronomy, in addition to scientific papers in American and foreign journals, and corresponded with scientists of Europe; apparently did not have access to telescope, and yet mathematical abilities made possible contributions of considerable insight and originality in domain of physical astronomy.

WORKS: "Chemical Researches in Animal and Vegetable Physiology," *Eclectic Medical Journal*, December 1850; "The Stability of Satellites Revolving in Small Orbits," *Proceedings of Am. Assoc. Advt. Sci.* 10 (1856): 111-13; *Popular Physical Astronomy* (Cincinnati, 1858). See *Roy. Soc. Cat.*, vol. 6 (also 6, additions), 8, 11. WORKS ABOUT: *DAB*,

19: 235-36 (DeLisle Stewart); *Popular Science Monthly* 15 (1879): 127-29, 556; ibid. 16 (1880): 125.

VENABLE, CHARLES SCOTT (April 19, 1827, "Longwood," Prince Edward County, Va.—August 11, 1900, Charlottesville, Va.). *Mathematics.* Son of Nathaniel E.—merchant, farmer, and legislator—and Mary Embra (Scott) Venable. Married Margaret Cantey McDowell, 1856; Mrs. Mary (Southall) Brown, 1876; five children by first marriage, one by second. EDUCATION: 1842, A.B., Hampden-Sydney College; 1845-1846 and 1847-1848, studied at Univ. of Virginia; 1852-1853, studied mathematics and astronomy, Berlin and Bonn. CAREER: At Hampden-Sydney College, tutor in mathematics (1843-1845) and professor of mathematics (1846-1856); 1856, appointed professor of natural philosophy, Univ. of Georgia; 1857-1860, professor of mathematics, South Carolina College (Univ. of South Carolina); 1860, appointed to commission to conduct eclipse observations in Labrador; 1861-1865, took part in military activities of Confederacy, from 1862 to 1865 as aide to Gen. Robert E. Lee, and achieved rank of lieutenant colonel; at Univ. of Virginia, professor of mathematics (1865-1896) and chairman of faculty (1870-1873 and 1886-1888); chairman of trustees, Miller Manual Labor School, Albemarle County, Va. SCIENTIFIC CONTRIBUTIONS: Noted especially as teacher of mathematics, and for promotion of science education at Univ. of Virginia. Motivated by realization of importance of applied sciences in well-being of South, he was largely responsible for establishment in 1867 of schools of applied chemistry and engineering at Univ. of Virginia; later, also played central role in beginnings of schools of practical astronomy, of biology and agriculture, and of natural history and geology in the university. Not an original contributor to mathematics, and a projected series of textbooks on college mathematics was not completed, although he did publish volumes on arithmetic and elementary algebra; of planned series on advanced mathematics, only translation and adaptation of Legendre's *Elements of Geometry* was published (1875).

WORKS: "Report of Prof. C. S. Venable on the Total Eclipse of July 18, 1860," *Annual Report of U.S. Coast Survey, 1860* (Washington, D.C. 1861), pp. 255-61. See *L. C. Cat. Prd. Cards 1942.* MSS: Univ. N.C.-SHC (*NUCMC* 60-707); Univ. Tex.-TA (*NUCMC* 71-456). WORKS ABOUT: *DAB*, 19: 245-46 (Charles W. Dabney); William M. Thornton, *Charles Scott Venable* ([Charlottesville, Va.?, 1901]), 15 pp.

VENABLE, FRANCIS PRESTON (1856-1934). *Chemistry.* Professor of chemistry and president, Univ. of North Carolina.

WORKS ABOUT: See *AMS*; *DAB*.

VERRILL, ADDISON EMERY (1839-1926). *Zoology.* Professor of zoology, Yale Univ.; curator of zoology department, Peabody Museum, Yale.

WORKS ABOUT: See *AMS*; *DAB*; *DSB*.

W

WACHSMUTH, CHARLES (September 13, 1829, Hanover, Germany—February 7, 1896, Burlington, Iowa). *Paleontology.* Son of Christian Wachsmuth, lawyer and German legislator. Married Bernandina Lorenz, 1855; no information on children. EDUCATION: Under father's direction, prepared for legal career until forced to abandon studies because of ill health. CAREER: Became associated with mercantile establishment in Hamburg, Germany, and in 1852 went to New York as company's agent; ca. 1854-1865, conducted grocery business at Burlington, Iowa; 1865, retired from commercial activities and thereafter, at Burlington, devoted efforts to paleontological studies and collecting; 1873, worked on collections with J. L. R. Agassiz [q.v.] at Cambridge, Mass. MEMBERSHIPS: Am. Assoc. Advt. Sci. and other societies. SCIENTIFIC CONTRIBUTIONS: At Burlington, began collection of fossils in attempt to restore failing health; during lifetime accumulated collection of rare crinoids from around the world, as well as an extensive library on the subject, and won international reputation as authority on these specimens; published earliest important descriptions of fossil finds in reports of Illinois Geological Survey; in 1873 collection was bought by Agassiz for Harvard. Thereafter, Wachsmuth began creation of second and enlarged collection, and with Frank Springer undertook project to examine and update descriptions of all known crinoids; result of this partnership was *Memoir* [see Works], with atlas of some 1,500 illustrations; Springer worked on the undertaking for some two decades after Wachsmuth's death.

WORKS: "Revision of the Palaeocrinoidea," *Proceedings of Ac. Nat. Scis. Philad.* for *1879, 1881, 1885, 1886*; and *The North American Crinoidea Camerata*, Museum of Comparative Zoology *Memoirs* 20 and 21 (Cambridge, Mass., 1897), both works with F. Springer. See list of chief works in *DAB*; bibliography in Calvin [below], p. 376. WORKS ABOUT: *DAB*, 19: 297-98 (Charles R. Keyes); Samuel Calvin, "Memoir," *Bulletin of Geol. Soc. Am.* 8 (1896-1897) : 374-76.

WADSWORTH, MARSHMAN EDWARD (1847-1921). *Geology.* President and professor of mineralogy, petrography, and geology, Michigan College of Mines; dean of school of mines and metallurgy and professor of mining and geology, Pennsylvania State Univ.; professor of mining geology and dean of School of Mines, Univ. of Pittsburgh.

WORKS ABOUT: See *AMS*; *NCAB*, 13 (1906) : 538; *World Who's Who in Science.*

WAGNER, WILLIAM (January 15, 1796, Philadelphia, Pa.—January 17, 1885, Philadelphia, Pa.). *Science benefactor*; *Paleontology.* Son of John, cloth merchant and importer, and Mary (Ritz) Baker Wagner. Married Caroline M. Say, 1824; Louisa Binney, 1841; no information on children. EDUCATION: 1808, graduated from Philadelphia Academy; ca. 1813, apprenticed to Philadelphia merchant and financier Stephen Girard, while continuing study of languages and mathematics. CAREER: During affiliation with Stephen Girard, in 1814 had charge of convoy carrying goods for safe storage at Reading, Pa., and from 1816 to 1818 served as assistant supercargo on Girard ship to Far East; 1818, ended employment with Girard and commenced independent business career; 1833, suffered substantial financial losses in mining of anthracite coal; 1840, having achieved business success, retired to country estate and pursued scientific studies, accumulating museum collection; 1847, began giving free lectures in home on geology, mineralogy, and conchology, and in 1852 moved lectures to larger quarters in public building; 1855, Wagner Free Institute of Science chartered by state of Pennsylvania, with Wagner himself teaching geological sciences; 1864, institute incorporated with board of trustees, Wagner serving until death as president; 1865, dedicated new building. SCIENTIFIC CONTRIBUTIONS: Chief contribution was founding and endowment of Wagner Free Institute of Science, which was granted right to confer degrees under terms of 1864 incorporation, and to which founder gave at least one-half million dollars in addition to museum collections; institute had faculty in all science fields, and lecture courses were free of charge. Wagner was interested in geology and mineralogy, and during youthful voyage of two years to Far East, started accumulation of minerals, shells, plants, and fossils, the beginnings of museum about which free institute was built; scientific work done chiefly in paleontology [see Works].

WORKS: "Descriptions of Five New Fossils, of the Older Pliocene Formation of Maryland and North Carolina," *Journal of Ac. Nat. Scis. Philad.* 8 (1838): 51-53, with one plate; also W. H. Dall, "Notes on the Paleontological Publications of Professor William Wagner," *Transactions of Wagner Free Institute of Science* 5 [pt. 2] (1898): 7-11, reproduces three plates prepared by Wagner about 1839, with Dall's notes. MSS: Wagner Free Institute of Science (*NUCMC* 66-1503). WORKS ABOUT: *DAB*, 19: 313-14 (Joseph Jackson).

WAILES, BENJAMIN LEONARD COVINGTON (August 1, 1797, Columbia County, Ga.—November 16, 1862, Washington [?], Miss.). *Natural history.* Son of Levin, surveyor and land agent, and Eleanor (Davis) Wailes. Married Rebecca S. M. Covington, 1820; ten children. EDUCATION: Attended Jefferson College (Washington, Miss. Territory). CAREER: 1814-1820, engaged in land surveying and held clerical positions in land offices, especially in Mississippi Territory and in Louisiana, and also served as assistant to Indian agent; after 1820, settled near, and later in, Washington, Miss., engaged chiefly in cotton planting and in managing personal property and plantations owned by family; at Jefferson College, trustee 1824-1862 (at death, president of board), librarian (1829-1836), and treasurer (1837-1854); 1825-1826, member of Mississippi state legislature; 1826-1835, register of land office, Washington, Miss.; 1834, elected a director, Agricultural Bank of Natchez; ca. 1835, cashier, West Feliciana Railroad Co.; 1852-1854, assistant professor of agriculture and geological sciences, Univ. of Mississippi, assigned to state survey. MEMBERSHIPS: Am. Assoc. Advt. Sci.; Ac. Nat. Scis. Philad. SCIENTIFIC CONTRIBUTIONS: In surveying and other activities achieved intimate knowledge of physical geography of old Southwest; amassed large personal collection of specimens of rocks, shells, fossils (including mammalian forms), animals, plants, and Indian relics, and contributed to other local collections, to Smithson. Instn., and to other naturalists; duties as assistant professor in university consisted of field work for biological and geological survey, and after resignation of director of survey, Wailes wrote report [see Works]. Best work done in accumulation of data and of specimens; not generally well informed regarding theoretical work, and virtues of state survey report were the publication of factual information rather than interpretation. From 1839 to 1843, president of Agricultural, Horticultural and Botanical Society of Jefferson College. Left manuscript on Indian archaeology.

WORKS: *Report on the Agriculture and Geology of Mississippi, Embracing a Sketch of the Social and Natural History of the State* ([Philadelphia], 1854), 371 pp. See Meisel, *Bibliog. Amer. Nat. Hist.* MSS: Mississippi Department of Archives and History (*NUCMC* 60-1548); Duke Univ. Library (*NUCMC* 63-132); also see Sydnor, *Gentleman* [below], pp. 307-10. WORKS ABOUT: *DAB*, 19: 315-16 (Charles S. Sydnor); Sydnor, *A Gentleman of the Old Natchez Region: B. C. L. Wailes* (Durham, N.C., 1938).

WALCOTT, CHARLES DOOLITTLE (1850-1927). *Geology*; *Paleontology*. Director, U.S. Geol. Surv.; secretary, Smithson. Instn.

MSS: Smithson. Instn. Archives (*NUCMC* 72-1260). WORKS ABOUT: See *AMS*; *DAB*; *DSB*.

WALKER, SEARS COOK (March 23, 1805, Wilmington, Mass.—January 30, 1853, Cincinnati, Ohio). *Astronomy*; *Mathematics*. Son of Benjamin, farmer, and Susanna (Cook) Walker; half-brother of E. O. Kendall [q.v.]. Apparently never married. EDUCATION: 1825, A.B., Harvard College; later studied medicine in Philadelphia. CAREER: 1825-1827, teacher in Boston area; 1827-1836, teacher in Philadelphia; 1836-1845, actuary to Pennsylvania Company for Insurance on Lives and Granting Annuities; ca. 1845, through investment and commercial losses, was left financially destitute; 1845-1847, held appointment at U.S. Naval Observatory, Washington; 1847-1853, director of longitude computations, office of U.S. Coast Survey; 1852, spent some time in mental asylum, and then went to Cincinnati to stay with brother. MEMBERSHIPS: Am. Phil. Soc. and other societies. SCIENTIFIC CONTRIBUTIONS: After about 1836, devoted spare time to astronomical observations and study, and in 1837 planned and helped establish one of first astronomical observatories in U.S., at Philadelphia Central High School; with these instruments, the first to be imported from Germany, made and published number of observations, and in 1843 published important paper on meteors [see Works]. At Naval Observatory undertook studies relating to Neptune; in early 1847 (some four months after planet's discovery) announced identity of Neptune and star observed by Lalande in 1795, and with this information was able to compute planet's orbit. Of equal importance with Neptune studies was role in devising and perfecting so-called American method for longitude determination, involving use of telegraph; among Walker's contributions to this technique (carried out with advice of A. D. Bache [q.v.]) were the introduction of telegraphing of transits of stars (first used in 1846) and the application of graphic means for registering intervals of time with time of observations; this latter apparatus was shown to have general usefulness in astronomical work, in addition to its value for longitude determinations.

WORKS: "Researches Concerning the Periodical Meteors of August and November," *Transactions of Am. Phil. Soc.*, n.s. 8 (1843): 87-140; "Researches Relative to the Planet Neptune," *Smithson. Contr. Knowl.* 2 (1851). See *Roy. Soc. Cat.*, vol. 6, which lists forty papers. MSS: Commonplace book in Univ. of Pennsylvania Library—Rare Book Collection. WORKS ABOUT: *DAB*, 19: 359-60 (George W. Littlehales); Benjamin A. Gould, Jr., commemorative address, *Proceedings of Am. Assoc. Advt. Sci.* 8 (1854): 18-45.

WALLER, ELWYN (1846-1919). *Chemistry*. Professor of analytical chemistry, School of Mines, Columbia Univ.; analytical and consulting chemist, New York City.

WORKS ABOUT: See *AMS*; *NCAB*, 13 (1906): 344; *Who Was Who in America*, vol. 1.

WALSH, BENJAMIN DANN (September 21, 1808, Clapton, London, England—November 18, 1869, Rock Island, Ill.). *Entomology*. Son of Benjamin Walsh; from family of some wealth. Married Rebecca Finn, ca. 1837; no children. EDUCATION: A.B. in 1831 and A.M. in 1834, Trinity College, Cambridge. CAREER: For several years in England, pursued writing interests, published separate works, and wrote for *Blackwood's Magazine*; 1838, came to U.S. and took up farming near Cambridge, Henry County, Ill.; 1850, moved to Rock Island, Ill., and there established lumber business in 1851; 1858, retired from business, gaining some income from real estate properties, and thereafter devoted chief efforts to entomology; ca. 1865-1867, associate editor and then editor, *Practical Entomologist* (established by Entomological Society of Philadelphia); 1867, appointed Illinois state entomologist; 1868, with C. V. Riley [q.v.], established *American Entomologist*. MEMBERSHIPS: Am. Assoc. Advt. Sci. SCIENTIFIC CONTRIBUTIONS: After retirement in 1858, began studies in entomology and thereafter contributed to agricultural newspapers in effort to acquaint farmers with knowledge of insect pests; wrote alone or as coauthor over 800 items, many of them brief notes in *American Entomologist*. Among significant contributions were his demonstration that planting practices of American farmers encouraged growth of insect population and his early advocacy of importation of foreign predators for control of imported pests. Efforts in economic entomology led to appointment as second state entomologist (after A. Fitch, [q.v.]); with urging of farmers and horticulturists, began this work before appointment was confirmed by legislature, and published a single report in 1867 *Transactions of Illinois State Horticultural Society*; in addition to papers in agricultural press, published about dozen scientific articles in *Proceedings of Bos. Soc. Nat. Hist.* and *Transactions of American Entomological Society* (1862-1866). Cambridge Univ. classmate of Charles Darwin and early advocate of his ideas on evolution. Collection of some 10,000 species destroyed in Chicago fire of 1871.

WORKS: Samuel Henshaw, *Bibliography of the More Important Contributions to American Economic Entomology* . . . (Washington, D.C., 1889), pp. 1-95. MSS: Smithson. Instn. Archives (*NUCMC* 74-983); letters from Darwin in Am. Phil Soc. Library; correspondence with H. Hagen [q.v.] in Harv.U.-MCZ; most papers destroyed in Chicago fire. WORKS ABOUT: *DAB*, 19: 388-89 (Leland O. Howard); Mallis, *Amer. Entomologists*, pp. 43-48.

WALTER, THOMAS (ca. 1740, Hampshire, England—January 17, 1789, [Berkeley County?], South Carolina). *Botany*. No information on parentage. Married Anne Lesesne, 1769; Ann Peyre, 1777; Dorothy Cooper, after 1780; three [four?] children by second marriage, one by third. EDUCATION: Although details are not known, apparently enjoyed sound and thorough educational preparation. CAREER: As young man, emigrated to South Carolina, came into possession of plantation on Santee River, probably in St. John's Parish (Berkeley County), and remained there for rest of life. SCIENTIFIC CONTRIBUTIONS: Undertook collection and description of flowering plants within fifty-mile radius of South Carolina home; manuscript catalogue dated 1787 was taken to England by botanical collector and friend, John Fraser, and was published at latter's expense under title *Flora Caroliniana*; this work, in Latin, included abbreviated descriptions of some 1,000 species constituting 435 genera and is distinguished as first reasonably complete flora covering a locality in eastern North America and using Linnaean binomial system of nomenclature; *Flora* included in excess of 200 new species and was based on Walter's extensive herbarium as well as Fraser's collections, both of which were taken to England along with manuscript; Walter herbarium, in state of some neglect, was acquired in 1863 by British Museum. Maintained botanical garden on estate, and there reportedly cultivated many species described in *Flora*; with Fraser, made unsuccessful attempt to introduce a Carolina grass into England. Published no work other than *Flora*.

WORKS: *Flora Caroliniana* (London, 1788), 263 pp. WORKS ABOUT: *DAB*, 19: 396-97 (William R. Maxon); Maxon, "Thomas Walter, Botanist," *Smithson. Misc. Coll.* 95, no. 8 (1936).

WARD, HENRY AUGUSTUS (1834-1906). *Natural science; Meteorites.* Founder and president, Ward's Natural Science Establishment (Rochester, N.Y.).

MSS: Univ. of Rochester Library (*NUCMC* 61-1278). WORKS ABOUT: See *AMS*; *DAB*; Roswell Ward, *Henry Augustus Ward: Museum Builder to America* (Rochester, N.Y., 1948).

WARD, LESTER FRANK (1841-1913). *Botany; Sociology.* Worked for U.S. Treasury Department; geologist and paleontologist, U.S. Geol. Surv.; professor of sociology, Brown Univ.

MSS: Lib. Cong. (*NUCMC* 62-4525); Brown Univ. Library (Hamer). WORKS ABOUT: See *AMS*; *DAB*.

WARDER, JOHN ASTON (January 19, 1812, Philadelphia, Pa.—July 14, 1883, North Bend, Ohio). *Forestry; Horticulture.* Son of Jeremiah, merchant and farmer, and Ann (Aston) Warder. Married Elizabeth Bowne Haines, 1836; seven children, including Robert B. Warder [q.v.]. EDUCATION: 1836, M.D., Jefferson Medical College (Philadelphia). CAREER: 1837-1855, medical practice, Cincinnati, Ohio; 1850-1853, editor, *Western Horticultural Review*; 1854, co-editor, *Horticultural Review and Botanical Magazine*; 1855, took up residence on farm at North Bend, Ohio; 1871-1876, member, Ohio State Board of Agriculture; 1873, appointed U.S. commissioner for forestry to Vienna World's Fair. MEMBERSHIPS: Am. Assoc. Advt. Sci. and other scientific and agricultural societies. SCIENTIFIC CONTRIBUTIONS: Early career devoted especially to medical practice, and also published one or two items on that subject. After giving up medicine, moved to farm and there carried out number of experiments on varieties and cultivation of fruits and contributed articles to horticultural and agricultural publications; some articles were prepared for, and published in annual reports of, U.S. Patent Office's division of agriculture; published book on apples (1867) and *Hedges and Evergreens* (1857). During final fifteen years of life was especially devoted to forestry; became leader in that field and served from 1875 to 1882 as first president of American Forestry Association; as commissioner to Vienna exposition prepared report on forests and forestry (1876), and also contributed articles to *American Journal of Forestry* during period of its publication, 1882-1883 [*Journal* was official publication of American Forestry Congress, which absorbed Forestry Association]; at time of death, was engaged in study of forestry of Northwest for U.S. Department of Agriculture.

WORKS: See Banks [below], p. 12. MSS: Banks makes reference to manuscripts [in private hands?]. WORKS ABOUT: *DAB*, 19: 444-45 (Russell H. Anderson); Laura S. V. Banks, "John Aston Warder," *American Forests* 73 (November 1967): 10-13, 66, 68.

WARDER, ROBERT BOWNE (March 28, 1848, Cincinnati, Ohio—July 23, 1905, Washington, D.C.). *Chemistry.* Son of John A. [q.v.]—physician, farmer, and forester—and Elizabeth Bowne (Haines) Warder. Married Gulielma M. Dorland, 1884; no information on children. EDUCATION: 1866, A.B., Earlham College (Indiana); 1874, S.B., Lawrence Scientific School, Harvard; 1874-1875, studied chemistry in Germany, at Giessen and Berlin. CAREER: 1866-1867, high school teacher, Mooresville, Ind.; 1869-1871, assistant in chemistry, Univ. of Illinois; 1871, associated with Ohio Geological Survey; 1872, assistant on Indiana Geological Survey; 1875-1879, professor of chemistry and physics, Univ. of Cincinnati; 1879-1880, professor of chemistry and physics, Haverford College; 1883-1887, professor of chemistry, Purdue Univ., and also Indiana state chemist; 1887-1905, professor of chemistry, Howard Univ. (Washington, D.C.). MEMBERSHIPS: German and American chemical societies; Am. Assoc. Advt. Sci.; and other societies. SCIENTIFIC CONTRIBUTIONS: During studies in Germany was especially interested in methods of teaching chemistry and in applications of chemistry in practical scientific work, and throughout career self-improvement as teacher was major concern. Early became interested in interrelations of physics and chemistry, and chief research was in physical chemistry, including early works on speed of saponification and atomic motion in liquid molecules; later wrote on theory of albuminoid ammonia, theories of geometric isomerism, and other topics. Especially noted as critic of work of other chemists, and later writings related particularly to studies of velocity of chemical reactions, using data generated by other researchers.

WORKS: "Evidence of Atomic Motion Within Liquid Molecules, as Based Upon the Speed of Chemical Action," *American Chemical Journal* 3 (1881-1882): 294-95; "The Speed of Saponification of Ethyl Acetate," ibid., pp. 340-49; "Dynamical Theory of Albuminoid Ammonia," ibid. 11 (1889): 365-78; "The Major Premise in Physical Chemistry," *Science* n.s. 2 (1895): 651-54; "Speed of Esterification, as Compared with Theory," *Journal of Physical Chemistry* 1 (1896-1897): 149-56. See *Roy. Soc. Cat.*, vols.

11, 19. WORKS ABOUT: *NCAB*, 13 (1906) : 571; obituary, *Science*, n.s. 23 (1906) : 195-97.

WARREN, CYRUS MOORS (January 15, 1824, West Dedham, Mass.—August 13, 1891, Manchester, Vt.). *Chemistry.* Son of Jesse, blacksmith and foundry owner, and Betsey (Jackson) Warren. Married Lydia Ross, 1849; seven children. EDUCATION: Attended country school, and engaged in private study; 1855, S.B., Lawrence Scientific School, Harvard; thereafter, for several years, studied chemistry in Paris, Germany, and London. CAREER: Engaged in teaching and farm work; 1847, joined brother in manufacture of tarred roofing at Cincinnati, Ohio, an enterprise in which other members of family also later became involved; success of business permitted period of advanced study of chemistry in U.S. and Europe; 1863, set up private chemical research laboratory, Boston; established Warren Chemical and Manufacturing Co. in Boston and Warren-Scharf Asphalt Paving Co. in New York, and served as president and treasurer; 1866-1868, professor of organic chemistry, Massachusetts Institute of Technology. MEMBERSHIPS: Am. Ac. Arts Scis. SCIENTIFIC CONTRIBUTIONS: Combined interest and abilities in business management, inventiveness and perfection of analytical techniques, and chemical research. Chief scientific work, done especially during mid-1860s, grew out of business concern with tars, family having gained monopoly of coal tar from gasworks, a product formerly treated as waste; devised process for "fractional condensation," and employed technique in studies of tars, isolating various component hydrocarbons; later, also performed pioneering studies on constituent hydrocarbons of Pennsylvania petroleum. In these important studies of volatile hydrocarbons, discovered new substances and helped to define better the characteristics of others, and also developed patentable techniques for fractional distillation, determination of vapor densities, analysis of organic compounds, and other processes.

WORKS: "On a Process of Fractional Condensation; Applicable to the Separation of Bodies Having Small Differences Between Their Boiling Points," *Memoirs of Am. Ac. Arts. Scis.* n.s. 9 (1867) : 121-34; "On the Volatile Hydrocarbons in Pennsylvania Petroleum" (1868), *Proceedings of Am. Ac. Arts Scis.* 27 (1891-1892) : 56-87. See list of works in memoir [below], pp. 402-3. WORKS ABOUT: *DAB*, 19: 471-72 (C. A. Browne) ; memoir, *Proceedings of Am. Ac. Arts Scis.* 27 (1891-1892) : 391-403.

WARREN, GOUVERNEUR KEMBLE (January 8, 1830, Cold Spring, N.Y.—August 8, 1882, Newport, R.I.). *Engineering*; *Geology.* Son of Sylvanus Warren, man of prominence in local affairs. Married Emily Forbes Chase, 1863; two children. EDUCATION: 1850, graduated, U.S. Military Academy, West Point. CAREER: 1850, assigned to U.S. Corps of Topographical Engineers; 1850-1854, served on engineering and surveying projects on Mississippi delta, falls of Ohio, and Rock Island and Des Moines rapids; 1854, assisted with preparation of map and reports for Pacific Railroad Surveys; 1855, chief topographical engineer of Sioux expedition; 1855-1859, in charge of reconnaissances and preparation of maps and reports for Dakota and Nebraska territories; 1859-1861, assistant professor of mathematics, West Point; 1861, became lieutenant colonel, New York Volunteers; 1863, chief topographical engineer, Army of Potomac; 1866-1870, chief engineer, surveys and improvements of upper Mississippi; 1869-1870, chief engineer, Rock Island bridge across Mississippi; 1870-1882, superintending engineer of surveys and improvements of various rivers and harbors on Atlantic Coast; 1879, promoted to lieutenant colonel, Corps of Engineers; in period after Civil War, duties also included engineering work on number of railroad, bridge, and other projects. MEMBERSHIPS: Natl. Ac. Scis. and other societies. SCIENTIFIC CONTRIBUTIONS: Assisted A. A. Humphreys [q.v.] with preparation of general report on Pacific Railroad Surveys and was given assignment by Humphreys to prepare general map, published with accompanying memoir as volume 11 of railroad survey reports (1859). After Civil War, prepared report on earlier explorations in Nebraska and Dakota (published in 1875) ; also published important reports on Minnesota River (1874), improvements along Fox and Wisconsin rivers (1876), and bridging of Mississippi between St. Paul and St. Louis (1878) ; also prepared number of other reports on engineering and surveying projects. In later writings —for example, Minnesota River report—introduced idea of southern elevation and northern subsidence of the continent.

WORKS: See chief works listed in *DAB*. MSS: N.Y. St. Lib. (Hamer). WORKS ABOUT: *DAB*, 19: 473-74 (William A. Ganoe) ; Henry L. Abbot, "Memoir," *Biog. Mems. of Natl. Ac. Scis.* 2 (1886) : 175-88. Also see Emerson G. Taylor, *Gouverneur Kemble Warren: The Life and Letters of an American Soldier* (Boston and New York, 1932).

WARREN, JOHN COLLINS (August 1, 1778, Boston, Mass.—May 4, 1856, Boston, Mass.). *Anatomy*; *Paleontology.* Son of John, surgeon and medical educator, and Abigail (Collins) Warren. Married Susan Powell Mason, 1803; Anne Winthrop, 1843; six children by first marriage. EDUCATION: 1797, A.B., Harvard College; 1798, began study of medicine with father; 1799-1802, studied medicine in London, Edinburgh, and Paris. HONORS: M.D., Harvard Univ., 1819. CAREER: 1802, began medical practice with father in Boston; assisted with anatomical dissections for father's lectures at Harvard Medical School; at Harvard Medical School, adjunct professor (1809-1815), Hersey professor of anatomy and surgery (1815-1847), and dean (1816-1819); 1821-1853, principal surgeon, Massachusetts General Hospital; 1824, appointed a consulting physician of Boston; 1837-1838 and 1851, traveled to Europe. MEMBERSHIPS: President, Bost. Soc. Nat. Hist., 1847-1856. SCIENTIFIC CONTRIBUTIONS: Chief interests during active professional career were in medicine and surgery. With James Jackson made lasting changes in medical practice and education in Boston: prepared *Pharmacopeia* (1808) for Massachusetts Medical Society, took leading role in founding of Massachusetts General Hospital and in removing Harvard Medical School from Cambridge to Boston (1815), and also helped to found *New England Journal of Medicine and Surgery* (1812). Became authority on tumors and in 1837 published chief medical work, *Surgical Observations on Tumours with Cases and Operations*; in 1846, performed first non-dental operation with use of ether anesthesia. In conjunction with duties as professor of anatomy, gathered important museum collection; in 1822, published *Comparative View of the Sensorial and Nervous Systems in Men and Animals,* including anatomical examination of crania of North American Indians. After retirement, became particularly interested in geology and paleontology and amassed large collection, purchasing mastodon bones found at Newburgh, N.Y.; anatomical description of these bones constituted chief scientific publication [see Works]. Carried out agricultural experiments on country estate.

WORKS: *The Mastodon Giganteus of North America*, 2d ed. (Boston, 1855), 260 pp. with 30 plates. See *Roy. Soc. Cat.*, vol. 6, MSS: Harv. U.-MS (*NUCMC* 62-3750); Harv. U.-UA (*NUCMC* 65-1290); Mass. Hist. Soc. (Hamer). WORKS ABOUT: *DAB*, 19: 480-81 (Henry R. Viets); [Jeffries Wyman], "Memoir," *Proceedings of Bos. Soc. Nat. Hist.* 6 (1856-1859): 73-83.

WARREN, JOSEPH WEATHERHEAD (1849-1916). *Physiology.* Professor of physiology, Bryn Mawr College; assistant to Pennsylvania commissioner of health.

WORKS ABOUT: See *AMS*; *Who Was Who in America*, vol. 1.

WATERHOUSE, BENJAMIN (March 4, 1754, Newport, R.I.—October 2, 1846, Cambridge, Mass.). *Natural history*; *Medicine.* Son of Timothy—tanner, cabinet maker, and judge of court of common pleas—and Hannah (Proud) Waterhouse. Married Elizabeth Oliver, 1788; Louisa Lee, 1819; six children by first marriage. EDUCATION: Ca. 1770-1775, apprenticed to Newport physician; 1775-1778, studied medicine with relative, Dr. John Fothergill, and with others in London and in Edinburgh; 1780, M.D., Univ. of Leyden, and for year thereafter studied history and law at Leyden. CAREER: 1782, established medical practice at Newport; at Harvard Univ., professor of theory and practice of physic (1783-1812) and lecturer in natural history (1788-1809); 1784-1786, lecturer in natural history, Brown Univ.; 1807-1809, physician to U.S. Marine Hospital, Charlestown, Mass.; 1812, removed from Harvard professorship on account of differences with colleagues, and thereafter continued medical practice at Cambridge; 1813-1821, held presidential appointment as medical superintendent for New England military posts. MEMBERSHIPS: Am. Ac. Arts Scis.; Am. Phil. Soc. SCIENTIFIC CONTRIBUTIONS: One of best-prepared physicians in U.S. at time, and as early as 1791, in lecture published as *The Rise, Progress, and Present State of Medicine*, showed appreciation of experimental work. In medical history, chiefly remembered for early advocacy of Edward Jenner's smallpox vaccination; in 1800 vaccinated own children, and later (1802) persuaded Boston board of health to carry out experiments that helped to overcome public opposition, emphasizing importance of use of pure vaccine. In natural history lectures at Brown and at Harvard was particularly interested in mineralogy and botany; established mineralogical collection at Harvard, and through contacts with English correspondents procured information and specimens in support of natural history study at Harvard; botanical lectures first published between 1804 and 1808 in *Monthly Anthology* and later in separate publication [see Works]. Published widely read work on effects of use of tobacco and alcohol. Motivated throughout career by desire to introduce European practices and interests into U.S.

WORKS: *A Prospect of Exterminating the Small Pox*, 2 pts. (Cambridge, Mass., 1800, 1802); *The Botanist* (Boston, 1811). MSS: Harv. U.-MS (*NUCMC* 62-3751). WORKS ABOUT: *DAB*, 19: 529-32 (Henry R. Viets); Josiah C. Trent, "Benjamin Waterhouse," *Journal of the History of Medicine* 1 (1946): 357-64. Also see John B. Blake, *Benjamin Waterhouse and the Introduction of Vaccination: A Reappraisal* (Philadelphia, 1957), 95 pp.

WATSON, JAMES CRAIG (January 28, 1838, near Fingal, Ontario, Canada—November 22, 1880, Madison, Wis). *Astronomy.* Son of William—farmer, schoolmaster, and factory worker—and Rebecca (Bacon) Watson. Married Annette Waite, 1860; no children. EDUCATION: 1857, A.B., Univ. of Michigan. HONORS: Lalande Prize, French Academy of Sciences, 1870. CAREER: Ca. 1851 [age thirteen], became factory engineer at Ann Arbor, Mich.; at

Univ. of Michigan, astronomical assistant to Prof. F. F. E. Brunnow [q.v.] (1858-1859), professor of astronomy and director of observatory (1859-1860 and 1863-1879), and professor of physics (1860-1863); took part in eclipse expeditions in 1869 to Iowa, 1870 to Sicily, and 1878 to Wyoming; 1874, headed expedition to observe transit of Venus in China, and on return trip to U.S. devoted several weeks to advising Egyptian army on conduct of geodetic survey; 1879, appointed director, Washburn Observatory, Univ. of Wisconsin; engaged in various business enterprises, including work as insurance actuary. MEMBERSHIPS: Natl. Ac. Scis. and other societies. SCIENTIFIC CONTRIBUTIONS: Before age twenty-one had published some fifteen papers in *Astronomical Journal*; early work dealt especially with observations and computations of comets and minor planets; during early 1860s, undertook considerable work in reduction of Washington zones, begun by B. A. Gould [q.v.], showing great skill in computation; between 1863 and 1877, discovered twenty-two asteroids, and devoted some ten years to preparation of charts of stars near ecliptic, an aid to discovery of asteroids. In 1868, published *Theoretical Astronomy* [see Works], used as textbook in U.S. and in Europe. In 1869, commenced work on lunar theory with B. Peirce [q.v.]. Became convinced of existence of Leverrier's planet Vulcan, and in 1878 undertook eclipse expedition to Wyoming that resulted in report of discovery of two intramercurial planets (one supposedly Vulcan); however, this discovery failed to be verified by later research. In 1878, published *Tables for the Calculation of Simple or Compound Interest*.

WORKS: *A Popular Treatise on Comets* (Philadelphia, 1861); *Theoretical Astronomy Relating to the Motions of the Heavenly Bodies Around the Sun . . .* (Philadelphia, 1868); "Discovery of [Two Intra-Mercurial Planets]," *Am. Journ. Sci.* 3d ser. 16 (1878): 230-33, 310-13. See *Roy. Soc. Cat.*, vols. 6, 8, 11. MSS: Univ. Mich.-MHC (*NUCMC* 65-627). WORKS ABOUT: *DAB*, 19: 543-44 (W. Carl Rufus); George C. Comstock, "Memoir," *Biog. Mems. of Natl. Ac. Scis.* 3 (1895): 43-57.

WATSON, SERENO (December 1, 1826, East Windsor Hill, Conn.—March 9, 1892, Cambridge, Mass.). *Botany*. Son of Henry, merchant and farmer, and Julia (Reed) Watson. Never married. EDUCATION: 1847, A.B., Yale College; for several years, studied medicine under physicians in New England, New York, and Illinois; 1866-1867, studied chemistry and mineralogy, Sheffield Scientific School, Yale. HONORS: Ph.D., Iowa College, 1878. CAREER: After college, spent several years as teacher in New England, Pennsylvania, New York, and Iowa, with intervening periods devoted to medical studies; 1856-1861, secretary, Planters' Insurance Co. (Greensboro, Ala.); 1861-1866, editorial work on *Journal of Education*, Hartford; 1867, went to California, became assistant on U.S. Geological Survey of Fortieth Parallel led by C. R. King [q.v.] and appointed survey's botanist in 1868; 1869, began study of botanical collections of King survey, at Yale Univ.; 1870, continued botanical studies with A. Gray [q.v.] at herbarium at Harvard Univ.; at Harvard, curator of Gray herbarium (1874-1892) and instructor in phytography (1881-1884); 1885, attempted botanical exploration in Guatemala, which was terminated by illness. MEMBERSHIPS: Natl. Ac. Scis. and other societies. SCIENTIFIC CONTRIBUTIONS: Active botanical interests did not begin until his early forties; botanical field work confined essentially to period with King exploration, and at that time collected extensively in Great Basin; results of this work appeared as volume 5 (1871) of reports of King survey and included important information on geography of plants, as well as significant taxonomic contributions. Thereafter, took up study of botany of California, and in this work benefited from prior efforts of W. H. Brewer [q.v.], who along with Gray assisted with volume 1 of resulting two-volume work, the first systematic flora of California [see Works]; completed one volume of *Bibliographical Index to North American Botany* (1878), worked on completion of L. Lesquereux and T. P. James [qq.v.], *Manual of the Mosses of North America* (1884), and with J. M. Coulter [q.v.] completed revision of Gray's *Manual of the Botany of Northern U.S.* (1889). Published series of "Contributions to American Botany," especially in *Proceedings of Am. Ac. Arts Scis.*, in which he drew upon extensive curatorial labors on specimens from western U.S. and Mexico and gave further results of his characteristically important and careful work on plant systematics and relations.

WORKS: *Botany of California*, 2 vols. (Cambridge, Mass., 1876, 1880). References to chief works and to bibliography in *DSB*. MSS: Harv. U.-GH (Hamer). WORKS ABOUT: *DAB*, 19: 547-48 (Benjamin L. Robinson); *DSB*, 14: 192-93 (Elizabeth N. Shor).

WEAD, CHARLES KASSON (1848-1925). *Physics*. Examiner, U.S. Patent Office.

MSS: Univ. Mich.-MHC (*NUCMC* 65-629). WORKS ABOUT: See *AMS*; *Who Was Who in America*, vol. 1.

WEBSTER, FRANCIS MARION (1849-1916). *Entomology*. Associated with agricultural experiment stations of Illinois, Indiana, and Ohio; in charge of Section on Cereal and Forage Crop Insects, U.S. Bureau of Entomology.

WORKS ABOUT: See *AMS*; *NCAB*, 13 (1906): 603-4; Mallis, *Amer. Entomologists*, pp. 92-94.

WEBSTER, JOHN WHITE (May 20, 1793, Boston, Mass.—August 30, 1850, Boston, Mass.). *Chemistry*. Son of Redford, successful apothecary, and Hannah (White) Webster. Married Harriet Fredrica Hickling, 1818; four children. EDUCATION: A.B. in 1811 and M.D. in 1815, Harvard Univ.; thereafter, studied medicine and surgery in London. CAREER: Inherited some wealth, which was eventually expended in living beyond means; for several years after completion of degree at Harvard Medical School, pursued medical studies in London and traveled; at Harvard Univ., lecturer on chemistry, mineralogy, and geology (1824-1826), adjunct professor of chemistry (1826-1827), and Erving professor of chemistry and mineralogy (1827-1850); 1824-1826, associate editor, *Boston Journal of Philosophy and the Arts*; 1850, convicted and hanged for murder of Dr. George Parkman, to whom Webster had been financially indebted. MEMBERSHIPS: Am. Ac. Arts Scis.; Am. Assoc. Advt. Sci. SCIENTIFIC CONTRIBUTIONS: Not greatly distinguished as researcher or as teacher of chemistry, but nonetheless did some worthwhile work, which was subsequently overshadowed by murder conviction. *Manual* [see Works], based on Harvard lectures and compiled from works of European scientists, was effective as textbook and kept up with advances in chemistry, the enlarged edition of 1839 giving full review of Berzelius's innovative system of element symbols, formulas, and equations; familiarity with current developments in science also shown in preparation of American edition of Justus Liebig's *Organic Chemistry in Its Applications to Agriculture and Physiology* (1841), with notes referring to American experimental work in agriculture. Among other duties at Harvard, during years 1827-1836 arranged some 26,000 mineralogical specimens in university collection; published several articles in *Am. Journ. Sci., Boston Journal Philosophy,* and foreign publications, on topics in geology, chemistry (including analysis of minerals and of meteor), magnetism, and electricity.

WORKS: *A Description of the Island of St. Michael, Comprising an Account of Its Geological Structure . . .* (Boston, 1821); *A Manual of Chemistry on the Basis of Professor Brande's* (Boston, 1826). See also *Roy. Soc. Cat.*, vol 6. MSS: Harv. U.-UA (*NUCMC* 65-1291); some items in E. N. Horsford [q.v.] Papers at Rens. Poly. Inst.-Arch. WORKS ABOUT: *DAB*, 19: 592-93 (Edmund L. Pearson), I. Bernard Cohen, *Some Early Tools of American Science . . . in Harvard University* (Cambridge, Mass., 1950); Guralnick, *Science and American College*, pp. 218-19.

WEIGHTMAN, WILLIAM (September 30, 1813, Waltham, Lincolnshire, England—August 25, 1904, Philadelphia, Pa.). *Chemistry*. Son of William and Anne (Farr) Weightman. Married Louise Stelwagon, 1841; three children. EDUCATION: No information. CAREER: 1829 [age sixteen], emigrated to U.S. and began work with Farr and Kunzi chemical manufacturing establishment in Philadelphia, in which uncle, John Farr, was partner; in 1836 company changed to Farr, Powers & Weightman, and in 1847 became Powers & Weightman; filled role as company chemist; 1878-1904, conducted company business affairs as well, and achieved wealth; invested in property and became largest real estate owner in Philadelphia; director of Philadelphia Trust Co., Northern Trust Co., and Commercial National Bank. MEMBERSHIPS: Franklin Institute; Philadelphia College of Pharmacy. SCIENTIFIC CONTRIBUTIONS: Chemical career intimately linked to manufacturing firm. Played chief role in establishment of company's international reputation for innovations in production processes and introduction of new chemicals, for which firm won Franklin Institute's gold medal; company was first to manufacture quinine sulphate, promoted substitution of alkaloids of cinchona, cinchonidine, and cinchonine for more expensive quinine, and introduced an improved means of production of citric acid in U.S.

WORKS: *On Chinoidine* (Philadelphia, 1849), 22 pp. WORKS ABOUT: *DAB*, 19: 607-8 (Andrew G. DuMez); Howard B. French, obituary, *Am. Journ. Pharm.* 77 (April 1905): 151-53.

WELLS, DAVID AMES (June 17, 1828, Springfield, Mass.—November 5, 1898, Norwich, Conn.). *Chemistry*; *Geology*; *Science editing and writing*. Son of James and Rebecca (Ames) Wells; mother was the daughter of successful paper manufacturer. Married Mary Dwight, 1860; Ellen Dwight, 1879; one child by first marriage. EDUCATION: 1847, A.B., Williams College; 1851, B.S., Lawrence Scientific School, Harvard. HONORS: D.C.L., Oxford Univ., 1874, and other honorary degrees. CAREER: 1847-1848, reporter, *Springfield Republican*; 1850-1866, conducted *Annual of Scientific Discovery*; 1851-1852, resident graduate at Lawrence Scientific School and lecturer on physics and chemistry at Groton Academy; 1857-1858, worked with G. P. Putnam and Co., publishers; 1858-1865, lived at Troy, N.Y.; 1861-1864, associate member, U.S. Sanitary Commission; 1865, took up residence at Norwich, Conn.; acquired wealth through marriage, publishing, and investments; 1865-1866, chairman, national revenue commission; 1866-1870, special U.S. commissioner of revenue; 1870, appointed chairman, New York tax commission; with home at Norwich, wrote and lectured on economic questions, participated in politics, and in 1876 and 1890 was unsuccessful candidate for Congress. MEMBERSHIPS: Am. Assoc. Advt. Sci. SCIENTIFIC CONTRIBUTIONS: During early years, invented device for folding newspapers, and made significant improvements in textile manufacture. In addition to *Annual*, produced several successful textbooks, including chemistry (1858), geology (1861), and natural philosophy (1863), and also *Yearbook of Agriculture* (1856). During early career enjoyed reputation as scientist, including agricultural work on soil problems, plant breeding, and the like; during 1850s published several articles on geological-chemical subjects, including paper on soil analysis for secretary of Ohio State Board of Agriculture, with several papers in *Am. Journ. Sci.* and in proceedings of Am. Assoc. Advt. Sci. and Bos. Soc. Nat. Hist. Best remembered for later work in economics, especially on tariffs, theory of money and currency question, and taxation.

WORKS: See Joyner [below], pp. 223-29, and *Roy. Soc. Cat.*, vol. 6. MSS: Springfield Public Library (Joyner; there also see pp. 222-23); Lib. Cong. (Hamer). WORKS ABOUT: *DAB*, 19: 637-38 (Broadus Mitchell); Fred B. Joyner, *David Ames Wells, Champion of Free Trade* (Cedar Rapids, Iowa, 1939).

WELLS, SAMUEL ROBERTS (April 4, 1820, West Hartford, Conn.—April 13, 1875, New York, N.Y.). *Phrenology*. Father pioneer in west central New York, apparently farmer. Married Charlotte Fowler, 1844; no children. EDUCATION: Attended district school; apprenticed to tanner and currier; studied medicine. CAREER: 1843, began work with phrenologists O. S. Fowler and L. N. Fowler [qq.v.] in New York; 1845, became partner in renamed firm, Fowlers and Wells, and took charge of publishing department; 1858-1860, accompanied L. N. Fowler on phrenological lecture tour throughout U.S. and Canada; 1860-1862, accompanied L. N. Fowler to England; ca. 1864, Fowlers having withdrawn, phrenological publishing firm became known as S. R. Wells. SCIENTIFIC CONTRIBUTIONS: In division of responsibilities with Fowler brothers, became business manager of firm that published books, as well as articles in *American Phrenological Journal*, on phrenology and on wide range of reform ideas, including education, penology, care of insane, temperance, antitobacco and antilacing views, vegetarianism, marriage and sex, animal magnetism and phreno-magnetism, hydropathy, women's rights, and child and vocational guidance; 1850-1862, edited *Water-Cure Journal*; 1863-1875, edited *American Phrenological Journal*; during 1850s, Wells or others compiled manuals on self-improvement, such as *How to Write, Talk, Behave, Do Business.* Also issued works on breeding and raising farm animals, on growing field crops, and on other practical subjects; published second, revised edition of Whitman's *Leaves of Grass*; also established Patent Agency; Wells wrote several works on phrenology and related subjects, being particularly interested in physiognomy. Wife Charlotte, sister of O. S. and L. N. Fowler, took over firm after husband's death.

WORKS: *The New Physiognomy; or, Signs of Character* (New York, 1867); *Wedlock; or, The Right Relations of the Sexes* (New York, 1869); *How to Read Character* (New York, 1870). WORKS ABOUT: Madeleine B. Stern, *Heads & Headlines: The Phrenological Fowlers* (Norman, Okla. 1971); *Appleton's CAB*, vol. 6: 431.

WELLS, WILLIAM CHARLES (May 24, 1757, Charleston, S.C.—September 18, 1817, London, England). *Physics*; *Physiology*. Son of Robert, printer and bookseller, and Mary Wells. Never married. EDUCATION: Ca. 1768 [age eleven], sent to school in Dumfries, Scotland; 1770, entered Univ. of Edinburgh; 1771-1774, apprenticed to physician-botanist Alexander Garden [q.v.] at Charleston; 1775-1778, attended lectures at medical school, Univ. of Edinburgh; for short period, attended Univ. of Leyden; 1780, M.D., Univ. of Edinburgh; 1788, licensed by Royal College of Physicians. HONORS: Rumford Medal, Roy. Soc. CAREER: At beginning of Revolutionary War, went as loyalist to Great Britain; 1779, surgeon, Scottish regiment in Dutch service in Europe; 1781-1784, physician at Charleston, S.C., and at St. Augustine, Fla.; 1784-1817, medical practice, London; 1795-1817, physician at St. Thomas' Hospital. MEMBERSHIPS: Roy. Soc. SCIENTIFIC CONTRIBUTIONS: Began serious scientific studies after establishing medical practice in London. 1792 work on vision [see Works] won him election to Roy. Soc.; later, did further work on optics and vision. Conducted other studies in physics and physiology; 1790-1810, published about dozen papers on medical topics that appeared in *Transactions of Society for Promotion of Medical and Chirurgical Knowledge.* Chief contribution to biological sciences was report (1813) on white woman having skin with brown splotches, in which idea of evolution by natural selection was introduced; in 1866 edition of *Origin of Species*, Darwin credited Wells with being first to suggest role of natural selection. Major work in physics dealt with phenomenon of dew (1814), presenting original theory of dew as atmospheric vapor condensed on cooled objects; related this conclusion to meteorological conditions; while this work was based on careful investigation and won Rumford

Medal, idea was not generally accepted for some seventy years.

WORKS: Chief works collected and republished after death as *Two Essays: One Upon Single Vision with Two Eyes; The Other on Dew . . . and an Account of a Female of the White Race of Mankind, Part of Whose Skin Resembles That of a Negro; With Some Observations on the Causes of the Differences in Colour and Form Between the White and Negro Races of Men* (London, 1818). WORKS ABOUT: *DAB*, 19: 644-45 (Richard H. Shryock); *DSB*, 14: 253-54 (William Dock).

WEST, BENJAMIN (March 1730, Rehoboth, Mass.—August 26, 1813, Providence, R.I.). *Astronomy*. Son of John West, farmer. Married Elizabeth Smith, 1753; eight children. EDUCATION: For three months, attended town school at Bristol, R.I.; self-educated through use of libraries of interested acquaintances. HONORS: M.A., Harvard College, 1770; LL.D., Brown Univ., 1792. CAREER: 1753, opened private school at Providence, R.I.; for some twenty years before Revolution, operated dry goods and book store in Providence; 1763, almanac prepared for that year was first of many issued in subsequent years; during Revolution, manufactured clothing at Providence for American troops; after war, again operated school at Providence; 1787-1788, professor of mathematics, Protestant Episcopal Academy, Philadelphia; at Brown Univ., professor of mathematics and astronomy (1788-1799) and also professor of natural philosophy (1798-1799); ca. 1799, opened school of navigation at Providence; 1802-1813, postmaster of Providence. MEMBERSHIPS: Am. Ac. Arts Scis. SCIENTIFIC CONTRIBUTIONS: Achievements began with calculations for almanac for 1763, published in Providence and, beginning with issue for 1764, known as *New England Almanack*; calculations won wide acclaim for accuracy, and West contributed to almanacs published for Boston, Providence, and Nova Scotia; bibliographical history of almanacs complicated by use of pseudonym (Isaac Bickerstaff) and conflicts arising from sale of calculations to rival publishers; contributed calculations for some 200 almanacs published from 1763 to early 1800s. As early as 1766, sent comet observations to J. Winthrop [q.v.] at Harvard; took important part in observations at Providence of 1769 transit of Venus, collaborating with J. Brown [q.v.], and West prepared the report on this project [see Works].

WORKS: *An Account of the Observation of Venus Upon the Sun . . . 1769, at Providence* (Providence, 1769), 22 pp.; "An Account of the Observations Made in Providence . . . of the Eclipse of the Sun . . . 1781," *Memoirs of Am. Ac. Arts Scis.* 1 (1785): 156-58; "On the Extraction of Roots," ibid., pp. 165-72. WORKS ABOUT: *DAB*, 20: 5-6 (Raymond C. Archibald); Shipton, *Biog. Sketches, Harvard*, 12 (1962): 220-26.

WETHERILL, CHARLES MAYER (November 4, 1825, Philadelphia, Pa.—March 5, 1871, Bethlehem, Pa.). *Chemistry*. Son of Charles, member of family of chemical manufacturers, and Margaretta (Mayer) Wetherill. Married Mary Benbridge, 1856; no information on children. EDUCATION: 1845, graduated, Univ. of Pennsylvania; 1845-1846, student in chemical laboratory of J. C. Booth and M. H. Boyé [qq.v.]; 1846-1847, studied chemistry in Paris; 1848, Ph.D., Univ. of Giessen. HONORS: M.D., New York Medical College, 1853. CAREER: 1849-1853, operated laboratory for chemical instruction and analysis, Philadelphia; 1853, appointed to prepare exhibit of Pennsylvania chemical and mineralogical products for New York Crystal Palace Exposition; thereafter, made mineralogical journey through North Central states; ca. 1856-1862, resided at Lafayette, Ind., where engaged in private research and writing; 1862-1863, first chemist in new U.S. Department of Agriculture; 1863-1866, chemist in Smithson. Instn.; 1866-1871, professor of chemistry, Lehigh Univ.; 1871, appointed director of chemical department, Univ. of Pennsylvania, but died before assuming duties. MEMBERSHIPS: Member Am. Phil. Soc. and other societies. SCIENTIFIC CONTRIBUTIONS: Chemical investigations in Philadelphia laboratory in early 1850s included study of minerals, illuminating gas, and asphalt, as well as of such biological products as royal jelly (of queen bee), foods, and adipocere, the latter attracting wide notice; studies at Lafayette, Ind., related chiefly to questions of local interest, of particular note being analysis of white sulphur water of that place and also work on *The Manufacture of Vinegar* (1860); during tenure in Washington, D.C., prepared *Report on the Chemical Analysis of Grapes* (1862), issued as first scientific bulletin of Department of Agriculture, and engaged in study of gunpowder under special assignment from president. Most elaborate chemical study related to condition of air in U.S. Capitol [see Works], prepared while at Smithson. Instn., where also studied variety of other theoretical and applied questions, including investigations of ammonium amalgam and of crystallization of sulfur and glass. Courses offered at Lehigh were considered models for other colleges and universities, and Wetherill published lecture outlines.

WORKS: *Warming and Ventilating the Capitol*, U.S. House Executive Document no. 100, 39th Congress, 1st session (Washington, D.C., 1866), 96 pp. Also see *Roy. Soc. Cat.*, vols. 6, 8, 12. MSS. Univ. of Pennsylvania (*DAB*). WORKS ABOUT: *DAB*, 20: 22-23 (C. A. Browne); Edgar F. Smith, "Charles

Mayer Wetherill," *Journ. Chem. Educ.* 6 (1929), in 6 pts.

WHEELER, GEORGE MONTAGUE (October 9, 1842, Hopkinton, Mass.—May 3, 1905, New York, N.Y.). *Topographical engineering.* Son of John and Miriam P. (Daniels) Wheeler. Married Lucy Blair; no information on children. EDUCATION: 1866, graduated U.S. Military Academy, West Point. CAREER: 1866, commissioned second lieutenant, U.S. Corps of Engineers, and assigned to survey duty in San Francisco Bay area; 1868-1871, engineer, staff of commanding general, Department of California; 1871-1879, director, geographical surveys of U.S. territory west of 100th meridian; 1879, promoted to rank of captain, and during same year survey organization was dissolved in favor of new U.S. Geol. Surv.; 1879-1888, engaged especially in supervision and preparation of survey reports, with periods of leave on account of illness; 1881-1883, as U.S. commissioner, Third International Geographical Congress and Exhibition (Venice), prepared report on that congress, and devoted some time to study of European governmental land and marine surveys; 1888, retired from army; 1890, given rank of major (as of 1888); last years spent in New York. MEMBERSHIPS: Am. Assoc. Advt. Sci. SCIENTIFIC CONTRIBUTIONS: As early as 1869, devoted efforts chiefly to survey and reconnaissance duty, especially in search of routes for troop transport. Great work of life was directorship of Surveys West of 100th Meridian; being survey under sponsorship of U.S. War Department, it emphasized astronomical, geodetic, and topographical work with production of maps of interest for military purposes; survey also encompassed concern for mineral, geological, botanical, zoological, and ethnological aspects of the area (as well as climate, water, and other natural features), employed geologists and naturalists, and collected large numbers of natural history specimens, deposited mainly in Smithson. Instn.; nevertheless, many scientists in 1879 opposed continued involvement of army in western explorations. By 1889, Wheeler survey had resulted in publication of some forty reports and 164 maps.

WORKS: *Report Upon U.S. Geographical Surveys West of the 100th Meridian,* 7 vols. (Washington, D.C., 1875-1889), of which vol. 1, *Geographical Report* (1889), was written by Wheeler. MSS: See references in Bartlett [below], especially Natl. Arch. WORKS ABOUT: *DAB,* 20: 47-48 (Thomas M. Spaulding); Richard A. Bartlett, *Great Surveys of the American West* (Norman, Okla., 1962), especially pp. 333-72.

WHIPPLE, AMIEL WEEKS (1816, Greenwich, Hampshire County, Mass.—May 7, 1863, Washington, D.C.). *Topographical engineering.* Son of David and Abigail (Pepper) Whipple. Married Eleanor Sherburne, 1843; at least one child. EDUCATION: For a time, attended Amherst College; 1841, graduated, U.S. Military Academy, West Point. CAREER: 1834, engaged in teaching at Concord, Mass.; 1841, assigned to U.S. Army Corps of Topographical Engineers; 1841-1843, engaged in engineering duty at Baltimore, New Orleans, and Portsmouth, N.H.; 1844-1849, assistant astronomer, survey of U.S. northeastern boundary; 1849-1853, assistant astronomer, U.S.-Mexico boundary survey; 1853-1856, chief engineer for railway survey and exploration near thirty-fifth parallel from Mississippi to Pacific; 1855, promoted captain, Topographical Engineers; 1856-1861, lighthouse engineer and also chief engineer for ship channels through St. Clair flats and St. Mary's River, facilitating larger shipping on Great Lakes; during Civil War, held engineering and military commands, including chief topographical engineer with Army of Potomac at the Battle of Bull Run; 1861, promoted to major, Topographical Engineers, and in 1863 to major general, U.S. Volunteers; 1863, fatally wounded at Battle of Chancellorsville. SCIENTIFIC CONTRIBUTIONS: Chief scientific labors performed in conjunction with army engineering duties; in *Whipple Report* [see Works], prepared in conjunction with U.S.-Mexico boundary survey, demonstrated personal interest in study of Indians; also prepared report for work on Pacific Railroad Surveys [see Works].

WORKS: *The Whipple Report: Journal of an Expedition from San Diego, California, to the Rio Grande . . . 1849* (Los Angeles, 1961), edited by E. I. Edwards, originally published in 1851 as U.S. Senate Document; *Report of Explorations for a Railway Route, Near the Thirty-fifth Parallel . . . ,* Pacific Railroad Surveys, vols. 3-4 (Washington, D.C., 1856), with Joseph C. Ives. See Meisel, *Bibliog. Amer. Nat. Hist.* MSS: Oklahoma Historical Society (Hamer). WORKS ABOUT: *DAB,* 20: 66-67 (Thomas M. Spaulding); Cullum, *Biographical Register, West Point,* 2: 65-66.

WHIPPLE, SQUIRE (September 16, 1804, Hardwick, Mass.—March 15, 1888, Albany, N.Y.). *Civil engineering.* Son of James, farmer and cotton mill owner, and Electa (Johnson) Whipple. Married Anna Case, 1837; apparently no children. EDUCATION: Attended Hartwick (N.Y.) and Fairfield (N.Y.) academies; 1830, A.B., Union College (Schenectady). CAREER: Before entering college in 1829, taught school; 1830-1832, civil engineering service with Baltimore and Ohio Railroad; engaged as civil engineer on canals, including Erie Canal, on New York and Erie Railroad, and on other projects, especially in New York state; manufacturer of surveying instruments; 1841, added bridge building to other activities; later, settled at Albany as small-scale instrument-maker. MEMBERSHIPS: Honorary member of American Society of Civil Engineers. SCIENTIFIC CONTRIBUTIONS: In addition to engineering and instrument making, also devoted efforts to invention and in 1840 devised apparatus for weighing canal boats; later built weighing lock scale used on Erie Canal at Utica. In 1841, patented bowstring iron truss bridge, and several years later invented Whipple's trapezoidal truss bridge, used for long-span bridges of wood or iron; before 1850, designed most iron bridges in U.S.; among notable early bridge engineering projects was structure of 146-foot span, built in 1852-1853 for Rensselaer and Saratoga Railroad, near West Troy, N.Y.; in 1872 patented plan for lift drawbridge designed for Erie Canal at Utica. Enduring reputation as American pioneer in bridge engineering rests chiefly on 1847 *Work on Bridge Building* [see Works], of which enlarged edition appeared in 1872; *Work* was original with Whipple, and in it engineering design was based on elementary principles and dealt effectively for first time with analysis of stresses in truss, introducing consideration of theoretical questions involved in framed structures. Published several papers in early volumes of *Transactions of American Society of Civil Engineers.*

WORKS: *A Work on Bridge Building, Consisting of Two Essays, the One Elementary and General, the Other Giving Original Plans, and Practical Details for Iron and Wooden Bridges* (Utica, N.Y., 1847), with appendix (Albany, 1869); *The Way to Happiness: Being an Essay on the Motives to Human Actions and the Fundamental Principles of Morality* (Utica, N.Y., 1847). See bibliography in Sayre [below], p. 27. WORKS ABOUT: *DAB*, 20: 70-71 (James K. French); Mortimer F. Sayre et al., *Squire Whipple . . .*, Union [College] Worthies no. 4 (Schenectady, N.Y., [1949]), 28 pp.

WHITE, CHARLES ABIATHAR (1826-1910). *Geology; Paleontology; Zoology.* Iowa state geologist; professor of geology, Iowa State Univ.; served with Wheeler, Powell, and Hayden surveys; curator, invertebrate fossils, U.S. Natl. Mus.; geologist and paleontologist, U.S. Geol. Surv.

MSS: Iowa State Department of History and Archives (*NUCMC* 62-4165). WORKS ABOUT: See *AMS*; *DAB.*

WHITE, ISRAEL CHARLES (1848-1927). *Geology.* Professor of geology, West Virginia Univ.; superintendent, West Virginia State Geological Survey.

MSS: West Virginia Univ. Library (*NUCMC* 60-270). WORKS ABOUT: See *AMS*; *DAB*; *DSB.*

WHITFIELD, ROBERT PARR (1828-1910). *Paleontology; Geology.* Curator of geology, Am. Mus. Nat. Hist.

WORKS ABOUT: See *AMS*; *DAB.*

WHITING, SARAH FRANCES (1847-1927). *Physics; Astronomy.* Professor of physics and physical astronomy, director of Whitin Observatory, Wellesley College.

WORKS ABOUT: See *AMS*; *Nota. Am. Wom.*

WHITMAN, CHARLES OTIS (1842-1910). *Zoology.* Professor of zoology, Univ. of Chicago; director of Marine Biological Laboratory, Woods Hole, Mass.

WORKS ABOUT: See *AMS*; *DAB*; *DSB.*

WHITNEY, JOSIAH DWIGHT (November 23, 1819, Northampton, Mass.—August 19, 1896, Lake Sunapee, N.H.). *Geology*; *Chemistry.* Son of Josiah Dwight, successful banker, and Sarah (Williston) Whitney. Married Louisa (Goddard) Howe, 1854; one child. EDUCATION: 1839, A.B., Yale College; for short time, studied chemistry with R. Hare [q.v.], Philadelphia; 1841, began study of law at Northampton; 1842-1845 and 1845-1847, studied geology and chemistry in Paris, Berlin, and Giessen. HONORS: Name given to Mt. Whitney, Calif., highest point in contiguous United States. CAREER: 1840-1841, assisted C. T. Jackson [q.v.] on geological survey of New Hampshire; 1847-ca. 1849, engaged in mineral survey of Lake Superior region, initially as assistant to Jackson; ca. 1850, began practice as mining consultant, first at Brookline and later at Cambridge, Mass.; 1855-1858, held title of professor of chemistry at Univ. of Iowa, assigned as chemist and mineralogist to Geological Survey of Iowa; during same period, also did some work for geological surveys of Illinois and Wisconsin; 1860-1874, director, Geological Survey of California, though after 1868 survey work was largely suspended because of lack of appropriation; at Harvard Univ., Sturgis-Hooper professor of geology (1865-1896) and dean of short-lived School of Mining and Practical Engineering (1868-1875). MEMBERSHIPS: Natl. Ac. Scis.; Geological Society of London; and other societies. SCIENTIFIC CONTRIBUTIONS: After Jackson departed from Michigan survey at end of first year, Whitney and J. W. Foster [q.v.] prepared final reports, issued in 1850 and 1851 as congressional documents. Extensive work as mining consultant contributed directly to publication entitled *Metallic Wealth* [see Works], a pioneering attempt at systematic study of ore deposits; work for Illinois and Wisconsin surveys related especially to study of zinc and lead deposits. State of California published only three volumes of final report on state survey, but later Whitney, at personal expense and with aid of

Harvard Museum of Comparative Zoology, published some of material accumulated during California survey.

WORKS: *The Metallic Wealth of the U.S.* (Philadelphia, 1854); *The Auriferous Gravels of the Sierra Nevada of California* (Cambridge, Mass., 1880); *The Climatic Changes of Later Geological Times: A Discussion Based on Observations Made in the Cordilleras of North America* (Cambridge, Mass., 1882). See also *DAB*, *DSB*, and Edwin T. Brewster, *Life and Letters of J. D. Whitney* (Boston and New York, 1909), pp. 387-400. MSS: Yale Univ. and Univ. of Calif.-Ban. (*DSB*). WORKS ABOUT: *DAB*, 20: 161-63 (George P. Merrill); *DSB*, 14: 315-16 (Gerald D. Nash).

WHITNEY, MARY WATSON (1847-1921). *Astronomy.* Professor of astronomy and director of observatory, Vassar College.

WORKS ABOUT: See *AMS; DAB; Nota. Am. Wom.*

WHITTLESEY, CHARLES (October 4, 1808, Southington, Conn.—October 18, 1886, Cleveland, Ohio). *Geology.* Son of Asaph—merchant, farmer, justice of peace, postmaster, and iron manufacturer—and Vesta (Hart) Whittlesey. Married Mary (Lyon) Morgan, 1858; no children. EDUCATION: Attended school at Tallmadge, Ohio; 1831, graduated, U.S. Military Academy, West Point; later, studied law. CAREER: 1831, assigned to U.S. Fifth Infantry; 1832, stationed at Fort Howard, Wis., and participated in Black Hawk War; 1832, resigned from army; established law practice at Cleveland and, 1836-1837, part owner and coeditor, *Whig and Herald;* 1837-1839, assistant geologist, Ohio survey; 1839-1840, engaged in examination of Indian mounds in Ohio; 1844, made agricultural survey of Hamilton County, Ohio; 1845, became geologist for private company in Detroit; 1847-1851, engaged in U.S. survey of Lake Superior and upper Mississippi region, associated with D. D. Owen [q.v.]; thereafter, active as mining engineer in Michigan, Wisconsin, and Minnesota; 1858-1860, associated with State Geological Survey of Wisconsin; 1861, became assistant quartermaster general on staff of Ohio governor, and later that year was appointed colonel of Ohio Twentieth Regiment and performed engineering and military duty; 1862, resigned from army and resumed geological studies of Lake Superior and upper Mississippi region. MEMBERSHIPS: Am. Assoc. Advt. Sci. SCIENTIFIC CONTRIBUTIONS: Author of some 200 items related to geology, archaeology, history, and religion and a few on topographical geology, with little or no interest in paleontology. Work for Ohio survey related especially to topography and geography, and with financial aid of Joseph Sullivant of Columbus continued study of Indian mounds thereafter; planned joint publication never appeared, although Ephraim G. Squier and E. H. Davis used notes in their *Ancient Monuments of the Mississippi Valley* (1848); through extensive explorations, became expert on geology and topography of Lake Superior and upper Mississippi and of Ohio, work that had particular significance for development of mineral resources. Publications included four articles in *Smithson. Contr. Knowl.* series: "Ancient Works in Ohio" (1852), "Fluctuations of Level in the North American Lakes" (1860), "Ancient Mining on the Shores of Lake Superior" (1863), and "Fresh Water Glacial Drift in the Northeastern States" (1866). During later years, especially interested in question of relations of science and religion.

WORKS: See above; also *L. C. Cat. Prd. Cards 1942* and *Roy. Soc. Cat.*, vols. 6, 8, 11, 14. MSS: Western Reserve Historical Society, Cleveland (*NUCMC* 75-1803). WORKS ABOUT: *Appleton's CAB*, 6: 496; C. C. Baldwin, "Col. Charles Whittlesey," *Magazine of Western History* 5 (1886-1887): 534-48.

WILDER, BURT GREEN (1841-1925). *Vertebrate morphology.* Professor of neurology and vertebrate zoology, Cornell Univ.

MSS: Corn. U.-CRH (*NUCMC* 62-3435). WORKS ABOUT: See *AMS; NCAB*, 4 (1897 [copyright 1891 and 1902]): 481; *World Who's Who in Science.*

WILEY, HARVEY WASHINGTON (1844-1930). *Chemistry.* Chief chemist, U.S. Department of Agriculture; professor of agricultural chemistry, George Washington Univ.

MSS: Lib. Cong. (*NUCMC* 62-2703); see also *DSB*. WORKS ABOUT: See *AMS; DAB; DSB.*

WILKES, CHARLES (April 3, 1798, New York, N.Y.—February 8, 1877, Washington, D.C.). *Exploration; Geophysics; Astronomy.* Son of John De Ponthieu, prosperous businessman, and Mary (Seton) Wilkes. Married Jane Renwick (sister of James Renwick [q.v.]), 1826; Mary H. (Lynch) Bolton, 1854; four children by first marriage, and one by second. EDUCATION: Attended private preparatory school; 1818, became midshipman and attended a naval school in Boston. HONORS: Founder's Medal, Royal Geographical Society of London, 1847. CAREER: 1815-1818, merchant marine service; 1818, appointed midshipman; engaged in sea duty for three years in Mediterranean and two years in South Pacific; 1826, promoted to lieutenant; spent long periods of time waiting for orders and on leave of absence; 1832-1833, duty on survey of Narragansett Bay; 1833, appointed superintendent, U.S. Depot of Charts and Instruments; 1836, went to Europe to procure instruments for projected exploring expedition; 1837-1838, engaged in survey of St. George's Bank and Savannah River; 1838-1842, commander, U.S. Exploring Expedition to the Pacific (Wilkes Expedition); 1843-1861, engaged in preparation of reports on expedition; 1861, was in command of *San Jacinto,* and removed by force Confederate commissioners on British ship (*Trent* affair); 1862-1863, commander of squadron in West Indies; 1864, court-martialed; 1866, commissioned rear admiral on retired list. MEMBERSHIPS: Am. Phil. Soc.; Am. Assoc. Advt. Sci. SCIENTIFIC CONTRIBUTIONS: During period of extended shore leave, worked with F. R. Hassler [q.v.] of Coast Survey; developed reputation for abilities in hydrography and geodesy, and during period before command of exploring expedition, erected and used private astronomical observatory. Expedition, accompanied by corps of scientists, explored coast of Antarctica, islands of Pacific, and coast of American Northwest; prepared general *Narrative* of expedition (1844) in five volumes, as well as volumes on *Meteorology* (1851) and *Hydrography* (1861; with two volumes of charts, 1858). Also prepared volume on *Physics,* with calculations based on astronomical, geographical, and meteorological observations, but it was never published.

WORKS: See Tyler [below] for references to publications and to manuscripts. MSS: Lib. Cong. (*NUCMC* 62-4650). WORKS ABOUT: *DAB,* 20: 216-18 (Charles O. Paullin); David B. Tyler, *The Wilkes Expedition,* Am. Phil. Soc. Memoirs 73 (Philadelphia, 1968). Also see Daniel M. Henderson, *Hidden Coasts: A Biography of Admiral Charles Wilkes* (New York, 1953), a popularization.

WILLIAMS, GEORGE HUNTINGTON (January 28, 1856, Utica, N.Y.—July 12, 1894, Utica, N.Y.). *Mineralogy; Petrology.* Son of Robert Stanton, wealthy businessman, and Abigail (Doolittle) Williams. Married Mary Clifton Wood, 1886; three children. EDUCATION: 1878, A.B., Amherst College; 1879-1882, studied in Germany at Göttingen and Heidelberg; 1882, Ph.D., Univ. of Heidelberg. CAREER: 1879, teacher of science, Utica Free Academy; at Johns Hopkins Univ., fellow by courtesy (1882-1883), associate in mineralogy (1883-1885), associate professor of mineralogy (1885-1889), and associate professor (1889-1891) and professor (1891-1894) of inorganic geology. MEMBERSHIPS: A vice-president, Geol. Soc. Am.; member, Geological Society of London and other societies. SCIENTIFIC CONTRIBUTIONS: Despite early death (age thirty-eight), made important contributions to American science through introduction and application of most advanced methods in microscopic petrography. Observations and microscopic study of greenstone schist in Michigan, based on field work there in summers of 1884 and 1885, was his most extensive and probably most important single publication [see Works]. Investigated especially the Piedmont region of Maryland, in close association with U.S. Geol. Surv.; chief work on Maryland geology was *The Gabbros and . . . Hornblende* [see Works], a valuable study of relationship and chemical nature of those rocks; also published articles entitled "Cortlandt Series of the Hudson" in *Am. Journ. Sci.,* shorter papers on mineralogy and geology, reviews of literature, and contributions to works of reference; textbook (*Elements* [see Works]) widely used in U.S. and England. Displayed notable ability in production of mechanical devices to facilitate petrographic studies.

WORKS: *The Gabbros and Associated Hornblende Rocks Occurring in the Neighborhood of Baltimore, Md.,* U.S. Geol. Surv. *Bulletin* no. 28 (Washington, D.C., 1886); *The Greenstone Schist Areas of the Menominee and Marquette Regions of Michigan,* U.S. Geol. Surv. *Bulletin* no. 62 (Washington, D.C., 1890); *Elements of Crystallography* (New York, 1890). Bibliography in Clark [below], pp. 437-40. WORKS ABOUT: *DAB,* 20: 263 (George P. Merrill); William B. Clark, "Memorial," *Bulletin of Geol. Soc. Am.* 6 (1894-1895): 432-40.

WILLIAMS, HENRY SHALER (1847-1918). *Geology; Paleontology.* Professor of geology and paleontology, Cornell Univ.; Silliman Professor, Yale Univ.

MSS: Corn. U.-CRH (Hamer). WORKS ABOUT: See *AMS; DAB; DSB.*

WILLIAMSON, HUGH (December 5, 1735, West Nottingham, Pa.—May 22, 1819, New York, N.Y.). *Astronomy; Climatology; Medicine.* Son of John W., wealthy clothier, and Mary (Davison) Williamson. Married Maria Apthorpe, 1789; two children. EDUCATION: 1757, A.B., College of Philadelphia (Univ. of Pennsylvania); 1759, studied theology in Connecticut and was licensed to preach; 1764, went to Europe and studied medicine at Edinburgh, London, and Utrecht; 1768, M.D., Univ. of Utrecht. HONORS: LL.D., Univ. of Leyden. CAREER: For period of time, active as preacher; 1760-1763, professor of mathematics, College of Philadelphia; following European medical studies, established practice in Philadelphia; 1773-1776, went to Europe to solicit aid for academy at Newark (Del.); while in England secured for B. Franklin [q.v.] the Hutchinson-Oliver letters from Massachusetts and then went to Holland; 1777, settled at Edenton, N.C., where established mercantile business and received appointment as surgeon general of state troops; 1782-1785 and 1788, member of North Carolina legislature; 1782-

1785 and 1787-1789, member of Continental Congress; 1787, member of Federal Convention; 1789-1793, North Carolina member of U.S. Congress; ca. 1793, moved to New York and there pursued scientific and literary interests. MEMBERSHIPS: Am. Phil. Soc. and other societies. SCIENTIFIC CONTRIBUTIONS: From early age had particular interest in mathematics. In 1769, appointed to commission to study transits of Venus and Mercury, and during same year observations of comet gave rise to theory that tail of comet was not fire, but rather sun's reflection in comet's atmosphere [see Works]. In England in mid-1770s, became friend of B. Franklin [q.v.] and collaborated with him on electrical experiments, publishing paper on electric eel in *Phil. Trans. of Roy. Soc.* (1775); wrote papers on medical subjects, including one of some significance on North Carolina fevers (1797); during later period of life in New York, wrote on such subjects as fascination of serpents (1807), native climate of pestilence (1810), iron rods as protection against lightning (1810), and navigable canals (1810). Most influential work was *Observations* [see Works], which was intended as answer to European critics of American climate and gave evidence of considerable research; also wrote two-volume *History of North Carolina* (1812).

WORKS: "An Essay on Comets," *Transactions of Am. Phil. Soc.* 1 (1771): 27-36 (app.); *Observations on the Climate in Different Parts of America* (New York, 1811). WORKS ABOUT: *DAB*, 20: 298-300 (J. G. De R. Hamilton); David C. Whitney, *Founders of Freedom in America* (Chicago, 1965), pp. 225-29.

WILLISTON, SAMUEL WENDELL (1852-1918). *Paleontology.* Professor of geology and anatomy and dean of Medical School, Univ. of Kansas; professor of paleontology, Univ. of Chicago.

MSS: See Shor [below]. WORKS ABOUT: See *AMS; DAB;* Elizabeth N. Shor, *Fossils and Flies: The Life of a Compleat Scientist, Samuel Wendell Williston* (Norman, Okla., 1971).

WILSON, ALEXANDER (July 6, 1766, Paisley, Scotland—August 23, 1813, Philadelphia, Pa.). *Ornithology.* Son of Alexander, weaver and smuggler, and Mary (McNab) Wilson. Never married. EDUCATION: Attended Paisley Grammar School, and began preparation for ministry under divinity student; 1779-1782, apprenticed to weaver; largely self-educated. CAREER: While resident in Scotland, worked as weaver and peddler, in 1790 published volume of poetry, and for a time was imprisoned in case involving libel and blackmail; 1794, went to Philadelphia and for some years engaged in teaching, 1796-1801, at Milestown, Pa., and 1802-1806, at Gray's Ferry, Pa.; 1807, employed by Philadelphia publisher Samuel Bradford as assistant editor on new edition of Abraham Rees's *Cyclopaedia*, and for remainder of life worked on *Ornithology*, published by Bradford; 1808, undertook journey from Maine to Georgia in search of subscribers for *Ornithology*, and in 1809 visited St. Augustine, Fla.; 1810, made journey down Ohio River and to New Orleans, in search of bird specimens and subscribers. MEMBERSHIPS: Am. Phil. Soc.; Ac. Nat. Scis. Philad. SCIENTIFIC CONTRIBUTIONS: Scientific reputation rests entirely on *Ornithology* [see Works], to which attention was devoted only during last decade of life. Project took definite form through association with W. Bartram [q.v.] while teaching at Gray's Ferry; ambition was to classify, describe, and picture all American bird species. Resulting volumes were superbly executed and set new standard for natural history publications in America, showing particular success in effective relations of well-written text and colored illustrations drawn by author; the work covered eastern U.S. north of Florida and contained paintings and descriptions of 264 species, including 48 new species of American birds; also included life histories for many of them. Completed eight volumes before death, and work was finished by friend G. Ord [q.v.].

WORKS: *American Ornithology*, 9 vols. (Philadelphia, 1808-1814). MSS: Ac. Nat. Scis. Philad. (*NUCMC* 66-93); Harv. U.-MCZ (Hamer). WORKS ABOUT: *DAB*, 20: 317-19 (Witmer Stone); *DSB*, 14: 415-18 (Robert Cantwell). Also see Cantwell, *Alexander Wilson, Naturalist and Pioneer* (Philadelphia, 1961).

WILSON, EDMUND BEECHER (1856-1939). *Biology; Embryology; Cytology.* Professor of zoology, Columbia Univ.

WORKS ABOUT: See *AMS; DAB,* supp. 2; *DSB.*

WILSON, WILLIAM POWELL (1844-1927). *Botany.* Professor of anatomy and physiology of plants, Univ. of Pennsylvania; organized and directed Philadelphia Commercial Museum.

WORKS ABOUT: See *AMS; Who Was Who in America,* vol. 1.

WINCHELL, ALEXANDER (December 31, 1824, Northeast, N.Y.—February 19, 1891, Ann Arbor, Mich.). *Geology.* Son of Horace, schoolteacher, and Caroline (McAllister) Winchell; brother of Newton H. Winchell [q.v.]. Married Julia Lines, 1849; six children. EDUCATION: 1842, began study at Amenia Seminary, Dutchess County, N.Y.; 1847, graduated, Wesleyan Univ. (Connecticut). CAREER: 1841-1842, taught school; 1847, appointed teacher of natural science, Pennington Male Seminary (New Jersey); thereafter, appointed teacher of natural history, Amenia Seminary; 1850, became head of academy of Newbern, Ala.; 1851, inaugurated Mesopotamia Female Seminary, Eutaw, Ala.; 1853, assumed presidency, Masonic Univ., Selma, Ala.; at Univ. of Michigan, professor of physics and engineering (1853-1855), professor of geology, zoology, and botany (1855-1873), and professor of geology and paleontology (1879-1891); 1859-1861 and 1869-1871, director, Michigan State Geological Survey; at Syracuse Univ., was chancellor (1873-1874) and appointed professor of geology, botany, and zoology (1874); 1875-1878, professor of geology and zoology, Vanderbilt Univ.; active as popular lecturer on science. MEMBERSHIPS: President, Geol. Soc. Am., 1891. SCIENTIFIC CONTRIBUTIONS: Especially noted as popular lecturer and writer on science, as promoter of geological studies, and as most active founder of Geol. Soc. Am. Published bibliography included 255 publications on wide variety of scientific subjects, especially geology, but also evolution and science and religion, astronomy, climatology, meteorology, and zoology. Of his scientific work, contributions to stratigraphy and paleontology had special significance, and majority of 304 species that he introduced were fossil forms; established basin shape of Michigan strata, and wrote on so-called Marshall group of strata, oil region of Michigan, and Archean rocks in Minnesota; in economic sphere, particularly remembered for work in relation to Saginaw Valley salt beds. Of writings on evolution and science and religion, *World Life* [see Works] was most ambitious and significant general work, encompassing both cosmology and geology.

WORKS: *World Life; or, Comparative Geology* (Chicago, 1883). Chief works listed in *DSB;* references to bibliographies in *DAB*. MSS: Minnesota Historical Society (*DAB*); Univ. Mich.-MHC (*NUCMC* 65-654). WORKS ABOUT: *DAB,* 20: 373-74 (George P. Merrill); *DSB,* 14: 439-40 (author not listed).

WINCHELL, NEWTON HORACE (1839-1914). *Geology.* Minnesota state geologist; professor of geology, Univ. of Minnesota; editor, *American Geologist.*

MSS: Minnesota Historical Society (*NUCMC* 70-1647). WORKS ABOUT: See *AMS; DAB; DSB.*

WINLOCK, JOSEPH (February 6, 1826, Shelby County, Ky.—June 11, 1875, Cambridge, Mass.). *Astronomy; Mathematics.* Son of Fielding, lawyer, and Nancy (Peyton) Winlock. Married Mary Isabella Lane, 1856; six children, including William C. Winlock [q.v.]. EDUCATION: 1845, graduated, Shelby College (Shelbyville, Ky.). CAREER: 1845, appointed professor of mathematics and astronomy, Shelby College; for *American Ephemeris and Nautical Almanac,* Cambridge, Mass., computer (1852-1857) and superintendent (1858-1859 and 1861-1866); 1857-1858, professor of mathematics, U.S. Naval Observatory; 1859-1861, head of department of mathematics, U.S. Naval Academy, Annapolis; at Harvard Univ., Phillips professor of astronomy and director of observatory (1866-1875) and professor of geodesy (1868-1875). MEMBERSHIPS: Natl. Ac. Scis. SCIENTIFIC CONTRIBUTIONS: B. Peirce [q.v.], whom he met at 1851 meeting of Am. Assoc. Advt. Sci. at Cincinnati, played crucial role in advancing Winlock's scientific career. Had use of refracting telescope from Shelby during early years in Cambridge, and with B. A. Gould [q.v.] carried out observations that appeared in Gould's *Astronomical Journal.* During years with *Nautical Almanac,* made number of significant contributions, of which tables of Mercury were of particular importance. As director of Harvard Observatory, chief interest was in matters of instrumentation, for which he drew upon notable mechanical and inventive abilities, with intention of increasing the work and research capacity of observatory; to observatory was added new meridian circle prepared by London firm, with advice of Winlock, based on four months of investigation in Europe; this instrument subsequently was used for determination of Harvard zone stars for Astronomische Gesellschaft, a project considered by Winlock to be of first importance; endeavored to procure quality spectroscopes for observatory, became actively interested in solar photography, installed apparatus to produce daily pictures of sun, and led U.S. Coast Survey eclipse expeditions to Kentucky (1869) and Spain (1870), making notable photographs of solar corona. Later began project to produce engravings of planets, lunar craters, sunspots, nebulae, and the like; outcome of research efforts, which appeared especially in observatory's *Annals,* were carried to completion by others after his early death.

WORKS: See bibliographical note in *DSB*. MSS: Harv. U.-UA. WORKS ABOUT: *DAB,* 20:389-90 (Margaret Harwood); *DSB,* 14: 448-49 (Deborah J. Warner); Bessie Z. Jones and Lyle Boyd, *The Harvard College Observatory: The First Four Directorships* . . . (Cambridge, Mass., 1971).

WINLOCK, WILLIAM CRAWFORD (March 27, 1859, Cambridge, Mass.—September 20, 1896, Bay Head, N.J.). *Astronomy.* Son of Joseph [q.v.], astronomer, and Isabella (Lane) Winlock. Married Alice (Broom) Munroe, 1883; three children. EDUCATION: 1880, A.B., Harvard College. CAREER: 1874 (age fifteen), assistant, U.S. Coast Survey; 1879 and 1880, aide, Harvard College Observatory; 1880-1889, assistant astronomer, U.S. Naval Observatory, Washington, D.C.; 1886-1896, professor of astronomy, Columbian Univ. (George Washington Univ.); at Smithson. Instn., appointed curator of international exchanges (1889) and also became assistant in charge of office (1891); 1889-1896, honorary curator of apparatus, U.S. Natl. Mus. MEMBERSHIPS: Secretary, Phil. Soc. Wash., 1887-1896; member, Astronomische Gesellschaft of Leipzig and other societies. SCIENTIFIC CONTRIBUTIONS: As Harvard undergraduate, prepared paper on solar spectrum [see Works, 1881]. At Naval Observatory, worked with transit circle, made in excess of 9,000 observations, and while there published noteworthy work on 1882 comet [see Works], based largely on personal observations with transit instrument and accompanied by own drawings; independent scientific publications dealt mainly with comets. During period 1885-1892, beginning while still at Naval Observatory, prepared annual review of "progress of astronomy" for Smithson. Instn.; after going to Smithsonian in 1889, administrative and other duties left little time for active scientific work, though official duties did require correspondence with leading international scientific personalities and institutions. Wrote for *Sidereal Messenger,* for time edited astronomical column for *Science,* and wrote popular articles on the subject for encyclopedia and for *Harper's Weekly.*

WORKS: "On the Group 'b' in the Solar Spectrum," *Proceedings of Am. Ac. Arts Scis.* 16 (1880-1881): 398-405; "Observations of the Great Comet of 1882," in U.S. Naval Observatory, *Washington Observations, 1880* (Washington, D.C., 1884), app. 1, 38 pp. See also *Roy. Soc. Cat.*, vol. 19. WORKS ABOUT: *NCAB*, 9 (1907 [copyright 1899 and 1907]): 267; Harvard College Class of 1880, *Secretary's Report,* no. 6 (1900), pp. 74-75; J. R. Eastman, "Obituary notice," *Bulletin of Phil. Soc. Wash.* 13 (1895-1899): 431-34.

WINTHROP, JOHN (December 19, 1714, Boston, Mass.—May 3, 1779, Cambridge, Mass.). *Astronomy*; *Physics*; *Mathematics.* Son of Adam, chief justice, and Anne (Wainwright) Winthrop. Married Rebecca Townsend, 1746; Hannah (Fayerweather) Tolman, 1756; five children by first marriage. EDUCATION: 1732, A.B., Harvard College. HONORS: LL.D., Univ. of Edinburgh in 1771 and Harvard in 1773. CAREER: 1738-1779, Hollis professor of mathematics and natural philosophy, Harvard College. MEMBERSHIPS: Roy. Soc.; Am. Phil. Soc. SCIENTIFIC CONTRIBUTIONS: Established at Harvard in 1746 the first American laboratory for experiments in physics, used for teaching laws of mechanics, heat, light, and motions of heavenly bodies according to Newtonian precepts, and also utilized laboratory for personal research. In 1751, introduced study of fluxions (differential and integral calculus) into Harvard curriculum. In 1739 commenced astronomical observations with study of sunspots, probably first such investigation in English colonies; in 1740, 1743, and 1769, carried out observations of transits of Mercury, and employed results in attempt to determine difference in longitude between Greenwich, England, and Cambridge, Mass.; studied earthquake of 1755 that struck New England, and in public lecture of 1759, in regard to Halley's comet of 1682, made first prediction of return of a comet. Chief astronomical work probably done in relation to transits of Venus in 1761 and 1769; in order to observe 1761 transit, led expedition to St. John's, Newfoundland; for some twenty years, carried out magnetic and meteorological observations; held first place among colonial astronomers, and his observations appeared especially in *Phil. Trans of Roy. Soc.*. A devotee of Newtonian science and a supporter of the theories of electricity advanced by B. Franklin [q.v.].

WORKS: *A Lecture on Earthquakes* (Boston, 1755); *Two Lectures on Comets* (Boston, 1759); *Relation of a Voyage from Boston to Newfoundland, for the Observation of the Transit of Venus, June 6, 1761* (Boston, 1761); *Two Lectures on the Parallax and Distance of the Sun, as Deducible from the Transit of Venus* (Boston, 1769). See *DSB* for list of works appearing in *Phil. Trans. of Roy. Soc.* MSS: Harv. U.-UA (*NUCMC* 65-1299). WORKS ABOUT: *DAB*, 20: 414-16 (Frederick E. Brasch); *DSB*, 14: 452-53 (G. L'E. Turner).

WINTHROP, JOHN, Jr. (February 12, 1606, Groton, Suffolk, England—April 5, 1676, Boston, Mass.). *Natural philosophy*; *Medicine.* Son of John, governor of Massachusetts Bay, and Mary (Forth) Winthrop. Married Martha Fones, 1631; Elizabeth Reade, 1635; one child by first marriage, seven by second. EDUCATION: Studied at Free Grammar School, Bury St. Edmunds; 1622, began two years of study at Trinity College, Dublin; 1625, admitted to Inner Temple. CAREER: Soon abandoned law; 1627, appointed secretary to naval captain; 1628-1629, traveled in Europe; 1631, followed father to Massachusetts Bay Colony; 1632-1649, assistant, Massachusetts Bay; 1633, founder of Ipswich, Mass.; 1634-1635, in England; 1636, went to Connecticut as governor; ca. 1636, again settled at Ipswich; by 1639, was residing at Salem; undertook manufacture of salt; 1641-1643, in England; 1644, began construction of ironworks at Lynn and Braintree, Mass.; 1646, took up residence in vicinity of present-day New London, Conn., and in 1648 became a magistrate there; later, moved to New Haven, and pursued interests in ironworks; 1657 and 1659-1676, was governor of Connecticut, and took up residence at Hartford; 1661-1663, in England, where procured charter uniting Connecticut and New Haven colonies; was landowner and inherited some means from mother, but various commercial ventures achieved little success. MEMBERSHIPS: Elected first colonial fellow, Roy. Soc. SCIENTIFIC CONTRIBUTIONS: Motivated by local requirements of colonists; in hopes of procuring exportable commodities, made search for mineral resources, examined possible iron deposits in Maine, New Hampshire, and Massachusetts, and engaged in production of salt, saltpeter, indigo, and other products, but achieved long-term success only with iron and graphite. Credited as a founder of industrial chemistry in America; from early years in England had interest in alchemy, although full details of this activity are not now known. Achieved wide reputation in New England for medical knowledge, and at death held reputation as region's chief medical authority, being particularly interested in chemical prescriptions. By 1660, had in operation what appears to be first large telescope in America, though observations were not of great consequence. Interested in wide range of scientific subjects, results now extant chiefly in correspondence; ranks as first notable student of science in British America.

WORKS: Correspondence on scientific and other subjects in *The Winthrop Papers* (Boston, 1929-). See bibliographical note in *DSB*. WORKS ABOUT: *DAB*, 20: 411-13 (James T. Adams); *DSB*, 14: 451-52 (Ronald S. Wilkinson). Also see Robert C. Black, *The Younger John Winthrop* (New York and London, 1966).

WISLIZENUS, FREDERICK ADOLPHUS (May 21, 1810, Königsee, Schwarzburg, Rudolstadt, Germany—September 22, 1889, St. Louis, Mo.). *Meteorology*; *Natural history*. Son of Christian Anton Wislizenus, minister. Married Lucy Crane, 1850; had children. EDUCATION: Studied at universities of Jena, Göttingen, and Würzburg; 1834, M.D., Univ. of Zurich; later, studied in Paris hospitals. CAREER: 1835, arrived in New York; 1836-1839, medical practice, Mascoutah, St. Clair County, Ill.; 1839, made journey to present state of Idaho; ca. 1840-1846, medical practice, St. Louis; 1846-1847, made private scientific expedition to Chihuahua, Mexico, was detained for time as prisoner of war, then returned to St. Louis as temporary surgeon in U.S. Army; 1850, married in Constantinople; 1852, settled permanently as physician in St. Louis after unsuccessful attempt to locate in California. MEMBERSHIPS: A founder of Academy of Science of St. Louis (1856). SCIENTIFIC CONTRIBUTIONS: At St. Louis in 1840 published German-language account of travels to Idaho during the previous year, which was presented by son in 1912 translation as *A Journey to the Rocky Mountains* and included descriptions of plants, animals, and other natural phenomena. In 1846-1847 expedition to Southwest, achieved first scientific exploration of much of region crossed, despite interruptions caused by political conditions; account of this later expedition [see Works] included meteorological observations and also three charts that showed route taken, geological formations, and topography of the area; a botanical appendix based on substantial collections of specimens was prepared by Wislizenus's partner, Dr. G. Englemann [q.v.]. Published papers on Indian graves, atmospheric electricity, matter and force, army worm in *Transactions of Academy of Science of St. Louis*, and published meteorological reports in that journal and elsewhere; during later years, wrote especially on medicine.

WORKS: *Memoirs of a Tour to Northern Mexico, Connected with Col. Doniphan's Expedition in 1846 and 1847* (Glorieta, N. Mex., 1969), originally published in 1848 as U.S. Senate Document. See bibliography in Schlueter [below]. WORKS ABOUT: *DAB*, 20: 430-31 (W. J. Ghent); Robert E. Schlueter, "Frederick Adolphus Wislizenus . . . ," *Isis* 28 (1938): 38-52.

WISTAR, CASPAR (September 13, 1761, Philadelphia, Pa.—January 22, 1818, Philadelphia, Pa.). *Anatomy*. Son of Richard, glass manufacturer, and Sarah (Wyatt) Wistar. Married Isabella Marshall, 1788; Elizabeth Mifflin, 1798; three children by second marriage. EDUCATION: 1782, bachelor of medicine, Univ. of State of Pennsylvania (later part of Univ. of Pennsylvania); 1783, began year of study in London; 1786, M.D., Univ. of Edinburgh. CAREER: 1787, returned to Philadelphia from medical study in Europe, and immediately was elected member of College of Physicians of Philadelphia; ca. 1787, became a physician of Philadelphia Dispensary; 1789, appointed professor of chemistry, College of Philadelphia (later part of Univ. of Pennsylvania); at Univ. of Pennsylvania, became adjunct professor of anatomy, surgery, and midwifery (1792), appointed professor of anatomy and midwifery (1808), and was professor of anatomy (1810-1818); 1793-1810, staff member, Philadelphia Hospital; until several years prior to death, carried on medical practice in addition to other activities. MEMBERSHIPS: President, Am. Phil. Soc., 1815-1818. SCIENTIFIC CONTRIBUTIONS: While in Edinburgh, elected by fellow students as president of their Royal Medical Society and of the Edinburgh Natural History Society. As teacher of anatomy in U.S., introduced use of large-scale models, and also provided student groups with specimens for study and identification; *System of Anatomy* [see Works], earliest American textbook in that field, was warmly praised and widely used; major contribution to anatomical knowledge was study of relations of ethmoid and sphenoid bones of skull [see Works]. During early years, published papers on evaporation in *Transactions of Am. Phil.*

Soc., and later in same journal also published studies of paleontological interest based on specimens from Big Bone Lick, Ky. During lifetime had weekly gatherings in home for members of Am. Phil Soc., which continue as Wistar Association.

WORKS: "Observations on Those Processes of the Ethmoid Bone, Which Originally Form the Sphenoidal Sinuses," *Transactions of Am. Phil Soc.*, n.s. 1 (1818): 371-75; *A System of Anatomy for the Use of Students of Medicine*, 2 vols. (Philadelphia, 1811). See bibliographical note in *DSB*. MSS: Am. Phil. Soc. (*NUCMC* 61-944); also Hist. Soc. Penn. and College of Physicians of Philadelphia (*DSB*). WORKS ABOUT: *DAB*, 20: 533-34 (Francis R. Packard); *DSB*, 14: 456-57 (Whitfield J. Bell, Jr.).

WITTHAUS, RUDOLPH AUGUST (1846-1915). *Chemistry*; *Toxicology*. Professor of chemistry and toxicology, Univ. of Vermont; professor of physiological chemistry and physics, Univ. of City of New York (New York Univ.); professor of chemistry, physics, and toxicology, Cornell Univ.

WORKS ABOUT: See *AMS*; *DAB*.

WOOD, EDWARD STICKNEY (April 28, 1846, Cambridge, Mass.—July 11, 1905, Pocasset, Mass.). *Chemistry*; *Medicine*. Son of Alfred, grocer, and Laura (Stickney) Wood. Married Irene Eldridge Hills, 1872; Elizabeth A. Richardson, 1883; one child by first marriage. EDUCATION: 1867, A.B., Harvard College; 1871, M.D., Harvard Medical School; 1872, spent six months studying physiological and medical chemistry at Berlin and Vienna. CAREER: At Harvard Medical School, assistant professor of chemistry (1871-1876) and professor of chemistry (1876-1905); 1873, appointed chemist, Massachusetts General Hospital (Boston); served on sanitary commissions for state of Massachusetts and city of Boston; served on committee for revision of 1880 edition of U.S. *Pharmacopoeia*. MEMBERSHIPS: Am. Ac. Arts Scis.; American Pharmaceutical Association. SCIENTIFIC CONTRIBUTIONS: At Harvard, taught one of earliest comprehensive courses in U.S. on chemistry in relation to medicine. In 1874, became member of commission to study sanitary conditions of rivers in Boston area; also reported on gas lighting for Boston. Among publications were studies of bloodstains and of arsenic poisoning; also prepared revision of K. T. L. Neubauer and J. Vogel's *Guide to the Qualitative and Quantitative Analysis of the Urine* (1879) and contributed to publications on medical jurisprudence; publications appeared chiefly in medical journals and works of reference and as public documents. Especially noted as a leading American expert in questions of chemistry in relation to legal matters and won renown for role as expert witness in murder trials.

WORKS: "Arsenic as a Domestic Poison," *Report of Massachusetts Board of Health* 5 (1884), Supp., pp. 211-67. Chief publications listed in Harvard College Class of 1867, *Secretary's Report*, nos. 5 (1876)—11 (1902). WORKS ABOUT: *DAB*, 20: 455-56 (Henry R. Viets); J. Collins Warren, memoir, *Proceedings of Am. Ac. Arts Scis.* 51 (1915-1916): 929-30.

WOODHOUSE, JAMES (November 17, 1770, Philadelphia, Pa.—June 4, 1809, Philadelphia, Pa.). *Chemistry*. Son of William, bookseller and stationer, and Anne (Martin) Woodhouse. Never married. EDUCATION: B.A. in 1787 and M.D. in 1792, Univ. of Pennsylvania. CAREER: 1795-1809, professor of chemistry, Univ. of Pennsylvania; 1802, visited England and France. MEMBERSHIPS: Founder, Chemical Society of Philadelphia (1792); member, Am. Phil. Soc. SCIENTIFIC CONTRIBUTIONS: Interest in chemistry attributed to influence of B. Rush [q.v.], and doctoral thesis itself dealt with medical and chemical qualities of vegetable products. By 1795, had small well-equipped laboratory and carried out experiments that tended to disprove phlogiston theory, even though J. Priestley [q.v.], that theory's advocate, was frequent laboratory visitor; among chemical studies were analysis of what proved to be basaltic rock formations, experiments on metals and nitric acid, investigation of chemistry and production of white starch, and study of manufacture of purified camphor; also studied anthracite and bituminous coal and bread making; in 1806, carried out experiments on nitrous oxide gas and its effects as an anaesthetic. Of special interest is experiment, carried out in 1808, that reportedly resulted in isolation of potassium by new procedure; this experiment also is of interest in relation to early study on fixation of nitrogen from the air and on synthesis of ammonia. Notable as teacher, made use of laboratory instruction, and produced *The Young Chemist's Pocket Companion* (1797), an early student aid to experimentation; edited and annotated several standard works in chemistry.

WORKS: "Account of an Experiment in Which Potash Calcined with Charcoal Took Fire on the Addition of Water, and Ammoniacal Gas Was Produced," *Nicholson's Journal of Natural Philosophy, Chemistry and the Arts* 21 (1808): 290-91. See *Roy. Soc. Cat.*, vol. 6. MSS: Univ. of Pennsylvania (Browne [below]). WORKS ABOUT: *DAB*, 20: 491-92 (Edgar F. Smith); C. A. Browne article and K. S. Love and P. H. Emmett article, both on Woodhouse's experiments on synthetic production of ammonia, in *Journ. Chem. Educ.* 9 (1932): 1744-50.

WOODHOUSE, SAMUEL WASHINGTON (June 27, 1821, Philadelphia, Pa.—October 23, 1904, Philadelphia, Pa.). *Ornithology*; *Natural history*. Son of Samuel, commodore in U.S. Navy, and H. Matilda Roberts Woodhouse. Married Sarah A. Peck, 1872; survived by two children. EDUCATION: Attended private classical schools of Philadelphia and West Haven, Conn.; 1847, M.D., Univ. of Pennsylvania. CAREER: For a period before entering upon medical studies, engaged in farming in Chester County, Pa.; ca. 1848, assistant resident physician, Philadelphia Hospital; 1849-1850, surgeon and naturalist, U.S. Army Corps of Topographical Engineers' survey of Creek and Cherokee boundary; 1851-1852, member of Zuni River expedition; 1853, surgeon to private exploring expedition to Central America; 1854-1856, surgeon, Fort Delaware; 1856, resigned from army service; 1859-1860, surgeon, Cope's line of packets between Philadelphia and Liverpool; during Civil War, resident physician, Eastern Penitentiary, Philadelphia; thereafter, said to have retired to private life. MEMBERSHIPS: Ac. Nat. Scis. Philad.; Am. Ornith. Un.; and other societies. SCIENTIFIC CONTRIBUTIONS: Chief scientific work done in connection with Creek and Cherokee boundary survey, initially under Lt. Lorenzo Sitgreaves, and with Zuni exploration, also under leadership of Lt. Sitgreaves; during first survey, accumulated important natural history specimens, though few new birds and mammals were discovered; Zuni expedition went from San Antonio, Tex., to San Francisco and passed through territory largely unexplored by naturalists; Zuni expedition resulted in six new birds and mammals, reported upon by Woodhouse in Sitgreaves's report, while collections of reptiles, fishes, and plants were described by others. Collections during Central American exploration were not large, and no report ever appeared. For number of years, lost contact with ornithological workers, but during last years interest was revived and in 1899 and 1903 attended congresses of Am. Ornith. Un.

WORKS: Papers appeared in *Proceedings of Ac. Nat. Scis. Philad.* 5 (1850-1851) and 6 (1852-1853); and in L. Sitgreaves, *Report of an Expedition Down the Zuni and Colorado Rivers* . . . Senate Executive Document (Washington, D.C., 1853), pp. 31-105. See Meisel, *Bibliog. Amer. Nat. Hist.* MSS: Ac. Nat. Scis. Philad. WORKS ABOUT: Witmer Stone, "S. W. Woodhouse," *Cassinia: Proceedings of the Delaware Valley Ornithological Club* 8 (1904): 1-5 (abbreviated in *Auk* 20 [1905]: 104-6).

WOODWARD, ROBERT SIMPSON (1849-1924). *Astronomy*; *Geodesy*; *Mathematics*; *Physics*. Professor of mechanics and mathematical physics and dean of science, Columbia Univ.; president, Carnegie Institution of Washington.

MSS: See *DSB*. WORKS ABOUT: See *AMS*; *DAB*; *DSB*.

WORTHEN, AMOS HENRY (October 31, 1813, Bradford, Vt.—May 6, 1888, Warsaw, Ill.). *Geology*. Son of Thomas, farmer, and Susannah (Adams) Worthen. Married Sarah B. Kimball, 1834; seven children. EDUCATION: Studied at Bradford Academy. CAREER: 1834, moved to Kentucky; ca. 1835-1836, teacher, Cumminsville, Ohio; 1836, moved to Warsaw, Ill., and there with brothers-in-law entered dry goods business; 1842-1844, lived in Boston, Mass.; 1844, returned to Warsaw and until 1855 continued mercantile business; 1853, employed for time as assistant with Illinois survey; 1855-1857, assistant to state geologist of Iowa; 1858-1888, Illinois state geologist; 1877, appointed curator, Illinois State Historical Library and Natural History Museum. MEMBERSHIPS: Am. Assoc. Advt. Sci.; Natl. Ac. Scis.; and other societies. SCIENTIFIC CONTRIBUTIONS: After move to Illinois, began to collect geological specimens from vicinity of home at Warsaw and engaged in correspondence and exchange of specimens with eastern scientists; prepared two chapters for report by J. Hall [q.v.] as state geologist of Iowa and aided Hall with construction of geological section along Mississippi River, also published in Iowa report. Chief work done as director of Illinois survey, for which he recruited an accomplished staff of scientific assistants. Worthen's own important studies related especially to Lower Carboniferous strata; prepared eight volumes of reports (last published posthumously), which covered all sections of state, and reviewed mineral resources; reports were particularly noted for contributions to paleontology, with descriptions of over 1,600 species of invertebrate and of vertebrate and botanical fossil forms, most of which were described there in published form for first time.

WORKS: *Geological Survey of Illinois*, 8 vols. in 9 (Springfield, Ill., 1866-1890). See bibliography of works in White [below], pp. 348-61, with tables of contents of reports. WORKS ABOUT: *DAB*, 20: 537-38 (Carey Croneis); Charles A. White, "Memoir," *Biog. Mems. of Natl. Ac. Scis.* 3 (1895): 339-62.

WRIGHT, ALBERT ALLEN (April 27, 1846, Oberlin, Ohio—April 2, 1905, Oberlin, Ohio). *Geology*; *Natural history*. Son of William Wheeler and Susan (Allen) Wright. Married Mary Lyon Bedortha, 1874; Mary P. B. Hill, 1891; survived by one child from first marriage and one from second. EDUCATION: 1865, A.B., Oberlin College; for two years, studied at Union Theological Seminary, N.Y.; 1870, degree in theology, Oberlin Seminary; 1875, Ph.B., School of Mines, Columbia Univ. CAREER: 1864, served with Ohio National Guard, in defense of Washington; ca. 1865-ca. 1867, teacher in Cleveland Institute; 1870-ca. 1872, professor of mathematics and natural science, Berea College (Kentucky); 1874-1905, professor of geology and natural history, Oberlin College; 1874, 1884, and 1893, employed on special research projects by Ohio Geological Survey; 1884-1885, traveled in Europe. MEMBERSHIPS: Geol. Soc. Am. SCIENTIFIC CONTRIBUTIONS: Chief labors devoted to teaching; for a time at Oberlin had charge of zoology and botany in addition to geology, and there established laboratory study by students; actively involved in practical scientific needs of local community, including plan and construction of exemplary sewer system and waterworks for city of Oberlin. As president of Ohio Academy of Sciences (and elsewhere), promoted topographical survey of Ohio, finally authorized by state in cooperation with U.S. Geol. Surv. In 1874 contributed to Ohio Geological Survey a report on lake ridges of Lorain County; in 1884 prepared report for survey on coal seams of Holmes County, and in 1893 wrote on "Ventral Armor of Dinichthys" (extinct placoderm fish) based on Oberlin collection; results of all these researches appeared in reports of the Ohio survey; geological studies related especially to Ohio, and while publications were limited in number, they were considered to have been well executed; geological publications relating to work outside Ohio included study of drift and glaciation in New Jersey; among other works was list of local flowering plants and ferns. At time of death engaged in study of thin sections of bryozoans.

WORKS: Bibliographies in both Wright and Wilder [below]. MSS: Oberlin College Archives. WORKS ABOUT: G. F. Wright, "A. A. Wright," *American Geologist* 36 (1905): 65-68; Frank A. Wilder, "Memoir," *Bulletin of Geol. Soc. of Am.* 17 (1906-1907): 687-90.

WRIGHT, ARTHUR WILLIAMS (1836-1915). *Experimental physics.* Professor of experimental physics and director of Sloane Physical Laboratory, Yale Univ.

MSS: Yale Univ. Library (*NUCMC* 74-1209). WORKS ABOUT: See *AMS*; *NCAB*, 13 (1906): 348; *World Who's Who in Science.*

WRIGHT, CHARLES (October 29, 1811, Wethersfield, Conn.—August 11, 1885, Wethersfield, Conn.). *Botanical exploration.* Son of James, carpenter and joiner, and Mary (Goodrich) Wright. Never married. EDUCATION: 1835, A.B., Yale College. CAREER: 1835, became teacher on plantation, Natchez, Miss.; 1837-1845, surveyor and teacher at various locations in eastern Texas, and began botanical explorations; 1845-1846, teacher in short-lived Rutersville College (Fayette County, Tex.); 1846-1847, private tutor in Rutersville, Tex.; 1847-1848, teacher at Austin; 1848-1849 (winter), worked at Harvard Univ. herbarium on plants previously sent to A. Gray [q.v.]; 1849, accompanied U.S. troops from San Antonio to El Paso; 1850, tutor, San Marcos, Tex.; 1850-1851, operated school at New Braunfels, Tex.; 1851-1852, botanist with U.S.-Mexico boundary survey; 1852, permanently departed from Texas; 1853-1856, botanist, North Pacific Surveying and Exploring Expedition, under John Rodgers and Cadwalader Ringgold; 1856-1867, chiefly engaged in botanical explorations in Cuba; 1868, acting curator, Harvard herbarium; 1871, botanical explorations in Santo Domingo; 1875-1876, librarian, Bussey Institution, Harvard; 1876-1885, resided at Wethersfield. SCIENTIFIC CONTRIBUTIONS: As one of most important botanical explorers of period, he collected number of new species and contributed valuable observations; collections gathered from various parts of world related to wide range of botanical groups and were prepared for publication by many botanists, especially A. Gray with whom he began correspondence in 1844. Started botanical explorations in Texas at time when flora of that region was little known; during 1849 expedition from San Antonio to El Paso, collected some 1,400 species, and later with Mexican boundary survey collected extensively in New Mexico and Arizona. Also collected extensively at Cape of Good Hope, Hong Kong, Loo Choo Islands (Ryuku), and Japan during North Pacific expedition; later did equally important work in survey of Cuban botany, and also explored flora around Wethersfield.

WORKS: A. Gray, "Plantae Wrightianae Texano-Neo-Mexicanae," *Smithson. Contr. Knowl.* 3 (1852) and 5 (1853). See Meisel, *Bibliog. Amer. Nat. Hist.*; *Roy. Soc. Cat.*, vols. 8, 12, 19. MSS: Harv. U.-GH (Hamer); Mo. Bot. Gard. (*DAB*). WORKS ABOUT: *DAB*, 20: 545-46 (Samuel W. Geiser); Geiser, *Naturalists of the Frontier* (Dallas, 1948), pp. 172-98.

WRIGHT, CHAUNCEY (September 20, 1830, Northampton, Mass.—September 11, 1875, Cambridge, Mass.). *Philosophy of science*; *Mathematics.* Son of Ansel, groceryman and deputy sheriff, and Elizabeth (Boleyn) Wright. Never married. EDUCATION: 1852, A.B., Harvard College. CAREER: 1852-1870, computer for *American Ephemeris and Nautical Almanac* office in Cambridge; 1862-1870, recording secretary and editor, Am. Ac. Arts Scis.; 1870, delivered lectures on psychology, Harvard Univ.; 1872, made trip to Europe; 1874-1875, instructor in mathematical physics, Harvard. MEMBERSHIPS: Am. Ac. Arts Scis. SCIENTIFIC CONTRIBUTIONS: As Harvard undergraduate, demonstrated special abilities and interest in science and mathematics, and in mathematical piecework for *Nautical Almanac* devised means for facilitating computations that greatly reduced time devoted to that work, making available more time for philosophical studies; contributed several articles to *Mathematical Monthly* (published at Cambridge). In 1864, began publication of philosophical series in *North American Review* and in period 1864-1869 became established as philosopher, while during period 1870-1875 chief works appeared; greatest influence on his thought was that of Darwin, and increasingly turned attention to questions of logic in relation to evolutionary theory; wrote on arrangement of leaves on plant, as an aspect of evolutionary thought; published important papers on evolution, including "Genesis of Species" (1871), in *North American Review.* Interest in Darwinian ideas related ultimately to larger concern with structure of scientific thought itself, and represented new type of American philosopher, with sound background in both philosophy and technical aspects of science per se. With exception of major philosophical essay, "Evolution of Self-Consciousness" (1873), writings were prepared in response to occasion rather than as part of planned program; chief influence exerted not through publications or teaching, but in philosophical discussions at Cambridge, of particular importance being relations with W. James and C. S. Peirce [qq.v.].

WORKS: *Philosophical Discussions* (New York, 1877), edited by C. E. Norton, a collection of Wright's essays and reviews. See also annotated bibliography in Madden [below], pp. 162-64. MSS: Harv. U.-HL (Madden [below]). WORKS ABOUT: *DAB,* 20: 547-48 (Ernest S. Bates); Edward H. Madden, *Chauncey Wright,* Great American Thinkers Series (New York, [1964]).

WURTZ, HENRY (June 5, 1828, Easton, Pa.—November 8, 1910, Brooklyn, N.Y.). *Chemistry.* Son of John J. and Ann (Novus) Wurts. Married; survived by five children. EDUCATION: 1848, A.B., Princeton Univ.; 1848-1849, said to have been enrolled at Lawrence Scientific School, Harvard [catalogue lists Jacob Henry Wurtz]. HONORS: Ph.D., Stevens Institute of Technology, 1877. CAREER: Ca. 1850, associated with laboratory of O. W. Gibbs [q.v.], New York; 1851, worked as assistant to B. Silliman, Jr. [q.v.], at Yale Univ.; 1854-1856, chemist and mineralogist, New Jersey State Geological Survey; 1857 (summer), explored geology of North Carolina; 1857-1858, lecturer on chemistry, medical college at Kingston, Ontario, Canada; 1858, appointed professor of chemistry and pharmacy, National Medical College of Washington, D.C. (later part of George Washington Univ.); during same period acted as chemical examiner in U.S. Patent Office; 1861, established private consulting laboratory, New York City; 1868-1871, editor, *American Gas Light Journal,* continuing private consulting business in laboratory at Hoboken, N.J.; 1876, appointed judge of exhibits and special examiner of ceramics, Philadelphia Centennial Exhibition; for some ten years, devoted efforts chiefly to developmental work relating to coal distillation processes; during later years, continued private chemical consultant business. MEMBERSHIPS: Am. Assoc. Advt. Sci. SCIENTIFIC CONTRIBUTIONS: In 1850, published papers on supposed new mineral and on New Jersey greensand as source of potash; employment with New Jersey survey yielded noteworthy analyses of water of Delaware River; later, in North Carolina, discovered deposits of cobalt and nickel, and while at Washington published result of studies on use of blowpipe (1859). Researches as consultant included study of extraction of precious metals from ores, and utilization of new mineral from Virginia; during period of editorship of *Gas Light Journal,* patented means for production of fuel gas (1869); studied river and water supplies of New Jersey (1873, 1874); and wrote on new processes in analysis of gas (1875). In 1876, published results of research on Japanese and Chinese porcelains, based on work done for Philadelphia exhibition, and during same year published suggestive essay on geometrical chemistry. Thereafter, researches related especially to processes for distillation of coal and of paraffin hydrocarbons and other chemical substances; received number of patents for this work.

WORKS: See references to chief works in *DAB*; also see *Roy. Soc. Cat.,* vols. 6, 8, 11, 19. MSS: N.Y. Pub. Lib. (Hamer). WORKS ABOUT: *DAB,* 20: 571-72 (C. A. Browne); *NCAB,* 7 (1897 [copyright 1892]): 519.

WYMAN, JEFFRIES (August 11, 1814, Chelmsford, Mass.—September 4, 1874, Bethlehem, N.H.). *Anatomy*; *Ethnology.* Son of Rufus, physician, and Ann (Morrill) Wyman. Married Adeline Wheelwright, 1850; Annie Williams Whitney, 1861; two children by first marriage, one by second. EDUCATION: A.B. in 1833 and M.D. in 1837, Harvard Univ.; 1841, studied anatomy, physiology, and zoology in Paris and London. CAREER: Ca. 1837, began medical practice at Boston; 1838-1840, demonstrator in anatomy at Harvard Medical School; at Lowell Institute (Boston), appointed curator (1840) and lectured on comparative anatomy and physiology (1840-1841); 1843-1848, professor of anatomy and physiology, Hampden-Sydney Medical College (Virginia); 1847-1874, Hersey professor of anatomy, Harvard Univ.; 1848-1849, Lowell Institute lecturer; 1856, made collecting journey to Surinam; 1857-1866, connected with private medical school, Cambridge, Mass.; 1858, accompanied expedition of J. M. Forbes to South America; 1866-1874, curator, Peabody Museum of American Archaeology and Ethnology, Harvard; during most of adult life, spent winters in warm climate, going often to Florida and to Europe. MEMBERSHIPS: President, Bos. Soc. Nat. Hist., 1856-1870; member, Natl. Ac. Scis. and other societies. SCIENTIFIC CONTRIBUTIONS: Early showed ability in preparation of natural history specimens and had talent as artist; scientific interests were furthered by aid from men of wealth. Came to hold reputation as foremost American authority in anatomy; with appointment to Hersey professorship, took up work at Cambridge, teaching anatomy and physiology to undergraduates, and there assembled notable anatomical collection. Undertook several collecting expeditions; during later years, as curator of archaeology and ethnology, also built up collections in that area as well. Author of some 175 fairly short papers, dealing chiefly with anatomy; wrote notable work on nervous system of frog (1853) and series of articles on anatomy of blind fish of Mammoth Cave (1843-1854); during years 1862-1867 carried out experiments on presence of organisms in boiled water. Chief work done on structure of gorilla [see Works] and earned international reputation for study of higher primate anatomy. During later years particularly interested in shell heaps in Maine, Massachusetts, and Florida, which were studied as evidence regarding their builders.

WORKS: "Notice of the External Characters, Habits, and Osteology of Troglodytes gorilla . . . ," *Boston Journal of Natural History* 5 (1845-1847): 417-42, with T. S. Savage [q.v.]. Bibliography in *Biog. Mems. of Natl. Ac. Scis.* 2 (1886): 118-26. MSS: Harv. U.-MS (*DSB*). WORKS ABOUT: *DAB*, 20: 583-84 (Hubert L. Clark); *DSB*, 14: 532-34 (A. Hunter Dupree).

X

XÁNTUS, JÁNOS (October 5, 1825, Csokonya, county of Somogy, Hungary—December 13, 1894, Budapest, Hungary). *Natural history collector*; *Ornithology*. Son of Ignácznak—solicitor, land agent, and steward for an estate—and Terézia Vandertich Xántus. Married Gabriella Doleschall, 1873; later, Ilona Steden; one child. EDUCATION: Attended law academy at Györ, Hungary; 1847, passed bar examination at Pest. CAREER: During 1848 war of independence, enlisted in Hungarian army; 1851, went to U.S.; at St. Louis (1851-1853), New Orleans (ca. 1853-ca. 1854), and New Buda, Decatur county, Iowa (ca. 1854-ca. 1855), but specific activities during these years are uncertain; 1855-1859, in U.S. Army, was stationed first at Fort Riley (Kansas), received rank of hospital steward (1857), then stationed at Fort Tejon, Calif. (1857-1859) ; 1859-1861, U.S. Coast Survey tide observer, Cape San Lucas (Lower California) ; 1861-1862, visited Hungary; 1862-1863, U.S. consul, Manzanillo, Mexico; 1864, left Mexico and returned to Hungary; 1866, became director, zoological garden at Pest; 1868, appointed collector with Austro-Hungarian East Asiatic Expedition; 1869-1871, continued Asian explorations independently; 1872, appointed keeper of ethnographical section and also served as biology librarian, National Museum at Budapest. MEMBERSHIPS: Ac. Nat. Scis. Philad. SCIENTIFIC CONTRIBUTIONS: Chief scientific work in U.S. done as collector at Fort Tejon and Cape San Lucas, Calif., under patronage of S. F. Baird [q.v.] and Smithson. Instn. Overall evaluation obscured by numerous fabrications written in letters to family and by publication of excerpts from travel and exploration reports by various Americans, translated from English by Xántus and presented as his own. While at Fort Riley, began sending specimens to Ac. Nat. Scis. Philad.; especially interested in birds, in California he collected full range of botanical and zoological specimens, with useful notes and drawings; during years on Cape San Lucas contributed 290 new species; in 1862 was commended by Baird for having assembled most complete natural history collections ever made by one person in America within a comparable span of time. Not trained in taxonomy, and so the few papers published under own name were perfected by Baird. Also carried out some explorations and collecting while in Mexico.

WORKS: Four short papers in *Proceedings of Ac. Nat. Scis. Philad.*, vols. 10 (1858) : 117; 11 (1859) : 189-93 and 297-99; and 12 (1860) : 568. See Madden [below], chap. 8, for detailed review of bibliographic complexities caused by Xántus's plagiarizings and fabrications, and chapter 9, for bibliography of works by other authors, based on his collections. MSS: Especially with Baird Papers, Smithson. Instn. (Madden). WORKS ABOUT: *DAB*, 20: 589-90 (Charles Feleky) [incorporates many traditional errors]; Henry M. Madden, *Xántus* . . . (Palo Alto, Calif., 1949).

Y

YANDELL, LUNSFORD PITTS (July 4, 1805, near Hartsville, Sumner County, Tenn.—February 4, 1878, Louisville, Ky.). *Paleontology.* Son of Wilson, physician, and Elizabeth (Pitts) Yandell. Married Susan Juliet Wendell, 1825; Eliza Bland, 1861; four children by first marriage. EDUCATION: Attended Bradley Academy (Murfreesboro, Tenn.); 1822-1823, studied medicine, Transylvania Univ. (Lexington, Ky.); 1825, M.D., Univ. of Maryland. CAREER: 1826, established medical practice, Murfreesboro; 1830, moved to Nashville, Tenn.; 1831-1837, professor of chemistry and pharmacy, Transylvania Univ.; 1832-1836, editor, *Transylvania Journal of Medicine and Associated Sciences*; at Louisville Medical Institute (after 1846, medical department, Univ. of Louisville), professor of chemistry and materia medica (1837-1849), professor of physiology and pathological anatomy (1849-1859), and dean; 1840-1855, coeditor, *Western Journal of Medicine and Surgery;* 1859, assumed professorship in Memphis Medical College; 1861, entered service as hospital surgeon, Confederate army; 1862, licensed to preach by Memphis Presbytery; 1864, ordained as minister of church at Dancyville, Tenn.; 1867-1878, medical practice, Louisville. MEMBERSHIPS: Ac. Nat. Scis. Philad.; Bos. Soc. Nat. Hist. SCIENTIFIC CONTRIBUTIONS: After move to Louisville in 1837, took up active study of natural history, especially collection and study of fossil forms from surrounding region; in 1847 published *Geology of Kentucky* with B. F. Shumard [q.v.] [see Works], and later published other works on paleontology. Active author in several fields and is credited with some 100 papers on medicine, geology, local history, biography, education, and religion; beginning in 1873, contributed series of popular articles to local publication, *Home and School,* on what fossils teach, blood and the brain, food and digestion, mind and body, physiology of sleep and dreams, birds, and other subjects. Various paleontological authors honored Yandell as discoverer, by giving his name to fossil specimens.

WORKS: *Contributions to the Geology of Kentucky* (Louisville, Ky., 1847), 36 pp., with B. F. Shumard; "On the Distribution of the Crinoidea in the Western States," *Proceedings of Am. Assoc. Advt. Sci.* 5 (1851): 229-35; "Description of a New Genus of Crinoidea," *Am. Journ. Sci.* 2d ser. 20 (1855): 135-37. MSS: Filson Club (Louisville, Ky.). WORKS ABOUT: DAB, 20: 596-97 (James M. Phalen); J. M. Toner, "Sketch," *Nashville Journal of Medicine,* n.s. 21 (1878): 71-79.

YARNALL, MORDECAI (April, 1816, near Urbana, Ohio—February 27, 1879, Washington, D.C.). *Astronomy.* No information on parentage or marital status. EDUCATION: Reportedly graduated from school of civil engineering, Bacon College (Kentucky). CAREER: 1837-1839, assistant professor of mathematics, Bacon College; 1839, appointed professor of mathematics, U.S. Navy, and from 1839 to 1852 served as instructor at sea; for period during 1840s, associated with O. M. Mitchel [q.v.] at Cincinnati Observatory; 1852-1878, assigned to U.S. Naval Observatory, Washington, D.C. MEMBERSHIPS: Am. Assoc. Advt. Sci. SCIENTIFIC CONTRIBUTIONS: Naval Observatory had begun observations of stars in mid-1840s, and Yarnall was given task of preparing new star catalogue; during years with observatory, conducted number of observations with transit instrument and meridian circle, and also carried out computations that reduced variant data to standard form; resulting *Catalogue* (1878) [see Works] earned author reputation as astronomer of note.

WORKS: *Catalogue of Stars Observed at the U.S. Naval Observatory During the Years 1845 to 1877, and prepared for publication by Prof. M. Yarnall,* 2d ed. (Washington, D.C., 1878). WORKS ABOUT: *Appleton's CAB,* vol. 7: 287; Marc Rothenberg, "The Educational and Intellectual Background of American Astronomers, 1825-1875" (Ph. D. diss., Bryn Mawr College, 1974), pp. 182-84.

YEATES, WILLIAM SMITH (December 15, 1856, Murfreesboro, N.C.—February 19, 1908, Atlanta, Ga.). *Geology*; *Mineralogy*. Son of Jesse J., lawyer and congressman, and Virginia (Scott) Yeates. Married Julia W. Moore, 1884; three children. EDUCATION: Studied at Randolph-Macon College; 1878, B.A., Emory and Henry College (Virginia). CAREER: 1879, distributing messenger, U.S. Commission on Fish; for year and a half, taught school; 1880-1881, clerk, Tenth U.S. Census, in division of fisheries statistics; 1881-1893, assistant curator in charge of minerals and gems, U.S. Natl. Mus. (Washington, D.C.); at Columbian Univ. (later, George Washington Univ.), instructor (1884-1887), professor of mineralogy (1887-1890), and professor of mineralogy and geology (1890-1893); 1893-1908, state geologist of Georgia. MEMBERSHIPS: Geol. Soc. Am.; Am. Chem. Soc.; and other societies. SCIENTIFIC CONTRIBUTIONS: Chiefly noted as scientific administrator, rather than researcher, and is credited with few original publications; at U.S. Natl. Mus., developed reputation for ability to prepare exhibits, and assembled the museum's mineral and gem displays for several world's fairs, including World's Columbian and Louisiana Purchase expositions. Through extensive museum work, became known for ability to identify minerals visually, often even pinpointing specimen's locality. As head of Georgia survey, directed attention primarily to questions of economic interest, and reports issued during term of office dealt especially with location of building stone, manganese, phosphates, ochres, coal, gold, and the like, and also touched on water power and subterranean water; established Geological Museum in Atlanta.

WORKS: See Nickles, *Geol. Lit. on N. Am.* WORKS ABOUT: *NCAB*, 13 (1906): 60; George P. Merrill, "Memoir," *Bulletin of Geol. Soc. Am.* 20 (1909-1910): 618-19; S. W. McCallie, "In Memoriam," *Bulletin of Geological Survey of Georgia* 19 (1909): 7-8.

YOUMANS, EDWARD LIVINGSTON (June 3, 1821, Coeymans, Albany County, N.Y.—January 18, 1887, New York, N.Y.). *Science journalism*; *Popularization of science*. Son of Vincent, mechanic and farmer, and Catherine (Scofield) Youmans; brother of William J. Youmans [q.v.]. Married Catherine E. (Newton) Lee, 1861; no children. EDUCATION: Formal education limited to elementary grades; largely self-educated. CAREER: At young age, developed ophthalmia and at times during early life was nearly blind and dependent on sister for reading and carrying out chemical experiments; 1851-1868, with improvement in eyesight, traveled widely in U.S. as lyceum lecturer on science; established association with House of Appleton, publishers, as writer, editor, and adviser on scientific publications; 1869-1870, editor, *Appleton's Journal of Popular Literature, Science, and Art*; 1871-1874, editor, "Scientific Miscellany" for *Galaxy*; 1871-1887, founder and editor, International Scientific Series published by Appleton; 1872-1887, founder and editor, *Popular Science Monthly* (Appleton). MEMBERSHIPS: Am. Assoc. Advt. Sci. SCIENTIFIC CONTRIBUTIONS: Through writing, lecturing, and editing, took the leading role in popularizing science in late nineteenth-century America and imbued American people with appreciation of scientific methods. A devotee of writings of Herbert Spencer, Youmans did much to introduce them into U.S. and contributed significantly to popular acceptance of evolution theory in this country; also helped introduce work of other leading scientists, and in International Scientific Series persuaded leading American, and especially European, scientists to write on their subjects for the public. Also attempted to reform teaching of science, and *A Class-Book of Chemistry* (1851) was widely popular and influential as school textbook and for popular reading; also wrote other popular works, including *Chemical Chart* (1855) to accompany *Class-Book*, *Handbook of Household Science* (1857), and others. Most lasting contribution was establishment of *Popular Science Monthly*, which exerted greater influence on spread of scientific culture than any other publication of the time; brother William continued as editor of *Monthly* after Edward's death.

WORKS: See footnote reference in Haar [below], especially p. 193. WORKS ABOUT: *DAB*, 20: 615-16 (Harry G. Good); Charles M. Haar, "E. L. Youmans: A Chapter in the Diffusion of Science in America," *Journal of History of Ideas* 9 (1948): 193-213. Also see John Fiske, *Edward Livingston Youmans, Interpreter of Science for the People*... (Freeport, N.Y., 1972; originally published in New York, 1894).

YOUMANS, WILLIAM JAY (October 14, 1838, Milton, near Saratoga, N.Y.—April 10, 1901, Mt. Vernon, N.Y.). *Science journalism; Popularization of science*. Son of Vincent, mechanic and farmer, and Catherine (Scofield) Youmans; brother of Edward L. Youmans [q.v.]. Married Celia Greene, 1866; four children. EDUCATION: Studied with Charles A. Joy, Columbia Univ. 1860-1861, attended Sheffield Scientific School, Yale; 1865, M.D., Univ. of City of New York (New York Univ.); ca. 1866, went to London, studied with Thomas Huxley. CAREER: Ca. 1869-1872, medical practice, Winona, Minn.; 1872-1900, associated with *Popular Science Monthly*, from 1887 to 1900 as editor-in-chief; 1900, *Monthly* was sold, retired to farm at Mt. Vernon, N.Y. MEMBERSHIPS: Am. Assoc. Advt. Sci. SCIENTIFIC CONTRIBUTIONS: Career in science largely influenced by older brother, Edward, who founded *Popular Sci-*

ence Monthly. Especially interested in physiology and chemistry; on return from European studies, at author's request, carried through the press Huxley's *Elements of Physiology and Hygiene: A Text-Book for Educational Institutions* (1868), adapted by Youmans to American needs, with addition of instructional aids and seven chapters on hygiene. Writings appeared especially in *Monthly,* with regular column on scientific subjects, and continued brother's great interest in promotion of evolutionary thought of Herbert Spencer; wrote series of sketches of American scientists, first published in the *Monthly* and later collected together in *Pioneers of Science in America* (1896); during period 1880-1900, prepared reviews of progress in chemistry, metallurgy, meteorology, and physiology for *Appleton's Annual Cyclopaedia.* Continued brother's devotion to popularization of science and science education.

WORKS: See above. WORKS ABOUT: *DAB,* 20: 616-17 (Harry G. Good); *NCAB,* 2 (1899 [copyright 1891 and 1899]): 466.

YOUNG, AARON (December 19, 1819, Wiscasset, Maine—January 13, 1898, Belmont, Mass.). *Botany.* Son of Aaron, surveyor of lumber and justice of peace, and Mary (Colburn) Young. Said never to have been married. EDUCATION: For two years, attended Bowdoin College; 1842-1843, studied at Jefferson Medical College (Philadelphia). CAREER: 1840-1841, assistant in chemistry, Bowdoin College; went to Philadelphia, and there sought without success to gain relief from total deafness; thereafter, returned to Maine and for year tried to establish practice as aurist; ca. 1843, became apothecary in drugstore, Bangor, Maine; 1847-1848, engaged in botanical survey for state of Maine; after 1850, engaged in series of activities, including practice as ear surgeon at Auburn, Lewiston, and Portland, Maine; peddled cure-all; 1852-1854, wrote and printed three short-lived newspapers, *Farmer and Mechanic, Pansophist,* and *Touchstone*; 1859, published *Franklin Journal of Aural Surgery and Rational Medicine,* Farmington, Maine; during Civil War, fled to New Brunswick, under pressure at Bangor on account of political views; 1863-1873, American consul at Rio Grande do Sul, Brazil; 1875, returned to Boston to practice and became member of Massachusetts Medical Society; details of last years uncertain. MEMBERSHIPS: Secretary, Bangor Natural History Society, 1839-1840. SCIENTIFIC CONTRIBUTIONS: In 1843, published "Flora of Bangor" in *Bangor Daily Whig and Courier*; initiated and carried out Botanical Survey of Maine, and with several associates was probably first to explore entire length of Mt. Katahdin; drew attention to extensive true alpine flora on its summit and made collection of most obvious specimens; also first to make systematic collection of marine algae of Maine, and surveyed hardwoods in York County; in 1848, submitted preliminary report on botanical survey, which was published in *Maine Farmer* (March-May 1848) and related chiefly to work on Mt. Katahdin; at same time, began preparation of exsiccatae of Maine flora, of which only one volume appeared [see Works]. With end of state support for survey, and with lack of subscribers for *Flora,* gave up active botanical studies. While in Brazil, collected for Smithson. Instn.

WORKS: *A Flora of Maine, Illustrated with Specimens from Nature* . . . (Bangor, Maine, 1848), 28 pp. MSS: Am. Phil. Soc. WORKS ABOUT: *DAB,* 20: 617-18 (Henry R. Viets); Arthur H. Norton, "Dr. A. Young, Jr. and the Botanical Survey of Maine," *Rhodora* 37 (January 1935): 1-16.

YOUNG, CHARLES AUGUSTUS (1834-1908). *Astronomy.* Professor of astronomy, Princeton Univ.

WORKS ABOUT: See *AMS*; *DAB*; *DSB.*

Appendixes

The following appendixes list names of the scientists according to certain biographical data, under the following general headings:

A. Year of Birth
B. Place of Birth
C. Education
D. Occupation
E. Fields of Science

Parentheses have been placed around the names of scientists who are subjects of minor sketches in this *Dictionary* (those whose names appeared in the *American Men of Science* volumes and who therefore are not written up fully in this work). The minor sketches do not list the scientists' place of birth or level of education. For purposes of making these appendixes complete and uniform in content, this information has been compiled for Appendix B (Place of Birth) and Appendix C (Education), using the entries for the scientists in *American Men of Science*. Supplementary sources have been consulted when necessary. Therefore, all five appendixes include all of the names in the *Dictionary*, even though the biographical data represented in Appendixes B and C do not appear in the minor biographical sketches in the main body of this work.

Dates following names in Appendixes B-E are the scientists' dates of birth.

Appendix A
YEAR OF BIRTH

Not Known
Freeman, Thomas (d. 1821)
Grew, Theophilus (d. 1759)
Lawson, John (d. 1711)
Meade, William (d. 1833)
More, Thomas (fl. 1670-1724)

1606
Winthrop, John, Jr.

1608
Josselyn, John (ca. 1608)

1613
Sherman, John

1639
Mather, Increase

1650
Banister, John

1658
Brattle, Thomas

1662
Mather, Cotton (1662/1663)

1673
Coleman, Benjamin

1674
Logan, James

1675
Dudley, Paul

1680
Boylston, Zabdiel

1683
Catesby, Mark

1688
Colden, Cadwallader
Robie, Thomas (1688/1689)

1691
Douglass, William (ca. 1691)

1692
Jones, Hugh (ca. 1692)

1694
Clayton, John

1699
Bartram, John

1700
Fry, Joshua (ca. 1700)

1702
Greenwood, Isaac

1703
Clap, Thomas

1704
Godfrey, Thomas

1705
Leeds, John

1706
Franklin, Benjamin

1708
Lining, John

1711
Kinnersley, Ebenezer
Mitchell, John

1714
Winthrop, John

1715
Chalmers, Lionel

1717
De Brahm, William G.

1720
Romans, Bernard (ca. 1720)

1722
Marshall, Humphry

1724
Colden, Jane

1726
Bowdoin, James

1730
Garden, Alexander
Hutchins, Thomas
West, Benjamin

1731
Banneker, Benjamin
Oliver, Andrew

1732
Rittenhouse, David

1733
Brown, Joseph
Priestley, Joseph

1735
Williamson, Hugh

1737
Biddle, Owen

1739
Bartram, William

1740
Walter, Thomas (ca. 1740)

1741
Kuhn, Adam

1742
Cutler, Manasseh
Peale, Charles W.

1743
Jefferson, Thomas
Patterson, Robert
Pike, Nicholas

1744
Jeffries, John (1744/1745)

1746
Michaux, André
Rush, Benjamin

1749
Dunbar, William
Madison, James
Melsheimer, Friedrich V.

1750
Daboll, Nathan

1751
Abbot, John

1752
Schöpf, Johann D.

1753
Minto, Walter
Mühlenberg, Gotthilf H. E.
Thompson, Benjamin
(Count Rumford)

1754
Ellicott, Andrew
Ramsay, Alexander (ca. 1754)
Waterhouse, Benjamin

1757
Wells, William C.

1759
Cooper, Thomas
Mansfield, Jared

1761
Wistar, Caspar

1762
Morey, Samuel

1763
Maclure, William
Peck, William D.
Saugrain de Vigni, Antoine F.

1764
Mitchill, Samuel L.

1765
Folger, Walter

1766
Barton, Benjamin S.
Wilson, Alexander

1769
Hayden, Horace H.

1770
Hassler, Ferdinand R.
Michaux, François A.
Shecut, John L. E. W.
Woodhouse, James

1771
Elliott, Stephen
Maclean, John
Mease, James

1773
Bowditch, Nathaniel
Harrison, John
Horsfield, Thomas

1774
Griscom, John
Pursh, Frederick

1775
Adrain, Robert

1776
Eaton, Amos
Gibbs, George
Lewis, Enoch
Troost, Gerard

1777
Bruce, Archibald

1778
Atwater, Caleb
Cotting, John R.
Lesueur, Charles A.
Warren, John C.

1779
Baldwin, William
Farrar, John
Silliman, Benjamin

1780
Cleaveland, Parker
Schweinitz, Lewis D. von

1781
Hare, Robert
Ord, George

1782
Cist, Jacob
Darlington, William
Guthrie, Samuel

1783
Gorham, John
Hildreth, Samuel P.
Rafinesque, Constantine S.

1784
Dewey, Chester
Gummere, John
LeConte, John E.

1785
Audubon, John J.
Beaumont, William
Drake, Daniel
Espy, James P.

1786
Barton, William P. C.
Bigelow, Jacob
Nicollet, Joseph N.
Nuttall, Thomas
Rush, James

1787
Patterson, Robert M.
Say, Thomas

1788
Abert, John J.
Cutbush, James

1789
Bond, William C.

Redfield, William C.
Taylor, Richard C.

1790

Bachman, John
Delafield, Joseph
Green, Jacob
Strong, Theodore
Teschemacher, James E.

1791

Mitchell, William
Olmsted, Denison
Treadwell, Daniel

1792

Bache, Franklin
De Kay, James E.
Lea, Isaac
Locke, John
Renwick, James, Sr.
Sartwell, Henry P.
Vanuxem, Lardner

1793

Dana, James F.
Hitchcock, Edward
Horner, William E.
Kirtland, Jared P.
Lincecum, Gideon
Mitchell, Elisha
Mitchell, John K.
Phelps, Almira H. L.
Schoolcraft, Henry R.
Van Rensselaer, Jeremiah
Webster, John W.

1794

Burritt, Elijah H.
Couper, James H.
Durand, Elias
Godman, John D.
Holbrook, John E.
Johnson, Walter R.
Michener, Ezra
Ruffin, Edmund
Short, Charles W.

1795

Dana, Samuel L.
Harris, Thaddeus W.
Percival, James G.

1796

Curley, James
Emmet, John P.
Harlan, Richard
Joslin, Benjamin F.
Thompson, Zadock
Torrey, John
Wagner, William

1797

Emerson, George B.
Ferguson, James
Henry, Joseph
Hentz, Nicholas M.
James, Edwin
Morris, Margaretta H.
Perrine, Henry
Pitcher, Zina
Ravenel, Edmund
Smyth, William
Wailes, Benjamin L. C.

1798

Beck, Lewis C.
Cooper, William (ca. 1798)
Dunglison, Robley
Eights, James
Griffith, Robert E.
Metcalfe, Samuel L.
Wilkes, Charles

1799

Caswell, Alexis
Emmons, Ebenezer
Graham, James D.
Keating, William H.
Morton, Samuel G.
Peale, Titian R.
Saxton, Joseph
Tyson, Philip T.

1801

Coan, Titus
Deane, James
Lindheimer, Ferdinand J.
Marsh, George P.
Seybert, Henry
Snell, Ebenezer S.
Twining, Alexander C.

1802

Carpenter, George W.
Christy, David
Rogers, James B.

1803

Binney, Amos
Bonaparte, Charles L.
Conrad, Timothy A.
Courtenay, Edward H.
James, Thomas P.
Morris, John G.
Paine, Robert T.
Sullivant, William S.

1804

Anthony, John G.
Bartlett, William H. C.
Browne, Daniel J.
Clark, Alvan
Mather, William W.
Rogers, William B.
Savage, Thomas S.
Shepard, Charles U.
Storer, David H.
Whipple, Squire

1805

Gould, Augustus A.
Jackson, Charles T.
Peter, Robert
Pickering, Charles
Tuomey, Michael
Walker, Sears C.
Yandell, Lunsford P.

1806

Alexander, Stephen
Bache, Alexander D.
Bridges, Robert
Coffin, James H.
Hayes, Augustus A.
Jackson, John B. S.
Johnston, John
Lawrence, George N.
Lesquereux, Leo
Mapes, James J.
Maury, Matthew F.
Taylor, Charlotte De B. S.

1807

Agassiz, J. Louis R.
Alger, Francis
Alter, David
Clemson, Thomas G.
Davis, Charles H.
Grimes, James S.
Guyot, Arnold H.
Owen, David D.
Plummer, John T.
Riddell, John L.

1808

Curtis, Moses A.
Fitz, Henry
Hallowell, Edward
Jacobs, Michael
Rogers, Henry D.
Walsh, Benjamin D.
Whittlesey, Charles

1809

Barnard, Frederick A. P.
Bland, Thomas
Buckley, Samuel B.
Chapman, Alvan W.
Engelmann, George
Fitch, Asa
Fowler, Orson S.
Gibbes, Robert W.
Houghton, Douglass
Hubbard, Oliver P.
Mitchel, Ormsby M.
Peirce, Benjamin
Townsend, John K.

1810

Booth, James C.
Brackenridge, William D.
Gibbes, Lewis R.
Gray, Asa
Humphreys, Andrew A.
McCay, Charles F.
Norton, William A.
Trautwine, John C.
Wislizenus, Frederick A.

1811

Bailey, Jacob W.
Batchelder, John M.
Brocklesby, John
Draper, John W.
Emory, William H.
Fowler, Lorenzo N.
Gilliss, James M.
Hall, James
Herrick, Edward C.
Lapham, Increase A.
Loomis, Elias
Wright, Charles

1812

Alexander, John H.
(Boyé, Martin H.)
Evans, John
Frazer, John F.
Haldeman, Samuel S.
Page, Charles G.
Warder, John A.

1813

Cassin, John
Dana, James D.
Glover, Townend
Kellogg, Albert
Lovering, Joseph
Peters, Christian H. F.
Rogers, Robert E.
Schubert, Ernst
Weightman, William
Worthen, Amos H.

1814

Adams, Charles B.
Brewer, Thomas M.
Kirkwood, Daniel
Lyman, Chester S.
McArthur, William P.
Ravenel, Henry W.
Schott, Arthur C. V.
Sylvester, James J.
Wyman, Jeffries

1815

Barnard, John G.
Cabot, Samuel, Jr.
Coffin, John H. C.
Foster, John W.
Holmes, Francis S.

1816

Bayma, Joseph
Douglas, Silas H.
Rutherfurd, Lewis M.
Sestini, Benedict
Silliman, Benjamin, Jr.
Whipple, Amiel W.
Yarnall, Mordecai

1817

Ayres, William O.
Ferrel, William
Hagen, Hermann A.
Hager, Albert D.
Meek, Fielding B.
Swallow, George C.
Tuckerman, Edward

1818

Behr, Hans Herman
Cook, George H.
Hill, Thomas
Horsford, Eben N.
Kendall, Ezra O.
LeConte, John
McCulloh, Richard S.
Mitchell, Maria
Smith, Hamilton L.
Smith, John L.
Tracy, Clarissa T.
Vaughan, Daniel (ca. 1818)

1819

Eustis, Henry L.
Jenks, John W. P.
Lane, Jonathan H.
Lesley, J. Peter
Lippincott, James S.
Nichols, James R.
Ravenel, St. Julien
Squibb, Edward R.
Whitney, Josiah D.
Young, Aaron

1820

Bent, Silas
Chase, Pliny E.
Chauvenet, William
Genth, Frederick A., Sr.
Marcy, Oliver
Morfit, Campbell
Peck, William G.
Shumard, Benjamin F.
(Swift, Lewis)
Wells, Samuel R.

1821

Brunnow, Franz F. E.
Kneeland, Samuel
Thurber, George
Woodhouse, Samuel W.
Youmans, Edward L.

1822

(Edwards, William H.)
(Gibbs, O. Wolcott)
Girard, Charles F.
Hough, Franklin B.
Hunt, Edward B.
Judd, Orange
Lintner, Joseph A.
Newberry, John S.
Norton, John P.
Porter, John A.
Porter, Thomas C.
Runkle, John D.
(Safford, James M.)
Shattuck, Lydia W.
Vasey, George

1823

Baird, Spencer F.
Blodget, Lorin
Elliott, Ezekiel B.
Gambel, William
Hubbard, Joseph S.
Kedzie, Robert C.
Lea, Mathew C.
LeConte, Joseph
Leidy, Joseph
Parry, Charles C.

Pourtalès, Louis F. de (1823/1824)
Salisbury, James H.
Stallo, Johann B.

1824

(Doremus, Robert O.)
Gould, Benjamin A., Jr.
Holder, Joseph B.
Marcou, Jules
Mohr, Charles T.
Tilghman, Richard A.
Trask, John B.
Warren, Cyrus M.
Winchell, Alexander

1825

(Blake, William P.)
Bond, George P.
Dalton, John C., Jr.
(Davidson, George)
Gattinger, Augustin
Hilgard, Julius E.
LeConte, John L.
Porcher, Francis P.
Stoddard, John F.
(Thomas, Cyrus)
Wetherill, Charles M.
Xántus, János

1826

Brooke, John M.
Clark, Henry J.
Hanks, Henry G.
Hunt, Thomas S.
Loomis, Mahlon
Meehan, Thomas
Miles, Manly
Puryear, Bennet
Schott, Charles A.
Watson, Sereno
(White, Charles A.)
Winlock, Joseph

1827

(Broadhead, Garland C.)
Burnett, Waldo I.
Cooke, Josiah P.
(Goessmann, Charles A.)
Jones, William L.
Kerr, Washington C.
Keyt, Alonzo T.
(Stearns, Robert E. C.)
Venable, Charles S.

1828

Boll, Jacob
(Brewer, William H.)
Carll, John F.
Collett, John
Hammond, William A.
(Lattimore, Samuel A.)
Osten Sacken, Carl R. R. von der
Pugh, Evan
Trowbridge, William P.
Wells, David A.
(Whitfield, Robert P.)
Wurtz, Henry

1829

Crocker, Lucretia
(Ellis, Job B.)
(Hall, Asaph)
Hayden, Ferdinand V.
(Mitchell, Silas W.)
Oliver, James E.
Orton, Edward F. B.
Tuttle, Charles W.
Wachsmuth, Charles

1830

Cooper, James G.
(Johnson, Samuel W.)
Mitchell, Henry
Newton, Hubert A.
Orton, James
Warren, Gouverneur K.
Wright, Chauncey

1831

(Abbot, Henry L.)
Bodley, Rachel L.
(Brush, George J.)
Engelmann, Henry
Marsh, Othniel C.
Masterman, Stillman
Nason, Henry B.
Rood, Ogden N.

1832

Clark, Alvan G.
Cutting, Hiram A.
Lemmon, John G.
(Mallet, John W.)
Prescott, Albert B.
Prime, Temple
Rogers, William A.
Stimpson, William
(Stockwell, John N.)
(Storer, Francis H.)

1833

(Beal, William J.)
Binney, William G.
(Brackett, Cyrus F.)
(Cooley, Leroy C.)
(Davenport, George E.)
(Hilgard, Eugene W.)
Lindenkohl. Adolph
Lyman, Theodore
(Peck, Charles H.)

1834

(Caldwell, George C.)
De Forest, Erastus L.
Eaton, Daniel C.
Furbish, Kate
(Langley, Samuel P.)
(Peirce, James M.)
Powell, John W.
(Ward, Henry A.)
(Young, Charles A.)

1835

(Agassiz, Alexander)
(Anthony, William A.)
(Barker, George F.)
Claypole, Edward W.
Collier, Peter
(Compton, Alfred G.)
(Elliot, Daniel G.)
(Fontaine, William M.)
Green, Francis M.
Kennicott, Robert
(Lyman, Benjamin S.)
(Newcomb, Simon)
(Uhler, Philip R.)

1836

Allen, Oscar D.
(Chandler, Charles F.)
(Eastman, John R.)
(Hague, James D.)
(Hinrichs, Gustavus D.)
(Hitchcock, Charles H.)
(Hough, George W.)
Mapes, Charles V.
Mason, James W.
Mayer, Alfred M.
(Morgan, Andrew P.)
Morton, Henry
Safford, Truman H.
Samuels, Edward A.
Smith, Erminnie A. P.
Steele, Joel D.
Strecker, Herman
(Wright, Arthur W.)

1837

Allen, Timothy F.
Bowser, Edward A.

(Dolbear, Amos E.)
Draper, Henry
(Frisby, Edgar)
(Gill, Theodore N.)
Harkness, William
(Pumpelly, Raphael)
(Scudder, Samuel H.)
(Seaman, William H.)
(Searle, Arthur)

1838

(Abbe, Cleveland)
(Allen, Joel A.)
Bradley, Frank H.
(Burnham, Sherburne W.)
(Cresson, Ezra T.)
(Fernald, Charles H.)
(Hill, George W.)
Hyatt, Alpheus
Marx, George
(Morley, Edward W.)
(Morse, Edward S.)
(Muir, John)
(Niles, William H.)
(Parish, Samuel B.)
Tryon, George W.
Watson, James C.
Youmans, William J.

1839

(Burrill, Thomas J.)
(Crafts, James M.)
Gabb, William M.
Gibbs, Josiah W.
(Goodale, George L.)
Michie, Peter S.
Packard, Alpheus S., Jr.
(Peirce, Charles S.)
(Putnam, Frederic W.)
(Rising, Willard B.)
Thompson, Almon H.
(Verrill, Addison E.)
(Winchell, Newton H.)

1840

(Bowditch, Henry P.)
(Brashear, John A.)
(Calvin, Samuel)
Cope, Edward D.
(Hague, Arnold)
Hartt, Charles F.
Horn, George H.
(Julien, Alexis A.)
(McClintock, Emory)
Sennett, George B.
(Snow, Francis H.)

1841

Allen, Harrison
(Draper, Daniel)
(Dutton, Clarence E.)
Eimbeck, William
(Emmons, Samuel F.)
Grote, Augustus R.
(Johnson, William W.)
(Mendenhall, Thomas C.)
Rice, Charles
(Sargent, Charles S.)
(Shaler, Nathaniel S.)
(Smith, Eugene A.)
(Stevenson, John J.)
Trumbull, Gurdon
(Ward, Lester F.)
(Wilder, Burt G.)

1842

Coues, Elliott
Drown, Thomas M.
(Dudley, Charles B.)
Gardiner, James T.
(James, William)
King, Clarence R.
(Ladd, George T.)
(Richards, Ellen H. S.)
(Sharples, Stephen P.)
Sturtevant, Edward L.
Wheeler, George M.
(Whitman, Charles O.)

1843

(Abbott, Charles C.)
(Babcock, Stephen M.)
Bolton, Henry C.
Brandegee, Townshend S.
(Chamberlin, Thomas C.)
Chester, Albert H.
(Doolittle, Charles L.)
(Dwight, Thomas)
(Emerson, Benjamin K.)
(Garman, Samuel)
(Gilbert, Grove K.)
(Greene, Edward L.)
(Hyde, Edward W.)
Riley, Charles V.
(Smith, Sidney I.)
Torrey, Bradford
(Trowbridge, John)

1844

(Atwater, Wilbur O.)
Babcock, James F.
Brandegee, Mary K. L. C.
(Brooks, William R.)
(Carhart, Henry S.)
(Cornwall, Henry B.)
(Eddy, Henry T.)
(Farlow, William G.)
(Forbes, Stephen A.)
(Frazer, Persifor)
(Koenig, George A.)
Platt, Franklin
Procter, John R.
Schwarz, Eugene A.
(Wiley, Harvey W.)
(Wilson, William P.)

1845

(Bessey, Charles E.)
(Dall, William H.)
(Hall, Christopher W.)
Newton, Henry
(Rice, William N.)
(Skinner, Aaron N.)

1846

(Bean, Tarleton H.)
(Boss, Lewis)
Caswell, John H.
(Chandler, Seth C.)
(Gannett, Henry)
(Hall, G. Stanley)
(Hay, Oliver P.)
(Holden, Edward S.)
(Holmes, William H.)
(Mixter, William G.)
(Pickering, Edward C.)
(Remsen, Ira)
(Todd, James E.)
(Waller, Elwyn)
(Witthaus, Rudolph A.)
Wood, Edward S.
Wright, Albert A.

1847

(Becker, George F.)
(Bell, Alexander G.)
(Clarke, Frank W.)
(Edison, Thomas A.)
(Emerton, James H.)
(Hagen, John G.)
Irving, Roland D.
(Jackson, Charles L.)
(Leffmann, Henry)
(Mark, Edward L.)
(Nipher, Francis E.)
(Sadtler, Samuel P.)
(Stevens, Walter L.)
(Stone, Ormond)
(Stringham, Washington I.)
(Wadsworth, Marshman E.)
(Whiting, Sarah F.)

(Whitney, Mary W.)
(Williams, Henry S.)

1848
(Bailey, Edgar H. S.)
(Blair, Andrew A.)
(Brooks, William K.)
Burgess, Edward
(Chartard, Thomas M.)
(Collins, Frank S.)
(Cross, Charles R.)
Harrington, Mark W.
(Hastings, Charles S.)
(Holland, William J.)
(MacBride, Thomas H.)
Martin, Henry N.
(Mellus, Edward L.)
(Morse, Harmon N.)
(Renouf, Edward)
Rowland, Henry A.
(Tuckerman, Alfred)
Warder, Robert B.
(Wead, Charles K.)
(White, Israel C.)

1849
Baker, Marcus
(Brush, Charles F.)
(Byerly, William E.)
(Clapp, Cornelia M.)
(Comstock, John H.)
(Comstock, Theodore B.)
(Dana, Edward S.)
(Dudley, William R.)
Hazen, Henry A.
Hill, Henry B.
(Lloyd, John U.)
(Munroe, Charles E.)
Ober, Frederick A.
(Spalding, Volney M.)
(Warren, Joseph W.)
(Webster, Francis M.)
(Woodward, Robert S.)

1850
(Arthur, Joseph C.)
(Beman, Wooster W.)
(Beyer, Henry G.)
(Blake, Francis)
(Branner, John C.)
(Crosby, William O.)
(Davis, William M.)
(Diller, Joseph S.)
(Fairchild, Herman L.)
(Hering, Daniel W.)
Hubbard, Henry G.
(Jenkins, Edward H.)
(Kellerman, William A.)
(Mabery, Charles F.)
(Nicholson, Henry H.)
(Phillips, Francis C.)
(Ridgway, Robert)
Shepard, James H.
(Shufeldt, Robert W.)
(Smith, William B.)
(Story, William E.)
(Thomas, Benjamin F.)
(Upham, Warren)
(Walcott, Charles D.)

1851
(Brewster, William)
(Coulter, John M.)
(Gage, Simon H.)
Goode, George B.
(Jordan, David S.)
Lilley, George
(McMurtrie, William)
(Rees, John K.)

1852
(Austen, Peter T.)
(Gooch, Frank A.)
McCalley, Henry
(Michelson, Albert A.)
(Minot, Charles S.)
Mitchell, James A.
Mott, Henry A., Jr.
(Rathbun, Richard)
(Russell, Israel C.)
(Stoddard, John T.)
(Williston, Samuel W.)

1853
(Flint, Albert S.)
(Halsted, George B.)
(Heilprin, Angelo)
(Hillebrand, William F.)
(McGee, William J.)
(Maschke, Heinrich)
(Michael, Arthur)
(Schaeberle, John M.)
(Underwood, Lucien M.)

1854
Ashburner, Charles A.
(Atkinson, George F.)
(Bailey, Solon I.)
(Comstock, Anna B.)
(Hart, Edward)
(Kingsley, John S.)
(Kinnicutt, Leonard P.)
(Merrill, George P.)
(Millspaugh, Charles F.)
(Nichols, Edward L.)
(Peirce, Benjamin O.)
(Smith, Edgar F.)
(Takamine, Jokichi)

1855
(Ashmead, William H.)
(Brigham, Albert P.)
(Comstock, George C.)
Craig, Thomas
(Elkin, William L.)
(Hall, Edwin H.)
(MacFarlane, John M.)

1856
(Barus, Carl)
Beecher, Charles E.
(Chittenden, Russell H.)
Good, Adolphus C.
(Penfield, Samuel L.)
(Venable, Francis P.)
Williams, George H.
(Wilson, Edmund B.)
Yeates, William S.

1857
(Barnard, Edward E.)
(Clarke, John M.)
Cory, Charles B.
Day, William C.
(Donaldson, Henry H.)
(Fleming, Williamina P. S.)
(Hollick, Charles A.)
(Howard, Leland O.)
(Iddings, Joseph P.)
Keeler, James E.
(Osborn, Henry F.)
(Pritchett, Henry S.)
(Trelease, William)
(Van Hise, Charles R.)

1858
(Barnes, Charles R.)
(Britton, Elizabeth G. K.)
(Campbell, Marius R.)
Eigenmann, Rosa S.
(Fine, Henry B.)
Hayden, Edward E.
(Hill, Robert T.)
(Howe, Herbert A.)
Knight, Wilbur C.
(Nutting, Charles C.)
(Pickering, William H.)
(Salisbury, Rollin D.)
(Scott, Charlotte A.)
(True, Frederick W.)

1859

Brace, De Witt B.
(Britton, Nathaniel L.)
(Frankforter, George B.)
(Howe, James L.)
(Jayne, Horace F.)
(Kemp, James F.)
(McMurrich, James P.)
Winlock, William C.

1860

(Clark, William B.)
(Frear, William)
(Knowlton, Frank H.)
(Lindgren, Waldemar)
(Maltby, Margaret E.)
(Pirsson, Louis V.)
(Prosser, Charles S.)
(Rydberg, Per A.)
(Taylor, Frank B.)

1861

Hatcher, John B.
Palmer, Arthur W.

1864

Schweinitz, Emil A. de (ca. 1864)

1867

Ewell, Ervin E.

Appendix B
PLACE OF BIRTH

NOT KNOWN

Christy, David 1802
Cooper, William ca. 1798
De Brahm, William G. 1717
Grew, Theophilus d. 1759
More, Thomas fl. 1670-1724

UNITED STATES

Alabama

McCalley, Henry 1852
(Smith, Eugene A. 1841)

Connecticut

Ayres, William O. 1817
(Bailey, Edgar H. S. 1848)
Beaumont, William 1785
(Beman, Wooster W. 1850)
Bradley, Frank H. 1838
Brandegee, Townshend S. 1843
Burritt, Elijah H. 1794
(Chittenden, Russell H. 1856)
Coan, Titus 1801
(Cornwall, Henry B. 1844)
Cutler, Manasseh 1742
Daboll, Nathan 1750
(Dana, Edward S. 1849)
De Forest, Erastus L. 1834
(Dolbear, Amos E. 1837)
(Dudley, William R. 1849)
(Dutton, Clarence E. 1841)
Gibbs, Josiah W. 1839
(Hall, Asaph 1829)
Hayden, Horace H. 1769
Herrick, Edward C. 1811
Hubbard, Joseph S. 1823
Hubbard, Oliver P. 1809
Hunt, Thomas S. 1826
Kellogg, Albert 1813
Kirtland, Jared P. 1793
Loomis, Elias 1811
Lyman, Chester S. 1814
Mansfield, Jared 1759
Mather, William W. 1804
Mitchell, Elisha 1793
Morey, Samuel 1762
Newberry, John S. 1822
Olmsted, Denison 1791
(Osborn, Henry F. 1857)
Peck, William G. 1820
Percival, James G. 1795
Phelps, Almira H. L. 1793
Redfield, William C. 1789
Rogers, William A. 1832
Rood, Ogden N. 1831
Savage, Thomas S. 1804
Silliman, Benjamin 1779
Silliman, Benjamin, Jr. 1816
Smith, Hamilton L. 1818
(True, Frederick W. 1858)
Trumbull, Gurdon 1841
Twining, Alexander C. 1801
Watson, Sereno 1826
Wells, Samuel R. 1820
Whittlesey, Charles 1808
(Wright, Arthur W. 1836)
Wright, Charles 1811

Delaware

Keating, William H. 1799
Lea, Isaac 1792
Squibb, Edward R. 1819

District of Columbia

Gilliss, James M. 1811
Hyatt, Alpheus 1838

Florida

Brooke, John M. 1826

Georgia

Couper, James H. 1794
Jones, William L. 1827
LeConte, John 1818
LeConte, Joseph 1823
Lincecum, Gideon 1793
(Stevens, Walter L. 1847)
Taylor, Charlotte De B. S. 1806
Wailes, Benjamin L. C. 1797

Hawaii

(Hillebrand, William F. 1853)

Illinois

(Chamberlin, Thomas C. 1843)
Eigenmann, Rosa S. 1858
(Forbes, Stephen A. 1844)
Harrington, Mark W. 1848
Hatcher, John B. 1861
(Howard, Leland O. 1857)
Keeler, James E. 1857
Knight, Wilbur C. 1858
Lilley, George 1851
(Mixter, William G. 1846)
(Nutting, Charles C. 1858)
(Ridgway, Robert 1850)
(Stone, Ormond 1847)
(Ward, Lester F. 1841)
(Wilson, Edmund B. 1856)

Indiana

(Barnes, Charles R. 1858)
Collett, John 1828
(Doolittle, Charles L. 1843)
Goode, George B. 1851
(Hay, Oliver P. 1846)
(Lattimore, Samuel A. 1828)
Meek, Fielding B. 1817
(Taylor, Frank B. 1860)
(Wiley, Harvey W. 1844)

Iowa

(Campbell, Marius R. 1858)
(McGee, William J. 1853)

Kentucky

(Blair, Andrew A. 1848)
Mitchel, Ormsby M. 1809
Procter, John R. 1844
(Shaler, Nathaniel S. 1841)
Short, Charles W. 1794
(Smith, William B. 1850)
Winlock, Joseph 1826

Louisiana

(Elkin, William L. 1855)
Kennicott, Robert 1835

Maine

Allen, Oscar D. 1836
Bond, William C. 1789
(Brackett, Cyrus F. 1833)
Coffin, John H. C. 1815
Emerson, George B. 1797
(Fernald, Charles H. 1838)
(Gannett, Henry 1846)
(Goodale, George L. 1839)
(Hall, Edwin H. 1855)
Johnston, John 1806
(Mabery, Charles F. 1850)
Masterman, Stillman 1831
(Mellus, Edward L. 1848)
(Merrill, George P. 1854)
(Morse, Edward S. 1838)
Oliver, James E. 1829
Packard, Alpheus S., Jr. 1839
(Smith, Sidney I. 1843)
Smyth, William 1797
Storer, David H. 1804
Swallow, George C. 1817
Tuttle, Charles W. 1829
(Verrill, Addison E. 1839)
(Wadsworth, Marshman E. 1847)
(Whitman, Charles O. 1842)
Young, Aaron 1819

Maryland

Alexander, John H. 1812
Banneker, Benjamin 1731
(Chatard, Thomas M. 1848)
Courtenay, Edward H. 1803
Emory, William H. 1811
Godman, John D. 1794
Hammond, William A. 1828
(Hering, Daniel W. 1850)
(Iddings, Joseph P. 1857)
Kirkwood, Daniel 1814
Leeds, John 1705
McCulloh, Richard S. 1818
Mayer, Alfred M. 1836
Peale, Charles W. 1742
Plummer, John T. 1807
Rogers, Robert E. 1813
Tyson, Philip T. 1799
(Uhler, Philip R. 1835)

Massachusetts

(Abbot, Henry L. 1831)
Adams, Charles B. 1814
Alger, Francis 1807
(Allen, Joel A. 1838)
Atwater, Caleb 1778
Babcock, James F. 1844
Bailey, Jacob W. 1811
(Barker, George F. 1835)
Barnard, Frederick A. P. 1809
Barnard, John G. 1815
Bigelow, Jacob 1786
Binney, Amos 1803
Binney, William G. 1833
(Blake, Francis 1850)
Bond, George P. 1825
(Bowditch, Henry P. 1840)
Bowditch, Nathaniel 1773
Bowdoin, James 1726
Boylston, Zabdiel 1680
Brattle, Thomas 1658
Brewer, Thomas M. 1814
(Brewster, William 1851)
Burgess, Edward 1848
Burnett, Waldo I. 1827
(Burrill, Thomas J. 1839)
Cabot, Samuel, Jr. 1815
(Caldwell, George C. 1834)
Caswell, Alexis 1799
(Chandler, Charles F. 1836)
(Chandler, Seth C. 1846)
Chapman, Alvan W. 1809
Chase, Pliny E. 1820
Clap, Thomas 1703
(Clapp, Cornelia M. 1849)
Clark, Alvan 1804
Clark, Alvan G. 1832
Clark, Henry J. 1826
(Clarke, Frank W. 1847)
Cleaveland, Parker 1780
Coffin, James H. 1806
(Collins, Frank S. 1848)
Colman, Benjamin 1673
Cooke, Josiah P. 1827
Cory, Charles B. 1857
Cotting, John R. 1778
(Crafts, James M. 1839)
Crocker, Lucretia 1829
Curtis, Moses A. 1808
(Dall, William H. 1845)
Dalton, John C., Jr. 1825
(Davenport, George E. 1833)
Davis, Charles H. 1807
Deane, James 1801
Dewey, Chester 1784
Dudley, Paul 1675
(Dwight, Thomas 1843)
(Eddy, Henry T. 1844)
(Emerton, James H. 1847)
Emmons, Ebenezer 1799
(Emmons, Samuel F. 1841)
Eustis, Henry L. 1819
(Farlow, William G. 1844)
Farrar, John 1779
Fitz, Henry 1808
(Flint, Albert S. 1853)
Folger, Walter 1765
Foster, John W. 1815
Franklin, Benjamin 1706
(Gooch, Frank A. 1852)
Gorham, John 1783
Gould, Benjamin A., Jr. 1824
Green, Francis M. 1835
Greenwood, Isaac 1702
Grimes, James S. 1807
Guthrie, Samuel 1782
(Hague, Arnold 1840)
(Hague, James D. 1836)
(Hall, G. Stanley 1846)
Hall, James 1811
Harris, Thaddeus W. 1795
Hayden, Edward E. 1858
Hayden, Ferdinand V. 1829
Hildreth, Samuel P. 1783
Hill, Henry B. 1849
(Hitchcock, Charles H. 1836)
Hitchcock, Edward 1793
Holder, Joseph B. 1824
(Howe, James L. 1859)
(Jackson, Charles L. 1847)
Jackson, Charles T. 1805
Jackson, John B. S. 1806
Jeffries, John 1744/1745
(Jenkins, Edward H. 1850)
Jenks, John W. P. 1819
Johnson, Walter R. 1794
Kendall, Ezra O. 1818
(Kinnicutt, Leonard P. 1854)
Kneeland, Samuel 1821
(Langley, Samuel P. 1834)
Lovering, Joseph 1813
(Lyman, Benjamin S. 1835)
Lyman, Theodore 1833
Marcy, Oliver 1820
Mather, Cotton 1662/1663
Mather, Increase 1639
(Minot, Charles S. 1852)
Mitchell, Henry 1830
Mitchell, Maria 1818

Mitchell, William 1791
(Munroe, Charles E. 1849)
Nason, Henry B. 1831
Nichols, James R. 1819
(Niles, William H. 1838)
Ober, Frederick A. 1849
Oliver, Andrew 1731
Page, Charles G. 1812
Paine, Robert T. 1803
Peck, William D. 1763
Peirce, Benjamin 1809
(Peirce, Benjamin O. 1854)
(Peirce, Charles S. 1839)
(Peirce, James M. 1834)
(Pickering, Edward C. 1846)
(Pickering, William H. 1858)
(Putnam, Frederic W. 1839)
(Rice, William N. 1845)
(Richards, Ellen H. S. 1842)
Riddell, John L. 1807
Robie, Thomas 1688/1689
Samuels, Edward A. 1836
(Sargent, Charles S. 1841)
Sartwell, Henry P. 1792
(Scudder, Samuel H. 1837)
(Skinner, Aaron N. 1845)
Snell, Ebenezer S. 1801
(Snow, Francis H. 1840)
(Stearns, Robert E. C. 1827)
Stimpson, William 1832
(Stockwell, John N. 1832)
(Stoddard, John T. 1852)
(Storer, Francis H. 1832)
(Story, William E. 1850)
Strong, Theodore 1790
Sturtevant, Edward L. 1842
Thompson, Benjamin
(Count Rumford) 1753
Torrey, Bradford 1843
Trask, John B. 1824
Treadwell, Daniel 1791
(Trowbridge, John 1843)
Tuckerman, Edward 1817
Walker, Sears C. 1805
Warren, Cyrus M. 1824
Warren, John C. 1778
(Warren, Joseph W. 1849)
Webster, John W. 1793
Wells, David A. 1828
West, Benjamin 1730
Wheeler, George M. 1842
Whipple, Amiel W. 1816
Whipple, Squire 1804
(White, Charles A. 1826)
Whitney, Josiah D. 1819
(Whitney, Mary W. 1847)
(Wilder, Burt G. 1841)
(Williston, Samuel W. 1852)
Winlock, William C. 1859
Winthrop, John 1714
Wood, Edward S. 1846
Wright, Chauncey 1830
Wyman, Jeffries 1814

Michigan

(Atkinson, George F. 1854)
Baker, Marcus 1849
(Beal, William J. 1833)
Eaton, Daniel C. 1834
Ewell, Ervin E. 1867
Hubbard, Henry G. 1850
(Hyde, Edward W. 1843)
Lemmon, John G. 1832
Shepard, James H. 1850
Trowbridge, William P. 1828
(Wilson, William P. 1844)
(Woodward, Robert S. 1849)

Missouri

Bent, Silas 1820
(Holden, Edward S. 1846)
McArthur, William P. 1814
Morfit, Campbell 1820
(Pritchett, Henry S. 1857)

New Hampshire

(Bailey, Solon I. 1854)
Batchelder, John M. 1811
Browne, Daniel J. 1804
Coues, Elliott 1842
Dana, James F. 1793
Dana, Samuel L. 1795
(Eastman, John R. 1836)
(Emerson, Benjamin K. 1843)
Evans, John 1812
Furbish, Kate 1834
Gould, Augustus A. 1805
Locke, John 1792
Pike, Nicholas 1743
Shattuck, Lydia W. 1822
Thompson, Almon H. 1839
(Upham, Warren 1850)
(Webster, Francis M. 1849)
(Young, Charles A. 1834)

New Jersey

(Abbott, Charles C. 1843)
Conrad, Timothy A. 1803
Cook, George H. 1818
Drake, Daniel 1785
Griscom, John 1774
(Halsted, George B. 1853)
Hill, Thomas 1818
Hutchins, Thomas 1730
LeConte, John E. 1784
(McMurtrie, William 1851)
(Morley, Edward W. 1838)
(Parish, Samuel B. 1838)
Perrine, Henry 1797

New York

(Abbe, Cleveland 1838)
Alexander, Stephen 1806
(Arthur, Joseph C. 1850)
(Atwater, Wilbur O. 1844)
(Austen, Peter T. 1852)
(Babcock, Stephen M. 1843)
Bachman, John 1790
Beck, Lewis C. 1798
(Becker, George F. 1847)
Beecher, Charles E. 1856
(Blake, William P. 1825)
Blodget, Lorin 1823
Bolton, Henry C. 1843
Brace, De Witt B. 1859
(Brewer, William H. 1828)
(Brigham, Albert P. 1855)
(Britton, Elizabeth G. K. 1858)
(Britton, Nathaniel L. 1859)
Bruce, Archibald 1777
(Brush, George J. 1831)
Buckley, Samuel B. 1809
(Carhart, Henry S. 1844)
Carll, John F. 1828
Caswell, John H. 1846
Chester, Albert H. 1843
(Clarke, John M. 1857)
Colden, Jane 1724
Collier, Peter 1835
(Comstock, Anna B. 1854)
(Cooley, Leroy C. 1833)
Cooper, James G. 1830
(Cross, Charles R. 1848)
Dana, James D. 1813
Delafield, Joseph 1790
(Donaldson, Henry H. 1857)
(Doremus, Robert O. 1824)
Douglass, Silas H. 1816
(Draper, Daniel 1841)
(Dudley, Charles B. 1842)
Eaton, Amos 1776
(Edwards, William H. 1822)
Eights, James 1798
(Elliot, Daniel G. 1835)
Elliott, Ezekiel B. 1823
(Ellis, Job B. 1829)
Fitch, Asa 1809
Fowler, Lorenzo N. 1811
Fowler, Orson S. 1809
(Gage, Simon H. 1851)
Gardiner, James T. 1842

(Gibbs, O. Wolcott 1822)
(Gilbert, Grove K. 1843)
(Gill, Theodore N. 1837)
Gray, Asa 1810
(Hastings, Charles S. 1848)
Henry, Joseph 1797
(Hill, George W. 1838)
(Hollick, Charles A. 1857)
Horsford, Eben N. 1818
Hough, Franklin B. 1822
(Hough, George W. 1836)
Houghton, Douglass 1809
(Howe, Herbert A. 1858)
Hunt, Edward B. 1822
Irving, Roland D. 1847
(James, William 1842)
(Johnson, Samuel W. 1830)
(Johnson, William W. 1841)
(Jordan, David S. 1851)
Judd, Orange 1822
(Julien, Alexis A. 1840)
Kedzie, Robert C. 1823
(Kemp, James F. 1859)
(Kingsley, John S. 1854)
Lane, Jonathan H. 1819
Lapham, Increase A. 1811
Lawrence, George N. 1806
LeConte, John L. 1825
Lintner, Joseph A. 1822
(Lloyd, John U. 1849)
Loomis, Mahlon 1826
Mapes, Charles V. 1836
Mapes, James J. 1806
(Mark, Edward L. 1847)
Marsh, Othniel C. 1831
Mason, James W. 1836
(Michael, Arthur 1853)
Miles, Manly 1826
(Millspaugh, Charles F. 1854)
Mitchill, Samuel L. 1764
Morton, Henry 1836
Mott, Henry A., Jr. 1852
Newton, Henry 1845
Newton, Hubert A. 1830
(Nipher, Francis E. 1847)
Norton, John P. 1822
Norton, William A. 1810
Orton, Edward F. B. 1829
Orton, James 1830
(Peck, Charles H. 1833)
(Penfield, Samuel L. 1856)
(Pirsson, Louis V. 1860)
Pitcher, Zina 1797
Porter, John A. 1822
Powell, John W. 1834
Prescott, Albert B. 1832
Prime, Temple 1832
(Prosser, Charles S. 1860)
(Pumpelly, Raphael 1837)
(Rathbun, Richard 1852)
(Rees, John K. 1851)
(Remsen, Ira 1846)
(Renouf, Edward 1848)
(Rising, Willard B. 1839)
Runkle, John D. 1822
(Russell, Israel C. 1852)
Rutherfurd, Lewis M. 1816
Salisbury, James H. 1823
Schoolcraft, Henry R. 1793
(Seaman, William H. 1837)
Sennett, George B. 1840
(Shufeldt, Robert W. 1850)
Smith, Erminnie A. P. 1836
(Spalding, Volney M. 1849)
Steele, Joel D. 1836
(Stevenson, John J. 1841)
Stoddard, John F. 1825
(Stringham, Washington I. 1847)
(Swift, Lewis 1820)
Torrey, John 1796
(Trelease, William 1857)
(Tuckerman, Alfred 1848)
(Underwood, Lucien M. 1853)
Van Rensselaer, Jeremiah 1793
(Walcott, Charles D. 1850)
(Waller, Elwyn 1846)
(Ward, Henry A. 1834)
Warren, Gouverneur, K. 1830
(Wead, Charles K. 1848)
(Whitfield, Robert P. 1828)
(Whiting, Sarah F. 1847)
Wilkes, Charles 1798
Williams, George H. 1856
(Williams, Henry S. 1847)
Winchell, Alexander 1824
(Winchell, Newton H. 1839)
(Witthaus, Rudolph A. 1846)
Youmans, Edward L. 1821
Youmans, William J. 1838

North Carolina

Kerr, Washington C. 1827
Schweinitz, Emil A. de ca. 1864
Yeates, William S. 1856

Ohio

(Barus, Carl 1856)
(Bessey, Charles E. 1845)
Bodley, Rachael L. 1831
(Brooks, William K. 1848)
(Brush, Charles F. 1849)
(Comstock, Theodore B. 1849)
(Crosby, William O. 1850)
Day, William C. 1857
(Edison, Thomas A. 1847)
(Frankforter, George B. 1859)
Hanks, Henry G. 1826
(Holmes, William H. 1846)
(Kellerman, William A. 1850)
Keyt, Alonzo T. 1827
(Ladd, George T. 1842)
(Maltby, Margaret E. 1860)
(Mendenhall, Thomas C. 1841)
(Morgan, Andrew P. 1836)
(Safford, James M. 1822)
Sullivant, William S. 1803)
(Thomas, Benjamin F. 1850)
(Todd, James E. 1846)
Warder, Robert B. 1848
Wright, Albert A. 1846
Yarnall, Mordecai 1816

Pennsylvania

Allen, Harrison 1841
Alter, David 1807
Ashburner, Charles A. 1854
(Ashmead, William H. 1855)
Bache, Alexander D. 1806
Bache, Franklin 1792
Baird, Spencer F. 1823
Baldwin, William 1779
Bartlett, William H. C. 1804
Barton, Benjamin S. 1766
Barton, William P. C. 1786
Bartram, John 1699
Bartram, William 1739
(Bean, Tarleton H. 1846)
Biddle, Owen 1737
Booth, James C. 1810
(Brashear, John A. 1840)
Bridges, Robert 1806
(Byerly, William E. 1849)
Carpenter, George W. 1802
Cassin, John 1813
Chauvenet, William 1820
Cist, Jacob 1782
Clemson, Thomas G. 1807
Cope, Edward D. 1840
Craig, Thomas 1855
(Cresson, Ezra T. 1838)
Cutbush, James 1788
Darlington, William 1782
(Davis, William M. 1850)
(Diller, Joseph S. 1850)
Drown, Thomas M. 1842
Ellicott, Andrew 1754
Espy, James P. 1785
(Fairchild, Herman L. 1850)
Ferrel, William 1817
(Fine, Henry B. 1858)

Frazer, John F. 1812
(Frazer, Persifor 1844)
(Frear, William 1860)
Gabb, William M. 1839
Gambel, William 1823
(Garman, Samuel 1843)
Godfrey, Thomas 1704
Good, Adolphus C. 1856
Green, Jacob 1790
Griffith, Robert E. 1798
Gummere, John 1784
Haldeman, Samuel S. 1812
Hallowell, Edward 1808
Hare, Robert 1781
Harlan, Richard 1796
Harrison, John 1773
(Hart, Edward 1854)
Horn, George H. 1840
Horsfield, Thomas 1773
Humphreys, Andrew A. 1810
Jacobs, Michael 1808
James, Thomas P. 1803
(Jayne, Horace F. 1859)
Kuhn, Adam 1741
Lea, Mathew C. 1823
(Leffmann, Henry 1847)
Leidy, Joseph 1823
Lesley, J. Peter 1819
Lewis, Enoch 1776
Lippincott, James S. 1819
McCay, Charles F. 1810
(McClintock, Emory 1840)
Marshall, Humphry 1722
Mease, James 1771
Michener, Ezra 1794
(Mitchell, Silas W. 1829)
Morris, John G. 1803
Morris, Margaretta H. 1797
Morton, Samuel G. 1799
Mühlenberg, Gotthilf H. E. 1753
Ord, George 1781
Patterson, Robert M. 1787
Peale, Titian R. 1799
(Phillips, Francis C. 1850)
Pickering, Charles 1805
Platt, Franklin 1844
Porter, Thomas C. 1822
Pugh, Evan 1828
Rittenhouse, David 1732
Rogers, Henry D. 1808
Rogers, James B. 1802
Rogers, William B. 1804
Rowland, Henry A. 1848
Rush, Benjamin 1746
Rush, James 1786
(Sadtler, Samuel P. 1847)
Saxton, Joseph 1799
Say, Thomas 1787
Schweinitz, Lewis D. von 1780
Seybert, Henry 1801
(Sharples, Stephen P. 1842)
Shumard, Benjamin F. 1820
(Smith, Edgar F. 1854)
Strecker, Herman 1836
Tilghman, Richard A. 1824
Townsend, John K. 1809
Tracy, Clarissa T. 1818
Trautwine, John C. 1810
Tryon, George W. 1838
Vanuxem, Lardner 1792
Wagner, William 1796
Warder, John A. 1812
Wetherill, Charles M. 1825
Williamson, Hugh 1735
Wistar, Caspar 1761
Woodhouse, James 1770
Woodhouse, Samuel W. 1821
Wurtz, Henry 1828

Rhode Island

Anthony, John G. 1804
(Anthony, William A. 1835)
(Boss, Lewis 1846)
Brown, Joseph 1733
Gibbs, George 1776
(Greene, Edward L. 1843)
Joslin, Benjamin F. 1796
King, Clarence R. 1842
Shepard, Charles U. 1804
Thurber, George 1821
Waterhouse, Benjamin 1754

South Carolina

Elliott, Stephen 1771
Gibbes, Lewis R. 1810
Gibbes, Robert W. 1809
Holbrook, John E. 1794
Holmes, Francis S. 1815
Porcher, Francis P. 1825
Ravenel, Edmund 1797
Ravenel, Henry W. 1814
Ravenel, St. Julien 1819
Shecut, John L. E. W. 1770
Smith, John L. 1818
Wells, William C. 1757

Tennessee

(Barnard, Edward E. 1857)
Brandegee, Mary K. L. C. 1844
(Branner, John C. 1850)
(Hill, Robert T. 1858)
(MacBride, Thomas H. 1848)
(Thomas, Cyrus 1825)
Yandell, Lunsford P. 1805

Vermont

Allen, Timothy F. 1837
(Burnham, Sherburne W. 1838)
(Clark, William B. 1860)
Cutting, Hiram A. 1832
Hager, Albert D. 1817
(Hall, Christopher W. 1845)
Hayes, Augustus A. 1806
James, Edwin 1797
(Knowlton, Frank H. 1860)
Marsh, George P. 1801
(Morse, Harmon N. 1848)
Safford, Truman H. 1836
Thompson, Zadock 1796
Worthen, Amos H. 1813

Virginia

Abert, John J. 1788
(Broadhead, Garland C. 1827)
Draper, Henry 1837
(Fontaine, William M. 1835)
Graham, James D. 1799
Horner, William E. 1793
Jefferson, Thomas 1743
Madison, James 1749
Maury, Matthew F. 1806
Metcalfe, Samuel L. 1798
Mitchell, John 1711
Mitchell, John K. 1793
Puryear, Bennet 1826
Ruffin, Edmund 1794
Venable, Charles S. 1827
(Venable, Francis P. 1856)
(White, Israel C. 1848)

Wisconsin

(Comstock, George C. 1855)
(Comstock, John H. 1849)
(Nicholson, Henry H. 1850)
(Salisbury, Rollin D. 1858)
(Van Hise, Charles R. 1857)

FOREIGN

Austria

(Hagen, John G. 1847)

Brazil

Glover, Townend 1813

Canada

Bowser, Edward A. 1837
Hartt, Charles F. 1840
(McMurrich, James P. 1859)
(Newcomb, Simon 1835)
Watson, James C. 1838

China

(Coulter, John M. 1851)

Denmark

(Boyé, Martin H. 1812)

England

Abbot, John 1751
Banister, John 1650
Bland, Thomas 1809
Brocklesby, John 1811
(Brooks, William R. 1844)
Catesby, Mark 1683
Claypole, Edward W. 1835
Clayton, John 1694
(Compton, Alfred G. 1835)
Cooper, Thomas 1759
(Davidson, George 1825)
Draper, John W. 1811
Dunglison, Robley 1798
(Frisby, Edgar 1837)
Fry, Joshua ca. 1700
Grote, Augustus R. 1841
Jones, Hugh ca. 1692
Josselyn, John ca. 1608
Kinnersley, Ebenezer 1711
Lawson, John d. 1711
Meehan, Thomas 1826
(Nichols, Edward L. 1854)
Nuttall, Thomas 1786
Palmer, Arthur W. 1861
Parry, Charles C. 1823
Peter, Robert 1805
Priestley, Joseph 1733
Renwick, James, Sr. 1792
Riley, Charles V. 1843
(Scott, Charlotte A. 1858)
(Searle, Arthur 1837)
Sherman, John 1613
Sylvester, James J. 1814
Taylor, Richard C. 1789
Teschemacher, James E. 1790
Vasey, George 1822
Walsh, Benjamin D. 1808
Walter, Thomas ca. 1740
Weightman, William 1813
Winthrop, John, Jr. 1606

France

Bonaparte, Charles L. 1803
Durand, Elias 1794
Girard, Charles F. 1822
Hentz, Nicholas M. 1797
Lesueur, Charles A. 1778
Marcou, Jules 1824
Michaux, André 1746
Michaux, François A. 1770
Nicollet, Joseph N. 1786
Saugrain de Vigni, Antoine F. 1763

Germany

Behr, Hans H. 1818
(Beyer, Henry G. 1850)
Brunnow, Franz F. E. 1821
Eimbeck, William 1841
Engelmann, George 1809
Engelmann, Henry 1831
Gattinger, Augustin 1825
Genth, Frederick A., Sr. 1820
(Goessmann, Charles, A. 1827)
Hagen, Hermann A. 1817
(Hilgard, Eugene W. 1833)
Hilgard, Julius E. 1825
(Hinrichs, Gustavus D. 1836)
(Koenig, George A. 1844)
Lindenkohl, Adolph 1833
Lindheimer, Ferdinand J. 1801
Marx, George 1838
(Maschke, Heinrich 1853)
Melsheimer, Friedrich V. 1749
(Michelson, Albert A. 1852)
Mohr, Charles T. 1824
Peters, Christian H. F. 1813
Pursh, Frederick 1774
Rice, Charles 1841
(Schaeberle, John M. 1853)
Schöpf, Johann D. 1752
Schott, Arthur C. V. 1814
Schott, Charles A. 1826
Schubert, Ernst 1813
Schwarz, Eugene A. 1844
Stallo, Johann B. 1823
Wachsmuth, Charles 1829
Wislizenus, Frederick A. 1810

Hungary

(Heilprin, Angelo 1853)
Xántus, János 1825

India

Hazen, Henry A. 1849

Ireland

Adrain, Robert 1775
Colden, Cadwallader 1688
Curley, James 1796
Emmet, John P. 1796
Freeman, Thomas d. 1821
Logan, James 1674
(Mallet, John W. 1832)
Martin, Henry N. 1848
Meade, William d. 1833
Mitchell, James A. 1852
Patterson, Robert 1743
Tuomey, Michael 1805
Vaughan, Daniel ca. 1818

Italy

Bayma, Joseph 1816
Sestini, Benedict 1816

Japan

(Takamine, Jokichi 1854)

Netherlands

Romans, Bernard ca. 1720
Troost, Gerard 1776

Portugal

De Kay, James E. 1792
(probable birthplace)

Russia

Osten, Sacken, Carl R. R. von der 1828

Scotland

(Bell, Alexander G. 1847)
Brackenridge, William D. 1810
(Calvin, Samuel 1840)
Chalmers, Lionel 1715
Douglass, William ca. 1691
Dunbar, William 1749
Ferguson, James 1797
(Fleming, Williamina P. S. 1857)
Garden, Alexander 1730
Harkness, William 1837
Lining, John 1708
(MacFarlane, John M. 1855)
Maclean, John 1771
Maclure, William 1763
Michie, Peter S. 1839
Minto, Walter 1753
(Muir, John 1838)
Owen, David D. 1807
Ramsay, Alexander ca. 1754
Wilson, Alexander 1766

Sweden

(Lindgren, Waldemar 1860)
(Rydberg, Per A. 1860)

Switzerland

(Agassiz, Alexander 1835)
Agassiz, J. Louis R. 1807
Boll, Jacob 1828
Guyot, Arnold H. 1807
Hassler, Ferdinand R. 1770
Lesquereux, Leo 1806
Pourtalès, Louis F. de 1823/1824

Turkey

Rafinesque, Constantine S. 1783

West Indies

Audubon, John J. 1785
(Holland, William J. 1848)

Appendix C
EDUCATION

Names generally are listed only by the highest level of education, with special weight given to education that culminated in a formal degree. However, some exceptions have been made. When a person received an A.B. degree and subsequently received an M.D., Ph.D., A.M., or professional science degree, the name will appear under both the bachelor's (Graduated from College) and the advanced degree. If a person received an MD. and a Ph.D., the name likewise appears under both degrees. In an instance in which an individual received an A.B. degree and subsequently studied science or medicine as a postgraduate activity not leading to an advanced degree, the name appears only under the heading Graduated from College. However, if a person attended college without receiving a degree and subsequently studied medicine or science without receiving a degree, the name appears only with the advanced training.

This appendix does not include a separate listing for persons who studied law, theology, and the like. There were instances in which such studies were pursued on a postgraduate level or after attendance at a college. If a person studied law or theology without attending a liberal arts college, the name appears in the Miscellaneous listing.

Names in this appendix are listed under the following headings:

1. Not Known
2. Miscellaneous
3. Attended College (no degree)
4. Graduated from College
5. Master of Arts Degree
6. Studied Medicine (no degree)
7. Medical Degree
8. Formal Study of Science, Engineering, or Agriculture (no degree)
9. Graduated from U.S. Military Academy (West Point)
10. Graduated from U.S. Naval Academy (Annapolis)
11. Professional Degree in Science or Engineering
12. Doctor of Philosophy or Doctor of Science Degree

1. NOT KNOWN

Adrain, Robert 1775
Batchelder, John M. 1811
Bent, Silas 1820
Biddle, Owen 1737
Bonaparte, Charles L. 1803
Brown, Joseph 1733
Browne, Daniel J. 1804
Catesby, Mark 1683
Christy, David 1802
Cooper, William ca. 1798
Cutbush, James 1788
De Brahm, William G. 1717
Dunbar, William 1749
Ferguson, James 1797
Fitz, Henry 1808
Freeman, Thomas d. 1821
Gibbs, George 1776
Grew, Theophilus d. 1759
Grimes, James S. 1807
Hutchins, Thomas 1730
Josselyn, John ca. 1608
Lawrence, George N. 1806
Lawson, John d. 1711
Leeds, John 1705
Meehan, Thomas 1826
More, Thomas fl. 1670-1724
(Morgan, Andrew P. 1836)
Morris, Margaretta H. 1797
Ord, George 1781
Osten Sacken, Carl R. R. von der 1828
Patterson, Robert 1743

Rice, Charles 1841
Romans, Bernard ca. 1720
Saugrain de Vigni, Antoine F. 1763
Schubert, Ernst 1813
Taylor, Richard C. 1789
Teschemacher, James E. 1790
Trautwine, John C. 1810
Trumbull, Gurdon 1841
Wachsmuth, Charles 1829
Walter, Thomas ca. 1740
Weightman, William 1813

2. MISCELLANEOUS

In this category are listed persons who neither attended college nor pursued a special professional course in science or medicine. Included are a variety of educational backgrounds, ranging from individuals who had no formal education to those graduating from public high schools or private academies, and encompassing persons who were apprenticed to various trades or crafts, those who had private tutors, individuals described as self-educated, those who attended only the common schools, and the like. Also included are persons who studied law and theology, in one or two instances resulting in a professional degree in one of those subjects (e.g., W. H. Seaman received a bachelor of law degree).

Abbot, John 1751
Alger, Francis 1807
Anthony, John G. 1804
(Ashmead, William H. 1855)
Audubon, John J. 1785
Banneker, Benjamin 1731
Bartram, John 1699
Bartram, William 1739
(Blake, Francis 1850)
Bland, Thomas 1809
Bond, William C. 1789
Bowditch, Nathaniel 1773
Brackenridge, William D. 1810
(Brashear, John A. 1840)
(Brewster, William 1851)
(Brooks, William R. 1844)
(Burnham, Sherburne W. 1838)
Carll, John F. 1828
Carpenter, George W. 1802
Cassin, John 1813
(Chandler, Seth C. 1846)
Cist, Jacob 1782
Clark, Alvan 1804
Clark, Alvan G. 1832
Coan, Titus 1801
Colden, Jane 1724
(Collins, Frank S. 1848)
Conrad, Timothy A. 1803
Cope, Edward D. 1840
(Cresson, Ezra T. 1838)
Daboll, Nathan 1750
(Dall, William H. 1845)
(Davenport, George E. 1833)
(Davidson, George 1825)
(Draper, Daniel 1841)
Durand, Elias 1794
(Edison, Thomas A. 1847)
Eigenmann, Rosa S. 1858
Ellicott, Andrew 1754
(Elliot, Daniel G. 1835)
(Emerton, James H. 1847)
(Fernald, Charles H. 1838)
(Fleming, Williamina P. S. 1857)
Folger, Walter 1765
Fowler, Lorenzo N. 1811
Franklin, Benjamin 1706
Furbish, Kate 1834
Gabb, William M. 1839
Gardiner, James T. 1842
(Gill, Theodore N. 1837)
Glover, Townend 1813
Godfrey, Thomas 1704
Green, Francis M. 1835
Griscom, John 1774
Grote, Augustus R. 1841
Gummere, John 1784
Hager, Albert D. 1817
Hanks, Henry G. 1826
Hare, Robert 1781
Harrison, John 1773
Hayes, Augustus A. 1806
Henry, Joseph 1797
Herrick, Edward C. 1811
Hilgard, Julius E. 1825
Hitchcock, Edward 1793
Holmes, Francis S. 1815
Hunt, Thomas S. 1826
Kendall, Ezra O. 1818
Kennicott, Robert 1835
Kinnersley, Ebenezer 1711
Kirkwood, Daniel 1814
(Langley, Samuel P. 1834)
Lapham, Increase A. 1811
Lea, Isaac 1792
Lea, Mathew C. 1823
Lesquereux, Leo 1806
Lesueur, Charles A. 1778
Lewis, Enoch 1776
Lintner, Joseph A. 1822
(Lloyd, John U. 1849)
Logan, James 1674
Loomis, Mahlon 1826
McArthur, William P. 1814
(McGee, William J. 1853)

Maclure, William 1763
Mapes, James J. 1806
Marshall, Humphry 1722
Masterman, Stillman 1831
Maury, Matthew F. 1806
Meek, Fielding B. 1817
(Mendenhall, Thomas C. 1841)
Michaux, André 1746
Mitchell, Henry 1830
Mitchell, Maria 1818
Mitchell, William 1791
Morey, Samuel 1762
Nuttall, Thomas 1786
Peale, Charles W. 1742
Peale, Titian R. 1799
Phelps, Almira H. L. 1793
Pourtalès, Louis F. de 1823/1824
Priestley, Joseph 1733
Pursh, Frederick 1774
Rafinesque, Constantine S. 1783
Redfield, William C. 1789
(Ridgway, Robert 1850)
Riley, Charles V. 1843
Rittenhouse, David 1732
Rogers, Henry D. 1808
Samuels, Edward A. 1836
Saxton, Joseph 1799
Say, Thomas 1787
Schweinitz, Lewis D. von 1780
(Seaman, William H. 1837)
Sennett, George B. 1840
Smith, Erminnie A. P. 1836
(Stearns, Robert E. C. 1827)
Stimpson, William 1832
(Stockwell, John N. 1832)
Strecker, Herman 1836
(Swift, Lewis 1820)
Taylor, Charlotte De B. S. 1806
(Thomas, Cyrus 1825)
Thompson, Benjamin
(Count Rumford) 1753
Thurber, George 1821
Torrey, Bradford 1843
Townsend, John K. 1809
Tracy, Clarissa T. 1818
Treadwell, Daniel 1791
Tryon, George W. 1838
Tuttle, Charles W. 1829
Tyson, Philip T. 1799
Vaughan, Daniel ca. 1818
Wagner, William 1796
(Walcott, Charles D. 1850)
West, Benjamin 1730
(Whitfield, Robert P. 1828)
Wilkes, Charles 1798
Wilson, Alexander 1766
Worthen, Amos H. 1813
Xántus, János 1825
Youmans, Edward L. 1821

3. ATTENDED COLLEGE

Here are listed persons known to have attended college without evidence of their having received a degree. In the case of J. Curley, J. G. Hagen, and B. Sestini, all Jesuits, their training is equated to attendance at college. H. Engelmann is said to have graduated from the University of Berlin, but the degree received is not known.

Bachman, John 1790
(Barnard, Edward E. 1857)
(Bell, Alexander G. 1847)
Blodget, Lorin 1823
Boll, Jacob 1828
(Broadhead, Garland C. 1827)
Burritt, Elijah H. 1794
(Calvin, Samuel 1840)
Chalmers, Lionel 1715
Clayton, John 1694
Cooper, Thomas 1759
Curley, James 1796
Engelmann, Henry 1831
Fry, Joshua ca. 1700
Gilliss, James M. 1811
(Hagen, John G. 1847)
Haldeman, Samuel S. 1812
(Hall, Asaph 1829)
Hassler, Ferdinand R. 1770
(Hinrichs, Gustavus D. 1836)
Lemmon, John G. 1832
Lindheimer, Ferdinand J. 1801
Lippincott, James S. 1819
Marcou, Jules 1824
Mayer, Alfred M. 1836
Melsheimer, Friedrich V. 1749
(Michael, Arthur 1853)
Minto, Walter 1753
Morfit, Campbell 1820
Mühlenberg, Gotthilf H. E. 1753
(Muir, John 1838)
Nicollet, Joseph N. 1786
Norton, John P. 1822
Platt, Franklin 1844
Powell, John W. 1834
(Rathbun, Richard 1852)
Rogers, William B. 1804
Ruffin, Edmund 1794
Schoolcraft, Henry R. 1793
Schwarz, Eugene A. 1844
Sestini, Benedict 1816
Sherman, John 1613
(Skinner, Aaron N. 1845)
Stallo, Johan B. 1823

(Taylor, Frank B. 1860)
(Uhler, Philip R. 1835)
Wailes, Benjamin L. C. 1797
Winthrop, John, Jr. 1606

4. GRADUATED FROM COLLEGE

For institutions represented by more than two graduates among the scientists, personal names are listed under the name of the college. For those persons receiving a degree from institutions represented by only one or two graduates, names appear under Other at the end of the list of colleges.

Amherst College
Adams, Charles B. 1814
Allen, Timothy F. 1837
Chapman, Alvan W. 1809
(Clark, William B. 1860)
(Clarke, John M. 1857)
Coffin, James H. 1806
(Emerson, Benjamin K. 1843)
Fowler, Orson S. 1809
(Goodale, George L. 1839)
(Hitchcock, Charles H. 1836)
(Holland, William J. 1848)
(Howe, James L. 1859)
(Kemp, James F. 1859)
(Morse, Harmon N. 1848)
Nason, Henry B. 1831
Shepard, Charles U. 1804
Snell, Ebenezer S. 1801
(Stoddard, John T. 1852)
Williams, George H. 1856

Bowdoin College
(Brackett, Cyrus F. 1833)
Coffin, John H. C. 1815
(Hall, Edwin H. 1855)
Johnston, John 1806
Packard, Alpheus S., Jr. 1839
Smyth, William 1797
Storer, David H. 1804
Sturtevant, Edward L. 1842
Swallow, George C. 1817
(Wadsworth, Marshman E. 1847)
(Whitman, Charles O. 1842)

Brown University
Binney, Amos 1803
Caswell, Alexis 1799
Holbrook, John E. 1794
Jenks, John W. P. 1819
Rogers, William A. 1832

Cambridge University
Martin, Henry N. 1848
Sylvester, James J. 1814
Walsh, Benjamin D. 1808

City College of New York
(Abbe, Cleveland 1838)
(Compton, Alfred G. 1835)
Mason, James W. 1836
Newton, Henry 1845
(Remsen, Ira 1846)

Columbia University
Bolton, Henry C. 1843
Bruce, Archibald 1777
Caswell, John H. 1846
(Cornwall, Henry B. 1844)
(Gibbs, O. Wolcott 1822)
LeConte, John E. 1784
(McClintock, Emory 1840)
(Rees, John K. 1851)
Renwick, James, Sr. 1792
(Witthaus, Rudolph A. 1846)

Dartmouth College
(Boss, Lewis 1846)
Cotting, John R. 1778
Hazen, Henry A. 1849
Marsh, George P. 1801
(Upham, Warren 1850)
(Young, Charles A. 1834)

Dickinson College
Baird, Spencer F. 1823
Bridges, Robert 1806
Morris, John G. 1803

University of Georgia
Jones, William L. 1827
LeConte, John 1818
LeConte, Joseph 1823

Hamilton College
Elliott, Ezekiel B. 1823
Orton, Edward F. B. 1829
(Rising, Willard B. 1839)

Hanover College
(Barnes, Charles R. 1858)
(Coulter, John M. 1851)
(Wiley, Harvey W. 1844)

Harvard College
(Agassiz, Alexander 1835)
(Becker, George F. 1847)
Bigelow, Jacob 1786
Binney, William G. 1833
Bond, George P. 1825
(Bowditch, Henry P. 1840)
Bowdoin, James 1726
Brattle, Thomas 1658
Brewer, Thomas M. 1814
Burgess, Edward 1848
(Byerly, William E. 1849)
Cabot, Samuel, Jr. 1815
Chase, Pliny E. 1820

Clap, Thomas 1703
Cleaveland, Parker 1780
Colman, Benjamin 1673
Cooke, Josiah P. 1827
Dalton, John C., Jr. 1825
Dana, James F. 1793
Dana, Samuel L. 1795
Davis, Charles H. 1807
Dudley, Paul 1675
(Dwight, Thomas 1843)
Emerson, George B. 1797
(Emmons, Samuel F. 1841)
Eustis, Henry L. 1819
(Farlow, William G. 1844)
Farrar, John 1779
(Flint, Albert S. 1853)
(Gooch, Frank A. 1852)
Gorham, John 1783
Gould, Augustus A. 1805
Gould, Benjamin A., Jr. 1824
Greenwood, Isaac 1702
Harris, Thaddeus W. 1795
Hill, Henry B. 1849
Hill, Thomas 1818
Hubbard, Henry G. 1850
(Jackson, Charles L. 1847)
Jackson, John B. S. 1806
Jeffries, John 1744/1745
Johnson, Walter R. 1794
Kneeland, Samuel 1821
Lovering, Joseph 1813
(Lyman, Benjamin S. 1835)
Lyman, Theodore 1833
Mapes, Charles V. 1836
Mather, Cotton 1662/1663
Mather, Increase 1639
Oliver, Andrew 1731
Oliver, James E. 1829
Page, Charles G. 1812
Paine, Robert T. 1803
Peck, William D. 1763
Peirce, Benjamin 1809
(Peirce, Benjamin O. 1854)
(Peirce, Charles S. 1839)
(Peirce, James M. 1834)
Pickering, Charles 1805
Pike, Nicholas 1743
Robie, Thomas 1688/1689
Safford, Truman H. 1836
(Sargent, Charles S. 1841)
(Searle, Arthur 1837)
(Story, William E. 1850)
(Stringham, Washington I. 1847)
(Tuckerman, Alfred 1848)
Walker, Sears C. 1805
(Waller, Elwyn 1846)
Warren, John C. 1778
(Warren, Joseph W. 1849)
Webster, John W. 1793
Winlock, William C. 1859
Winthrop, John 1714
Wood, Edward S. 1846
Wright, Chauncey 1830
Wyman, Jeffries 1814

University of Michigan

Baker, Marcus 1849
(Beal, William J. 1833)
(Beman, Wooster W. 1850)
Harrington, Mark W. 1848
(Mark, Edward L. 1847)
(Spalding, Volney M. 1849)
Watson, James C. 1838
(Winchell, Newton H. 1839)

New York University

Clark, Henry J. 1826
(Doremus, Robert O. 1824)
(Parish, Samuel B. 1838)
(Stevenson, John J. 1841)

Oberlin College

Hayden, Ferdinand V. 1829
Kedzie, Robert C. 1823
(Maltby, Margaret E. 1860)
(Todd, James E. 1846)
Wright, Albert A. 1846

University of Pennsylvania

Bache, Franklin 1792
Booth, James C. 1810
Frazer, John F. 1812
(Frazer, Persifor 1844)
Green, Jacob 1790
Hallowell, Edward 1808
(Jayne, Horace F. 1859)
Keating, William H. 1799
Lesley, J. Peter 1819
Mease, James 1771
(Mitchell, Silas W. 1829)
Morton, Henry 1836
Patterson, Robert M. 1787
Tilghman, Richard A. 1824
Wetherill, Charles M. 1825
Williamson, Hugh 1735
Woodhouse, James 1770

Princeton University

Barton, William P. C. 1786
(Fine, Henry B. 1858)
(Halsted, George B. 1853)
McCulloh, Richard S. 1818
(Osborn, Henry F. 1857)
Rood, Ogden N. 1831
Rush, Benjamin 1746

Rush, James 1786
Wurtz, Henry 1828

University of South Carolina

Gibbes, Lewis R. 1810
Gibbes, Robert W. 1809
Porcher, Francis P. 1825
Ravenel, Henry W. 1814
(Stevens, Walter L. 1847)

Union College

Alexander, Stephen 1806
Beck, Lewis C. 1798
(Cooley, Leroy C. 1833)
(Ellis, Job B. 1829)
Hough, Franklin B. 1822
(Hough, George W. 1836)
Joslin, Benjamin F. 1796
(Julien, Alexis A. 1840)
Parry, Charles C. 1823
(Peck, Charles H. 1833)
Tuckerman, Edward 1817
Whipple, Squire 1804

Wesleyan University

(Atwater, Wilbur O. 1844)
Buckley, Samuel B. 1809
(Carhart, Henry S. 1844)
Goode, George B. 1851
Judd, Orange 1822
Marcy, Oliver 1820
(Rice, William N. 1845)
Winchell, Alexander 1824

Williams College

Atwater, Caleb 1778
(Brooks, William K. 1848)
Curtis, Moses A. 1808
Dewey, Chester 1784
Eaton, Amos 1776
(Edwards, William H. 1822)
Emmons, Ebenezer 1799
(Hall, G. Stanley 1846)
(Kingsley, John S. 1854)
(Morley, Edward W. 1838)
Orton, James 1830
Rutherfurd, Lewis M. 1816
(Scudder, Samuel H. 1837)
(Snow, Francis H. 1840)
Wells, David A. 1828

Yale College

Ayres, William O. 1817
Barnard, Frederick A. P. 1809
Bradley, Frank H. 1838
Brocklesby, John 1811
Chauvenet, William 1820
Collier, Peter 1835
Couper, James H. 1794
Cutler, Manasseh 1742
(Dana, Edward S. 1849)
Dana, James D. 1813
De Forest, Erastus L. 1834
Delafield, Joseph 1790
(Donaldson, Henry H. 1857)
(Dudley, Charles B. 1842)
(Dutton, Clarence E. 1841)
Eaton, Daniel C. 1834
(Eddy, Henry T. 1844)
Elliott, Stephen 1771
Gibbs, Josiah W. 1839
Hubbard, Joseph S. 1823
Hubbard, Oliver P. 1809
(Jenkins, Edward H. 1850)
(Johnson, William W. 1841)
Lane, Jonathan H. 1819
Loomis, Elias 1811
Lyman, Chester S. 1814
Mansfield, Jared 1759
Marsh, Othniel C. 1831
Mitchell, Elisha 1793
Newton, Hubert A. 1830
Olmsted, Denison 1791
Percival, James G. 1795
Porter, John A. 1822
Savage, Thomas S. 1804
Silliman, Benjamin 1779
Silliman, Benjamin, Jr. 1816
Smith, Hamilton L. 1818
Strong, Theodore 1790
Sullivant, William S. 1803
Twining, Alexander C. 1801
Van Rensselaer, Jeremiah 1793
Watson, Sereno 1826
Whitney, Josiah D. 1819
(Wright, Arthur W. 1836)
Wright, Charles 1811

Other

Alexander, John H. 1812
(Babcock, Stephen M. 1843)
(Bailey, Solon I. 1854)
Banister, John 1650
Bayma, Joseph 1816
Bodley, Rachel L. 1831
(Boyé, Martin H. 1812)
Brace, De Witt B. 1859
(Brigham, Albert P. 1855)
(Britton, Elizabeth G. K. 1858)
(Burrill, Thomas J. 1839)
(Chamberlin, Thomas C. 1843)
(Chatard, Thomas M. 1848)
Claypole, Edward W. 1835
Colden, Cadwallader 1688
Collett, John 1828
Coues, Elliott 1842
Crocker, Lucretia 1829

Day, William C. 1857
(Dolbear, Amos E. 1837)
Espy, James P. 1785
Ferrel, William 1817
(Fontaine, William M. 1835)
(Frankforter, George B. 1859)
(Frear, William 1860)
(Frisby, Edgar 1837)
(Gilbert, Grove K. 1843)
Good, Adolphus C. 1856
(Hall, Christopher W. 1845)
Harkness, William 1837
Hartt, Charles F. 1840
(Hay, Oliver P. 1846)
(Hill, George W. 1838)
Houghton, Douglass 1809
(Howe, Herbert A. 1858)
Jacobs, Michael 1808
James, Edwin 1797
Jefferson, Thomas 1743
Jones, Hugh ca. 1692
Keeler, James E. 1857
Kerr, Washington C. 1827
(Ladd, George T. 1842)
(Lattimore, Samuel A. 1828)
LeConte, John L. 1825
(MacBride, Thomas H. 1848)
McCay, Charles F. 1810
(McMurrich, James P. 1859)
Madison, James 1749
(Mallet, John W. 1832)
Newberry, John S. 1822
(Nicholson, Henry H. 1850)
(Nutting, Charles C. 1858)
Porter, Thomas C. 1822
(Pritchett, Henry S. 1857)
Puryear, Bennet 1826
(Richards, Ellen H. S. 1842)
Riddell, John L. 1807
(Sadtler, Samuel P. 1847)
(Safford, James M. 1822)
Schweinitz, Emil A. de ca. 1864
Shattuck, Lydia 1822
Short, Charles W. 1794
(Smith, Eugene A. 1841)
(Smith, William B. 1850)
Steele, Joel D. 1836
Stoddard, John F. 1825
(Stone, Ormond 1847)
Thompson, Zadock 1796
Venable, Charles S. 1827
(Venable, Francis P. 1856)
(Ward, Lester F. 1841)
Warder, Robert B. 1848
(Wead, Charles K. 1848)
(White, Israel C. 1848)
(Whiting, Sarah F. 1847)
(Whitney, Mary W. 1847)
Winlock, Joseph 1826
Yeates, William S. 1856

5. MASTER OF ARTS DEGREE

(Bailey, Solon I. 1854)
(Brigham, Albert P. 1855)
(Flint, Albert S. 1853)
(Nicholson, Henry H. 1850)
(Nipher, Francis E. 1847)

6. STUDIED MEDICINE

There is no evidence that persons in this group received the medical degree.

Barton, Benjamin S. 1766
Beaumont, William 1785
Boylston, Zabdiel 1680
Cutting, Hiram A. 1832
Douglas, Silas H. 1816
Eights, James 1798
Gattinger, Augustin 1825
Guthrie, Samuel 1782
Hayden, Horace H. 1769
Hentz, Nicholas M. 1797
Hildreth, Samuel P. 1783
Holder, Joseph B. 1824
James, Thomas P. 1803
Lincecum, Gideon 1793
Lining, John 1708
Maclean, John 1771
Michaux, François A. 1770
Mitchell, John 1711
Perrine, Henry 1797
Ramsay, Alexander ca. 1754
Sartwell, Henry P. 1792
Shecut, John L. E. W. 1770
Trask, John B. 1824
Wells, Samuel R. 1820
Young, Aaron 1819

7. MEDICAL DEGREE

(Abbott, Charles C. 1843)
Agassiz, J. Louis R. 1807
Allen, Harrison 1841
Allen, Timothy F. 1837
Alter, David 1807
Ayres, William O. 1817
Bache, Franklin 1792
Baldwin, William 1779
(Barker, George F. 1835)
Barton, William P. C. 1786
(Bean, Tarleton H. 1846)
Beck, Lewis C. 1798

Behr, Hans H. 1818
(Beyer, Henry G. 1850)
Bigelow, Jacob 1786
Binney, Amos 1803
(Bowditch, Henry P. 1840)
(Boyé, Martin H. 1812)
(Brackett, Cyrus F. 1833)
Brandegee, Mary K. L. C. 1844)
Brewer, Thomas M. 1814
Bridges, Robert 1806
Bruce, Archibald 1777
Burnett, Waldo I. 1827
Cabot, Samuel, Jr. 1815
Chapman, Alvan W. 1809
Collett, John 1828
Cooper, James G. 1830
Coues, Elliott 1842
Dalton, John C., Jr. 1825
Dana, James F. 1793
Dana, Samuel L. 1795
Darlington, William 1782
Deane, James 1801
De Kay, James E. 1792
(Doremus, Robert O. 1824)
Douglass, William ca. 1691
Drake, Daniel 1785
Draper, Henry 1837
Draper, John W. 1811
Drown, Thomas M. 1842
Dunglison, Robley 1798
(Dwight, Thomas 1843)
Emmet, John P. 1796
Engelmann, George 1809
Evans, John 1812
(Farlow, William G. 1844)
Fitch, Asa 1809
Gambel, William 1823
Garden, Alexander 1730
Gibbes, Lewis R. 1810
Gibbes, Robert W. 1809
(Gibbes, O. Wolcott 1822)
Girard, Charles F. 1822
Godman, John D. 1794
(Goodale, George L. 1839)
Gorham, John 1783
Gould, Augustus A. 1805
Gray, Asa 1810
Grith, Robert E. 1798
Hagen, Hermann A. 1817
Hallowell, Edward 1808
Hammond, William A. 1828
Harkness, William 1837
Harlan, Richard 1796
Harris, Thaddeus W. 1795
Hayden, Ferdinand V. 1829
Holbrook, John E. 1794
Horn, George H. 1840
Horner, William E. 1793
Horsfield, Thomas 1773
Hough, Franklin B. 1822
Jackson, Charles T. 1805
Jackson, John B. S. 1806
(James, William 1842)
(Jayne, Horace F. 1859)
Jeffries, John 1744/1745
Jones, William L. 1827
(Jordan, David S. 1851)
Joslin, Benjamin F. 1796
Kedzie, Robert C. 1823
Kellogg, Albert 1813
Keyt, Alonzo T. 1827
Kirtland, Jared P. 1793
Kneeland, Samuel 1821
Kuhn, Adam 1741
LeConte, John 1818
LeConte, John L. 1825
LeConte, Joseph 1823
(Leffmann, Henry 1847)
Leidy, Joseph 1823
Locke, John 1792
(Mallet, John W. 1832)
Marx, George 1838
Meade, William d. 1833
Mease, James 1771
(Mellus, Edward L. 1848)
Metcalfe, Samuel L. 1798
Michener, Ezra 1794
Miles, Manly 1826
(Millspaugh, Charles F. 1854)
Mitchell, John K. 1793
(Mitchell, Silas W. 1829)
Mitchill, Samuel L. 1764
Morton, Samuel G. 1799
Newberry, John S. 1822
Nichols, James R. 1819
Owen, David D. 1807
Packard, Alpheus S., Jr. 1839
Page, Charles G. 1812
Parry, Charles C. 1823
Patterson, Robert M. 1787
Percival, James G. 1795
Peter, Robert 1805
Pickering, Charles 1805
Pitcher, Zina 1797
Plummer, John T. 1807
Porcher, Francis P. 1825
Prescott, Albert B. 1832
Ravenel, Edmund 1797
Ravenel, St. Julien 1819
(Remsen, Ira 1846)
Riddell, John L. 1807
Rogers, James B. 1802
Rogers, Robert E. 1813
Rush, Benjamin 1746

Rush, James 1786
Salisbury, James H. 1823
Savage, Thomas S. 1804
Schöpf, Johann D. 1752
Short, Charles W. 1794
(Shufeldt, Robert W. 1850)
Shumard, Benjamin F. 1820
Smith, John L. 1818
Squibb, Edward R. 1819
Storer, David H. 1804
Sturtevant, Edward L. 1842
Swallow, George C. 1817
Torrey, John 1796
Troost, Gerard 1776
Van Rensselaer, Jeremiah 1793
Vasey, George 1822
Warder, John A. 1812
(Warren, Joseph W. 1849)
Waterhouse, Benjamin 1754
Webster, John W. 1793
Wells, William C. 1757
(White, Charles A. 1826)
(Wilder, Burt G. 1841)
(Wiley, Harvey W. 1844)
Williamson, Hugh 1735
(Williston, Samuel W. 1852)
Wislizenus, Frederick A. 1810
Wistar, Caspar 1761
(Witthaus, Rudolph A. 1846)
Wood, Edward S. 1846
Woodhouse, James 1770
Woodhouse, Samuel W. 1821
Wyman, Jeffries 1814
Yandell, Lunsford P. 1805
Youmans, William J. 1838

8. FORMAL STUDY OF SCIENCE, ENGINEERING, OR AGRICULTURE

Persons in this group pursued a course of study in science or science-related subjects at a college, scientific school, or polytechnic school. There is no evidence that they received professional science degrees as a result of these studies. However, several did graduate from European polytechnic schools.

(Allen, Joel A. 1838)
Babcock, James F. 1844
(Boyé, Martin H. 1812)
(Campbell, Marius R. 1858)
Clemson, Thomas G. 1807
Cory, Charles B. 1857
Eimbeck, William 1841
(Hague, James D. 1836)
(Heilprin, Angelo 1853)
(Johnson, Samuel W. 1830)
Lindenkohl, Adolph 1833
Mitchell, James A. 1852
Mohr, Charles T. 1824
(Morse, Edward S. 1838)
Ober, Frederick A. 1849
Prime, Temple 1832
Procter, John R. 1844
(Pumpelly, Raphael 1837)
Schott, Arthur C. V. 1814
Schott, Charles A. 1826
Seybert, Henry 1801
Smith, Erminnie A. P. 1836
Thompson, Almon H. 1839
Vanuxem, Lardner 1792
(Ward, Henry A. 1834)

9. GRADUATED U.S. MILITARY ACADEMY (WEST POINT)

(Abbot, Henry L. 1831)
Abert, John J. 1788
Bache, Alexander D. 1806
Bailey, Jacob W. 1811
Barnard, John G. 1815
Bartlett, William H. C. 1804
Courtenay, Edward H. 1803
Emory, William H. 1811
Eustis, Henry L. 1819
Graham, James D. 1799
(Holden, Edward S. 1846)
Humphreys, Andrew A. 1810
Hunt, Edward B. 1822
Mather, William W. 1804
Michie, Peter S. 1839
Mitchel, Ormsby M. 1809
Norton, William A. 1810
Peck, William G. 1820
Trowbridge, William P. 1828
Warren, Gouverneur K. 1830
Wheeler, George M. 1842
Whipple, Amiel W. 1816
Whittlesey, Charles 1808

10. GRADUATED U.S. NAVAL ACADEMY (ANNAPOLIS)

(Blair, Andrew A. 1848)
Brooke, John M. 1826
Hayden, Edward E. 1858
(Michelson, Albert A. 1852)

11. PROFESSIONAL DEGREE IN SCIENCE OR ENGINEERING (B.S., M.S., PH.B., C.E., M.E., B.N.S.)

(Abbe, Cleveland 1838)
(Agassiz, Alexander 1835)
(Anthony, William A. 1835)
Ashburner, Charles A. 1854

(Atkinson, George F. 1854)
(Barker, George F. 1835)
(Beal, William J. 1833)
(Bean, Tarleton H. 1846)
(Bessey, Charles E. 1845)
(Blake, William P. 1825)
Bowser, Edward A. 1837
Brandegee, Townshend S. 1843
(Brewer, William H. 1828)
(Brush, Charles F. 1849)
(Brush, George J. 1831)
Clark, Henry J. 1826
(Clarke, Frank W. 1847)
Claypole, Edward W. 1835
(Comstock, Anna B. 1854)
(Comstock, George C. 1855)
(Comstock, John H. 1849)
Cook, George H. 1818
(Crafts, James M. 1839)
(Crosby, William O. 1850)
(Cross, Charles R. 1848)
(Davis, William M. 1850)
De Forest, Erastus L. 1834
(Diller, Joseph S. 1850)
(Dolbear, Amos E. 1837)
(Doolittle, Charles L. 1843)
(Dudley, William R. 1849)
(Eastman, John R. 1836)
Eaton, Daniel C. 1834
Emmons, Ebenezer 1799
Ewell, Ervin E. 1867
(Fairchild, Herman L. 1850)
Fitch, Asa 1809
Foster, John W. 1815
(Gage, Simon H. 1851)
(Gannett, Henry 1846)
(Garman, Samuel 1843)
(Greene, Edward L. 1843)
(Hague, Arnold 1840)
Hall, James 1811
Hatcher, John B. 1861
(Hering, Daniel W. 1850)
(Hill, Robert T. 1858)
(Holden, Edward S. 1846)
(Holmes, William H. 1846)
Horsford, Eben N. 1818
(Howard, Leland O. 1857)
Hyatt, Alpheus 1838
(Hyde, Edward W. 1843)
(Iddings, Joseph P. 1857)
Irving, Roland D. 1847
Jones, William L. 1827
(Jordan, David S. 1851)
(Kemp, James F. 1859)
King, Clarence R. 1842
LeConte, Joseph 1823
(Lindgren, Waldemar 1860)
Lyman, Theodore 1833
McCalley, Henry 1852
(Mixter, William G. 1846)
(Newcomb, Simon 1835)
(Niles, William H. 1838)
Packard, Alpheus S., Jr. 1839
(Peirce, Charles S. 1839)
(Penfield, Samuel L. 1856)
(Pickering, Edward C. 1846)
(Pickering, William H. 1858)
(Pirsson, Louis V. 1860)
(Prosser, Charles S. 1860)
(Putnam, Frederic W. 1839)
(Richards, Ellen H. S. 1842)
Rowland, Henry A. 1848
Runkle, John D. 1822
(Russell, Israel C. 1852)
Salisbury, James H. 1823
(Salisbury, Rollin D. 1858)
(Schaeberle, John M. 1853)
(Scudder, Samuel H. 1837)
(Sharples, Stephen P. 1842)
Shepard, James H. 1850
(Smith, Sidney I. 1843)
(Storer, Francis H. 1832)
(Takamine, Jokichi 1854)
(True, Frederick W. 1858)
Tuomey, Michael 1805
(Verrill, Addison E. 1839)
Warder, Robert B. 1848
Warren, Cyrus M. 1824
(Webster, Francis M. 1849)
Wells, David A. 1828
(Wilder, Burt G. 1841)
(Wiley, Harvey W. 1844)
(Woodward, Robert S. 1849)
Wright, Albert A. 1846
Yarnall, Mordecai 1816

12. DOCTOR OF PHILOSOPHY OR DOCTOR OF SCIENCE DEGREE

Agassiz, J. Louis R. 1807
Allen, Oscar D. 1836
(Arthur, Joseph C. 1850)
(Atwater, Wilbur O. 1844)
(Austen, Peter T. 1852)
(Babcock, Stephen M. 1843)
(Bailey, Edgar H. S. 1848)
(Barnes, Charles R. 1858)
(Barus, Carl 1856)
(Becker, George F. 1847)
Beecher, Charles E. 1856
(Beyer, Henry G. 1850)
Bolton, Henry C. 1843
Brace, De Witt B. 1859
(Branner, John C. 1850)

(Britton, Nathaniel L. 1859)
(Brooks, William K. 1848)
Brunnow, Franz F. E. 1821
(Byerly, William E. 1849)
(Caldwell, George C. 1834)
(Chandler, Charles F. 1836)
(Chatard, Thomas M. 1848)
Chester, Albert H. 1843
(Chittenden, Russell H. 1856)
(Clapp, Cornelia M. 1849)
(Clark, William B. 1860)
Collier, Peter 1835
(Comstock, Theodore B. 1849)
(Cornwall, Henry B. 1844)
(Coulter, John M. 1851)
Craig, Thomas 1855
(Dana, Edward S. 1849)
Day, William C. 1857
(Donaldson, Henry H. 1857)
(Dudley, Charles B. 1842)
(Eddy, Henry T. 1844)
(Elkin, William L. 1855)
(Emerson, Benjamin K. 1843)
(Fine, Henry B. 1858)
(Forbes, Stephen A. 1844)
(Frankforter, George B. 1859)
(Frazer, Persifor 1844)
(Frear, William 1860)
Genth, Frederick A., Sr. 1820
Gibbs, Josiah W. 1839
(Goessmann, Charles A. 1827)
(Gooch, Frank A. 1852)
Gould, Benjamin A., Jr. 1824
Guyot, Arnold H. 1807
(Hall, Edwin H. 1855)
(Hall, G. Stanley 1846)
(Halsted, George B. 1853)
(Hart, Edward 1854)
(Hastings, Charles S. 1848)
(Hay, Oliver P. 1846)
(Hilgard, Eugene W. 1833)
(Hillebrand, William F. 1853)
(Hollick, Charles A. 1857)
(Howe, Herbert H. 1858)
(Howe, James L. 1859)
(Jenkins, Edward H. 1850)
(Kellerman, William A. 1850)
(Kingsley, John S. 1854)
(Kinnicutt, Leonard P. 1854)
Knight, Wilbur C. 1858
(Knowlton, Frank H. 1860)
(Koenig, George A. 1844)
Lilley, George 1851
(Mabery, Charles F. 1850)
(MacFarlane, John M. 1855)
(McMurrich, James P. 1859)
(McMurtrie, William 1851)
(Mallet, John W. 1832)
(Maltby, Margaret E. 1860)
(Mark, Edward L. 1847)
Martin, Henry N. 1848
(Maschke, Heinrich 1853)
(Merrill, George P. 1854)
(Minot, Charles S. 1852)
(Morse, Harmon N. 1848)
Mott, Henry A., Jr. 1852
(Munroe, Charles E. 1849)
Nason, Henry B. 1831
Newton, Henry 1845
(Nichols, Edward L. 1854)
(Osborn, Henry F. 1857)
Palmer, Arthur W. 1861
(Peirce, Benjamin O. 1854)
Peters, Christian H. F. 1813
(Phillips, Francis C. 1850)
(Pritchett, Henry S. 1857)
Pugh, Evan 1828
(Rees, John K. 1851)
(Remsen, Ira 1846)
(Renouf, Edward 1848)
(Rice, William N. 1845)
(Rising, Willard B. 1839)
(Rydberg, Per A. 1860)
(Sadtler, Samuel P. 1847)
Schweinitz, Emil A. de ca. 1864
(Scott, Charlotte A. 1858)
(Shaler, Nathaniel S. 1841)
(Smith, Edgar F. 1854)
(Smith, Eugene A. 1841)
(Smith, William B. 1850)
(Spalding, Volney M. 1849)
(Stevenson, John J. 1841)
(Stoddard, John T. 1852)
(Story, William E. 1850)
(Stringham, Washington I. 1847)
(Thomas, Benjamin F. 1850)
(Trelease, William 1857)
(Trowbridge, John 1843)
(Tuckerman, Alfred 1848)
(Underwood, Lucien M. 1853)
(Van Hise, Charles R. 1857)
(Venable, Francis P. 1856)
(Wadsworth, Marshman E. 1847)
(Waller, Elwyn 1846)
Wetherill, Charles M. 1825
(White, Israel C. 1848)
(Whitman, Charles O. 1842)
Williams, George H. 1856
(Williams, Henry S. 1847)
(Williston, Samuel W. 1852)
(Wilson, Edmund B. 1856)
(Wilson, William P. 1844)
(Wright, Arthur W. 1836)

Appendix D OCCUPATION

The occupations or activities of the scientists are grouped into four general categories:

I. Not Known
II. Non-Science
III. Science-Related
IV. Science

A number of problems are encountered in any classification of occupations, and this listing is no exception. To the usual difficulties, here is encountered the problem of trying to designate what was and was not science. Just as the scheme itself in many ways is arbitrary, the placement of persons in these categories also is accompanied by some uncertainties. A recurring question is whether to define a person's occupation in terms of what he or she did (the activity) or to designate the occupation in relation to the place where the activity was practiced (that is, the employer). This scheme incorporates elements of both these approaches. It is hoped that the notes given in defining the various headings or categories, and the several cross-references, can help to overcome the inevitable difficulties.

Very frequently, an individual was engaged in more than one occupation (activity) and these sometimes were carried out in more than one institutional setting. In this appendix, names are listed under a maximum of three headings, and therefore these listings represent only the individual's chief occupational activities and not his or her whole career. Consequently, the lists do *not* include the name of every person who was ever engaged in a particular activity or who was ever employed in a particular institution.

Names in this appendix are listed under the following headings:

I. Not Known
II. Non-Science Employment or Activity
  1. Administrator/Clerk
  2. Agriculture
  3. Artist
  4. Business
  5. Clergyman
  6. College President
  7. Craftsman (see III-5)
  8. Lawyer
  9. Lecturer, Public (see IV-6)
  10. Librarian
  11. Literary Work (non-science)
  12. Military Service
  13. Professor (non-science)
  14. Public Official (elective and appointive)
  15. Teacher/Educator
  16. Tradesman
  17. Wealth (Independent)
III. Science-Related Employment or Activity
  1. Actuary
  2. Artist (see II-3)
  3. Business (see II-4)
  4. Horticulturist/Gardener
  5. Instrument Maker/Craftsman
  6. Inventor
  7. Medicine, Practitioner
  8. Medicine, Professor
  9. Pharmacist
  10. Surveyor/Explorer
IV. Science Employment or Activity
  1. Consultant/Engineer/Industrial Scientist
  2. Government Science, Foreign
  3. Government Science, States
  4. Government Science, U.S. (federal)
  5. Independent Study
  6. Lecturer, Public
  7. Literary Work
  8. Museum Curator/Associate
  9. Natural History Exploration and Collection
  10. Observatory Director/Staff Member
  11. Professor (science)
  12. Research Foundation Administrator/Associate
  13. Teacher/Educator of Science

I. NOT KNOWN

Bland, Thomas 1809
(Davenport, George E. 1833)
Eights, James 1798
Grote, Augustus R. 1841
Masterman, Stillman 1831
(Mellus, Edward L. 1848)
(Morgan, Andrew P. 1836)
Tyson, Philip T. 1799

II. NON-SCIENCE EMPLOYMENT OR ACTIVITY

II-1. Administrator/Clerk

Includes persons in private and public service (that is, government, business and industry, societies and philanthropic organizations, private foundations, and educational institutions).

Bent, Silas 1820
Blodget, Lorin 1823
Christy, David 1802
Cist, Jacob 1782
Clayton, John 1694
(Collins, Frank S. 1848)
Hager, Albert D. 1817
Herrick, Edward C. 1811
Hough, Franklin B. 1822
(Pritchett, Henry S. 1857)
Schoolcraft, Henry R. 1793
Torrey, Bradford 1843
(Upham, Warren 1850)
(Ward, Lester F. 1841)
Watson, Sereno 1826

II-2. Agriculture (farmer, planter)

Abbot, John 1751
Banister, John 1650
Banneker, Benjamin 1731
Buckley, Samuel B. 1809
Clemson, Thomas G. 1807
Collett, John 1828
Couper, James H. 1794
Dunbar, William 1749
Elliott, Stephen 1771
Holmes, Francis S. 1815
James, Edwin 1797
Lippincott, James S. 1819
Marshall, Humphry 1722
(Parish, Samuel B. 1838)
Peck, William D. 1763
Ravenel, Edmund 1797
Ravenel, Henry W. 1814
Ruffin, Edmund 1794
Sturtevant, Edward L. 1842
Wailes, Benjamin L. C. 1797
Walsh, Benjamin D. 1808
Walter, Thomas ca. 1740

II-3. Artist

In addition to painters and sculptors, this category also includes a designer of museum models. Most of the persons in this group used their artistic talents in their scientific work.

Abbot, John 1751
Audubon, John J. 1785
Clark, Alvan 1804
(Emerton, James H. 1847)
Lesueur, Charles A. 1778
Peale, Charles W. 1742
Strecker, Herman 1836
Trumbull, Gurdon 1841

II-4. Business

Under the subheading Manufacturer are several persons involved in production of chemicals, drugs, and fertilizers. Among the activities included under the general business category (Other) are persons active as merchants, bankers, investors, wholesale druggists, printers and publishers, storekeepers, and the like.

Manufacturer

Alger, Francis 1807
Brown, Joseph 1733
(Brush, Charles F. 1849)
Guthrie, Samuel 1782
Hanks, Henry G. 1826
Hare, Robert 1781
Harrison, John 1773
Holmes, Francis S. 1815
Horsford, Eben N. 1818
(Lloyd, John U. 1849)
Mapes, Charles V. 1836
Mapes, James J. 1806
Nichols, James R. 1819
Ravenel, St. Julien 1819
Sennett, George B. 1840
Squibb, Edward R. 1819
Treadwell, Daniel 1791
Tryon, George W. 1838
Warren, Cyrus M. 1824
Weightman, William 1813
West, Benjamin 1730
Winthrop, John, Jr. 1606

Other

(Agassiz, Alexander 1835)
Alexander, John H. 1812
Anthony, John G. 1804
Audubon, John J. 1785
Biddle, Owen 1737
Binney, Amos 1803
Binney, William G. 1833
Bowdoin, James 1726
Brattle, Thomas 1658
Carpenter, George W. 1802
Cassin, John 1813
Caswell, John H. 1846
Chase, Pliny E. 1820
Cist, Jacob 1782
Conrad, Timothy A. 1803
(Cresson, Ezra T. 1838)
Cutler, Manasseh 1742
Cutting, Hiram A. 1832
Darlington, William 1782
(Edwards, William H. 1822)
Elliott, Stephen 1771
Franklin, Benjamin 1706
Gardiner, James T. 1842
Gibbes, Robert W. 1809
Hubbard, Henry G. 1850
James, Thomas P. 1803
Keating, William H. 1799
Lapham, Increase A. 1811
Lawrence, George N. 1806
Lea, Isaac 1792
Lincecum, Gideon 1793
Lintner, Joseph A. 1822
McCay, Charles F. 1810
Maclure, William 1763
Marx, George 1838
Mitchell, William 1791
Morey, Samuel 1762
Nichols, James R. 1819
Ord, George 1781
Prime, Temple 1832
Rafinesque, Constantine S. 1783
Redfield, William C. 1789
Strecker, Herman 1836
Sullivant, William S. 1803
Teschemacher, James E. 1790
Wachsmuth, Charles 1829
Wagner, William 1796
Walsh, Benjamin D. 1808
Williamson, Hugh 1735
Worthen, Amos H. 1813

II-5. Clergyman

Bachman, John 1790
Banister, John 1650
Bayma, Joseph 1816
(Brigham, Albert P. 1855)
Clap, Thomas 1703
Coan, Titus 1801
Colman, Benjamin 1673
Cotting, John R. 1778
Curley, James 1796
Curtis, Moses A. 1808
Cutler, Manasseh 1742
Good, Adolphus C. 1856
(Hagen, John G. 1847)
Hill, Thomas 1818
Jones, Hugh ca. 1692
Madison, James 1749
Mather, Cotton 1662/1663
Mather, Increase 1639
Melsheimer, Friedrich V. 1749
Morris, John G. 1803
Mühlenberg, Gotthilf H. E. 1753
Orton, James 1830
Priestley, Joseph 1733
Savage, Thomas S. 1804
Schweinitz, Lewis D. von 1780
Sestini, Benedict 1816
Sherman, John 1613

II-6. College President

Barnard, Frederick A. P. 1809
Bayma, Joseph 1816
Clap, Thomas 1703
(Comstock, Theodore B. 1849)
Cooper, Thomas 1759
Drown, Thomas M. 1842
(Eddy, Henry T. 1844)
(Hall, G. Stanley 1846)
Hill, Thomas 1818
Hitchcock, Edward 1793
(Holden, Edward S. 1846)
(Jordan, David S. 1851)
LeConte, John 1818
Lilley, George 1851
Madison, James 1749
Mather, Increase 1639
(Mendenhall, Thomas C. 1841)
Morton, Henry 1836
Orton, Edward F. B. 1829
(Pritchett, Henry S. 1857)
Pugh, Evan 1828
(Remsen, Ira 1846)
Rogers, William B. 1804
Runkle, John D. 1822
Stoddard, John F. 1825
(Van Hise, Charles R. 1857)
(Venable, Francis P. 1856)
(Wadsworth, Marshman E. 1847)

II-7. Craftsman

See III-15, below, which lists general craftsmen with makers of scientific instruments. In some instances a single individual was involved in both activities.

II-8. Lawyer

Atwater, Caleb 1778
Bland, Thomas 1809
Cooper, Thomas 1759
(Edwards, William H. 1822)
Folger, Walter 1765
Foster, John W. 1815
Grimes, James S. 1807
Keating, William H. 1799
Marsh, George P. 1801
Paine, Robert T. 1803
Rutherfurd, Lewis M. 1816
Stallo, Johann B. 1823
(Thomas, Cyrus 1825)
Tuttle, Charles W. 1829
Whittlesey, Charles 1808

II-9. Lecturer, Public

See IV-6, below, for persons active as lecturers on scientific and non-scientific subjects.

II-10. Librarian

Includes F. W. True, librarian at the Smithsonian Institution.

Harris, Thaddeus W. 1795
Herrick, Edward C. 1811
(Holden, Edward S. 1846)
Mitchell, Maria 1818
(True, Frederick W. 1858)
(Tuckerman, Alfred 1848)
(Uhler, Philip R. 1835)

II-11. Literary Work

Includes persons active as authors, editors, journalists, and publishers.

Blodget, Lorin 1823
Brewer, Thomas M. 1814
Browne, Daniel J. 1804
Burrit, Elijah H. 1794
Christy, David 1802
Gibbes, Robert W. 1809
Lewis, Enoch 1776
Lindheimer, Ferdinand J. 1801
(Mitchell, Silas W. 1829)
Percival, James G. 1795
Samuels, Edward A. 1836
Steele, Joel D. 1836
Torrey, Bradford 1843

II-12. Military Service

Includes persons active in regular army and naval service in the United States and in foreign countries. The names of persons teaching science or engaged in other scientific work in the army or navy are listed elsewhere.

Bent, Silas 1820
Brooke, John M. 1826
Davis, Charles H. 1807
(Dutton, Clarence E. 1841)
Emory, William H. 1811
Gilliss, James M. 1811
Green, Francis M. 1835
Hayden, Edward E. 1858
Hutchins, Thomas 1730
McArthur, William P. 1814
Mather, William W. 1804
Maury, Matthew F. 1806
Wilkes, Charles 1798

II-13. Professor (non-science subjects)

Haldeman, Samuel S. 1812
Kinnersley, Ebenezer 1711
(Ward, Lester F. 1841)

II-14. Public Official (elective and appointive)

Includes persons holding public office in local, state, and federal service, as well as the diplomatic or foreign service. Several held office with foreign governments.

Atwater, Caleb 1778
Bowdoin, James 1726
Colden, Cadwallader 1688
Darlington, William 1782
Dudley, Paul 1675
Elliott, Stephen 1771
Franklin, Benjamin 1706
Fry, Joshua ca. 1700
Jefferson, Thomas 1743
Lapham, Increase A. 1811
Leeds, John 1705
Logan, James 1674
Marsh, George P. 1801
Mitchill, Samuel L. 1764
Oliver, Andrew 1731
Osten Sacken, Carl R. R. von der 1828
Perrine, Henry 1797
Pike, Nicholas 1743
Pitcher, Zina 1797

Procter, John R. 1844
Stallo, Johann B. 1823
Thompson, Benjamin
(Count Rumford) 1753
Wells, David A. 1828
Williamson, Hugh 1835
Winthrop, John, Jr. 1606
Young, Aaron 1819

II-15. Teacher/Educator

Persons in this category taught at the precollege level. Also see IV-13.

Abbot, John 1751
Ayres, William O. 1817
Burritt, Elijah H. 1794
Chase, Pliny E. 1820
Coffin, James H. 1806
Cotting, John R. 1778
Crocker, Lucretia 1829
Dewey, Chester 1784
(Ellis, Job B. 1829)
Emerson, George B. 1797
Ferrel, William 1817
Griscom, John 1774
Gummere, John 1784
Hentz, Nicholas M. 1797
Hubbard, Oliver P. 1809
Jenks, John W. P. 1819
Johnson, Walter R. 1794
Lewis, Enoch 1776
Locke, John 1792
Mansfield, Jared 1759
Mitchell, William 1791
Phelps, Almira H. L. 1793
Pike, Nicholas 1743
Steele, Joel D. 1836
Stoddard, John F. 1825
Thompson, Zadock 1796
Tracy, Clarissa T. 1818
Tuomey, Michael 1805
Vaughan, Daniel ca. 1818
Walker, Sears C. 1805
Wilson, Alexander 1766
Wright, Charles 1811

II-16. Tradesman

Activities represented here were glazier, carpenter, miller, and saddle maker.

Godfrey, Thomas 1704
Hager, Albert D. 1817
Marshall, Humphry 1722
Redfield, William C. 1789

II-17. Wealth (Independent)

Persons in this category generally inherited wealth.

Binney, William G. 1833
Brattle, Thomas 1658
Cooper, William ca. 1798
Cory, Charles B. 1857
De Forest, Erastus L. 1834
Furbish, Kate 1834
Gibbs, George 1776
Lea, Mathew C. 1823
LeConte, John E. 1784
LeConte, John L. 1825
Lyman, Theodore 1833
Oliver, Andrew 1731
Prime, Temple 1832
Seybert, Henry 1801

III. SCIENCE-RELATED EMPLOYMENT OR ACTIVITY

III-1. Actuary

Included is E. B. Elliott, actuary to the U.S. government.

Bartlett, William H. C. 1804
Bowditch, Nathaniel 1773
Elliott, Ezekiel B. 1823
McCay, Charles F. 1810
(McClintock, Emory 1840)
Mason, James W. 1836
Sylvester, James J. 1814
Walker, Sears C. 1805

III-2. Artist

See II-3, above, which includes persons who used their artistic talents for scientific purposes.

III-3. Business

See II-4, above (especially Manufacturer), which includes some persons engaged in businesses more or less directly related to science (for example, chemical and drug manufacture).

III-4. Horticulturist/Gardener

Bartram, John 1699
Bartram, William 1739
Brackenridge, William D. 1810
Meehan, Thomas 1826
Pursh, Frederick 1774

III-5. Instrument Maker/Craftsman

Includes persons active in making telescopes and other scientific instruments and employed as locksmith, watch and clock maker, and yacht designer.

Bond, William C. 1789
(Brashear, John A. 1840)
Burgess, Edward 1848
Clark, Alvan 1804
Clark, Alvan G. 1832
Fitz, Henry 1808
Folger, Walter 1765
Lesquereux, Leo 1806
Rittenhouse, David 1732
Saxton, Joseph 1799
Whipple, Squire 1804

III-6. Inventor

(Bell, Alexander G. 1847)
(Blake, Francis 1850)
(Brush, Charles F. 1849)
(Edison, Thomas A. 1847)
Treadwell, Daniel 1791

III-7. Medicine, Practitioner

Includes several persons who were dentists; also includes persons employed as physicians or surgeons in the army or navy.

Allen, Harrison 1841
Allen, Timothy F. 1837
Alter, David 1807
Ayres, William O. 1817
Bache, Franklin 1792
Baldwin, William 1779
Barton, Benjamin S. 1766
Barton, William P. C. 1786
Beaumont, William 1785
Beck, Lewis C. 1798
Behr, Hans H. 1818
(Beyer, Henry G. 1850)
Bigelow, Jacob 1786
Boylston, Zabdiel 1680
Bridges, Robert 1806
Bruce, Archibald 1777
Cabot, Samuel, Jr. 1815
Chalmers, Lionel 1715
Chapman, Alvan W. 1809
Cooper, James G. 1830
Coues, Elliott 1842
Cutting, Hiram A. 1832
Dana, James F. 1793
Darlington, William 1782
Deane, James 1801
Douglas, Silas H. 1816
Douglass, William ca. 1691
Drake, Daniel 1785
Dunglison, Robley 1798
Emmet, John P. 1796
Emmons, Ebenezer 1799
Engelmann, George 1809
Fitch, Asa 1809
Garden, Alexander 1730
Gattinger, Augustin 1825
Gibbes, Robert W. 1809
Girard, Charles F. 1822
Godman, John D. 1794
Gorham, John 1783
Gould, Augustus A. 1805
Griffith, Robert E. 1798
Guthrie, Samuel 1782
Hagen, Hermann A. 1817
Hallowell, Edward 1808
Hammond, William A. 1828
Harlan, Richard 1796
Harris, Thaddeus W. 1795
Hayden, Horace H. 1769
Hildreth, Samuel P. 1783
Holbrook, John E. 1794
Holder, Joseph B. 1824
Horn, George H. 1840
Horner, William E. 1793
Horsfield, Thomas 1773
Hough, Franklin B. 1822
Houghton, Douglass 1809
Jackson, John B. S. 1806
James, Edwin 1797
Jeffries, John 1744/1745
Joslin, Benjamin F. 1796
Josselyn, John ca. 1608
Kedzie, Robert C. 1823
Kellogg, Albert 1813
Keyt, Alonzo T. 1827
Kirtland, Jared P. 1793
Kneeland, Samuel 1821
Kuhn, Adam 1741
Lincecum, Gideon 1793
Lining, John 1708
Loomis, Mahlon 1826
Maclean, John 1771
Meade, William d. 1833
Mease, James 1771
Metcalfe, Samuel L. 1798
Michener, Ezra 1794
Mitchell, John 1711
Mitchell, John K. 1793
(Mitchell, Silas W. 1829)
Mitchill, Samuel L. 1764
Morton, Samuel G. 1799
Newberry, John S. 1822
Parry, Charles C. 1823
Perrine, Henry 1797

Pickering, Charles 1805
Pitcher, Zina 1797
Plummer, John T. 1807
Porcher, Francis P. 1825
Ramsay, Alexander ca. 1754
Ravenel, Edmund 1797
Ravenel, St. Julien 1819
Riddell, John L. 1807
Robie, Thomas 1688/1689
Rush, Benjamin 1746
Rush, James 1786
Salisbury, James H. 1823
Sartwell, Henry P. 1792
Saugrain de Vigni, Antoine F. 1763
Schöpf, Johann D. 1752
Shecut, John L. E. W. 1770
Short, Charles W. 1794
(Shufeldt, Robert W. 1850)
Shumard, Benjamin F. 1820
Squibb, Edward R. 1819
Storer, David H. 1804
Trask, John B. 1824
Van Rensselaer, Jeremiah 1793
Vasey, George 1822
Warder, John A. 1812
Warren, John C. 1778
Waterhouse, Benjamin 1754
Wells, William C. 1757
Williamson, Hugh 1735
Wislizenus, Frederick A. 1810
Wistar, Caspar 1761
Woodhouse, Samuel W. 1821
Wyman, Jeffries 1814
Yandell, Lunsford P. 1805
Young, Aaron 1819

III-8. Medicine, Professor

Allen, Timothy F. 1837
Ayres, William O. 1817
Barton, Benjamin S. 1766
Drake, Daniel 1785
Dunglison, Robley 1798
Godman, John D. 1794
Griffith, Robert E. 1798
Hammond, William A. 1828
Hayden, Horace H. 1769
Holbrook, John E. 1794
Kirtland, Jared P. 1793
Kuhn, Adam 1741
Mitchell, John K. 1793
Porcher, Francis P. 1825
Rush, Benjamin 1746
Storer, David H. 1804
Warren, John C. 1778
Waterhouse, Benjamin 1754

III-9. Pharmacist

Wholesale druggists and drug manufacturers are listed under Business (II-4, above).

Boll, Jacob 1828
Boylston, Zabdiel 1680
Carpenter, George W. 1802
Cutbush, James 1788
Durand, Elias 1794
Mohr, Charles T. 1824
Rice, Charles 1841
Thurber, George 1821
Troost, Gerard 1776

III-10. Surveyor/Explorer

Includes several persons also known as engineers and persons employed as surveyors by the government (but not including U.S. Army engineers).

Banneker, Benjamin 1731
Brandegee, Townshend S. 1843
De Brahm, William G. 1717
Dunbar, William 1749
Ellicott, Andrew 1754
Ferguson, James 1797
Freeman, Thomas d. 1821
(Heilprin, Angelo 1853)
Lawson, John d. 1711
Leeds, John 1705
(Muir, John 1838)
Romans, Bernard ca. 1720
Taylor, Richard C. 1789

IV. SCIENCE EMPLOYMENT OR ACTIVITY

IV-1. Consultant/Engineer/Industrial Scientist

Persons in this category were active as scientists in industry, as private consultants, as civil and mining engineers, as patent agents, in private geological work, and the like. Chemical work is especially prevalent, as well as work in mining, geology, and engineering.

Ashburner, Charles A. 1854
(Austen, Peter T. 1852)
Babcock, James F. 1844
Batchelder, John M. 1811
(Blair, Andrew A. 1848)
Booth, James C. 1810
Bradley, Frank H. 1838
Browne, Daniel J. 1804
Carll, John F. 1828
(Chatard, Thomas M. 1848)
Chester, Albert H. 1843
Clemson, Thomas G. 1807

(Comstock, Theodore B. 1849)
Courtenay, Edward H. 1803
(Crosby, William O. 1850)
Dana, Samuel L. 1795
(Dudley, Charles B. 1842)
Eimbeck, William 1841
Engelmann, Henry 1831
(Frazer, Persifor 1844)
Genth, Frederick A., Sr. 1820
(Hague, James D. 1836)
Hayes, Augustus A. 1806
Hill, Henry B. 1849
(Hill, Robert T. 1858)
(Hinrichs, Gustavus D. 1836)
Hunt, Thomas S. 1826
Jackson, Charles T. 1805
Johnson, Walter R. 1794
Keating, William H. 1799
King, Clarence R. 1842
Knight, Wilbur C. 1858
Lesley J. Peter 1819
(Lyman, Benjamin S. 1835)
(McMurtrie, William 1851)
Mapes, James J. 1806
Morfit, Campbell 1820
Mott, Henry A., Jr. 1852
(Nicholson, Henry H. 1850)
Page, Charles G. 1812
Platt, Franklin 1844
Redfield, William C. 1789
Romans, Bernard ca. 1720
(Sadtler, Samuel P. 1847)
(Sharples, Stephen P. 1842)
(Takamine, Jokichi 1854)
Taylor, Richard C. 1789
Tilghman, Richard A. 1824
Trautwine, John C. 1810
Trowbridge, William P. 1828
Twining, Alexander C. 1801
(Waller, Elwyn 1846)
Whipple, Squire 1804
Whitney, Josiah D. 1819
Whittlesey, Charles 1808
Wurtz, Henry 1828

IV-2. Government Science, Foreign

Includes Americans who did scientific work for foreign governments, as well as foreign-born persons who did such work in their native country.

Gabb, William M. 1839
Hartt, Charles F. 1840
Hunt, Thomas S. 1826
(Lyman, Benjamin S. 1835)
(Takamine, Jokichi 1854)

IV-3. Government Science, States

Persons in this category worked especially in relation to state geological, natural history and topographical surveys. Also included are individuals who worked as state botanists or entomologists, with state agricultural experiment stations and the like, and also on public health. Museum curators generally are listed elsewhere (see IV-8, below).

Adams, Charles B. 1814
Alexander, John H. 1812
(Arthur, Joseph C. 1850)
Ashburner, Charles A. 1854
Babcock, James F. 1844
(Babcock, Stephen M. 1843)
(Bean, Tarleton H. 1846)
Beck, Lewis C. 1798
(Bessey, Charles E. 1845)
(Blake, William P. 1825)
(Boyé, Martin H. 1812)
Bradley, Frank H. 1838
Buckley, Samuel B. 1809
(Calvin, Samuel 1840)
Carll, John F. 1828
(Clark, William B. 1860)
(Clarke, John M. 1857)
Collett, John 1828
Collier, Peter 1835
Conrad, Timothy A. 1803
Cook, George H. 1818
Cotting, John R. 1778
Cutting, Hiram A. 1832
De Kay, James E. 1792
Emmons, Ebenezer 1799
Fitch, Asa 1809
(Forbes, Stephen A. 1844)
Foster, John W. 1815
(Frazer, Persifor 1844)
(Frear, William 1860)
Gabb, William M. 1839
Gardiner, James T. 1842
Hager, Albert D. 1817
Hall, James 1811
Hanks, Henry G. 1826
(Hitchcock, Charles H. 1836)
Hitchcock, Edward 1793
Houghton, Douglass 1809
Jackson, Charles T. 1805
(Jenkins, Edward H. 1850)
(Johnson, Samuel W. 1830)
Kedzie, Robert C. 1823
Kerr, Washington C. 1827
Lesley, J. Peter 1819
Lintner, Joseph A. 1822
(Lyman, Benjamin S. 1835)
McCalley, Henry 1852
Mather, William W. 1804

Meek, Fielding B. 1817
Newberry, John S. 1822
Olmsted, Denison 1791
Orton, Edward F. B. 1829
Owen, David D. 1807
Percival, James G. 1795
Platt, Franklin 1844
Procter, John R. 1844
(Pumpelly, Raphael 1837)
Riley, Charles V. 1843
Rogers, Henry D. 1808
Rogers, James B. 1802
Rogers, Robert E. 1813
Rogers, William B. 1804
(Safford, James M. 1822)
Samuels, Edward A. 1836
Shepard, James H. 1850
Shumard, Benjamin F. 1820
(Smith, Eugene A. 1841)
Sturtevant, Edward L. 1842
Swallow, George C. 1817
Thompson, Zadock 1796
(Todd, James E. 1846)
Troost, Gerard 1776
Tuomey, Michael 1805
Tyson, Philip T. 1799
(Upham, Warren 1850)
Vanuxem, Lardner 1792
(Warren, Joseph W. 1849)
(Webster, Francis M. 1849)
(White, Israel C. 1848)
Whitney, Josiah D. 1819
(Winchell, Newton H. 1839)
Worthen, Amos H. 1813
Yeates, William S. 1856

IV-4. Government Science, U.S. (federal)

For U.S. Naval Observatory, see IV-10, below; for Smithsonian Institution, see IV-12, below; also see III-1 and III-10, above.

Agriculture Department

Browne, Daniel J. 1804
Collier, Peter 1835
Ewell, Ervin E. 1867
Glover, Townend 1813
Hough, Franklin B. 1822
(Howard, Leland O. 1857)
Hubbard, Henry G. 1850
(McMurtrie, William 1851)
Marx, George 1838
Riley, Charles V. 1843
Schwarz, Eugene A. 1844
Schweinitz, Emil A. de ca. 1864
Vasey, George 1822
(Webster, Francis M. 1849)
Wetherill, Charles M. 1825
(Wiley, Harvey W. 1844)

Army Engineers

(Abbot, Henry L. 1831)
Abert, John J. 1788
Barnard, John G. 1815
Emory, William H. 1811
Graham, James D. 1799
Humphreys, Andrew A. 1810
Hunt, Edward B. 1822
Le Conte, John E. 1784
Mansfield, Jared 1759
Schott, Arthur C. V. 1814
Warren, Gouverneur K. 1830
Wheeler, George M. 1842
Whipple, Amiel W. 1816

Bureau of American Ethnology

(Holmes, William H. 1846)
(McGee, William J. 1853)
Powell, John W. 1834
(Thomas, Cyrus 1825)

Coast Survey

Bache, Alexander D. 1806
Baker, Marcus 1849
Batchelder, John M. 1811
(Blake, Francis 1850)
Bowser, Edward A. 1837
Davidson, George 1825)
Eimbeck, William 1841
Ferguson, James 1797
Ferrel, William 1817
Gould, Benjamin A., Jr. 1824
Hassler, Ferdinand R. 1770
Hilgard, Julius E. 1825
Lane, Jonathan H. 1819
Lindenkohl, Adolph 1833
McArthur, William P. 1814
(Mendenhall, Thomas C. 1841)
Mitchell, Henry 1830
Peirce, Benjamin 1809
(Peirce, Charles S. 1839)
Pourtalès, Louis F. de 1823/1824
Saxton, Joseph 1799
Schott, Charles A. 1826
Trowbridge, William P. 1828
Walker, Sears C. 1805

Geological and Topographical Surveys Pre-1879

Cope, Edward D. 1840
(Emmons, Samuel F. 1841)
Engelmann, Henry 1831
Evans, John 1812

(Frazer, Persifor 1844)
Gardiner, James T. 1842
(Gilbert, Grove K. 1843)
(Hague, Arnold 1840)
(Hague, James D. 1836)
Hayden, Ferdinand V. 1829
King, Clarence R. 1842
Marcou, Jules 1824
Meek, Fielding B. 1817
Newton, Henry 1845
Owen, David D. 1807
Powell, John W. 1834
Thompson, Almon H. 1839
Wheeler, George M. 1842
Whittlesey, Charles 1808

Geological Survey Post-1879

Baker, Marcus 1849
(Becker, George F. 1847)
(Campbell, Marius R. 1858)
(Chamberlin, Thomas C. 1843)
(Chatard, Thomas M. 1848)
(Clark, William B. 1860)
(Clarke, Frank W. 1847)
(Dall, William H. 1845)
(Diller, Joseph S. 1850)
(Dutton, Clarence E. 1841)
(Emerson, Benjamin K. 1843)
(Emmons, Samuel F. 1841)
(Gannett, Henry 1846)
(Gilbert, Grove K. 1843)
(Hague, Arnold 1840)
(Hall, Christopher W. 1845)
Hayden, Ferdinand V. 1829
Hill, Robert T. 1858)
(Hillebrand, William F. 1853)
(Holmes, William H. 1846)
(Iddings, Joseph P. 1857)
Irving, Roland D. 1847
King, Clarence R. 1842
(Knowlton, Frank H. 1860)
(Lindgren, Waldemar 1860)
(McGee, William J. 1853)
Marsh, Othniel C. 1831
(Osborn, Henry F. 1857)
Powell, John W. 1834
(Pumpelly, Raphael 1837)
(Russell, Israel C. 1852)
(Scudder, Samuel H. 1837)
(Shaler, Nathaniel S. 1841)
(Stearns, Robert E. C. 1827)
(Taylor, Frank B. 1860)
Thompson, Almon H. 1839
(Upham, Warren 1850)
(Van Hise, Charles R. 1857)
(Walcott, Charles D. 1850)
(Ward, Lester F. 1841)
(White, Charles A. 1826)

Nautical Almanac

Coffin, John H. C. 1815
Davis, Charles H. 1807
Ferrel, William 1817
(Hill, George W. 1838)
Mitchell, Maria 1818
(Newcomb, Simon 1835)
Oliver, James E. 1829
Runkle, John D. 1822
Schubert, Ernst 1813
Winlock, Joseph 1826
Wright, Chauncey 1830

Patent Office

Lane, Jonathan H. 1819
Page, Charles G. 1812
Peale, Titian R. 1799
(Seaman, William H. 1837)
(Wead, Charles K. 1848)

Weather Bureau

(Abbe, Cleveland 1838)
Harrington, Mark W. 1848
Hazen, Henry A. 1849

Other

Blodget, Lorin 1823
Booth, James C. 1810
Brackenridge, William D. 1810
Dana, James D. 1813
Delafield, Joseph 1790
Espy, James P. 1785
Hayden, Edward E. 1858
(Hillebrand, William F. 1853)
Hutchins, Thomas 1730
(McGee, William J. 1853)
Patterson, Robert 1743
Patterson, Robert M. 1787
Pickering, Charles 1805
(Rathbun, Richard 1852)
Rittenhouse, David 1732
Torrey, John 1796
Wilkes, Charles 1798

IV-5. Independent Study

Individuals listed here pursued study or investigation (chiefly on scientific topics) independently of any obvious or persistent occupational position. Such study was based on wealth or separate income, various or miscellaneous pecuniary positions, and the like.

(Abbott, Charles C. 1843)
Bolton, Henry C. 1843
Bonaparte, Charles L. 1803

Brandegee, Mary K. L. C. 1844
(Britton, Elizabeth G. K. 1858)
Brown, Joseph 1733
Burnett, Waldo I. 1827
Catesby, Mark 1683
Colden, Jane 1724
Conrad, Timothy A. 1803
Cooper, William ca. 1798
Coues, Elliott 1842
De Forest, Erastus L. 1834
Delafield, Joseph 1790
Eigenmann, Rosa S. 1858
Espy, James P. 1785
Fowler, Orson S. 1809
Furbish, Kate 1834
Glover, Townend 1813
Haldeman, Samuel S. 1812
Hayes, Augustus A. 1806
Lapham, Increase A. 1811
Lea, Mathew C. 1823
Lesquereux, Leo 1806
Lippincott, James S. 1819
Maclure, William 1763
Marcou, Jules 1824
(Mellus, Edward L. 1848)
Metcalfe, Samuel L. 1798
Morris, Margaretta H. 1797
Osten Sacken, Carl R. R. von der 1828
Paine, Robert T. 1803
(Peirce, Charles S. 1839)
Pickering, Charles 1805
(Pickering, William H. 1858)
Rush, James 1786
Rutherfurd, Lewis M. 1816
(Scudder, Samuel H. 1837)
Smith, Erminnie A. P. 1836
Smith, John L. 1818
(Stockwell, John N. 1832)
Sullivant, William S. 1803
Taylor, Charlotte De B. S. 1806
Tryon, George W. 1838
Wachsmuth, Charles 1829
Walsh, Benjamin D. 1808
Warder, John A. 1812
Wetherill, Charles M. 1825
Wilson, Alexander 1766
Wright, Chauncey 1830

IV-6. Lecturer, Public

Includes persons active as lecturers on scientific subjects and lecturers on non-scientific subjects as well.

Fowler, Lorenzo N. 1811
Fowler, Orson S. 1809
Grimes, James S. 1807
Kinnersley, Ebenezer 1711
Ramsay, Alexander ca. 1754
Youmans, Edward L. 1821

IV-7. Literary Work

Persons in this category engaged in editorial and related labors on scientific, medical, agricultural, and similar publications. Also included are persons who published and calculated almanacs.

(Chamberlin, Thomas C. 1843)
(Chandler, Seth C. 1846)
Cope, Edward D. 1840
(Coulter, John M. 1851)
Daboll, Nathan 1750
(Dana, Edward S. 1849)
Dana, James D. 1813
(Emerton, James H. 1847)
Godfrey, Thomas 1704
Gould, Benjamin A., Jr. 1824
Harrington, Mark W. 1848
Judd, Orange 1822
Mapes, James J. 1806
Meehan, Thomas 1826
Nichols, James R. 1819
Packard, Alpheus S., Jr. 1839
Robie, Thomas 1688/1689
Ruffin, Edmund 1794
Silliman, Benjamin 1779
Silliman, Benjamin, Jr. 1816
Thurber, George 1821
Trask, John B. 1824
Wells, David A. 1828
Wells, Samuel R. 1820
West, Benjamin 1730
(Winchell, Newton H. 1839)
Yandell, Lunsford P. 1805
Youmans, Edward L. 1821
Youmans, William J. 1838

IV-8. Museum Curator/Associate

Included in this category, in addition to persons in private, university, or society museums, are persons in state museums and foreign museums. Also included are persons who worked in botanical gardens. Persons who worked in the U.S. National Museum are listed with the Smithsonian Institution (see IV-12, below).

(Agassiz, Alexander 1835)
(Allen, Joel A. 1838)
Anthony, John G. 1804
(Bean, Tarleton H. 1846)
Beecher, Charles E. 1856
Brandegee, Mary K. L. C. 1844
(Brewster, William 1851)

(Britton, Elizabeth G. K. 1858)
(Britton, Nathaniel L. 1859)
(Clarke, John M. 1857)
Cory, Charles B. 1857
(Elliot, Daniel G. 1835)
(Garman, Samuel 1843)
Grote, Augustus R. 1841
Hagen, Hermann A. 1817
Hall, James 1811
Hatcher, John B. 1861
(Hay, Oliver P. 1846)
Holder, Joseph B. 1824
(Holland, William J. 1848)
(Hollick, Charles A. 1857)
(Holmes, William H. 1846)
Horsfield, Thomas 1773
Hyatt, Alpheus 1838
Jenks, John W. P. 1819
(Julien, Alexis A. 1840)
Kennicott, Robert 1835
Lyman, Theodore 1833
(Millspaugh, Charles F. 1854)
(Morse, Edward S. 1838)
Nuttall, Thomas 1786
(Osborn, Henry F. 1857)
Peale, Charles W. 1742
Peale, Titian R. 1799
(Peck, Charles H. 1833)
Pourtalès, Louis F. de 1823/1824
(Putnam, Frederic W. 1839)
(Rydberg, Per A. 1860)
(Scudder, Samuel H. 1837)
Stimpson, William 1832
(Trelease, William 1857)
(Verrill, Addison E. 1839)
Watson, Sereno 1826
(Whitfield, Robert P. 1828)
(Wilson, William P. 1844)
Wyman, Jeffries 1814
Xántus, János, 1825

IV-9. Natural History Exploration and Collection

Abbot, John 1751
Bartram, John 1699
Bartram, William 1739
Boll, Jacob 1828
Brandegee, Townshend S. 1843
Buckley, Samuel B. 1809
Catesby, Mark 1683
Cooper, James G. 1830
Eights, James 1798
(Ellis, Job B. 1829)
Gambel, William 1823
Hatcher, John B. 1861
Horsfield, Thomas 1773
Josselyn, John ca. 1608
Kennicott, Robert 1835
LeConte, John L. 1825
Lemmon, John G. 1832
Lesueur, Charles A. 1778
Lindheimer, Ferdinand J. 1801
Michaux, André 1746
Michaux, François A. 1770
More, Thomas fl. 1670-1724
Nicollet, Joseph N. 1786
Nuttall, Thomas 1786
Ober, Frederick A. 1849
Orton, James 1830
Parry, Charles C. 1823
Peale, Titian R. 1799
Pursh, Frederick 1774
Ravenel, Henry W. 1814
Say, Thomas 1787
Schott, Arthur C. V. 1814
Townsend, John K. 1809
(Ward, Henry A. 1834)
Wright, Charles 1811
Xántus, János 1825

IV-10. Observatory Director/Staff Member

Included in this category are persons working for the U.S. Naval Observatory (listed separately) and for private and university or college observatories. Also included are several persons connected with foreign observatories (B. A. Gould, J. N. Nicollet, E. Schubert).

Naval Observatory

Coffin, John H. C. 1815
Davis, Charles H. 1807
(Eastman, John R. 1836)
Ferguson, James 1797
Gilliss, James M. 1811
(Hall, Asaph 1829)
Harkness, William 1837
Hayden, Edward E. 1858
Hubbard, Joseph S. 1823
Masterman, Stillman 1831
Maury, Matthew F. 1806
(Newcomb, Simon 1835)
(Skinner, Aaron N. 1845)
Winlock, William C. 1859
Yarnall, Mordecai 1816

Other

(Barnard, Edward E. 1857)
Bond, George P. 1825
Bond, William C. 1789
(Boss, Lewis 1846)
(Brooks, William R. 1844)
Brunnow, Franz F. E. 1821
(Burnham, Sherburne W. 1838)

(Comstock, George C. 1855)
(Doolittle, Charles L. 1843)
(Draper, Daniel 1841)
(Elkin, William L. 1855)
(Fleming, Williamina P. S. 1857)
(Flint, Albert S. 1853)
Gould, Benjamin A., Jr. 1834
(Hagen, John G. 1847)
(Holden, Edward S. 1846)
(Hough, George W. 1836)
(Howe, Herbert A. 1858)
Keeler, James E. 1857
(Langley, Samuel P. 1834)
Mitchel, Ormsby M. 1809
Nicollet, Joseph N. 1786
(Pickering, Edward C. 1846)
(Pickering, William H. 1858)
(Rees, John K. 1851)
Rogers, William A. 1832
Safford, Truman H. 1836
(Schaeberle, John M. 1853)
Schubert, Ernst 1813
(Stone, Ormond 1847)
(Swift, Lewis 1820)
Tuttle, Charles W. 1829
Watson, James C. 1838
(Whiting, Sarah F. 1847)
(Whitney, Mary W. 1847)
Winlock, Joseph 1826

IV-11. Professor (science)

This category includes persons who were professors of science in colleges and universities, in medical and pharmaceutical schools, and in such noncollegiate institutions as the Academy of Natural Sciences of Philadelphia. For professors of medicine, see III-8, above.

For institutions represented by more than two professors, personal names are listed under the name of the institution. All other names are listed under Other at the end of the alphabetical list of colleges or universities. (Since persons in this appendix are listed under as many as three headings, a single name may appear under more than one institution and also appear in the Other category.)

University of Alabama

Barnard, Frederick A. P. 1809
(Smith, Eugene A. 1841)
Tuomey, Michael 1805

Amherst College

Adams, Charles B. 1814
(Emerson, Benjamin K. 1843)
Hitchcock, Edward 1793
Shepard, Charles U. 1804
Snell, Ebenezer S. 1801
Tuckerman, Edward 1817

Brown University

(Barus, Carl 1856)
Caswell, Alexis 1799
Jenks, John W. P. 1819
Packard, Alpheus S., Jr. 1839
West, Benjamin 1730

University of California

(Blake, William P. 1825)
(Davidson, George 1825)
(Greene, Edward L. 1843)
(Hilgard, Eugene W. 1833)
LeConte, John 1818
LeConte, Joseph 1823
(Rising, Willard B. 1839)
(Stringham, Washington I. 1847)

University of Chicago

(Barnard, Edward E. 1857)
(Barnes, Charles R. 1858)
(Burnham, Sherburne W. 1838)
(Chamberlin, Thomas C. 1843)
(Coulter, John M. 1851)
(Donaldson, Henry H. 1857)
(Iddings, Joseph P. 1857)
(Maschke, Heinrich 1853)
(Michelson, Albert A. 1852)
(Salisbury, Rollin D. 1858)
(Whitman, Charles O. 1842)
(Williston, Samuel W. 1852)

City College of New York

(Compton, Alfred G. 1835)
(Doremus, Robert O. 1824)
(Gibbs, O. Wolcott 1822)
Mason, James W. 1836

College of Physicians and Surgeons of New York

Bruce, Archibald 1777
Dalton, John C., Jr. 1825
Mitchill, Samuel L. 1764
Torrey, John 1796

Columbia University

Adrain, Robert 1775
(Chandler, Charles F. 1836)
(Kemp, James F. 1859)
McCulloh, Richard S. 1818
(Maltby, Margaret E. 1860)
Newberry, John S. 1822
(Osborn, Henry F. 1857)
Peck, William G. 1820

(Rees, John K. 1851)
Renwick, James, Sr. 1792
Rood, Ogden N. 1831
Trowbridge, William P. 1828
(Underwood, Lucien M. 1853)
(Waller, Elwyn 1846)
(Wilson, Edmund B. 1856)
(Woodward, Robert S. 1849)

Cornell University

(Anthony, William A. 1835)
(Atkinson, George F. 1854)
(Caldwell, George C. 1834)
(Comstock, Anna B. 1854)
(Comstock, John H. 1849)
(Gage, Simon H. 1851)
Hartt, Charles F. 1840
(Nichols, Edward L. 1854)
Oliver, James E. 1829
(Wilder, Burt G. 1841)
(Williams, Henry S. 1847)
(Witthaus, Rudolph A. 1846)

Dartmouth College

Dana, James F. 1793
(Hitchcock, Charles H. 1836)
Hubbard, Oliver P. 1809

George Washington University

(Gill, Theodore N. 1837)
(Munroe, Charles E. 1849)
Schweinitz, Emil A. de ca. 1864
(Wiley, Harvey W. 1844)
Winlock, William C. 1859
Yeates, William S. 1856

Harvard University

Agassiz, J. Louis R. 1807
(Bailey, Solon I. 1854)
Bigelow, Jacob 1786
(Bowditch, Henry P. 1840)
(Byerly, William E. 1849)
Cooke, Josiah P. 1827
(Davis, William M. 1850)
(Dwight, Thomas 1843)
Eustis, Henry L. 1819
(Farlow, William G. 1844)
Farrar, John 1779
(Gibbs, O. Wolcott 1822)
(Goodale, George L. 1839)
Gorham, John 1783
Gray, Asa 1810
Greenwood, Isaac 1702
(Hall, Edwin H. 1855)
Hill, Henry B. 1849
Horsford, Eben N. 1818
(Jackson, Charles L. 1847)
Jackson, John B. S. 1806
(James, William 1842)
Lovering, Joseph 1813
(Mark, Edward L. 1847)
(Michael, Arthur 1853)
(Minot, Charles S. 1852)
Peck, William D. 1763
Peirce, Benjamin 1809
(Peirce, Benjamin O. 1854)
(Peirce, James M. 1834)
(Pickering, Edward C. 1846)
(Pumpelly, Raphael 1837)
(Putnam, Frederic W. 1839)
Robie, Thomas 1688/1689
(Sargent, Charles S. 1841)
(Searle, Arthur 1837)
(Shaler, Nathaniel S. 1841)
(Storer, Francis H. 1832)
Treadwell, Daniel 1791
(Trowbridge, John 1843)
Webster, John W. 1793
Whitney, Josiah D. 1819
Winlock, Joseph 1826
Winthrop, John 1714
Wood, Edward S. 1846
Wyman, Jeffries 1814

University of Illinois

(Burrill, Thomas J. 1839)
(Forbes, Stephen A. 1844)
(Kingsley, John S. 1854)
Palmer, Arthur W. 1861
(Trelease, William 1857)

State University of Iowa

(Calvin, Samuel 1840)
(Hinrichs, Gustavus D. 1836)
(MacBride, Thomas H. 1848)
(Nutting, Charles C. 1858)
(White, Charles A. 1826)

Johns Hopkins University

(Brooks, William K. 1848)
(Clark, William B. 1860)
Craig, Thomas 1855
Martin, Henry N. 1848
(Morse, Harmon N. 1848)
(Newcomb, Simon 1835)
(Remsen, Ira 1846)
(Renouf, Edward 1848)
Rowland, Henry A. 1848
Sylvester, James J. 1814
Williams, George H. 1856

University of Kansas

(Bailey, Edgar H. S. 1848)
(Snow, Francis H. 1840)

(Todd, James E. 1846)
(Williston, Samuel W. 1852)

Lafayette College
Coffin, James H. 1806
Drown, Thomas M. 1842
(Hart, Edward 1854)
Porter, Thomas C. 1822

Massachusetts Institute of Technology
(Crafts, James M. 1839)
(Crosby, William O. 1850)
(Cross, Charles R. 1848)
Drown, Thomas M. 1842
Hunt, Thomas S. 1826
Hyatt, Alpheus 1838
Kneeland, Samuel 1821
(Lindgren, Waldemar 1860)
(Niles, William H. 1838)
(Richards, Ellen H. S. 1842)
Rogers, William B. 1804
Runkle, John D. 1822

University of Massachusetts
Clark, Henry J. 1826
(Fernald, Charles H. 1838)
(Goessmann, Charles A. 1827)

Michigan State University
(Beal, William J. 1833)
Kedzie, Robert C. 1823
Miles, Manly 1826

University of Michigan
(Beman, Wooster W. 1850)
Brunnow, Franz F. E. 1821
(Carhart, Henry S. 1844)
Douglas, Silas H. 1816
Harrington, Mark W. 1848
Houghton, Douglass 1809
(McMurrich, James P. 1859)
Prescott, Albert B. 1832
(Russell, Israel C. 1852)
(Spalding, Volney M. 1849)
Watson, James C. 1838
Winchell, Alexander 1824

University of Minnesota
(Eddy, Henry T. 1844)
(Frankforter, George B. 1859)
(Hall, Christopher W. 1845)
(Winchell, Newton H. 1839)

University of Nebraska
(Bessey, Charles E. 1845)
Brace, De Witt B. 1859
(Nicholson, Henry H. 1850)

New York University
Draper, Henry 1837
Draper, John W. 1811
(Hering, Daniel W. 1850)
Joslin, Benjamin F. 1796
Loomis, Elias 1811
(Stevenson, John J. 1841)
(Witthaus, Rudolph A. 1846)

Northwestern University
(Carhart, Henry S. 1844)
(Hough, George W. 1836)
Marcy, Oliver 1820

Ohio State University
(Kellerman, William A. 1850)
Orton, Edward F. B. 1829
(Prosser, Charles S. 1860)
(Thomas, Benjamin F. 1850)

Pennsylvania State University
(Frear, William 1860)
Pugh, Evan 1828
(Wadsworth, Marshman E. 1847)

University of Pennsylvania
Adrian, Robert 1775
Allen, Harrison 1841
Bache, Alexander D. 1806
(Barker, George F. 1835)
Barton, William P. C. 1786
Cope, Edward D. 1840
(Doolittle, Charles L. 1843)
Frazer, John F. 1812
Genth, Frederick A., Sr. 1820
Grew, Theophilus d. 1759
Hare, Robert 1781
Horner, William E. 1793
(Jayne, Horace F. 1859)
Kendall, Ezra O. 1818
(Koenig, George A. 1844)
Kuhn, Adam 1741
Leidy, Joseph 1823
Lesley, J. Peter 1819
(MacFarlane, John M. 1855)
Patterson, Robert 1743
Patterson, Robert M. 1787
Rogers, Henry D. 1808
Rogers, James B. 1802
Rogers, Robert E. 1813
(Sadtler, Samuel P. 1847)
Say, Thomas 1787
(Smith, Edgar F. 1854)
(Wilson, William P. 1844)
Wistar, Caspar 1761
Woodhouse, James 1770

Princeton University
Alexander, Stephen 1806
(Brackett, Cyrus F. 1833)
(Cornwall, Henry B. 1844)
(Fine, Henry B. 1858)
Guyot, Arnold H. 1807
Henry, Joseph 1797
Maclean, John 1771
Minto, Walter 1753
Torrey, John 1796
(Young, Charles A. 1834)

Rensselaer Polytechnic Institute
(Clarke, John M. 1857)
Eaton, Amos 1776
Nason, Henry B. 1831

University of Rochester
Dewey, Chester 1784
(Fairchild, Herman L. 1850)
(Lattimore, Samuel A. 1828)

Rutgers University
(Austen, Peter T. 1852)
Beck, Lewis C. 1798
Bowser, Edward A. 1837
Chester, Albert H. 1843
Cook ,George H. 1818
Strong, Theodore 1790

University of South Carolina
Cooper, Thomas 1759
LeConte, John 1818
LeConte, Joseph 1823
Vanuxem, Lardner 1792

Stanford University
(Branner, John C. 1850)
(Dudley, William R. 1849)
(Parish, Samuel B. 1838)

Transylvania University
Peter, Robert 1805
Rafinesque, Constantine S. 1783
Short, Charles W. 1794

Tufts University
(Dolbear, Amos E. 1837)
(Kingsley, John S. 1854)
(Michael, Arthur 1853)

U.S. Military Academy, West Point
Bailey, Jacob W. 1811
Bartlett, William H. C. 1804
Courtenay, Edward H. 1803
Cutbush, James 1788
Ellicott, Andrew 1754
Mansfield, Jared 1759
Michie, Peter S. 1839

U.S. Naval Academy, Annapolis
Chauvenet, William 1820
Coffin, John H. C. 1815
(Johnson, William W. 1841)
(Munroe, Charles E. 1849)

Vassar College
(Cooley, Leroy C. 1833)
Mitchell, Maria 1818
Orton, James 1830
(Whitney, Mary W. 1847)

University of Vermont
Collier, Peter 1835
Thompson, Zadock 1796
(Witthaus, Rudolph A. 1846)

University of Virginia
Courtenay, Edward H. 1803
Emmet, John P. 1796
(Fontaine, William M. 1835)
(Mallet, John W. 1832)
(Stone, Ormond 1847)
Venable, Charles S. 1827

Washington and Lee University
(Howe, James L. 1859)
McCulloh, Richard S. 1818
(Stevens, Walter L. 1847)

Washington University
Chauvenet, William 1820
(Nipher, Francis E. 1847)
(Trelease, William 1857)

Wesleyan University
(Atwater, Wilbur O. 1844)
Johnston, John 1806
(Rice, William N. 1845)

College of William and Mary
Fry, Joshua ca. 1700
Jones, Hugh ca. 1692
Madison, James 1749

Williams College
Dewey, Chester 1784
Emmons, Ebenezer 1799
Safford, Truman H. 1836

University of Wisconsin

(Babcock, Stephen M. 1843)
(Barnes, Charles R. 1858)
(Comstock, George C. 1855)
Irving, Roland D. 1847
(Van Hise, Charles R. 1857)

Yale University

Allen, Oscar D. 1836
Beecher, Charles E. 1856
(Brewer, William H. 1828)
(Brush, George J. 1831)
(Chittenden, Russell H. 1856)
(Dana, Edward S. 1849)
Dana, James D. 1813
Eaton, Daniel C. 1834
Gibbs, Josiah W. 1839
(Gooch, Frank A. 1852)
(Hastings, Charles S. 1848)
(Johnson, Samuel W. 1830)
(Ladd, George T. 1842)
Loomis, Elias 1811
Lyman, Chester S. 1814
Marsh, Othniel C. 1831
(Mixter, William G. 1846)
Newton, Hubert A. 1830
Norton, John P. 1822
Norton, William A. 1810
Olmsted, Denison 1791
(Penfield, Samuel L. 1856)
(Pirsson, Louis V. 1860)
Porter, John A. 1822
Silliman, Benjamin 1779
Silliman, Benjamin, Jr. 1816
(Smith, Sidney I. 1843)
(Verrill, Addison E. 1839)
(Williams, Henry S. 1847)
(Wright, Arthur W. 1836)

Other

Adams, Charles B. 1814
Alexander, John H. 1812
(Anthony, William A. 1835)
(Arthur, Joseph C. 1850)
Babcock, James F. 1844
Bache, Franklin 1792
Bayma, Joseph 1816
Behr, Hans H. 1818
(Blake, William P. 1825)
Bodley, Rachel L. 1831
Bolton, Henry C. 1843
Bradley, Frank H. 1838
Bridges, Robert 1806
(Brigham, Albert P. 1855)
(Broadhead, Garland C. 1827)
Brocklesby, John 1811
Brooke, John M. 1826
(Brooks, William R. 1844)
Brunnow, Franz F. E. 1821
Chester, Albert H. 1843
(Clapp, Cornelia M. 1849)
Claypole, Edward W. 1835
Cleaveland, Parker 1780
Curley, James 1796
Day, William C. 1857
(Eddy, Henry T. 1844)
(Emerson, Benjamin K. 1843)
(Frisby, Edgar 1837)
Gibbes, Lewis R. 1810
Green, Jacob 1790
(Greene, Edward L. 1843)
(Hall, G. Stanley 1846)
(Halsted, George B. 1853)
(Hay, Oliver P. 1846)
(Heilprin, Angelo 1853)
(Heinrichs, Gustavus D. 1836)
Holmes, Francis S. 1815
(Howe, Herbert A. 1858)
(Hyde, Edward W. 1843)
Jacobs, Michael 1808
Jones, William L. 1827
(Jordan, David S. 1851)
Joslin, Benjamin F. 1796
Kerr, Washington C. 1827
(Kinnicutt, Leonard P. 1854)
Kirkwood, Daniel 1814
Knight, Wilbur C. 1858
(Koenig, George A. 1844)
(Langley, Samuel P. 1834)
(Leffman, Henry 1847)
Lilley, George 1851
(Lloyd, John U. 1849)
Locke, John 1792
(Mabery, Charles F. 1850)
McCay, Charles F. 1810
McCulloh, Richard S. 1818
(McMurrich, James P. 1859)
Mather, William W. 1804
Mayer, Alfred M. 1836
Mitchel, Ormsby M. 1809
Mitchell, Elisha 1793
Mitchell, James A. 1852
(Morley, Edward W. 1838)
(Niles, William H. 1838)
Norton, William A. 1810
Peters, Christian H. F. 1813
(Phillips, Francis C. 1850)
Porter, Thomas C. 1822
Puryear, Bennet 1826
Ravenel, Edmund 1797
Riddell, John L. 1807
Rogers, Henry D. 1808
Rogers, William A. 1832
(Sadtler, Samuel P. 1847)

(Safford, James M. 1822)
(Scott, Charlotte A. 1858)
(Seaman, William H. 1837)
Sestini, Benedict 1816
(Sharples, Stephen P. 1842)
Shattuck, Lydia W. 1822
Shepard, Charles U. 1804
Shepard, James H. 1850
Short, Charles W. 1794
(Skinner, Aaron N. 1845)
Smith, Hamilton L. 1818
Smith, John L. 1818
(Smith, William B. 1850)
Smyth, William 1797
(Stockwell, John N. 1832)
(Stoddard, John T. 1852)
(Story, William E. 1850)
Swallow, George C. 1817
Sylvester, James J. 1814
(Todd, James E. 1846)
Tracy, Clarissa T. 1818
Troost, Gerard 1776
Twining, Alexander C. 1801
Vaughan, Daniel ca. 1818
Venable, Charles S. 1827
(Venable, Francis P. 1856)
(Wadsworth, Marshman E. 1847)
Warder, Robert B. 1848
(Warren, Joseph W. 1849)
Wetherill, Charles M. 1825
(White, Israel C. 1848)
(Whiting, Sarah F. 1847)
Wright, Albert A. 1846
Yandell, Lunsford P. 1805

IV-12. Research Foundation Administrator/Associate

In this category, persons associated with the Smithsonian Institution are listed separately, and this group includes curators in the U.S. National Museum (see also II-10, above). Among the institutions represented under the Other heading are the Wistar Institute of Philadelphia, the Carnegie Institution of Washington, and the Franklin Institute.

Smithsonian Institution

(Ashmead, William H. 1855)
Baird, Spencer F. 1823
(Gill, Theodore N. 1837)
Girard, Charles F. 1822
Goode, George B. 1851
(Greene, Edward L. 1843)
Henry, Joseph 1797
(Langley, Samuel P. 1834)
Meek, Fielding B. 1817
(Merrill, George P. 1854)
(Rathbun, Richard 1852)
(Ridgway, Robert 1850)
(Stearns, Robert E. C. 1827)
(True, Frederick W. 1858)
(Walcott, Charles D. 1850)
(White, Charles A. 1826)
Winlock, William C. 1859
Yeates, William S. 1856

Other

(Donaldson, Henry H. 1857)
(Hay, Oliver P. 1846)
(Mark, Edward L. 1847)
Morton, Henry 1836
(Spalding, Volney M. 1849)
(Whitman, Charles O. 1842)
(Woodward, Robert S. 1849)

IV-13. Teacher/Educator of Science

Persons in this category taught at the precollege level.

(Boyé, Martin H. 1812)
Crocker, Lucretia 1829
Daboll, Nathan 1750
Espy, James P. 1785
Greenwood, Isaac 1702
Griscom, John 1774
Gummere, John 1784
Johnson, Walter R. 1794
Lewis, Enoch 1776
Marcy, Oliver 1820
Phelps, Almira H. L. 1793
Pike, Nicholas 1743
Steele, Joel D. 1836
Vaughan, Daniel ca. 1818
Wagner, William 1796

Appendix E
FIELDS OF SCIENCE

This appendix lists the names of scientists according to the subjects or fields of science in which they worked. An individual name may appear in more than one category, and no limit was imposed on the number of times a single name could appear. In practice, however, names generally do not appear more than three times.

The headings ordinarily are of a general rather than a specific nature (for example, Zoology, rather than Entomology). A topical arrangement of the headings is used in order to bring together related subjects. Names in this appendix are listed under the following headings:

1. General Science
2. Mathematics
3. Physical Sciences
4. Chemistry
5. Physics
6. Engineering
7. Technology and Applied Science
8. Instrumentation and Invention
9. Microscopy
10. Astronomy
11. Meteorology
12. Geophysics
13. Oceanography
14. Hydrography and Hydrology
15. Geodesy
16. Geography
17. Topographical Engineering and Surveying
18. Exploration and Travels
19. Cartography
20. Natural History and Nature Study
21. Geology
22. Mineralogy
23. Paleontology
24. Biology (life forms and processes)
25. Zoology
26. Medicine
27. Psychology and Phrenology
28. Botany
29. Agriculture
30. Humanistic Aspects of Science
31. Science Administration and Education
32. Promotion and Communication of Scientific Work

1. GENERAL SCIENCE

Mather, Cotton 1662/1663
Mease, James 1771

2. MATHEMATICS

This category also includes several persons engaged in work in arithmetic, statistics, and actuarial science.

Adrain, Robert 1775
Alexander, John H. 1812
Baker, Marcus 1849
Banneker, Benjamin 1731
Barnard, Frederick A. P. 1809
Barnard, John G. 1815
Bartlett, William H. C. 1804
Bayma, Joseph 1816
(Becker, George F. 1847)
(Beman, Wooster W. 1850)
Bowditch, Nathaniel 1773
Bowser, Edward A. 1837
(Byerly, William E. 1849)
Chauvenet, William 1820
Coffin, James H. 1806
Coffin, John H. C. 1815
Courtenay, Edward H. 1803
Craig, Thomas 1855
Daboll, Nathan 1750
De Forest, Erastus L. 1834
(Doolittle, Charles L. 1843)
Dunbar, William 1749
(Eastman, John R. 1836)

(Eddy, Henry T. 1844)
Ellicott, Andrew 1754
Elliott, Ezekiel B. 1823
Farrar, John 1779
(Fine, Henry B. 1858)
Folger, Walter 1765
(Frisby, Edgar 1837)
Fry, Joshua ca. 1700
(Gannett, Henry 1846)
Gibbs, Josiah W. 1839
Godfrey, Thomas 1704
Greenwood, Isaac 1702
Grew, Theophilus d. 1759
Gummere, John 1784
(Hagen, John G. 1847)
(Halsted, George B. 1853)
Hassler, Ferdinand R. 1770
(Hill, George W. 1838)
Hill, Thomas 1818
(Hyde, Edward W. 1843)
(Johnson, William W. 1841)
Jones, Hugh ca. 1692
Kendall, Ezra O. 1818
Leeds, John 1705
Lewis, Enoch 1776
Lilley, George 1851
Loomis, Elias 1811
Lovering, Joseph 1813
McCay, Charles F. 1810
(McClintock, Emory 1840)
Mansfield, Jared 1759
(Maschke, Heinrich 1853)
Mason, James W. 1836
Minto, Walter 1753
Newton, Hubert A. 1830
Nicollet, Joseph N. 1786
Oliver, James E. 1829
Patterson, Robert 1743
Peck, William G. 1820
Peirce, Benjamin 1809
(Peirce, Benjamin O. 1854)
(Peirce, James M. 1834)
Pike, Nicholas 1743
Runkle, John D. 1822
Safford, Truman H. 1836
(Scott, Charlotte A. 1858)
Sestini, Benedict 1816
Sherman, John 1613
(Smith, William B. 1850)
Smyth, William 1797
Snell, Ebenezer S. 1801
Stoddard, John F. 1825
(Story, William E. 1850)
(Stringham, Washington I. 1847)
Strong, Theodore 1790
Sylvester, James J. 1814
Thompson, Zadock 1796
Vaughan, Daniel ca. 1818
Venable, Charles S. 1827
Walker, Sears C. 1805
Winlock, Joseph 1826
Winthrop, John 1714
(Woodward, Robert S. 1849)
Wright, Chauncey 1830

3. PHYSICAL SCIENCES

This category includes individuals engaged in general work in physics, chemistry, astronomy, and related subjects.

Caswell, Alexis 1799
Chase, Pliny E. 1820
Saugrain de Vigni, Antoine F. 1763

4. CHEMISTRY

This category also includes persons engaged in work in pharmacology, in agricultural, geological, industrial, and sanitary chemistry, and also in some aspects of metallurgy and physiological chemistry.

Allen, Oscar D. 1836
(Atwater, Wilbur O. 1844)
(Austen, Peter T. 1852)
Babcock, James F. 1844
(Babcock, Stephen M. 1843)
Bache, Franklin 1792
(Bailey, Edgar H. S. 1848)
(Barker, George F. 1835)
Beck, Lewis C. 1798
(Blair, Andrew A. 1848)
Bodley, Rachel L. 1831
Bolton, Henry C. 1843
Booth, James C. 1810
(Boyé, Martin H. 1812)
(Brewer, William H. 1828)
Bridges, Robert 1806
(Caldwell, George C. 1834)
Carpenter, George W. 1802
(Chandler, Charles F. 1836)
(Chatard, Thomas M. 1848)
Chester, Albert H. 1843
(Clarke, Frank W. 1847)
Clemson, Thomas G. 1807
Collier, Peter 1835
Cooke, Josiah P. 1827
Cooper, Thomas 1759
(Cornwall, Henry B. 1844)
Cotting, John R. 1778
(Crafts, James M. 1839)
Cutbush, James 1788
Dana, James F. 1793
Dana, Samuel L. 1795

Day, William C. 1857
(Doremus, Robert O. 1824)
Douglas, Silas H. 1816
Draper, John W. 1811
Drown, Thomas M. 1842
(Dudley, Charles B. 1842)
Emmet, John P. 1796
Ewell, Ervin E. 1867
(Frankforter, George B. 1859)
Frazer, John F. 1812
(Frazer, Persifor 1844)
(Frear, William 1860)
Genth, Frederick A., Sr. 1820
Gibbes, Lewis R. 1810
(Gibbs, O. Wolcott 1822)
(Goessmann, Charles A. 1827)
(Gooch, Frank A. 1852)
Gorham, John 1783
Green, Jacob 1790
Griscom, John 1774
Guthrie, Samuel 1782
Hanks, Henry G. 1826
Hare, Robert 1781
Harrison, John 1773
(Hart, Edward 1854)
Hayes, Augustus A. 1806
(Hilgard, Eugene W. 1833)
Hill, Henry B. 1849
(Hillebrand, William F. 1853)
(Hinrichs, Gustavus D. 1836)
Horsford, Eben N. 1818
(Howe, James L. 1859)
Hunt, Thomas S. 1826
(Jackson, Charles L. 1847
Jackson, Charles T. 1805
(Jenkins, Edward H. 1850)
(Johnson, Samuel W. 1830)
Johnson, Walter R. 1794
Johnston, John 1806
Jones, William L. 1827
Kedzie, Robert C. 1823
(Kinnicutt, Leonard P. 1854)
(Koenig, George A. 1844)
(Lattimore, Samuel A. 1828)
Lea, Mathew C. 1823
(Leffmann, Henry 1847)
(Lloyd, John U. 1849)
(Mabery, Charles F. 1850)
McCulloh, Richard S. 1818
Maclean, John 1771
(McMurtrie, William 1851)
(Mallet, John W. 1832)
(Maltby, Margaret E. 1860)
Mapes, Charles V. 1836
Mapes, James J. 1806
Metcalfe, Samuel L. 1798
(Michael, Arthur 1853)
Mitchell, John 1711
Mitchell, John K. 1793
Mitchill, Samuel L. 1764
(Mixter, William G. 1846)
Morey, Samuel 1762
Morfit, Campbell 1820
(Morley, Edward W. 1838)
(Morse, Harmon N. 1848)
Morton, Henry 1836
Mott, Henry A., Jr. 1852
(Munroe, Charles E. 1849)
Nason, Henry B. 1831
Nichols, James R. 1819
(Nicholson, Henry H. 1850)
Norton, John P. 1822
Palmer, Arthur W. 1861
Patterson, Robert M. 1787
Peter, Robert 1805
(Phillips, Francis C. 1850)
Plummer, John T. 1807
Porter, John A., 1822
Prescott, Albert B. 1832
Priestley, Joseph 1733
Pugh, Evan 1828
Puryear, Bennet 1826
Ravenel, St. Julien 1819
(Remsen, Ira 1846)
(Renouf, Edward 1848)
Rice, Charles 1841
(Richards, Ellen H. S. 1842)
(Rising, Willard B. 1839)
Rogers, James B. 1802
Rogers, Robert E. 1813
Ruffin, Edmund 1794
Rush, Benjamin 1746
(Sadtler, Samuel P. 1847)
(Safford, James M. 1822)
(Seaman, William H. 1837)
(Sharples, Stephen P. 1842)
Shepard, James H. 1850
Silliman, Benjamin 1779
Silliman, Benjamin, Jr. 1816
(Smith, Edgar F. 1854)
Smith, John L. 1818
Squibb, Edward R. 1819
(Stoddard, John T. 1852)
(Storer, Francis H. 1832)
(Takamine, Jokichi 1854)
Tilghman, Richard A. 1824
Torrey, John 1796
(Tuckerman, Alfred 1848)
Tyson, Philip T. 1799
Vaughan, Daniel ca. 1818
(Venable, Francis P. 1856)
(Waller, Elwyn 1846)
Warder, Robert B. 1848
Warren, Cyrus M. 1824

Webster, John W. 1793
Weightman, William 1813
Wells, David A. 1828
Wetherill, Charles M. 1825
Whitney, Josiah D. 1819
(Wiley, Harvey W. 1844)
(Witthaus, Rudolph A. 1846)
Wood, Edward S. 1846
Woodhouse, James 1770
Wurtz, Henry 1828

5. PHYSICS

This category also includes persons engaged in work in natural philosophy, mechanics, electricity, and electrical engineering.

(Abbot, Henry L. 1831)
Alter, David 1807
(Anthony, William A. 1835)
Bache, Alexander D. 1806
(Barker, George F. 1835)
Barnard, Frederick A. P. 1809
Barnard, John G. 1815
(Barus, Carl 1856)
Batchelder, John M. 1811
Bayma, Joseph 1816
(Becker, George F. 1847)
(Bell, Alexander G. 1847)
(Blake, Francis 1850)
Bowdoin, James 1726
(Boyé, Martin H. 1812)
Brace, De Witt B. 1859
(Brackett, Cyrus F. 1833)
(Brashear, John A. 1840)
Brooke, John M. 1826
(Brooks, William R. 1844)
Brown, Joseph 1733
(Brush, Charles F. 1849)
(Carhart, Henry S. 1844)
Colden, Cadwallader 1688
(Compton, Alfred G. 1835)
(Cooley, Leroy C. 1833)
(Cross, Charles R. 1848)
(Dolbear, Amos E. 1837)
(Doremus, Robert O. 1824)
(Dutton, Clarence E. 1841)
(Eddy, Henry T. 1844)
(Edison, Thomas A. 1847)
Elliott, Ezekiel B. 1823
Farrar, John 1779
Franklin, Benjamin 1706
Gibbs, Josiah W. 1839
Greenwood, Isaac 1702
Grimes, James S. 1807
(Hall, Edwin H. 1855)
(Hastings, Charles S. 1848)
Henry, Joseph 1797
(Hering, Daniel W. 1850)
Hunt, Edward B. 1822
Johnson, Walter R. 1794
Johnston, John 1806
Joslin, Benjamin F. 1796
Kinnersley, Ebenezer 1711
Lane, Jonathan H. 1819
LeConte, John 1818
Lining, John 1708
Locke, John 1792
Logan, James 1674
Loomis, Mahlon 1826
Lovering, Joseph 1813
McCulloh, Richard S. 1818
Madison, James 1749
(Maltby, Margaret E. 1860)
Mansfield, Jared 1759
Mayer, Alfred M. 1836
(Mendenhall, Thomas C. 1841)
(Michelson, Albert A. 1852)
Michie, Peter S. 1839
Morey, Samuel 1762
(Morley, Edward W. 1838)
Morton, Henry 1836
(Nichols, Edward L. 1854)
(Nipher, Francis E. 1847)
Norton, William A. 1810
Olmsted, Denison 1791
Page, Charles G. 1812
Patterson, Robert M. 1787
(Peirce, Benjamin O. 1854)
(Peirce, Charles S. 1839)
Priestley, Joseph 1733
Renwick, James, Sr. 1792
Rittenhouse, David 1732
Rogers, William A. 1832
Rood, Ogden N. 1831
Rowland, Henry A. 1848
Snell, Ebenezer S. 1801
(Stevens, Walter L. 1847)
(Thomas, Benjamin F. 1850)
Thompson, Benjamin (Count Rumford) 1753
Treadwell, Daniel 1791
(Trowbridge, John 1843)
(Wead, Charles K. 1848)
Wells, William C. 1757
(Whiting, Sarah F. 1847)
Winthrop, John 1714
Winthrop, John, Jr. 1606
(Woodward, Robert S. 1849)
(Wright, Arthur W. 1836)

6. ENGINEERING

This category includes persons engaged in work in civil, hydraulic, military, and mining engineering and also some who worked in topographical engineering.

(Abbot, Henry L. 1831)
Barnard, John G. 1815
Batchelder, John M. 1811
Clemson, Thomas G. 1807
(Comstock, Theodore B. 1849)
De Brahm, William G. 1717
(Eddy, Henry T. 1844)
(Emmons, Samuel F. 1841)
Eustis, Henry L. 1819
Freeman, Thomas d. 1821
(Hague, James D. 1836)
Humphreys, Andrew A. 1810
Hutchins, Thomas 1730
Irving, Roland D. 1847
(Lyman, Benjamin S. 1835)
Michie, Peter S. 1839
Newton, Henry 1845
Renwick, James, Sr. 1792
Romans, Bernard ca. 1720
Trautwine, John C. 1810
Trowbridge, William P. 1828
Twining, Alexander C. 1801
Warren, Gouverneur K. 1830
Whipple, Squire 1804

7. TECHNOLOGY AND APPLIED SCIENCE

Alexander, John H. 1812
Bigelow, Jacob 1786
(Dudley, Charles B. 1842)
Godfrey, Thomas 1704

8. INSTRUMENTATION AND INVENTION

This category also includes J. Jeffries, who did work in aeronautics.

Babcock, James F. 1844
Batchelder, John M. 1811
(Bell, Alexander G. 1847)
(Blake, Francis 1850)
Brooke, John M. 1826
(Brush, Charles F. 1849)
Clark, Alvan 1804
Clark, Alvan G. 1832
(Edison, Thomas A. 1847)
Fitz, Henry 1808
Folger, Walter 1765
Jeffries, John 1744/1745
Morey, Samuel 1762
Riddell, John L. 1807
Rittenhouse, David 1732
Saugrain de Vigni, Antoine F. 1763
Saxton, Joseph 1799
Treadwell, Daniel 1791
Trowbridge, William P. 1828
Twining, Alexander C. 1801

9. MICROSCOPY

Brocklesby, John 1811
Riddell, John L. 1807
Salisbury, James H. 1823
Smith, Hamilton L. 1818

10. ASTRONOMY

This category also includes persons engaged in work in astrophysics.

(Abbe, Cleveland 1838)
Alexander, Stephen 1806
(Bailey, Solon I. 1854)
Banneker, Benjamin 1731
(Barnard, Edward E. 1857)
Barnard, Frederick A. P. 1809
Bartlett, William H. C. 1804
Biddle, Owen 1737
Bond, George P. 1825
Bond, William C. 1789
(Boss, Lewis 1846)
Bowditch, Nathaniel 1773
Bowdoin, James 1726
(Brashear, John A. 1840)
Brattle, Thomas 1658
Brooke, John M. 1826
(Brooks, William R. 1844)
Brown, Joseph 1733
Brunnow, Franz F. E. 1821
(Burnham, Sherburne W. 1838)
Burritt, Elijah H. 1794
Caswell, Alexis 1799
(Chandler, Seth C. 1846)
Chauvenet, William 1820
Clap, Thomas 1703
Clark, Alvan 1804
Clark, Alvan G. 1832
(Compton, Alfred G. 1835)
(Comstock, George C. 1855)
Curley, James 1796
(Davidson, George 1825)
Davis, Charles H. 1807
(Dolbear, Amos E. 1837)
(Doolittle, Charles L. 1843)
Draper, Henry 1837
Dunbar, William 1749
(Eastman, John R. 1836)
(Elkin, William L. 1855)

Emory, William H. 1811
Farrar, John 1779
Ferguson, James 1797
(Fleming, Williamina P. S. 1857)
(Flint, Albert S. 1853)
Folger, Walter 1765
Freeman, Thomas d. 1821
(Frisby, Edgar 1837)
Gibbes, Lewis R. 1810
Gilliss, James M. 1811
Gould, Benjamin A., Jr. 1824
Grew, Theophilus d. 1759
(Hagen, John G. 1847)
(Hall, Asaph 1829)
Harkness, William 1837
Harrington, Mark W. 1848
Herrick, Edward C. 1811
(Hill, George W. 1838)
Hill, Thomas 1818
(Holden, Edward S. 1846)
(Hough, George W. 1836)
(Howe, Herbert A. 1858)
Hubbard, Joseph S. 1823
Keeler, James E. 1857
Kendall, Ezra O. 1818
Kirkwood, Daniel 1814
(Langley, Samuel P. 1834)
Leeds, John 1705
Loomis, Elias 1811
Lyman, Chester S. 1814
Madison, James 1749
Masterman, Stillman 1831
Mather, Increase 1639
Mitchel, Ormsby M. 1809
Mitchell, Maria 1818
Mitchell, William 1791
(Newcomb, Simon 1835)
Newton, Hubert A. 1830
Nicollet, Joseph N. 1786
Norton, William A. 1810
Oliver, Andrew 1731
Olmsted, Denison 1791
Paine, Robert T. 1803
Peirce, Benjamin 1809
Peters, Christian H. F. 1813
(Pickering, Edward C. 1846)
(Pickering, William H. 1858)
(Pritchett, Henry S. 1857)
(Rees, John K. 1851)
Rittenhouse, David 1732
Robie, Thomas 1688/1689
Rogers, William A. 1832
Rutherfurd, Lewis M. 1816
Safford, Truman H. 1836
(Schaeberle, John M. 1853)
Schubert, Ernst 1813
(Searle, Arthur 1837)
Sestini, Benedict 1816
(Skinner, Aaron N. 1845)
Smith, Hamilton L. 1818
(Stockwell, John N. 1832)
(Stone, Ormond 1847)
(Swift, Lewis 1820)
Tuttle, Charles W. 1829
Twining, Alexander C. 1801
Vaughan, Daniel ca. 1818
Walker, Sears C. 1805
Watson, James C. 1838
West, Benjamin 1730
(Whiting, Sarah F. 1847)
(Whitney, Mary W. 1847)
Wilkes, Charles 1798
Williamson, Hugh 1735
Winlock, Joseph 1826
Winlock, William C. 1859
Winthrop, John 1714
(Woodward, Robert S. 1849)
Yarnall, Mordecai 1816
(Young, Charles A. 1834)

11. METEOROLOGY

This category also includes persons engaged in work in climatology.

(Abbe, Cleveland 1838)
Blodget, Lorin 1823
Brocklesby, John 1811
Chalmers, Lionel 1715
Coffin, James H. 1806
(Draper, Daniel 1841)
Dunbar, William 1749
Engelmann, George 1809
Espy, James P. 1785
Ferrel, William 1817
Franklin, Benjamin 1706
Harrington, Mark W. 1848
Hayden, Edward E. 1858
Hazen, Henry A. 1849
Jacobs, Michael 1808
Jeffries, John 1744/1745
Joslin, Benjamin F. 1796
Lapham, Increase A. 1811
Lining, John 1708
Lippincott, James S. 1819
Loomis, Elias 1811
Masterman, Stillman 1831
Maury, Matthew F. 1806
Oliver, Andrew 1731
Redfield, William C. 1789
Robie, Thomas 1688/1689
Williamson, Hugh 1735
Wislizenus, Frederick A. 1810

12. GEOPHYSICS

Bache, Alexander D. 1806
Frazner, John F. 1812
Schott, Charles A. 1826
Trowbridge, William P. 1828
Wilkes, Charles 1798

13. OCEANOGRAPHY

(Agassiz, Alexander 1835)
Bent, Silas 1820
Franklin, Benjamin 1706
Lindenkohl, Adolph 1833
Maury, Matthew F. 1806
Pourtalès, Louis F. de 1823/1824

14. HYDROGRAPHY AND HYDROLOGY

Davis, Charles H. 1807
De Brahm, William G. 1717
Green, Francis M. 1835
McArthur, William P. 1814
(McGee, William J. 1853)
Mitchell, Henry 1830

15. GEODESY

(Davidson, George 1825)
Delafield, Joseph 1790
Eimbeck, William 1841
Hassler, Ferdinand R. 1770
Hilgard, Julius E. 1825
(Pritchett, Henry S. 1857)
(Rees, John K. 1851)
(Woodward, Robert S. 1849)

16. GEOGRAPHY

This category also includes G. P. Marsh, who wrote on the relations of man and environment.

Baker, Marcus 1849
(Brigham, Albert P. 1855)
(Davidson, George 1825)
(Davis, William M. 1850)
(Gannett, Henry 1846)
(Gilbert, Grove K. 1843)
Guyot, Arnold H. 1807
(Heilprin, Angelo 1853)
(Hill, Robert T. 1858)
Hutchins, Thomas 1730
Marsh, George P. 1801
(Niles, William H. 1838)
(Russell, Israel C. 1852)

17. TOPOGRAPHICAL ENGINEERING AND SURVEYING

Abert, John J. 1788
De Brahm, William G. 1717
Ellicott, Andrew 1754
Emory, William H. 1811
Fry, Joshua ca. 1700
Gardiner, James T. 1842
Graham, James D. 1799
(Lyman, Benjamin S. 1835)
Thompson, Almon H. 1839
Wheeler, George M. 1842
Whipple, Amiel W. 1816

18. EXPLORATION AND TRAVELS

Freeman, Thomas d. 1821
Lawson, John d. 1711
Nicollet, Joseph N. 1786
Orton, James 1830
Schöpf, Johann D. 1752
Wilkes, Charles 1798

19. CARTOGRAPHY

Lindenkohl, Adolph 1833
Mitchell, John 1711
Romans, Bernard ca. 1720

20. NATURAL HISTORY AND NATURE STUDY

Atwater, Caleb 1778
Bachman, John 1790
Bland, Thomas 1809
Boll, Jacob 1828
Buckley, Samuel B. 1809
Catesby, Mark 1683
Cist, Jacob 1782
Colman, Benjamin 1673
(Comstock, Anna B. 1854)
Cooper, William ca. 1798
Couper, James H. 1794
Cutting, Hiram A. 1832
Douglass, William ca. 1691
Drake, Daniel 1785
Dudley, Paul 1675
Eights, James 1798
Garden, Alexander 1730
Gibbes, Lewis R. 1810
Godman, John D. 1794
Good, Adolphus C. 1856
Hildreth, Samuel P. 1783
Holmes, Francis S. 1815
Horsfield, Thomas 1773
James, Edwin 1797

Jefferson, Thomas 1743
Josselyn, John ca. 1608
Kennicott, Robert 1835
Lapham, Increase A. 1811
Lawson, John d. 1711
LeConte, John 1818
LeConte, John E. 1784
LeConte, Joseph 1823
Lincecum, Gideon 1793
Michener, Ezra 1794
Mitchell, Elisha 1793
Mitchell, John 1711
Mitchill, Samuel L. 1764
More, Thomas fl. 1670-1724
Morton, Samuel G. 1799
(Muir, John 1838)
Orton, James 1830
Peale, Charles W. 1742
Peale, Titian R. 1799
Peck, William D. 1763
Pickering, Charles 1805
Pitcher, Zina 1797
Plummer, John T. 1807
(Putnam, Frederic W. 1839)
Rafinesque, Constantine S. 1783
Romans, Bernard ca. 1720
(Safford, James M. 1822)
Samuels, Edward A. 1836
Saugrain de Vigni, Antoine F. 1763
Schöpf, Johann D. 1752
Schott, Arthur C. V. 1814
Thompson, Zadock 1796
Van Rensselaer, Jeremiah 1793
Wailes, Benjamin L. C. 1797
(Ward, Henry A. 1834)
Waterhouse, Benjamin 1754
Wislizenus, Frederick A. 1810
Woodhouse, Samuel W. 1821
Wright, Albert A. 1846
Xántus, János 1825

21. GEOLOGY

This category also includes persons engaged in work relating to some aspects of metallurgy and in collection and study of meteorites.

Agassiz, J. Louis R. 1807
Ashburner, Charles A. 1854
(Becker, George F. 1847)
(Blake, William P. 1825)
Boll, Jacob 1828
(Boyé, Martin H. 1812)
Bradley, Frank H. 1838
(Branner, John C. 1850)
(Brewer, William H. 1828)
(Brigham, Albert P. 1855)
(Broadhead. Garland C. 1827)
(Calvin, Samuel 1840)
(Campbell, Marius R. 1858)
Carll, John F. 1828
(Chamberlin, Thomas C. 1843)
Christy, David 1802
Cist, Jacob 1782
(Clark, William B. 1860)
(Clarke, John M. 1857)
Claypole, Edward W. 1835
Clemson, Thomas G. 1807
Coan, Titus 1801
Collett, John 1828
(Comstock, Theodore B. 1849)
Cook, George H. 1818
Cooper, James G. 1830
Cotting, John R. 1778
(Crosby, William O. 1850)
Cutting, Hiram A. 1832
Dana, James D. 1813
(Davis, William M. 1850)
Deane, James 1801
(Diller, Joseph S. 1850)
(Dutton, Clarence E. 1841)
Eaton, Amos 1776
Eights, James 1798
(Emerson, Benjamin K. 1843)
Emmons, Ebenezer 1799
(Emmons, Samuel F. 1841)
Engelmann, Henry 1831
Evans, John 1812
(Fairchild, Herman L. 1850)
Foster, John W. 1815
(Frazer, Persifor 1844)
Gabb, William M. 1839
Gibbes, Robert W. 1809
(Gilbert, Grove K. 1843)
Guyot, Arnold H. 1807
Hager, Albert D. 1817
(Hague, Arnold 1840)
(Hague, James D. 1836)
(Hall, Christopher W. 1845)
Hall, James 1811
Hanks, Henry G. 1826
Hartt, Charles F. 1840
Hayden, Ferdinand V. 1829
Hayden, Horace H. 1769
(Heilprin, Angelo 1853)
(Hilgard, Eugene W. 1833)
(Hill, Robert T. 1858)
(Hitchcock, Charles H. 1836)
Hitchcock, Edward 1763
(Hollick, Charles A. 1857)
Holmes, Francis S. 1815
(Holmes, William H. 1846)
Houghton, Douglass 1809

Hubbard, Oliver P. 1809
Hunt, Thomas S. 1826
(Iddings, Joseph P. 1857)
Irving, Roland D. 1847
Jackson, Charles T. 1805
Johnson, Walter R. 1794
Jones, William L. 1827
(Julien, Alexis A. 1840)
(Kemp, James F. 1859)
Kerr, Washington C. 1827
King, Clarence R. 1842
Knight, Wilbur C. 1858
LeConte, Joseph 1823
Lesley, J. Peter 1819
(Lindgren, Waldemar 1860)
Locke, John 1792
(Lyman, Benjamin S. 1835)
Lyman, Chester S. 1814
(MacBride, Thomas H. 1848)
McCalley, Henry 1852
(McGee, William J. 1853)
Maclure, William 1763
Madison, James 1749
Marcou, Jules 1824
Marcy, Oliver 1820
Mather, William W. 1804
Meek, Fielding B. 1817
(Merrill, George P. 1854)
Mitchell, James A. 1852
Mitchill, Samuel L. 1764
(Muir, John 1838)
Newberry, John S. 1822
Newton, Henry 1845
(Niles, William H. 1838)
Olmsted, Denison 1791
Orton, Edward F. B. 1829
Owen, David D. 1807
Percival, James G. 1795
(Pirsson, Louis V. 1860)
Platt, Franklin 1844
Powell, John W. 1834
Procter, John R. 1844
(Prosser, Charles S. 1860)
(Pumpelly, Raphael 1837)
(Rice, William N. 1845)
Rogers, Henry D. 1808
Rogers, William B. 1804
(Russell, Israel C. 1852)
(Safford, James M. 1822)
(Salisbury, Rollin D. 1858)
Schoolcraft, Henry R. 1793
Schott, Arthur C. V. 1814
(Shaler, Nathaniel S. 1841)
Shumard, Benjamin F. 1820
Silliman, Benjamin 1779
Silliman, Benjamin, Jr. 1816
Smith, Erminnie A. P. 1836
(Smith, Eugene A. 1841)
(Stevenson, John J. 1841)
Swallow, George C. 1817
(Taylor, Frank B. 1860)
Taylor, Richard C. 1789
Teschemacher, James E. 1790
(Todd, James E. 1846)
Trask, John B. 1824
Troost, Gerard 1776
Tuomey, Michael 1805
Tyson, Philip T. 1799
(Uhler, Philip R. 1835)
(Upham, Warren 1850)
(Van Hise, Charles R. 1857)
Van Rensselaer, Jeremiah 1793
Vanuxem, Lardner 1792
(Wadsworth, Marshman E. 1847)
(Walcott, Charles D. 1850)
(Ward, Henry A. 1834)
Warren, Gouverneur K. 1830
Wells, David A. 1828
(White, Charles A. 1826)
(White, Israel C. 1848)
(Whitfield, Robert P. 1828)
Whitney, Josiah D. 1819
Whittlesey, Charles 1808
(Williams, Henry S. 1847)
Winchell, Alexander 1824
(Winchell, Newton H. 1839)
Worthen, Amos H. 1813
Wright, Albert A. 1846
Yeates, William S. 1856

22. MINERALOGY

This category also includes persons engaged in work in petrology and petrography.

Alger, Francis 1807
Beck, Lewis C. 1798
(Blake, William P. 1825)
Bruce, Archibald 1777
(Brush, George J. 1831)
Carpenter, George W. 1802
Caswell, John H. 1846
Chester, Albert H. 1843
Cleaveland, Parker 1780
(Cornwall, Henry B. 1844)
(Dana, Edward S. 1849)
Dana, James F. 1793
Dana, Samuel L. 1795
Delafield, Joseph 1790
(Diller, Joseph S. 1850)
Genth, Frederick A., Sr. 1820
Gibbs, George 1776
(Hall, Christopher W. 1845)

Hubbard, Oliver P. 1809
(Iddings, Joseph P. 1857)
Jackson, Charles T. 1805
(Julien, Alexis A. 1840)
Keating, William H. 1799
(Koenig, George A. 1844)
Meade, William d. 1833
(Nicholson, Henry H. 1850)
(Penfield, Samuel L. 1856)
Seybert, Henry 1801
Shepard, Charles U. 1804
Silliman, Benjamin 1779
Smith, John L. 1818
Teschemacher, James E. 1790
Troost, Gerard 1776
Williams, George H. 1856
Yeates, William S. 1856

23. PALEONTOLOGY

Agassiz, J. Louis R. 1807
Beecher, Charles E. 1856
(Calvin, Samuel 1840)
(Clarke, John M. 1857)
Conrad, Timothy A. 1803
Cope, Edward D. 1840
(Dall, William H. 1845)
(Fontaine, William M. 1835)
Gabb, William M. 1839
Gibbes, Robert W. 1809
Hall, James 1811
Hatcher, John B. 1861
(Hay, Oliver P. 1846)
(Heilprin, Angelo 1853)
(Holland, William J. 1848)
(Hollick, Charles A. 1857)
Hyatt, Alpheus 1838
(Knowlton, Frank H. 1860)
Leidy, Joseph 1823
Lesquereux, Leo 1806
Marcou, Jules 1824
Marsh, Othniel C. 1831
Meek, Fielding B. 1817
Newberry, John S. 1822
(Osborn, Henry F. 1857)
(Prosser, Charles S. 1860)
Redfield, William C. 1789
Shumard, Benjamin F. 1820
Wachsmuth, Charles 1829
Wagner, William 1796
(Walcott, Charles D. 1850)
Warren, John C. 1778
(White, Charles A. 1826)
(Whitfield, Robert P. 1828)
(Williams, Henry S. 1847)
(Williston, Samuel W. 1852)
Yandell, Lunsford P. 1805

24. BIOLOGY (LIFE FORMS AND PROCESSES)

Persons in this category were engaged in work in anatomy, biochemistry, cytology, embryology, histology, morphology, neurology, physiology, and aspects of physiological chemistry, toxicology, and the like.

Allen, Harrison 1841
Beaumont, William 1785
(Beyer, Henry G. 1850)
(Bowditch, Henry P. 1840)
Burnett, Waldo I. 1827
(Chittenden, Russell H. 1856)
Dalton, John C., Jr. 1825
(Donaldson, Henry H. 1857)
Dunglison, Robley 1798
(Dwight, Thomas 1843)
(Gage, Simon H. 1851)
(Gill, Theodore N. 1837)
Godman, John D. 1794
Hammond, William A. 1828
Harlan, Richard 1796
Horner, William E. 1793
Jackson, John B. S. 1806
(Jayne, Horace F. 1859)
Keyt, Alonzo T. 1827
(Kingsley, John S. 1854)
LeConte, Joseph 1823
Lining, John 1708
(McMurrich, James P. 1859)
Martin, Henry N. 1848
(Mellus, Edward L. 1848)
(Minot, Charles S. 1852)
Mitchell, John K. 1793
(Mitchell, Silas W. 1829)
(Osborn, Henry F. 1857)
Ramsay, Alexander ca. 1754
(Shufeldt, Robert W. 1850)
Schweinitz, Emil A. de ca. 1864
(Smith, Sidney I. 1843)
Vaughan, Daniel ca. 1818
Warren, John C. 1778
(Warren, Joseph W. 1849)
Wells, William C. 1757
(Wilder, Burt G. 1841)
(Wilson, Edmund B. 1856)
Wistar, Caspar 1761
(Witthaus, Rudolph A. 1846)
Wyman, Jeffries 1814

25. ZOOLOGY

Abbot, John 1751
(Abbott, Charles C. 1843)
Adams, Charles B. 1814

(Agassiz, Alexander 1835)
Agassiz, J. Louis R. 1807
(Allen, Joel A. 1838)
Anthony, John G. 1804
(Ashmead, William H. 1855)
Audubon, John J. 1785
Ayres, William O. 1817
Bachman, John 1790
Baird, Spencer F. 1823
Banister, John 1650
Barton, Benjamin S. 1766
Bartram, William 1739
(Bean, Tarleton H. 1846)
Behr, Hans H. 1818
Binney, Amos 1803
Binney, William G. 1833
Bland, Thomas 1809
Bonaparte, Charles L. 1803
Brewer, Thomas M. 1814
(Brewster, William 1851)
(Brooks, William K. 1848)
Burgess, Edward 1848
Cabot, Samuel, Jr. 1815
Cassin, John 1813
(Clapp, Cornelia M. 1849)
Clark, Henry J. 1826
(Comstock, Anna B. 1854)
(Comstock, John H. 1849)
Conrad, Timothy A. 1803
Cooper, James G. 1830
Cope, Edward D. 1840
Cory, Charles B. 1857
Coues, Elliott 1842
(Cresson, Ezra T. 1838)
(Dall, William H. 1845)
Dana, James D. 1813
De Kay, James E. 1792
(Edwards, William H. 1822)
Eigenmann, Rosa S. 1858
(Elliot, Daniel G. 1835)
(Emerton, James H. 1847)
(Fernald, Charles H. 1838)
Fitch, Asa 1809
(Forbes, Stephen A. 1844)
Gambel, William 1823
(Garman, Samuel 1843)
(Gill, Theodore N. 1837)
Girard, Charles F. 1822
Glover, Townend 1813
Goode, George B. 1851
Gould, Augustus A. 1805
Green, Jacob 1790
Griffith, Robert E. 1798
Grote, Augustus R. 1841
Hagen, Hermann A. 1817
Haldeman, Samuel S. 1812
Hallowell, Edward 1808
Harlan, Richard 1796
Harris, Thaddeus W. 1795
Hentz, Nicholas M. 1797
Herrick, Edward C. 1811
Holbrook, John E. 1794
Holder, Joseph B. 1824
(Holland, William J. 1848)
Horn, George H. 1840
(Howard, Leland O. 1857)
Hubbard, Henry G. 1850
Hyatt, Alpheus 1838
(Jayne, Horace F. 1859)
Jenks, John W. P. 1819
(Jordan, David S. 1851)
Kennicott, Robert 1835
(Kingsley, John S. 1854)
Kirtland, Jared P. 1793
Kneeland, Samuel 1821
(Knowlton, Frank H. 1860)
Lawrence, George N. 1806
Lea, Isaac 1792
LeConte, John E. 1784
LeConte, John L. 1825
Leidy, Joseph 1823
Lesueur, Charles A. 1778
Lintner, Joseph A. 1822
Lyman, Theodore 1833
(Mark, Edward L. 1847)
Marx, George 1838
Melsheimer, Friedrich V. 1749
Miles, Manly 1826
Morris, John G. 1803
Morris, Margaretta H. 1797
(Morse, Edward S. 1838)
Nuttall, Thomas 1786
(Nutting, Charles C. 1858)
Ober, Frederick A. 1849
Ord, George 1781
Osten Sacken, Carl R. R. von der 1828
Packard, Alpheus S., Jr. 1839
Prime, Temple 1832
(Rathbun, Richard 1852)
Ravenel, Edmund 1797
(Ridgway, Robert 1850)
Riley, Charles V. 1843
Samuels, Edward A. 1836
Savage, Thomas S. 1804
Say, Thomas 1787
Schwarz, Eugene A. 1844
(Scudder, Samuel H. 1837)
Sennett, George B. 1840
(Shufeldt, Robert W. 1850)
(Snow, Francis H. 1840)
(Stearns, Robert E. C. 1827)
Stimpson, William 1832
Storer, David H. 1804

Strecker, Herman 1836
Taylor, Charlotte De B. S. 1806
(Thomas, Cyrus 1825)
Torrey, Bradford 1843
Townsend, John K. 1809
(True, Frederick W. 1858)
Trumbull, Gurdon 1841
Tryon, George W. 1838
(Uhler, Philip R. 1835)
(Verrill, Addison E. 1839)
Walsh, Benjamin D. 1808
(Webster, Francis M. 1849)
(White, Charles A. 1826)
(Whitman, Charles O. 1842)
Wilson, Alexander 1766
Woodhouse, Samuel W. 1821
Xántus, János 1825

26. MEDICINE

Boylston, Zabdiel 1680
Chalmers, Lionel 1715
Colden, Cadwallader 1688
Colman, Benjamin 1673
(Doremus, Robert O. 1824)
Douglass, William ca. 1691
Drake, Daniel 1785
Joslin, Benjamin F. 1796
Kuhn, Adam 1741
Mather, Cotton 1662/1663
Mease, James 1771
Mitchell, John 1711
(Mitchell, Silas W. 1829)
Robie, Thomas 1688/1689
Rush, Benjamin 1746
Salisbury, James H. 1823
Waterhouse Benjamin 1754
Williamson, Hugh 1735
Winthrop, John, Jr. 1606
Wood, Edward 1846

27. PSYCHOLOGY AND PHRENOLOGY

Fowler, Lorenzo N. 1811
Fowler, Orson S. 1809
Grimes, James S. 1807
(Hall, G. Stanley 1846)
(James, William 1842)
(Ladd, George T. 1842)
Rush, Benjamin 1746
Rush, James 1786
Wells, Samuel R. 1820

28. BOTANY

This category also includes persons engaged in work in bacteriology, forestry, horticulture, phytogeography, and plant physiology.

Allen, Timothy F. 1837
(Arthur, Joseph C. 1850)
(Atkinson, George F. 1854)
Bailey, Jacob W. 1811
Baldwin, William 1779
Banister, John 1650
(Barnes, Charles R. 1858)
Barton, Benjamin S. 1766
Barton, William P. C. 1786
Bartram, John 1699
Bartram, William 1739
(Beal, William J. 1833)
Behr, Hans H. 1818
(Bessey, Charles E. 1845)
Bigelow, Jacob 1786
Bodley, Rachel L. 1831
Brackenridge, William D. 1810
Brandegee, Mary K. L. C. 1844
Brandegee, Townshend S. 1843
(Brewer, William H. 1828)
Bridges, Robert 1806
(Britton, Elizabeth G. K. 1858)
(Britton, Nathaniel L. 1859)
Buckley, Samuel B. 1809
(Burrill, Thomas J. 1839)
Catesby, Mark 1683
Chapman, Alvan W. 1809
Clark, Henry J. 1826
Clayton, John 1694
Colden, Cadwallader 1688
Colden, Jane 1724
(Collins, Frank S. 1848)
Cooper, James G. 1830
(Coulter, John M. 1851)
Curtis, Moses A. 1808
Cutler, Manasseh 1742
Darlington, William 1782
(Davenport, George E. 1833)
Dewey, Chester 1784
(Dudley, William R. 1849)
Durand, Elias 1794
Eaton, Amos 1776
Eaton, Daniel C. 1834
Elliott, Stephen 1771
(Ellis, Job B. 1829)
Emerson, George B. 1797
Engelmann, George 1809
(Farlow, William G. 1844)
Furbish, Kate 1834
Garden, Alexander 1730
Gattinger, Augustin 1825
(Goodale, George L. 1839)
Gray, Asa 1810
Green, Jacob 1790
(Greene, Edward L. 1843)
Griffith, Robert E. 1798
Hough, Franklin B. 1822

James, Edwin 1797
James, Thomas P. 1803
(Kellerman, William A. 1850)
Kellogg, Albert 1813
Kuhn, Adam 1741
Lemmon, John G. 1832
Lesquereux, Leo 1806
Lindheimer, Ferdinand J. 1801
Lippincott, James S. 1819
Locke, John 1792
Logan, James 1674
(MacBride, Thomas H. 1848)
(MacFarlane, John M. 1855)
Marshall, Humphry 1722
Meehan, Thomas 1826
Michaux, André 1746
Michaux, François A. 1770
(Millspaugh, Charles F. 1854)
Mohr, Charles T. 1824
(Morgan, Andrew P. 1836)
Mühlenberg, Gotthilf H. E. 1753
Nuttall, Thomas 1786
(Parish, Samuel B. 1838)
Parry, Charles C. 1823
(Peck, Charles H. 1833)
Perrine, Henry 1797
Porcher, Francis P. 1825
Porter, Thomas C. 1822
Pursh, Frederick 1774
Ravenel, Henry W. 1814
Riddell, John L. 1807
(Rydberg, Per A. 1860)
(Sargent, Charles S. 1841)
Sartwell, Henry P. 1792
Schweinitz, Lewis D. von 1780
Shattuck, Lydia W. 1822
Shecut, John L. E. W. 1770
Short, Charles W. 1794
(Spalding, Volney M. 1849)
(Stearns, Robert E. C. 1827)
Sturtevant, Edward L. 1842
Sullivant, William S. 1803
Teschemacher, James E. 1790
Thurber, George 1821
Torrey, John 1796
Tracy, Clarissa T. 1818
(Trelease, William 1857)
Tuckerman, Edward 1817
(Underwood, Lucien M. 1853)
Vasey, George 1822
Walter, Thomas ca. 1740
(Ward, Lester F. 1841)
Warder, John A. 1812
Watson, Sereno 1826
(Wilson, William P. 1844)
Wright, Charles 1811
Young, Aaron 1819

29. AGRICULTURE

(Brewer, William H. 1828)
Browne, Daniel J. 1804
Couper, James H. 1794
Jones, William L. 1827
Judd, Orange 1822
Mapes, James J. 1806
Miles, Manly 1826
Puryear, Bennet 1826
(Stearns, Robert E. C. 1827)
Sturtevant, Edward L. 1842

30. HUMANISTIC ASPECTS OF SCIENCE

This category includes persons engaged in the study of science and religion, logic, and philosophy of science.

Draper, John W. 1811
(Halsted, George B. 1853)
Hill, Thomas 1818
(Peirce, Charles S. 1839)
Stallo, Johann B. 1823
Wright, Chauncey 1830

31. SCIENCE ADMINISTRATION AND EDUCATION

This category also includes persons engaged in work relating to medical and engineering education.

Baird, Spencer F. 1823
Crocker, Lucretia 1829
Dunglison, Robley 1798
Goode, George B. 1851
Henry, Joseph 1797
Peale, Charles W. 1742
Rogers, William B. 1804
Runkle, John D. 1822

32. PROMOTION AND COMMUNICATION OF SCIENTIFIC WORK

This category includes persons engaged in the promotion and popularization of science, editorial and related work (including the preparation of textbooks and of illustrations), and patrons of science.

(Comstock, Anna B. 1854)
Gibbs, George 1776
Humphreys, Andrew A. 1810
Mather, Increase 1639
Phelps, Almira H. L. 1793
Steele, Joel D. 1836
Wagner, William 1796
Wells, David A. 1828
Youmans, Edward L. 1821
Youmans, William J. 1838

INDEX

Note: Page references for subject biographies are in boldface. Page references that are repeated, or followed by a number in parentheses, indicate that the indexed item appears in more than one biographical entry on the same page.

About the Author

CLARK A. ELLIOTT is Associate Curator at the Harvard University Archives. He has published in *American Archivist* and *Social Studies of Science*, among other journals.